GOVERNORS STATE
W9-BFU-674
3 1611 00223 9264

Thieme

Color Atlas of Biochemistry

Second edition, revised and enlarged

Jan Koolman
Professor
Philipps University Marburg
Institute of Physiologic Chemistry
Marburg, Germany

Klaus-Heinrich Roehm
Professor
Philipps University Marburg
Institute of Physiologic Chemistry
Marburg, Germany

215 color plates by Juergen Wirth

Thieme
Stuttgart · New York

Library of Congress Cataloging-in-Publication Data

This book is an authorized and updated translation of the 3rd German edition published and copyrighted 2003 by Georg Thieme Verlag, Stuttgart, Germany. Title of the German edition: Taschenatlas der Biochemie

Illustrator: Juergen Wirth, Professor of Visual Communication, University of Applied Sciences, Darmstadt, Germany

Translator: Michael Robertson, BA DPhil, Augsburg, Germany

1st Dutch edition 2004
1st English edition 1996
1st French edition 1994
2nd French edition 1999
3rd French edition 2004
1st German edition 1994
2nd German edition 1997
1st Greek edition 1999
1st Indonesian edition 2002
1st Italian edition 1997
1st Japanese edition 1996
1st Portuguese edition 2004
1st Russian edition 2000
1st Spanish edition 2004

Rüdigerstrasse 14, 70469 Stuttgart, Germany
http://www.thieme.de
Thieme New York, 333 Seventh Avenue, New York, NY 10001 USA
http://www.thieme.com

Cover design: Cyclus, Stuttgart
Cover drawing: CAP · cAMP bound to DNA
Typesetting by primustype Hurler GmbH, Notzingen
Printed in Germany by Appl, Wemding

ISBN 3-13-100372-3 (GTV)
ISBN 1-58890-247-1 (TNY)

About the Authors

Jan Koolman (left) was born in Lübeck, Germany, and grew up with the sea wind blowing off the Baltic. The high school he attended in the Hanseatic city of Lübeck was one that focused on providing a classical education, which left its mark on him. From 1963 to 1969, he studied biochemistry at the University of Tübingen. He then took his doctorate (in the discipline of chemistry) at the University of Marburg, under the supervision of biochemist Peter Karlson. In Marburg, he began to study the biochemistry of insects and other invertebrates. He took his postdoctoral degree in 1977 in the field of human medicine, and was appointed Honorary Professor in 1984. His field of study today is biochemical endocrinology. His other interests include educational methods in biochemistry. He is currently Dean of Studies in the Department of Medicine in Marburg; he is married to an art teacher.

Klaus-Heinrich Röhm (right) comes from Stuttgart, Germany. After graduating from the School of Protestant Theology in Urach —another institution specializing in classical studies—and following a period working in the field of physics, he took a diploma in biochemistry at the University of Tübingen, where the two authors first met. Since 1970, he has also worked in the Department of Medicine at the University of Marburg. He took his doctorate under the supervision of Friedhelm Schneider, and his postdoctoral degree in 1980 was in the Department of Chemistry. He has been an Honorary Professor since 1986. His research group is concerned with the structure and function of enzymes involved in amino acid metabolism. He is married to a biologist and has two children.

Jürgen Wirth (center) studied in Berlin and at the College of Design in Offenbach, Germany. His studies focused on free graphics and illustration, and his diploma topic was "The development and function of scientific illustration." From 1963 to 1977, Jürgen Wirth was involved in designing the exhibition space in the Senckenberg Museum of Natural History in Frankfurt am Main, while at the same time working as a freelance associate with several publishing companies, providing illustrations for schoolbooks, non-fiction titles, and scientific publications. He has received several awards for book illustration and design. In 1978, he was appointed to a professorship at the College of Design in Schwäbisch Gmünd, Germany, and in 1986 he became Professor of Design at the Academy of Design in Darmstadt, Germany. His specialist fields include scientific graphics/information graphics and illustration methods. He is married and has three children.

Preface

Biochemistry is a dynamic, rapidly growing field, and the goal of this color atlas is to illustrate this fact visually. The precise boundaries between biochemistry and related fields, such as cell biology, anatomy, physiology, genetics, and pharmacology, are difficult to define and, in many cases, arbitrary. This overlap is not coincidental. The object being studied is often the same—a nerve cell or a mitochondrion, for example—and only the point of view differs.

For a considerable period of its history, biochemistry was strongly influenced by chemistry and concentrated on investigating metabolic conversions and energy transfers. Explaining the composition, structure, and metabolism of biologically important molecules has always been in the foreground. However, new aspects inherited from biochemistry's other parent, the biological sciences, are now increasingly being added: the relationship between chemical structure and biological function, the pathways of information transfer, observance of the ways in which biomolecules are spatially and temporally distributed in cells and organisms, and an awareness of evolution as a biochemical process. These new aspects of biochemistry are bound to become more and more important.

Owing to space limitations, we have concentrated here on the biochemistry of humans and mammals, although the biochemistry of other animals, plants, and microorganisms is no less interesting. In selecting the material for this book, we have put the emphasis on subjects relevant to students of human medicine. The main purpose of the atlas is to serve as an overview and to provide visual information quickly and efficiently. Referring to textbooks can easily fill any gaps. For readers encountering biochemistry for the first time, some of the plates may look rather complex. It must be emphasized, therefore, that the atlas is not intended as a substitute for a comprehensive textbook of biochemistry.

As the subject matter is often difficult to visualize, symbols, models, and other graphic elements had to be found that make complicated phenomena appear tangible. The graphics were designed conservatively, the aim being to avoid illustrations that might look too spectacular or exaggerated. Our goal was to achieve a visual and aesthetic way of representing scientific facts that would be simple and at the same time effective for teaching purposes. Use of graphics software helped to maintain consistency in the use of shapes, colors, dimensions, and labels, in particular. Formulae and other repetitive elements and structures could be handled easily and precisely with the assistance of the computer.

Color-coding has been used throughout to aid the reader, and the key to this is given in two special color plates on the front and rear inside covers. For example, in molecular models each of the more important atoms has a particular color: gray for carbon, white for hydrogen, blue for nitrogen, red for oxygen, and so on. The different classes of biomolecules are also distinguished by color: proteins are always shown in brown tones, carbohydrates in violet, lipids in yellow, DNA in blue, and RNA in green. In addition, specific symbols are used for the important coenzymes, such as ATP and NAD^+. The compartments in which biochemical processes take place are color-coded as well. For example, the cytoplasm is shown in yellow, while the extracellular space is shaded in blue. Arrows indicating a chemical reaction are always black and those representing a transport process are gray.

In terms of the visual clarity of its presentation, biochemistry has still to catch up with anatomy and physiology. In this book, we sometimes use simplified ball-and-stick models instead of the classical chemical formulae. In addition, a number of compounds are represented by space-filling models. In these cases, we have tried to be as realistic as possible. The models of small molecules are based on conformations calculated by computer-based molecular modeling. In illustrating macromolecules, we used structural infor-

mation obtained by X-ray crystallography that is stored in the Protein Data Bank. In naming enzymes, we have followed the official nomenclature recommended by the IUBMB. For quick identification, EC numbers (in italics) are included with enzyme names.
To help students assess the relevance of the material (while preparing for an examination, for example), we have included symbols on the text pages next to the section headings to indicate how important each topic is. A filled circle stands for "basic knowledge," a half-filled circle indicates "standard knowledge," and an empty circle stands for "in-depth knowledge." Of course, this classification only reflects our subjective views.
This second edition was carefully revised and a significant number of new plates were added to cover new developments.

We are grateful to many readers for their comments and valuable criticisms during the preparation of this book. Of course, we would also welcome further comments and suggestions from our readers.

August 2004

Jan Koolman,
Klaus-Heinrich Röhm
Marburg

Jürgen Wirth
Darmstadt

Contents

Key to color-coding:
see front and rear inside covers

Introduction

This paperback atlas is intended for students of medicine and the biological sciences. It provides an introduction to biochemistry, but with its modular structure it can also be used as a reference book for more detailed information. The 216 color plates provide knowledge in the field of biochemistry, accompanied by detailed information in the text on the facing page. The degree of difficulty of the subject-matter is indicated by symbols in the text:

● stands for "basic biochemical knowledge"
◑ indicates "standard biochemical knowledge"
○ means "specialist biochemical knowledge."

Some general rules used in the structure of the illustrations are summed up in two *explanatory plates* inside the front and back covers. Keywords, definitions, explanations of unfamiliar concepts and chemical formulas can be found using the *index*. The book starts with a few **basics** in biochemistry (pp. 2–33). There is a brief explanation of the concepts and principles of chemistry (pp. 2–15). These include the periodic table of the elements, chemical bonds, the general rules governing molecular structure, and the structures of important classes of compounds. Several basic concepts of *physical chemistry* are also essential for an understanding of biochemical processes. Pages 16–33 therefore discuss the various forms of energy and their interconversion, reaction kinetics and catalysis, the properties of water, acids and bases, and redox processes.

These basic concepts are followed by a section on the structure of the important biomolecules (pp. 34–87). This part of the book is arranged according to the different classes of metabolites. It discusses carbohydrates, lipids, amino acids, peptides and proteins, nucleotides, and nucleic acids.

The next part presents the reactions involved in the interconversion of these compounds—the part of biochemistry that is commonly referred to as **metabolism** (pp. 88–195). The section starts with a discussion of the enzymes and coenzymes, and discusses the mechanisms of metabolic regulation and the so-called *energy metabolism*. After this, the central metabolic pathways are presented, once again arranged according to the class of metabolite (pp. 150–195).

The second half of the book begins with a discussion of the functional compartments within the cell, the **cellular organelles** (pp. 196–235). This is followed on pp. 236–265 by the current field of **molecular genetics** (*molecular biology*). A further extensive section is devoted to the biochemistry of individual **tissues and organs** (pp. 266–359). Here, it has only been possible to focus on the most important organs and organ systems—the digestive system, blood, liver, kidneys, muscles, connective and supportive tissues, and the brain.

Other topics include the biochemistry of **nutrition** (pp. 360–369), the structure and function of important **hormones** (pp. 370–393), and **growth and development** (pp. 394–405).

The paperback atlas concludes with a series of schematic **metabolic "charts"** (pp. 407–419). These plates, which are not accompanied by explanatory text apart from a brief introduction on p. 406, show simplified versions of the most important synthetic and degradative pathways. The charts are mainly intended for reference, but they can also be used to review previously learned material. The enzymes catalyzing the various reactions are only indicated by their EC numbers. Their names can be found in the systematically arranged and annotated enzyme list (pp. 420–430).

Periodic table

A. Biologically important elements ◑

There are 81 stable elements in nature. Fifteen of these are present in all living things, and a further 8–10 are only found in particular organisms. The illustration shows the first half of the **periodic table**, containing all of the biologically important elements. In addition to physical and chemical data, it also provides information about the distribution of the elements in the living world and their abundance in the human body. The laws of atomic structure underlying the periodic table are discussed in chemistry textbooks.

More than 99% of the atoms in animals' bodies are accounted for by just four elements—hydrogen (H), oxygen (O), carbon (C) and nitrogen (N). Hydrogen and oxygen are the constituents of water, which alone makes up 60–70% of cell mass (see p. 196). Together with carbon and nitrogen, hydrogen and oxygen are also the major constituents of the **organic compounds** on which most living processes depend. Many biomolecules also contain sulfur (S) or phosphorus (P). The above **macroelements** are essential for all organisms.

A second biologically important group of elements, which together represent only about 0.5% of the body mass, are present almost exclusively in the form of **inorganic ions.** This group includes the *alkali metals* sodium (Na) and potassium (K), and the *alkaline earth metals* magnesium (Mg) and calcium (Ca). The halogen *chlorine* (Cl) is also always ionized in the cell. All other elements important for life are present in such small quantities that they are referred to as **trace elements**. These include transition metals such as iron (Fe), zinc (Zn), copper (Cu), cobalt (Co) and manganese (Mn). A few *nonmetals,* such as iodine (I) and selenium (Se), can also be classed as essential trace elements.

B. Electron configurations: examples ○

The chemical properties of atoms and the types of bond they form with each other are determined by their electron shells. The **electron configurations** of the elements are therefore also shown in Fig. **A**. Fig. **B** explains the symbols and abbreviations used. More detailed discussions of the subject are available in chemistry textbooks.

The possible states of electrons are called **orbitals**. These are indicated by what is known as the principal quantum number and by a letter—s, p, or d. The orbitals are filled one by one as the number of electrons increases. Each orbital can hold a maximum of two electrons, which must have oppositely directed "spins." Fig. **A** shows the distribution of the electrons among the orbitals for each of the elements. For example, the six electrons of carbon (**B1**) occupy the 1s orbital, the 2s orbital, and two 2p orbitals. A filled 1s orbital has the same electron configuration as the noble gas helium (He). This region of the electron shell of carbon is therefore abbreviated as "He" in Fig. **A**. Below this, the numbers of electrons in each of the other filled orbitals (2s and 2p in the case of carbon) are shown on the right margin. For example, the electron shell of chlorine (**B2**) consists of that of neon (Ne) and seven additional electrons in 3s and 3p orbitals. In iron (**B3**), a transition metal of the first series, electrons occupy the 4s orbital even though the 3d orbitals are still partly empty. Many reactions of the transition metals involve empty d orbitals—e. g., redox reactions or the formation of complexes with bases.

Particularly stable electron arrangements arise when the outermost shell is fully occupied with eight electrons (the "**octet rule**"). This applies, for example, to the noble gases, as well as to ions such as Cl^- ($3s^2 3p^6$) and Na^+ ($2s^2 2p^6$). It is only in the cases of hydrogen and helium that two electrons are already sufficient to fill the outermost 1s orbital.

A. Biologically important elements

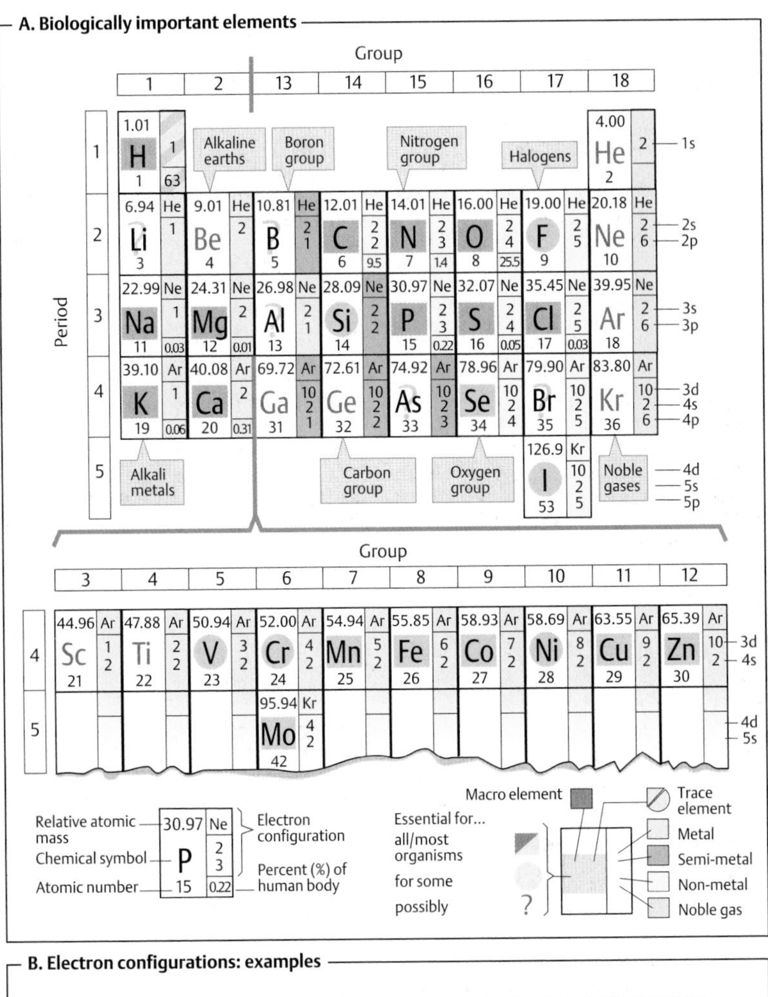

B. Electron configurations: examples

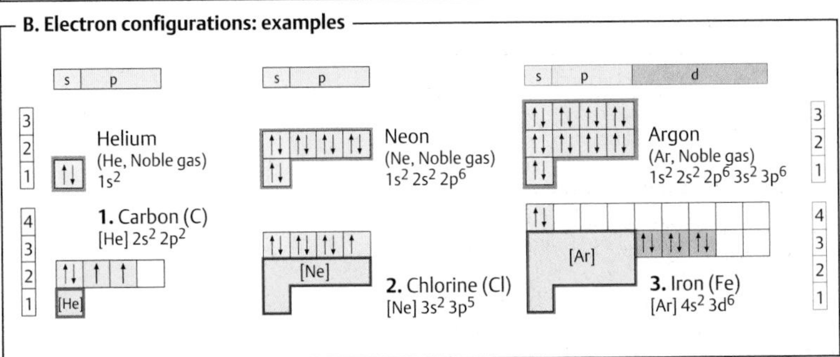

Bonds

A. Orbital hybridization and chemical bonding ○

Stable, covalent bonds between nonmetal atoms are produced when orbitals (see p. 2) of the two atoms form **molecular orbitals** that are occupied by one electron from each of the atoms. Thus, the four bonding electrons of the carbon atom occupy 2s and 2p atomic orbitals (**1a**). The 2s orbital is spherical in shape, while the three 2p orbitals are shaped like dumbbells arranged along the x, y, and z axes. It might therefore be assumed that carbon atoms should form at least *two different* types of molecular orbital. However, this is not normally the case. The reason is an effect known as **orbital hybridization**. Combination of the s orbital and the three p orbitals of carbon gives rise to four equivalent, tetrahedrally arranged sp^3 atomic orbitals (**sp^3 hybridization**). When these overlap with the 1s orbitals of H atoms, four equivalent σ-molecular orbitals (**1b**) are formed. For this reason, carbon is capable of forming four bonds—i. e., it has a valency of four. Single bonds between nonmetal atoms arise in the same way as the four σ or **single bonds** in methane (CH_4). For example, the hydrogen phosphate ion (HPO_4^{2-}) and the ammonium ion (NH_4^+) are also tetrahedral in structure (**1c**).

A second common type of orbital hybridization involves the 2s orbital and only *two* of the three 2p orbitals (2a). This process is therefore referred to as **sp^2 hybridization.** The result is three equivalent sp^2 hybrid orbitals lying in one plane at an angle of 120° to one another. The remaining $2p_x$ orbital is oriented perpendicular to this plane. In contrast to their sp^3 counterparts, sp^2-hybridized atoms form two *different* types of bond when they combine into molecular orbitals (**2b**). The three sp^2 orbitals enter into σ bonds, as described above. In addition, the electrons in the two $2p_x$ orbitals, known as **π electrons**, combine to give an additional, elongated π molecular orbital, which is located above and below the plane of the σ bonds. Bonds of this type are called **double bonds.** They consist of a σ bond and a π bond, and arise only when both of the atoms involved are capable of sp^2 hybridization. In contrast to single bonds, double bonds are not freely rotatable, since rotation would distort the π-molecular orbital. This is why all of the atoms lie in one plane (**2c**); in addition, *cis–trans* isomerism arises in such cases (see p. 8). Double bonds that are common in biomolecules are C=C and C=O. C=N double bonds are found in aldimines (Schiff bases, see p. 178).

B. Resonance ◑

Many molecules that have several double bonds are much less reactive than might be expected. The reason for this is that the double bonds in these structures cannot be localized unequivocally. Their π orbitals are not confined to the space between the double-bonded atoms, but form a shared, extended **π-molecular orbital.** Structures with this property are referred to as **resonance hybrids**, because it is impossible to describe their actual bonding structure using standard formulas. One can either use what are known as **resonance structures**—i. e., idealized configurations in which π electrons are assigned to specific atoms (cf. pp. 32 and 66, for example)—or one can use dashed lines as in Fig. **B** to suggest the extent of the delocalized orbitals. (Details are discussed in chemistry textbooks.)

Resonance-stabilized systems include carboxylate groups, as in *formate*; aliphatic hydrocarbons with conjugated double bonds, such as *1,3-butadiene*; and the systems known as **aromatic ring systems.** The best-known aromatic compound is *benzene,* which has six delocalized π electrons in its ring. Extended resonance systems with 10 or more π electrons absorb light within the visible spectrum and are therefore *colored.* This group includes the aliphatic carotenoids (see p. 132), for example, as well as the heme group, in which 18 π electrons occupy an extended molecular orbital (see p. 106).

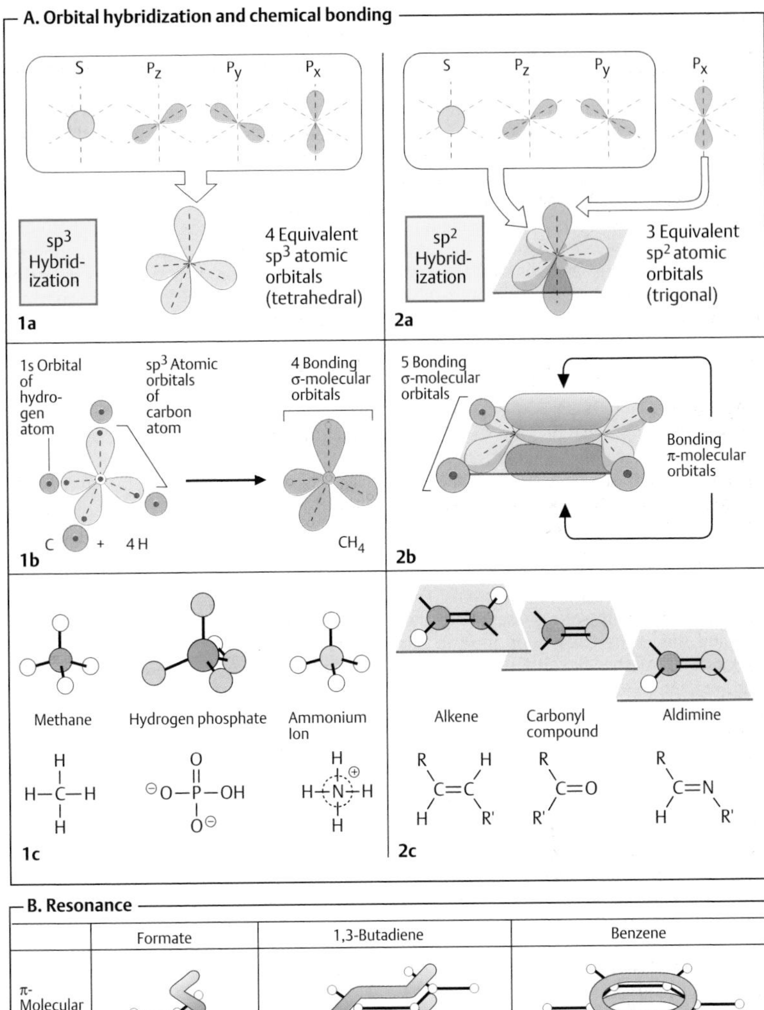

B. Resonance

	Formate	1,3-Butadiene	Benzene
π-Molecular orbitals			
Formula			

Molecular structure

The physical and chemical behavior of molecules is largely determined by their **constitution** (the type and number of the atoms they contain and their bonding). Structural formulas can therefore be used to predict not only the chemical reactivity of a molecule, but also its size and shape, and to some extent its conformation (the spatial arrangement of the atoms). Some data providing the basis for such predictions are summarized here and on the facing page. In addition, L-dihydroxyphenylalanine (L-dopa; see p. 352), is used as an example to show the way in which molecules are illustrated in this book.

A. Molecule illustrations ◑

In traditional two-dimensional **structural formulas** (**A1**), atoms are represented as letter symbols and electron *pairs* are shown as lines. Lines between two atomic symbols symbolize two **bonding electrons** (see p. 4), and all of the other lines represent **free electron pairs**, such as those that occur in O and N atoms. Free electrons are usually not represented explicitly (and this is the convention used in this book as well). Dashed or continuous circles or arcs are used to emphasize delocalized electrons.

Ball-and-stick models (**A2**) are used to illustrate the spatial structure of molecules. Atoms are represented as colored balls (for the color coding, see the inside front cover) and bonds (including multiple bonds) as gray cylinders. Although the relative bond lengths and angles correspond to actual conditions, the size at which the atoms are represented is too small to make the model more comprehensible.

Space-filling **van der Waals models** (**A3**) are useful for illustrating the actual shape and size of molecules. These models represent atoms as truncated balls. Their effective extent is determined by what is known as the van der Waals radius. This is calculated from the energetically most favorable distance between atoms that are not chemically bonded to one another.

B. Bond lengths and angles ○

Atomic radii and distances are now usually expressed in picometers (pm; 1 pm = 10^{-12} m). The old angstrom unit (Å, Å = 100 pm) is now obsolete. The length of single bonds approximately corresponds to the sum of what are known as the **covalent radii** of the atoms involved (see inside front cover). Double bonds are around 10–20% shorter than single bonds. In sp^3-hybridized atoms, the angle between the individual bonds is approx. 110°; in sp^2-hybridized atoms it is approx. 120°.

C. Bond polarity ○

Depending on the position of the element in the periodic table (see p. 2), atoms have different **electronegativity**—i. e., a different tendency to take up extra electrons. The values given in **C2** are on a scale between 2 and 4. The higher the value, the more electronegative the atom. When two atoms with very different electronegativities are bound to one another, the bonding electrons are drawn toward the more electronegative atom, and the **bond** is **polarized.** The atoms involved then carry positive or negative partial charges. In **C1**, the van der Waals surface is colored according to the different charge conditions (red = negative, blue = positive). Oxygen is the most strongly electronegative of the biochemically important elements, with C=O double bonds being especially highly polar.

D. Hydrogen bonds ◑

The **hydrogen bond**, a special type of noncovalent bond, is extremely important in biochemistry. In this type of bond, hydrogen atoms of OH, NH, or SH groups (known as hydrogen bond **donors**) interact with free electrons of **acceptor** atoms (for example, O, N, or S). The bonding energies of hydrogen bonds (10–40 kJ · mol^{-1}) are much lower than those of covalent bonds (approx. 400 kJ · mol^{-1}). However, as hydrogen bonds can be very numerous in proteins and DNA, they play a key role in the stabilization of these molecules (see pp. 68, 84). The importance of hydrogen bonds for the properties of water is discussed on p. 26.

A. Molecule illustrations

Chiral center

1. Formula illustration

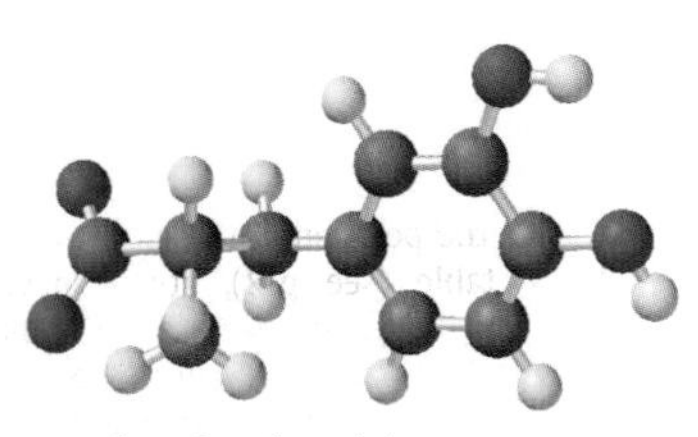

2. Ball- and-stick model

3. Van der Waals model

B. Bond lengths and angles

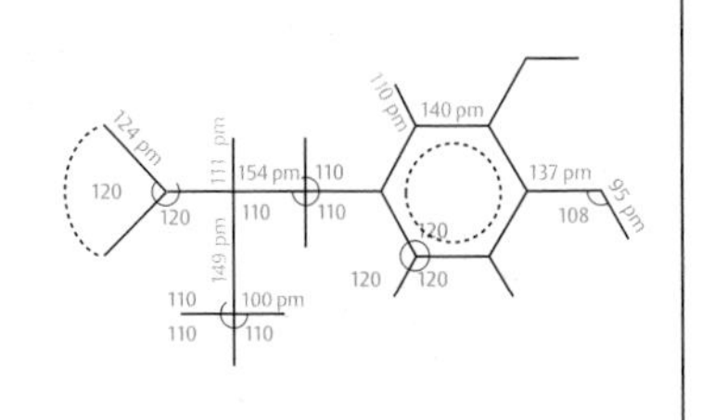

C. Bond polarity

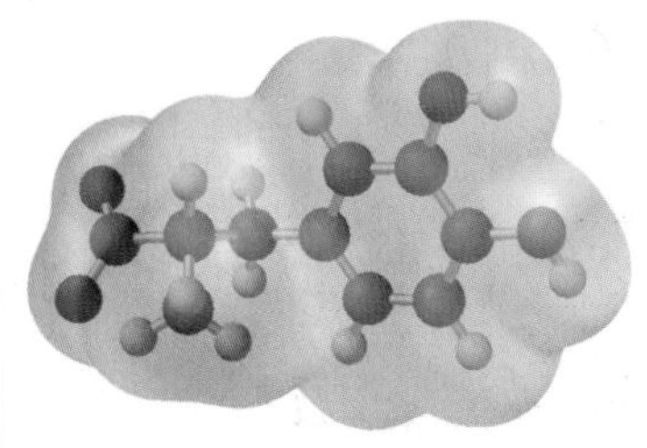

Positive Neutral Negative

1. Partial charges in L-dopa

0.9	2.1	2.5	3.0	3.5	4.0
Na	H	C	N	O	F

1 2 3 4

Increasing electronegativity

2. Electronegativities

D. Hydrogen bonds

Acid Base	Donor Acceptor	Dissociated acid Protonated base
A—H |B	A—H······|B	⊖A| H—B⊕
Initial state	Hydrogen bond	Complete reaction

1. Principle

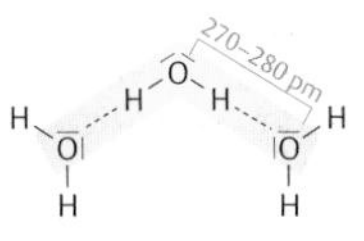

Water

Proteins

DNA

2. Examples

Isomerism

Isomers are molecules with the same composition (i. e. the same molecular formula), but with different chemical and physical properties. If isomers differ in the way in which their atoms are bonded in the molecule, they are described as **structural isomers** (cf. citric acid and isocitric acid, **D**). Other forms of isomerism are based on different arrangements of the substituents of bonds (**A**, **B**) or on the presence of chiral centers in the molecule (**C**).

A. *cis–trans* isomers ◑

Double bonds *are not freely rotatable* (see p. 4). If double-bonded atoms have different substituents, there are two possible orientations for these groups. In **fumaric acid**, an intermediate of the tricarboxylic acid cycle (see p. 136), the carboxy groups lie on *different* sides of the double bond (*trans* or *E* position). In its isomer **maleic acid**, which is not produced in metabolic processes, the carboxy groups lie on the *same* side of the bond (*cis* or *Z* position). *Cis–trans* isomers (**geometric isomers**) have different chemical and physical properties—e. g., their melting points (Fp.) and pK_a values. They can only be interconverted by chemical reactions.

In lipid metabolism, *cis–trans* isomerism is particularly important. For example, double bonds in natural fatty acids (see p. 48) usually have a *cis* configuration. By contrast, unsaturated intermediates of β oxidation have a *trans* configuration. This makes the breakdown of unsaturated fatty acids more complicated (see p. 166). Light-induced *cis–trans* isomerization of retinal is of central importance in the visual cycle (see p. 358).

B. Conformation ◑

Molecular forms that arise as a result of rotation around freely rotatable bonds are known as **conformers.** Even small molecules can have different conformations in solution. In the two conformations of **succinic acid** illustrated opposite, the atoms are arranged in a similar way to fumaric acid and maleic acid. Both forms are possible, although conformation 1 is more favorable due to the greater distance between the COOH groups and therefore occurs more frequently. Biologically active macromolecules such as proteins or nucleic acids usually have well-defined ("native") conformations, which are stabilized by interactions in the molecule (see p. 74).

C. Optical isomers ◑

Another type of isomerism arises when a molecule contains a **chiral center** or is chiral as a whole. Chirality (from the Greek *cheir,* hand) leads to the appearance of structures that behave like image and mirror-image and that cannot be superimposed ("mirror" isomers). The most frequent cause of chiral behavior is the presence of an asymmetric C atom—i. e., an atom with four *different* substituents. Then there are two forms **(enantiomers)** with different **configurations.** Usually, the two enantiomers of a molecule are designated as L and D forms. Clear classification of the configuration is made possible by the *R/S system* (see chemistry textbooks).

Enantiomers have very similar chemical properties, but they rotate polarized light in opposite directions (**optical activity**, see pp. 36, 58). The same applies to the enantiomers of **lactic acid.** The dextrorotatory L-lactic acid occurs in animal muscle and blood, while the D form produced by microorganisms is found in milk products, for example (see p. 148). The Fischer projection is often used to represent the formulas for chiral centers (cf. p. 58).

D. The aconitase reaction ○

Enzymes usually function *stereospecifically.* In chiral substrates, they only accept one of the enantiomers, and the reaction products are usually also sterically uniform. *Aconitate hydratase* (aconitase) catalyzes the conversion of citric acid into the constitution isomer isocitric acid (see p. 136). Although citric acid is not chiral, aconitase only forms one of the four possible isomeric forms of isocitric acid (*2R,3S*-isocitric acid). The intermediate of the reaction, the unsaturated tricarboxylic acid *aconitate,* only occurs in the *cis* form in the reaction. The *trans* form of aconitate is found as a constituent of certain plants.

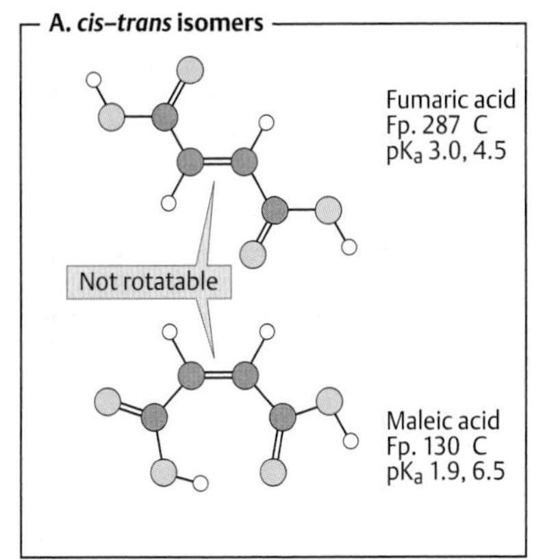

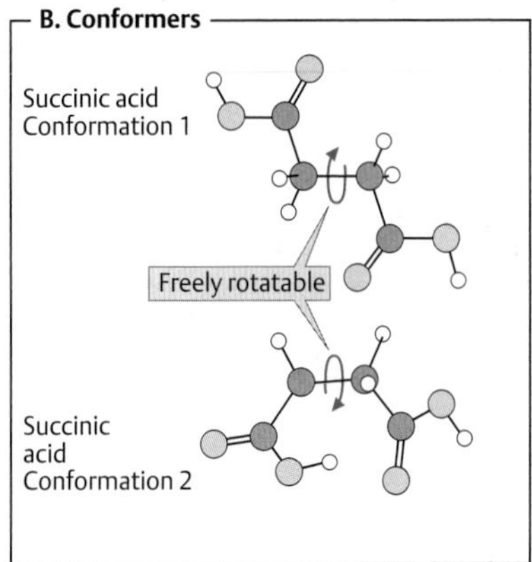

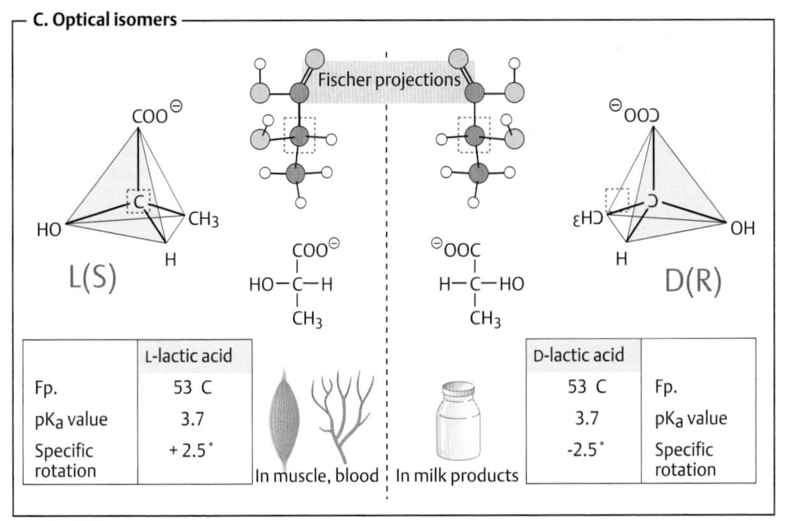

	L-lactic acid
Fp.	53 C
pKa value	3.7
Specific rotation	+ 2.5°

D-lactic acid	
53 C	Fp.
3.7	pKa value
-2.5°	Specific rotation

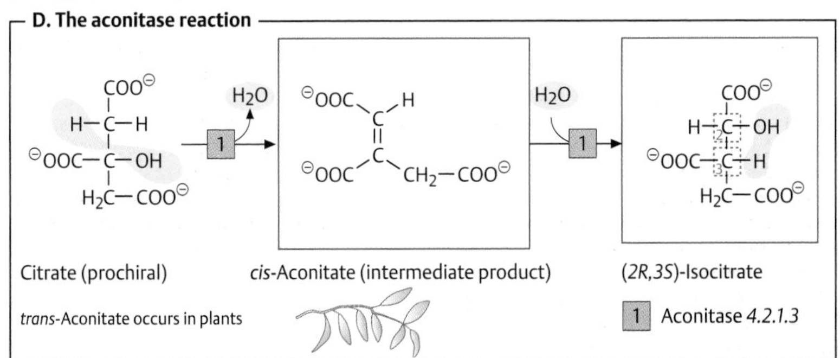

Biomolecules I

A. Important classes of compounds ●

Most biomolecules are derivatives of simple compounds of the non-metals oxygen (O), hydrogen (H), nitrogen (N), sulfur (S), and phosphorus (P). The biochemically important oxygen, nitrogen, and sulfur compounds can be formally derived from their compounds with hydrogen (i.e., H_2O, NH_3, and H_2S). In biological systems, phosphorus is found almost exclusively in derivatives of phosphoric acid, H_3PO_4.

If one or more of the hydrogen atoms of a non-metal hydride are replaced formally with another group, R—e.g., alkyl residues—then derived compounds of the type R-XH_{n-1}, R-XH_{n-2}-R, etc., are obtained. In this way, **alcohols** (R-OH) and **ethers** (R-O-R) are derived from water (H_2O); primary **amines** (R-NH_2), secondary amines (R-NH-R) and tertiary amines (R-N-R′R″) amines are obtained from ammonia (NH_3); and **thiols** (R-SH) and **thioethers** (R-S-R′) arise from hydrogen sulfide (H_2S). Polar groups such as -OH and -NH_2 are found as substituents in many organic compounds. As such groups are much more reactive than the hydrocarbon structures to which they are attached, they are referred to as **functional groups.**

New functional groups can arise as a result of **oxidation** of the compounds mentioned above. For example, the oxidation of a thiol yields a **disulfide** (R-S-S-R). Double oxidation of a primary alcohol (R-CH_2-OH) gives rise initially to an **aldehyde** (R-C(O)-H), and then to a **carboxylic acid** (R-C(O)-OH). In contrast, the oxidation of a secondary alcohol yields a **ketone** (R-C(O)-R). The carbonyl group (C=O) is characteristic of aldehydes and ketones.

The addition of an amine to the carbonyl group of an aldehyde yields—after removal of water—an **aldimine** (not shown; see p. 178). Aldimines are intermediates in amino acid metabolism (see p. 178) and serve to bond aldehydes to amino groups in proteins (see p. 62, for example). The addition of an alcohol to the carbonyl group of an aldehyde yields a **hemiacetal** (R-O-C(H)OH-R). The cyclic forms of sugars are well-known examples of hemiacetals (see p. 36). The oxidation of hemiacetals produces carboxylic acid esters.

Very important compounds are the **carboxylic acids** and their derivatives, which can be formally obtained by exchanging the OH group for another group. In fact, derivatives of this type are formed by nucleophilic substitutions of activated intermediate compounds and the release of water (see p. 14). **Carboxylic acid esters** (R-O-CO-R′) arise from carboxylic acids and alcohols. This group includes the fats, for example (see p. 48). Similarly, a carboxylic acid and a thiol yield a **thioester** (R-S-CO-R′). Thioesters play an extremely important role in carboxylic acid metabolism. The best-known compound of this type is acetyl-coenzyme A (see p. 12).

Carboxylic acids and primary amines react to form **carboxylic acid amides** (R-NH-CO-R′). The amino acid constituents of peptides and proteins are linked by carboxylic acid amide bonds, which are therefore also known as peptide bonds (see p. 66).

Phosphoric acid, H_3PO_4, is a tribasic (three-protic) acid—i.e., it contains three hydroxyl groups able to donate H^+ ions. At least one of these three groups is fully dissociated under normal physiological conditions, while the other two can react with alcohols. The resulting products are phosphoric acid monoesters (R-O-P(O)O-OH) and diesters (R-O-P(O)O-O-R′). **Phosphoric acid monoesters** are found in carbohydrate metabolism, for example (see p. 36), whereas **phosphoric acid diester** bonds occur in phospholipids (see p. 50) and nucleic acids (see p. 82).

Compounds of one acid with another are referred to as **acid anhydrides.** A particularly large amount of energy is required for the formation of an acid—anhydride bond. Phosphoric anhydride bonds therefore play a central role in the storage and release of chemical energy in the cell (see p. 122). Mixed anhydrides between carboxylic acids and phosphoric acid are also very important "energy-rich metabolites" in cellular metabolism.

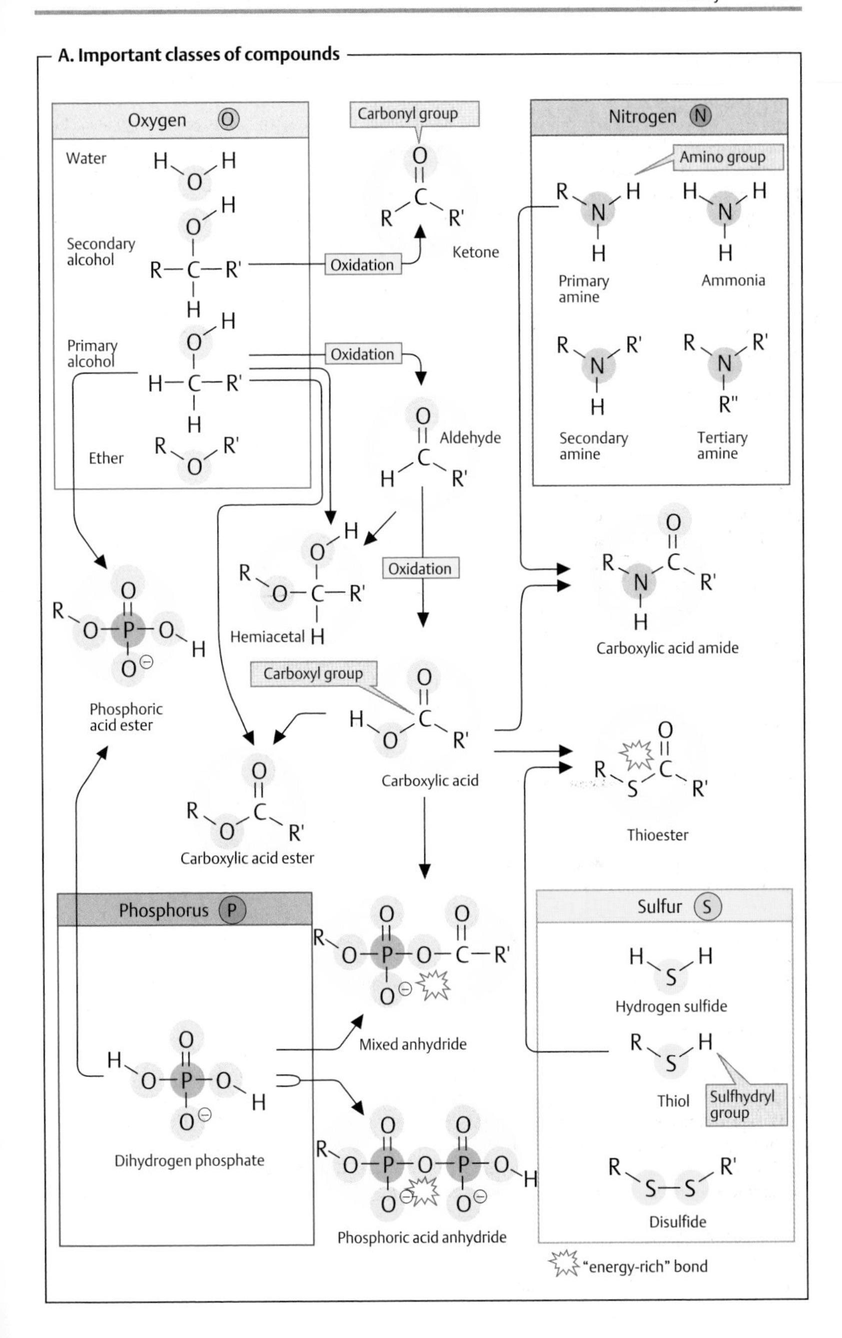
A. Important classes of compounds
Oxygen
Water
Secondary alcohol
Primary alcohol
Ether
Carbonyl group
Ketone
Oxidation
Oxidation
Aldehyde
Hemiacetal
Oxidation
Carboxyl group
Carboxylic acid
Carboxylic acid ester
Phosphoric acid ester
Nitrogen
Amino group
Primary amine
Ammonia
Secondary amine
Tertiary amine
Carboxylic acid amide
Thioester
Phosphorus
Dihydrogen phosphate
Mixed anhydride
Phosphoric acid anhydride
Sulfur
Hydrogen sulfide
Thiol
Sulfhydryl group
Disulfide
"energy-rich" bond

Biomolecules II

Many biomolecules are made up of smaller units in a modular fashion, and they can be broken down into these units again. The construction of these molecules usually takes place through condensation reactions involving the removal of water. Conversely, their breakdown functions in a hydrolytic fashion—i. e., as a result of water uptake. The page opposite illustrates this modular principle using the example of an important coenzyme.

A. Acetyl CoA ◑

Coenzyme A (see also p. 106) is a nucleotide with a complex structure (see p. 80). It serves to activate residues of carboxylic acids (acyl residues). Bonding of the carboxy group of the carboxylic acid with the thiol group of the coenzyme creates a **thioester bond** (-S-CO-R; see p. 10) in which the **acyl residue** has a **high chemical potential.** It can therefore be transferred to other molecules in exergonic reactions. This fact plays an important role in lipid metabolism in particular (see pp. 162 ff.), as well as in two reactions of the tricarboxylic acid cycle (see p. 136).

As discussed on p. 16, the **group transfer potential** can be expressed quantitatively as the change in free enthalpy (ΔG) during hydrolysis of the compound concerned. This is an arbitrary determination, but it provides important indications of the chemical energy stored in such a group. In the case of acetyl-CoA, the reaction to be considered is:

Acetyl CoA + H_2O → acetate + CoA

In standard conditions and at pH 7, the change in the chemical potential G (ΔG^0, see p. 18) in this reaction amounts to –32 $kJ \cdot mol^{-1}$ and it is therefore as high as the ΔG^0 of ATP hydrolysis (see p. 18). In addition to the "energy-rich" **thioester bond**, acetyl-CoA also has seven other hydrolyzable bonds with different degrees of stability. These bonds, and the fragments that arise when they are hydrolyzed, will be discussed here in sequence.

(1) The reactive thiol group of coenzyme A is located in the part of the molecule that is derived from **cysteamine.** Cysteamine is a *biogenic amine* (see p. 62) formed by decarboxylation of the amino acid cysteine.

(2) The amino group of cysteamine is bound to the carboxy group of another biogenic amine via an **acid amide bond** (-CO-NH-). β-**Alanine** arises through decarboxylation of the amino acid aspartate, but it can also be formed by breakdown of pyrimidine bases (see p. 186).

(3) Another **acid amide bond** (-CO-NH-) creates the compound for the next constituent, **pantoinate.** This compound contains a *chiral center* and can therefore appear in two enantiomeric forms (see p. 8). In natural coenzyme A, only one of the two forms is found, the (*R*)-pantoinate. Human metabolism is not capable of producing pantoinate itself, and it therefore has to take up a compound of β-alanine and pantoinate—**pantothenate** ("pantothenic acid")—in the form of a vitamin in food (see p. 366).

(4) The hydroxy group at C-4 of pantoinate is bound to a **phosphate** residue by an **ester bond.**

The section of the molecule discussed so far represents a functional unit. In the cell, it is produced from pantothenate. The molecule also occurs in a protein-bound form as **4′-phosphopantetheine** in the enzyme *fatty acid synthase* (see p. 168). In coenzyme A, however, it is bound to 3′,5′-adenosine diphosphate.

(5) When two phosphate residues bond, they do not form an ester, but an "energy-rich" **phosphoric acid anhydride bond,** as also occurs in other nucleoside phosphates. By contrast, (6) and (7) are ester bonds again.

(8) The base **adenine** is bound to C-1 of **ribose** by an **N-glycosidic** bond (see p. 36). In addition to C-2 to C-4, C-1 of ribose also represents a *chiral* center. The *β-configuration* is usually found in nucleotides.

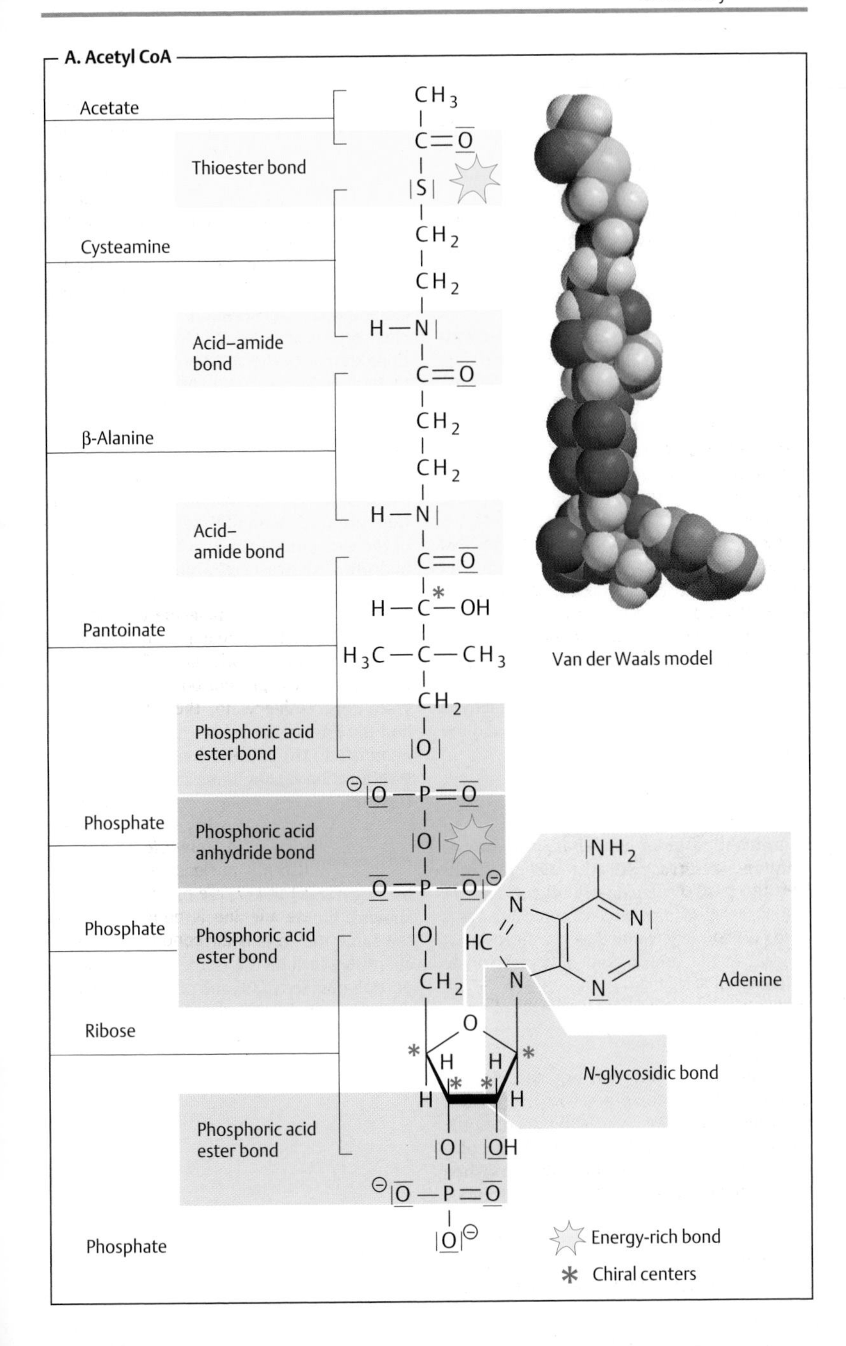
A. Acetyl CoA
Acetate
Thioester bond
Cysteamine
Acid–amide bond
β-Alanine
Acid–amide bond
Pantoinate
Van der Waals model
Phosphoric acid ester bond
Phosphate
Phosphoric acid anhydride bond
Phosphate
Phosphoric acid ester bond
Adenine
Ribose
N-glycosidic bond
Phosphoric acid ester bond
Phosphate
Energy-rich bond
Chiral centers

Chemical reactions

Chemical reactions are processes in which electrons or groups of atoms are taken up into molecules, exchanged between molecules, or shifted within molecules. Illustrated here are the most important types of reaction in organic chemistry, using simple examples. Electron shifts are indicated by red arrows.

A. Redox reactions ◑

In redox reactions (see also p. 32), **electrons** are **transferred** from one molecule (the reducing agent) to another (the oxidizing agent). One or two protons are often also transferred in the process, but the decisive criterion for the presence of a redox reaction is the electron transfer. The reducing agent is oxidized during the reaction, and the oxidizing agent is reduced.

Fig. **A** shows the oxidation of an alcohol into an aldehyde (**1**) and the reduction of the aldehyde to alcohol (**2**). In the process, one *hydride ion* is transferred (two electrons and one proton; see p. 32), which moves to the oxidizing agent A in reaction **1**. The superfluous proton is bound by the catalytic effect of a base B. In the reduction of the aldehyde (**2**), A-H serves as the reducing agent and the acid H-B is involved as the catalyst.

B. Acid–base reactions ◑

In contrast to redox reactions, only **proton transfer** takes place in acid–base reactions (see also p. 30). When an acid dissociates (**1**), water serves as a proton acceptor (i. e., as a base). Conversely, water has the function of an acid in the protonation of a carboxylate anion (**2**).

C. Additions/eliminations ◑

A reaction in which atoms or molecules are taken up by a multiple bond is described as **addition**. The converse of addition—i. e., the removal of groups with the formation of a double bond, is termed **elimination**. When water is added to an alkene (**1a**), a proton is first transferred to the alkene. The unstable carbenium cation that occurs as an intermediate initially takes up water (not shown), before the separation of a proton produces alcohol (**1b**). The elimination of water from the alcohol (**2**, dehydration) is also catalyzed by an acid and passes via the same intermediate as the addition reaction.

D. Nucleophilic substitutions ◑

A reaction in which one functional group (see p. 10) is replaced by another is termed **substitution**. Depending on the process involved, a distinction is made between nucleophilic and electrophilic substitution reactions (see chemistry textbooks). Nucleophilic substitutions start with the addition of one molecule to another, followed by elimination of the so-called *leaving group*.

The hydrolysis of an ester to alcohol and acid (**1**) and the esterification of a carboxylic acid with an alcohol (**2**) are shown here as an example of the S_N2 mechanism. Both reactions are made easier by the marked polarity of the C=O double bond. In the form of ester hydrolysis shown here, a proton is removed from a water molecule by the catalytic effect of the base B. The resulting strongly nucleophilic OH^- ion attacks the positively charged carbonyl C of the ester (**1a**), and an unstable sp^3-hybridized transition state is produced. From this, either water is eliminated (**2b**) and the ester re-forms, or the alcohol ROH is eliminated (**1b**) and the free acid results. In esterification (**2**), the same steps take place in reverse.

Further information

In **rearrangements** (isomerizations, not shown), groups are shifted within one and the same molecule. Examples of this in biochemistry include the isomerization of sugar phosphates (see p. 36) and of methylmalonyl-CoA to succinyl CoA (see p. 166).

A. Redox reactions

Alcohol

Aldehyde

B. Acid–base reactions

Acid

Anion

C. Additions/eliminations

Alkene

Carbonium ion

Alcohol

D. Nucleophilic substitutions

Ester

Transitional state

Alcohol

Carboxylic acid

Energetics

To obtain a better understanding of the processes involved in energy storage and conversion in living cells, it may be useful first to recall the physical basis for these processes.

A. Forms of work ●

There is essentially no difference between work and energy. Both are measured in **joule** (J = 1 N · m). An outdated unit is the **calorie** (1 cal = 4.187 J). **Energy is defined as the ability of a system to perform work.** There are many different forms of energy—e. g., mechanical, chemical, and radiation energy.

A system is capable of performing work when matter is moving along a potential gradient. This abstract definition is best understood by an example involving mechanical work (**A1**). Due to the earth's gravitational pull, the mechanical potential energy of an object is the greater the further the object is away from the center of the earth. A **potential difference** (ΔP) therefore exists between a higher location and a lower one. In a waterfall, the water spontaneously follows this potential gradient and, in doing so, is able to perform work—e. g., turning a mill.

Work and energy consist of two quantities: an **intensity** factor, which is a measure of the potential difference—i. e., the "driving force" of the process—(here it is the height difference) and a **capacity factor**, which is a measure of the quantity of the substance being transported (here it is the weight of the water). In the case of electrical work (**A2**), the intensity factor is the voltage—i. e., the electrical potential difference between the source of the electrical current and the "ground," while the capacity factor is the amount of charge that is flowing.

Chemical work and chemical energy are defined in an analogous way. The intensity factor here is the **chemical potential** of a molecule or combination of molecules. This is stated as **free enthalpy** *G* (also known as "Gibbs free energy"). When molecules spontaneously react with one another, the result is products at lower potential. The difference in the chemical potentials of the educts and products (the **change in free enthalpy, ΔG**) is a measure of the "driving force" of the reaction. The capacity factor in chemical work is the amount of matter reacting (in mol). Although absolute values for free enthalpy G cannot be determined, ΔG can be calculated from the equilibrium constant of the reaction (see p. 18).

B. Energetics and the course of processes ●

Everyday experience shows that water never flows uphill *spontaneously*. Whether a particular process can occur spontaneously or not depends on whether the potential difference between the final and the initial state, $\Delta P = P_2 - P_1$, is positive or negative. If P_2 is smaller than P_1, then ΔP will be negative, and the process will take place and perform work. Processes of this type are called **exergonic** (**B1**). If there is no potential difference, then the system is in **equilibrium** (**B2**). In the case of **endergonic** processes, ΔP is positive (**B3**). Processes of this type do *not* proceed spontaneously.

Forcing endergonic processes to take place requires the use of the principle of **energetic coupling**. This effect can be illustrated by a mechanical analogy (**B4**). When two masses M_1 and M_2 are connected by a rope, M_1 will move upward even though this part of the process is endergonic. The *sum* of the two potential differences ($\Delta P_{eff} = \Delta P_1 + \Delta P_2$) is the determining factor in coupled processes. When ΔP_{eff} is negative, the entire process can proceed.

Energetic coupling makes it possible to convert different forms of work and energy into one another. For example, in a flashlight, an exergonic chemical reaction provides an electrical voltage that can then be used for the endergonic generation of light energy. In the luminescent organs of various animals, it is a chemical reaction that produces the light. In the musculature (see p. 336), chemical energy is converted into mechanical work and heat energy. A form of storage for chemical energy that is used in all forms of life is **adenosine triphosphate** (ATP; see p. 122). Endergonic processes are usually driven by coupling to the strongly exergonic breakdown of ATP (see p. 122).

A. Forms of work

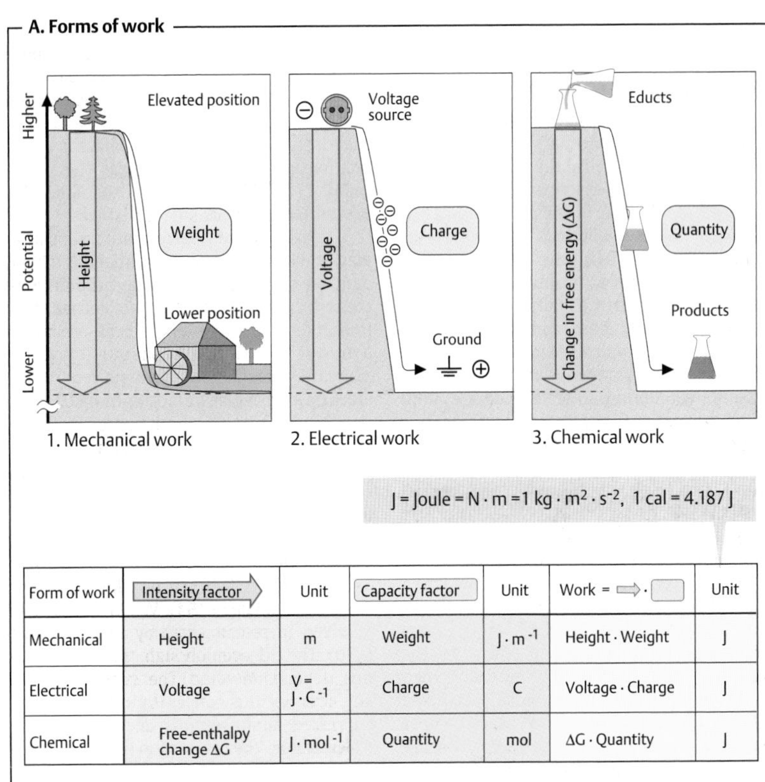

Form of work	Intensity factor	Unit	Capacity factor	Unit	Work = ⇨ · ▭	Unit
Mechanical	Height	m	Weight	$J \cdot m^{-1}$	Height · Weight	J
Electrical	Voltage	V = $J \cdot C^{-1}$	Charge	C	Voltage · Charge	J
Chemical	Free-enthalpy change ΔG	$J \cdot mol^{-1}$	Quantity	mol	ΔG · Quantity	J

B. Energetics and the course of processes

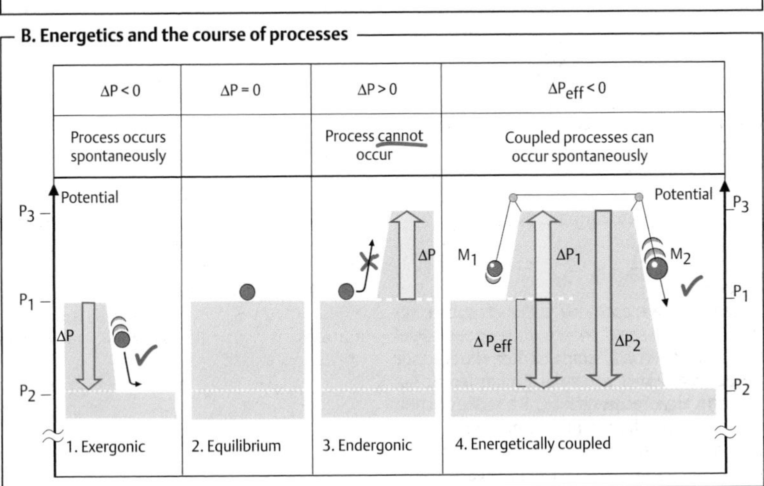

Equilibriums

A. Group transfer reactions ◑

Every chemical reaction reaches after a time a **state of equilibrium** in which the forward and back reactions proceed at the same speed. The **law of mass action** describes the concentrations of the educts (A, B) and products (C, D) *in equilibrium*. The **equilibrium constant K** is directly related to ΔG^0, the change in free enthalpy *G* involved in the reaction (see p. 16) under standard conditions ($\Delta G^0 = -R \cdot T \cdot \ln K$). For any given concentrations, the lower equation applies. At $\Delta G < 0$, the reaction proceeds spontaneously for as long as it takes for equilibrium to be reached (i. e., until $\Delta G = 0$). At $\Delta G > 0$, a *spontaneous* reaction is no longer possible (endergonic case; see p. 16). In biochemistry, ΔG is usually related to pH 7, and this is indicated by the "prime" symbol ($\Delta G^{0\prime}$ or $\Delta G'$).

As examples, we can look at two group transfer reactions (on the right). In ATP (see p. 122), the terminal phosphate residue is at a high chemical potential. Its transfer to water (reaction **a**, below) is therefore strongly **exergonic**. The equilibrium of the reaction ($\Delta G = 0$; see p. 122) is only reached when more than 99.9% of the originally available ATP has been hydrolyzed. ATP and similar compounds have a high **group transfer potential** for phosphate residues. Quantitatively, this is expressed as the **ΔG of hydrolysis** ($\Delta G^{0\prime} = -32\ kJ \cdot mol^{-1}$; see p. 122).

In contrast, the **endergonic** transfer of ammonia (NH_3) to glutamate (Glu, reaction **b**, $\Delta G^{0\prime} = +14\ kJ \cdot mol^{-1}$) reaches equilibrium so quickly that only minimal amounts of the product glutamine (Gln) can be formed in this way. The synthesis of glutamine from these preliminary stages is only possible through **energetic coupling** (see pp. 16, 124).

B. Redox reactions ◑

The course of electron transfer reactions (redox reactions, see p. 14) also follows the law of mass action. For a single redox system (see p. 32), the Nernst equation applies (top). The **electron transfer potential** of a redox system (i. e., its tendency to give off or take up electrons) is given by its **redox potential E** (in standard conditions, E^0 or $E^{0\prime}$). The *lower* the redox potential of a system is, the *higher* the chemical potential of the transferred electrons. To describe reactions between two redox systems, ΔE—the difference between the two systems' redox potentials—is usually used instead of ΔG. ΔG and ΔE have a simple relationship, but opposite signs (below). A redox reaction proceeds spontaneously when $\Delta E > 0$, i. e. $\Delta G < 0$.

The right side of the illustration shows the way in which the redox potential E is dependent on the composition (the proportion of the reduced form as a %) in two biochemically important redox systems (pyruvate/lactate and $NAD^+/NADH+H^+$; see pp. 98, 104). In the standard state (both systems reduced to 50%), electron transfer from lactate to NAD^+ is *not* possible, because ΔE is negative ($\Delta E = -0.13$ V, red arrow). By contrast, transfer can proceed successfully if the pyruvate/lactate system is reduced to 98% and $NAD^+/NADH$ is 98% oxidized (green arrow, $\Delta E = +0.08$ V).

C. Acid–base reactions ◑

Pairs of *conjugated* acids and bases are always involved in proton exchange reactions (see p. 30). The dissociation state of an acid–base pair depends on the H^+ concentration. Usually, it is not this concentration itself that is expressed, but its negative decadic logarithm, the **pH value**. The connection between the pH value and the dissociation state is described by the *Henderson–Hasselbalch equation* (below). As a measure of the **proton transfer potential** of an acid–base pair, its **pK_a value** is used—the negative logarithm of the acid constant K_a (where "a" stands for acid).

The *stronger* an acid is, the *lower* its pK_a value. The acid of the pair with the lower pK_a value (the stronger acid—in this case acetic acid, CH_3COOH) can protonate (green arrow) the base of the pair with the higher pK_a (in this case NH_3), while ammonium acetate (NH_4^+ and CH_3COO^-) only forms very little CH_3COOH and NH_3.

A. Group transfer reactions

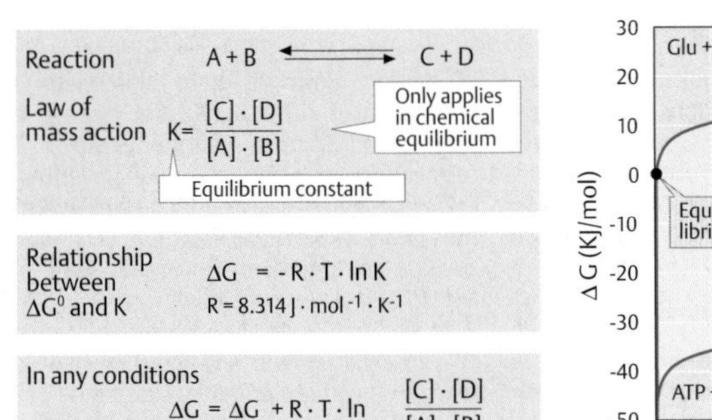

In any conditions

$$\Delta G = \Delta G + R \cdot T \cdot \ln \frac{[C] \cdot [D]}{[A] \cdot [B]}$$

Measure of group transfer potential

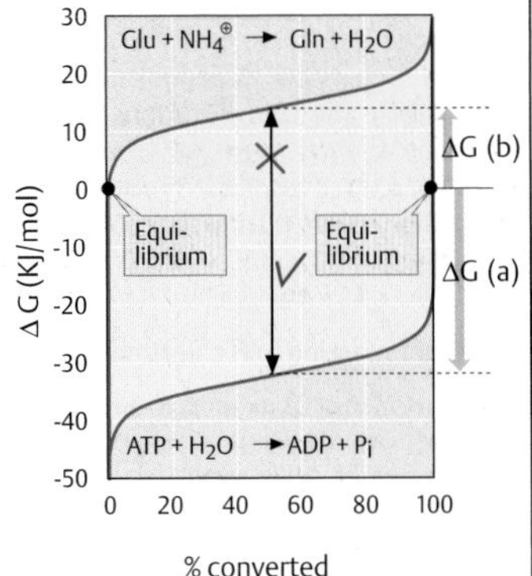

B. Redox reactions

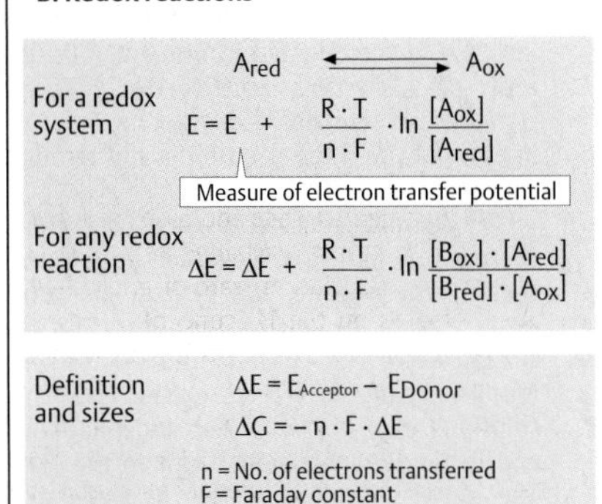

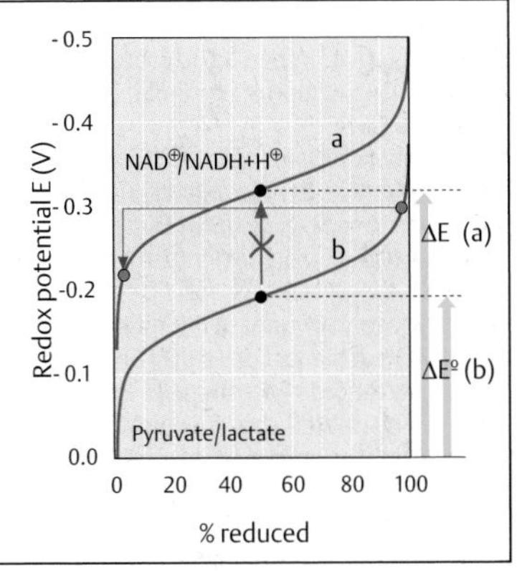

C. Acid–base reactions

Standard reaction

$$HA + H_2O \rightleftharpoons A^{\ominus} + H_3O^{\oplus}$$

Law of mass action

$$K = \frac{[A^{\ominus}] \cdot [H_3O^{\oplus}]}{[HA] \cdot [H_2O]}$$

Simplified

$$K_a = \frac{[A^{\ominus}] \cdot [H^{\oplus}]}{[HA]}$$

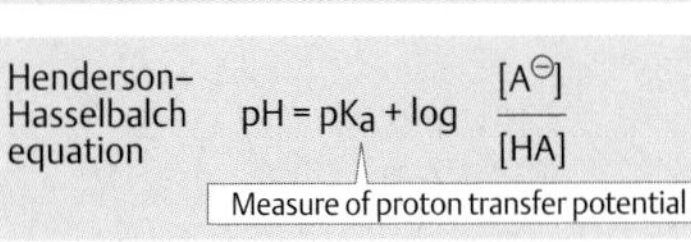

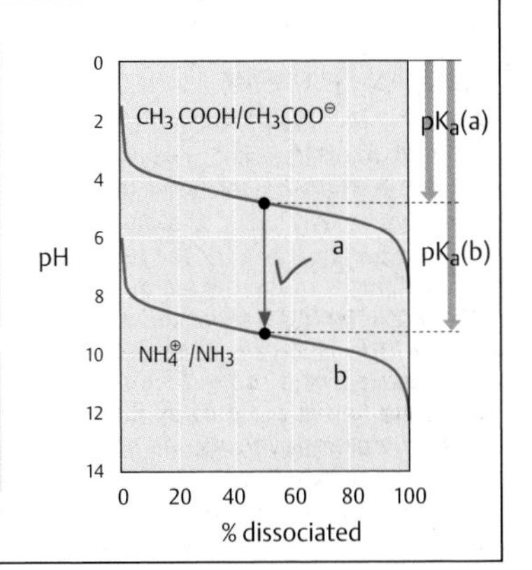

Enthalpy and entropy

The change in the free enthalpy of a chemical reaction (i. e., its ΔG) depends on a number of factors—e. g., the concentrations of the reactants and the temperature (see p. 18). Two further factors associated with molecular changes occurring during the reaction are discussed here.

A. Heat of reaction and calorimetry ◑

All chemical reactions involve heat exchange. Reactions that release heat are called **exothermic**, and those that consume heat are called **endothermic**. Heat exchange is measured as the enthalpy change ΔH (the heat of reaction). This corresponds to the heat exchange at constant pressure. In exothermic reactions, the system *loses* heat, and ΔH is negative. When the reaction is endothermic, the system gains heat, and ΔH becomes positive.

In many reactions, ΔH and ΔG are similar in magnitude (see **B1**, for example). This fact is used to estimate the caloric content of foods. In living organisms, nutrients are usually oxidized by oxygen to CO_2 and H_2O (see p. 112). The maximum amount of chemical work supplied by a particular foodstuff (i. e., the ΔG for the oxidation of the utilizable constituents) can be estimated by burning a weighed amount in a **calorimeter** in an oxygen atmosphere. The heat of the reaction increases the water temperature in the calorimeter. The reaction heat can then be calculated from the temperature difference ΔT.

B. Enthalpy and entropy ◑

The reaction enthalpy ΔH and the change in free enthalpy ΔG are not always of the same magnitude. There are even reactions that occur spontaneously ($\Delta G < 0$) even though they are endothermic ($\Delta H > 0$). The reason for this is that changes in the degree of order of the system also strongly affect the progress of a reaction. This change is measured as the **entropy change (ΔS)**.

Entropy is a physical value that describes the **degree of order of a system**. The *lower* the degree of order, the larger the entropy. Thus, when a process leads to increase in disorder—and everyday experience shows that this is the normal state of affairs—ΔS is positive for this process. An increase in the order in a system ($\Delta S < 0$) always requires an input of energy. Both of these statements are consequences of an important natural law, the Second Law of Thermodynamics. The connection between changes in enthalpy and entropy is described quantitatively by the **Gibbs–Helmholtz equation** ($\Delta G = \Delta H - T \cdot \Delta S$). The following examples will help explain these relationships.

In the *knall-gas* **(oxyhydrogen) reaction** (**1**), gaseous oxygen and gaseous hydrogen react to form liquid water. Like many redox reactions, this reaction is strongly exothermic (i. e., $\Delta H < 0$). However, during the reaction, the degree of order increases. The total number of molecules is reduced by one-third, and a more highly ordered liquid is formed from freely moving gas molecules. As a result of the increase in the degree of order ($\Delta S < 0$), the term $-T \cdot \Delta S$ becomes positive. However, this is more than compensated for by the decrease in enthalpy, and the reaction is still strongly exergonic ($\Delta G < 0$).

The **dissolution of salt in water** (**2**) is endothermic ($\Delta H > 0$)—i. e., the liquid cools. Nevertheless, the process still occurs spontaneously, since the degree of order in the system *decreases*. The Na^+ and Cl^- ions are initially rigidly fixed in a crystal lattice. In solution, they move about independently and in random directions through the fluid. The decrease in order ($\Delta S > 0$) leads to a negative $-T \cdot \Delta S$ term, which compensates for the positive ΔH term and results in a negative ΔG term overall. Processes of this type are described as being **entropy-driven**. The folding of proteins (see p. 74) and the formation of ordered lipid structures in water (see p. 28) are also mainly entropy-driven.

A. Heat of reaction and calorimetry

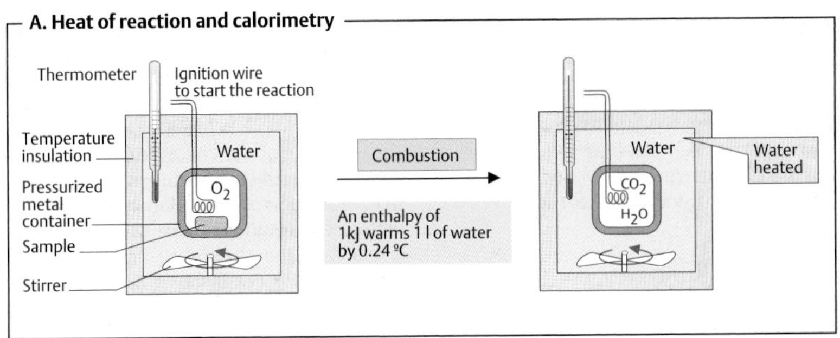

B. Enthalpy and entropy

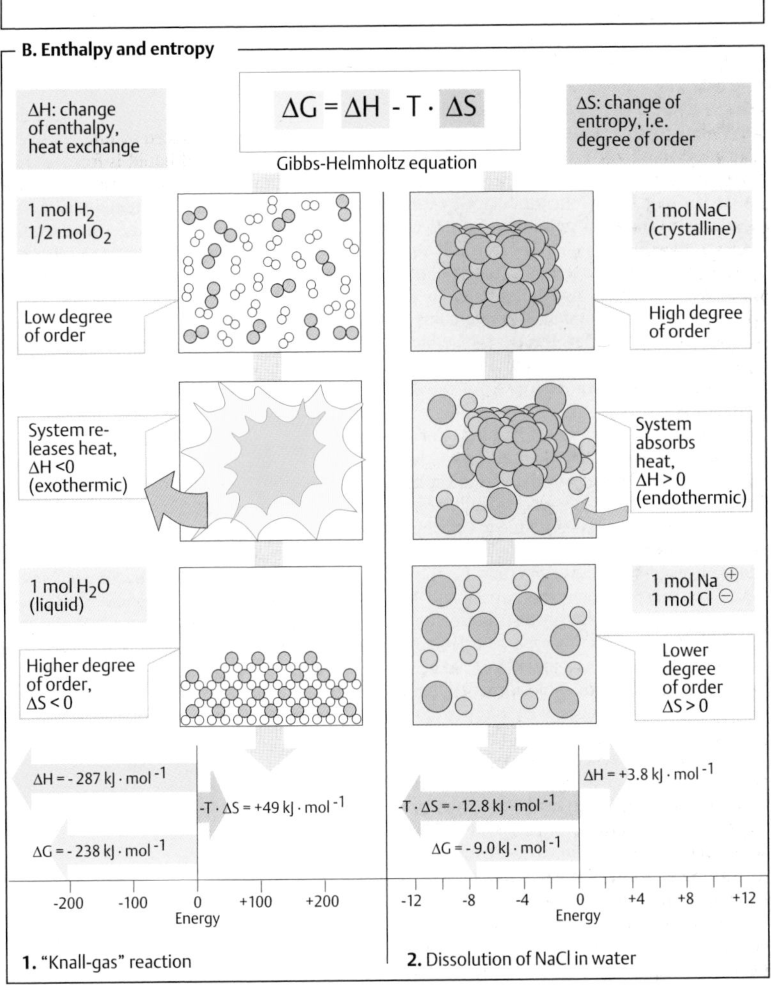

1. "Knall-gas" reaction

2. Dissolution of NaCl in water

Reaction kinetics

The change in free enthalpy ΔG in a reaction indicates whether or not the reaction can take place spontaneously in given conditions and how much work it can perform (see p. 18). However, it does not tell us anything about the *rate* of the reaction—i. e., its **kinetics**.

A. Activation energy ◑

Most organic chemical reactions (with the exception of acid–base reactions) proceed only very slowly, regardless of the value of ΔG. The reason for the slow reaction rate is that the molecules that react—the educts—have to have a certain minimum energy before they can enter the reaction. This is best understood with the help of an energy diagram (**1**) of the simplest possible reaction A → B. The educt A and the product B are each at a specific **chemical potential** (G_e and G_p, respectively). The change in the free enthalpy of the reaction, ΔG, corresponds to the difference between these two potentials. To be converted into B, A first has to overcome a potential energy barrier, the peak of which, G_a, lies well above G_e. The potential difference $G_a - G_e$ is the **activation energy E_a** of the reaction (in $kJ \cdot mol^{-1}$).

The fact that A can be converted into B at all is because the potential G_e only represents the average potential of all the molecules. Individual molecules may occasionally reach much higher potentials—e. g., due to collisions with other molecules. When the increase in energy thus gained is greater than E_a, these molecules can overcome the barrier and be converted into B. The energy distribution for a group of molecules of this type, as calculated from a simple model, is shown in (**2**) and (**3**). $\Delta n/n$ is the fraction of molecules that have reached or exceeded energy E (in kJ per mol). At 27 °C, for example, approximately 10% of the molecules have energies > 6 $kJ \cdot mol^{-1}$. The typical activation energies of chemical reactions are much higher. The course of the energy function at energies of around 50 $kJ \cdot mol^{-1}$ is shown in (**3**). Statistically, at 27 °C only two out of 10^9 molecules reach this energy. At 37 °C, the figure is already four. This is the basis for the long-familiar "Q_{10} law"—a rule of thumb that states that the speed of biological processes approximately doubles with an increase in temperature of 10 °C.

B. Reaction rate ◑

The velocity v of a chemical reaction is determined experimentally by observing the change in the concentration of an educt or product over time. In the example shown (again a reaction of the A → B type), 3 mmol of the educt A is converted per second and 3 mmol of the product B is formed per second in one liter of the solution. This corresponds to a rate of

$$v = 3\ mM \cdot s^{-1} = 3 \cdot 10^{-3}\ mol \cdot L^{-1} \cdot s^{-1}$$

C. Reaction order ◑

Reaction rates are influenced not only by the activation energy and the temperature, but also by the concentrations of the reactants. When there is only one educt, A (**1**), v is proportional to the concentration [A] of this substance, and a **first-order reaction** is involved. When *two* educts, A and B, react with one another (**2**), it is a **second order reaction** (shown on the right). In this case, the rate v is proportional to the *product* of the educt concentrations (12 mM^2 at the top, 24 mM^2 in the middle, and 36 mM^2 at the bottom). The proportionality factors k and k′ are the **rate constants** of the reaction. They are *not* dependent on the reaction concentrations, but depend on the external conditions for the reaction, such as temperature.

In **B**, only the kinetics of simple irreversible reactions is shown. More complicated cases, such as reaction with three or more reversible steps, can usually be broken down into first-order or second-order partial reactions and described using the corresponding equations (for an example, see the Michaelis–Menten reaction, p. 92).

A. Activation energy

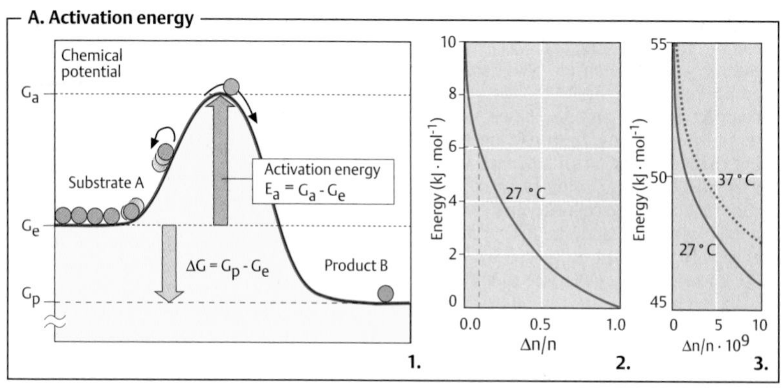

B. Reaction rate

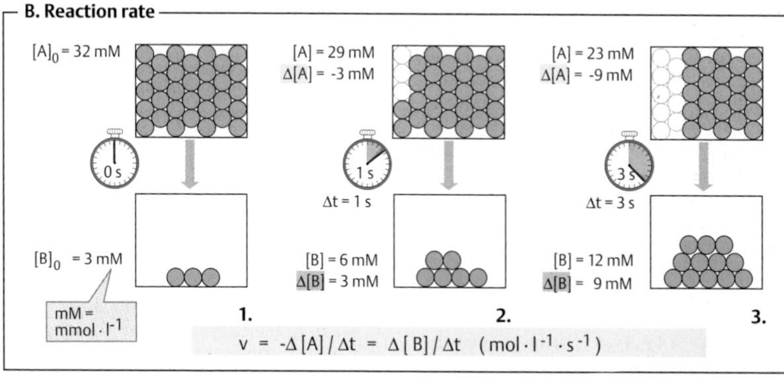

C. Reaction order

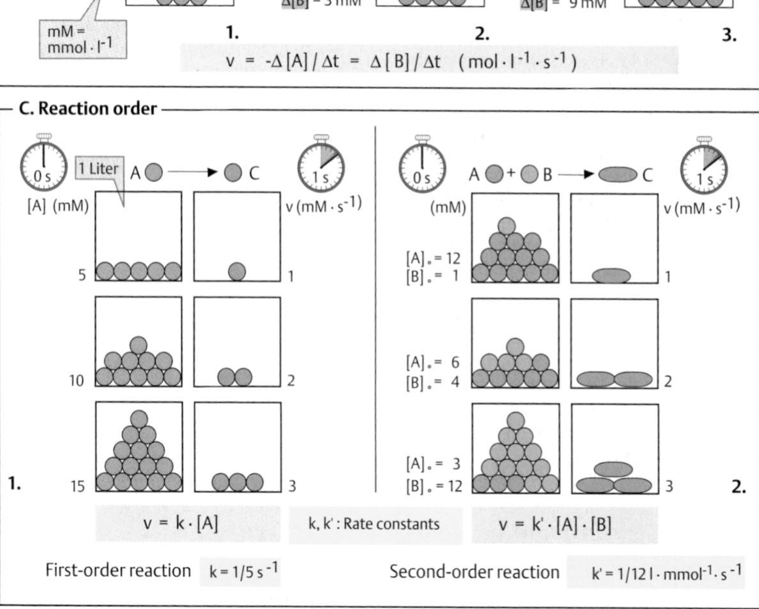

Catalysis

Catalysts are substances that accelerate chemical reactions without themselves being consumed in the process. Since catalysts emerge from the catalyzed reaction without being changed, even small amounts are usually sufficient to cause a powerful acceleration of the reaction. In the cell, **enzymes** (see p. 88) generally serve as catalysts. A few chemical changes are catalyzed by special RNA molecules, known as *ribozymes* (see p. 246).

A. Catalysis: principle ●

The reason for the slow rates of most reactions involving organic substances is the high **activation energy** (see p. 22) that the reacting molecules have to reach before they can react. In aqueous solution, a large proportion of the activation energy is required to remove the hydration shells surrounding the educts. During the course of a reaction, resonance-stabilized structures (see p. 4) are often temporarily suspended; this also requires energy. The highest point on the reaction coordinates corresponds to an energetically unfavorable **transition state** of this type (**1**).

A catalyst creates a new pathway for the reaction (**2**). When all of the transition states arising have a lower activation energy than that of the uncatalyzed reaction, the reaction will proceed more rapidly along the alternative pathway, even when the number of intermediates is greater. Since the starting points and end points are the same in both routes, the change in the enthalpy ΔG of the reaction is not influenced by the catalyst. Catalysts—including enzymes—are in principle *not* capable of altering the equilibrium state of the catalyzed reaction.

The often-heard statement that "a catalyst reduces the activation energy of a reaction" is not strictly correct, since a *completely different* reaction takes place in the presence of a catalyst than in uncatalyzed conditions. However, its activation energy is lower than in the uncatalyzed reaction.

B. Catalysis of H_2O_2 – breakdown by iodide ○

As a simple example of a catalyzed reaction, we can look at the disproportionation of hydrogen peroxide (H_2O_2) into oxygen and water. In the uncatalyzed reaction (at the top), an H_2O_2 molecule initially decays into H_2O and atomic oxygen (O), which then reacts with a second H_2O_2 molecule to form water and molecular oxygen (O_2). The activation energy E_a required for this reaction is relatively high, at 75 kJ · mol^{-1}. In the presence of **iodide** (I^-) as a catalyst, the reaction takes a different course (bottom). The intermediate arising in this case is hypoiodide (OI^-), which also forms H_2O and O_2 with another H_2O_2 molecule. In this step, the I^- ion is released and can once again take part in the reaction. The lower activation energy of the reaction catalyzed by iodide (E_a = 56 kJ · mol^{-1}) causes acceleration of the reaction by a factor of 2000, as the reaction rate depends exponentially on E_a ($v \sim e^{-Ea/R \cdot T}$).

Free metal ions such as iron (Fe) and platinum (Pt) are also effective catalysts for the breakdown of H_2O_2. **Catalase** (see p. 284), an enzyme that protects cells against the toxic effects of hydrogen peroxide (see p. 284), is much more catalytically effective still. In the enzyme-catalyzed disproportionation, H_2O_2 is bound to the enzyme's heme group, where it is quickly converted to atomic oxygen and water, supported by amino acid residues of the enzyme protein. The oxygen atom is temporarily bound to the central iron atom of the heme group, and then transferred from there to the second H_2O_2 molecule. The activation energy of the enzyme-catalyzed reaction is only 23 kJ · mol^{-1}, which in comparison with the uncatalyzed reaction leads to acceleration by a factor of $1.3 \cdot 10^9$.

Catalase is one of the most efficient enzymes there are. A single molecule can convert up to 10^8 (a hundred million) H_2O_2 molecules per second.

A. Catalysis: principle

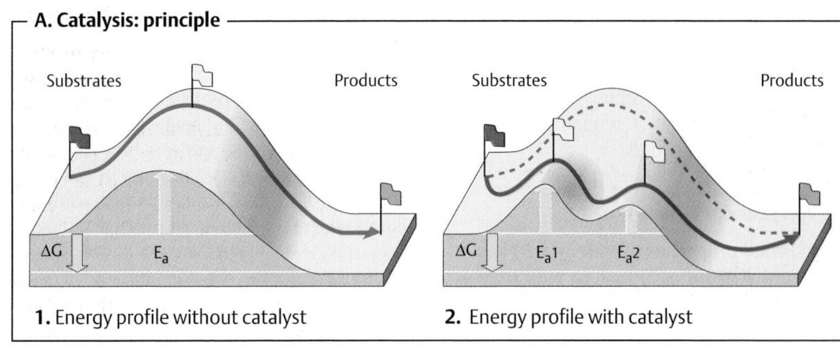

1. Energy profile without catalyst

2. Energy profile with catalyst

B. Catalysis of H_2O_2 – breakdown by iodide

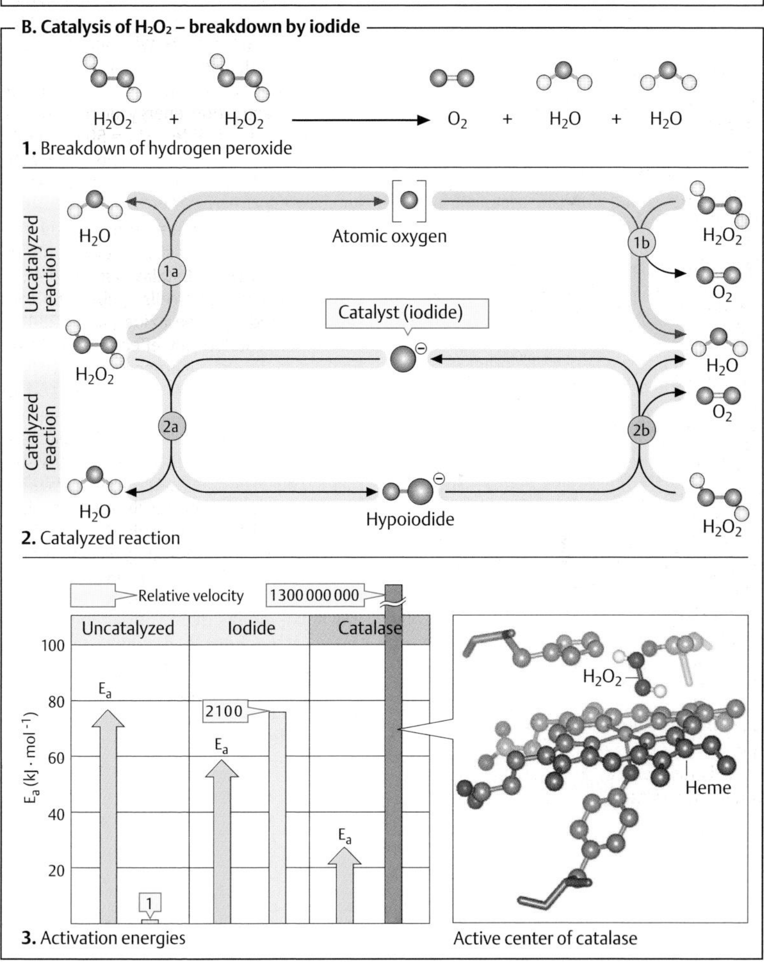

1. Breakdown of hydrogen peroxide

2. Catalyzed reaction

3. Activation energies

Active center of catalase

Water as a solvent

Life as we know it evolved in water and is still absolutely dependent on it. The properties of water are therefore of fundamental importance to all living things.

A. Water and methane ◑

The special properties of **water (H_2O)** become apparent when it is compared with **methane (CH_4)**. The two molecules have a similar mass and size. Nevertheless, the boiling point of water is more than 250 °C above that of methane. At temperatures on the earth's surface, water is liquid, whereas methane is gaseous. The high boiling point of water results from its high vaporization enthalpy, which in turn is due to the fact that the density of the electrons within the molecule is unevenly distributed. Two corners of the tetrahedrally-shaped water molecule are occupied by unshared electrons (green), and the other two by hydrogen atoms. As a result, the H–O–H bond has an angled shape. In addition, the O–H bonds are polarized due to the high electronegativity of oxygen (see p. 6). One side of the molecule carries a partial charge (δ) of about –0.6 units, whereas the other is correspondingly positively charged. The spatial separation of the positive and negative charges gives the molecule the properties of an **electrical dipole**. Water molecules are therefore attracted to one another like tiny magnets, and are also connected by hydrogen bonds (**B**) (see p. 6). When liquid water vaporizes, a large amount of energy has to be expended to disrupt these interactions. By contrast, methane molecules are not dipolar, and therefore interact with one another only weakly. This is why liquid methane vaporizes at very low temperatures.

B. Structure of water and ice ◑

The dipolar nature of water molecules favors the formation of **hydrogen bonds** (see p. 6). Each molecule can act either as a donor or an acceptor of H bonds, and many molecules in liquid water are therefore connected by H bonds (**1**). The bonds are in a state of constant fluctuation. Tetrahedral networks of molecules, known as water "clusters," often arise. As the temperature decreases, the proportion of water clusters increases until the water begins to crystallize. Under normal atmospheric pressure, this occurs at 0 °C. In **ice**, most of the water molecules are fixed in a **hexagonal lattice** (**3**). Since the distance between the individual molecules in the frozen state is on average greater than in the liquid state, the density of ice is lower than that of liquid water. This fact is of immense biological importance—it means, for example, that in winter, ice forms on the surface of open stretches of water first, and the water rarely freezes to the bottom.

C. Hydration ◑

In contrast to most other liquids, water is an excellent **solvent for ions.** In the electrical field of cations and anions, the dipolar water molecules arrange themselves in a regular fashion corresponding to the charge of the ion. They form **hydration shells** and shield the central ion from oppositely charged ions. Metal ions are therefore often present as hexahydrates ($[Me(H_2O)_6{}^{2+}]$, on the right). In the inner hydration sphere of this type of ion, the water molecules are practically immobilized and follow the central ion. Water has a high dielectric constant of 78—i. e., the electrostatic attraction force between ions is reduced to 1/78 by the solvent. Electrically charged groups in organic molecules (e. g., carboxylate, phosphate, and ammonium groups) are also well hydrated and contribute to water solubility. Neutral molecules with several hydroxy groups, such as glycerol (on the left) or sugars, are also easily soluble, because they can form H bonds with water molecules. The higher the proportion of polar functional groups there is in a molecule, the more water-soluble (**hydrophilic**) it is. By contrast, molecules that consist exclusively or mainly of hydrocarbons are poorly soluble or insoluble in water. These compounds are called **hydrophobic** (see p. 28).

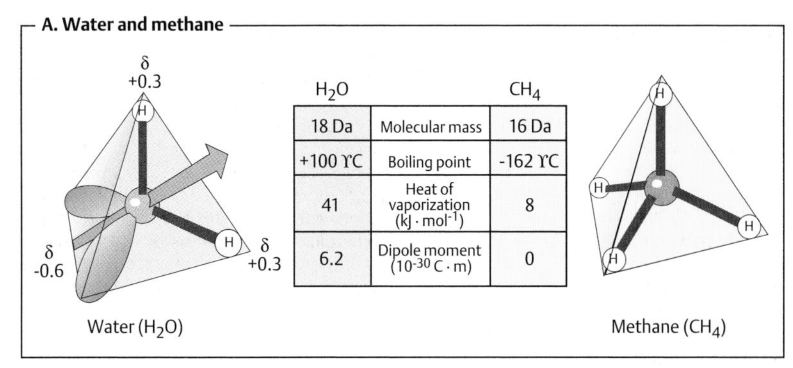

H_2O		CH_4
18 Da	Molecular mass	16 Da
+100 ϒC	Boiling point	-162 ϒC
41	Heat of vaporization ($kJ \cdot mol^{-1}$)	8
6.2	Dipole moment (10^{-30} C · m)	0

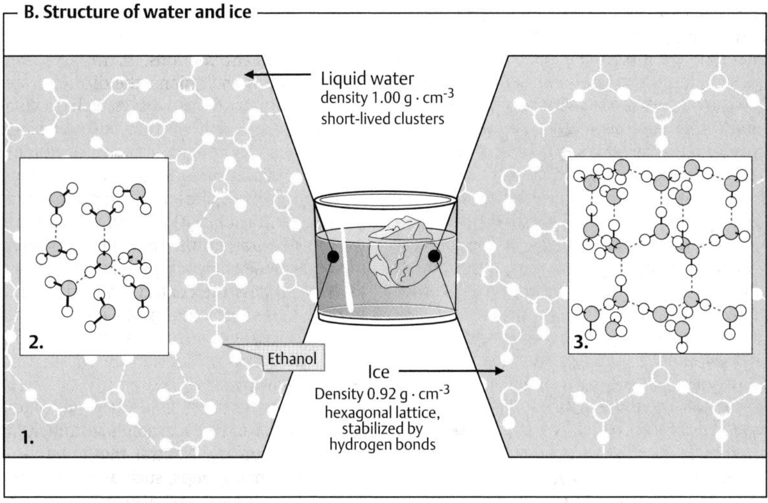

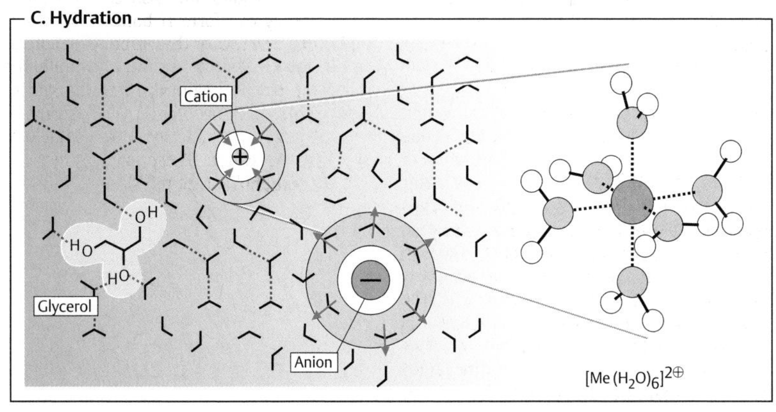

Hydrophobic interactions

Water is an excellent solvent for ions and for substances that contain polarized bonds (see p. 20). Substances of this type are referred to as **polar** or **hydrophilic** ("water-loving"). In contrast, substances that consist mainly of hydrocarbon structures dissolve only poorly in water. Such substances are said to be **apolar** or **hydrophobic**.

A. Solubility of methane ○

To understand the reasons for the poor water solubility of hydrocarbons, it is useful first to examine the energetics (see p. 16) of the processes involved. In (**1**), the individual terms of the Gibbs–Helmholtz equation (see p. 20) for the simplest compound of this type, **methane**, are shown (see p. 4). As can be seen, the transition from gaseous methane to water is actually exothermic ($\Delta H^0 < 0$). Nevertheless, the change in the free enthalpy ΔG^0 is positive (the process is endergonic), because the entropy term $T \cdot \Delta S^0$ has a strongly positive value. The entropy change in the process (ΔS^0) is evidently negative—i. e., a solution of methane in water has a *higher* degree of order than either water or gaseous methane. One reason for this is that the methane molecules are less mobile when surrounded by water. More importantly, however, the water around the apolar molecules forms cage-like **"clathrate" structures**, which—as in ice—are stabilized by H bonds. This strongly increases the degree of order in the water—and the more so the larger the area of surface contact between the water and the apolar phase.

B. The "oil drop effect" ◑

The spontaneous separation of oil and water, a familiar observation in everyday life, is due to the energetically unfavorable formation of clathrate structures. When a mixture of water and oil is firmly shaken, lots of tiny oil drops form to begin with, but these quickly coalesce spontaneously to form larger drops—the two phases separate. A larger drop has a smaller surface area than several small drops with the same volume. Separation therefore reduces the area of surface contact between the water and the oil, and consequently also the extent of clathrate formation. The ΔS for this process is therefore positive (the *dis*order in the water increases), and the negative term $-T \cdot \Delta S$ makes the separation process exergonic ($\Delta G < 0$), so that it proceeds spontaneously.

C. Arrangements of amphipathic substances in water ◑

Molecules that contain both polar *and* apolar groups are called **amphipathic** or amphiphilic. This group includes soaps (see p. 48), phospholipids (see p. 50), and bile acids (see p. 56).

As a result of the "oil drop effect" amphipathic substances in water tend to arrange themselves in such a way as to minimize the area of surface contact between the apolar regions of the molecule and water. On water surfaces, they usually form single-layer **films** (top) in which the polar "head groups" face toward the water. **Soap bubbles** (right) consist of double films, with a thin layer of water enclosed between them. In water, depending on their concentration, amphipathic compounds form **micelles**—i. e., spherical aggregates with their head groups facing toward the outside, or extended bilayered **double membranes**. Most biological membranes are assembled according to this principle (see p. 214). Closed hollow membrane sacs are known as **vesicles**. This type of structure serves to transport substances within cells and in the blood (see p. 278).

The separation of oil and water (**B**) can be prevented by adding a strongly amphipathic substance. During shaking, a more or less stable **emulsion** then forms, in which the surface of the oil drops is occupied by amphipathic molecules that provide it with polar properties externally. The emulsification of fats in food by bile acids and phospholipids is a vital precondition for the digestion of fats (see p. 314).

A. Solubility of methane

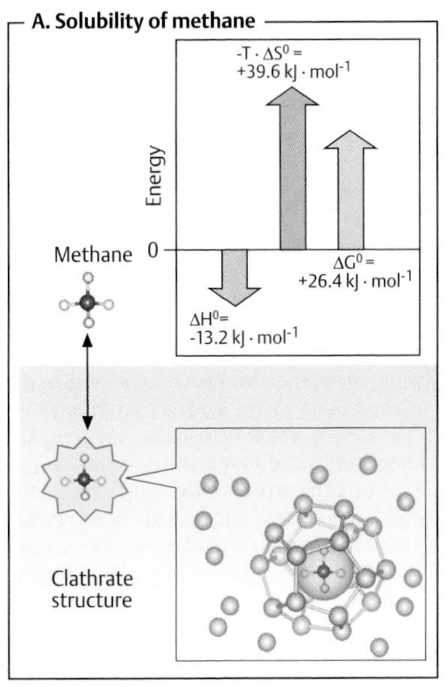

B. The "oil drop effect"

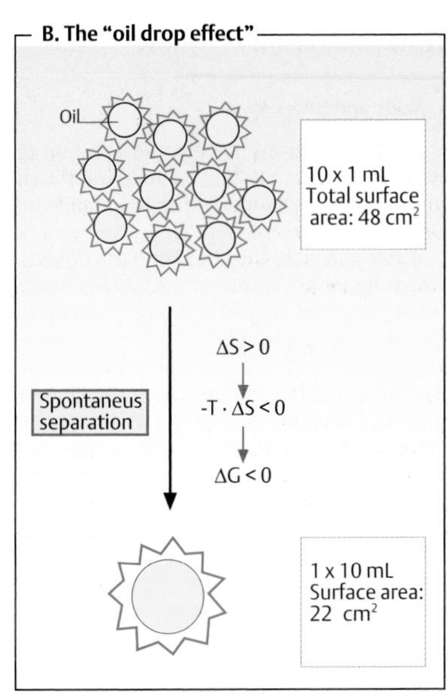

C. Arrangements of amphipathic substances in water

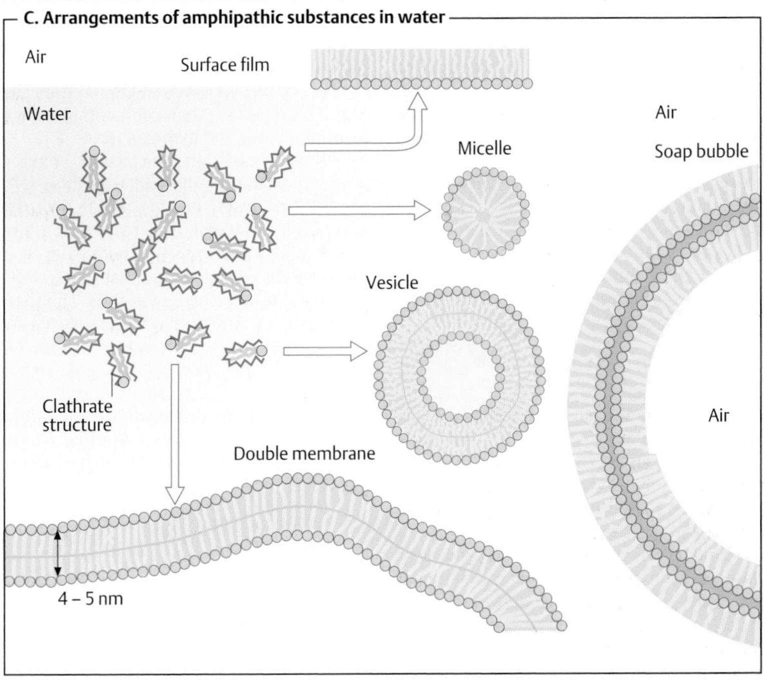

Acids and bases

A. Acids and bases ●

In general, **acids** are defined as substances that can donate hydrogen ions (protons), while **bases** are compounds that accept protons.

Water enhances the acidic or basic properties of dissolved substances, as water itself can act as either an acid or a base. For example, when **hydrogen chloride** (HCl) is in aqueous solution, it donates protons to the solvent (**1**). This results in the formation of chloride ions (Cl^-) and protonated water molecules (**hydronium ions, H_3O+**, usually simply referred to as H^+). The proton exchange between HCl and water is virtually quantitative: in water, HCl behaves as a *very strong acid* with a negative pK_a value (see p. 18).

Bases such as **ammonia** (NH_3) take over protons from water molecules. As a result of this, **hydroxyl ions** (OH^-) and positively charged ammonium ions (NH_4^+, **3**) form. Hydronium and hydroxyl ions, like other ions, exist in water in hydrated rather than free form (see p. 26).

Acid–base reactions always involve *pairs* of **acids** and the associated **conjugated bases** (see p. 18). The stronger the acid or base, the *weaker* the conjugate base or acid, respectively. For example, the very strongly acidic hydrogen chloride belongs to the very weakly basic chloride ion (**1**). The weakly acidic ammonium ion is conjugated with the moderately strong base ammonia (**3**).

The equilibrium constant K for the acid–base reaction between H_2O molecules (**2**) is very small. At 25 °C,

$$K = [H^+] \cdot [OH^-] / [H_2O] = 2 \cdot 10^{-16}\ mol \cdot L^{-1}$$

In pure water, the concentration $[H_2O]$ is practically constant at 55 $mol \cdot L^{-1}$. Substituting this value into the equation, it gives:

$$K_w = [H^+] \cdot [OH^-] = 1 \cdot 10^{-14}\ mol \cdot L^{-1}$$

The product $[H^+] \cdot [OH^-]$—the **ion product** of water—is constant even when additional acid–base pairs are dissolved in the water. At 25 °C, pure water contains H^+ and OH^- at concentrations of $1 \cdot 10^{-7}\ mol \cdot L^{-1}$ each; it is **neutral** and has a pH value of exactly 7.

B. pH values in the organism ◑

pH values in the cell and in the extracellular fluid are kept constant within narrow limits. In the blood, the pH value normally ranges only between 7.35 and 7.45 (see p. 288). This corresponds to a maximum change in the H^+ concentration of ca. 30%. The pH value of cytoplasm is slightly lower than that of blood, at 7.0–7.3. In lysosomes (see p. 234; pH 4.5–5.5), the H^+ concentration is several hundred times higher than in the cytoplasm. In the lumen of the gastrointestinal tract, which forms part of the outside world relative to the organism, and in the body's excretion products, the pH values are more variable. Extreme values are found in the stomach (ca. 2) and in the small bowel (> 8). Since the kidney can excrete either acids or bases, depending on the state of the metabolism, the pH of urine has a particularly wide range of variation (4.8–7.5).

C. Buffers ●

Short-term pH changes in the organism are cushioned by **buffer systems**. These are mixtures of a weak acid, HB, with its conjugate base, B^-, or of a weak base with its conjugate acid. This type of system can neutralize both hydronium ions and hydroxyl ions.

In the first case (left), the base (B^-) binds a large proportion of the added protons (H^+) and HB and water are formed. If hydroxyl ions (OH^-) are added, they react with HB to give B^- and water (right). In both cases, it is primarily the $[HB]/[B^-]$ ratio that shifts, while the pH value only changes slightly. The **titration curve** (top) shows that buffer systems are most effective at the pH values that correspond to the pK_a value of the acid. This is where the curve is at its steepest, so that the pH change, ΔpH, is at its smallest with a given increase Δc in $[H^+]$ or $[OH^-]$. In other words, the **buffer capacity** Δc/ ΔpH is highest at the pK_a value.

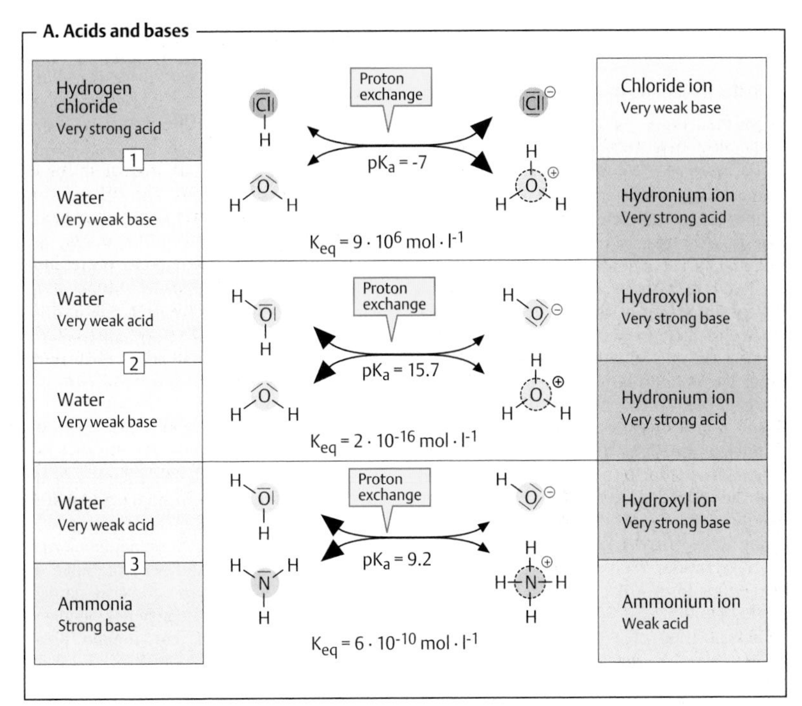
A. Acids and bases
Hydrogen chloride
Very strong acid
1
Water
Very weak base
Proton exchange
pKa = -7
Keq = 9 · 10^6 mol · l^-1
Chloride ion
Very weak base
Hydronium ion
Very strong acid
Water
Very weak acid
2
Water
Very weak base
Proton exchange
pKa = 15.7
Keq = 2 · 10^-16 mol · l^-1
Hydroxyl ion
Very strong base
Hydronium ion
Very strong acid
Water
Very weak acid
3
Ammonia
Strong base
Proton exchange
pKa = 9.2
Keq = 6 · 10^-10 mol · l^-1
Hydroxyl ion
Very strong base
Ammonium ion
Weak acid

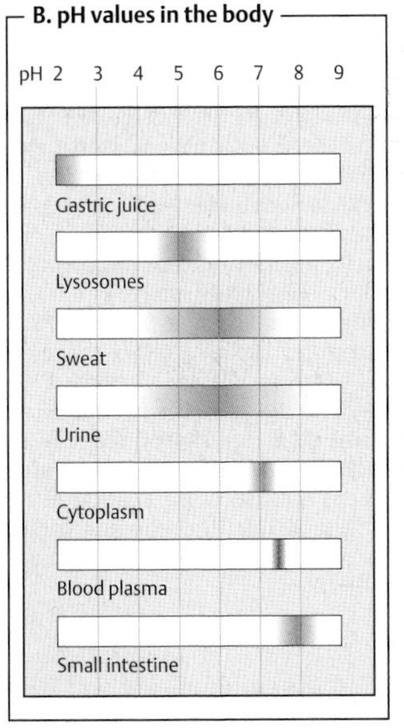
B. pH values in the body
pH 2 3 4 5 6 7 8 9
Gastric juice
Lysosomes
Sweat
Urine
Cytoplasm
Blood plasma
Small intestine

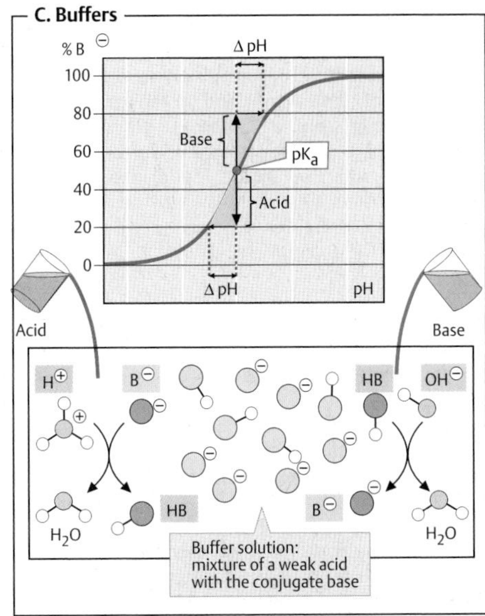
C. Buffers
% B
100
80
60
40
20
0
Δ pH
Base
pKa
Acid
Δ pH
pH
Acid
Base
H⊕
B⊖
HB
OH⊖
H2O
HB
B⊖
H2O
Buffer solution:
mixture of a weak acid
with the conjugate base

Redox processes

A. Redox reactions ●

Redox reactions are chemical changes in which electrons are transferred from one reaction partner to another (**1**; see also p. 18). Like acid–base reactions (see p. 30), redox reactions always involve *pairs* of compounds. A pair of this type is referred to as a **redox system** (**2**). The essential difference between the two components of a redox system is the number of electrons they contain. The more electronrich component is called the **reduced form** of the compound concerned, while the other one is referred to as the **oxidized form**. The reduced form of one system (the **reducing agent**) donates electrons to the oxidized form of another one (the **oxidizing agent**). In the process, the reducing agent becomes oxidized and the oxidizing agent is reduced (**3**). Any given reducing agent can reduce only certain other redox systems. On the basis of this type of observation, redox systems can be arranged to form what are known as **redox series** (**4**).

The position of a system within one of these series is established by its **redox potential E** (see p. 18). The redox potential has a sign; it can be more negative or more positive than a reference potential arbitrarily set at zero (the normal potential of the system [$2\ H^+/H_2$]). In addition, E depends on the concentrations of the reactants and on the reaction conditions (see p. 18). In redox series (**4**), the systems are arranged according to their increasing redox potentials. Spontaneous electron transfers are only possible if the redox potential of the donor is *more negative* than that of the acceptor (see p. 18).

B. Reduction equivalents ◑

In redox reactions, protons (H^+) are often transferred along with electrons (e^-), or protons may be released. The combinations of electrons and protons that occur in redox processes are summed up in the term **reduction equivalents**. For example, the combination $1\ e^-/1\ H^+$ corresponds to a hydrogen *atom*, while $2\ e^-$ and $2\ H^+$ together produce a hydrogen *molecule*. However, this does not mean that atomic or molecular hydrogen is actually transferred from one molecule to the other (see below). Only the combination $2\ e^-/1\ H^+$, the **hydride ion**, is transferred as a unit.

C. Biological redox systems ◑

In the cell, redox reactions are catalyzed by enzymes, which work together with soluble or bound redox cofactors.

Some of these factors contain **metal ions** as redox-active components. In these cases, it is usually single electrons that are transferred, with the metal ion changing its valency. Unpaired electrons often occur in this process, but these are located in d orbitals (see p. 2) and are therefore less dangerous than single electrons in non-metal atoms ("free radicals"; see below).

We can only show here a few examples from the many organic redox systems that are found. In the complete reduction of the **flavin coenzymes** FMN and FAD (see p. 104), $2\ e^-$ and $2\ H^+$ are transferred. This occurs in two separate steps, with a *semiquinone radical* appearing as an intermediate. Since organic radicals of this type can cause damage to biomolecules, flavin coenzymes never occur freely in solution, but remain firmly bound in the interior of proteins.

In the reduction or oxidation of **quinone/quinol systems**, free radicals also appear as intermediate steps, but these are less reactive than flavin radicals. Vitamin E, another quinone-type redox system (see p. 104), even functions as a radical scavenger, by delocalizing unpaired electrons so effectively that they can no longer react with other molecules.

The **pyridine nucleotides** NAD^+ and $NADP^+$ always function in unbound form. The oxidized forms contain an aromatic nicotinamide ring in which the positive charge is delocalized. The right-hand example of the two *resonance structures* shown contains an electron-poor, positively charged C atom at the *para* position to nitrogen. If a **hydride ion** is added at this point (see above), the reduced forms NADH or NADPH arise. No radical intermediate steps occur. Because a proton is released at the same time, the reduced pyridine nucleotide coenzymes are correctly expressed as $NAD(P)H{+}H^+$.

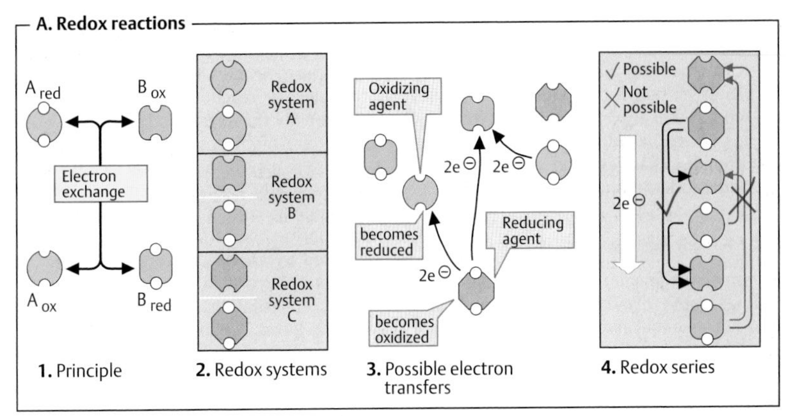
A. Redox reactions
A red
B ox
Electron exchange
A ox
B red
Redox system A
Redox system B
Redox system C
Oxidizing agent
becomes reduced
Reducing agent
becomes oxidized
2e⊖
Possible
Not possible
1. Principle
2. Redox systems
3. Possible electron transfers
4. Redox series

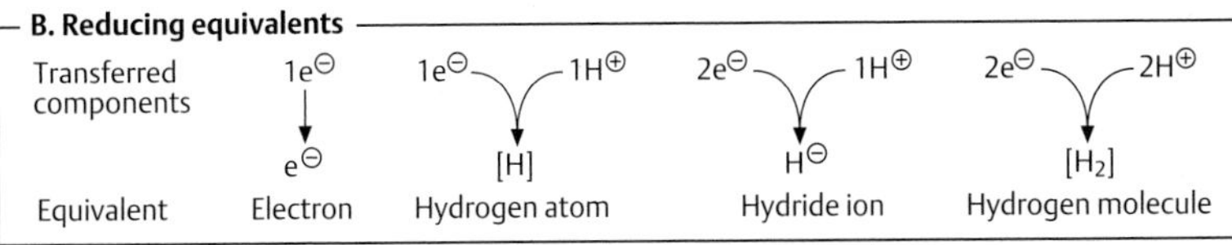
B. Reducing equivalents
Transferred components
1e⊖
1H⊕
2e⊖
2H⊕
e⊖
[H]
H⊖
[H2]
Equivalent
Electron
Hydrogen atom
Hydride ion
Hydrogen molecule

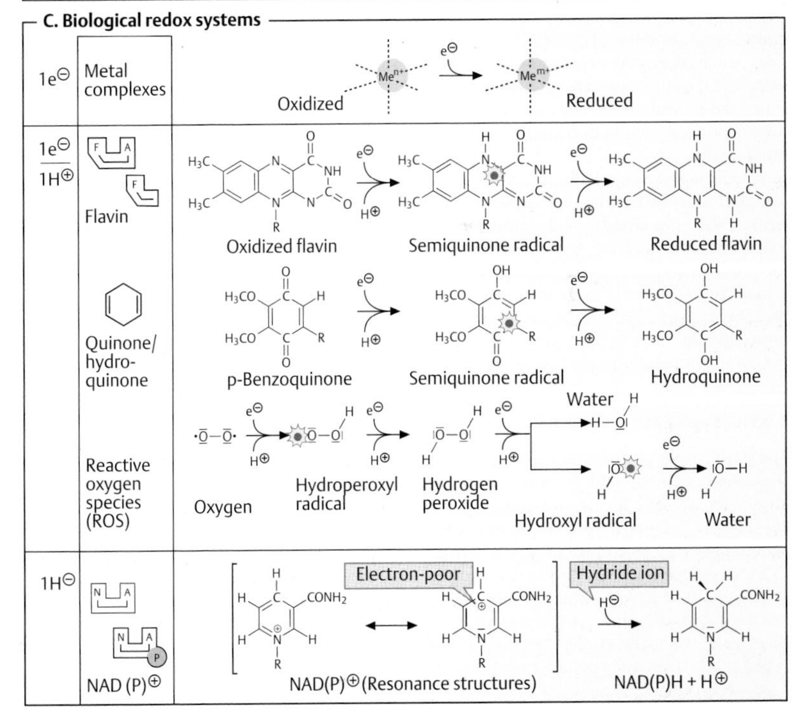
C. Biological redox systems
1e⊖
Metal complexes
Oxidized
Reduced
1e⊖ / 1H⊕
Flavin
Oxidized flavin
Semiquinone radical
Reduced flavin
Quinone/ hydro- quinone
p-Benzoquinone
Semiquinone radical
Hydroquinone
Reactive oxygen species (ROS)
Oxygen
Hydroperoxyl radical
Hydrogen peroxide
Water
Hydroxyl radical
Water
1H⊖
NAD (P)⊕
Electron-poor
Hydride ion
NAD(P)⊕ (Resonance structures)
NAD(P)H + H⊕

Overview

The **carbohydrates** are a group of naturally occurring carbonyl compounds (aldehydes or ketones) that also contain several hydroxyl groups. The carbohydrates include **single sugars (monosaccharides)** and their polymers, the **oligosaccharides** and **polysaccharides**.

A. Carbohydrates: overview ●

Polymeric carbohydrates–above all starch, as well as some disaccharides–are important (but not essential) **components of food** (see p. 360). In the gut, they are broken down into monosaccharides and resorbed in this form (see p. 272). The form in which carbohydrates are distributed by the blood of vertebrates is *glucose* ("blood sugar"). This is taken up by the cells and either broken down to obtain energy (glycolysis) or converted into other metabolites (see pp. 150–159). Several organs (particularly the liver and muscles) store *glycogen* as a polymeric **reserve carbohydrate** (right; see p. 156). The glycogen molecules are covalently bound to a protein, *glycogenin*. Polysaccharides are used by many organisms as **building materials**. For example, the cell walls of bacteria contain *murein* as a stabilizing component (see p. 40), while in plants *cellulose* and other polysaccharides fulfill this role (see p. 42). Oligomeric or polymeric carbohydrates are often covalently bound to lipids or proteins. The **glycolipids** and **glycoproteins** formed in this way are found, for example, in cell membranes (center). Glycoproteins also occur in the blood in solute form (plasma proteins; see p. 276) and, as components of *proteoglycans*, form important constituents of the intercellular substance (see p. 346).

B. Monosaccharides: structure ◑

The most important natural monosaccharide, **D-glucose**, is an aliphatic aldehyde with six C atoms, five of which carry a hydroxyl group (**1**). Since C atoms 2 to 5 represent chiral centers (see p. 8), there are 15 further isomeric *aldohexoses* in addition to D-glucose, although only a few of these are important in nature (see p. 38). Most natural monosaccharides have the same configuration at C-5 as D-glyceraldehyde–they belong to the D **series**.

The open-chained form of glucose shown in (**1**) is found in neutral solution in less than 0.1 % of the molecules. The reason for this is an intramolecular reaction in which one of the OH groups of the sugar is added to the aldehyde group of the *same* molecule (**2**). This gives rise to a cyclic **hemiacetal** (see p. 10). In aldohexoses, the hydroxy group at C-5 reacts preferentially, and a six-membered pyran ring is formed. Sugars that contain this ring are called **pyranoses**. By contrast, if the OH group at C-4 reacts, a five-part furan ring is formed. In solution, *pyranose* forms and *furanose* forms are present in equilibrium with each other and with the open-chained form, while in glucose polymers only the pyranose form occurs.

The **Haworth projection** (**2**) is usually used to depict sugars in the cyclic form, with the ring being shown in perspective as viewed from above. Depending on the configuration, the substituents of the chiral C atoms are then found above or below the ring. OH groups that lie on the *right* in the Fischer projection (**1**) appear *under* the ring level in the Haworth projection, while those on the *left* appear *above* it.

As a result of hemiacetal formation, an additional chiral center arises at C-1, which can be present in both possible configurations (anomers) (see p. 8). To emphasize this, the corresponding bonds are shown here using wavy lines.

The Haworth formula does not take account of the fact that the pyran ring is not plain, but usually has a *chair conformation*. In **B3**, two frequent conformations of D-glucopyranose are shown as ball-and-stick models. In the $^1C_4$ conformation (bottom), most of the OH groups appear vertical to the ring level, as in the Haworth projection (**axial** or **a** position). In the slightly more stable $^4C_1$ conformation (top), the OH groups take the **equatorial** or **e** position. At room temperature, each form can change into the other, as well as into other conformations.

A. Carbohydrates: overview

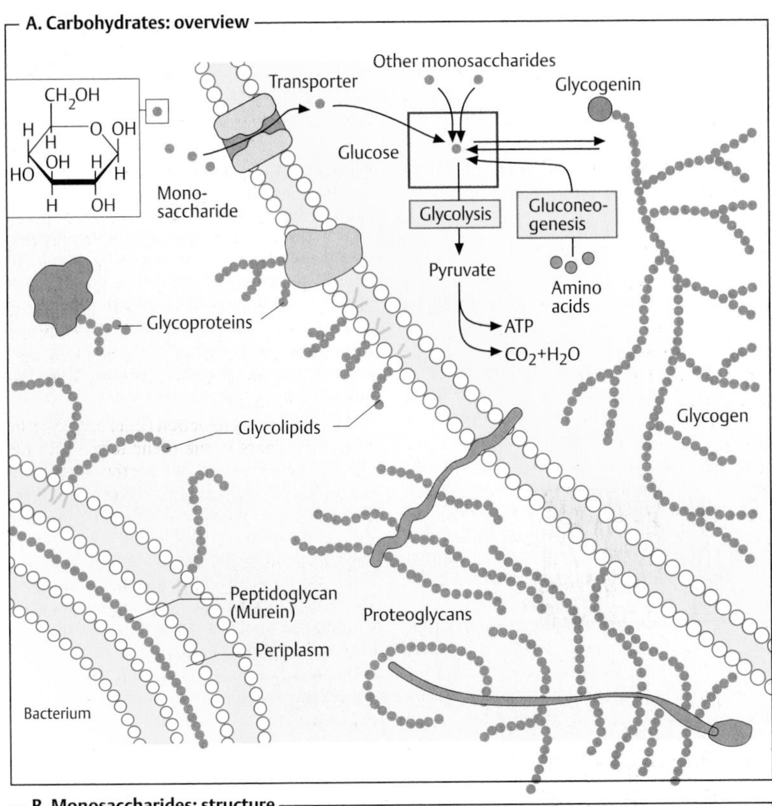

B. Monosaccharides: structure

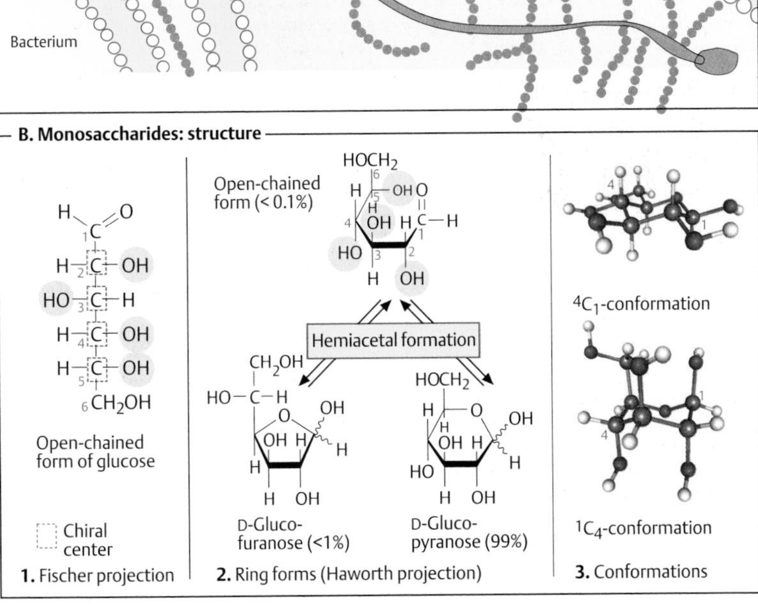

Chemistry of sugars

A. Reactions of the monosaccharides ◑

The sugars (monosaccharides) occur in the metabolism in many forms (derivatives). Only a few important conversion reactions are discussed here, using **D**-glucose as an example.

1. Mutarotation. In the cyclic form, as opposed to the open-chain form, aldoses have a chiral center at C-1 (see p. 34). The corresponding isomeric forms are called **anomers**. In the β-anomer (center left), the OH group at C-1 (the anomeric OH group) and the CH_2OH group lie on the *same* side of the ring. In the α-anomer (right), they are on different sides. The reaction that interconverts anomers into each other is known as *mutarotation* (**B**).

2. Glycoside formation. When the anomeric OH group of a sugar reacts with an alcohol, with elimination of water, it yields an *O-glycoside* (in the case shown, α -methylglucoside). The glycosidic bond is not a normal ether bond, because the OH group at C-1 has a hemiacetal quality. Oligosaccharides and polysaccharides also contain *O*-glycosidic bonds. Reaction of the anomeric OH group with an NH_2 or NH group yields an *N-glycoside* (not shown). *N*-glycosidic bonds occur in nucleotides (see p. 80) and in glycoproteins (see p. 44), for example.

3. Reduction and oxidation. Reduction of the anomeric center at C-1 of glucose (2) produces the sugar alcohol *sorbitol*. Oxidation of the aldehyde group at C-1 gives the intramolecular ester (lactone) of *gluconic acid* (a glyconic acid). Phosphorylated gluconolactone is an intermediate of the pentose phosphate pathway (see p. 152). When glucose is oxidized at C-6, *glucuronic acid* (a glycuronic acid) is formed. The strongly polar glucuronic acid plays an important role in biotransformations in the liver (see pp. 194, 316).

4. Epimerization. In weakly alkaline solutions, glucose is in equilibrium with the ketohexose D-*fructose* and the aldohexose D-*mannose*, via an enediol intermediate (not shown). The only difference between glucose and mannose is the configuration at C-2. Pairs of sugars of this type are referred to as *epimers*, and their interconversion is called *epimerization*.

5. Esterification. The hydroxyl groups of monosaccharides can form *esters* with acids. In metabolism, phosphoric acid esters such as *glucose 6-phosphate* and *glucose 1-phosphate* (6) are particularly important.

B. Polarimetry, mutarotation ○

Sugar solutions can be analyzed by **polarimetry**, a method based on the interaction between chiral centers and linearly polarized light—i. e., light that oscillates in only *one* plane. It can be produced by passing normal light through a special filter (a **polarizer**). A second polarizing filter of the same type (the **analyzer**), placed behind the first, only lets the polarized light pass through when the polarizer and the analyzer are in alignment. In this case, the field of view appears bright when one looks through the analyzer (**1**). Solutions of chiral substances rotate the plane of polarized light by an angle α either to the left or to the right. When a solution of this type is placed between the polarizer and the analyzer, the field of view appears darker (**2**). The angle of rotation, α, is determined by turning the analyzer until the field of view becomes bright again (**3**). A solution's **optical rotation** depends on the type of chiral compound, its concentration, and the thickness of the layer of the solution. This method makes it possible to determine the sugar content of wines, for example.

Certain procedures make it possible to obtain the α and β anomers of glucose in pure form. A 1-molar solution of α-D-glucose has a rotation value $[\alpha]_D$ of +112°, while a corresponding solution of β-D-glucose has a value of +19°. These values change spontaneously, however, and after a certain time reach the same end point of +52°. The reason for this is that, in solution, **mutarotation** leads to an equilibrium between the α and β forms in which, independently of the starting conditions, 62% of the molecules are present in the β form and 38% in the α form.

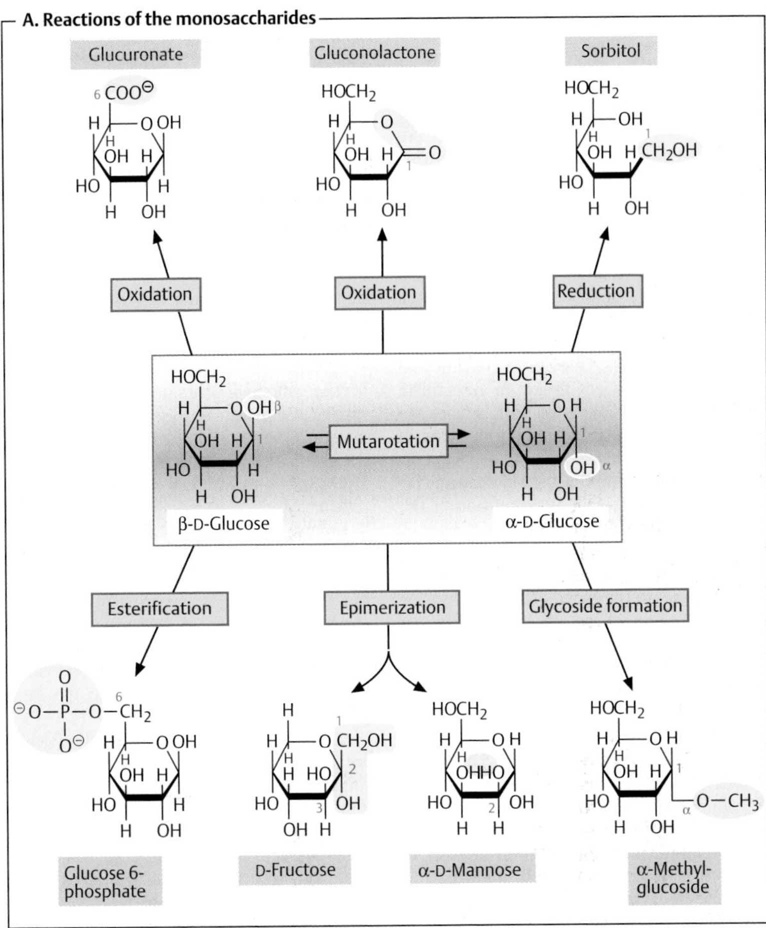
A. Reactions of the monosaccharides
Glucuronate
Gluconolactone
Sorbitol
Oxidation
Oxidation
Reduction
Mutarotation
β-D-Glucose
α-D-Glucose
Esterification
Epimerization
Glycoside formation
Glucose 6-phosphate
D-Fructose
α-D-Mannose
α-Methyl-glucoside

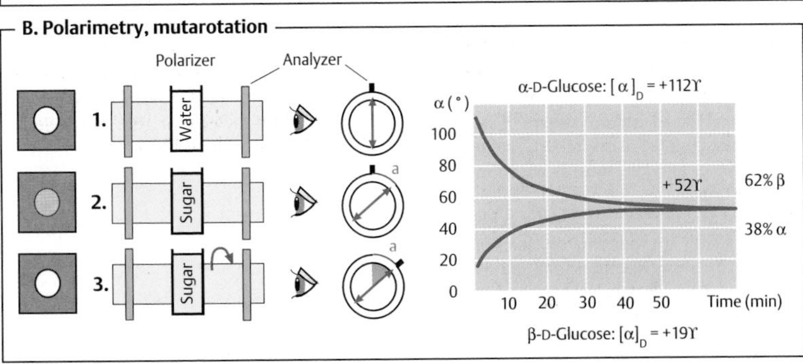
B. Polarimetry, mutarotation
Polarizer
Analyzer
Water
Sugar
Sugar
α-D-Glucose: [α]D = +112°
+ 52°
62% β
38% α
Time (min)
β-D-Glucose: [α]D = +19°

Monosaccharides and disaccharides

A. Important monosaccharides ◑

Only the most important of the large number of naturally occurring **monosaccharides** are mentioned here. They are classified according to the number of C atoms (into pentoses, hexoses, etc.) and according to the chemical nature of the carbonyl function into aldoses and ketoses.

The best-known **aldopentose** (**1**), D-*ribose*, is a component of RNA and of nucleotide coenzymes and is widely distributed. In these compounds, ribose always exists in the furanose form (see p. 34). Like ribose, D-*xylose* and L-*arabinose* are rarely found in free form. However, large amounts of both sugars are found as constituents of polysaccharides in the walls of plant cells (see p. 42).

The most important of the **aldohexoses** (**1**) is D-glucose. A substantial proportion of the biomass is accounted for by glucose polymers, above all cellulose and starch. Free D-glucose is found in plant juices ("grape sugar") and as "blood sugar" in the blood of higher animals. As a constituent of lactose (milk sugar), D-galactose is part of the human diet. Together with D-mannose, galactose is also found in glycolipids and glycoproteins (see p. 44).

Phosphoric acid esters of the **ketopentose** D-*ribulose* (**2**) are intermediates in the pentose phosphate pathway (see p. 152) and in photosynthesis (see p. 128). The most widely distributed of the **ketohexoses** is D-*fructose*. In free form, it is present in fruit juices and in honey. Bound fructose is found in sucrose (**B**) and plant polysaccharides (e. g., inulin).

In the **deoxyaldoses** (**3**), an OH group is replaced by a hydrogen atom. In addition to *2-deoxy-D-ribose*, a component of DNA (see p. 84) that is reduced at C-2, L-fucose is shown as another example of these. Fucose, a sugar in the λ series (see p. 34) is reduced at C-6.

The **acetylated amino sugars** *N-acetyl-D-glucosamine* and *N-acetyl-D-Galactosamine* (**4**) are often encountered as components of glycoproteins.

N-acetylneuraminic acid (sialic acid, **5**), is a characteristic component of glycoproteins. Other **acidic monosaccharides** such as D-*glucuronic acid*, D-*galacturonic acid*, and *liduronic acid*, are typical constituents of the glycosaminoglycans found in connective tissue.

Sugar alcohols (**6**) such as *sorbitol* and *mannitol* do not play an important role in animal metabolism.

B. Disaccharides ◑

When the anomeric hydroxyl group of one monosaccharide is bound glycosidically with one of the OH groups of another, a **disaccharide** is formed. As in all glycosides, the glycosidic bond does *not* allow mutarotation. Since this type of bond is formed stereospecifically by enzymes in natural disaccharides, they are only found in *one* of the possible configurations (α or β).

Maltose (**1**) occurs as a breakdown product of the starches contained in malt ("malt sugar"; see p. 148) and as an intermediate in intestinal digestion. In maltose, the anomeric OH group of one glucose molecule has an α-glycosidic bond with C-4 in a second glucose residue.

Lactose ("milk sugar," **2**) is the most important carbohydrate in the milk of mammals. Cow's milk contains 4.5% lactose, while human milk contains up to 7.5%. In lactose, the anomeric OH group of galactose forms a β-glycosidic bond with C-4 of a glucose. The lactose molecule is consequently elongated, and both of its pyran rings lie in the same plane.

Sucrose (**3**) serves in plants as the form in which carbohydrates are transported, and as a soluble carbohydrate reserve. Humans value it because of its intensely sweet taste. Sources used for sucrose are plants that contain particularly high amounts of it, such as sugar cane and sugar beet (*cane sugar, beet sugar*). Enzymatic hydrolysis of sucrose-containing flower nectar in the digestive tract of bees—catalyzed by the enzyme *invertase*—produces **honey**, a mixture of glucose and fructose. In sucrose, the two anomeric OH groups of glucose and fructose have a glycosidic bond; sucrose is therefore one of the non-reducing sugars.

A. Important monosaccharides

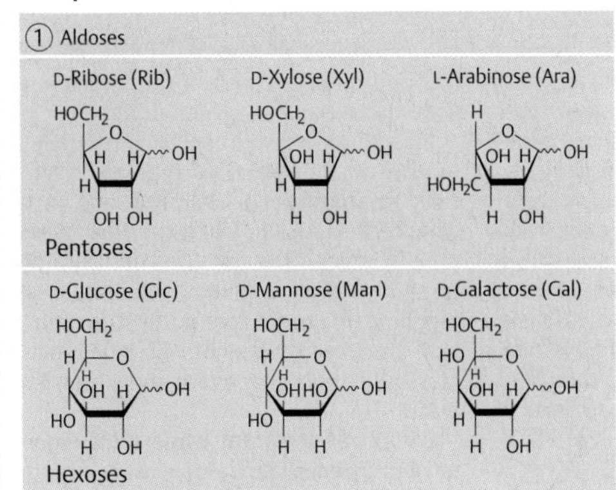

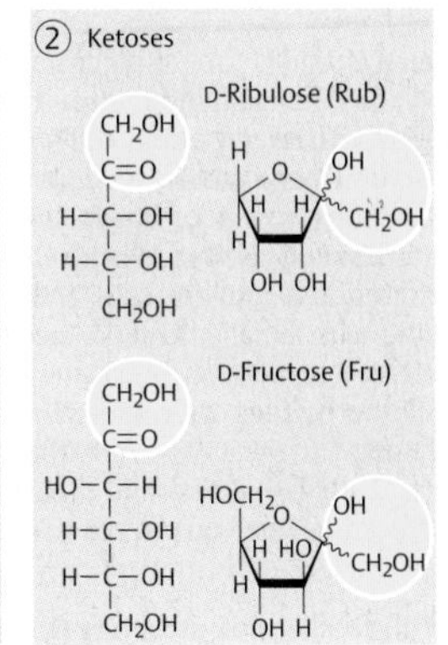

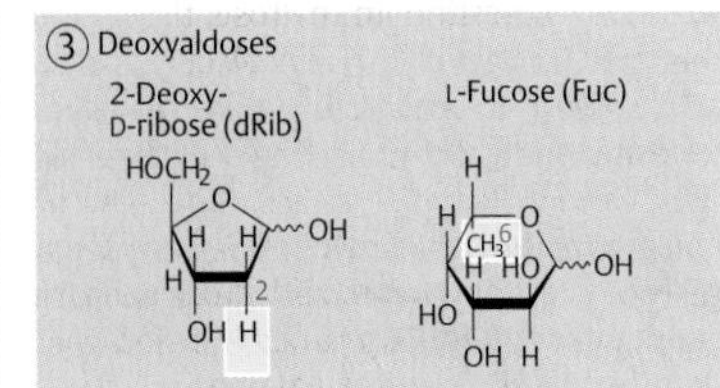

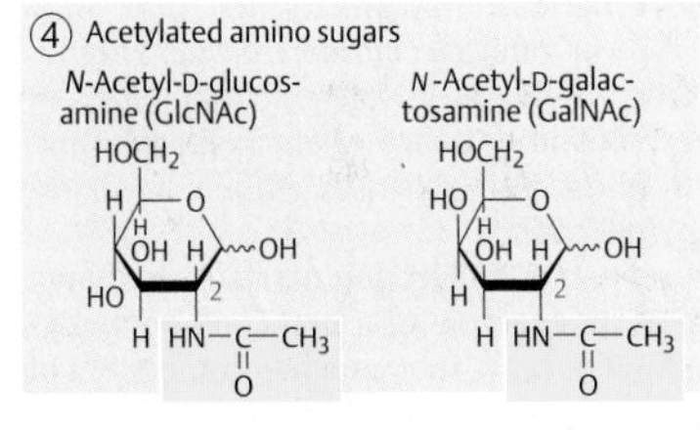

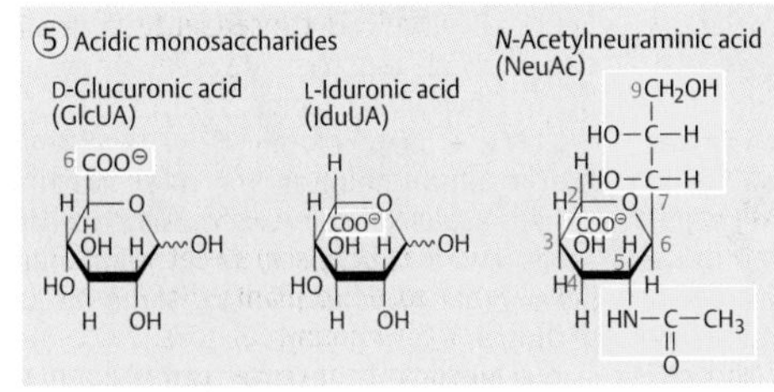

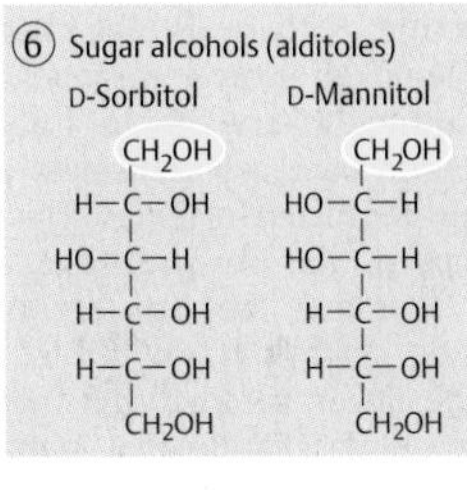

B. Disaccharides

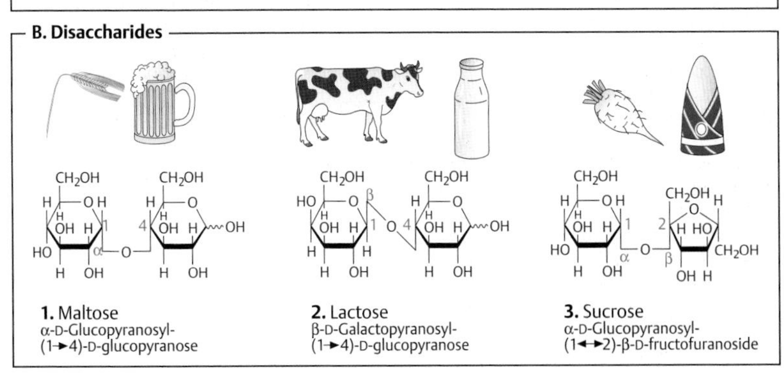

1. Maltose
α-D-Glucopyranosyl-(1→4)-D-glucopyranose

2. Lactose
β-D-Galactopyranosyl-(1→4)-D-glucopyranose

3. Sucrose
α-D-Glucopyranosyl-(1↔2)-β-D-fructofuranoside

Polysaccharides: overview

Polysaccharides are ubiquitous in nature. They can be classified into three separate groups, based on their different *functions*. **Structural polysaccharides** provide mechanical stability to cells, organs, and organisms. **Waterbinding polysaccharides** are strongly hydrated and prevent cells and tissues from drying out. Finally, **reserve polysaccharides** serve as carbohydrate stores that release monosaccharides as required. Due to their polymeric nature, reserve carbohydrates are osmotically less active, and they can therefore be stored in large quantities within the cell.

A. Polysaccharides: structure ◑

Polysaccharides that are formed from only *one* type of monosaccharide are called **homoglycans**, while those formed from different sugar constituents are called **heteroglycans**. Both forms can exist as either linear or branched chains.

A section of a **glycogen** molecule is shown here as an example of a branched homoglycan. Amylopectin, the branched component of vegetable starch (see p. 42), has a very similar structure. Both molecules mainly consist of α1→4-linked glucose residues. In glycogen, on average every 8th to 10th residue carries –via an α1→6 bond–another 1,4-linked chain of glucose residues. This gives rise to branched, tree-like structures, which in animal glycogen are covalently bound to a protein, *glycogenin* (see p. 156).

The linear heteroglycan **murein**, a structural polysaccharide that stabilizes the cell walls of bacteria, has a more complex structure. Only a short segment of this thread-like molecule is shown here. In murein, two different components, both β1→4-linked, alternate: *N-acetylglucosamine* (GlcNAc) and *N-acetylmuraminic acid* (MurNAc), a lactic acid ether of *N*-acetylglucosamine. *Peptides* are bound to the carboxyl group of the lactyl groups, and attach the individual strands of murein to each other to form a three-dimensional network (not shown). Synthesis of the network-forming peptides in murein is inhibited by *penicillin* (see p. 254).

B. Important polysaccharides ◑

The table gives an overview of the composition and make-up both of the glycans mentioned above and of several more.

In addition to murein, bacterial polysaccharides include **dextrans**–glucose polymers that are mostly α1→6-linked and α1→3-branched. In water, dextrans form viscous slimes or gels that are used for chromatographic separation of macromolecules after chemical treatment (see p. 78). Dextrans are also used as components of blood plasma substitutes (plasma expanders) and foodstuffs.

Carbohydrates from algae (e. g., **agarose** and **carrageenan**) can also be used to produce gels. Agarose has been used in microbiology for more than 100 years to reinforce culture media ("agar-agar"). Algal polysaccharides are also added to cosmetics and ready-made foods to modify the consistency of these products.

The **starches**, the most important vegetable reserve carbohydrate and polysaccharides from plant cell walls, are discussed in greater detail on the following page. **Inulin**, a fructose polymer, is used as a starch substitute in diabetics' dietary products (see p. 160). In addition, it serves as a test substance for measuring renal clearance (see p. 322).

Chitin, a homopolymer from β1→4-linked *N*-acetylglucosamine, is the most important structural substance in insect and crustacean shells, and is thus the most common animal polysaccharide. It also occurs in the cell wall of fungi.

Glycogen, the reserve carbohydrate of higher animals, is stored in the liver and musculature in particular (**A**, see pp. 156, 336). The formation and breakdown of glycogen are subject to complex regulation by hormones and other factors (see p. 120).

A. Polysaccharides: structure

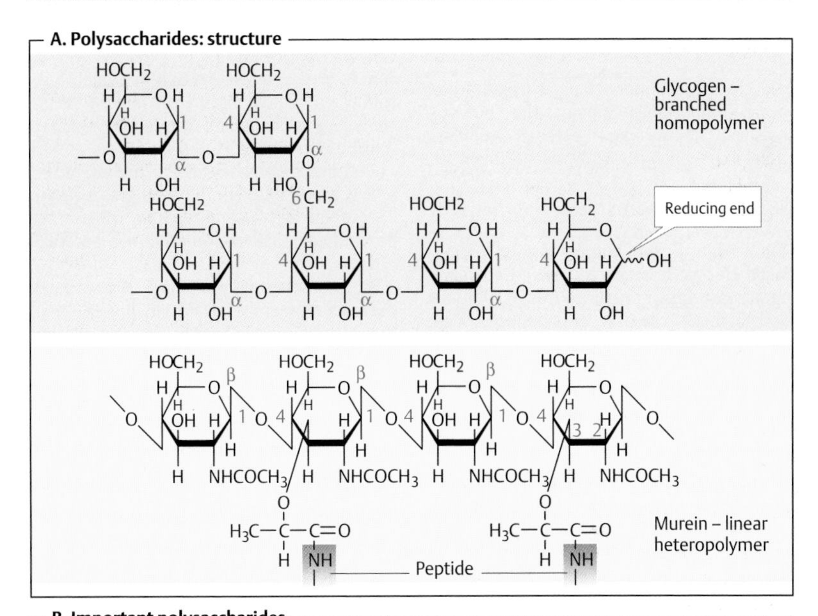

B. Important polysaccharides

Poly-saccharide	Mono-saccharide 1	Mono-saccharide 2	Linkage	Branch-ing	Occurrence	Function
Bacteria						
Murein	D-GlcNAc	D-MurNAc[1]	β1→4	—	Cell wall	SC
Dextran	D-Glc	—	α1→6	α1→3	Slime	WB
Plants						
Agarose	D-Gal	L-aGal[2]	β1→4	β1→3	Red algae (agar)	WB
Carrageenan	D-Gal	—	β1→3	α1→4	Red algae	WB
Cellulose	D-Glc	—	β1→4	—	Cell wall	SC
Xyloglucan	D-Glc	D-Xyl (D-Gal, L-Fuc)	β1→4	β1→6 (β1→2)	Cell wall (Hemicellulose)	SC SC
Arabinan	L-Ara	—	α1→5	α1→3	Cell wall (pectin)	SC
Amylose	D-Glc	—	α1→4	—	Amyloplasts	RC
Amylopectin	D-Glc	—	α1→4	α 1→6	Amyloplasts	RC
Inulin	D-Fru	—	β2→1	—	Storage cells	RC
Animals						
Chitin	D-GlcNAc	—	β1→4	—	Insects, crabs	SK
Glycogen	D-Glc	—	α1→4	α1→6	Liver, muscle	RK
Hyaluronic acid	D-GlcUA	D-GlcNAc	β1→4 β1→3	—	Connective tissue	SK,WB

SC= structural carbohydrate, RC= reserve carbohydrate, WB = water-binding carbohydrate; [1] *N*-acetylmuramic acid, [2] 3,6-anhydrogalactose

Plant polysaccharides

Two glucose polymers of plant origin are of special importance among the polysaccharides: β1→4-linked polymer **cellulose** and **starch**, which is mostly α1→4-linked.

A. Cellulose ◑

Cellulose, a linear homoglycan of β1→4-linked glucose residues, is the *most abundant organic substance* in nature. Almost half of the total biomass consists of cellulose. Some 40–50% of plant *cell walls* are formed by cellulose. The proportion of cellulose in *cotton fibers*, an important raw material, is 98%. Cellulose molecules can contain more than 10^4 glucose residues (mass $1\text{–}2 \cdot 10^6$ Da) and can reach lengths of 6–8 μm.

Naturally occurring cellulose is *extremely mechanically stable* and is highly *resistant* to chemical and enzymatic hydrolysis. These properties are due to the conformation of the molecules and their supramolecular organization. The unbranched β1→4 linkage results in linear chains that are stabilized by hydrogen bonds within the chain and between neighboring chains (**1**). Already during biosynthesis, 50–100 cellulose molecules associate to form an **elementary fibril** with a diameter of 4 nm. About 20 such elementary fibrils then form a **microfibril** (**2**), which is readily visible with the electron microscope.

Cellulose microfibrils make up the basic framework of the **primary wall** of young plant cells (**3**), where they form a complex network with other polysaccharides. The linking polysaccharides include **hemicellulose**, which is a mixture of predominantly neutral heteroglycans (xylans, xyloglucans, arabinogalactans, etc.). Hemicellulose associates with the cellulose fibrils via noncovalent interactions. These complexes are connected by neutral and acidic **pectins**, which typically contain galacturonic acid. Finally, a collagen-related protein, **extensin**, is also involved in the formation of primary walls.

In the higher animals, including humans, cellulose is **indigestible**, but important as **roughage** (see p. 273). Many herbivores (e. g., the ruminants) have symbiotic unicellular organisms in their digestive tracts that break down cellulose and make it digestible by the host.

B. Starch ◑

Starch, a **reserve polysaccharide** widely distributed in plants, is the *most important carbohydrate in the human diet.* In plants, starch is present in the chloroplasts in leaves, as well as in fruits, seeds, and tubers. The starch content is especially high in cereal grains (up to 75% of the dry weight), potato tubers (approximately 65%), and in other plant storage organs.

In these plant organs, starch is present in the form of microscopically small granules in special organelles known as **amyloplasts.** *Starch granules* are virtually insoluble in cold water, but swell dramatically when the water is heated. Some 15–25% of the starch goes into solution in colloidal form when the mixture is subjected to prolonged boiling. This proportion is called amylose ("soluble starch").

Amylose consists of *unbranched* α1→4-linked chains of 200–300 glucose residues. Due the α configuration at C-1, these chains form a *helix* with 6–8 residues per turn (**1**). The blue coloring that soluble starch takes on when iodine is added (the "iodine–starch reaction") is caused by the presence of these helices—the iodine atoms form chains inside the amylose helix, and in this largely nonaqueous environment take on a deep blue color. Highly branched polysaccharides turn brown or reddishbrown in the presence of iodine.

Unlike amylose, **amylopectin**, which is practically insoluble, is *branched.* On average, one in 20–25 glucose residues is linked to another chain via an α1→6 bond. This leads to an extended tree-like structure, which—like amylose—contains only *one* anomeric OH group (a "reducing end"). Amylopectin molecules can contain hundreds of thousands of glucose residues; their mass can be more than 10^8 Da.

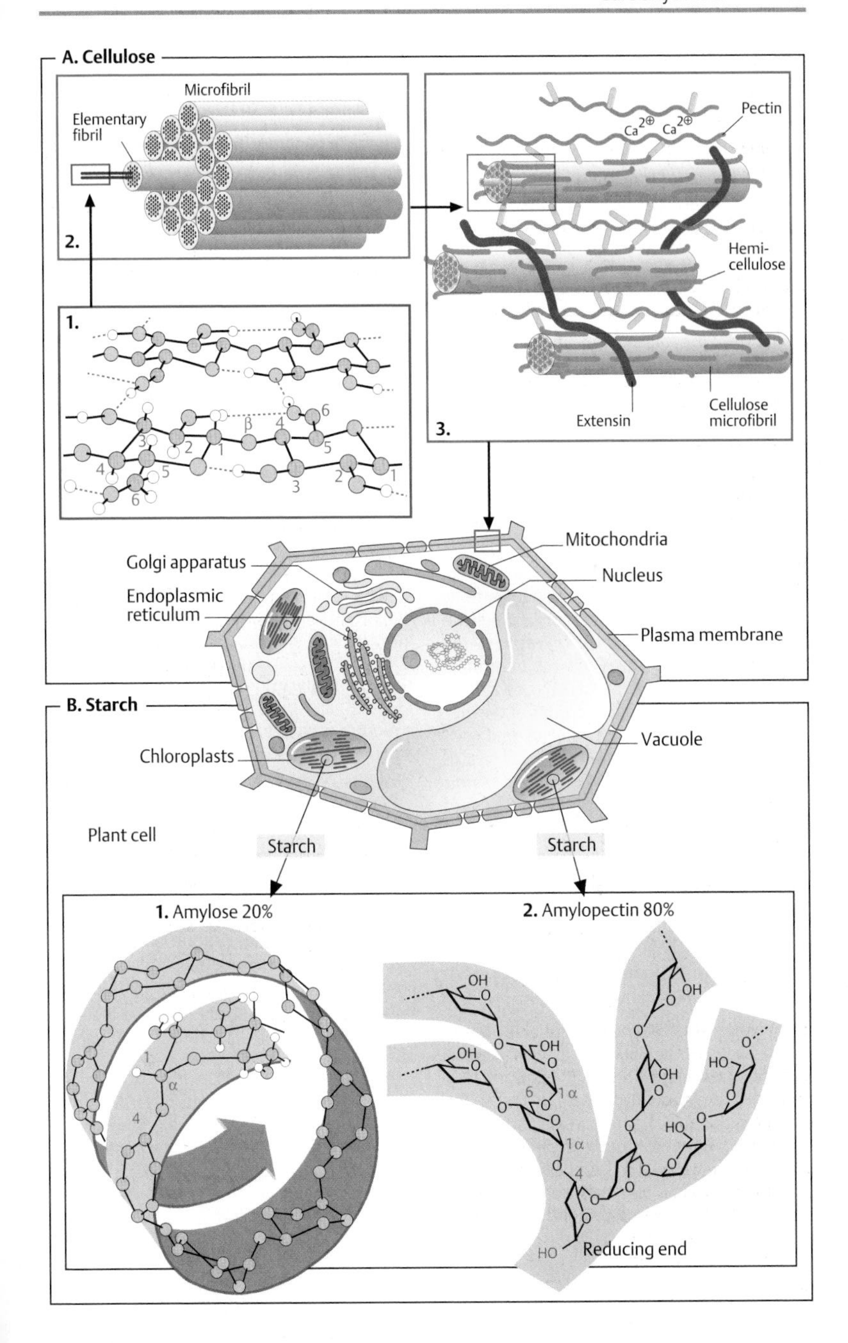
A. Cellulose
Microfibril
Elementary fibril
2.
1.
Pectin
Hemi-cellulose
Extensin
Cellulose microfibril
3.
Mitochondria
Golgi apparatus
Nucleus
Endoplasmic reticulum
Plasma membrane
B. Starch
Chloroplasts
Vacuole
Plant cell
Starch
Starch
1. Amylose 20%
2. Amylopectin 80%
Reducing end

Glycosaminoglycans and glycoproteins

A. Hyaluronic acid ○

As constituents of proteoglycans (see p. 346), the glycosaminoglycans—a group of acidic heteropolysaccharides—are important structural elements of the extracellular matrix.

Glycosaminoglycans contain *amino sugars* as well as *glucuronic acid* and *iduronic acid* as characteristic components (see p. 38). In addition, most polysaccharides in this group are esterified to varying extents by sulfuric acid, increasing their acidic quality. Glycosaminoglycans can be found in free form, or as components of proteoglycans throughout the organism.

Hyaluronic acid, an unesterified glycosaminoglycan with a relatively simple structure, consists of disaccharide units in which *N-acetylglucosamine* and *glucuronic acid* are alternately $\beta 1 \rightarrow 4$-linked and $\beta 1 \rightarrow 3$-linked. Due to the unusual $\beta 1 \rightarrow 3$ linkage, hyaluronic acid molecules–which may contain several thousand monosaccharide residues—are coiled like a helix. Three disaccharide units form each turn of the helix. The outwardfacing hydrophilic carboxylate groups of the glucuronic acid residues are able to bind Ca^{2+} ions. The **strong hydration** of these groups enables hyaluronic acid and other glycosaminoglycans to bind water up to 10 000 times their own volume in gel form. This is the function which hyaluronic acid has in the *vitreous body* of the eye, which contains approximately 1% hyaluronic acid and 98% water.

B. Oligosaccharide in immunoglobulin G (IgG) ○

Many proteins on the surface of the plasma membrane, and the majority of secreted proteins, contain oligosaccharide residues that are post-translationally added to the endoplasmic reticulum and in the Golgi apparatus (see p. 230). By contrast, cytoplasmic proteins are rarely glycosylated. **Glycoproteins** can contain more than 50% carbohydrate; however, the proportion of protein is generally much greater.

As an example of the carbohydrate component of a glycoprotein, the structure of one of the oligosaccharide chains of immunoglobulin G (IgG; see p. 300) is shown here. The oligosaccharide has an *N-glycosidic* link to the amide group of an asparagine residue in the F_c part of the protein. Its function is not known.

Like all *N*-linked carbohydrates, the oligosaccharide in IgG contains a T-shaped **core structure** consisting of two *N-acetylglucosamines* and three *mannose* residues (shown in violet). In addition, in this case the structure contains two further *N-acetylglucosamine* residues, as well as a *fucose* residue and a *galactose* residue. Glycoproteins show many different types of branching. In this case, we not only have $\beta 1 \rightarrow 4$ linkage, but also $\beta 1 \rightarrow 2$, $\alpha 1 \rightarrow 3$, and $\alpha 1 \rightarrow 6$ bonds.

C. Glycoproteins: forms ◑

On the cell surface of certain glycoproteins, **O-glycosidic** links are found between the carbohydrate part and a serine or threonine residue, instead of *N*-glycosidic links to asparagine residues. This type of link is less common than the ***N*-glycosidic** one.

There are two types of oligosaccharide structure with *N*-glycosidic links, which arise through two different biosynthetic pathways. During glycosylation in the ER, the protein is initially linked to an oligosaccharide, which in addition to the core structure contains six further mannose residues and three terminal glucose residues (see p. 230). The simpler from of oligosaccharide (the **mannose-rich** type) is produced when only the glucose residues are cleaved from the primary product, and no additional residues are added. In other cases, the mannose residues that are located outside the core structure are also removed and replaced by other sugars. This produces oligosaccharides such as those shown on the right (the **complex type**). At the external end of the structure, glycoproteins of the complex type often contain *N*-acetylneuraminic acid residues, which give the oligosaccharide components negative charges.

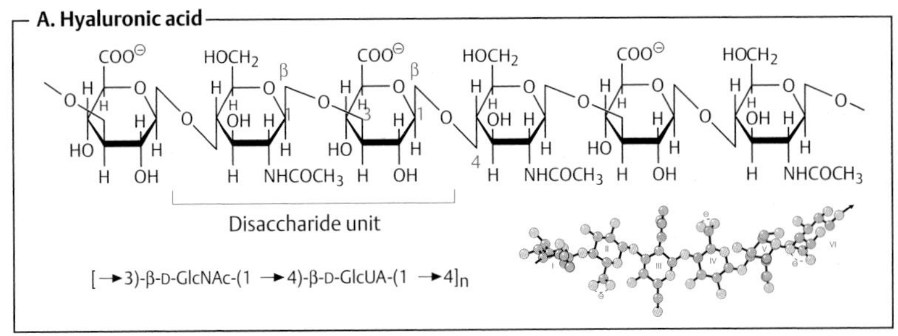
A. Hyaluronic acid
Disaccharide unit
[→3)-β-D-GlcNAc-(1 →4)-β-D-GlcUA-(1 →4]n

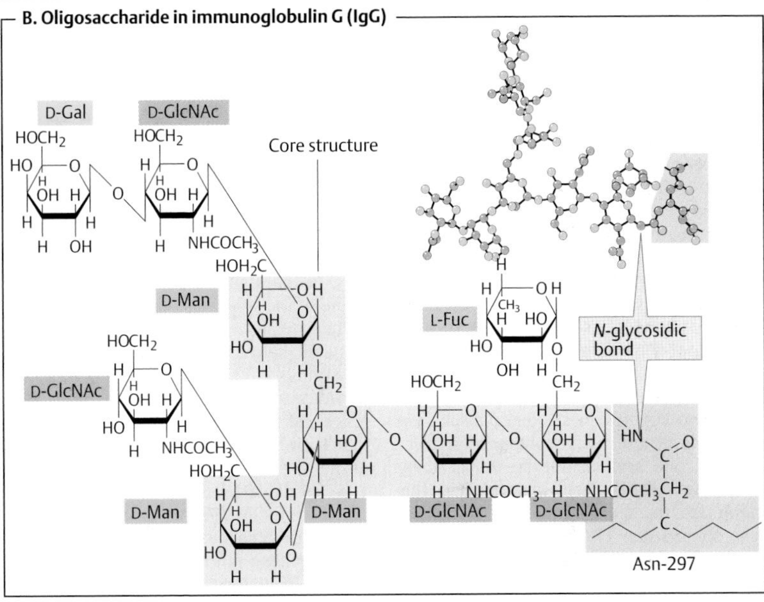
B. Oligosaccharide in immunoglobulin G (IgG)
D-Gal
D-GlcNAc
Core structure
D-Man
L-Fuc
N-glycosidic bond
D-GlcNAc
D-Man
D-Man
D-GlcNAc
D-GlcNAc
Asn-297

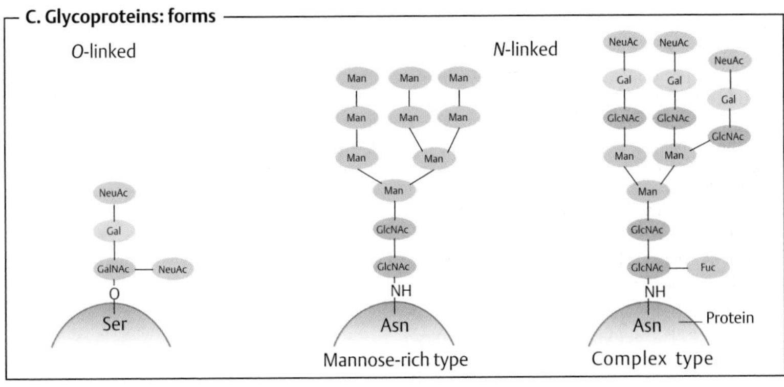
C. Glycoproteins: forms
O-linked
N-linked
NeuAc
Gal
GalNAc
NeuAc
Ser
Man
GlcNAc
NH
Asn
Mannose-rich type
Fuc
Protein
Complex type

Overview

A. Classification ●

The **lipids** are a large and heterogeneous group of substances of biological origin that are easily dissolved in organic solvents such as methanol, acetone, chloroform, and benzene. By contrast, they are either insoluble or only poorly soluble in water. Their low water solubility is due to a lack of polarizing atoms such as O, N, S, and P (see p. 6).

Lipids can be classified into substances that are either *hydrolyzable*– i. e., able to undergo hydrolytic cleavage—or *nonhydrolyzable*. Only a few examples of the many lipids known can be mentioned here. The individual classes of lipids are discussed in more detail in the following pages.

Hydrolyzable lipids (components shown in brackets). The simple **esters** include the *fats* (triacylglycerol; one glycerol + three acyl residues); the *waxes* (one fatty alcohol + one acyl residue); and the *sterol esters* (one sterol + one acyl residue). The **phospholipids** are esters with more complex structures. Their characteristic component is a phosphate residue. The phospholipids include the *phosphatidic acids* (one glycerol + two acyl residues + one phosphate) and the *phosphatides* (one glycerol + two acyl residues + one phosphate + one amino alcohol). In the **sphingolipids**, glycerol and one acyl residue are replaced by sphingosine. Particularly important in this group are the sugar-containing **glycolipids** (one sphingosine + one fatty acid + sugar). The *cerebrosides* (one sphingosine + one fatty acid + one sugar) and *gangliosides* (one sphingosine + one fatty acid + several different sugars, including neuraminic acid) are representatives of this group.

The components of the hydrolyzable lipids are linked to one another by **ester bonds**. They are easily broken down either enzymatically or chemically.

Non-hydrolyzable lipids. The **hydrocarbons** include the *alkanes* and *carotenoids*. The **lipid alcohols** are also not hydrolyzable. They include long-chained *alkanols* and cyclic *sterols* such as cholesterol, and *steroids* such as estradiol and testosterone. The most important **acids** among the lipids are *fatty acids*. The *eicosanoids* also belong to this group; these are derivatives of the polyunsaturated fatty acid arachidonic acid (see p. 390).

B. Biological roles ●

1. Fuel. Lipids are an important source of energy in the diet. In quantitative terms, they represent the principal energy reserve in animals. Neutral fats in particular are stored in specialized cells, known as *adipocytes*. Fatty acids are released from these again as needed, and these are then oxidized in the mitochondria to form water and carbon dioxide, with oxygen being consumed. This process also gives rise to reduced coenzymes, which are used for ATP production in the respiratory chain (see p. 140).

2. Nutrients. Amphipathic lipids are used by cells to build membranes (see p. 214). Typical membrane lipids include phospholipids, glycolipids, and cholesterol. Fats are only weakly amphiphilic and are therefore not suitable as membrane components.

3. Insulation. Lipids are excellent insulators. In the higher animals, neutral fats are found in the subcutaneous tissue and around various organs, where they serve as mechanical and thermal insulators. As the principal constituent of cell membranes, lipids also insulate cells from their environment mechanically and electrically. The impermeability of lipid membranes to ions allows the formation of the membrane potential (see p. 126).

4. Special tasks. Some lipids have adopted special roles in the body. Steroids, eicosanoids, and some metabolites of phospholipids have *signaling functions*. They serve as hormones, mediators, and second messengers (see p. 370). Other lipids form *anchors* to attach proteins to membranes (see p. 214). The lipids also produce *cofactors for enzymatic reactions*—e. g., vitamin K (see p. 52) and ubiquinone (see p. 104). The carotenoid retinal, a light-sensitive lipid, is of central importance in the *process of vision* (see p. 358).

Several lipids are not formed independently in the human body. These substances, as **essential fatty acids** and **fat-soluble vitamins**, are indispensable components of nutrition (see pp. 364ff.)

A. Classification

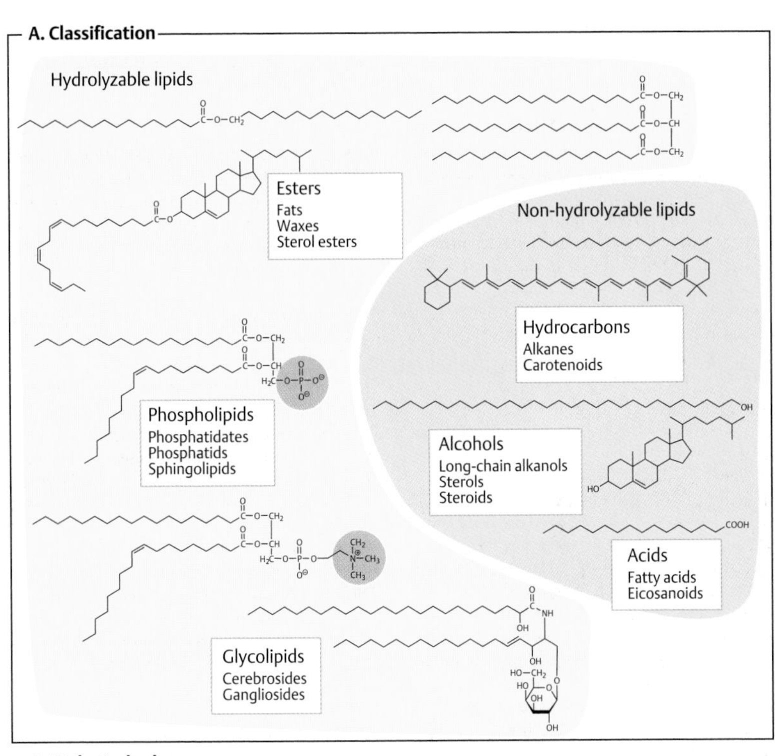

B. Biological roles

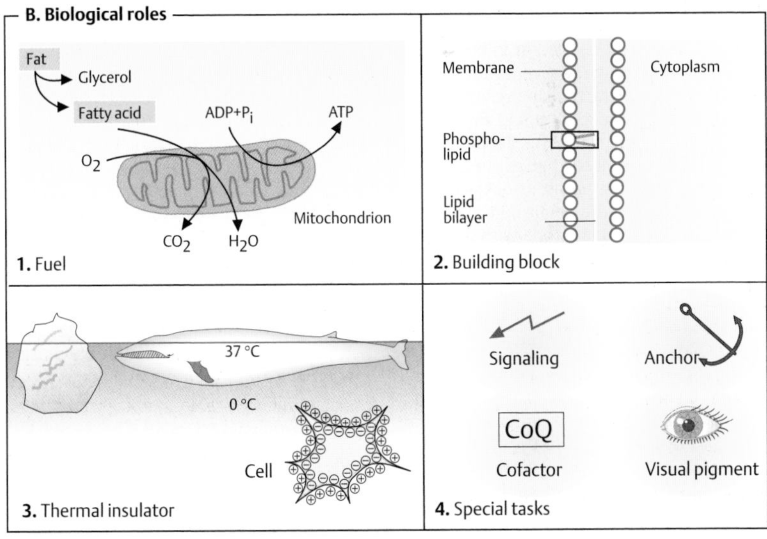

Fatty acids and fats

A. Carboxylic acids ◑

The naturally occurring **fatty acids** are carboxylic acids with unbranched hydrocarbon chains of 4–24 carbon atoms. They are present in all organisms as components of fats and membrane lipids. In these compounds, they are esterified with alcohols (glycerol, sphingosine, or cholesterol). However, fatty acids are also found in small amounts in unesterified form. In this case, they are known as *free fatty acids* (FFAs). As free fatty acids have strongly amphipathic properties (see p. 28), they are usually present in protein-bound forms.

The table lists the full series of aliphatic carboxylic acids that are found in plants and animals. In higher plants and animals, unbranched, longchain fatty acids with either 16 or 18 carbon atoms are the most common—e. g., palmitic and stearic acid. The number of carbon atoms in the longer, natural fatty acids is always even. This is because they are biosynthesized from C_2 building blocks (see p. 168).

Some fatty acids contain one or more isolated *double bonds*, and are therefore *"unsaturated."* Common **unsaturated fatty acids** include oleic acid and linoleic acid. Of the two possible *cis–trans* isomers (see p. 8), usually only the *cis* forms are found in natural lipids. Branched fatty acids only occur in bacteria. A shorthand notation with several numbers is used for precise characterization of the structure of fatty acids—e g., 18:2;9,12 for linoleic acid. The first figure stands for the number of C atoms, while the second gives the number of double bonds. The positions of the double bonds follow after the semicolon. As usual, numbering starts at the carbon with the highest oxidation state (i. e., the carboxyl group corresponds to C-1). Greek letters are also commonly used (α = C-2; β = C-3; ω = the last carbon, ω-3 = the third last carbon).

Essential fatty acids are fatty acids that have to be supplied in the diet. Without exception, these are all polyunsaturated fatty acids: the C_{20} fatty acid *arachidonic acid* (20:4;5,8,11,14) and the two C_{18} acids *linoleic acid* (18:2;9,12) and *linolenic acid* (18:3;9,12,15). The animal organism requires arachidonic acid to synthesize eicosanoids (see p. 390). As the organism is capable of elongating fatty acids by adding C_2 units, but is not able to introduce double bonds into the end sections of fatty acids (after C-9), arachidonic acid has to be supplied with the diet. Linoleic and linolenic acid can be converted into arachidonic acid by elongation, and they can therefore replace arachidonic acid in the diet.

B. Structure of fats ◑

Fats are esters of the trivalent alcohol *glycerol* with three fatty acids. When a single fatty acid is esterified with glycerol, the product is referred to as a *monoacylglycerol* (fatty acid residue = acyl residue).

Formally, esterification with additional fatty acids leads to *diacylglycerol* and ultimately to *triacylglycerol*, the actual fat (formerly termed "triglyceride"). As triacylglycerols are uncharged, they are also referred to as *neutral fats*. The carbon atoms of glycerol are not usually equivalent in fats. They are distinguished by their *"sn"* number, where sn stands for "stereospecific numbering."

The three acyl residues of a fat molecule may differ in terms of their chain length and the number of double bonds they contain. This results in a large number of possible combinations of individual fat molecules. When extracted from biological materials, fats always represent mixtures of very similar compounds, which differ in their fatty acid residues. A chiral center can arise at the middle C atom (*sn* -C-2) of a triacylglycerol if the two external fatty acids are different. The monoacylglycerols and diacylglycerols shown here are also chiral compounds. Nutritional fats contain palmitic, stearic, oleic acid, and linoleic acid particularly often. Unsaturated fatty acids are usually found at the central C atom of glycerol.

The length of the fatty acid residues and the number of their double bonds affect the melting point of the fats. The shorter the fatty acid residues and the more double bonds they contain, the lower their melting points.

A. Carboxylic acids

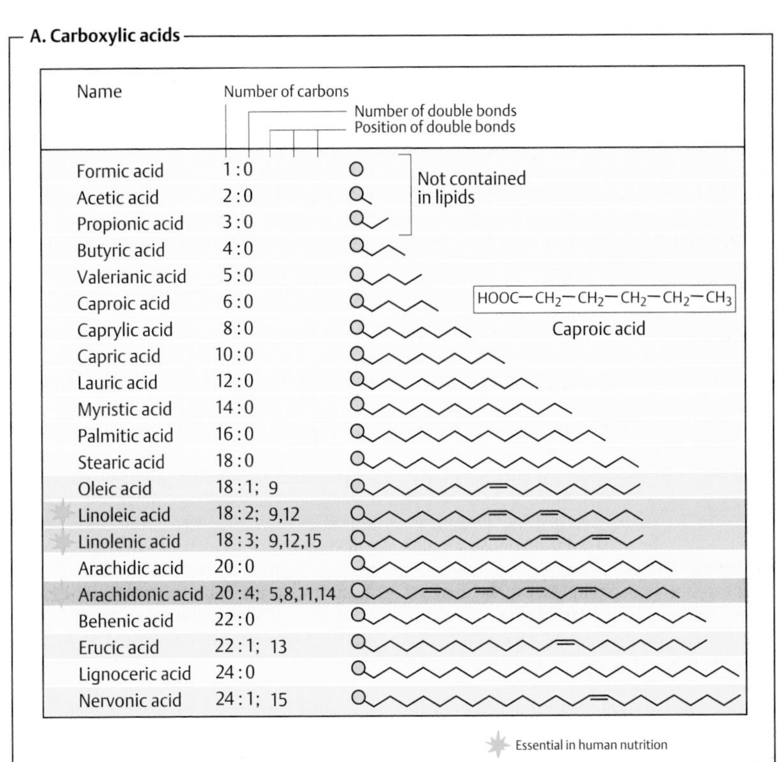

B. Structure of fats

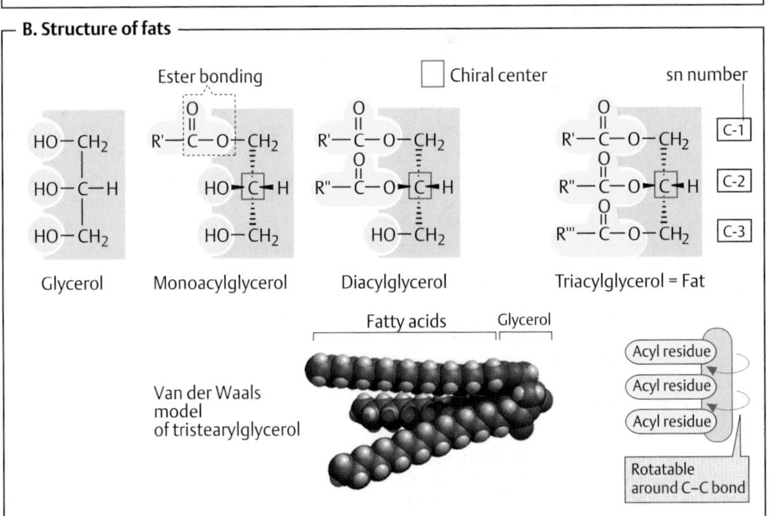

Phospholipids and glycolipids

A. Structure of phospholipids and glycolipids ◑

Fats (triacylglycerol, **1**) are esters of glycerol with three fatty acids (see p. 48). Within the cell, they mainly occur as fat droplets. In the blood, they are transported in the hydrophobic interior of lipoproteins (see p. 278).

Phospholipids (**2**) are the main constituents of biological membranes (see pp. 214–217). Their common feature is a phosphate residue that is esterified with the hydroxyl group at C-3 of glycerol. Due to this residue, phospholipids have at least one negative charge at a neutral pH.

Phosphatidates (anions of the phosphatidic acids), the simplest phospholipids, are phosphate esters of diacylglycerol. They are important intermediates in the biosynthesis of fats and phospholipids (see p. 170). Phosphatidates can also be released from phospholipids by phospholipases.

The other phospholipids can be derived from phosphatidates (residue = phosphatidyl). Their phosphate residues are esterified with the hydroxyl group of an amino alcohol (*choline*, *ethanolamine*, or *serine*) or with the cyclohexane derivative *myo-inositol*. *Phosphatidylcholine* is shown here as an example of this type of compound. When two phosphatidyl residues are linked with one glycerol, the result is *cardiolipin* (not shown), a phospholipid that is characteristic of the inner mitochondrial membrane. *Lysophospholipids* arise from phospholipids by enzymatic cleavage of an acyl residue. The hemolytic effect of bee and snake venoms is due in part to this reaction.

Phosphatidylcholine (lecithin) is the most abundant phospholipid in membranes. *Phosphatidylethanolamine* (cephalin) has an ethanolamine residue instead of choline, and *phosphatidylserine* has a serine residue. In *phosphatidylinositol*, phosphatidate is esterified with the sugarlike cyclic polyalcohol *myo*-inositol. A doubly phosphorylated derivative of this phospholipid, phosphatidylinositol 4,5-bisphosphate, is a special component of membranes, which, by enzymatic cleavage, can give rise to two *second messengers*, diacylglycerol (DAG) and inositol 1,4,5trisphosphate ($InsP_3$; see p. 386).

Some phospholipids carry additional charges, in addition to the negative charge at the phosphate residue. In phosphatidylcholine and phosphatidylethanolamine, the *N*-atom of the amino alcohol is positively charged. As a whole, these two phosphatides therefore appear to be neutral. In contrast, phosphatidylserine—with one additional positive charge and one additional negative charge in the serine residue—and phosphatidylinositol (with no additional charge) have a negative net charge, due to the phosphate residue.

Sphingolipids (**3**), which are found in large quantities in the membranes of nerve cells in the brain and in neural tissues, have a slightly different structure from the other membrane lipids discussed so far. In sphingolipids, *sphingosine*, an amino alcohol with an unsaturated alkyl side chain, replaces glycerol and one of the acyl residues. When sphingosine forms an amide bond to a fatty acid, the compound is called *ceramide* (**3**). This is the precursor of the sphingolipids. *Sphingomyelin* (2)—the most important sphingolipid—has an additional phosphate residue with a choline group attached to it on the sphingosine, in addition to the fatty acid.

Glycolipids (**3**) are present in all tissues on the outer surface of the plasma membrane. They consist of sphingosine, a fatty acid, and an oligosaccharide residue, which can sometimes be quite large. The phosphate residue typical of phospholipids is absent. *Galactosylceramide* and *glucosylceramide* (known as cerebroside) are simple representatives of this group. Cerebrosides in which the sugar is esterified with sulfuric acid are known as *sulfatides*. *Gangliosides* are the most complex glycolipids. They constitute a large family of membrane lipids with receptor functions that are as yet largely unknown. A characteristic component of many gangliosides is *N*-acetylneuraminic acid (sialic acid; see p. 38).

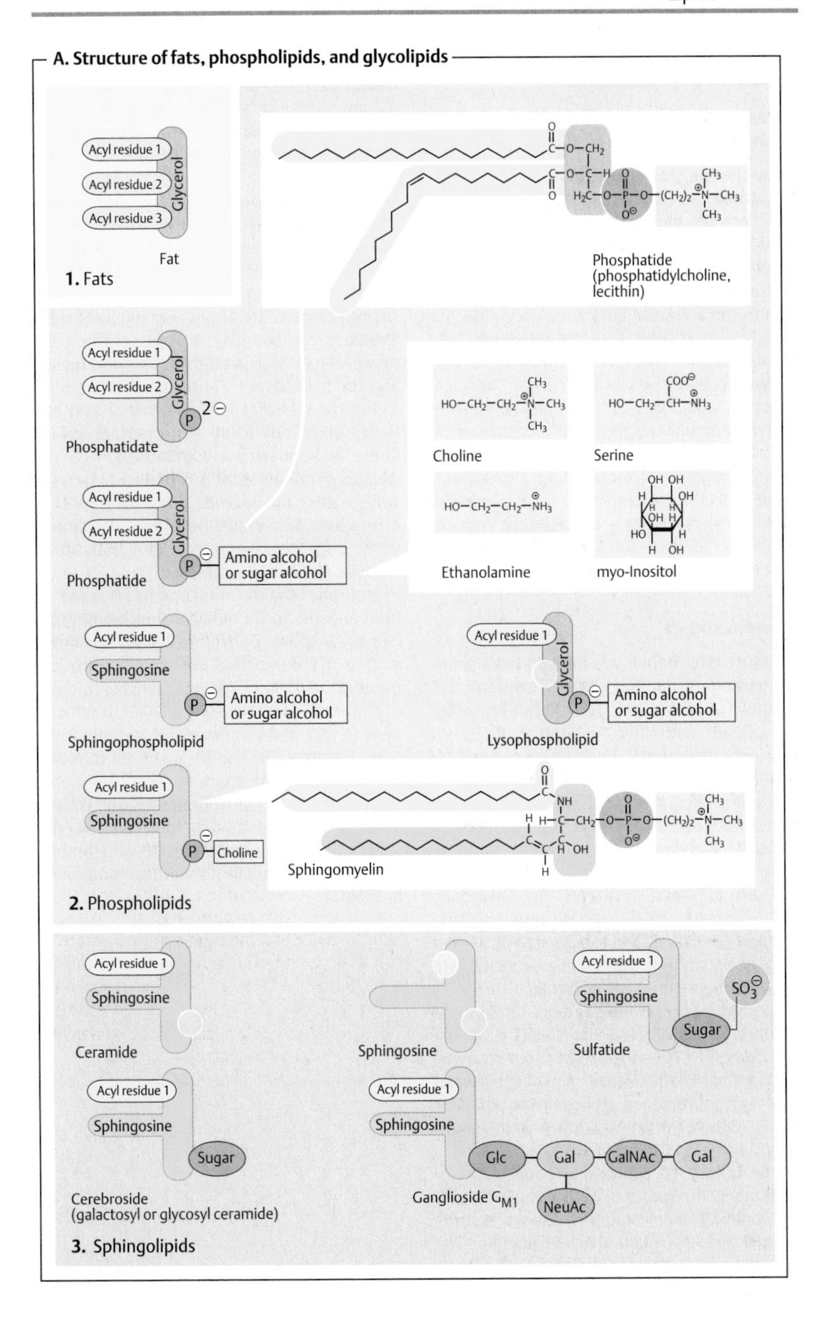
A. Structure of fats, phospholipids, and glycolipids
Acyl residue 1
Acyl residue 2
Acyl residue 3
Glycerol
Fat
1. Fats
Phosphatide
(phosphatidylcholine, lecithin)
Phosphatidate
Phosphatide
Amino alcohol or sugar alcohol
Choline
Serine
Ethanolamine
myo-Inositol
Sphingosine
Sphingophospholipid
Lysophospholipid
Choline
Sphingomyelin
2. Phospholipids
Ceramide
Sulfatide
Sugar
Cerebroside
(galactosyl or glycosyl ceramide)
Ganglioside G_{M1}
Glc
Gal
GalNAc
NeuAc
3. Sphingolipids

Isoprenoids

A. Activated acetic acid as a component of lipids ◑

Although the lipids found in plant and animal organisms occur in many different forms, they are all closely related biogenetically; they are all derived from **acetyl-CoA**, the "activated acetic acid" (see pp. 12, 110).

1. One major pathway leads from acetyl-CoA to the activated fatty acids (**acyl-CoA**; for details, see p. 168). *Fats, phospholipids,* and *glycolipids* are synthesized from these, and fatty acid derivatives in particular are formed. Quantitatively, this is the most important pathway in animals and most plants.

2. The second pathway leads from acetyl-CoA to isopentenyl diphosphate (*"active isoprene"*), the basic component for the **isoprenoids**. Its biosynthesis is discussed in connection with biosynthesis of the isoprenoid, cholesterol (see p. 172).

B. Isoprenoids ◑

Formally, isoprenoids are derived from a single common building block, isoprene (2-methyl-1,3-butadiene), a methyl-branched compound with five C atoms. Activated isoprene, *isopentenyl diphosphate,* is used by plants and animals to biosynthesize linear and cyclic oligomers and polymers. For the isoprenoids listed here—which only represent a small selection—the number of isoprene units (I) is shown.

From activated isoprene, the metabolic pathway leads via dimerization to activated *geraniol* (I = 2) and then to activated *farnesol* (I = 3). At this point, the pathway divides into two. Further extension of farnesol leads to chains with increasing numbers of isoprene units—e. g., *phytol* (I = 4), *dolichol* (I = 14–24), and *rubber* (I = 700–5000). The other pathway involves a "head-to-head" linkage between two farnesol residues, giving rise to *squalene* (I = 6), which, in turn, is converted to *cholesterol* (I = 6) and the other *steroids.*

The ability to synthesize particular isoprenoids is limited to a few species of plants and animals. For example, rubber is only formed by a few plant species, including the rubber tree (*Hevea brasiliensis*). Several isoprenoids that are required by animals for metabolism, but cannot be produced by them independently, are vitamins; this group includes *vitamins A, D, E,* and *K.* Due to its structure and function, vitamin D is now usually classified as a steroid hormone (see pp. 56, 330).

Isoprene metabolism in plants is very complex. Plants can synthesize many types of aromatic substances and volatile oils from isoprenoids. Examples include *menthol* (I= 2), *camphor* (I = 2), and *citronellal* (I = 2). These C_{10} compounds are also called *monoterpenes.* Similarly, compounds consisting of three isoprene units (I = 3) are termed *sesquiterpenes,* and the steroids (I = 6) are called *triterpenes.*

Isoprenoids that have hormonal and signaling functions form an important group. These include *steroid hormones* (I = 6) and *retinoate* (the anion of retinoic acid; I = 3) in vertebrates, and *juvenile hormone* (I = 3) in arthropods. Some plant hormones also belong to the isoprenoids—e. g., the cytokinins, abscisic acid, and brassinosteroids.

Isoprene chains are sometimes used as lipid anchors to fix molecules to membranes (see p. 214). Chlorophyll has a *phytyl* residue (I = 4) as a lipid anchor. Coenzymes with isoprenoid anchors of various lengths include *ubiquinone* (coenzyme Q; I = 6–10), *plastoquinone* (I = 9), and *menaquinone* (vitamin K; I = 4–6). Proteins can also be anchored to membranes by *isoprenylation.*

In some cases, an isoprene residue is used as an element to modify molecules chemically. One example of this is *N'*-isopentenyl-AMP, which occurs as a modified component in tRNA.

A. Activated acetic acid as a component of lipids

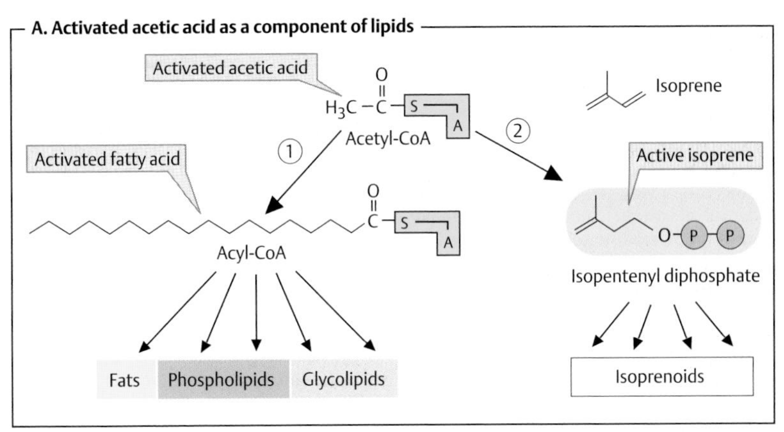

B. Isoprenoids

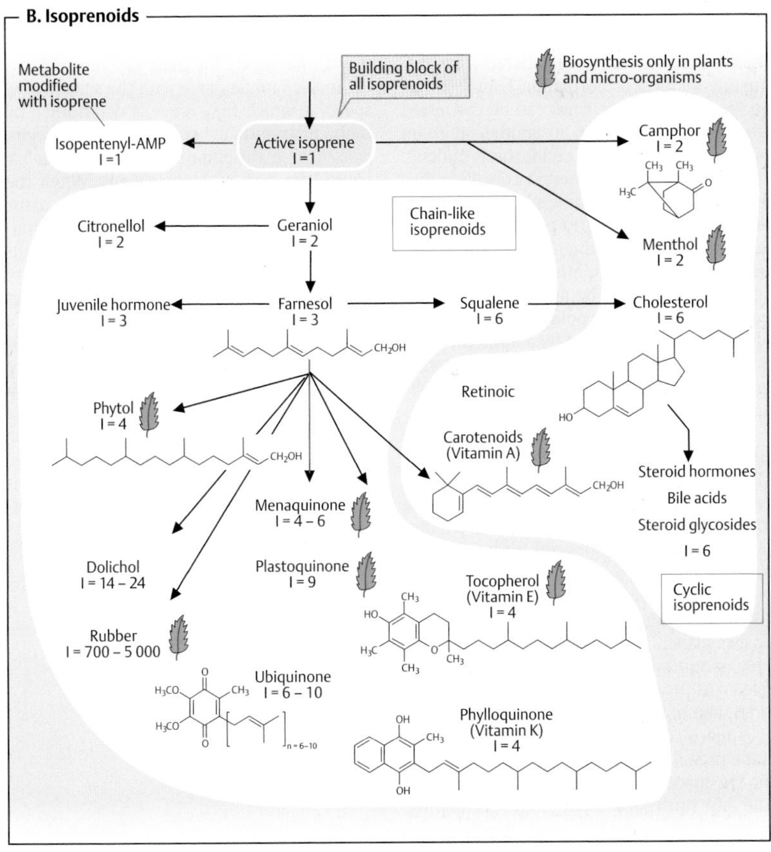

Steroid structure

A. Steroid building blocks ◑

Common to all of the steroids is a molecular core structure consisting of four saturated rings, known as *gonane*. At the end of the steroid core, many steroids also carry a side chain, as seen in *cholestane*, the basic component of the *sterols* (steroid alcohols).

B. Spatial structure ○

The four rings of the steroids are distinguished using the letters A, B, C, and D. Due to the tetrahedral arrangement of the single carbon bonds, the rings are not flat, but puckered. Various *ring conformations* are known by the terms "chair," "boat," and "twisted" (not shown). The *chair* and *boat* conformations are common. Fivemembered rings frequently adopt a conformation referred to as an "envelope". Some rings can be converted from one conformation to another at room temperature, but with steroids this is difficult.

Substituents of the steroid core lie either approximately in the same plane as the ring (e = *equatorial*) or nearly perpendicular to it (a = *axial*). In threedimensional representations, substituents pointing toward the observer are indicated by an unbroken line (β position), while bonds pointing into the plane of the page are indicated by a dashed line (α position). The so-called *angular* methyl groups at C-10 and C-13 of the steroids always adopt the β position.

Neighboring rings can lie in the same plane (*trans*; **2**) or at an angle to one another (*cis*; **1**). This depends on the positions of the substituents of the shared ring carbons, which can be arranged either *cis* or *trans* to the angular methyl group at C-10. The substituents of steroid that lie at the points of intersection of the individual rings are usually in *trans* position. As a whole, the core of most steroids is more or less planar, and looks like a flat disk. The only exceptions to this are the ecdysteroids, bile acids (in which A:B is *cis*), cardiac glycosides, and toad toxins.

A more realistic impression of the three-dimensional structure of steroids is provided by the space-filling model of *cholesterol* (**3**). The four rings form a fairly rigid scaffolding, onto which the much more mobile side chain is attached.

Steroids are relatively apolar (hydrophobic). Some polar groups—e.g., hydroxyl and oxo groups—give them amphipathic properties. This characteristic is especially pronounced with the bile acids (see p. 314).

C. Thin-layer chromatography ○

Thin-layer chromatography (TLC) is a powerful, mainly analytic, technique for rapidly separating lipids and other small molecules such as amino acids, nucleotides, vitamins, and drugs. The *sample* being analyzed is applied to a *plate* made of glass, aluminum, or plastic, which is covered with a thin layer of silica gel or other material (**1**). The plate is then placed in a chromatography chamber that contains some *solvent*. Drawn by capillary forces, the solvent moves up the plate (**2**). The substances in the sample move with the solvent. The speed at which they move is determined by their distribution between the *stationary phase* (the hydrophilic silica), and the *mobile phase* (the hydrophobic solvent). When the solvent reaches the top edge of the plate, the chromatography is stopped. After evaporation of the solvent, the separated substances can be made visible using appropriate staining methods or with physical processes (e.g., ultraviolet light) (**3**). The movement of a substance in a given TLC system is expressed as its R_f value. In this way, compounds that are not known can be identified by comparison with reference substances.

A process in which the polarity of the stationary and mobile phases is reversed—i.e., the stationary phase is apolar and the solvent is polar—is known as "reversed-phase thin-layer chromatography" (RP-TLC).

A. Steroid building blocks

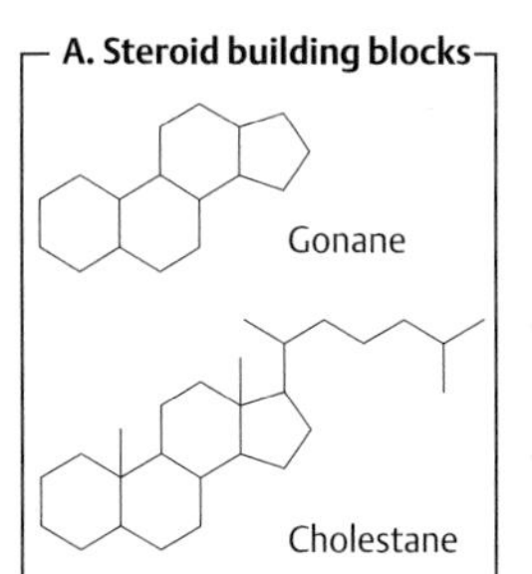

B. 3D structure

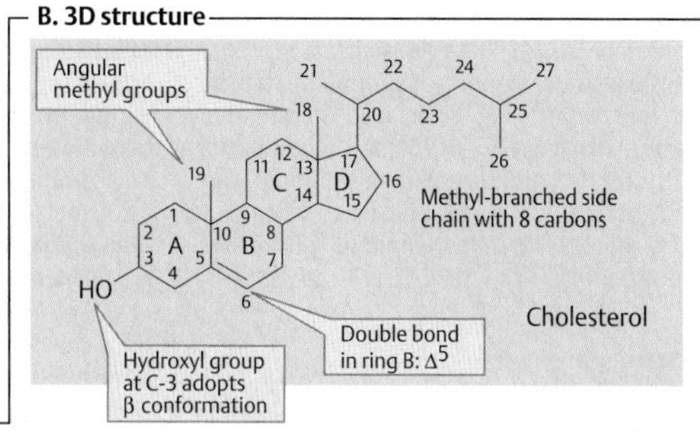

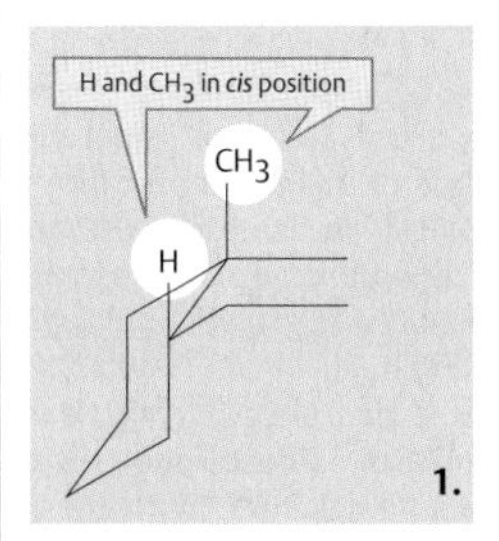

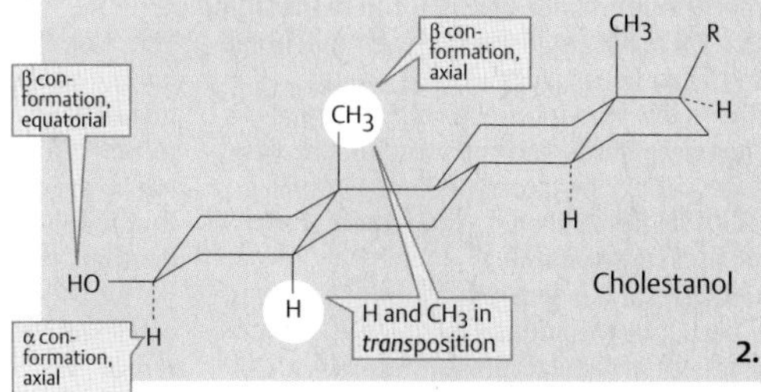

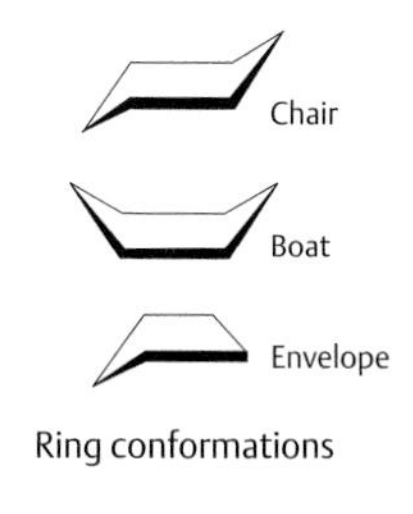

Ring conformations

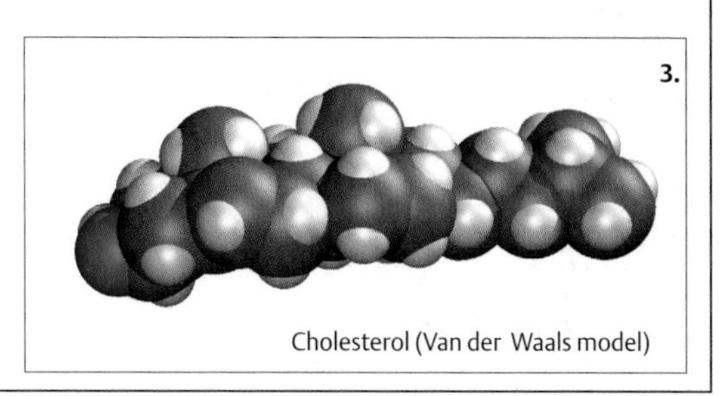

Cholesterol (Van der Waals model)

C. Thin-layer chromatography

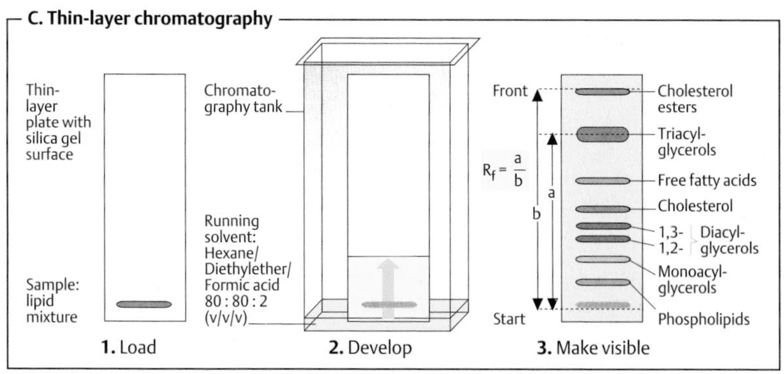

1. Load **2.** Develop **3.** Make visible

Steroids: overview

The three most important groups of steroids are the *sterols*, *bile acids*, and *steroid hormones*. Particularly in plants, compounds with steroid structures are also found that are notable for their pharmacological effects—steroid alkaloids, digitalis glycosides, and saponins.

A. Sterols ◑

Sterols are *steroid alcohols*. They have a β-positioned hydroxyl group at C-3 and one or more double bonds in ring B and in the side chain. There are no further oxygen functions, as in the carbonyl and carboxyl groups.

The most important sterol in animals is **cholesterol**. Plants and microorganisms have a wide variety of closely related sterols instead of cholesterol—e. g., **ergosterol**, **β-sitosterol**, and **stigmasterol**.

Cholesterol is present in all animal tissues, and particularly in neural tissue. It is a major constituent of cellular membranes, in which it regulates fluidity (see p. 216). The storage and transport forms of cholesterol are its esters with fatty acids. In lipoproteins, cholesterol and its fatty acid esters are associated with other lipids (see p. 278). Cholesterol is a constituent of the bile and is therefore found in many gallstones. Its biosynthesis, metabolism, and transport are discussed elsewhere (see pp. 172, 312).

Cholesterol-rich lipoproteins of the LDL type are particularly important in the development of arteriosclerosis, in which the arterial walls are altered in connection with an excess plasma cholesterol level. In terms of dietary physiology, it is important that plant foodstuffs are low in cholesterol. By contrast, animal foods can contain large amounts of cholesterol—particularly butter, egg yolk, meat, liver, and brain.

B. Bile acids ◑

Bile acids are synthesized from cholesterol in the liver (see p. 314). Their structures can therefore be derived from that of cholesterol. Characteristic for the bile acids is a side chain shortened by three C atoms in which the last carbon atom is oxidized to a carboxyl group. The double bond in ring B is reduced and rings A and B are in *cis* position relative to each other (see p. 54). One to three hydroxyl groups (in α position) are found in the steroid core at positions 3, 7, and 12. Bile acids keep bile cholesterol in a soluble state as micelles and promote the digestion of lipids in the intestine (see p. 270). **Cholic acid** and **chenodeoxycholic acid** are *primary bile acids* that are formed by the liver. Their dehydroxylation at C-7 by microorganisms from the intestinal flora gives rise to the *secondary bile acids* **lithocholic acid** and **deoxycholic acid**.

C. Steroid hormones ◑

The conversion of cholesterol to *steroid hormones* (see p. 376) is of minor importance quantitatively, but of major importance in terms of physiology. The steroid hormones are a group of lipophilic signal substances that regulate metabolism, growth, and reproduction (see p. 374).

Humans have six steroid hormones: **progesterone**, **cortisol**, **aldosterone**, **testosterone**, **estradiol**, and **calcitriol**. With the exception of calcitriol, these steroids have either no side chain or only a short side one consisting of two carbons. Characteristic for most of them is an oxo group at C-3, conjugated with a double bond between C-4 and C-5 of ring A. Differences occur in rings C and D. Estradiol is aromatic in ring A, and its hydroxyl group at C-3 is therefore phenolic. Calcitriol differs from other vertebrate steroid hormones; it still contains the complete carbon framework of cholesterol, but lightdependent opening of ring B turns it into what is termed a "secosteroid" (a steroid with an open ring).

Ecdysone is the steroid hormone of the arthropods. It can be regarded as an early form of the steroid hormones. Steroid hormones with signaling functions also occur in plants.

A. Sterols

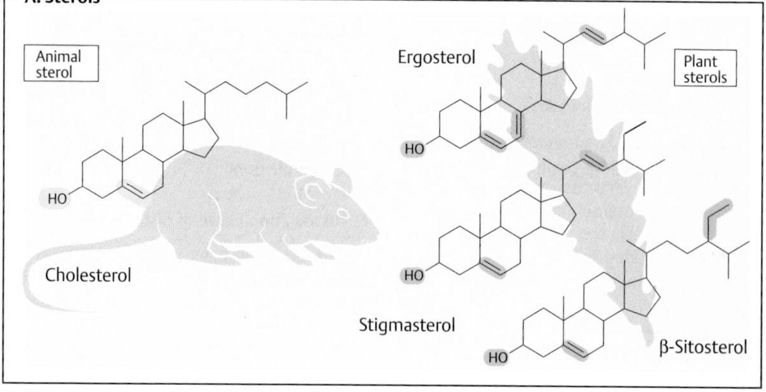

Animal sterol

HO

Cholesterol

Ergosterol

Plant sterols

HO

HO

Stigmasterol

HO

β-Sitosterol

B. Bile acids

OH

O

OH

HO

OH

H

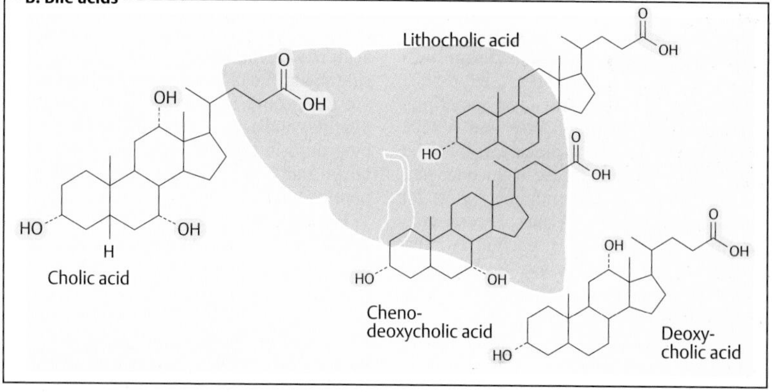

Cholic acid

Lithocholic acid

O

OH

HO

O

OH

HO

OH

Cheno-deoxycholic acid

OH

O

OH

HO

Deoxy-cholic acid

C. Steroid hormones

CH_2OH

C=O

HO

OH

O

Cortisol

CH_2OH

OHC

C=O

HO

O

Aldosterone

OH

O

Testosterone

OH

HO

Estradiol

CH_3

C=O

O

Progesterone

25

OH

CH_2

3

1

HO

OH

Calcitriol

OH

OH

HO

HO

OH

O

Ecdysone

Molting hormone of insects, spiders and crabs

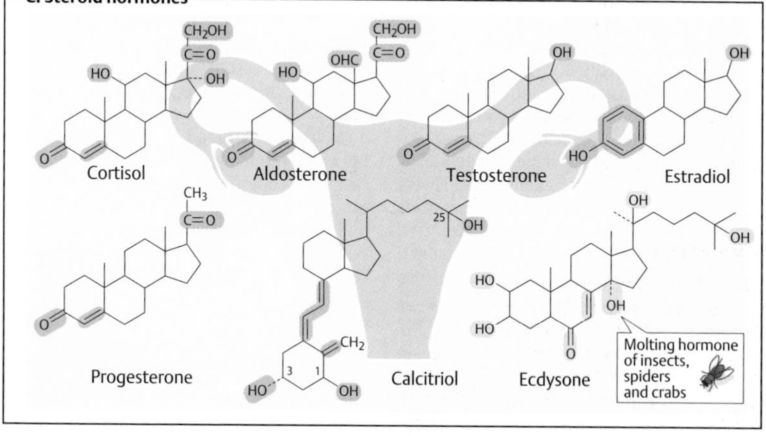

Amino acids: chemistry and properties

A. Amino acids: functions ●

The amino acids (2-aminocarboxylic acids) fulfill various functions in the organism. Above all, they serve as the **components of peptides and proteins**. Only the 20 *proteinogenic amino acids* (see p. 60) are included in the genetic code and therefore regularly found in proteins. Some of these amino acids undergo further (post-translational) change following their incorporation into proteins (see p. 62). Amino acids or their derivatives are also form components of **lipids**—e. g., serine in phospholipids and glycine in bile salts. Several amino acids function as **neurotransmitters** themselves (see p. 352), while others are precursors of neurotransmitters, mediators, or hormones (see p. 380). Amino acids are important (and sometimes essential) components of food (see p. 360). Specific amino acids form **precursors** for other metabolites—e. g., for glucose in gluconeogenesis, for purine and pyrimidine bases, for heme, and for other molecules. Several non-proteinogenic amino acids function as intermediates in the synthesis and breakdown of proteinogenic amino acids (see p. 412) and in the urea cycle (see p. 182).

B. Optical activity ◑

The natural amino acids are mainly α-amino acids, in contrast to β-amino acids such as β-alanine and taurine. Most α-amino acids have four different substituents at C-2 (Cα). The α atom therefore represents a *chiral center*—i. e., there are two different **enantiomers** (L- and D-amino acids; see p. 8). Among the proteinogenic amino acids, only glycine is *not* chiral (R = H). In nature, it is almost exclusively **L-amino acids** that are found. **D**-Amino acids occur in bacteria—e. g., in murein (see p. 40)—and in peptide antibiotics. In animal metabolism, D-Amino acids would disturb the enzymatic reactions of L-amino acids and they are therefore broken down in the liver by the enzyme *D-amino acid oxidase*.

The **Fischer projection** (center) is used to present the formulas for chiral centers in biomolecules. It is derived from their three-dimensional structure as follows: firstly, the tetrahedron is rotated in such a way that the most oxidized group (the carboxylate group) is at the top. Rotation is then continued until the line connecting line COO^- and R (red) is level with the page. In L-amino acids, the NH_3^+ group is then on the left, while in **D**-amino acids it is on the right.

C. Dissociation curve of histidine ◑

All amino acids have at least two ionizable groups, and their net charge therefore depends on the pH value. The COOH groups at the α-C atom have pK_a values of between 1.8 and 2.8 and are therefore more acidic than simple monocarboxylic acids. The basicity of the α-amino function also varies, with pK_a values of between 8.8 and 10.6, depending on the amino acid. Acidic and basic amino acids have additional ionizable groups in their side chain. The pK_a values of these side chains are listed on p. 60. The electrical charges of peptides and proteins are mainly determined by groups in the side chains, as most α-carboxyl and α-amino functions are linked to peptide bonds (see p. 66).

Histidine can be used here as an example of the pH-dependence of the net charge of an amino acid. In addition to the carboxyl group and the amino group at the α-C atom with pK_a values of 1.8 and 9.2, respectively, histidine also has an imidazole residue in its side chain with a pK_a value of 6.0. As the pH increases, the net charge (the sum of the positive and negative charges) therefore changes from +2 to –1. At pH 7.6, the net charge is zero, even though the molecule contains two almost completely ionized groups in these conditions. This pH value is called the **isoelectric point**.

At its isoelectric point, histidine is said to be **zwitterionic**, as it has both anionic *and* cationic properties. Most other amino acids are also zwitterionic at neutral pH. Peptides and proteins also have isoelectric points, which can vary widely depending on the composition of the amino acids.

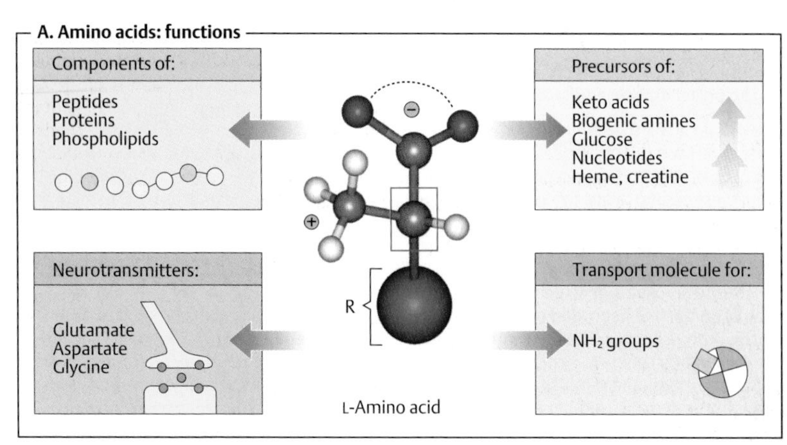
A. Amino acids: functions
Components of:
Peptides
Proteins
Phospholipids
Precursors of:
Keto acids
Biogenic amines
Glucose
Nucleotides
Heme, creatine
Neurotransmitters:
Glutamate
Aspartate
Glycine
Transport molecule for:
NH_2 groups
R
L-Amino acid

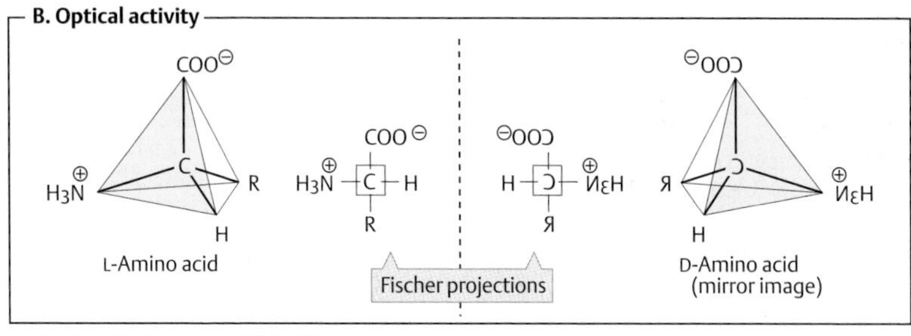
B. Optical activity
$COO^{\ominus}$
$H_3\overset{\oplus}{N}$
C
R
H
L-Amino acid
$COO^{\ominus}$
$H_3\overset{\oplus}{N}$ — C — H
R
Fischer projections
D-Amino acid
(mirror image)

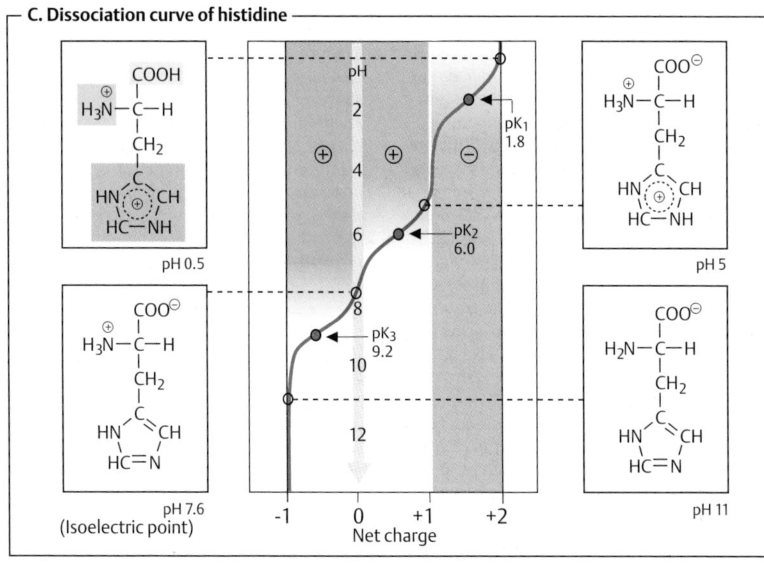
C. Dissociation curve of histidine
COOH
$H_3\overset{\oplus}{N}$—C—H
CH_2
C
HN CH
HC—NH
pH 0.5
$COO^{\ominus}$
$H_3\overset{\oplus}{N}$—C—H
CH_2
C
HN CH
HC=N
pH 7.6
(Isoelectric point)
pH
2
4
6
8
10
12
⊕ ⊕ ⊖
pK_1
1.8
pK_2
6.0
pK_3
9.2
-1
0
+1
+2
Net charge
$COO^{\ominus}$
$H_3\overset{\oplus}{N}$—C—H
CH_2
C
HN CH
HC—NH
pH 5
$COO^{\ominus}$
H_2N—C—H
CH_2
C
HN CH
HC=N
pH 11

Proteinogenic amino acids

A. The proteinogenic amino acids ◑

The amino acids that are included in the genetic code (see p. 248) are described as "proteinogenic." With a few exceptions (see p. 58), only these amino acids can be incorporated into proteins through *translation*. Only the side chains of the 20 proteinogenic amino acids are shown here. Their classification is based on the chemical structure of the side chains, on the one hand, and on their polarity on the other (see p. 6). The literature includes several slightly different systems for classifying amino acids, and details may differ from those in the system used here.

For each amino acid, the illustration names:

- *Membership of structural classes* I–VII (see below; e. g., III and VI for histidine)
- Name and abbreviation, formed from the first three letters of the name (e. g., histidine, His)
- The *one-letter symbol* introduced to save space in the electronic processing of sequence data (H for histidine)
- A quantitative *value for the polarity* of the side chain (bottom left; 10.3 for histidine). The more positive this value is, the *more polar* the amino acid is.

In addition, the polarity of the side chains is indicated by color. It increases from yellow, through light and dark green, to bluish green. For ionizing side chains, the corresponding pK_a values are also given (red numbers).

The **aliphatic** amino acids (class I) include *glycine, alanine, valine, leucine,* and *isoleucine*. These amino acids do not contain heteroatoms (N, O, or S) in their side chains and do not contain a ring system. Their side chains are markedly apolar. Together with threonine (see below), valine, leucine, and isoleucine form the group of *branched-chain amino acids*. The **sulfurcontaining amino acids** *cysteine* and *methionine* (class II), are also apolar. However, in the case of cysteine, this only applies to the undissociated state. Due to its ability to form disulfide bonds, cysteine plays an important role in the stabilization of proteins (see p. 72). Two cysteine residues linked by a disulfide bridge are referred to as *cystine* (not shown).

The **aromatic amino acids** (class III) contain resonancestabilized rings. In this group, only *phenylalanine* has strongly apolar properties. *Tyrosine* and *tryptophan* are moderately polar, and *histidine* is even strongly polar. The imidazole ring of histidine is already protonated at weakly acidic pH values. Histidine, which is only aromatic in protonated form (see p. 58), can therefore also be classified as a basic amino acid. Tyrosine and tryptophan show strong light absorption at wavelengths of 250–300 nm.

The **neutral** amino acids (class IV) have hydroxyl groups (*serine, threonine*) or amide groups (*asparagine, glutamine*). Despite their nonionic nature, the amide groups of asparagine and glutamine are markedly polar.

The carboxyl groups in the side chains of the **acidic** amino acids *aspartic acid* and *glutamic acid* (class V) are almost completely ionized at physiological pH values. The side chains of the **basic** amino acids *lysine* and *arginine* are also fully ionized—i. e., positively charged—at neutral pH. Arginine, with its positively charge guanidinium group, is particularly strongly basic, and therefore extremely polar.

Proline (VII) is a special case. Together with the α-C atom and the α-NH_2 group, its side chain forms a fivemembered ring. Its nitrogen atom is only weakly basic and is not protonated at physiological pH. Due to its ring structure, proline causes *bending of the peptide chain* in proteins (this is important in collagen, for example; see p. 70).

Several proteinogenic amino acids cannot be synthesized by the human organism, and therefore have to be supplied from the diet. These **essential amino acids** (see p. 360) are marked with a star in the illustration. Histidine and possibly also arginine are essential for infants and small children.

A. The proteinogenic amino acids

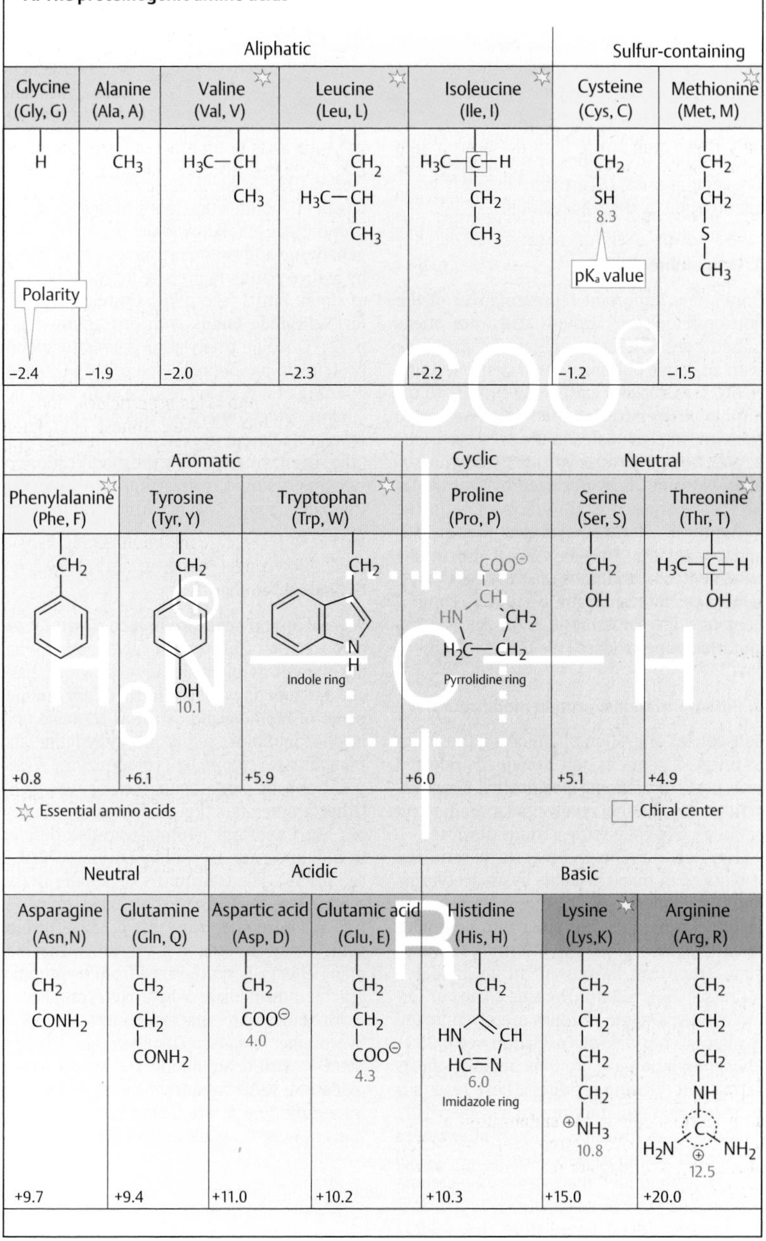

	Aliphatic				Sulfur-containing	
Glycine (Gly, G)	Alanine (Ala, A)	Valine ☆ (Val, V)	Leucine ☆ (Leu, L)	Isoleucine ☆ (Ile, I)	Cysteine (Cys, C)	Methionine ☆ (Met, M)
H	CH_3	$H_3C-CH-CH_3$	$CH_2-CH(CH_3)-CH_3$	H_3C-[C]$-H$, CH_2-CH_3	CH_2-SH (8.3, pK_a value)	$CH_2-CH_2-S-CH_3$
Polarity −2.4	−1.9	−2.0	−2.3	−2.2	−1.2	−1.5

Aromatic			Cyclic	Neutral	
Phenylalanine ☆ (Phe, F)	Tyrosine (Tyr, Y)	Tryptophan ☆ (Trp, W)	Proline (Pro, P)	Serine (Ser, S)	Threonine ☆ (Thr, T)
CH_2–phenyl	CH_2–phenyl–OH (10.1)	CH_2–indole (N–H); Indole ring	$COO^{\ominus}$, CH, HN, CH_2, H_2C-CH_2; Pyrrolidine ring	CH_2-OH	H_3C-[C]$-H$, OH
+0.8	+6.1	+5.9	+6.0	+5.1	+4.9

☆ Essential amino acids

[ ] Chiral center

Neutral		Acidic		Basic		
Asparagine (Asn,N)	Glutamine (Gln, Q)	Aspartic acid (Asp, D)	Glutamic acid (Glu, E)	Histidine (His, H)	Lysine ☆ (Lys,K)	Arginine (Arg, R)
CH_2-CONH_2	$CH_2-CH_2-CONH_2$	$CH_2-COO^{\ominus}$ (4.0)	$CH_2-CH_2-COO^{\ominus}$ (4.3)	CH_2–imidazole (HN, CH, HC=N) (6.0); Imidazole ring	$CH_2-CH_2-CH_2-CH_2-NH_3^{\oplus}$ (10.8)	$CH_2-CH_2-CH_2-NH-C(NH_2)(H_2N)^{\oplus}$ (12.5)
+9.7	+9.4	+11.0	+10.2	+10.3	+15.0	+20.0

$H_3N^{\oplus}-C-H$, $COO^{\ominus}$, R

Non-proteinogenic amino acids

In addition to the 20 proteinogenic amino acids (see p. 60), there are also many more compounds of the same type in nature. These arise during metabolic reactions (**A**) or as a result of enzymatic modifications of amino acid residues in peptides or proteins (**B**). The "biogenic amines" (**C**) are synthesized from α-amino acids by decarboxylation.

A. Rare amino acids ○

Only a few important representatives of the non-proteinogenic amino acids are mentioned here. The basic amino acid **ornithine** is an analogue of lysine with a shortened side chain. Transfer of a carbamoyl residue to ornithine yields **citrulline**. Both of these amino acids are intermediates in the urea cycle (see p. 182). **Dopa** (an acronym of 3,4-**d**ihydr**o**xy**p**henyl**a**lanine) is synthesized by hydroxylation of tyrosine. It is an intermediate in the biosynthesis of catecholamines (see p. 352) and of melanin. It is in clinical use in the treatment of *Parkinson's disease*. **Selenocysteine**, a cysteine analogue, occurs as a component of a few proteins—e. g., in the enzyme glutathione peroxidase (see p. 284).

B. Post-translational protein modification ◑

Subsequent alteration of amino acid residues in finished peptides and proteins is referred to as *post-translational modification*. These reactions usually only involve polar amino acid residues, and they serve various purposes.

The free α-amino group at the *N*-terminus is blocked in many proteins by an acetyl residue or a longer acyl residue (**acylation**). *N*-terminal glutamate can cyclize into a pyroglutamate residue, while the *C*-terminal carboxylate group can be present in an amidated form (see TSH, p. 380). The side chains of serine and asparagine residues are often linked to oligosaccharides (**glycosylation**, see p. 230). **Phosphorylation** of proteins mainly affects serine and tyrosine residues. These reactions have mainly regulatory functions (see p. 114). Aspartate and histidine residues of enzymes are sometimes phosphorylated, too. A special modification of glutamate residues, γ-**carboxylation**, is found in coagulation factors. It is essential for blood coagulation (see p. 290).

The ε-amino group of lysine residues is subject to a particularly large number of modifications. Its **acetylation** (or deacetylation) is an important mechanism for controlling genetic activity (see p. 244). Many coenzymes and cofactors are covalently linked to lysine residues. These include biotin (see p. 108), lipoic acid (see p. 106), and pyridoxal phosphate (see p. 108), as well as retinal (see p. 358). Covalent modification with **ubiquitin** marks proteins for breakdown (see p. 176). In collagen, lysine and proline residues are modified by **hydroxylation** to prepare for the formation of stable fibrils (see p. 70). Cysteine residues form **disulfide bonds** with one another (see p. 72). Cysteine **prenylation** serves to anchor proteins in membranes (see p. 214). Covalent bonding of a cysteine residue with heme occurs in cytochrome *c*. Flavins are sometimes covalently bound to cysteine or histidine residues of enzymes. Among the modifications of tyrosine residues, conversion into iodinated **thyroxine** (see p. 374) is particularly interesting.

C. Biogenic amines ◑

Several amino acids are broken down by *decarboxylation*. This reaction gives rise to what are known as biogenic amines, which have various functions. Some of them are **components of biomolecules**, such as *ethanolamine* in phospholipids (see p. 50). *Cysteamine* and *β-alanine* are components of coenzyme A (see p. 12) and of pantetheine (see pp. 108, 168). Other amines function as signaling substances. An important **neurotransmitter** derived from glutamate is γ-aminobutyrate (GABA, see p. 356). The transmitter *dopamine* is also a precursor for the catecholamines epinephrine and norepinephrine (see p. 352). The biogenic amine *serotonin*, a substance that has many effects, is synthesized from tryptophan via the intermediate 5-hydroxytryptophan.

Monamines are inactivated into aldehydes by *amine oxidase* (monoamine oxidase, "MAO") with deamination and simultaneous oxidation. MAO inhibitors therefore play an important role in pharmacological interventions in neurotransmitter metabolism.

A. Rare amino acids

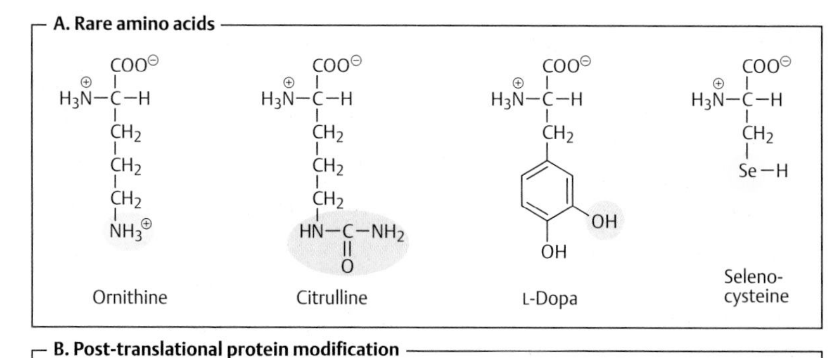

B. Post-translational protein modification

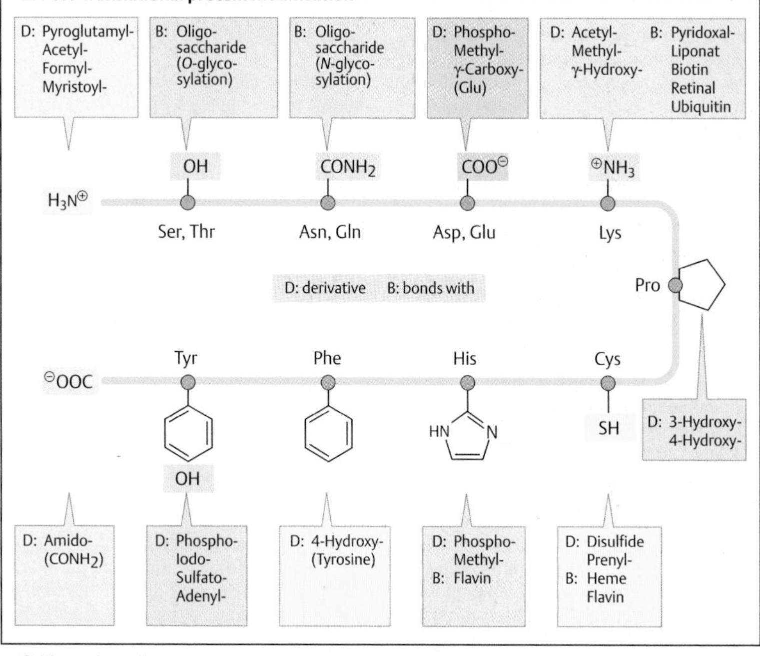

C. Biogenic amines

Amino acid	Amine	Function
Serine	Ethanol-amine	Glutamate
Cysteine	Cysteamine	Component of coenzyme A
Threonine	Amino-propanol	Component of vitamin B_{12}
Aspartate	β-Alanine	Component of coenzyme A

Amino acid	Amine	Function
Glutamate	γ-Amino-butyrate	Neurotrans-mitter (GABA)
Histidine	Histamine	Mediator, neuro-transmitter
Dopa	Dopamine	Neurotransmitter
5-Hydroxy-tryptophan	Serotonin	Mediator, neuro-transmitter

Peptides and proteins: overview

A. Proteins ◑

When amino acids are linked together by acid–amide bonds, linear macromolecules (peptides) are produced. Those containing more than ca. 100 amino acid residues are described as **proteins** (polypeptides). Every organism contains thousands of different proteins, which have a variety of functions. At a magnification of ca. 1.5 million, the semi-schematic illustration shows the structures of a few intra and extracellular proteins, giving an impression of their variety. The functions of proteins can be classified as follows.

Establishment and maintenance of structure. Structural proteins are responsible for the *shape and stability* of cells and tissues. A small part of a **collagen** molecule is shown as an example (right; see p. 70). The complete molecule is 1.5 · 300 nm in size, and at the magnification used here it would be as long as three pages of the book. **Histones** are also structural proteins. They organize the arrangement of DNA in chromatin. The basic components of chromatin, the *nucleosomes* (top right; see p. 218) consist of an octameric complex of histones, around which the DNA is coiled.

Transport. A wellknown transport protein is **hemoglobin** in the erythrocytes (bottom left). It is responsible for the transport of oxygen and carbon dioxide between the lungs and tissues (see p. 282). The blood plasma also contains many other proteins with transport functions. **Prealbumin** (transthyretin; middle), for example, transports the thyroid hormones thyroxin and triiodothyronine. **Ion channels** and other integral membrane proteins (see p. 220) facilitate the transport of ions and metabolites across biological membranes.

Protection and defense. The immune system protects the body from pathogens and foreign substances. An important component of this system is **immunoglobulin G** (bottom left; see p. 300). The molecule shown here is bound to an erythrocyte by complex formation with surface glycolipids (see p. 292).

Control and regulation. In biochemical signal chains, proteins function as signaling substances (hormones) and as hormone receptors. The complex between the growth hormone **somatotropin** and its **receptor** is shown here as an example (middle). Here, the extracellular domains of two receptor molecules here bind one molecule of the hormone. This binding activates the cytoplasmic domains of the complex, leading to further conduction of the signal to the interior of the cell (see p. 384). The small peptide hormone **insulin** is discussed in detail elsewhere (see pp. 76, 160). DNA-binding proteins (*transcription factors;* see p. 118) are decisively involved in regulating the metabolism and in differentiation processes. The structure and function of the **catabolite activator protein** (top left) and similar bacterial transcription factors have been particularly well investigated.

Catalysis. *Enzymes,* with more than 2000 known representatives, are the largest group of proteins in terms of numbers (see p. 88). The smallest enzymes have molecular masses of 10–15 kDa. Intermediatesized enzymes, such as **alcohol dehydrogenase** (top left) are around 100–200 kDa, and the largest—including **glutamine synthetase** with its 12 monomers (top right)—can reach more than 500 kDa.

Movement. The interaction between actin and myosin is responsible for muscle contraction and cell movement (see p. 332). **Myosin** (right), with a length of over 150 nm, is among the largest proteins there are. Actin filaments (**F-actin**) arise due to the polymerization of relatively small protein subunits (G-actin). Along with other proteins, **tropomyosin,** which is associated with F-actin, controls contraction.

Storage. Plants contain special **storage proteins**, which are also important for human nutrition (not shown). In animals, *muscle proteins* constitute a nutrient reserve that can be mobilized in emergencies.

A. Proteins

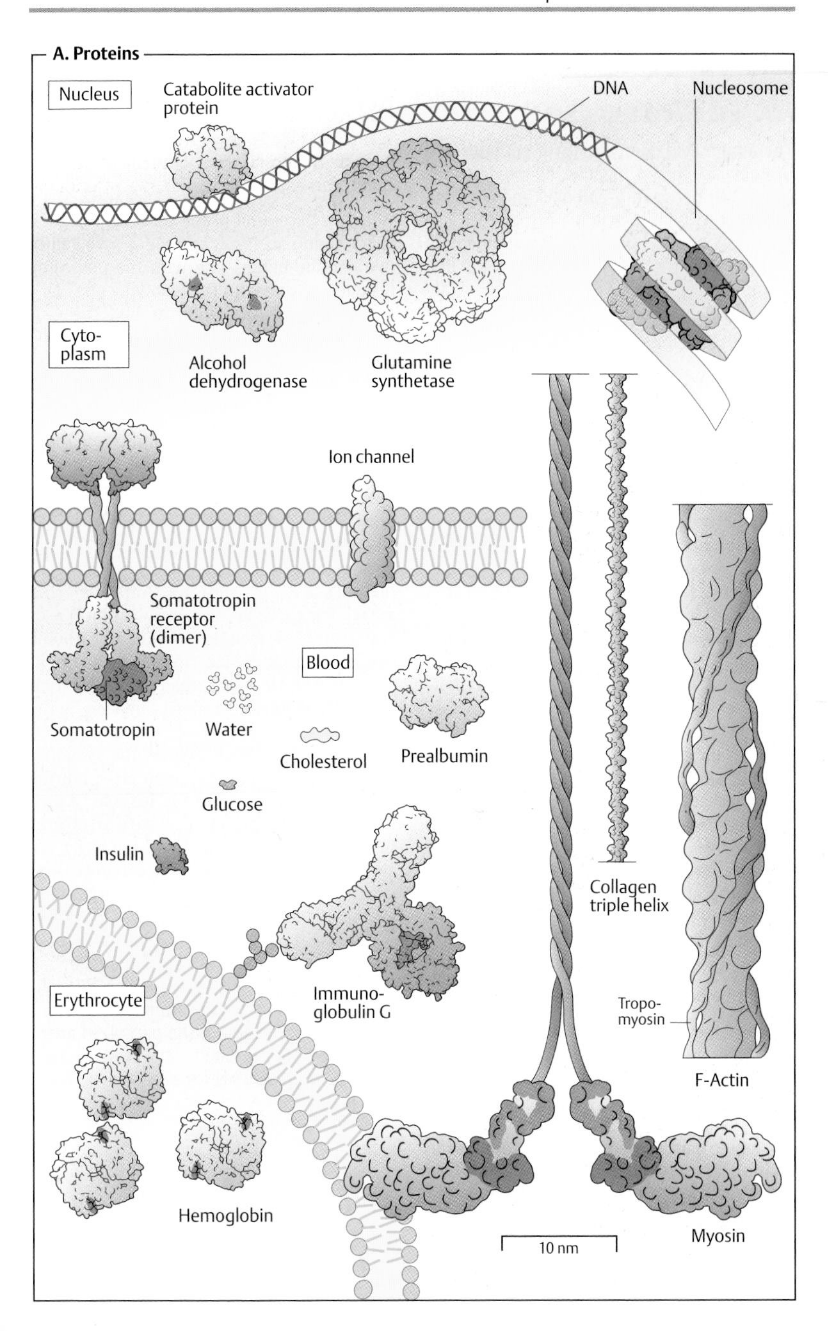
Nucleus
Catabolite activator protein
DNA
Nucleosome
Cyto-plasm
Alcohol dehydrogenase
Glutamine synthetase
Ion channel
Somatotropin receptor (dimer)
Somatotropin
Water
Blood
Cholesterol
Prealbumin
Glucose
Insulin
Collagen triple helix
Immuno-globulin G
Erythrocyte
Tropo-myosin
F-Actin
Hemoglobin
10 nm
Myosin

Peptide bonds

A. Peptide bond ◑

The amino acid components of peptides and proteins are linked together by *amide* bonds (see p. 60) between α-carboxyl and α-amino groups. This type of bonding is therefore also known as **peptide bonding**. In the **dipeptide** shown here, the serine residue has a free ammonium group, while the carboxylate group in alanine is free. Since the amino acid with the free NH_3^+ group is named first, the peptide is known as **seryl alanine**, or in abbreviated form Ser-Ala or SA.

B. Resonance ◑

Like all acid–amide bonds, the peptide bond is **stabilized by resonance** (see p. 4). In the conventional notation (top right) it is represented as a combination of a C=O double bond with a C–N single bond. However, a C=N double bond with charges at O and N could also be written (middle). Both of these are only extreme cases of electron distribution, known as *resonance structures*. In reality, the π electrons are *delocalized* throughout all the atoms (bottom). As a mesomeric system, the peptide bond is *planar*. Rotation around the C–N bond would only be possible at the expense of large amounts of energy, and the bond is therefore *not freely rotatable*. Rotations are only possible around the single bonds marked with arrows. The state of these is expressed using the angles ϕ and ψ (see **D**). The plane in which the atoms of the peptide bond lie is highlighted in light blue here and on the following pages.

C. Peptide nomenclature ◑

Peptide chains have a *direction* and therefore two different ends. The amino terminus (**N terminus**) of a peptide has a free ammonium group, while the carboxy terminus (**C terminus**) is formed by the carboxylate group of the last amino acid. In peptides and proteins, the amino acid components are usually linked in linear fashion. To express the **sequence** of a peptide, it is therefore sufficient to combine the three-letter or single-letter abbreviations for the amino acid residues (see p. 60). This sequence always starts at the N terminus. For example, the peptide hormone *angiotensin II* (see p. 330) has the sequence Asp-Arg-Val-Tyr-Ile-His-Pro-Phe, or DRVYIHPF.

D. Conformational space of the peptide chain ○

With the exception of the terminal residues, every amino acid in a peptide is involved in *two* peptide bonds (one with the preceding residue and one with the following one). Due to the restricted rotation around the C–N bond, rotations are only possible around the N–C_α and C_α–C bonds (**2**). As mentioned above, these rotations are described by the dihedral angles ϕ (phi) and ψ (psi). The angle describes rotation around the N–C_α bond; ψ describes rotation around C_α–C—i. e., the position of the subsequent bond.

For steric reasons, only specific combinations of the dihedral angles are possible. These relationships can be illustrated clearly by a so-called *ϕ/ψ diagram* (**1**). Most combinations of ϕ and ψ are sterically “forbidden” (red areas). For example, the combination ϕ = 0° and ψ = 180° (**4**) would place the two carbonyl oxygen atoms less than 115 pm apart—i. e., at a distance much smaller than the sum of their van der Waals radii (see p. 6). Similarly, in the case of ϕ = 180° and ψ = 0° (**5**), the two NH hydrogen atoms would collide. The combinations located within the green areas are the only ones that are sterically feasible (e. g., **2** and **3**). The important secondary structures that are discussed in the following pages are also located in these areas. The conformations located in the yellow areas are energetically less favorable, but still possible.

The ϕ/ψ diagram (also known as a **Ramachandran plot**) was developed from modeling studies of small peptides. However, the conformations of most of the amino acids in proteins are also located in the permitted areas. The corresponding data for the small protein, insulin (see p. 76), are represented by black dots in **1**.

A. Peptide bonds

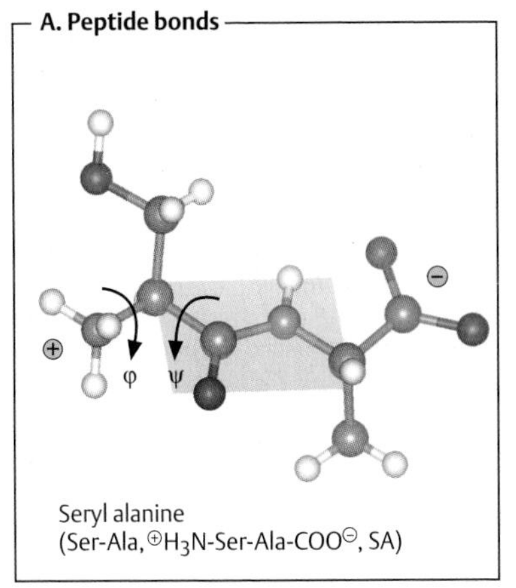

B. Resonance

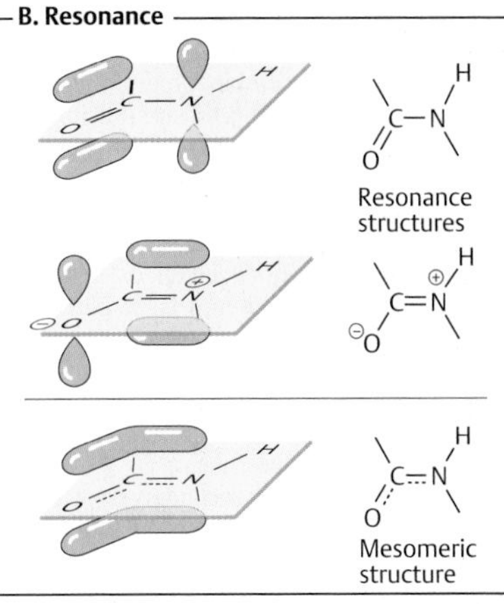

C. Peptide nomenclature

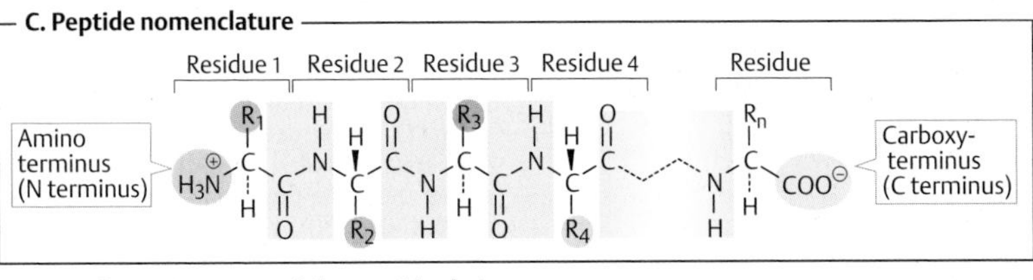

D. Conformation space of the peptide chain

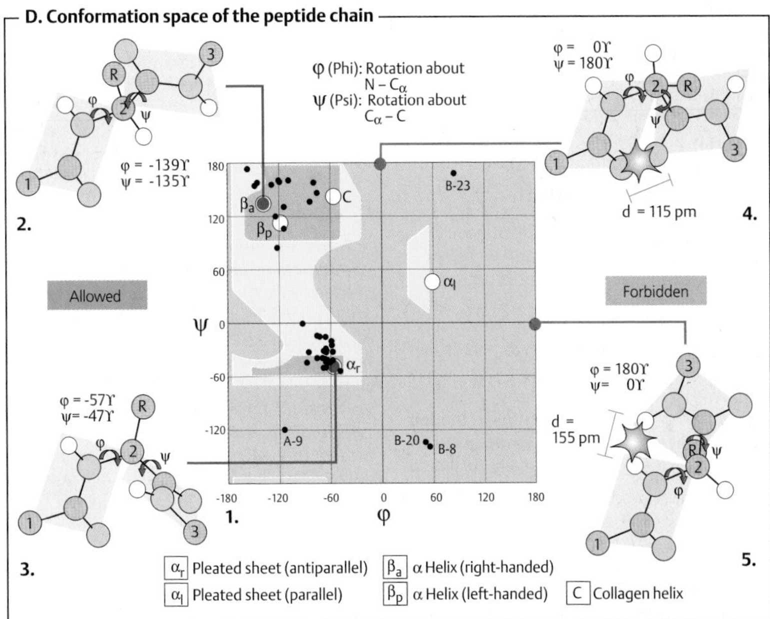

Secondary structures

In proteins, specific combinations of the dihedral angles ϕ and ψ (see p. 66) are much more common than others. When several successive residues adopt one of these conformations, defined **secondary structures** arise, which are stabilized by hydrogen bonds either within the peptide chain or between neighboring chains. When a large part of a protein takes on a defined secondary structure, the protein often forms mechanically stable filaments or fibers. **Structural proteins** of this type (see p. 70) usually have characteristic amino acid compositions.

The most important secondary structural elements of proteins are discussed here first. The illustrations only show the course of the peptide chain; the side chains are omitted. To make the course of the chains clearer, the levels of the peptide bonds are shown as blue planes. The dihedral angles of the structures shown here are also marked in diagram D1 on p. 67.

A. α-Helix ◑

The **right-handed** α-helix (α_R) is one of the most common secondary structures. In this conformation, the peptide chain is wound like a screw. Each turn of the screw (the screw axis in shown in orange) covers approximately 3.6 amino acid residues. The *pitch* of the screw (i. e., the smallest distance between two equivalent points) is 0.54 nm. α-Helices are stabilized by almost linear *hydrogen bonds* between the NH and CO groups of residues, which are four positions apart from each another in the sequence (indicated by red dots; see p. 6). In longer helices, most amino acid residues thus enter into *two* H bonds. Apolar or amphipathic α-helices with five to seven turns often serve to anchor proteins in biological membranes (*transmembrane helices;* see p. 214).

The mirror image of the α_R helix, the **left-handed α-helix** (α_L), is rarely found in nature, although it would be energetically “permissible.”

B. Collagen helix ◑

Another type of helix occurs in the collagens, which are important constituents of the connectivetissue matrix (see pp. 70, 344). The **collagen helix** is **left-handed**, and with a pitch of 0.96 nm and 3.3 residues per turn, it is steeper than the α-helix. In contrast to the α-helix, H bonds are not possible *within* the collagen helix. However, the conformation is stabilized by the association of three helices to form a righthanded **collagen triple helix** (see p. 70).

C. Pleated-sheet structures ◑

Two additional, almost stretched, conformations of the peptide chain are known as β **pleated sheets**, as the peptide planes are arranged like a regularly folded sheet of paper. Again, H bonds can only form between *neighboring chains* (“strands”) in pleated sheets. When the two strands run in opposite directions (**1**), the structure is referred to as an **antiparallel pleated sheet** (β_a). When they run in the same direction (**2**), it is a **parallel pleated sheet** (β_p). In both cases, the α-C atoms occupy the highest and lowest points in the structure, and the side chains point alternately straight up or straight down (see p. 71 C). The β_a structure, with its almost linear H bonds, is energetically more favorable. In extended pleated sheets, the individual strands of the sheet are usually not parallel, but twisted relative to one another (see p. 74).

D. β Turns ◑

β **Turns** are often found at sites where the peptide chain changes direction. These are sections in which four amino acid residues are arranged in such a way that the course of the chain reverses by about 180° into the opposite direction. The two turns shown (types I and II) are particularly frequent. Both are stabilized by hydrogen bonds between residues 1 and 4. β Turns are often located between the individual strands of antiparallel pleated sheets, or between strands of pleated sheets and α helices.

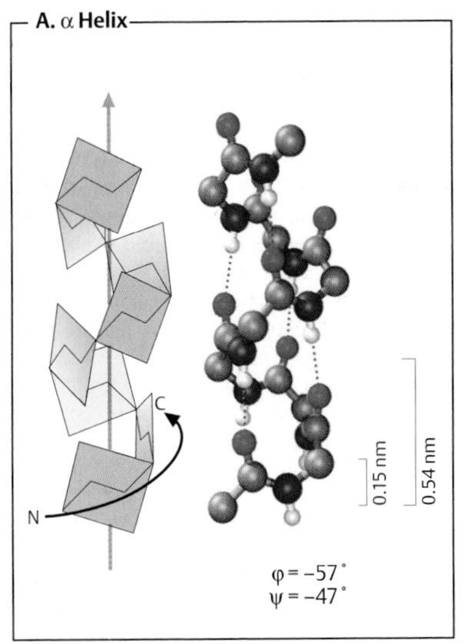
A. α Helix
N
C
0.15 nm
0.54 nm
φ = –57°
ψ = –47°

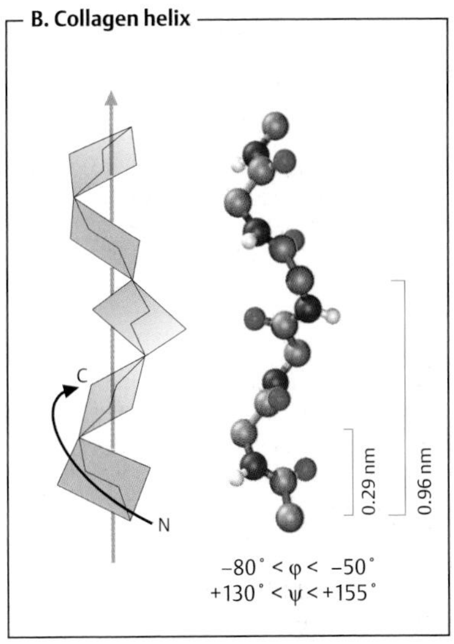
B. Collagen helix
C
N
0.29 nm
0.96 nm
–80° < φ < –50°
+130° < ψ < +155°

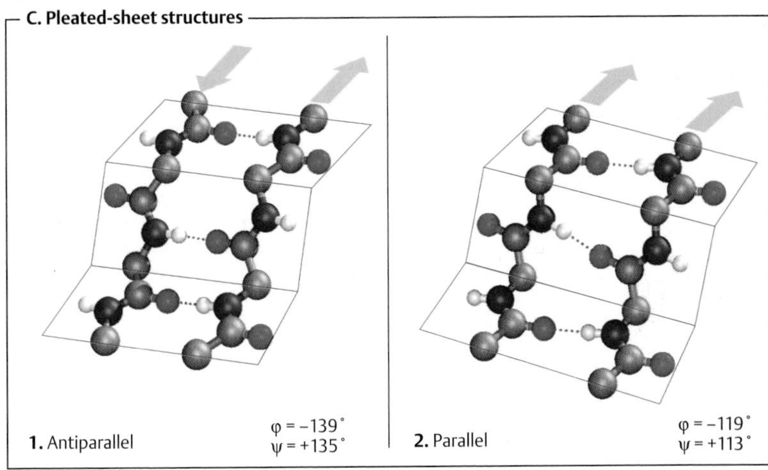
C. Pleated-sheet structures
1. Antiparallel
φ = –139°
ψ = +135°
2. Parallel
φ = –119°
ψ = +113°

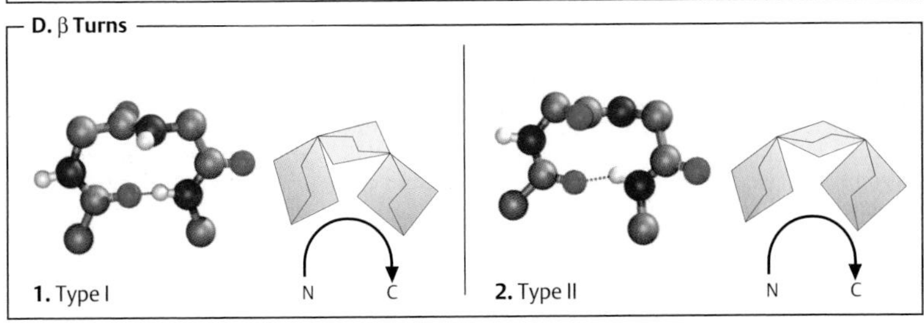
D. β Turns
1. Type I
N
C
2. Type II
N
C

Structural proteins

The **structural proteins** give extracellular structures mechanical stability, and are involved in the structure of the cytoskeleton (see p. 204). Most of these proteins contain a high percentage of specific secondary structures (see p. 68). For this reason, the amino acid composition of many structural proteins is also characteristic (see below).

A. α Keratin ○

α-**Keratin** is a structural protein that predominantly consists of α helices. Hair (wool), feathers, nails, claws and the hooves of animals consist largely of keratin. It is also an important component of the cytoskeleton (cytokeratin), where it appears in intermediate filaments (see p. 204).

In the keratins, large parts of the peptide chain show right-handed α-helical coiling. Two chains each form a left-handed **superhelix**, as is also seen in myosin (see p. 65). The superhelical keratin dimers join to form tetramers, and these aggregate further to form **protofilaments**, with a diameter of 3 nm. Finally, eight protofilaments then form an **intermediate filament**, with a diameter of 10 nm (see p. 204).

Similar keratin filaments are found in **hair**. In a single wool fiber with a diameter of about 20 μm, millions of filaments are bundled together within dead cells. The individual keratin helices are cross-linked and stabilized by numerous disulfide bonds (see p. 72). This fact is exploited in the *perming* of hair. Initially, the disulfide bonds of hair keratin are disrupted by reduction with thiol compounds (see p. 8). The hair is then styled in the desired shape and heat-dried. In the process, new disulfide bonds are formed by oxidation, which maintain the hairstyle for some time.

B. Collagen ◑

Collagen is the quantitatively most important protein in mammals, making up about 25% of the total protein. There are many different types of collagen, particularly in connective tissue. Collagen has an unusual amino acid composition. Approximately one-third of the amino acids are *glycine* (Gly), about 10% *proline* (Pro), and 10% *hydroxyproline* (Hyp). The two latter amino acids are only formed during collagen biosynthesis as a result of *posttranslational modification* (see p. 344).

The triplet Gly-X-Y (**2**) is constantly repeated in the sequence of collagen, with the X position often being occupied by Pro and the Y position by Hyp. The reason for this is that collagen is largely present as a **triple helix** made up of three individual collagen helices (**1**). In triple helices, every third residue lies on the inside of the molecule, where for steric reasons there is only room for glycine residues (**3**; the glycine residues are shown in yellow). Only a small section of a triple helix is illustrated here. The complete collagen molecule is approximately 300 nm long.

C. Silk fibroin ○

Silk is produced from the spun threads from silkworms (the larvae of the moth *Bombyx mori* and related species). The main protein in silk, **fibroin**, consists of antiparallel *pleated sheet structures* arranged one on top of the other in numerous layers (**1**). Since the amino acid side chains in pleated sheets point either straight up or straight down (see p. 68), only compact side chains fit between the layers. In fact, more than 80% of fibroin consists of glycine, alanine, and serine, the three amino acids with the shortest side chains. A typical repetitive amino acid sequence is *(Gly-Ala-Gly-Ala-Gly-Ser)*. The individual pleated sheet layers in fibroin are found to lie alternately 0.35 nm and 0.57 nm apart. In the first case, only glycine residues (R = H) are opposed to one another. The slightly greater distance of 0.57 nm results from repulsion forces between the side chains of alanine and serine residues (**2**).

A. α-Keratin

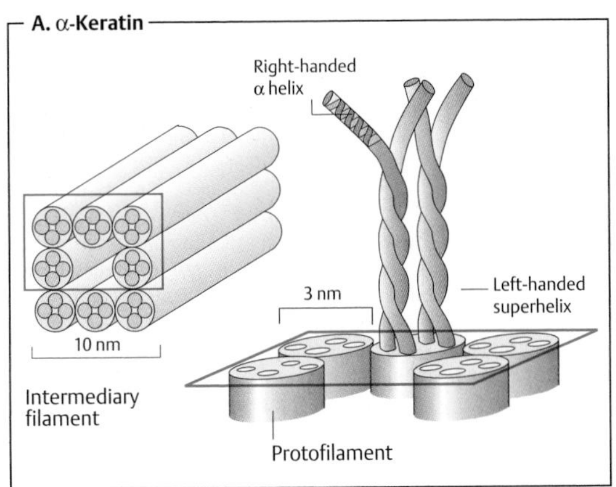

C. Silk fibroin

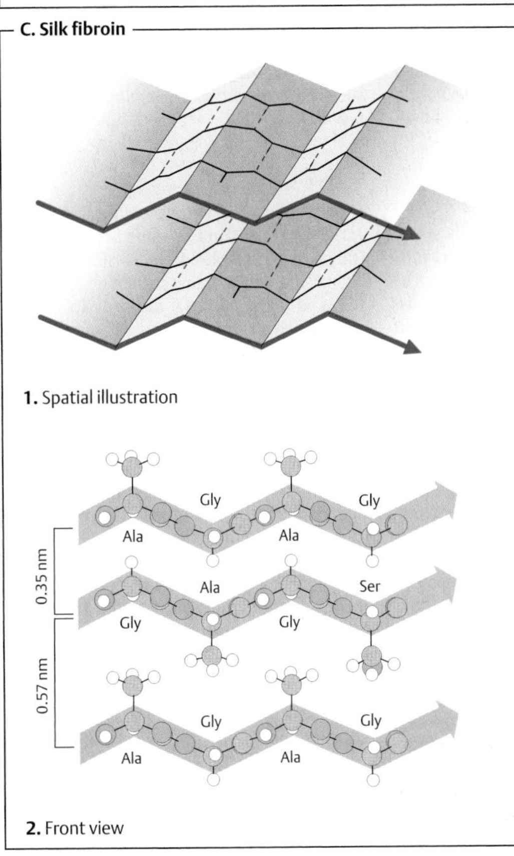

1. Spatial illustration

2. Front view

B. Collagen

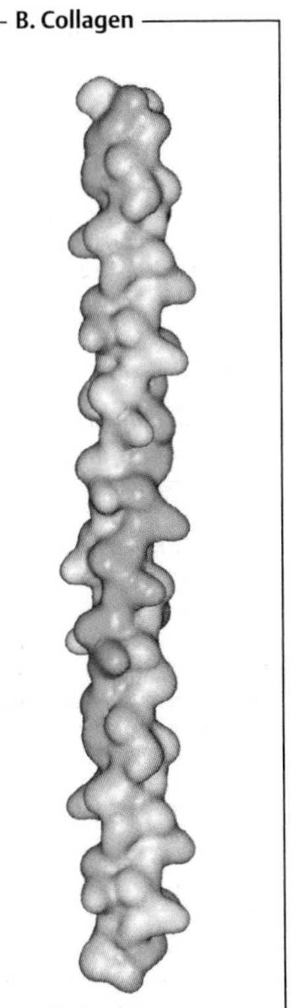

1. Triple helix (section)

Gly	Arg	Hyp
Gly	Gln	Arg
Gly	Pro	Hyp
Gly	Pro	Gln
Gly	Ala	Arg →
Gly	X	Y

2. Typical sequence

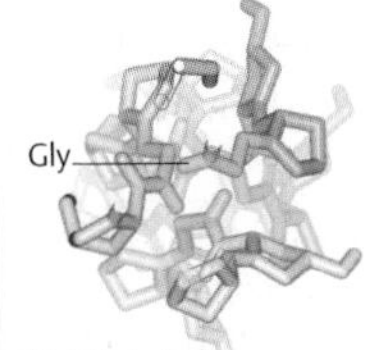

3. Triple helix (view from above)

Globular proteins

Soluble proteins have a more complex structure than the fibrous, completely insoluble structural proteins. The shape of soluble proteins is more or less spherical (globular). In their biologically active form, **globular proteins** have a defined spatial structure (the **native conformation**). If this structure is destroyed (**denaturation**; see p. 74), not only does the biological effect disappear, but the protein also usually precipitates in insoluble form. This happens, for example, when eggs are boiled; the proteins dissolved in the egg white are denatured by the heat and produce the solid egg white.

To illustrate protein conformations in a clear (but extremely simplified) way, *Richardson diagrams* are often used. In these diagrams, α-helices are symbolized by red cylinders or spirals and strands of pleated sheets by green arrows. Less structured areas of the chain, including the β-turns, are shown as sections of gray tubing.

A. Conformation-stabilizing interactions ◑

The native conformation of proteins is stabilized by a number of different interactions. Among these, only the **disulfide bonds** (**B**) represent covalent bonds. **Hydrogen bonds**, which can form inside secondary structures, as well as between more distant residues, are involved in all proteins (see p. 6). Many proteins are also stabilized by **complex formation** with metal ions (see pp. 76, 342, and 378, for example). The **hydrophobic effect** is particularly important for protein stability. In globular proteins, most hydrophobic amino acid residues are arranged in the interior of the structure in the native conformation, while the polar amino acids are mainly found on the surface (see pp. 28, 76).

B. Disulfide bonds ◑

Disulfide bonds arise when the SH groups of two cysteine residues are covalently linked as a dithiol by oxidation. Bonds of this type are only found (with a few exceptions) in extracellular proteins, because in the interior of the cell *glutathione* (see p. 284) and other reducing compounds are present in such high concentrations that disulfides would be reductively cleaved again. The small plant protein *crambin* (46 amino acids) contains three disulfide bonds and is therefore very stable. The high degree of stability of insulin (see p. 76) has a similar reason.

C. Protein dynamics ◑

The conformations of globular proteins are not rigid, but can change dramatically on binding of ligands or in contact with other proteins. For example, the enzyme *adenylate kinase* (see p. 336) has a mobile domain (domain = independently folded partial structure), which folds shut after binding of the substrate (yellow). The larger domain (bottom) also markedly alters its conformation. There are large numbers of **allosteric proteins** of this type. This group includes, for example, *hemoglobin* (see p. 280), *calmodulin* (see p. 386), and many allosteric enzymes such as *aspartate carbamoyltransferase* (see p. 116).

D. Folding patterns ○

The globular proteins show a high degree of variability in folding of their peptide chains. Only a few examples are shown here. Purely helically folded proteins such as *myoglobin* (**1**; see p. 74, heme yellow) are rare. In general, pleated sheet and helical elements exist alongside each other. In the hormone-binding domain of the *estrogen receptor* (**2**; see p. 378), a small, two-stranded pleated sheet functions as a "cover" for the hormone binding site (estradiol yellow). In *flavodoxin*, a small flavoprotein with a redox function (**3**; FMN yellow), a fan-shaped, pleated sheet made up of five parallel strands forms the core of the molecule. The conformation of the β subunit of the G-protein *transducin* (**4**; see pp. 224, 358) is very unusual. Seven pleated sheets form a large, symmetrical "β propeller." The *N*-terminal section of the protein contains one long and one short helix.

A. Conformation-stabilizing interactions

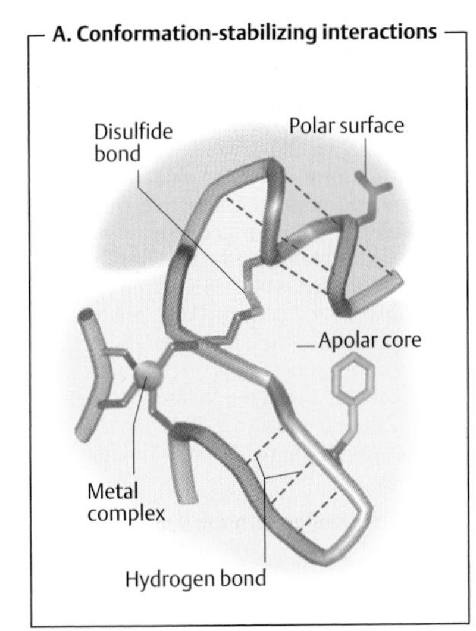

B. Disulfide bonds

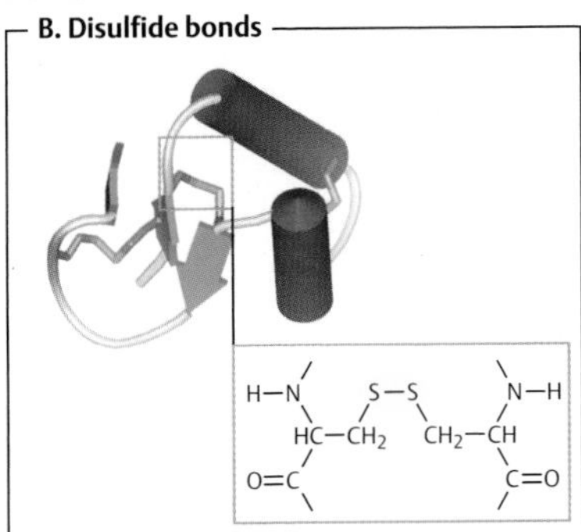

C. Protein dynamics

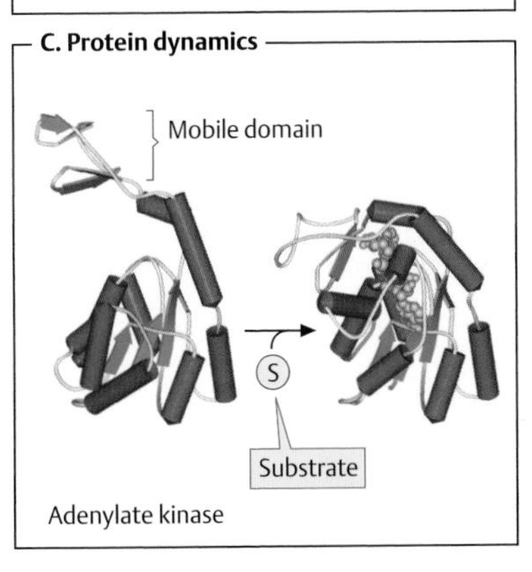

Adenylate kinase

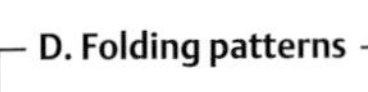

D. Folding patterns

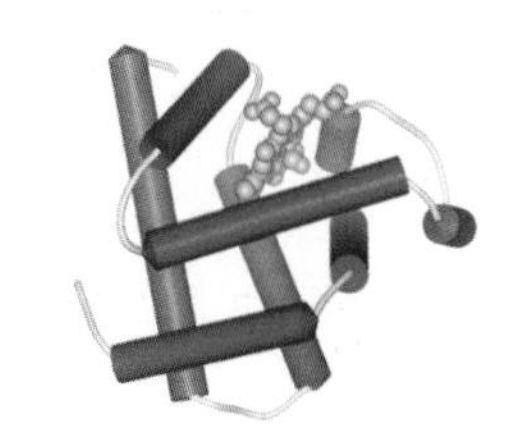

1. Myoglobin

2. Estrogen receptor (domain)

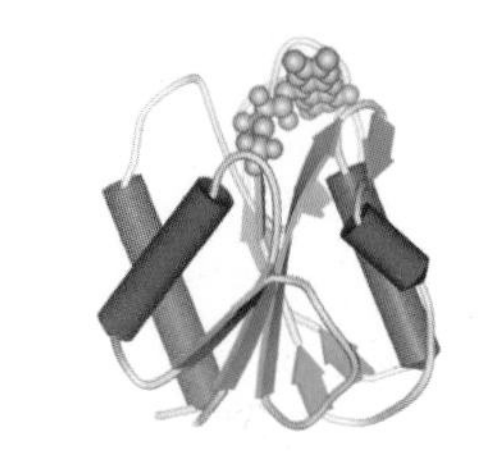

3. Flavodoxin

4. Transducin (β subunit)

Protein folding

Information about the biologically active (*native*) conformation of proteins is already encoded in their amino acid sequences. The native forms of many proteins arise spontaneously in the test tube and within a few minutes. Nevertheless, there are special auxiliary proteins (chaperonines) that support the folding of other proteins in the conditions present within the cell (see p. 232). An important goal of biochemistry is to understand the laws governing **protein folding**. This would make it possible to predict the conformation of a protein from the easily accessible DNA sequence (see p. 260).

A. Folding and denaturation of ribonuclease A ◑

The **folding** of proteins to the native form is favored under physiological conditions. The native conformation is lost, as the result of **denaturation**, at extreme pH values, at high temperatures, and in the presence of organic solvents, detergents, and other denaturing substances, such as urea.

The fact that a denatured protein can spontaneously return to its native conformation was demonstrated for the first time with **ribonuclease**, a digestive enzyme (see p. 266) consisting of 124 amino acids. In the native form (top right), there are extensive pleated sheet structures and three α helices. The eight cysteine residues of the protein are forming four disulfide bonds. Residues His-12, Lys-41 and His-119 (pink) are particularly important for catalysis. Together with additional amino acids, they form the enzyme's *active center.*

The disulfide bonds can be reductively cleaved by *thiols* (e.g., mercaptoethanol, HO-CH_2-CH_2-SH). If *urea* at a high concentration is also added, the protein unfolds completely. In this form (left), it is up to 35 nm long. Polar (green) and apolar (yellow) side chains are distributed randomly. The denatured enzyme is completely inactive, because the catalytically important amino acids (pink) are too far away from each other to be able to interact with each other and with the substrate.

When the urea and thiol are removed by dialysis (see p. 78), secondary and tertiary structures develop again spontaneously. The cysteine residues thus return to a sufficiently close spatial vicinity that disulfide bonds can once again form under the oxidative effect of atmospheric oxygen. The active center also reestablishes itself. In comparison with the denatured protein, the native form is astonishingly compact, at 4.5 · 2.5 nm. In this state, the apolar side chains (yellow) predominate in the interior of the protein, while the polar residues are mainly found on the surface. This distribution is due to the "hydrophobic effect" (see p. 28), and it makes a vital contribution to the stability of the native conformation (**B**).

B. Energetics of protein folding ○

The **energetics** of protein folding are not at present satisfactorily understood. Only a simplified model is discussed here. The conformation of a molecule is stable in any given conditions if the change in its free enthalpy during folding (ΔG_{fold}) is negative (see p. 16). The magnitude of the folding enthalpy is affected by several factors. The main factor working *against* folding is the strong increase in the ordering of the molecule involved. As discussed on p. 20, this leads to a negative change in entropy of ΔS_{conf} and therefore to a strongly positive entropy term $-T \cdot \Delta S$ (violet arrow). By contrast, the covalent and non-covalent bonds in the interior of the protein have a *stabilizing* influence. For this reason, the change in folding enthalpy ΔH_{fold} is negative (red arrow). A third factor is the change in the system's entropy due to the hydrophobic effect. During folding, the degree of order in the *surrounding water* decreases—i. e., ΔS_{water} is positive and therefore $-T \cdot \Delta S$ is negative (blue arrow). When the sum of these effects is negative (green arrow), the protein folds spontaneously into its native conformation.

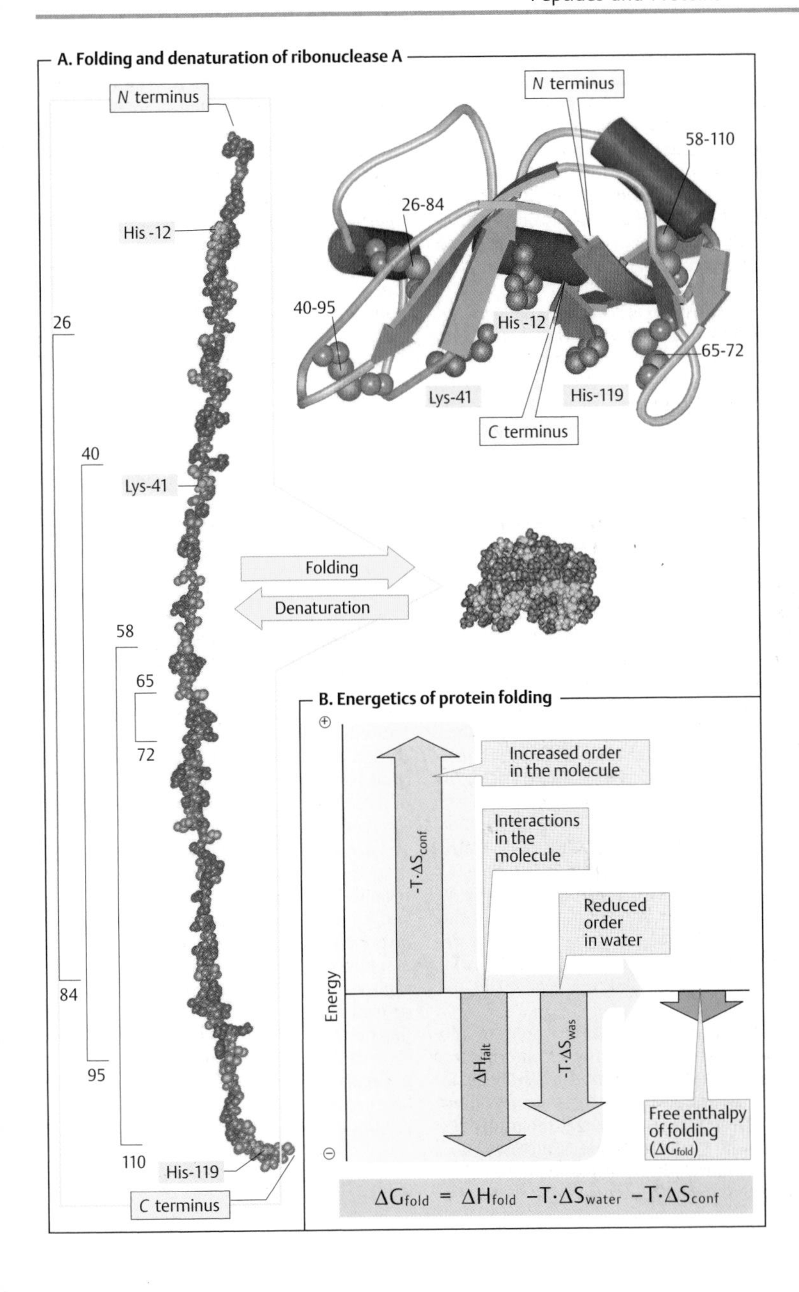
A. Folding and denaturation of ribonuclease A
N terminus
His -12
26
40
Lys-41
58
65
72
84
95
110
His-119
C terminus
N terminus
58-110
26-84
40-95
His -12
65-72
Lys-41
His-119
C terminus
Folding
Denaturation
B. Energetics of protein folding
Increased order in the molecule
Interactions in the molecule
Reduced order in water
-T·ΔS_conf
Energy
ΔH_falt
-T·ΔS_was
Free enthalpy of folding (ΔG_fold)
ΔG_fold = ΔH_fold −T·ΔS_water −T·ΔS_conf

Molecular models: insulin

The opposite page presents models of insulin, a small protein. The biosynthesis and function of this important hormone are discussed elsewhere in this book (pp. 160, 388). Monomeric insulin consists of 51 amino acids, and with a molecular mass of 5.5 kDa it is only half the size of the smallest enzymes. Nevertheless, it has the typical properties of a globular protein.

Large quantities of pure insulin are required for the treatment of *diabetes mellitus* (see p. 160). The annual requirement for insulin is over 500 kg in a country the size of Germany. Formerly, the hormone had to be obtained from the pancreas of slaughtered animals in a complicated and expensive procedure. **Human insulin**, which is produced by *overexpression* in genetically engineered bacteria, is now mainly used (see p. 262).

A. Structure of insulin ○

There are various different structural levels in proteins, and these can be briefly discussed again here using the example of insulin.

The **primary structure** of a protein is its amino acid sequence. During the biosynthesis of insulin in the pancreas, a continuous peptide chain with 84 residues is first synthesized—*proinsulin* (see p. 160). After folding of the molecule, the three disulfide bonds are first formed, and residues 31 to 63 are then proteolytically cleaved releasing the so-called *C peptide*. The molecule that is left over (**1**) now consists of two peptide chains, the *A chain* (21 residues, shown in yellow) and the *B chain* (30 residues, orange). One of the disulfide bonds is located inside the A chain, and the two others link the two chains together.

Secondary structures are regions of the peptide chain with a defined conformation (see p. 68) that are stabilized by H-bonds. In insulin (**2**), the α-helical areas are predominant, making up 57% of the molecule; 6% consists of β-pleated-sheet structures, and 10% of β-turns, while the remainder (27%) cannot be assigned to any of the secondary structures.

The three-dimensional conformation of a protein, made up of secondary structural elements and unordered sections, is referred to as the **tertiary structure**. In insulin, it is compact and wedge-shaped (**B**). The tip of the wedge is formed by the B chain, which changes its direction at this point.

Quaternary structure. Due to non-covalent interactions, many proteins assemble to form symmetrical complexes (oligomers). The individual components of oligomeric proteins (usually 2–12) are termed *subunits* or *monomers.* Insulin also forms quaternary structures. In the blood, it is partly present as a dimer. In addition, there are also hexamers stabilized by Zn^{2+} ions (light blue) (**3**), which represent the form in which insulin is stored in the pancreas (see p. 160).

B. Insulin (monomer) ○

The van der Waals model of monomeric insulin (**1**) once again shows the wedge-shaped tertiary structure formed by the two chains together. In the second model (**3**, bottom), the side chains of polar amino acids are shown in blue, while apolar residues are yellow or pink. This model emphasizes the importance of the "hydrophobic effect" for protein folding (see p. 74). In insulin as well, most hydrophobic side chains are located on the inside of the molecule, while the hydrophilic residues are located on the surface. Apparently in contradiction to this rule, several apolar side chains (pink) are found on the surface. However, all of these residues are involved in hydrophobic interactions that stabilize the dimeric and hexameric forms of insulin.

In the third model (**2**, right), the colored residues are those that are located on the surface and occur *invariably* (red) or *almost invariably* (orange) in all known insulins. It is assumed that amino acid residues that are not replaced by other residues during the course of evolution are essential for the protein's function. In the case of insulin, almost all of these residues are located on one side of the molecule. They are probably involved in the binding of the hormone to its receptor (see p. 224).

A. Structure of insulin

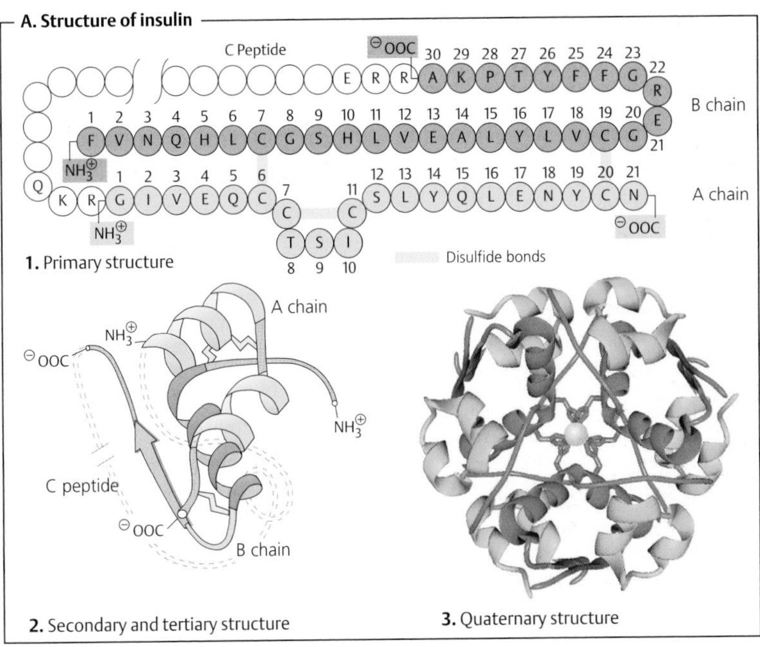

B. Insulin (monomer)

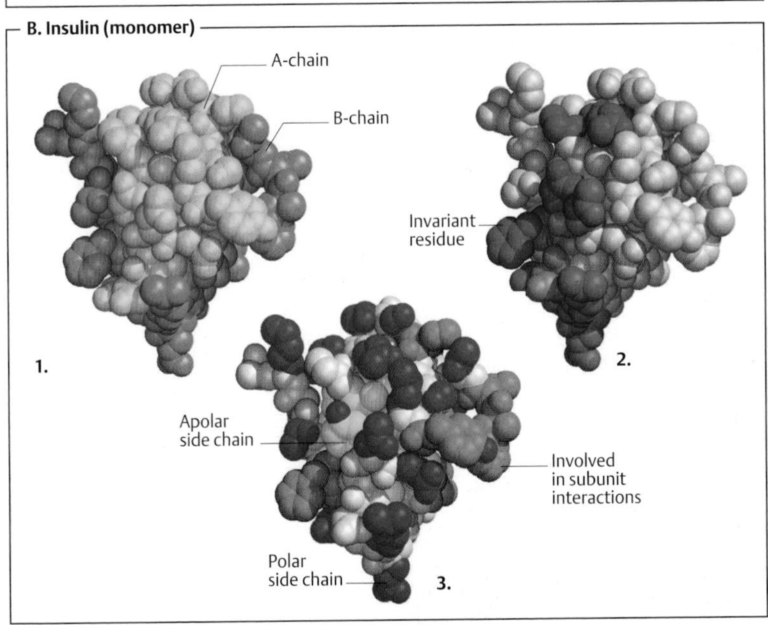

Isolation and analysis of proteins

Purified proteins are nowadays required for a wide variety of applications in research, medicine, and biotechnology. Since the globular proteins in particular are very unstable (see p. 72), purification is carried out at low temperatures (0–5 °C) and particularly gentle separation processes are used. A few of the methods of purifying and characterizing proteins are discussed on this page.

A. Salt precipitation ○

The solubility of proteins is strongly dependent on the salt concentration (*ionic strength*) of the medium. Proteins are usually poorly soluble in pure water. Their solubility increases as the ionic strength increases, because more and more of the well-hydrated anorganic ions (blue circles) are bound to the protein's surface, preventing aggregation of the molecules (**salting in**). At very high ionic strengths, the salt withdraws the hydrate water from the proteins and thus leads to aggregation and precipitation of the molecules (**salting out**). For this reason, adding salts such as ammonium sulfate $(NH_4)_2SO_4$ makes it possible to separate proteins from a mixture according to their degree of solubility (fractionation).

B. Dialysis ○

Dialysis is used to remove lower-molecular components from protein solutions, or to exchange the medium. Dialysis is based on the fact that due to their size, protein molecules are unable to pass through the pores of a **semipermeable membrane**, while lower-molecular substances distribute themselves evenly between the inner and outer spaces over time. After repeated exchanging of the external solution, the conditions inside the *dialysis tube* (salt concentration, pH, etc.) will be the same as in the surrounding solution.

C. Gel filtration ○

Gel permeation chromatography ("gel filtration") separates proteins according to their size and shape. This is done using a *chromatography column*, which is filled with spherical *gel particles* (diameter 10–500 μm) of polymeric material (shown schematically in **1a**). The insides of the particles are traversed by channels that have defined diameters. A protein mixture is then introduced at the upper end of the column (**1b**) and *elution* is carried out by passing a buffer solution through the column. Large protein molecules (red) are unable to penetrate the particles, and therefore pass through the column quickly. Medium-sized (green) and small particles (blue) are delayed for longer or shorter periods (**1c**). The proteins can be collected separately from the effluent (*eluate*) (**2**). Their elution volume V_e depends mainly on their molecular mass (**3**).

D. SDS gel electrophoresis ○

The most commonly used procedure for checking the purity of proteins is sodium dodecyl sulfate polyacrylamide gel electrophoresis (SDS-PAGE). In electrophoresis, molecules move in an electrical field (see p. 276). Normally, the speed of their movement depends on three factors—their size, their shape, and their electrical charge.

In SDS-PAGE, the protein mixture is treated in such a way that only the molecules' mass affects their movement. This is achieved by adding *sodium dodecyl sulfate* ($C_{12}H_{25}$-OSO_3Na), the sulfuric acid ester of lauryl alcohol (dodecyl alcohol). SDS is a *detergent* with strongly amphipathic properties (see p. 28). It separates oligomeric proteins into their subunits and denatures them. SDS molecules bind to the unfolded peptide chains (ca. 0.4 g SDS / g protein) and give them a strongly negative charge. To achieve complete denaturation, thiols are also added in order to cleave the disulfide bonds (**1**).

Following electrophoresis, which is carried out in a vertically arranged gel of polymeric acrylamide (**2**), the separated proteins are made visible by staining. In example (**3**), the following were separated: **a**) a cell extract with hundreds of different proteins, **b**) a protein purified from this, and **c**) a mixture of proteins with known masses.

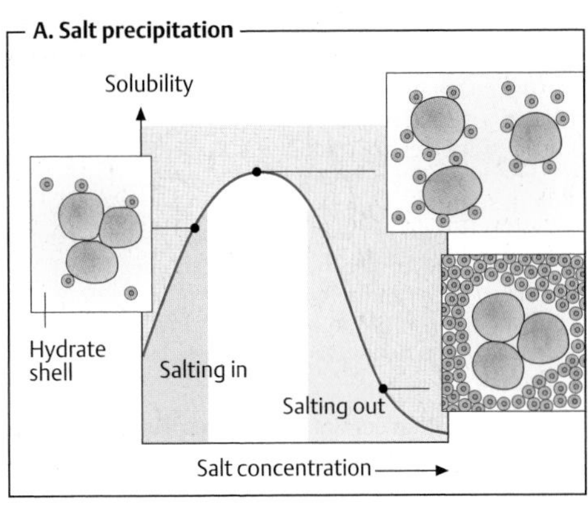
A. Salt precipitation
Solubility
Hydrate shell
Salting in
Salting out
Salt concentration

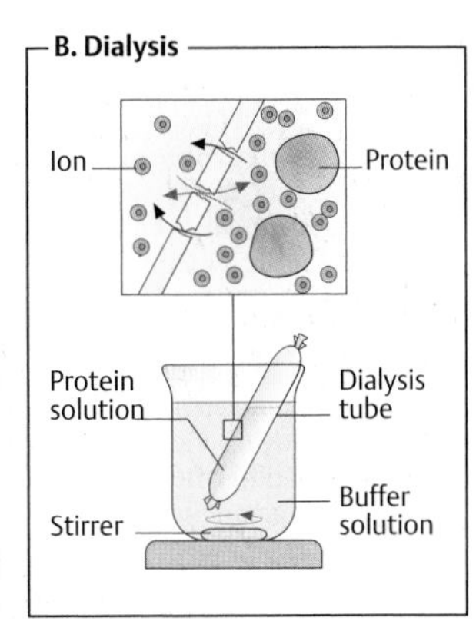
B. Dialysis
Ion
Protein
Protein solution
Dialysis tube
Buffer solution
Stirrer

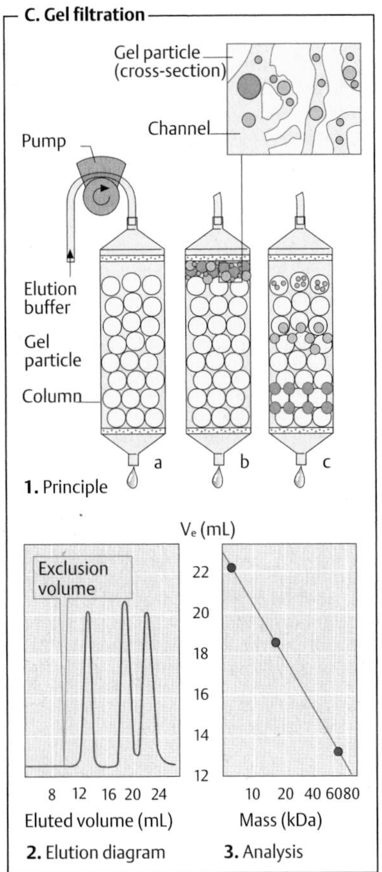
C. Gel filtration
Gel particle (cross-section)
Channel
Pump
Elution buffer
Gel particle
Column
a
b
c
1. Principle
Exclusion volume
8 12 16 20 24
Eluted volume (mL)
2. Elution diagram
Ve (mL)
22
20
18
16
14
12
10 20 40 60 80
Mass (kDa)
3. Analysis

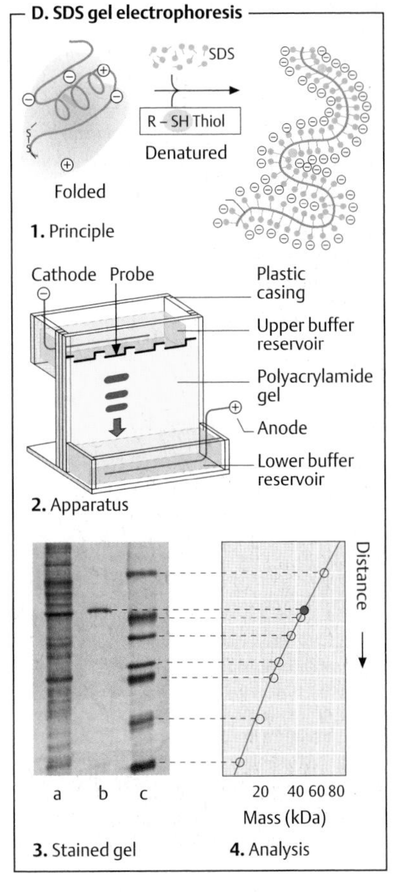
D. SDS gel electrophoresis
SDS
R – SH Thiol
Denatured
Folded
1. Principle
Cathode
Probe
Plastic casing
Upper buffer reservoir
Polyacrylamide gel
Anode
Lower buffer reservoir
2. Apparatus
a
b
c
3. Stained gel
Distance
20 40 60 80
Mass (kDa)
4. Analysis

Bases and nucleotides

The nucleic acids play a central role in the storage and expression of genetic information (see p. 236). They are divided into two major classes: **deoxyribonucleic acid (DNA)** functions solely in information storage, while **ribonucleic acids (RNAs)** are involved in most steps of gene expression and protein biosynthesis. All nucleic acids are made up from **nucleotide components**, which in turn consist of a *base*, a *sugar*, and a *phosphate residue*. DNA and RNA differ from one another in the type of the sugar and in one of the bases that they contain.

A. Nucleic acid bases ◑

The bases that occur in nucleic acids are *aromatic* heterocyclic compounds derived from either **pyrimidine** or **purine**. Five of these bases are the main components of nucleic acids in all living creatures. The purine bases **adenine** (abbreviation Ade, *not* "A"!) and **guanine** (Gua) and the pyrimidine base **cytosine** (Cyt) are present in both RNA *and* DNA. In contrast, **uracil** (Ura) is only found in RNA. In DNA, **uracil** is replaced by thymine (Thy), the 5-methyl derivative of uracil. 5-methylcytosine also occurs in small amounts in the DNA of the higher animals. A large number of other modified bases occur in tRNA (see p. 82) and in other types of RNA.

B. Nucleosides, nucleotides ◑

When a nucleic acid base is N-glycosidically linked to ribose or 2-deoxyribose (see p. 38), it yields a **nucleoside**. The nucleoside **adenosine** (abbreviation: A) is formed in this way from adenine and ribose, for example. The corresponding derivatives of the other bases are called *guanosine* (G), *uridine* (U), *thymidine* (T) and *cytidine* (C). When the sugar component is 2-deoxyribose, the product is a **deoxyribonucleoside**—e. g., 2′-deoxyadenosine (dA, not shown). In the cell, the 5′OH group of the sugar component of the nucleoside is usually esterified with phosphoric acid. 2′-Deoxythymidine (dT) therefore gives rise to **2′-deoxythymidine-5′-monophosphate (dTMP)**, one of the components of DNA (**2**). If the 5′phosphate residue is linked via an acid–anhydride bond to additional phosphate residues, it yields nucleoside diphosphates and triphosphates—e. g., ADP and ATP, which are important coenzymes in energy metabolism (see p. 106). All of these nucleoside phosphates are classified as **nucleotides**.

In nucleosides and nucleotides, the pentose residues are present in the furanose form (see p. 34). The sugars and bases are linked by an *N*-**glycosidic bond** between the C-1 of the sugar and either the N-9 of the purine ring or N-1 of the pyrimidine ring. This bond always adopts the β-configuration.

C. Oligonucleotides, polynucleotides ◑

Phosphoric acid molecules can form acid–anhydride bonds with each other. It is therefore possible for two nucleotides to be linked via the phosphate residues. This gives rise to *dinucleotides with a phosphoric acid–anhydride* structure. This group includes the coenzymes $NAD(P)^+$ and CoA, as well as the flavin derivative **FAD** (**1**; see p. 104).

If the phosphate residue of a nucleotide reacts with the 3′-OH group of a second nucleotide, the result is a *dinucleotide with a phosphoric acid diester structure*. Dinucleotides of this type have a free phosphate residue at the *5′ end* and a free OH group at the *3′ end*. They can therefore be extended with additional mononucleotides by adding further phosphoric acid diester bonds. This is the way in which **oligonucleotides**, and ultimately **polynucleotides**, are synthesized.

Polynucleotides consisting of ribonucleotide components are called **ribonucleic acid (RNA)**, while those consisting of deoxyribonucleotide monomers are called **deoxyribonucleic acid (DNA**; see p. 84). To describe the structure of polynucleotides, the abbreviations for the *nucleoside* components are written from left to right *in the 5′→3′* direction. The position of the phosphate residue is also sometimes indicated by a "p". In this way, the structure of the RNA segment shown Fig. **2** can be abbreviated as ..pUpG.. or simply as ..UG.. .

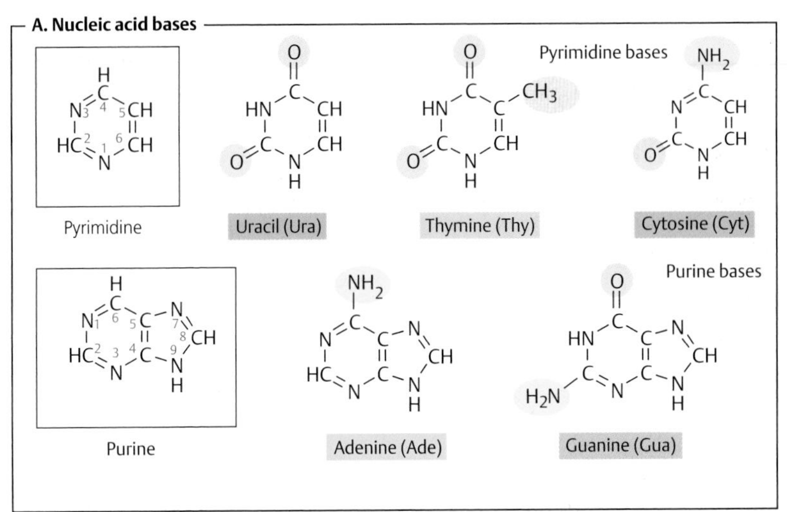
A. Nucleic acid bases
Pyrimidine bases
Pyrimidine
Uracil (Ura)
Thymine (Thy)
Cytosine (Cyt)
Purine bases
Purine
Adenine (Ade)
Guanine (Gua)

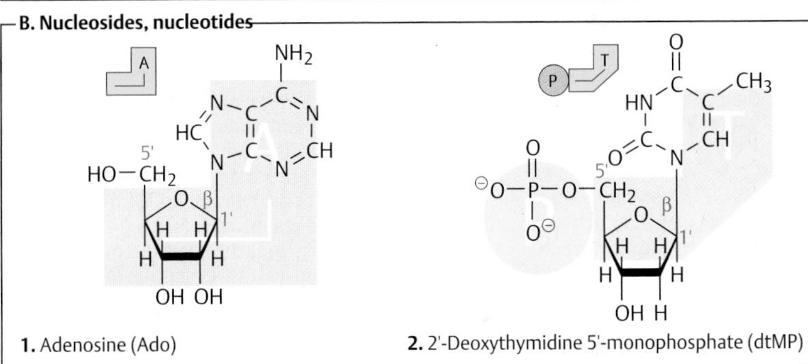
B. Nucleosides, nucleotides
1. Adenosine (Ado)
2. 2'-Deoxythymidine 5'-monophosphate (dtMP)

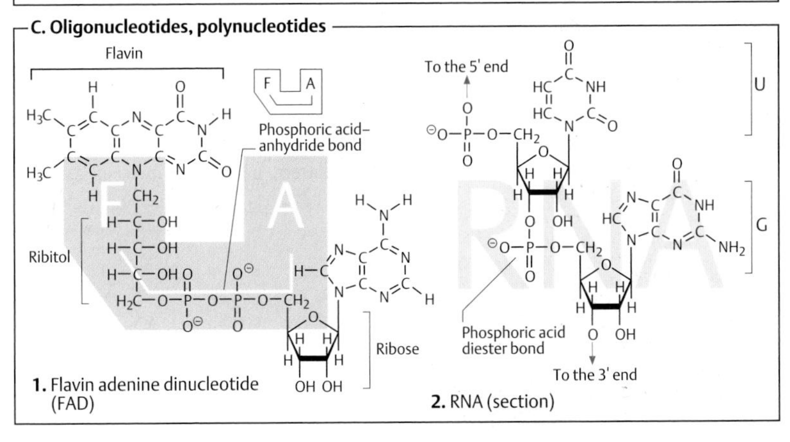
C. Oligonucleotides, polynucleotides
Flavin
Phosphoric acid–anhydride bond
Ribitol
Ribose
1. Flavin adenine dinucleotide (FAD)
To the 5' end
U
G
Phosphoric acid diester bond
To the 3' end
2. RNA (section)

RNA

Ribonucleic acids (RNAs) are polymers consisting of nucleoside phosphate components that are linked by phosphoric acid diester bonds (see p. 80). The bases the contain are mainly uracil, cytosine, adenine, and guanine, but many unusual and modified bases are also found in RNAs (**B**).

A. Ribonucleic acids (RNAs) ◑

RNAs are involved in all the individual steps of gene expression and protein biosynthesis (see pp. 242–253). The properties of the most important forms of RNA are summarized in the table. The schematic diagram also gives an idea of the secondary structure of these molecules.

In contrast to DNA, RNAs do not form extended double helices. In RNAs, the base pairs (see p. 84) usually only extend over a few residues. For this reason, substructures often arise that have a finger shape or clover-leaf shape in two-dimensional representations. In these, the paired stem regions are linked by loops. Large RNAs such as ribosomal 16S-rRNA (center) contain numerous "stem and loop" regions of this type. These sections are again folded three-dimensionally—i. e., like proteins, RNAs have a tertiary structure (see p. 86). However, tertiary structures are only known of small RNAs, mainly tRNAs. The diagrams in Fig. **B** and on p. 86 show that the "clover-leaf" structure is not recognizable in a three-dimensional representation.

Cellular RNAs vary widely in their size, structure, and lifespan. The great majority of them are ribosomal RNA **(rRNA)**, which in several forms is a structural and functional component of *ribosomes* (see p. 250). Ribosomal RNA is produced from DNA by transcription in the nucleolus, and it is processed there and assembled with proteins to form ribosome subunits (see pp. 208, 242). The bacterial 16S-rRNA shown in Fig. **A**, with 1542 nucleotides (nt), is a component of the small ribosomae subunit, while the much smaller 5S-rRNA (118 nt) is located in the large subunit.

Messenger RNAs **(mRNAs)** transfer genetic information from the cell nucleus to the cytoplasm. The primary transcripts are substantially modified while still in the nucleus (mRNA maturation; see p. 246). Since mRNAs have to be read codon by codon in the ribosome, they must not form a stable tertiary structure. This is ensured in part by the attachment of *RNA-binding proteins,* which prevent base pairing. Due to the varying amounts of information that they carry, the lengths of mRNAs also vary widely. Their lifespan is usually short, as they are quickly broken down after translation.

Small nuclear RNAs **(snRNAs)** are involved in the splicing of mRNA precursors (see p. 246). They associate with numerous proteins to form "spliceosomes."

B. Transfer RNA ($tRNA^{Phe}$) ◑

The transfer RNAs **(tRNAs)** function during translation (see p. 250) as links between the nucleic acids and proteins. They are small RNA molecules consisting of 70–90 nucleotides, which "recognize" specific mRNA codons with their *anticodons* through base pairing. At the same time, at their 3′ end (sequence .. CCA-3′) they carry the amino acid that is assigned to the relevant mRNA codon according to the genetic code (see p. 248).

The base sequence and the tertiary structure of the yeast tRNA specific for phenylalanine ($tRNA^{Phe}$) is typical of all tRNAs. The molecule (see also p. 86) contains a high proportion of unusual and modified components (shaded in dark green in Fig. **1**). These include *pseudouridine* (Ψ), *dihydrouridine* (D), *thymidine* (T), which otherwise only occurs in DNA, and many methylated nucleotides such as *7-methylguanidine* (m^7G) and—in the anticodon—*2′-O-methylguanidine* (m^2G). Numerous base pairs, sometimes deviating from the usual pattern, stabilize the molecule's conformation (**2**).

A. Ribonucleic acids (RNAs)

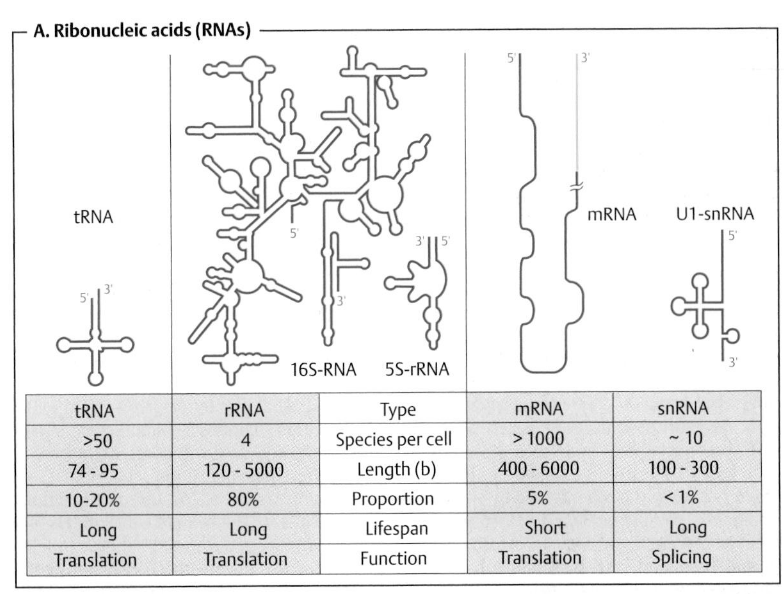

tRNA	rRNA	Type	mRNA	snRNA
>50	4	Species per cell	> 1000	~ 10
74 - 95	120 - 5000	Length (b)	400 - 6000	100 - 300
10-20%	80%	Proportion	5%	< 1%
Long	Long	Lifespan	Short	Long
Translation	Translation	Function	Translation	Splicing

B. Transfer RNA ($tRNA^{Phe}$)

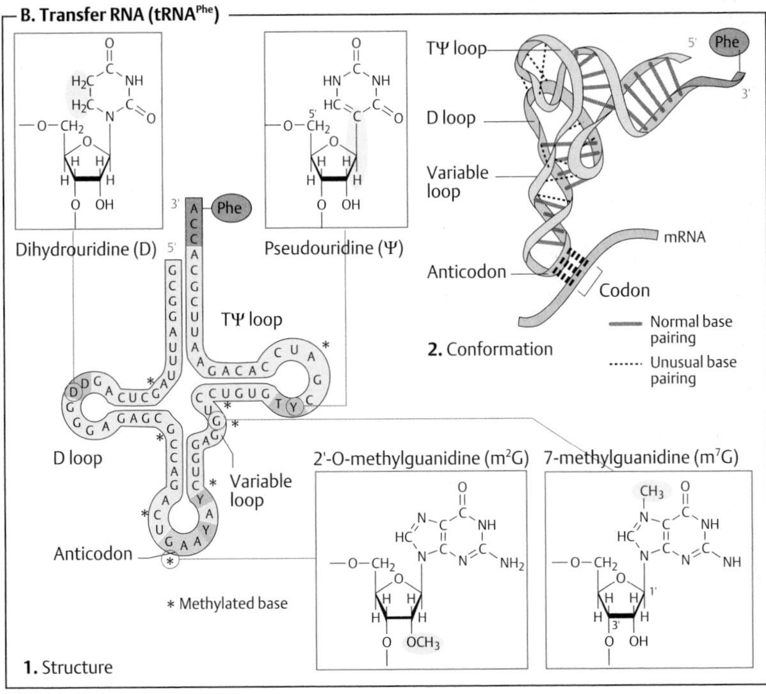

DNA

A. DNA: structure ●

Like RNAs (see p. 82), deoxyribonucleic acids (DNAs) are polymeric molecules consisting of nucleotide building blocks. Instead of ribose, however, DNA contains 2′-deoxyribose, and the *uracil* base in RNA is replaced by *thymine*. The spatial structure of the two molecules also differs (see p. 86).

The first evidence of the special structure of DNA was the observation that the amounts of adenine and thymine are almost equal in every type of DNA. The same applies to guanine and cytosine. The model of DNA structure formulated in 1953 explains these *constant base ratios:* intact DNA consists of *two* polydeoxynucleotide molecules ("strands"). Each base in one strand is linked to a *complementary* base in the other strand by H-bonds. Adenine is complementary to thymine, and guanine is complementary to cytosine. One purine base and one pyrimidine base are thus involved in each **base pair**.

The complementarity of A with T and of G with C can be understood by considering the H bonds that are possible between the different bases. Potential donors (see p. 6) are amino groups (Ade, Cyt, Gua) and ring NH groups. Possible acceptors are carbonyl oxygen atoms (Thy, Cyt, Gua) and ring nitrogen atoms. *Two* linear and therefore highly stable bonds can thus be formed in A–T pairs, and *three* in G–C pairs.

Base pairings of this type are only possible, however, when the *polarity* of the two strands differs—i. e., when they run in opposite directions (see p. 80). In addition, the two strands have to be intertwined to form a **double helix**. Due to steric hindrance by the 2′-OH groups of the ribose residues, RNA is unable to form a double helix. The structure of RNA is therefore less regular than that of DNA (see p. 82).

The conformation of DNA that predominates within the cell (known as **B-DNA**) is shown schematically in Fig. **A2** and as a van der Waals model in Fig. **B1**. In the schematic diagram (**A2**), the deoxyribose–phosphate "backbone" is shown as a ribbon. The bases (indicated by lines) are located on the inside of the **double helix**. This area of DNA is therefore apolar. By contrast, the molecule's surface is polar and negatively charged, due to the sugar and phosphate residues in the backbone. Along the whole length of the DNA molecule, there are two depressions—referred to as the "minor groove" and the "major groove"—that lie between the strands.

B. Coding of genetic information ●

In all living cells, DNA serves to **store genetic information**. Specific segments of DNA ("genes") are transcribed as needed into RNAs, which either carry out structural or catalytic tasks themselves or provide the basis for synthesizing proteins (see p. 82). In the latter case, the DNA codes for the primary structure of proteins. The "language" used in this process has four letters (A, G, C, and T). All of the words ("codons") contain three letters ("triplets"), and each triplet stands for one of the 20 proteinogenic amino acids.

The two strands of DNA are not functionally equivalent. The **template strand** (the (–) strand or "codogenic strand," shown in light gray in Fig. **1**) is the one that is read during the synthesis of RNA (transcription; see p. 242). Its sequence is complementary to the RNA formed. The **sense strand** (the (+) strand or "coding strand," shown in color in Figs. **1** and **2**) has the *same sequence as the RNA*, except that T is exchanged for U. By convention, it is agreed that gene sequences are expressed by reading the sequence of the sense strand in the 5′→3′ direction. Using the genetic code (see p. 248), in this case the protein sequence (**3**) is obtained directly in the reading direction usual for proteins—i. e., from the *N* terminus to the *C* terminus.

A. DNA: structure

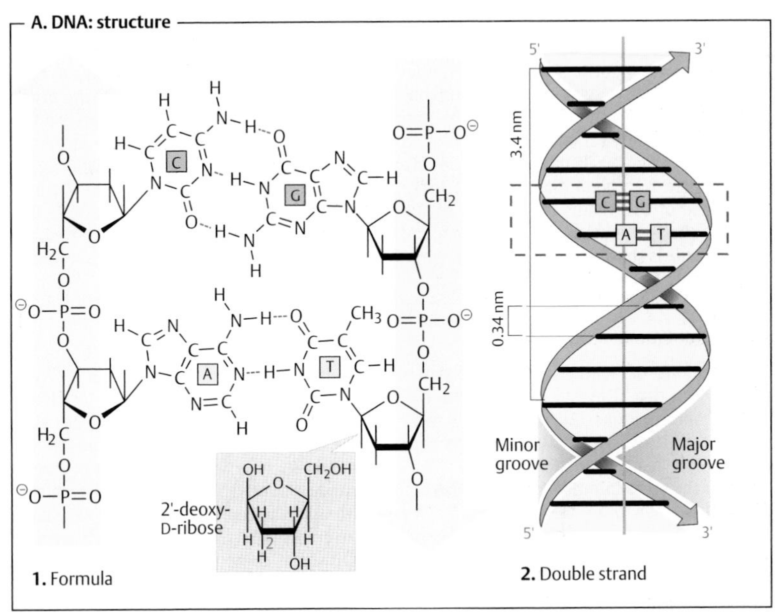

1. Formula

2. Double strand

B. Coding of genetic information

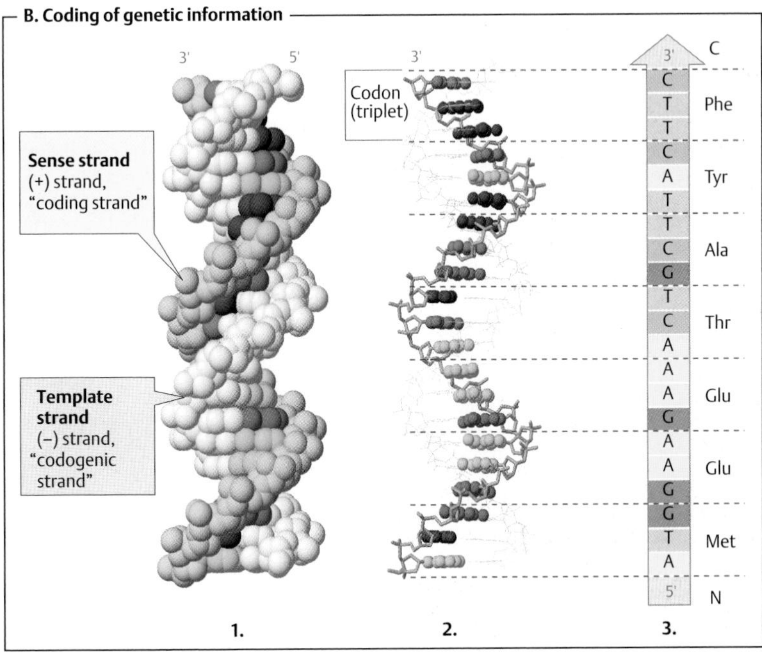

1.

2.

3.

Molecular models: DNA and RNA

The illustration opposite shows selected nucleic acid molecules. Fig. **A** shows various conformations of DNA, and Fig. **B** shows the spatial structures of two small RNA molecules. In both, the van der Waals models (see p. 6) are accompanied by ribbon diagrams that make the course of the chains clear. In all of the models, the polynucleotide "backbone" of the molecule is shown in a darker color, while the bases are lighter.

A. DNA: conformation ○

Investigations of synthetic DNA molecules have shown that DNA can adopt several different conformations. All of the DNA segments shown consist of 21 base pairs (bp) and have the same sequence.

By far the most common form is **B-DNA** (**2**). As discussed on p. 84, this consists of two antiparallel polydeoxynucleotide strands intertwined with one another to form a **right-handed double helix**. The "backbone" of these strands is formed by deoxyribose and phosphate residues linked by phosphoric acid diester bonds.

In the B conformation, the aromatic rings of the nucleobases are stacked at a distance of 0.34 nm almost at right angles to the axis of the helix. Each base is rotated relative to the preceding one by an angle of 35°. A complete turn of the double helix (360°) therefore contains around 10 base pairs (abbreviation: bp), i. e., the *pitch* of the helix is 3.4 nm. Between the backbones of the two individual strands there are two grooves with different widths. The *major groove* is visible at the top and bottom, while the narrower *minor groove* is seen in the middle. DNA-binding proteins and transcription factors (see pp. 118, 244) usually enter into interactions in the area of the major groove, with its more easily accessible bases.

In certain conditions, DNA can adopt the **A conformation** (**1**). In this arrangement, the double helix is still right-handed, but the bases are no longer arranged at right angles to the axis of the helix, as in the B form. As can be seen, the A conformation is more compact than the other two conformations. The minor groove almost completely disappears, and the major groove is narrower than in the B form. A-DNA arises when B-DNA is dehydrated. It probably does not occur in the cell.

In the **Z-conformation** (**3**), which can occur within GC-rich regions of B-DNA, the organization of the nucleotides is completely different. In this case, the helix is *left-handed*, and the backbone adopts a characteristic *zig-zag* conformation (hence "Z-DNA"). The Z double helix has a smaller pitch than B-DNA. DNA segments in the Z conformation probably have physiological significance, but details are not yet known.

B. RNA ○

RNA molecules are unable to form extended double helices, and are therefore less highly ordered than DNA molecules. Nevertheless, they have defined secondary and tertiary structures, and a large proportion of the nucleotide components enter into base pairings with other nucleotides. The examples shown here are **5S-rRNA** (see p. 242), which occurs as a structural component in ribosomes, and a **tRNA** molecule from yeast (see p. 82) that is specific for phenylalanine.

Both molecules are folded in such a way that the 3′ end and the 5′ end are close together. As in DNA, most of the bases are located in the inside of the structures, while the much more polar "backbone" is turned outwards. An exception to this is seen in the three bases of the *anticodon* of the tRNA (pink), which have to interact with mRNA and therefore lie on the surface of the molecule. The bases of the conserved CCA triplet at the 3′ end (red) also jut outward. During amino acid activation (see p. 248), they are recognized and bound by the ligases.

A. DNA: conformation

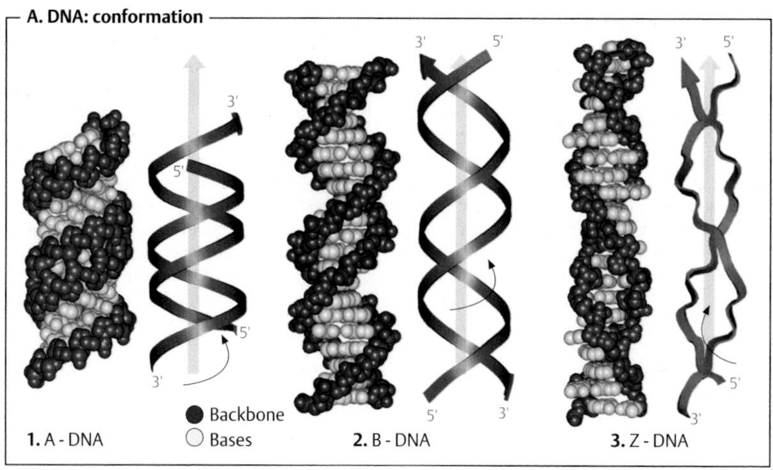

B. RNA

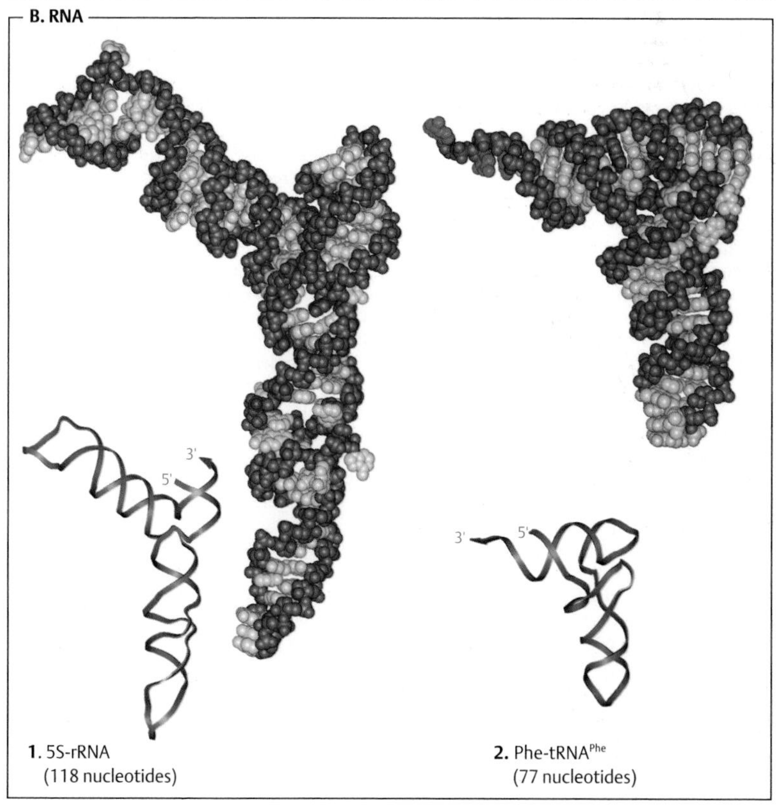

Enzymes: basics

Enzymes are **biological catalysts**—i. e., substances of biological origin that accelerate chemical reactions (see p. 24). The orderly course of metabolic processes is only possible because each cell is equipped with its own genetically determined set of enzymes. It is only this that allows coordinated sequences of reactions (**metabolic pathways**; see p. 112). Enzymes are also involved in many regulatory mechanisms that allow the metabolism to adapt to changing conditions (see p. 114). Almost all enzymes are **proteins**. However, there are also catalytically active ribonucleic acids, the *"ribozymes"* (see pp. 246, 252).

A. Enzymatic activity ●

The catalytic action of an enzyme, its **activity**, is measured by determining the **increase in the reaction rate** under precisely defined conditions—i. e., the difference between the turnover (violet) of the catalyzed reaction (orange) and uncatalyzed reaction (yellow) in a specific time interval. Normally, reaction rates are expressed as the *change in concentration per unit of time* ($mol \cdot l^{-1} \cdot s^{-1}$; see p. 22). Since the catalytic activity of an enzyme is independent of the volume, the unit used for enzymes is usually *turnover per unit time*, expressed in **katal** (kat, $mol \cdot s^{-1}$). However, the **international unit U** is still more commonly used (μmol turnover $\cdot min^{-1}$; 1 U = 16.7 nkat).

B. Reaction and substrate specificity ●

The action of enzymes is usually very *specific*. This applies not only to the type of reaction being catalyzed (**reaction specificity**), but also to the nature of the reactants ("substrates") that are involved (**substrate specificity**; see p. 94). In Fig. **B**, this is illustrated schematically using a bond-breaking enzyme as an example. Highly specific enzymes (type A, top) catalyze the cleavage of only *one* type of bond, and only when the structure of the substrate is the correct one. Other enzymes (type B, middle) have narrow reaction specificity, but broad substrate specificity. Type C enzymes (with low reaction specificity *and* low substrate specificity, bottom) are very rare.

C. Enzyme classes ◑

More than 2000 different enzymes are currently known. A system of *classification* has been developed that takes into account both their *reaction specificity* and their *substrate specificity*. Each enzyme is entered in the *Enzyme Catalogue* with a four-digit Enzyme Commission number (**EC number**). The first digit indicates membership of one of the six **major classes**. The next two indicate subclasses and subsubclasses. The last digit indicates where the enzyme belongs in the subsubclass. For example, lactate dehydrogenase (see pp. 98–101) has the EC number *1.1.1.27* (class 1, oxidoreductases; subclass 1.1, CH–OH group as electron *donor*; sub-subclass 1.1.1, $NAD(P)^+$ as electron *acceptor*).

Enzymes with similar reaction specificities are grouped into each of the six major classes:

The **oxidoreductases** (*class 1*) catalyze the transfer of reducing equivalents from one redox system to another.

The **transferases** (*class 2*) catalyze the transfer of other groups from one molecule to another. Oxidoreductases and transferases generally require coenzymes (see pp. 104ff.).

The **hydrolases** (*class 3*) are also involved in group transfer, but the acceptor is always a *water molecule*.

Lyases (*class 4*, often also referred to as "synthases") catalyze reactions involving either the cleavage or formation of chemical bonds, with double bonds either arising or disappearing.

The **isomerases** (*class 5*) move groups within a molecule, without changing the gross composition of the substrate.

The ligation reactions catalyzed by **ligases** ("synthetases," *class 6*) are energy-dependent and are therefore always coupled to the hydrolysis of nucleoside triphosphates.

In addition to the enzyme name, we also usually give its EC number. The *annotated enzyme list* (pp. 420ff.) includes all of the enzymes mentioned in this book, classified according to the Enzyme Catalog system.

A. Enzymatic activity

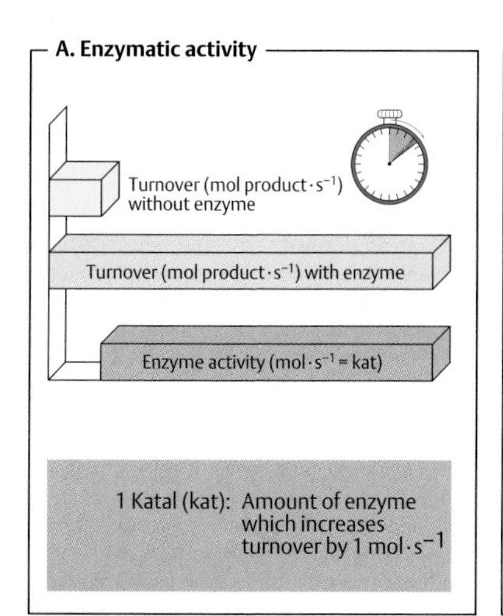

B. Reaction and substrate specificity

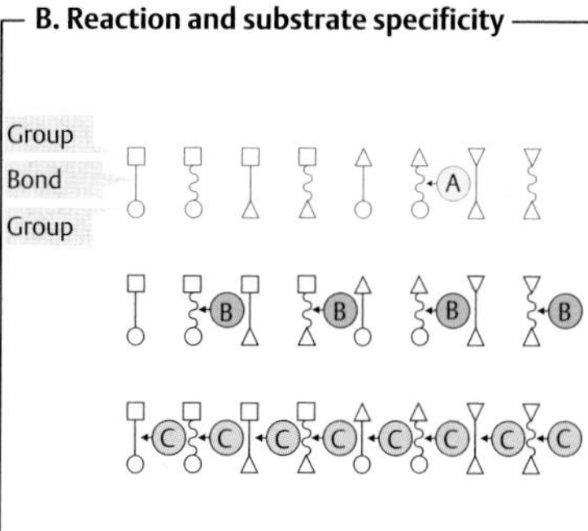

	Reaction specificity	Substrate specificity
Ⓐ	High	High
Ⓑ	High	Low
Ⓒ	Low	Low

C. The enzyme classes

Class	Reaction type	Important subclasses
1 Oxidoreductases	○ = Reduction equivalent A_{red} + B_{ox} ⇌ A_{ox} + B_{red}	Dehydrogenases Oxidases, peroxidases Reductases Monooxygenases Dioxygenases
2 Transferases	A–B + C ⇌ A + B–C	C_1-Transferases Glycosyltransferases Aminotransferases Phosphotransferases
3 Hydrolases	A–B + H_2O ⇌ A–H + B–OH	Esterases Glycosidases Peptidases Amidases
4 Lyases ("synthases")	A + B ⇌ A–B	C-C-Lyases C-O-Lyases C-N-Lyases C-S-Lyases
5 Isomerases	A ⇌ Iso-A	Epimerases *cis trans* Isomerases Intramolecular transferases
6 Ligases ("synthetases")	X = A, G, U, C B + A + XTP ⇌ A–B + XDP + P	C-C-Ligases C-O-Ligases C-N-Ligases C-S-Ligases

Enzyme catalysis

Enzymes are extremely effective **catalysts**. They can increase the rate of a catalyzed reaction by a factor of 10^{12} or more. To grasp the mechanisms involved in enzyme catalysis, we can start by looking at the course of an uncatalyzed reaction more closely.

A. Uncatalyzed reaction ○

The reaction A + B → C + D is used as an example. In solution, **reactants A and B** are surrounded by a shell of water molecules (the *hydration shell*), and they move in random directions due to thermal agitation. They can only react with each other if they collide in a favorable orientation. This is not very probable, and therefore only occurs rarely. Before conversion into the products C + D, the **collision complex A-B** has to pass through a **transition state**, the formation of which usually requires a large amount of **activation energy, E_a** (see p. 22). Since only a few A–B complexes can produce this amount of energy, a productive transition state arises even less often than a collision complex. In solution, a large proportion of the activation energy is required for the *removal of the hydration shells* between A and B. However, charge displacements and other *chemical processes* within the reactants also play a role. As a result of these limitations, conversion only happens occasionally in the absence of a catalyst, and the reaction rate v is low, even when the reaction is thermodynamically possible—i. e., when $\Delta G < 0$ (see p. 18).

B. Enzyme-catalyzed reaction ◑

Shown here is a *sequential mechanism* in which substrates A and B are bound and products C and D are released, in that order. Another possible reaction sequence, known as the *"ping-pong mechanism,"* is discussed on p. 94.

Enzymes are able to bind the reactants (their *substrates*) specifically at the **active center**. In the process, the substrates are oriented in relation to each other in such a way that they take on the *optimal orientation* for the formation of the transition state (**1**–**3**). The **proximity and orientation of the substrates** therefore strongly increase the likelihood that *productive* A–B complexes will arise. In addition, binding of the substrates results in removal of their hydration shells. As a result of the **exclusion of water**, very different conditions apply in the active center of the enzyme during catalysis than in solution (**3**–**5**). A third important factor is the **stabilization of the transition state** as a result of interactions between the amino acid residues of the protein and the substrate (**4**). This further reduces the activation energy needed to create the transition state. Many enzymes also take up groups from the substrates or transfer them to the substrates during catalysis.

Proton transfers are particularly common. This **acid–base catalysis** by enzymes is much more effective than the exchange of protons between acids and bases in solution. In many cases, chemical groups are temporarily bound covalently to the amino acid residues of the enzyme or to coenzymes during the catalytic cycle. This effect is referred to as **covalent catalysis** (see the transaminases, for example; p. 178). The principles of enzyme catalysis sketched out here are discussed in greater detail on p. 100 using the example of lactate dehydrogenase.

C. Principles of enzyme catalysis ◑

Although it is difficult to provide quantitative estimates of the contributions made by individual catalytic effects, it is now thought that the enzyme's **stabilization of the transition state** is the most important factor. It is not tight binding of the *substrate* that is important, therefore—this would increase the activation energy required by the reaction, rather than reducing it—but rather the binding of the transition state. This conclusion is supported by the very high affinity of many enzymes for analogues of the transition state (see p. 96). A simple mechanical analogy may help clarify this (right). To transfer the metal balls (the reactants) from location EA (the substrate state) via the higher-energy transition state to EP (the product state), the magnet (the catalyst) has to be orientated in such a way that its attractive force acts on the transition state (bottom) rather than on EA (top).

A. Uncatalyzed reaction

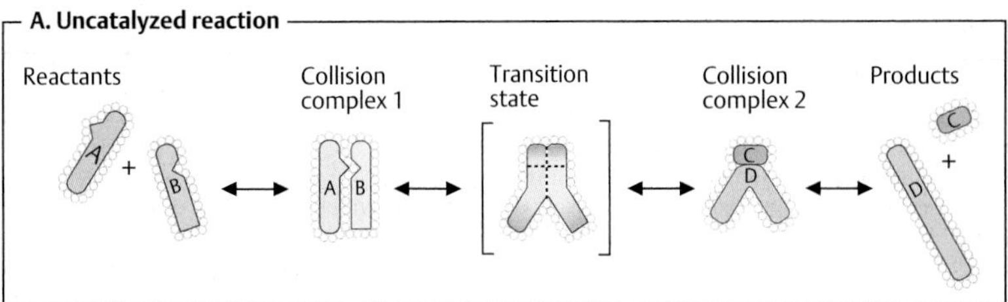

B. Enzyme-catalyzed reaction

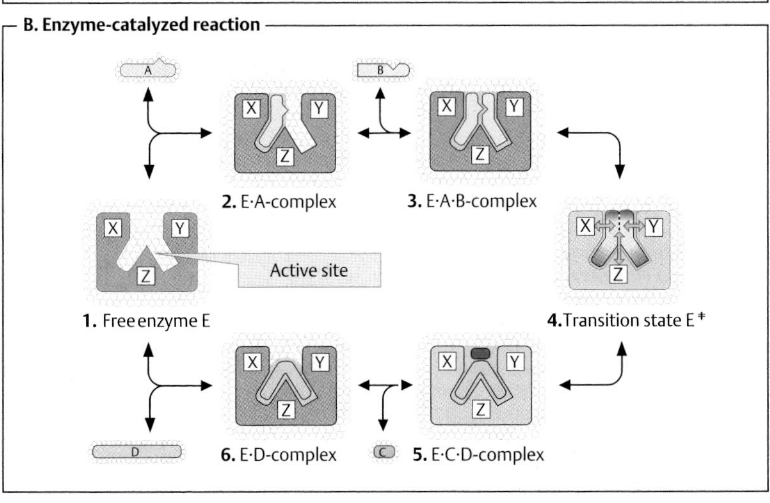

C. Principles of enzyme catalysis

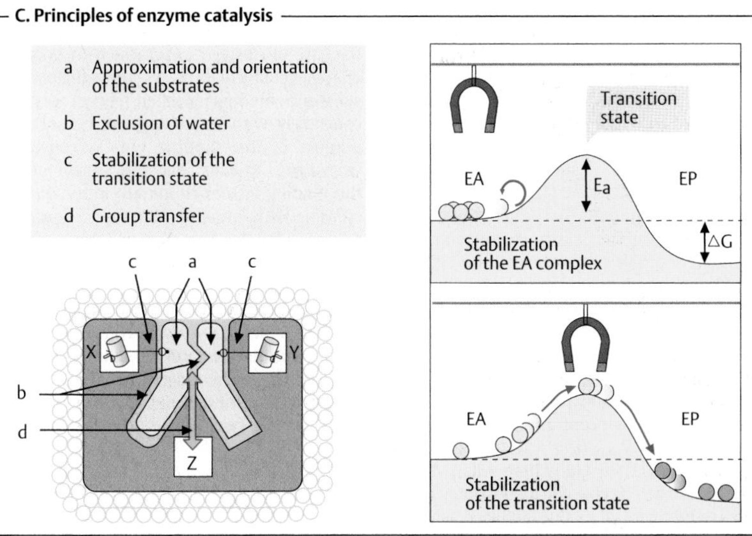

Enzyme kinetics I

The **kinetics** of enzyme-catalyzed reactions (i. e., the dependence of the reaction rate on the reaction conditions) is mainly determined by the *properties of the catalyst.* It is therefore more complex than the kinetics of an uncatalyzed reaction (see p. 22). Here we discuss these issues using the example of a simple first-order reaction (see p. 22)

A. Michaelis–Menten kinetics ◑

In the absence of an enzyme, the reaction rate v is proportional to the concentration of substance A (top). The constant k is the *rate constant* of the uncatalyzed reaction. Like all catalysts, the enzyme E (total concentration $[E]_t$) creates a new reaction pathway. Initially, A is bound to E (partial reaction 1, left). If this reaction is in chemical equilibrium, then with the help of the law of mass action—and taking into account the fact that $[E]_t = [E] + [EA]$—one can express the concentration [EA] of the *enzyme–substrate* complex as a function of [A] (left). The **Michaelis constant** K_m thus describes the state of equilibrium of the reaction. In addition, we know that $k_{cat} > k$—in other words, enzyme-bound substrate reacts to B much faster than A alone (partial reaction 2, right). k_{cat}, the enzyme's **turnover number**, corresponds to the number of substrate molecules converted by one enzyme molecule per second. Like the conversion A → B, the formation of B from EA is a first-order reaction—i. e., $v = k \cdot [EA]$ applies. When this equation is combined with the expression already derived for EA, the result is the **Michaelis–Menten equation.**

In addition to the *variables* v and [A], the equation also contains two *parameters* that do not depend on the substrate concentration [A], but describe properties of the enzyme itself: the product $k_{cat} \cdot [E]_g$ is the limiting value for the reaction rate at a very high [A], the **maximum velocity V_{max}** of the reaction (recommended abbreviation: V). The **Michaelis constant** K_m characterizes the *affinity* of the enzyme for a substrate. It corresponds to the substrate concentration at which v reaches half of V_{max} (if $v = V_{max}/2$, then $[A]/(K_m + [A]) = 1/2$, i. e. [A] is then $= K_m$). A *high* affinity of the enzyme for a substrate therefore leads to a *low* K_m value, and vice versa. Of the two enzymes whose *substrate saturation curves* are shown in diagram 1, enzyme 2 has the higher affinity for A [$K_m = 1\ \text{mmol} \cdot l^{-1}$]; V_{max}, by contrast, is much lower than with enzyme 1.

Since v approaches V *asymptotically* with increasing values of [A], it is difficult to obtain reliable values for V_{max}—and thus for K_m as well—from diagrams plotting v against [A]. To get around this, the Michaelis–Menten equation can be arranged in such a way that the measured points lie on a *straight line.* In the **Lineweaver–Burk plot (2)**, 1/v is plotted against 1/[A]. The intersections of the line of best fit with the axes then produce $1/V_{max}$ and $-1/K_m$. This type of diagram is very clear, but for practical purposes it is less suitable for determining V_{max} and K_m. Calculation methods using personal computers are faster and more objective.

B. Isosteric and allosteric enzymes ◑

Many enzymes can occur in various *conformations* (see p. 72), which have different catalytic properties and whose proportion of the total number of enzyme molecules is influenced by substrates and other ligands (see pp. 116 and 280, for example). **Allosteric enzymes** of this type, which are usually present in oligomeric form, can be recognized by their S-shaped (*sigmoidal*) saturation curves, which cannot be described using the Michaelis model. In the case of isosteric enzymes (with only *one* enzyme conformation, **1**), the efficiency of substrate binding (dashed curve) declines constantly with increasing [A], because the number of free binding sites is constantly decreasing. In most allosteric enzymes (**2**), the binding efficiency initially rises with increasing [A], because the free enzyme is present in a low-affinity conformation (square symbols), which is gradually converted into a higher-affinity form (round symbols) as a result of binding with A. It is only at high [A] values that a lack of free binding sites becomes noticeable and the binding strength decreases again. In other words, the affinity of allosteric enzymes is not constant, but depends on the type and concentration of the ligand.

A. Michaelis Menten kinetics

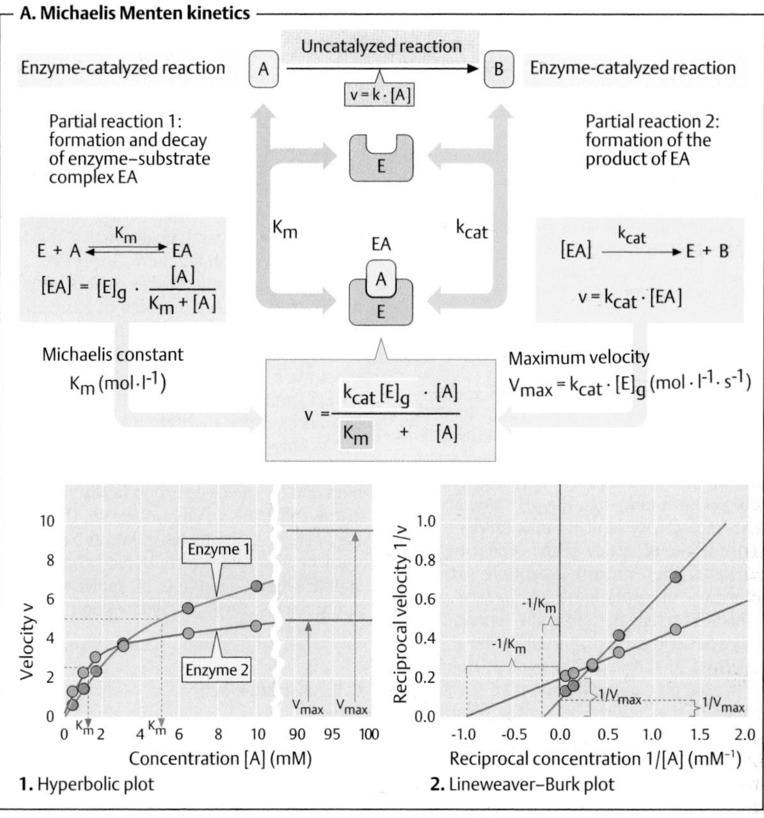

1. Hyperbolic plot

2. Lineweaver–Burk plot

B. Isosteric and allosteric enzymes

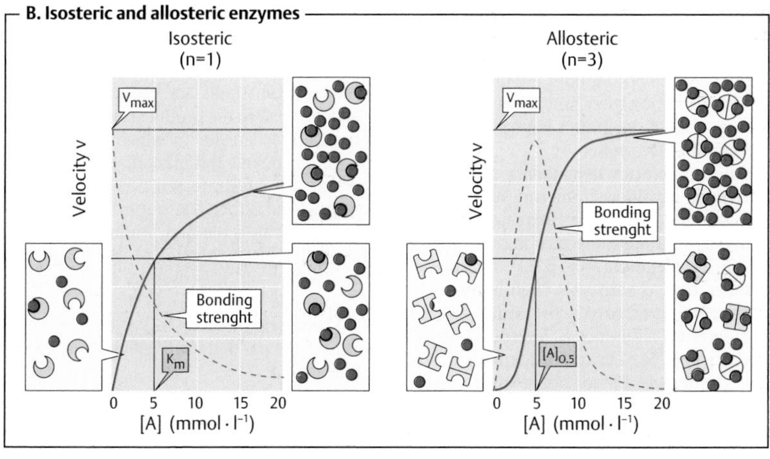

Enzyme kinetics II

The catalytic properties of enzymes, and consequently their activity (see p. 90), are influenced by numerous factors, which all have to be optimized and controlled if activity measurements are to be carried out in a useful and reproducible fashion. These factors include physical quantities (temperature, pressure), the chemical properties of the solution (pH value, ionic strength), and the concentrations of the relevant substrates, cofactors, and inhibitors.

A. pH and temperature dependency of enzyme activity ◑

The effect of enzymes is strongly dependent on the pH value (see p. 30). When the activity is plotted against pH, a *bell-shaped curve* is usually obtained (**1**). With animal enzymes, the **pH optimum**—i. e., the pH value at which enzyme activity is at its maximum—is often close to the pH value of the cells (i. e., pH 7). However, there are also exceptions to this. For example, the proteinase *pepsin* (see p. 270), which is active in the acidic gastric lumen, has a pH optimum of 2, while other enzymes (at least in the test tube) are at their most active at pH values higher than 9. The bell shape of the activity–pH profile results from the fact that amino acid residues with ionizable groups in the side chain are essential for catalysis. In example (**1**), these are a basic group B ($pK_a = 8$), which has to be protonated in order to become active, and a second acidic amino acid AH ($pK_a = 6$), which is only active in a dissociated state. At the optimum pH of 7, around 90% of both groups are present in the active form; at higher and lower values, one or the other of the groups increasingly passes into the inactive state.

The **temperature dependency** of enzymatic activity is usually asymmetric. With increasing temperature, the increased thermal movement of the molecules initially leads to a rate acceleration (see p. 22). At a certain temperature, the enzyme then becomes unstable, and its activity is lost within a narrow temperature difference as a result of denaturation (see p. 74). The optimal temperatures of the enzymes in higher organisms rarely exceed 50 °C, while enzymes from thermophilic bacteria found in hot springs, for instance, may still be active at 100 °C.

B. Substrate specificity ◑

Enzymes "recognize" their substrates in a highly specific way (see p. 88). It is only the marked **substrate specificity** of the enzymes that makes a regulated metabolism possible. This principle can be illustrated using the example of the two closely related proteinases *trypsin* and *chymotrypsin*. Both belong to the group of serine proteinases and contain the same "triad" of catalytically active residues (Asp–His–Ser, shown here in green; see p. 176). Trypsin selectively cleaves peptide bonds on the C-terminal side of basic amino acids (lysine and arginine), while chymotrypsin is specific for hydrophobic residues. The substrate binding "pockets" of both enzymes have a similar structure, but their amino acid sequences differ slightly. In trypsin, a negatively charged aspartate residue (Asp-189, red) is arranged in such a way that it can bind and fix the basic group in the side chain of the substrate. In chymotrypsin, the "binding pocket" is slightly narrower, and it is lined with neutral and hydrophobic residues that stabilize the side chains of apolar substrate amino acids through hydrophobic interactions (see p. 28).

C. Bisubstrate kinetics ○

Almost all enzymes—in contrast to the simplified description given on p. 92—have more than one substrate or product. On the other hand, it is rare for more than two substrates to be bound *simultaneously*. In bisubstrate reactions of the type $A + B \rightarrow C + D$, a number of reaction sequences are possible. In addition to the *sequential mechanisms* (see p. 90), in which all substrates are bound in a specific sequence before the product is released, there are also mechanisms in which the first substrate A is bound and immediately cleaved. A part of this substrate remains bound to the enzyme, and is then transferred to the second substrate B after the first product C has been released. This is known as the **ping-pong mechanism**, and it is used by *transaminases*, for example (see p. 178). In the Lineweaver–Burk plot (right; see p. 92), it can be recognized in the parallel shifting of the lines when [B] is varied.

A. pH and temperature dependency of enzyme activity

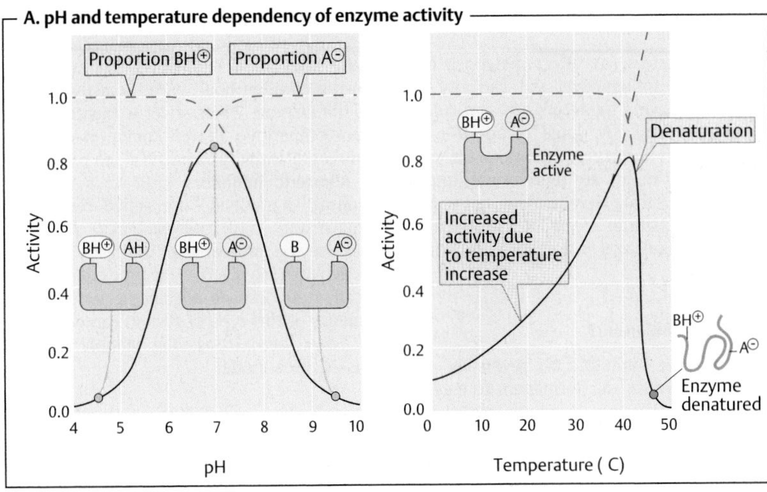

B. Substrate specificity

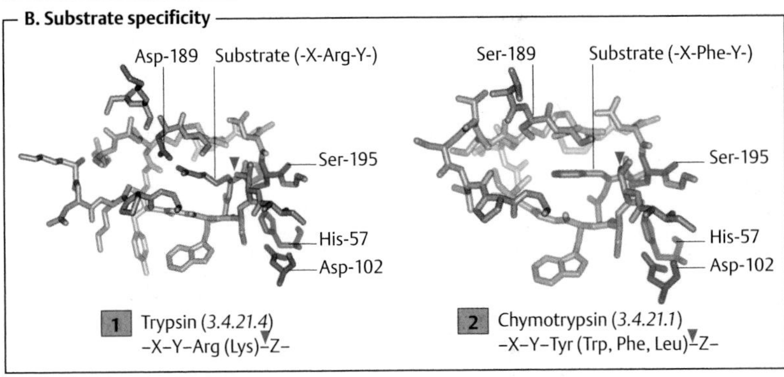

C. Bisubstrate kinetics

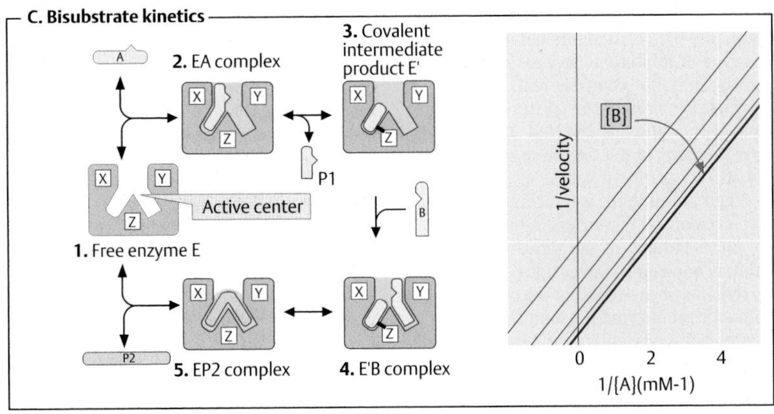

Inhibitors

Many substances can affect metabolic processes by influencing the activity of enzymes. **Enzyme inhibitors** are particularly important here. A large proportion of **medicines** act as enzyme inhibitors. Enzyme-kinetic experiments are therefore an important aspect of drug development and testing procedures. Natural *metabolites* are also involved in regulatory processes as inhibitors (see p. 114).

A. Types of inhibitor ◑

Most enzyme inhibitors act **reversibly**—i. e., they do not cause any permanent changes in the enzyme. However, there are also **irreversible** inhibitors that permanently modify the target enzyme. The mechanism of action of an inhibitor—its **inhibition type**—can be determined by comparing the kinetics (see p. 92) of the inhibited and uninhibited reactions (**B**). This makes it possible to distinguish *competitive inhibitors* (left) from *noncompetitive inhibitors* (right), for example. *Allosteric inhibition* is particularly important for metabolic regulation (see below).

Substrate analogs (**2**) have properties similar to those of one of the substrates of the target enzyme. They are bound by the enzyme, but cannot be converted further and therefore *reversibly* block some of the enzyme molecules present. A *higher* substrate concentration is therefore needed to achieve a half-maximum rate; the Michaelis constant K_m increases (**B**). High concentrations of the substrate displace the inhibitor again. The maximum rate V_{max} is therefore not influenced by this type of inhibition. Because the substrate and the inhibitor compete with one another for the *same* binding site on the enzyme, this type of inhibition is referred to as **competitive. Analogs of the transition state** (**3**) usually also act competitively.

When an inhibitor interacts with a group that is important for enzyme activity, but does not affect binding of the substrate, the inhibition is **non-competitive** (right). In this case, K_m remains unchanged, but the concentration of functional enzyme $[E]_t$, and thus V_{max}, decrease. Non-competitive inhibitors generally act irreversibly, by modifying functional groups of the target enzyme (**4**).

"Suicide substrates" (**5**) are substrate analogs that also contain a reactive group. Initially, they bind reversibly, and then they form a covalent bond with the active center of the enzyme. Their effect is therefore also non-competitive. A well-known example of this is the antibiotic *penicillin* (see p. 254).

Allosteric inhibitors bind to a separate binding site outside the active center (**6**). This results in a *conformational change* in the enzyme protein that indirectly reduces its activity (see p. 116). Allosteric effects practically only occur in *oligomeric enzymes.* The kinetics of this type of system can no longer be described using the simple Michaelis–Menten model.

B. Inhibition kinetics ◑

In addition to the Lineweaver–Burk plot (see p. 92), the *Eadie–Hofstee plot* is also commonly used. In this case, the velocity v is plotted against v /[A]. In this type of plot, V_{max} corresponds to the intersection of the approximation lines with the v axis, while K_m is derived from the gradient of the lines. Competitive and non-competitive inhibitors are also easily distinguishable in the Eadie–Hofstee plot. As mentioned earlier, **competitive** inhibitors only influence K_m, and not V_{max}. The lines obtained in the absence and presence of an inhibitor therefore intersect on the ordinate. **Non-competitive inhibitors** produce lines that have the same slope (K_m unchanged) but intersect with the ordinate at a lower level. Another type of inhibitor, not shown here, in which V_{max} and K_m are reduced by the same factor, is referred to as **uncompetitive.** Inhibitors with purely uncompetitive effects are rare. A possible explanation for this type of inhibition is selective binding of the inhibitor to the EA complex.

Allosteric enzymes shift the target enzyme's saturation curve to the left (see p. 92). In Eadie–Hofstee and Lineweaver–Burk plots (see p. 92), allosteric enzymes are recognizable because they produce curved lines (not shown).

A. Types of inhibitor

Competitive

C I C C
a

C I C C
b

2. Substrate analogs

C I C C

3. Transition state analog

1. Uninhibited

C C C

Allosteric

I

v
[A]

6.

Noncompetitive

I C C C

4. Modifying reagent

C I C C
a

C I C C
b

5. “Suicide substrate”

B. Kinetics of inhibition

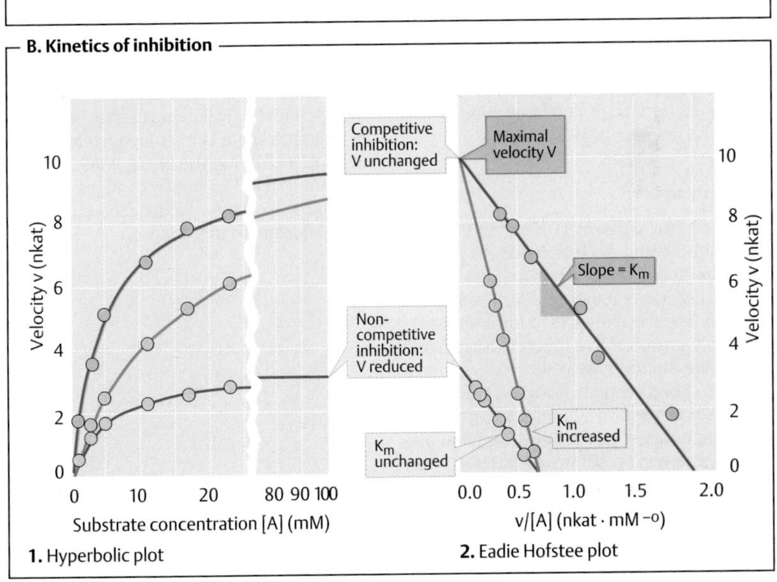

1. Hyperbolic plot

2. Eadie Hofstee plot

Lactate dehydrogenase: structure

Lactate dehydrogenase (LDH, *EC 1.1.1.27*) is discussed in some detail here and on the next page as an example of the structure and function of an enzyme.

A. Lactate dehydrogenase: structure ○

The active form of lactate dehydrogenase (mass 144 kDa) is a **tetramer** consisting of four subunits (**1**). Each monomer is formed by a peptide chain of 334 amino acids (36 kDa). In the tetramer, the subunits occupy *equivalent positions* (**1**); each monomer has an active center. Depending on metabolic conditions, LDH catalyzes NADH-dependent reduction of pyruvate to lactate, or NAD^+-dependent oxidation of lactate to pyruvate (see p. 18).

The active center of an LDH subunit is shown schematically in Fig. **2**. The peptide backbone is shown as a light blue tube. Also shown are the substrate *lactate* (red), the coenzyme *NAD^+* (yellow), and three amino acid side chains (Arg-109, Arg-171, and His-195; green), which are directly involved in the catalysis. A *peptide loop* (pink) formed by amino acid residues 98–111 is also shown. In the absence of substrate and coenzyme, this partial structure is open and allows access to the substrate binding site (not shown). In the enzyme · lactate · NAD^+ complex shown, the peptide loop closes the active center. The catalytic cycle of lactate dehydrogenase is discussed on the next page.

B. Isoenzymes ◑

There are two *different* LDH subunits in the organism—M and H—which have a slightly different amino acid sequence and consequently different catalytic properties. As these two subunits can associate to form tetramers randomly, a total of five different **isoenzymes** of LDH are found in the body.

Fig. **1** shows sections from the amino acid sequences of the two subunits, using the single-letter notation (see p. 60). A common precursor gene was probably duplicated at some point in evolution. The two genes then continued to develop further independently of each other through mutation and selection.

The differences in sequence between the M and H subunits are mainly *conservative*—i. e., both residues are of the same type, e. g. glycine (G) and alanine (A), or arginine (R) and lysine (K). Non-conservative exchanges are less frequent—e. g., lysine (K) for glutamine (Q), or threonine (T) for glutamic acid (E). Overall, the H subunit contains more acidic and fewer basic residues than the M form, and it therefore has a more strongly negative charge. This fact is exploited to separate the isoenzymes using electrophoresis (**2**; see pp. 78, 276). The isoenzyme LDH-1, consisting of four H subunits, migrates fastest, and the M_4 isoenzyme is slowest.

The separation and analysis of isoenzymes in blood samples is important in the diagnosis of certain diseases. Normally, only small amounts of enzyme activity are found in serum. When an organ is damaged, intracellular enzymes enter the blood and can be demonstrated in it (**serum enzyme diagnosis**). The total activity of an enzyme reflects the severity of the damage, while the type of isoenzyme found in the blood provides evidence of the site of cellular injury, since each of the genes is expressed in the various organs at different levels. For example, the liver and skeletal muscles mainly produce M subunits of lactate dehydrogenase (M for muscle), while the brain and cardiac muscle mainly express H subunits (H for heart). In consequence, each organ has a characteristic *isoenzyme pattern* (**3**). Following cardiac infarction, for example, there is a strong increase in the amount of LDH-1 in the blood, while the concentration of LDH-5 hardly changes. The isoenzymes of *creatine kinase* (see p. 336) are also of diagnostic importance.

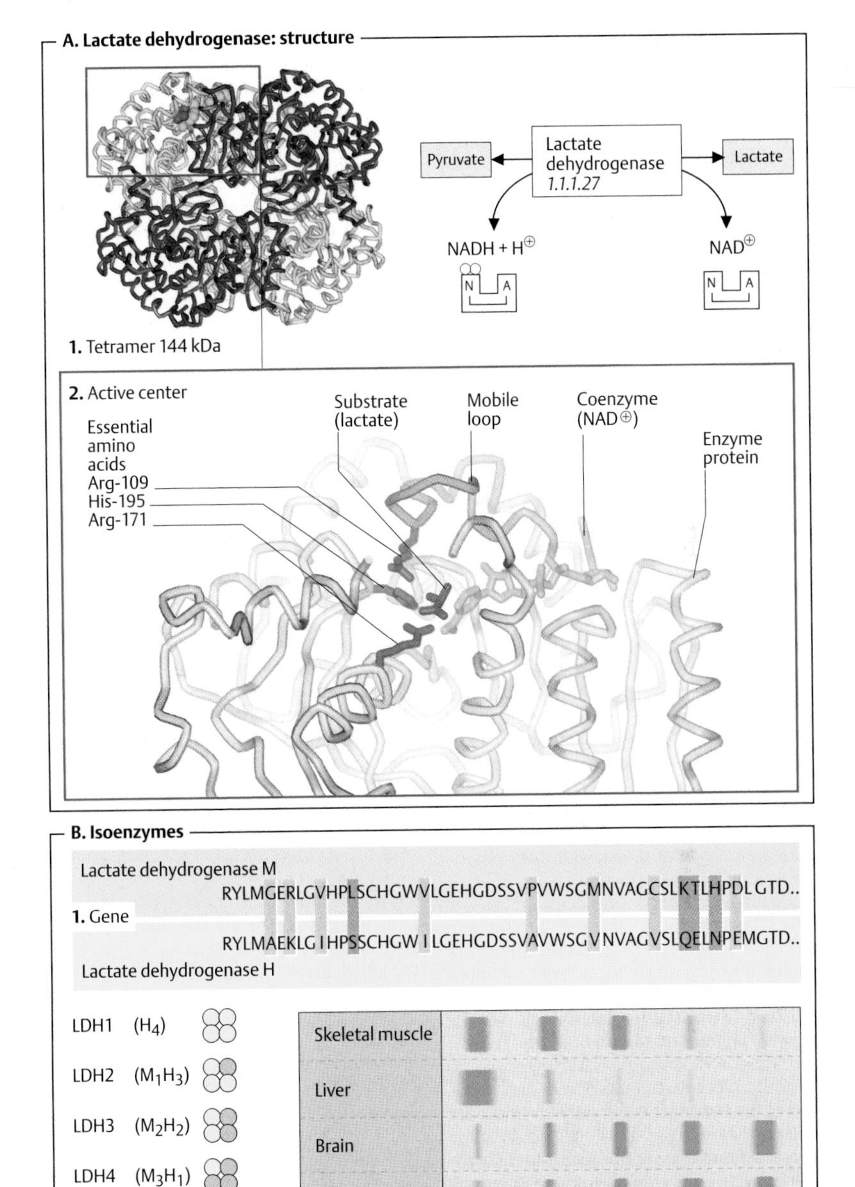
A. Lactate dehydrogenase: structure
Pyruvate
Lactate dehydrogenase
1.1.1.27
Lactate
NADH + H⊕
NAD⊕
N
A
1. Tetramer 144 kDa
2. Active center
Essential amino acids
Arg-109
His-195
Arg-171
Substrate (lactate)
Mobile loop
Coenzyme (NAD⊕)
Enzyme protein
B. Isoenzymes
Lactate dehydrogenase M
RYLMGERLGVHPLSCHGWVLGEHGDSSVPVWSGMNVAGCSLKTLHPDLGTD..
1. Gene
RYLMAEKLGIHPSSCHGWILGEHGDSSVAVWSGVNVAGVSLQELNPEMGTD..
Lactate dehydrogenase H
LDH1 (H4)
LDH2 (M1H3)
LDH3 (M2H2)
LDH4 (M3H1)
LDH5 (M4)
2. Forms
Skeletal muscle
Liver
Brain
Cardiac muscle
LDH5
LDH4
LDH3
LDH2
LDH1
3. Separation by gel electrophoresis

Lactate dehydrogenase: mechanism

The principles of enzyme catalysis discussed on p. 90 can be illustrated using the reaction mechanism of lactate dehydrogenase (LDH) as an example.

A. Lactate dehydrogenase: catalytic cycle ○

LDH catalyzes the transfer of hydride ions (see p. 32) from lactate to NAD^+ or from NADH to pyruvate.

$$\text{L-lactate} + NAD^+ \leftrightarrow \text{pyruvate} + NADH + H^+$$

The equilibrium of the reaction strongly favors lactate *formation*. At high concentrations of lactate and NAD^+, however, oxidation of lactate to pyruvate is also possible (see p. 18). LDH catalyzes the reaction in *both* directions, but—like all enzymes—it has *no* effect on chemical equilibrium.

As the reaction is reversible, the catalytic process can be represented as a closed loop. The **catalytic cycle** of LDH is reduced to six "snapshots" here. Intermediate steps in catalysis such as those shown here are extremely short-lived and therefore difficult to detect. Their existence was deduced indirectly from a large number of experimental findings—e. g., kinetic and binding measurements.

Many amino acid residues play a role in the **active center** of LDH. They can mediate the binding of the substrate and coenzyme, or take part in one of the steps in the catalytic cycle directly. Only the side chains of three particularly important residues are shown here. The positively charged guanidinium group of **arginine-171** binds the carboxylate group of the substrate by electrostatic interaction. The imidazole group of **histidine-195** is involved in acid–base catalysis, and the side chain of **arginine-109** is important for the stabilization of the transition state. In contrast to His-195, which changes its charge during catalysis, the two essential arginine residues are constantly protonated. In addition to these three residues, the **peptide loop 98–111** mentioned on p. 98 is also shown here schematically (red). Its function consists of closing the active center after binding of the substrate and coenzyme, so that water molecules are largely excluded during the electron transfer.

We can now look at the **partial reactions** involved in LDH-catalyzed pyruvate reduction.

In the free enzyme, His195 is protonated (**1**). This form of the enzyme is therefore described as $E \cdot H^+$. The coenzyme NADH is bound first (**2**), followed by pyruvate (**3**). It is important that the carbonyl group of the pyruvate in the enzyme and the active site in the nicotinamide ring of the coenzyme should have a fairly optimal position in relation to each other, and that this orientation should become fixed (*proximity and orientation of the substrates*). The 98–111 loop now closes over the active center. This produces a marked decrease in polarity, which makes it easier to achieve the **transition state** (**4**; *water exclusion*). In the transition state, a hydride ion, H^- (see p. 32), is transferred from the coenzyme to the carbonyl carbon (*group transfer*). The transient—and energetically unfavorable—negative charge on the oxygen that occurs here is stabilized by electrostatic interaction with Arg-109 (*stabilization of the transition state*). At the same time, a proton from His-195 is transferred to this oxygen atom (*group transfer*), giving rise to the enzyme-bound products lactate and NAD^+ (**5**). After the loop opens, lactate dissociates from the enzyme, and the temporarily uncharged imidazole group in His-195 again binds a proton from the surrounding water (**6**). Finally, the oxidized coenzyme NAD^+ is released, and the initial state (**1**) is restored. As the diagram shows, the proton that appears in the reaction equation ($NADH + H^+$) is not bound together with NADH, but after release of the lactate—i. e., between steps (**5**) and (**6**) of the *previous* cycle.

Exactly the same steps occur during the oxidation of lactate to pyruvate, but in the opposite direction. As mentioned earlier, the direction which the reaction takes depends not on the enzyme, but on the equilibrium state—i. e., on the concentrations of all the reactants and the pH value (see p. 18).

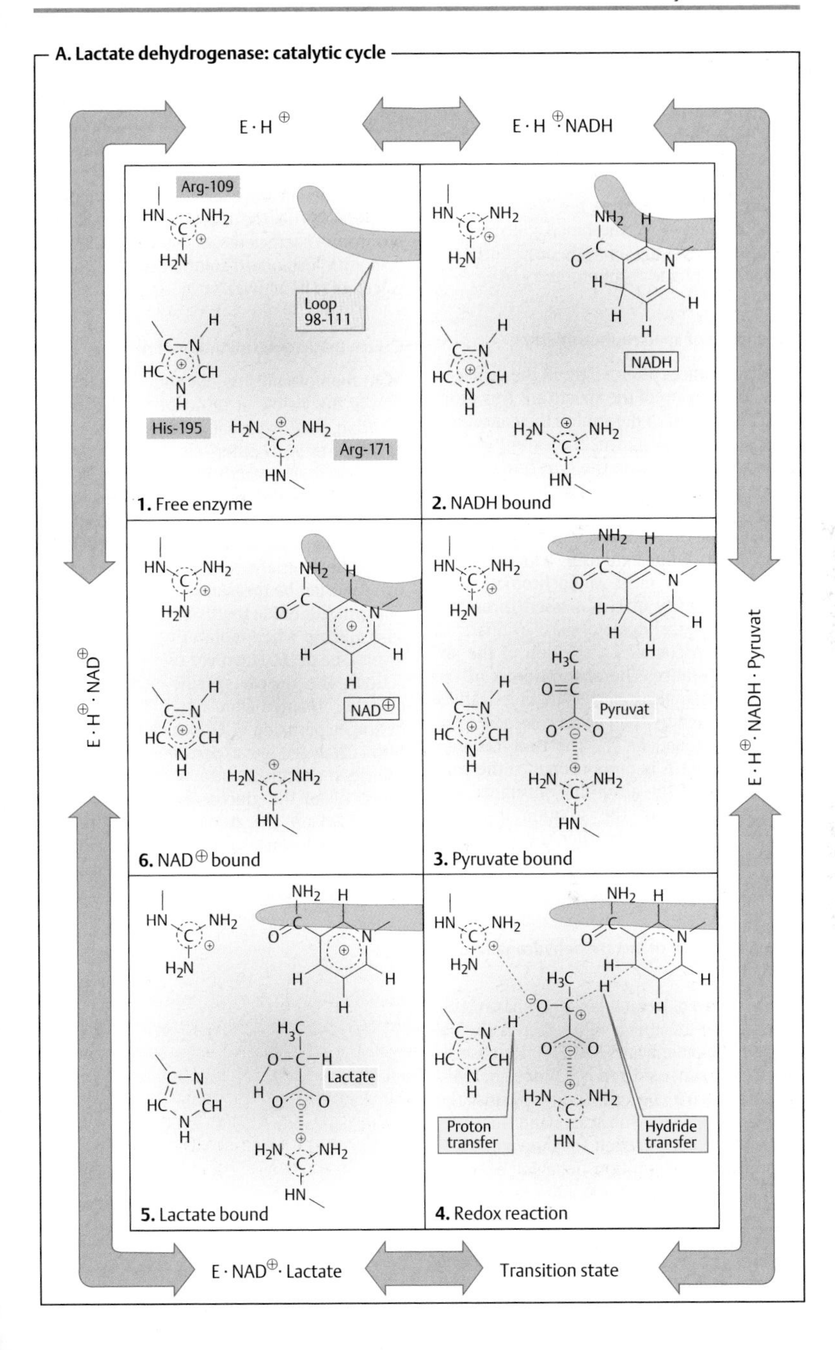
A. Lactate dehydrogenase: catalytic cycle
E · H⊕
E · H⊕ · NADH
E · H⊕ · NADH · Pyruvat
Transition state
E · NAD⊕ · Lactate
E · H⊕ · NAD⊕
Arg-109
Loop 98-111
His-195
Arg-171
1. Free enzyme
NADH
2. NADH bound
Pyruvat
3. Pyruvate bound
Proton transfer
Hydride transfer
4. Redox reaction
Lactate
5. Lactate bound
NAD⊕
6. NAD⊕ bound

Enzymatic analysis

Enzymes play an important role in *biochemical analysis*. In biological material—e. g., in body fluids—even tiny quantities of an enzyme can be detected by measuring its catalytic activity. However, enzymes are also used as *reagents* to determine the concentrations of metabolites—e. g., the blood glucose level (**C**). Most enzymatic analysis procedures use the method of spectrophotometry (**A**).

A. Principle of spectrophotometry ○

Many substances *absorb* light in the visible or ultraviolet region of the spectrum. This property can be used to determine the concentration of such a substance. The extent of light absorption depends on the type and concentration of the substance and on the wavelength of the light used. **Monochromatic light**—i. e., light with a defined wavelength isolated from white light using a monochromator—is therefore used. Monochromatic light with an intensity of I_0 is passed through a rectangular vessel made of glass or quartz (a *cuvet*), which contains a solution of the absorbing substance. The **absorption A** of the solution (often also referred to as its *extinction*) is defined as the *negative decadic logarithm of the quotient* I/I_0. The **Beer–Lambert law** states that A is proportional to the concentration c of the absorbing substance and the thickness d of the solution it passes through. As mentioned earlier, the **absorption coefficient** ε depends on the type of substance and the wavelength.

B. Measurement of lactate dehydrogenase activity ○

Measurement of lactate dehydrogenase (LDH) activity takes advantage of the fact that while the reduced coenzyme NADH + H^+ absorbs light at 340 nm, oxidized NAD^+ does not. *Absorption spectra* (i. e., plots of A against the wavelength) for the substrates and the coenzymes of the LDH reaction are shown in Fig. **1**. Differences in absorption behavior between NAD^+ and NADH between 300 and 400 nm result from changes in the nicotinamide ring during oxidation or reduction (see p. 32). To measure the activity, a solution containing lactate and NAD^+ is placed in a cuvet, and absorption is recorded at a *constant wavelength* of 340 nm. The uncatalyzed LDH reaction is very slow. It is only after addition of the enzyme that measurable quantities of NADH are formed and absorption increases. Since according to the Beer–Lambert law the rate of the increase in absorption $\Delta A/\Delta t$ is proportional to the reaction rate $\Delta c/\Delta t$. The absorption coefficient ε at 340 nm or comparison with a standard solution can be used to calculate LDH activity.

C. Enzymatic determination of glucose ○

Most biomolecules do not show any absorption in the visible or ultraviolet spectrum. In addition, they are usually present in the form of mixtures with other—similar—compounds that would also react to a chemical test procedure. These two problems can be avoided by using an appropriate enzyme to produce a colored dye selectively from the metabolite that is being analyzed. The absorption of the dye can then be measured.

A procedure (**1**) that is often used to measure glucose when monitoring blood glucose levels (see p. 160) involves two successive reactions. The glucose-specific enzyme *glucose oxidase* (obtained from fungi) first produces hydrogen peroxide, H_2O_2, which in the second step—catalyzed by a *peroxidase*—oxidizes a colorless precursor into a green dye (**2**). When all of the glucose in the sample has been used up, the amount of dye formed—which can be measured on the basis of its light absorption—is equivalent to the quantity of glucose originally present.

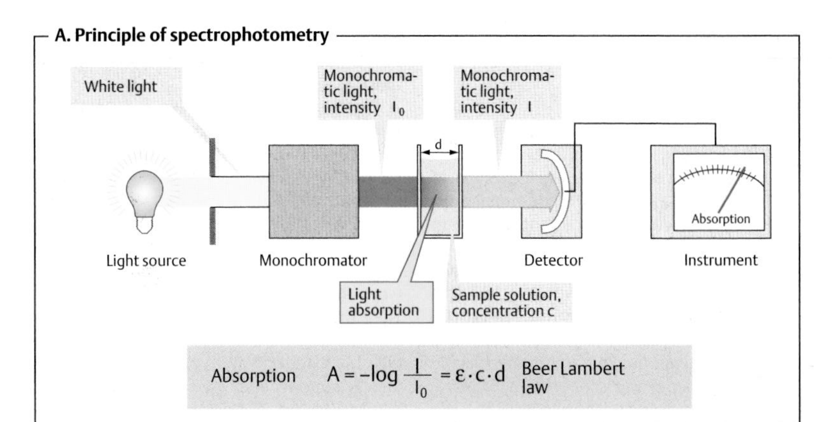
A. Principle of spectrophotometry
White light
Monochromatic light, intensity I0
Monochromatic light, intensity I
d
Absorption
Light source
Monochromator
Detector
Instrument
Light absorption
Sample solution, concentration c
Absorption A = –log I/I0 = ε·c·d
Beer Lambert law

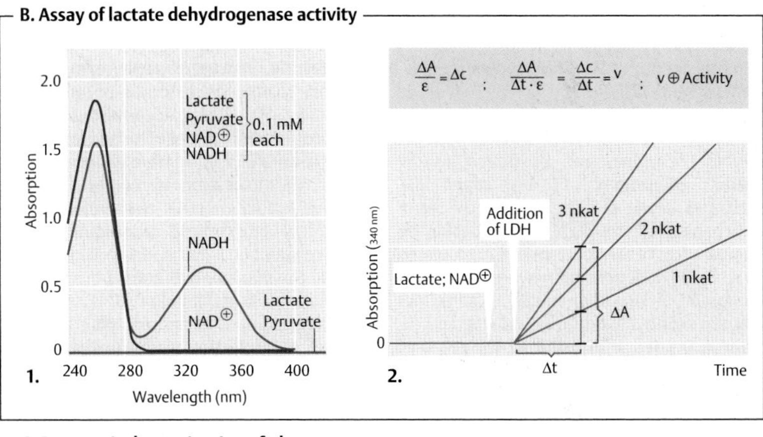
B. Assay of lactate dehydrogenase activity
2.0
1.5
1.0
0.5
0
Absorption
Lactate
Pyruvate
NAD⊕
NADH
0.1 mM each
NADH
NAD⊕
Lactate
Pyruvate
1.
240
280
320
360
400
Wavelength (nm)
ΔA/ε = Δc ; ΔA/(Δt·ε) = Δc/Δt = v ; v ⊕ Activity
Absorption (340 nm)
Addition of LDH
3 nkat
2 nkat
1 nkat
Lactate; NAD⊕
ΔA
0
Δt
2.
Time

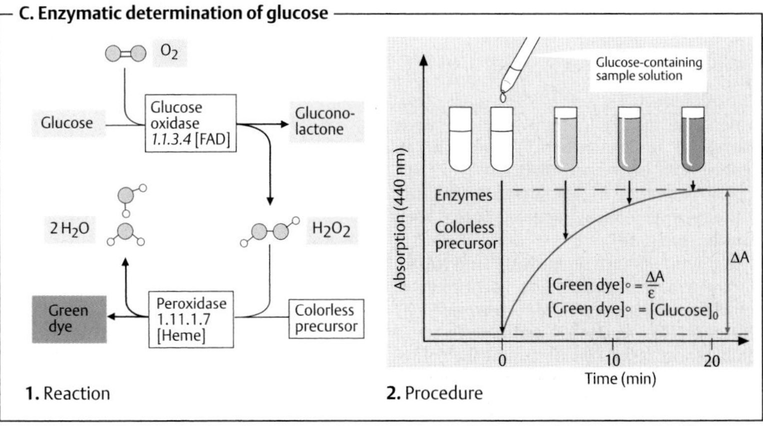
C. Enzymatic determination of glucose
O2
Glucose
Glucose oxidase 1.1.3.4 [FAD]
Glucono-lactone
2 H2O
H2O2
Green dye
Peroxidase 1.11.1.7 [Heme]
Colorless precursor
1. Reaction
Glucose-containing sample solution
Absorption (440 nm)
Enzymes
Colorless precursor
ΔA
[Green dye]o = ΔA/ε
[Green dye]o = [Glucose]0
0
10
20
Time (min)
2. Procedure

Coenzymes 1

A. Coenzymes: definitions ●

In many enzyme-catalyzed reactions, electrons or groups of atoms are transferred from one substrate to another. This type of reaction always also involves additional molecules, which temporarily accept the group being transferred. Helper molecules of this type are called **coenzymes**. As they are not catalytically active themselves, the less frequently used term *"cosubstrate"* would be more appropriate. In contrast to substrates for which a given enzyme is usually specific (see p. 88), coenzymes cooperate with many enzymes of varying substrate specificity. We have rather arbitrarily divided the coenzymes here into group-transferring and redox coenzymes. Strictly speaking, redox coenzymes also transfer groups—namely, reducing equivalents (see p. 32).

Depending on the type of interaction with the enzyme, a distinction is made between soluble coenzymes and prosthetic groups. **Soluble coenzymes** (**1**) are *bound like substrates* during a reaction, undergo a chemical change, and are then *released again*. The original form of the coenzyme is regenerated by a second, independent reaction. **Prosthetic groups** (**2**), on the other hand, are coenzymes that are *tightly bound to the enzyme* and remain associated with it during the reaction. The part of the substrate bound by the coenzyme is later transferred to another substrate or coenzyme of the *same* enzyme (not shown in Fig. **2**).

B. Redox coenzymes 1 ◑

All oxidoreductases (see p. 88) require coenzymes. The most important of these redox coenzymes are shown here. They can act in soluble form (S) or prosthetically (P). Their normal potentials $E^{0'}$ are shown in addition to the type of reducing equivalent that they transfer (see p. 18).

The pyridine nucleotides **NAD⁺** and **NADP⁺** (**1**) are widely distributed as coenzymes of dehydrogenases. They transport *hydride ions* ($2e^-$ and 1 H^+; see p. 32) and *always* act in soluble form. NAD^+ transfers reducing equivalents from catabolic pathways to the respiratory chain and thus contributes to energy metabolism. In contrast, reduced $NADP^+$ is the most important *reductant* involved in biosynthesis (see p. 112).

The flavin coenzymes **FMN** and **FAD** (**2**, **3**) contain *flavin* (isoalloxazine) as a redox-active group. This is a three-membered, *N*-containing ring system that can accept a maximum of two electrons and two protons during reduction. FMN carries the phosphorylated sugar alcohol *ribitol* at the flavin ring. FAD arises from FMN through bonding with AMP. The two coenzymes are functionally similar. They are found in *dehydrogenases, oxidases,* and *monooxygenases*. In contrast to the pyridine nucleotides, flavin reactions give rise to *radical intermediates* (see p. 32). To prevent damage to cell components, the flavins always remain bound as prosthetic groups in the enzyme protein.

The role of **ubiquinone** (coenzyme Q, **4**) in transferring reducing equivalents in the respiratory chain is discussed on p. 140. During reduction, the *quinone* is converted into the *hydroquinone* (ubiquinol). The isoprenoid side chain of ubiquinone can have various lengths. It holds the molecule in the membrane, where it is freely mobile. Similar coenzymes are also found in photosynthesis (plastoquinone; see p. 132). *Vitamins E and K* (see p. 52) also belong to the quinone/hydroquinone systems.

L-Ascorbic acid (vitamin C, **5**) is a powerful reducing agent. As an **antioxidant**, it provides nonspecific protection against oxidative damage (see p. 284), but it is also an essential **cofactor** for various monooxygenases and dioxygenases. Ascorbic acid is involved in the hydroxylation of proline and lysine residues during the biosynthesis of collagen (see p. 344), in the synthesis of catecholamines (see p. 352) and bile acids (see p. 314), as well as in the breakdown of tyrosine (see p. 415). The reduced form of the coenzyme is a relatively strong acid and forms salts, the **ascorbates**. The oxidized form is known as **dehydroascorbic acid**. The stimulation of the immune system caused by ascorbic acid has not yet been fully explained.

A. Coenzymes: definitions

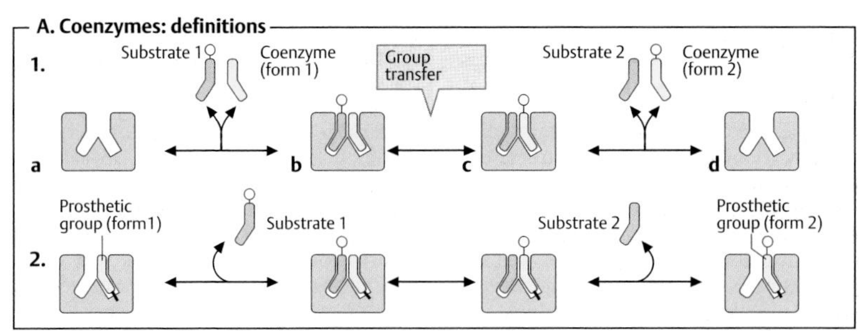

B. Redox coenzymes

Coenzyme	Oxidized form	Reduced form	Type	Transferred	$E^{0\prime}$ (V)
1. NAD(P)$^{\oplus}$ (ox., red.)			L	$H^{\ominus}$	−0.32
2. Flavin mononucleotide (FMN) (ox., red.)		Ribitol (Rit)	P	2[H]	−0.3 to +0.2
3. Flavin adenine dinucleotide (FAD) (ox., red.)		Ribitol	P	2[H]	−0.3 to +0.2
4. Ubiquinone (coenzym Q)		$[\ ]_{6-10}$	L	2[H]	−0 to +0.2
5. Ascorbic acid			L	2[H]	+0.1

Coenzymes 2

A. Redox coenzymes 2 ◑

In **lipoic acid** (**6**), an intramolecular *disulfide bond* functions as a redox-active structure. As a result of reduction, it is converted into the corresponding *dithiol*. As a prosthetic group, lipoic acid is usually covalently bound to a lysine residue (R) of the enzyme, and it is then referred to as **lipoamide**. Lipoamide is mainly involved in oxidative decarboxylation of 2-oxo acids (see p. 134). The peptide coenzyme **glutathione** is a similar disulfide/dithiol system (not shown; see p. 284).

Iron–sulfur clusters (**7**) occur as prosthetic groups in oxidoreductases, but they are also found in lyases—e. g., *aconitase* (see p. 136) and other enzymes. Iron–sulfur clusters consist of 2–4 iron ions that are coordinated with cysteine residues of the protein (-SR) and with anorganic sulfide ions (S). Structures of this type are only stable in the interior of proteins. Depending on the number of iron and sulfide ions, distinctions are made between $[Fe_2S_2]$, $[Fe_3S_4]$, and $[Fe_4S_4]$ clusters. These structures are particularly numerous in the respiratory chain (see p. 140), and they are found in all complexes except complex IV.

Heme coenzymes (**8**) with redox functions exist in the *respiratory chain* (see p. 140), in *photosynthesis* (see p. 128), and in *monooxygenases* and *peroxidases* (see p. 24). Heme-containing proteins with redox functions are also referred to as **cytochromes**. In cytochromes, in contrast to hemoglobin and myoglobin, the iron changes its valence (usually between +2 and +3). There are several classes of heme (a, b, and c), which have different types of substituent $-R_1$ to $-R_3$. Hemoglobin, myoglobin, and the heme enzymes contain heme b. Two types of heme a are found in cytochrome *c* oxidase (see p. 132), while heme c mainly occurs in cytochrome *c*, where it is covalently bound with cysteine residues of the protein part via thioester bonds.

B. Group-transferring coenzymes 1 ◑

The **nucleoside phosphates** (**1**) are not only *precursors* for nucleic acid biosynthesis; many of them also have coenzyme functions. They serve for *energy conservation*, and as a result of *energetic coupling* (see p. 124) also allow endergonic processes to proceed. Metabolites are often made more reactive ("activated") as a result of the transfer of phosphate residues (*phosphorylation*). Bonding with nucleoside diphosphate residues (mainly UDP and CDP) provides activated precursors for polysaccharides and lipids (see p. 110). Endergonic formation of bonds by *ligases* (enzyme class 6) also depends on nucleoside triphosphates.

Acyl residues are usually activated by transfer to **coenzyme A** (**2**). In coenzyme A (see p. 12), *pantetheine* is linked to 3′-phospho-ADP by a phosphoric acid anhydride bond. Pantetheine consists of three components connected by amide bonds—*pantoic acid*, *β-alanine*, and *cysteamine*. The latter two components are biogenic amines formed by the decarboxylation of aspartate and cysteine, respectively. The compound formed from pantoic acid and β–alanine (*pantothenic acid*) has vitamin-like characteristics for humans (see p. 368). Reactions between the thiol group of the cysteamine residue and carboxylic acids give rise to **thioesters**, such as acetyl CoA. This reaction is strongly endergonic, and it is therefore coupled to exergonic processes. Thioesters represent the *activated form of carboxylic acids*, because acyl residues of this type have a high chemical potential and are easily transferred to other molecules. This property is often exploited in metabolism.

Thiamine diphosphate (TPP, **3**), in cooperation with enzymes, is able to activate aldehydes or ketones as *hydroxyalkyl groups* and then to pass them on to other molecules. This type of transfer is important in the transketolase reaction, for example (see p. 152). Hydroxyalkyl residues also arise in the decarboxylation of oxo acids. In this case, they are released as aldehydes or transferred to lipoamide residues of 2-oxoacid dehydrogenases (see p. 134). The functional component of TPP is the sulfur- and nitrogen-containing *thiazole ring*.

A. Redox coenzymes 2

Coenzyme	Oxidized form	Reduced form	Type	Transferred	$E^{0\prime}$
6. Lipoamide (ox. / red.)			P	2[H]	−0.29
7. Iron–sulfur cluster	$[Fe_2S_2]^{n+}$	$[Fe_4S_4]^{m+}$	P	$1e^{\ominus}$	−0.6 to +0.5
8. Heme (ox. 3+ / red. 2+)	Fe^{3+}	Fe^{2+}	P	$1e^{\ominus}$	0 to +0.5

B. Group-transferring coenzymes 1

Coenzyme (symbol)	Free form	Charged form	Group(s) transferred	Important enzymes
1. Nucleoside phosphates			P; B-Rib; B-Rib-P; B-Rib-PP	Phosphotransferases; Nucleotidyltransferases (2.7.n.n); Ligases (6.n.n.n)
2. Coenzyme A			Acyl residues	Acyltransferases (2.3.n.n); CoA transferases (2.8.3.n)
3. Thiamine diphosphate (TPP)			Hydroxyalkyl residues	Decarboxylases (4.1.1.n); Oxoacid dehydrogenases (1.2.4.n); Transketolase (2.2.1.1)

Coenzymes 3

A. Group-transferring coenzymes 2 ◑

Pyridoxal phosphate (**4**) is the most important coenzyme in amino acid metabolism. Its role in *transamination* reactions is discussed in detail on p. 178. Pyridoxal phosphate is also involved in other reactions involving amino acids, such as *decarboxylations* and *dehydrations*. The aldehyde form of pyridoxal phosphate shown here (left) is not generally found in free form. In the absence of substrates, the aldehyde group is covalently bound to the ε-amino group of a lysine residue as *aldimine* ("Schiff's base"). **Pyridoxamine phosphate** (right) is an intermediate of transamination reactions. It reverts to the aldehyde form by reacting with 2-oxoacids (see p. 178).

Biotin (**5**) is the coenzyme of the *carboxylases*. Like pyridoxal phosphate, it has an amide-type bond via the carboxyl group with a lysine residue of the carboxylase. This bond is catalyzed by a specific enzyme. Using ATP, biotin reacts with hydrogen carbonate (HCO_3^-) to form *N*-**carboxybiotin** . From this activated form, *carbon dioxide* (CO_2) is then transferred to other molecules, into which a carboxyl group is introduced in this way. Examples of biotindependent reactions of this type include the formation of oxaloacetic acid from pyruvate (see p. 154) and the synthesis of malonyl-CoA from acetyl-CoA (see p. 162).

Tetrahydrofolate (THF, **6**) is a coenzyme that can transfer *C_1 residues* in different oxidation states. THF arises from the vitamin *folic acid* (see p. 366) by double hydrogenation of the heterocyclic pterin ring. The C_1 units being transferred are bound to N-5, N-10, or both nitrogen atoms. The most important derivatives are:

a) N^5**-formyl-THF** and N^{10}**-formyl-THF**, in which the formyl residue has the oxidation state of a carboxylic acid;

b) N^5**-methylene-THF**, with a C_1 residue in the oxidation state of an aldehyde; and

c) N^5**-methyl-THF**, in which the methyl group has the oxidation state of an alcohol.

C_1 units transferred by THF play a role in the synthesis of methionine (see p. 412), purine nucleotides (see p. 188), and dTMP (see p. 190), for example. Due to the central role of THF derivatives in the biosynthesis of DNA precursors, the enzymes involved in THF metabolism are primary targets for cytostatic drugs (see p. 402).

The **cobalamins** (**7**) are the chemically most complex form of coenzyme. They also represent the only natural substances that contain the transition metal *cobalt* (Co) as an essential component. Higher organisms are unable to synthesize cobalamins themselves, and are therefore dependent on a supply of **vitamin B_{12}** synthesized by bacteria (see p. 368).

The central component of the cobalamins is the **corrin** ring, a member of the tetrapyrroles, at the center of which the cobalt ion is located. The end of one of the side chains of the ring carries a nucleotide with the unusual base *dimethylbenzimidazole.* The ligands for the metal ion are the four N atoms of the pyrrole ring, a nitrogen from dimethylbenzimidazole, and a **group X**, which is organometallically bound—i. e., *mainly covalently.*

In **methylcobalamin**, X is a methyl group. This compound functions as a coenzyme for several *methyltransferases,* and among other things is involved in the synthesis of methionine from homocysteine (see p. 418). However, in human metabolism, in which methionine is an essential amino acid, this reaction does not occur.

Adenosylcobalamin (coenzyme B_{12}) carries a covalently bound adenosyl residue at the metal atom. This is a coenzyme of various *isomerases,* which catalyze rearrangements following a radical mechanism. The radical arises here through *homolytic cleavage* of the bond between the metal and the adenosyl group. The most important reaction of this type in animal metabolism is the rearrangement of methylmalonyl-CoA to form succinyl-CoA, which completes the breakdown of odd-numbered fatty acids and of the branched amino acids valine and isoleucine (see pp. 166 and 414).

A. Group-transferring coenzymes 2

Coenzyme	Free form	Charged form	Group(s) transferred	Important enzymes
4. Pyridoxal phosphate  PLP	O=C–H, HO, CH_2O–(P), H_3C, $\oplus$ N–H, H	$\oplus$ NH_3, H–C–H, HO, CH_2O–(P), H_3C, $\oplus$ N–H, H	Amino group Amino acid residues	Transaminases (2.6.1.n) Many lyases (4.n.n.n)
5. Biotin B	O, H–N, C, NH, H_2C, S, CH, H, N, C, O	O, O, $\ominus$O–C, N, C, NH, H_2C, S, CH, H, N, C, O	$[CO_2]$	Carboxylases (6.4.1.n)
4. Pyridoxal phosphate  THF	H_2N, N, N, H, N, H, H, H, N, CH_2, H, HN, H, $COO^\ominus$, C–N–C–H, O, CH_2, CH_2, $COO^\ominus$ a) H, N, H, H, H, 5, N, CH_2, C, HN–, O, H 10 b) H, N, H, H, H, N, H, CH_2, N–, O=C, H c) H, N, H, H, H, N, CH_2, HC$\oplus$N– d) H, N, H, H, H, N, CH_2, H_2C–N– e) H, N, H, H, H, N, CH_2, H_3C, HN–		C_1 groups a) N^5-Formyl b) N^{10}-Formyl c) N^5N^{10}-Methenyl d) N^5N^{10}-Methylene e) N^5N^{10}-Methyl	C_1 transferases (2.1.n.n)
7. Cobalamin coenzymes	H_2N, C=O, H_2N, C=O, H_3C, H_3C, H_3C, X, O, C, NH_2, H_2N, C, O, N, N–$Co^{2\oplus}$, N, CH_3, NH_2, C, O, HN–C, H_3C, N, H_2C, O, CH_3, CH_3, CH_3, H_3C–CH, N, CH_3, O, $\ominus$O–P=O, N, H_3C, O=C, NH_2, O, HO, O, $HOCH_2$		X = Adenosyl- X = Methyl-	Mutases *(5.4.n.n)* Methyltransferases *(2.1.1.n.)*

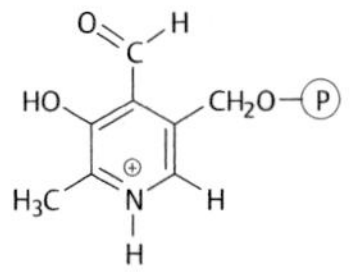

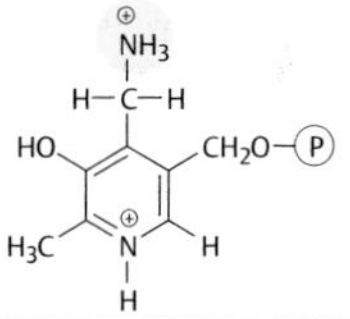

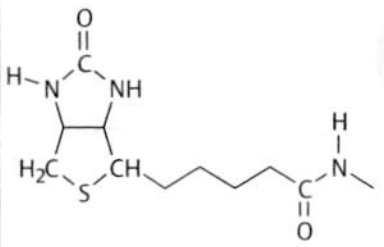

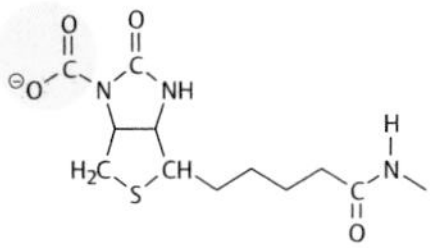

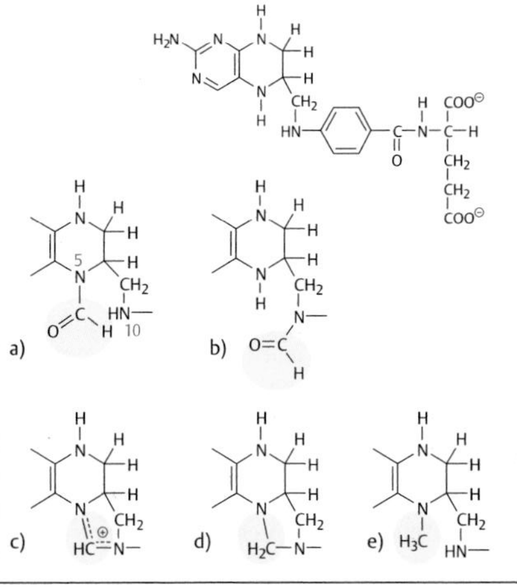

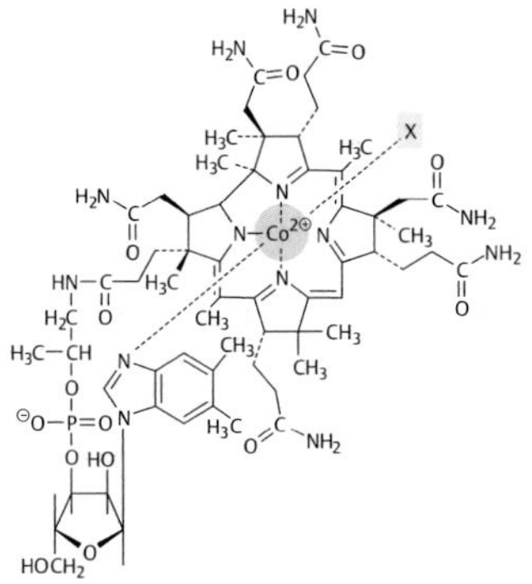

Activated metabolites

Many coenzymes (see pp. 104ff.) serve to *activate* molecules or groups that are poorly reactive. Activation consists of the formation of reactive intermediate compounds in which the group concerned is located at a higher chemical potential and can therefore be transferred to other molecules in an exergonic reaction (see p. 124). Acetyl-CoA is an example of this type of compound (see p. 12).

ATP and the other **nucleoside triphosphate coenzymes** not only transfer phosphate residues, but also provide the nucleotide components for this type of activation reaction. On this page, we discuss metabolites or groups that are activated in the metabolism by bonding with nucleosides or nucleotides. Intermediates of this type are mainly found in the metabolism of complex carbohydrates and lipids.

A. Activated metabolites ◑

1. Uridine diphosphate glucose (UDPglucose)

The inclusion of glucose residues into polymers such as glycogen or starches is an endergonic process. The activation of the **glucose** building blocks that is required for this takes places in several steps, in which two ATPs are used per glucose. After the phosphorylation of free glucose, glucose 6-phosphate is isomerized to glucose 1-phosphate (**a**), reaction with UTP (**b**) then gives rise to UDPglucose, in which the anomeric OH group at C-1 of the sugar is bound with phosphate. This "energy-rich" compound (an acetal phosphate) allows exergonic transfer of glucose residues to glycogen (**c**; see pp. 156, 408) or other acceptors.

2. Cytidine diphosphate choline (CDPcholine)

The amino alcohol **choline** is activated for inclusion in phospholipids following a similar principle (see p. 170). Choline is first phosphorylated by ATP to form choline phosphate (**a**), which by reaction with CTP and cleavage of diphosphate, then becomes CDPcholine. In contrast to (**1**), it is not choline that is transferred from CDPcholine, but rather choline phosphate, which with diacylglycerol yields phosphatidylcholine (lecithin).

3. Phosphoadenosine phosphosulfate (PAPS)

Sulfate residues occur as strongly polar groups in various biomolecules—e.g., in *glycosaminoglycans* (see p. 346) and *conjugates* of steroid hormones and xenobiotics (see p. 316). In the synthesis of the "activated sulfate" PAPS, ATP first reacts with anorganic sulfate to form adenosine phosphosulfate (APS, **a**). This intermediate already contains the "energy-rich" mixed anhydride bond between phosphoric acid and sulfuric acid. In the second step, the 3′-OH group of APS is phosphorylated, with ATP being used again. After transfer of the sulfate residue to OH groups (**c**), adenosine-3′,5′-bisphosphate remains.

4. *S*-adenosyl methionine (SAM)

The coenzyme *tetrahydrofolate* (THF) is the main agent by which C_1 fragments are transferred in the metabolism. THF can bind this type of group in various oxidation states and pass it on (see p. 108). In addition, there is "activated methyl," in the form of *S*-adenosyl methionine (SAM). SAM is involved in many **methylation reactions**—e.g., in creatine synthesis (see p. 336), the conversion of norepinephrine into epinephrine (see p. 352), the inactivation of norepinephrine by methylation of a phenolic OH group (see p. 316), and in the formation of the active form of the cytostatic drug 6-mercaptopurine (see p. 402).

SAM is derived from degradation of the proteinogenic amino acid **methionine**, to which the adenosyl residue of an ATP molecule is transferred. After release of the activated methyl group, *S*-adenosyl homocysteine (SAH) is left over. This can be converted back into methionine in two further steps. Firstly, cleavage of the adenosine residue gives rise to the non-proteinogenic amino acid **homocysteine**, to which a methyl group is transferred once again with the help of N^5-methyl-THF (see p. 418). Alternatively, homocysteine can also be broken down into propionyl-CoA.

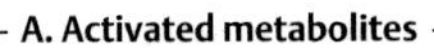

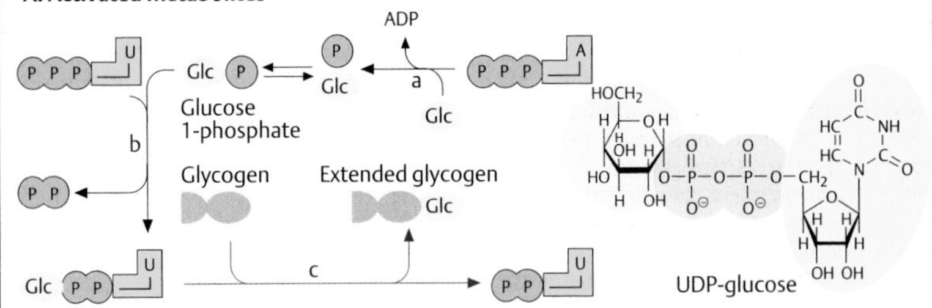

1. Uridine diphosphate glucose (UDP-glucose)

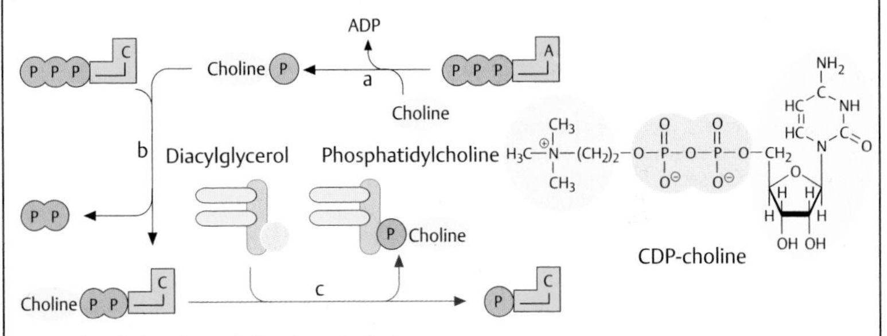

2. Cytidine diphosphate choline (CDPcholine)

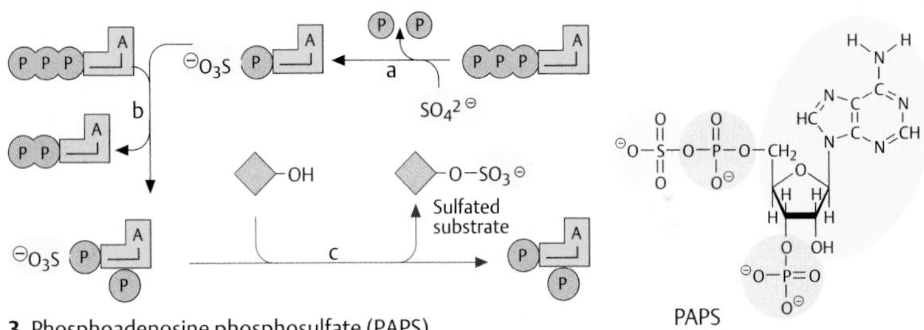

3. Phosphoadenosine phosphosulfate (PAPS)

Methionine
a
Homocysteine
b
N^5-Methyl-THF
Adenosine
d
c
Methylated substrate
S-adenosyl-homocysteine
SAM

4. *S*-adenosyl methionine (SAM)

Intermediary metabolism

Hundreds of chemical reactions are constantly taking place in every cell, and taken together these are referred to as the **metabolism**. The chemical compounds involved in this are known as **metabolites**. Outside of the cell, almost all of the chemical changes in metabolites would only take place very slowly and without any specific direction. By contrast, organized sequences of chemical reactions with a high rate of throughput, known as **metabolic pathways**, become possible through the existence of specific **enzymes** (see p. 88).

A. Intermediary metabolism: overview ●

A number of central metabolic pathways are common to most cells and organisms. These pathways, which serve for synthesis, degradation, and interconversion of important metabolites, and also for energy conservation, are referred to as the **intermediary metabolism**.

In order to survive, all cells constantly require organic and inorganic *nutrients*, as well as *chemical energy*, which is mainly derived from ATP (see below). Depending on the way in which these needs are satisfied, organisms can be classified into autotrophic and heterotrophic groups. The **autotrophs**, which include plants and many microorganisms, can synthesize organic molecules from inorganic precursors (CO_2). An autotrophic lifestyle is possible through **photosynthesis**, for example (see p. 128). The **heterotrophs**—e.g., animals and fungi—depend on organic substances supplied in their diet. The schema shown on this page provides an overview of animal metabolism.

The polymeric substances contained in the diet (proteins, carbohydrates, and nucleic acids—top) cannot be used by the organism directly. Digestive processes first have to degrade them to monomers (amino acids, sugars, nucleotides). These are then mostly broken down by **catabolic pathways** (pink arrows) into smaller fragments. The metabolites produced in this way (generally referred to as the "metabolite pool") are then either used to obtain energy through further catabolic conversion, or are built up again into more complex molecules by **anabolic pathways** (blue arrows). Of the numerous metabolites in the pool, only three particularly important representatives—pyruvate, acetyl-CoA, and glycerol—are shown here. These molecules represent connecting links between the metabolism of proteins, carbohydrates, and lipids. The metabolite pool also includes the intermediates of the tricarboxylic acid cycle (**6**). This cyclic pathway has both catabolic and anabolic functions—i. e., it is **amphibolic** (violet; see p. 138).

Waste products from the degradation of organic substances in animal metabolism include *carbon dioxide* (CO_2), *water* (H_2O), and *ammonia* (NH_3). In mammals, the toxic substance ammonia is incorporated into *urea* and excreted in this form (see p. 182).

The most important form of storage for chemical energy in all cells is **adenosine triphosphate** (ATP, see p. 122). ATP *synthesis* requires energy—i. e., the reaction is *endergonic*. Conversely, cleavage of ATP into ADP and phosphate releases energy. *Exergonic* hydrolysis of ATP, as a result of **energetic coupling** (see p. 16), makes energy-dependent (*endergonic*) processes possible. For example, most anabolic pathways, as well as movement and transport processes, are energy-dependent.

The most important pathway for the synthesis of ATP is **oxidative phosphorylation** (see p. 122). In this process, catabolic pathways first form reduced cofactors ($NADH+H^+$, QH_2, $ETFH_2$). Electrons are then transferred from these compounds to oxygen. This strongly exergonic process is catalyzed by the **respiratory chain** and used indirectly for the ATP synthesis (see p. 140). In *anaerobic conditions*—i. e., in the absence of oxygen—most organisms can fall back on ATP that arises in glycolysis (**3**). This less efficient type of ATP synthesis is referred to as **fermentation** (see p. 146).

While NADH exclusively supplies oxidative phosphorylation, $NADPH+H^+$—a very similar coenzyme—is the reducing agent for anabolic pathways. $NADPH + H^+$ is mainly formed in the pentose phosphate pathway (PPP, **1**; see p. 152).

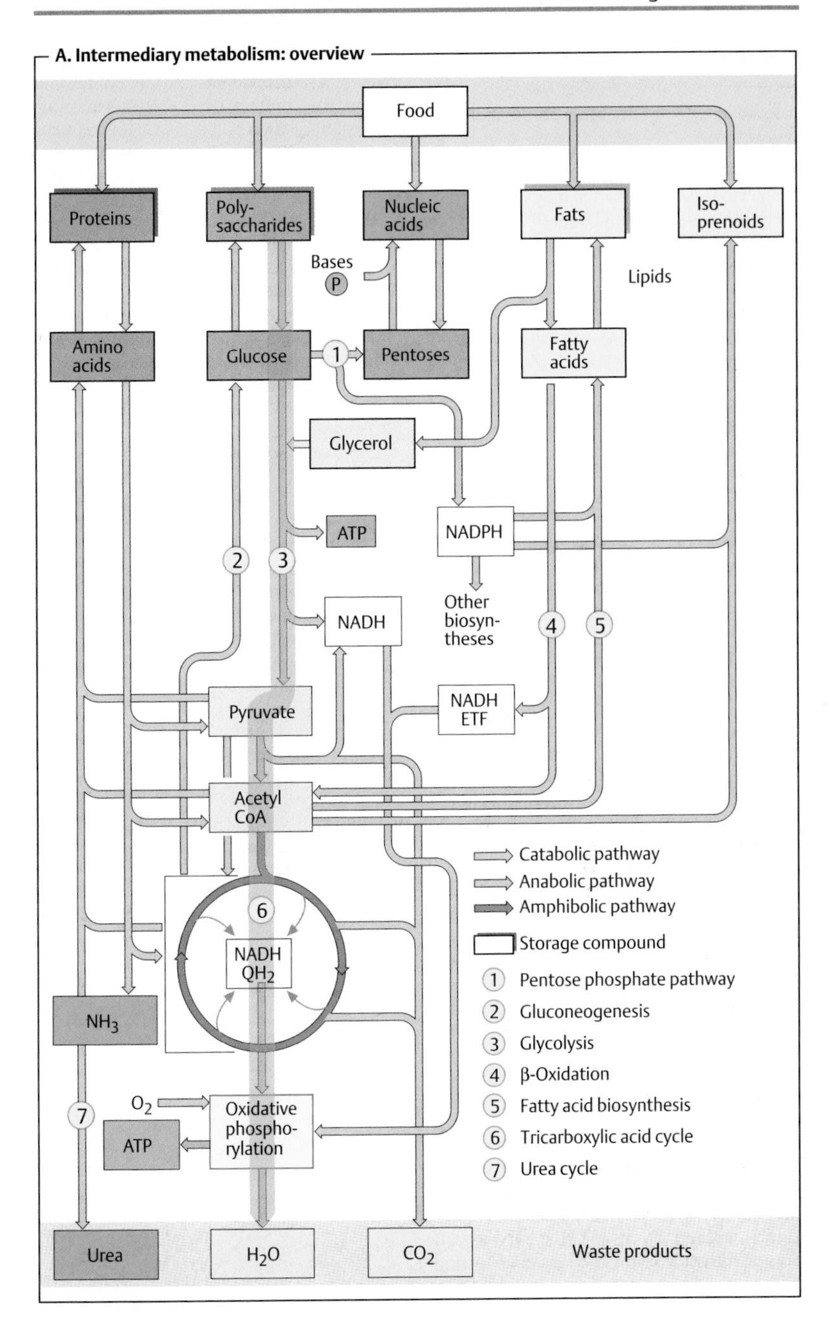
A. Intermediary metabolism: overview
Food
Proteins
Poly-saccharides
Nucleic acids
Fats
Iso-prenoids
Bases
P
Lipids
Amino acids
Glucose
1
Pentoses
Fatty acids
Glycerol
ATP
NADPH
2
3
NADH
Other biosyn-theses
4
5
NADH ETF
Pyruvate
Acetyl CoA
6
NADH QH2
NH3
7
O2
ATP
Oxidative phospho-rylation
Urea
H2O
CO2
Waste products
Catabolic pathway
Anabolic pathway
Amphibolic pathway
Storage compound
1 Pentose phosphate pathway
2 Gluconeogenesis
3 Glycolysis
4 β-Oxidation
5 Fatty acid biosynthesis
6 Tricarboxylic acid cycle
7 Urea cycle

Regulatory mechanisms

A. Fundamental mechanisms of metabolic regulation ◑

The activities of all metabolic pathways are subject to precise regulation in order to adjust the synthesis and degradation of metabolites to physiological requirements. An overview of the regulatory mechanisms is presented here. Further details are shown on pp. 116ff.

Metabolite flow along a metabolic pathway is mainly determined by the activities of the **enzymes** involved (see p. 88). To regulate the pathway, it is sufficient to change the activity of the enzyme that catalyzes the *slowest* step in the reaction chain. Most metabolic pathways have **key enzymes** of this type on which the regulatory mechanisms operate. The activity of key enzymes is regulated at three independent levels:

Transcriptional control. Here, Biosynthesis of the enzyme protein is influenced at the genetic level (**1**). Interventions in enzyme synthesis mainly affect synthesis of the corresponding mRNA—i. e., *transcription* of the gene coding for the enzyme. The term "transcriptional control" is therefore used (see pp. 118, 244). This mechanism is mediated by *regulatory proteins* (transcription factors) that act directly on DNA. The genes have a special regulatory segment for this purpose, known as the *promoter* region, which contains binding sites (control elements) for regulatory proteins. The activity of these proteins is, in turn, affected by metabolites or hormones. When synthesis of a protein is increased by transcriptional control, the process is referred to as **induction**; when it is reduced or suppressed, it is referred to as **repression**. Induction and repression processes take some time and are therefore not immediately effective.

Interconversion of key enzymes (**2**) takes effect considerably faster than transcriptional control. In this case, the enzyme is already present at its site of effect, but it is initially still inactive. It is only when needed that it is converted into the catalytically active form, after signaling and mediation from second messengers (see p. 120) through an *activating enzyme* (E_1). If the metabolic pathway is no longer required, an *inactivating enzyme* (E_2) returns the key enzyme to its inactive resting state.

Interconversion processes in most cases involve **ATP-dependent phosphorylation** of the enzyme protein by a *protein* kinase or **dephosphorylation** of it by a *protein phosphatase* (see p. 120). The phosphorylated form of the key enzyme is usually the more active one, but the reverse may also occur.

Modulation by ligands. An important variable that regulates flow through a metabolic pathway is **precursor availability** (metabolite A in the case shown here). The availability of precursor A increases along with the activity of the metabolic pathways that form A (**3**) and it decreases with increasing activity of other pathways that also consume A (**4**). Transport from one cell compartment to another can also restrict the availability of A.

Coenzyme availability can also often have a limiting effect (**5**). If the coenzyme is regenerated by a second, independent metabolic pathway, the speed of the second pathway can limit that of the first one. For example, glycolysis and the tricarboxylic acid cycle are mainly regulated by the availability of NAD^+ (see p. 146). Since NAD^+ is regenerated by the respiratory chain, the latter indirectly controls the breakdown of glucose and fatty acids (respiratory control, see p. 144).

Finally, the activity of key enzymes can be regulated by *ligands* (substrates, products, coenzymes, or other effectors), which as *allosteric effectors* do not bind at the active center itself, but at another site in the enzyme, thereby modulating enzyme activity (**6**; see p. 116). Key enzymes are often inhibited by immediate reaction products, by end products of the reaction chain concerned (*"feedback" inhibition*), or by metabolites from completely different metabolic pathways. The precursors for a reaction chain can stimulate their own utilization through enzyme activation.

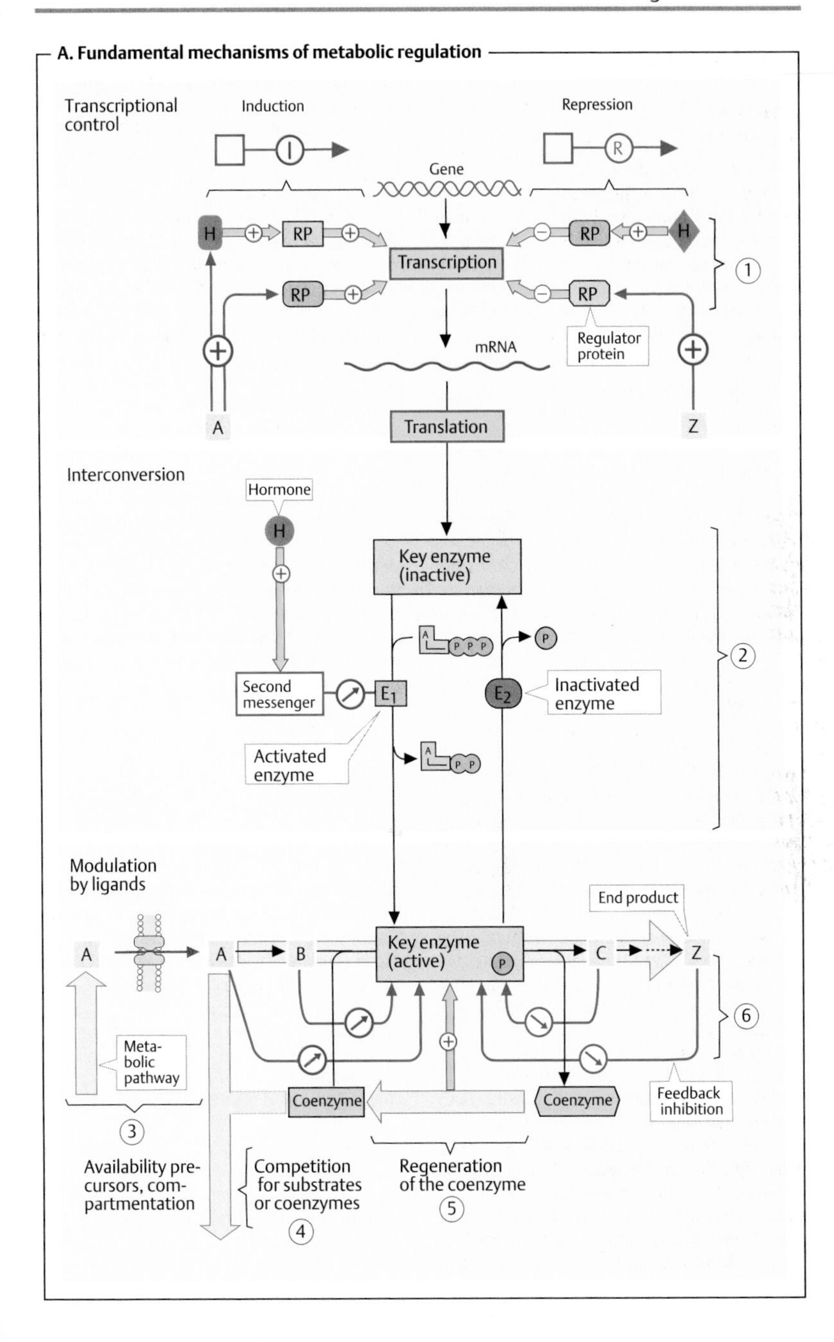
A. Fundamental mechanisms of metabolic regulation
Transcriptional control
Induction
Repression
Gene
H
RP
Transcription
RP
Regulator protein
mRNA
A
Translation
Z
Interconversion
Hormone
H
Key enzyme (inactive)
Second messenger
E1
E2
Inactivated enzyme
Activated enzyme
Modulation by ligands
End product
A
A
B
Key enzyme (active)
C
Z
Metabolic pathway
Coenzyme
Coenzyme
Feedback inhibition
Availability precursors, compartmentation
Competition for substrates or coenzymes
Regeneration of the coenzyme
1
2
3
4
5
6

Allosteric regulation

The regulation of **aspartate carbamoyltransferase** (ACTase), a key enzyme of pyrimidine biosynthesis (see p. 188) is discussed here as an example of allosteric regulation of enzyme activity. Allosteric effects are mediated by the substrate itself or by inhibitors and activators (*allosteric effectors*, see p. 114). The latter bind at special sites outside the active center, producing a conformational change in the enzyme protein and thus indirectly lead to an alteration in its activity.

A. Aspartate carbamoyltransferase: reaction ○

ACTase catalyzes the transfer of a carbamoyl residue from carbamoyl phosphate to the amino group of L-aspartate. The *N*-carbamoyl L-aspartate formed in this way already contains all of the atoms of the later pyrimidine ring (see p. 188). The ACTase of the bacterium *Escherichia coli* is inhibited by cytidine triphosphate (CTP), an end product of the anabolic metabolism of pyrimidines, and is activated by the precursor ATP.

B. Kinetics ◑

In contrast to the kinetics of isosteric (normal) enzymes, allosteric enzymes such as ACTase have **sigmoidal** (S-shaped) **substrate saturation curves** (see p. 92). In allosteric systems, the enzyme's affinity to the substrate is not constant, but depends on the substrate concentration [A]. Instead of the Michaelis constant K_m (see p. 92), the *substrate concentration at half-maximal rate* ($[A]_{0.5}$) is given. The sigmoidal character of the curve is described by the **Hill coefficient h.** In isosteric systems, h = 1, and h increases with increasing sigmoidicity.

Depending on the enzyme, *allosteric effectors* can influence the maximum rate V_{max}, the semi-saturation concentration $[A]_{0.5}$, and the Hill coefficient h. If it is mainly V_{max} that is changed, the term **"V system"** is used. Much more common are **"K systems"**, in which allosteric effects only influence $[A]_{0.5}$ and h.

The K type also includes ACTase. The inhibitor CTP in this case leads to *right-shifting* of the curve, with an increase in $[A]_{0.5}$ and h (curve II). By contrast, the activator ATP causes a *left shift;* it reduces both $[A]_{0.5}$ and h (curve III). This type of allosteric effect was first observed in *hemoglobin* (see p. 280), which can be regarded as an "honorary" enzyme.

C. R and T states ○

Allosteric enzymes are almost always *oligomers* with 2–12 subunits. ACTase consists of six catalytic subunits (blue) and six regulatory subunits (yellow). The latter bind the allosteric effectors CTP and ATP. Like hemoglobin, ACTase can also be present in two conformations—the less active **T state** (for "tense") and the more active **R state** (for "relaxed"). Substrates and effectors influence the equilibrium between the two states, and thereby give rise to sigmoidal saturation behavior. With increasing aspartate concentration, the equilibrium is shifted more and more toward the R form. ATP also stabilizes the R conformation by binding to the regulatory subunits. By contrast, binding of CTP to the same sites promotes a transition to the T state. In the case of ACTase, the structural differences between the R and T conformations are particularly dramatic. In T → R conversion, the catalytic subunits separate from one another by 1.2 nm, and the subunits also rotate around the axis of symmetry. The conformations of the subunits themselves change only slightly, however.

D. Structure of a dimer ○

The subunits of ACTase each consist of two *domains*—i. e., independently folded partial structures. The *N*-terminal domain of the regulatory subunit (right) mediates interaction with CTP or ATP (green). A second, Zn^{2+}-containing domain (Zn^{2+} shown in light blue) establishes contact with the neighboring catalytic subunit. Between the two domains of the catalytic subunit lies the active center, which is occupied here by two substrate analogs (red).

A. Aspartate carbamoyltransferase: reaction

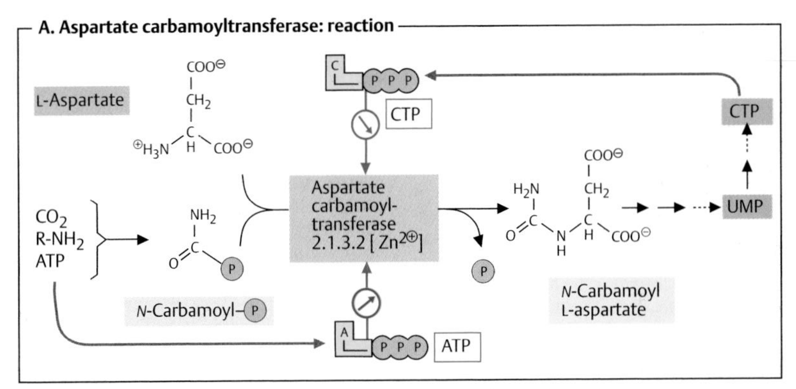

B. Kinetics

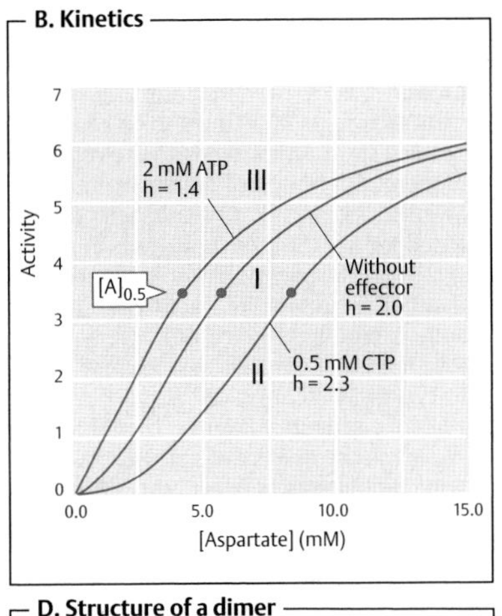

C. R and T conformation

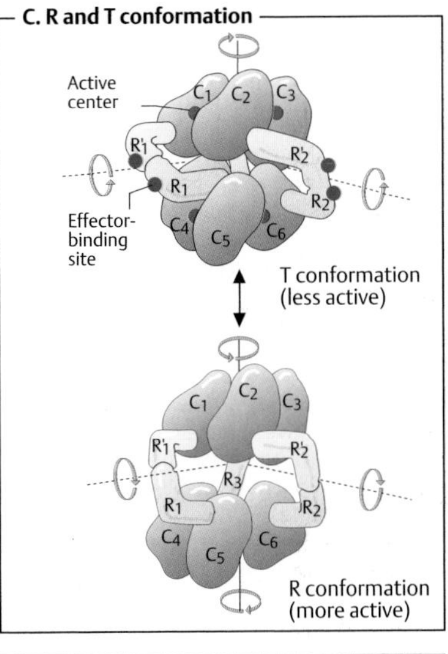

D. Structure of a dimer

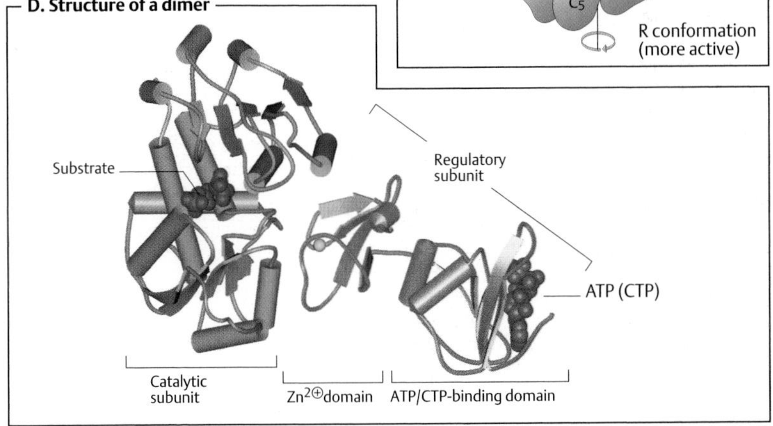

Transcription control

A. Functioning of regulatory proteins ◑

Regulatory proteins (transcription factors) are involved in controlling gene expression in all cells. These regulatory proteins bind to specific DNA sequences and thereby activate or inhibit the transcription of genes **(Transcription control)**. The effects of transcription factors are usually reversible and are often controlled by *ligands* or by *interconversion.*

The nomenclature for transcription factors is confusing. Depending on their mode of action, various terms are in use both for the proteins themselves and for the DNA sequences to which they bind. If a factor blocks transcription, it is referred to as a **repressor**; otherwise, it is called an **inducer.** DNA sequences to which regulatory proteins bind are referred to as **control elements.** In prokaryotes, control elements that serve as binding sites for RNA polymerases are called **promoters,** whereas repressor-binding sequences are usually called **operators.** Control elements that bind activating factors are termed **enhancers,** while elements that bind inhibiting factors are known as **silencers.**

The numerous regulatory proteins that are known can be classified into four different groups (**1**–**4**), based on their mechanisms of action. **Negative gene regulation**—i. e., switching off of the gene concerned—is carried out by **repressors.** Some repressors only bind to DNA (**1a**) in the absence of specific ligands (L). In this case, the complex between the repressor and the ligand loses its ability to bind to the DNA, and the promoter region becomes accesible for binding of RNA polymerase (**1b**). It is often the free repressor that does not bind to the DNA, so that transcription is only blocked in the presence of the ligand (**2a**, **2b**). A distinction between two different types of **positive gene regulation** can be made in the same way. If it is only the free inducer that binds, then transcription is inhibited by the appropriate ligand (**3**). Conversely, many **inducers** only become active when they have bound a ligand (**4**). This group includes the receptors for steroid hormones, for example (see p. 378).

B. Lactose operon ○

The well-investigated **lactose operon** of the bacterium *Escherichial coli* can be used here as an example of transcriptional control. The *lac* operon is a DNA sequence that is simultaneously subject to negative and positive control. The operon contains the *structural genes* for three proteins that are required for the utilization of lactose (one transporter and two enzymes), as well as *control elements* that serve to regulate the operon.

Since lactose is converted to glucose in the cell, there is no point in expressing the genes if glucose is already available. And indeed, the genes are in fact only transcribed when *glucose is absent* and *lactose is present* (**3**). This is achieved by interaction between two regulatory proteins. In the absence of lactose, the ***lac* repressor** blocks the promoter region (**2**). When lactose is available, it is converted into *allolactose*, which binds to the repressor and thereby detaches it from the operator (**3**). However, this is still not sufficient for the transcription of the structural genes. For binding of the RNA polymerase to take place, an *inducer*—the **catabolite activator protein (CAP)**—is required, which only binds to the DNA when it is present as a complex with 3,5′-cyclo-AMP (cAMP; see p. 386). cAMP, a signal for nutrient deficiency, is only formed by *E. coli* in the *absence* of glucose.

The interaction between the CAP–cAMP complex and DNA is shown in Fig. **4**. Each subunit of the dimeric inducer (yellow or orange) binds one molecule of cAMP (red). Contact with the DNA (blue) is mediated by two "recognition helices" that interact with the major groove of the DNA. The bending of the DNA strand caused by CAP has functional significance.

Transcription control is much more complex in eukaryotes (see p. 244). The number of transcription factors involved is larger, and in addition the gene activity is influenced by the state of the chromatin (see p. 238).

A. Functions of regulatory proteins

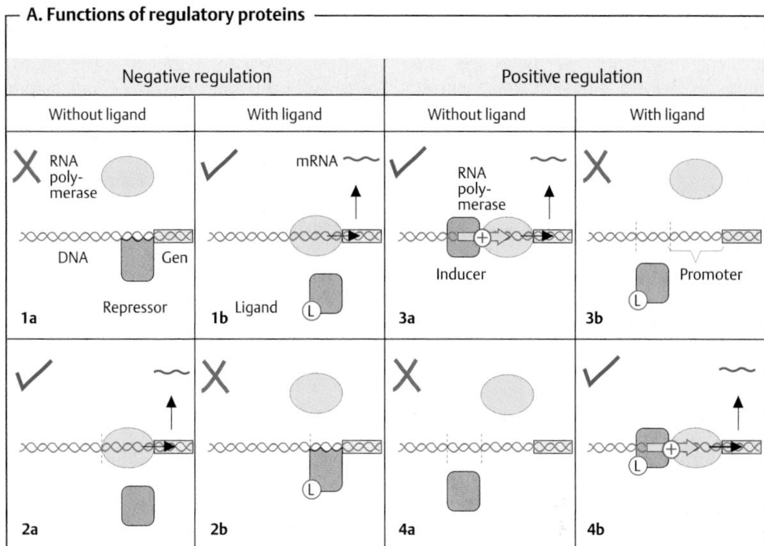

B. Lactose operon

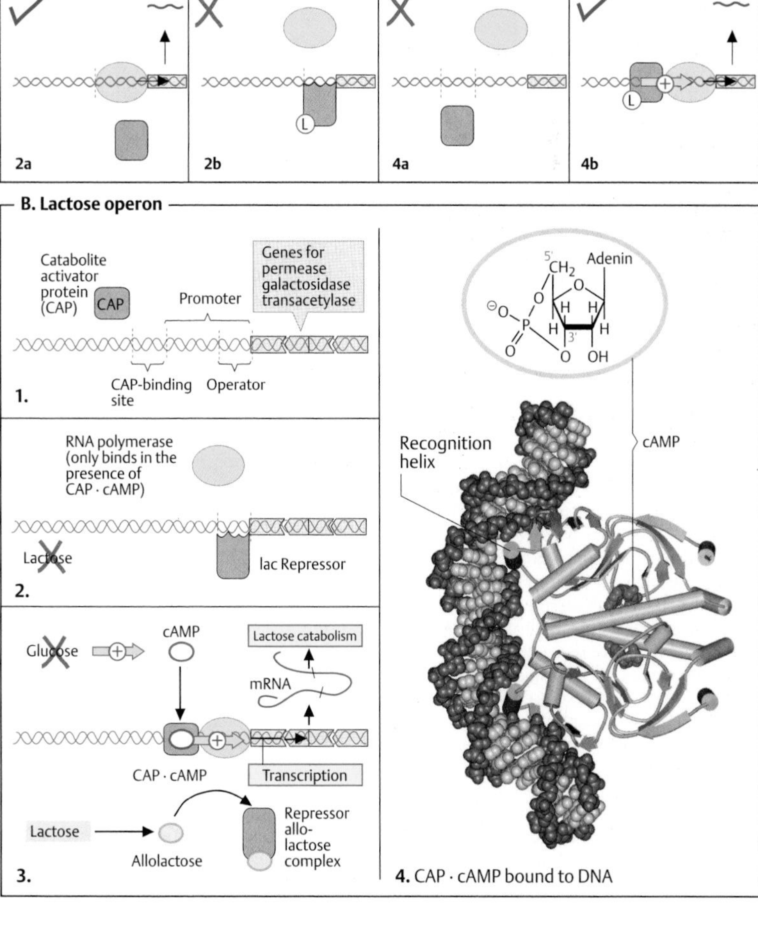

4. CAP · cAMP bound to DNA

Hormonal control

In higher organisms, metabolic and other processes (growth, differentiation, control of the internal environment) are controlled by **hormones** (see pp. 370ff.)

A. Principles of hormone action ●

Depending on the type of hormone, hormone signals are transmitted to the target cells in different ways. Apolar (lipophilic) hormones penetrate the cell and act in the cell nucleus, while polar (hydrophilic) hormones act on the external cell membrane.

Lipophilic hormones, which include the steroid hormones, thyroxine, and retinoic acid, bind to a specific *receptor protein* inside their target cells. The complex formed by the hormone and the receptor then influences *transcription* of specific genes in the cell nucleus (see pp. 118, 244). The group of **hydrophilic hormones** (see p. 380) consists of hormones derived from amino acids, as well as peptide hormones and proteohormones. Their *receptors* are located in the plasma membrane. Binding of the hormone to this type of receptor triggers a signal that is transmitted to the interior of the cell, where it controls the processes that allow the hormone signal to take effect (**signal transduction**; see pp. 384ff.)

B. Hormonal regulation of glucose metabolism in the liver ◑

The liver plays a major role in glucose homeostasis in the organism (see p. 310). If glucose deficiency arises, the liver releases glucose into the blood, and when blood sugar levels are high, it takes glucose up from the blood and converts it into different metabolites. Several hormones from both groups are involved in controlling these processes. A very simplified version of the way in which they work is presented here. **Glycogen** is the form in which glucose is stored in the liver and muscles. The rate of glycogen synthesis is determined by *glycogen synthase* (bottom right), while its breakdown is catalyzed by *glycogen phosphorylase* (bottom left).

Regulation by interconversion (bottom). If the blood glucose level falls, the peptide hormone **glucagon** is released. This activates glycogen breakdown, releasing glucose, and at the same time inhibits glycogen synthesis. Glucagon binds to receptors in the plasma membrane (bottom left) and, with mediation by a G-protein (see p. 386), activates the enzyme *adenylate cyclase*, which forms the *second messenger* 3,5′-cyclo-AMP (**cAMP**) from ATP. cAMP binds to another enzyme, *protein kinase A* (PK-A), and activates it. PK-A has several points of attack. Through *phosphorylation*, it converts the active form of *glycogen synthase* into the inactive form, thereby terminating the synthesis of glycogen. Secondly, it activates another protein kinase (not shown), which ultimately converts the inactive form of *glycogen phosphorylase* into the active form through phosphorylation. The active phosphorylase releases glucose 1-phosphate from glycogen, which after conversion into glucose 6-phosphate supplies free glucose. In addition, via an inhibitor (I) of protein phosphatase (PP), active PK-A inhibits inactivation of glycogen phosphorylase. When the cAMP level falls again, *phosphoprotein phosphatases* become active, which dephosphorylate the various phosphoproteins in the cascade described, and thereby arrest glycogen breakdown and re-start glycogen synthesis. Activation and inactivation of proteins through phosphorylation or dephosphorylation is referred to as **interconversion**.

In contrast to glucagon, the peptide hormone **insulin** (see p. 76) increases glycogen synthesis and inhibits glycogen breakdown. Via several intermediates, it inhibits protein kinase GSK-3 (bottom right; for details, see p. 388) and thereby prevents inactivation of glycogen synthase. In addition, insulin reduces the cAMP level by activating *cAMP phosphodiesterase* (PDE).

Regulation by transcriptional control (top). If the liver's glycogen reserves have been exhausted, the steroid hormone **cortisol** maintains glucose release by initiating the conversion of amino acids into glucose (*gluconeogenesis*; see p. 154). In the cell nucleus, the complex of cortisol and its receptor (see p. 378) binds to the promoter regions of various key enzymes of gluconeogenesis and leads to their transcription. The active enzymes are produced through translation of the mRNA formed. Control of the transcription of the gluconeogenesis enzyme *PEP carboxykinase* is discussed on p. 244.

A. Principles of hormone action

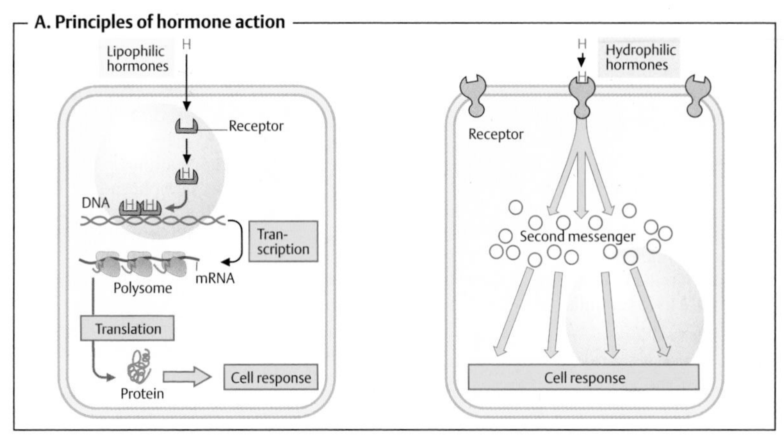

B. Hormonal regulation of glucose metabolism in the liver

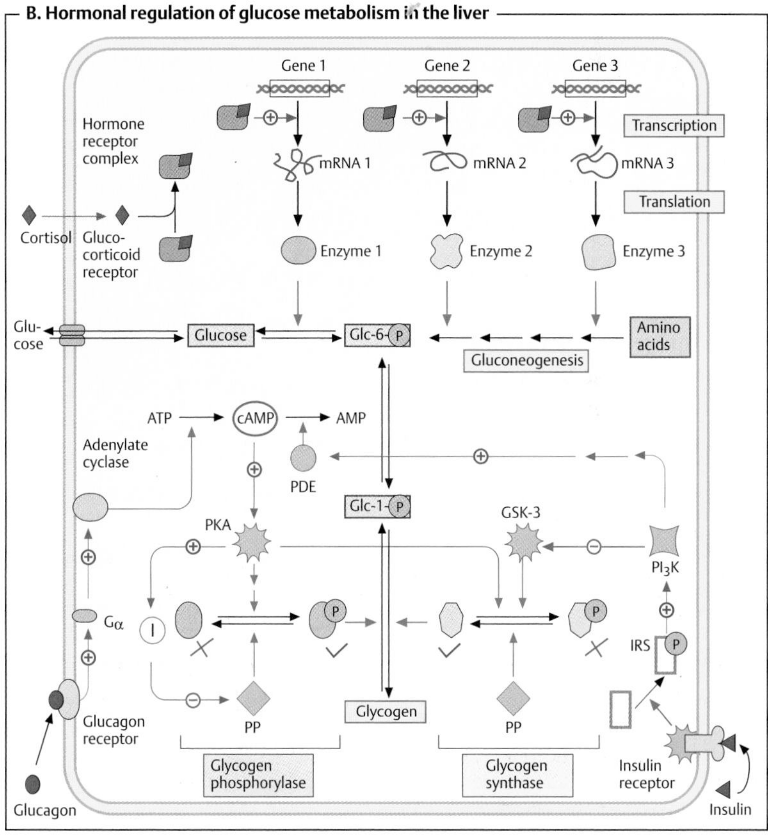

ATP

The nucleotide coenzyme **adenosine triphosphate** (ATP) is the most important **form of chemical energy** in all cells. Cleavage of ATP is strongly exergonic. The energy this provides (ΔG; see p. 16) is used to drive endergonic processes (such as biosynthesis and movement and transport processes) through *energetic coupling* (see p. 124). The other *nucleoside triphosphate coenzymes* (GTP, CTP, and UTP) have similar chemical properties to ATP, but they are used for different tasks in metabolism (see p. 110).

A. ATP: structure ◑

In ATP, a chain of three phosphate residues is linked to the 5′-OH group of the nucleoside adenosine (see p. 80). These phosphate residues are termed α, β, and γ. The α phosphate is bound to ribose by a *phosphoric acid ester bond*. The linkages *between* the three phosphate residues, on the other hand, involve much more unstable *phosphoric acid anhydride bonds*. The active coenzyme is in fact generally a complex of ATP with an Mg^{2+} ion, which is coordinatively bound to the α and β phosphates ($Mg^{2+} \cdot ATP^{4-}$). However, the term "ATP" is usually used for the sake of simplicity.

B. Hydrolysis energies ◑

The formula for phosphate residues shown in Fig. **A**, with single and double bonds, is not an accurate representation of the actual charge distribution. In ATP, the oxygen atoms of all three phosphate residues have similarly strong negative charges (orange), while the phosphorus atoms represent centers of positive charge. One of the reasons for the instability of phosphoric anhydride bonds is the *repulsion between these negatively charged oxygen atoms*, which is partly relieved by cleavage of a phosphate residue. In addition, the free phosphate anion formed by hydrolysis of ATP is *better hydrated* and *more strongly resonance-stabilized* than the corresponding residue in ATP. This also contributes to the strongly exergonic character of ATP hydrolysis.

In *standard conditions*, the change in free enthalpy $\Delta G^{0\prime}$ (see p. 18) that occurs in the hydrolysis of phosphoric acid anhydride bonds amounts to –30 to –35 $kJ \cdot mol^{-1}$ at pH 7. The particular anhydride bond of ATP that is cleaved only has a minor influence on $\Delta G^{0\prime}$ (**1**–**2**). Even the hydrolysis of diphosphate (also known as pyrophosphate; **4**) still yields more than –30 $kJ \cdot mol^{-1}$. By contrast, cleavage of the ester bond between ribose and phosphate only provides –9 $kJ \cdot mol^{-1}$ (**3**).

In the cell, the ΔG of ATP hydrolysis is substantially larger, because the concentrations of ATP, ADP and P_i are much lower than in standard conditions and there is an excess of ATP over ADP (see p. 18). The pH value and Mg^{2+} concentration also affect the value of ΔG. The *physiological energy yield* of ATP hydrolysis to ADP and anorganic phosphate (P_i) is probably around –50 $kJ \cdot mol^{-1}$.

C. Types of ATP formation ●

Only a few compounds contain phosphate residues with a group transfer potential (see p. 18) that is high enough to transfer them to ADP and thus allow **ATP synthesis**. Processes that raise anorganic phosphate to this type of high potential are called **substrate level phosphorylations** (see p. 124). Reactions of this type take place in glycolysis (see p. 150) and in the tricarboxylic acid cycle (see p. 136). Another "energy-rich" phosphate compound is *creatine phosphate*, which is formed from ATP in muscle and can regenerate ATP as needed (see p. 336).

Most cellular ATP does not arise in the way described above (i. e., by transfer of phosphate residues from organic molecules to ADP), but rather by **oxidative phosphorylation**. This process takes place in mitochondria (or as light-driven phosphorylation in chloroplasts) and is energetically coupled to a proton gradient over a membrane. These H^+ gradients are established by electron transport chains and are used by the enzyme *ATP synthase* as a source of energy for direct linking of anorganic phosphate to ADP. In contrast to substrate level phosphorylation, oxidative phosphorylation requires the presence of oxygen (i. e., *aerobic* conditions).

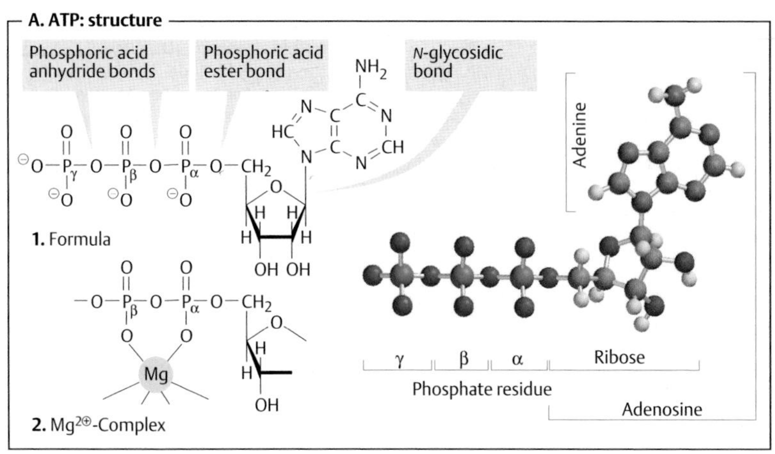
A. ATP: structure
Phosphoric acid anhydride bonds
Phosphoric acid ester bond
N-glycosidic bond
1. Formula
2. Mg2⊕-Complex
Adenine
γ
β
α
Ribose
Phosphate residue
Adenosine

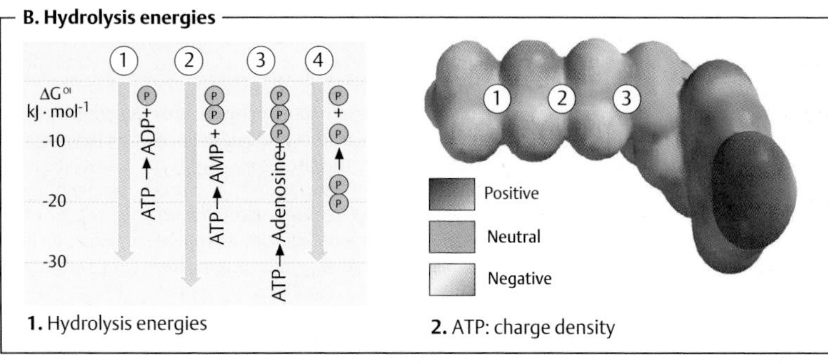
B. Hydrolysis energies
ΔG°' kJ · mol-1
-10
-20
-30
ATP → ADP+ P
ATP → AMP + P P
ATP → Adenosine+ P P P
P P → P + P
1. Hydrolysis energies
Positive
Neutral
Negative
2. ATP: charge density

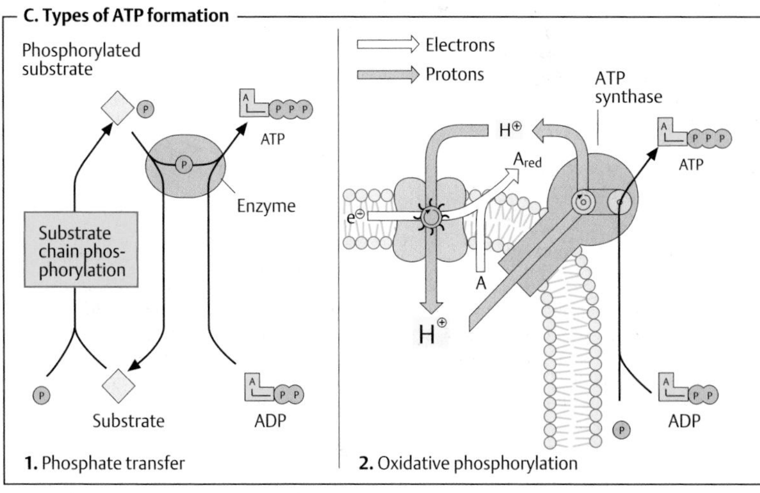
C. Types of ATP formation
Phosphorylated substrate
ATP
Enzyme
Substrate chain phos-phorylation
Substrate
ADP
1. Phosphate transfer
Electrons
Protons
ATP synthase
H⊕
Ared
e⊖
A
H⊕
ATP
ADP
2. Oxidative phosphorylation

Energetic coupling

The cell stores chemical energy in the form of "energy-rich" metabolites. The most important metabolite of this type is adenosine triphosphate (**ATP**), which drives a large number of energy-dependent reactions via **energetic coupling** (see p. 16).

A. Energetic coupling ◑

The change in free enthalpy ΔG^0 during hydrolysis (see p. 18) has been arbitrarily selected as a measure of the group transfer potential of "energy-rich" compounds. However, this does not mean that ATP is in fact hydrolyzed in energetically coupled reactions. If ATP hydrolysis and an endergonic process were simply allowed to run alongside each other, the hydrolysis would only produce heat, without influencing the endergonic process. For coupling, the two reactions have to be linked in such a way that a *common intermediate* arises. This connection is illustrated here using the example of the **glutamine synthetase reaction.**

Direct transfer of NH_3 to glutamate is endergonic ($\Delta G^{0'} = +14\ kJ \cdot mol^{-1}$; see p. 18), and can therefore not take place. In the cell, the reaction is divided into two exergonic steps. First, the γ-phophate residue is transferred from ATP to glutamate. This gives rise to an "energy-rich" *mixed acid anhydride.* In the second step, the phosphate residue from the intermediate is substituted by NH_3, and glutamine and free phosphate are produced. The energy balance of the reaction as a whole ($\Delta G^{0'} = -17\ kJ \cdot mol^{-1}$) is the sum of the changes in free enthalpy of direct glutamine synthesis ($\Delta G^{0'} = 14\ kJ \cdot mol^{-1}$) plus ATP hydrolysis ($\Delta G^{0'} = -31\ kJ \cdot mol^{-1}$), although ATP has not been hydrolyzed at all.

B. Substrate-level phosphorylation ◑

As mentioned earlier (see p. 122), there are a few metabolites that transfer phosphate to ADP in an exergonic reaction and can therefore form ATP. In ATP synthesis, anorganic phosphate or phosphate bound in an ester-like fashion is transferred to bonds with a high phosphate transfer potential. Reactions of this type are termed *"substrate-level phosphorylations,"* as they represent individual steps within metabolic pathways.

In the **glyceraldehyde 3-phosphate dehydrogenation** reaction, a step involved in glycolysis (**1**; see also **C**), the aldehyde group in glyceraldehyde 3-phosphate is oxidized into a carboxyl group. During the reaction, an anorganic phosphate is also introduced into the product, producing a mixed acid anhydride—1,3-bisphosphoglycerate. **Phosphopyruvate hydratase** ("enolase", **2**) catalyzes the elimination of water from 2-phosphoglycerate. In the *enol phosphate* formed (phosphoe*nol* pyruvate), the phosphate residue—in contrast to 2-phosphoglycerate—is at an extremely high potential ($\Delta G^{0'}$ of hydrolysis: $-62\ kJ \cdot mol^{-1}$). A third reaction of this type is the formation of succinyl phosphate, which occurs in the tricarboxylic acid cycle as an individual step in the **succinyl CoA ligase** reaction. Here again, anorganic phosphate is introduced into a mixed acid anhydride bond to be transferred from there to GDP. Succinyl phosphate is only an intermediate here, and is not released by the enzyme.

In the literature, the term "substrate level phosphorylation" is used inconsistently. Some authors use it to refer to reactions in which *anorganic* phosphate is raised to a high potential, while others use it for the subsequent reactions, in which ATP or GTP is formed from the energy-rich intermediates.

C. Glyceraldehyde-3-phosphate dehydrogenase ○

The reaction catalyzed during glycolysis by *glyceraldehyde-3-phosphate dehydrogenase* (GADPH) is shown here in detail. Initially, the SH group of a cysteine residue of the enzyme is added to the carbonyl group of glyceraldehyde 3-phosphate (**a**). This intermediate is oxidized by NAD^+ into an "energy-rich" thioester (**b**). In the third step (**c**), anorganic phosphate displaces the thiol, and the mixed anhydride *1,3-bisphosphoglycerate* arises. In this bond, the phosphate residue is at a high enough potential for it to be transferred to ADP in the next step (not shown; see p. 150).

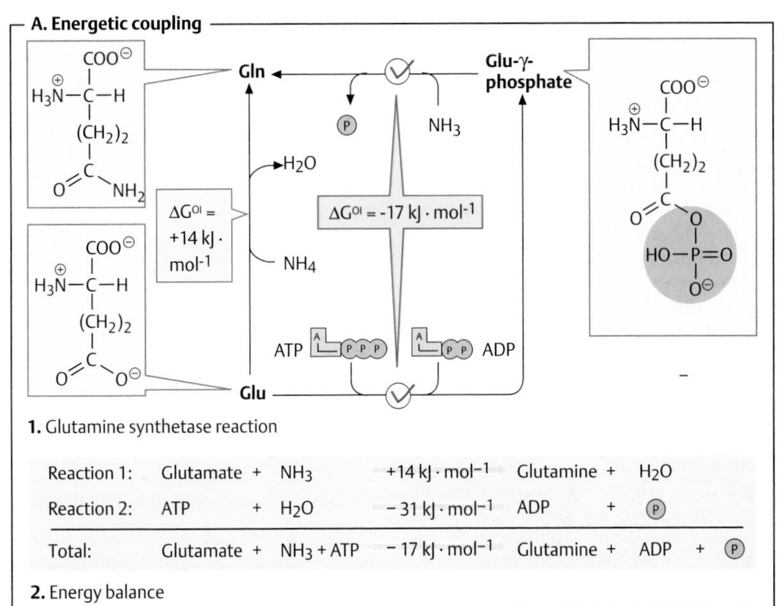
A. Energetic coupling
Gln
Glu-γ-phosphate
P
NH3
H2O
ΔG°' = +14 kJ · mol-1
ΔG°' = -17 kJ · mol-1
NH4
ATP
ADP
Glu
1. Glutamine synthetase reaction
Reaction 1: Glutamate + NH3 +14 kJ · mol-1 Glutamine + H2O
Reaction 2: ATP + H2O – 31 kJ · mol-1 ADP + P
Total: Glutamate + NH3 + ATP – 17 kJ · mol-1 Glutamine + ADP + P
2. Energy balance

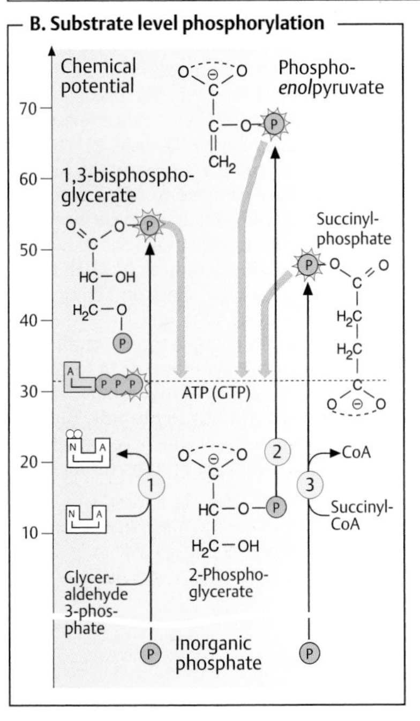
B. Substrate level phosphorylation
Chemical potential
70
60
50
40
30
20
10
Phospho-enolpyruvate
1,3-bisphospho-glycerate
Succinyl-phosphate
ATP (GTP)
CoA
Succinyl-CoA
Glycer-aldehyde 3-phos-phate
2-Phospho-glycerate
Inorganic phosphate

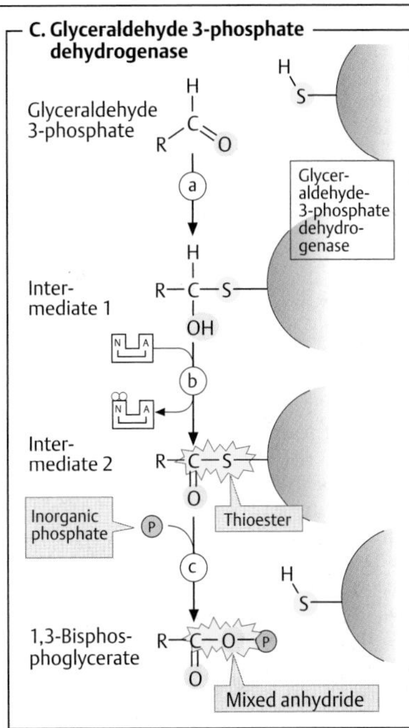
C. Glyceraldehyde 3-phosphate dehydrogenase
Glyceraldehyde 3-phosphate
Glycer-aldehyde-3-phosphate dehydro-genase
Inter-mediate 1
Inter-mediate 2
Thioester
Inorganic phosphate
1,3-Bisphos-phoglycerate
Mixed anhydride

Energy conservation at membranes

Metabolic energy can be stored not only in the form of "energy-rich" bonds (see p. 122), but also by separating electric charges from each other using an insulating layer to prevent them from redistributing. In the field of technology, this type of system would be called a *condenser*. Using the same principle, energy is also stored ("conserved") at cell membranes. The membrane functions as an insulator; electrically charged atoms and molecules (*ions*) function as charges.

A. Electrochemical gradient ◑

Although artificial lipid membranes are almost impermeable to ions, biological membranes contain **ion channels** that selectively allow individual ion types to pass through (see p. 222). Whether an ion can cross this type of membrane, and if so in which direction, depends on the **electrochemical gradient**—i. e., on the concentrations of the ion on each side of the membrane (the *concentration gradient*) and on the *difference* in the electrical potential between the interior and exterior, the **membrane potential.**

The membrane potential of resting cells (**resting potential**; see p. 350) is –0.05 to –0.09 V—i. e., there is an excess negative charge on the inner side of the plasma membrane. The main contributors to the resting potential are the two cations Na^+ and K^+, as well as Cl^- and organic anions (**1**). Data on the concentrations of these ions outside and inside animal cells, and permeability coefficients, are shown in the table (**2**).

The behavior of an ion type is described quantitatively by the **Nernst equation** (**3**). $\Delta\psi_G$ is the membrane potential (in volts, V) at which there is *no* net transport of the ion concerned across the membrane (**equilibrium potential**). The factor R·T/F·n has a value of 0.026 V for monovalent ions at 25 °C. Thus, for K^+, the table (**2**) gives an equilibrium potential of ca. –0.09 V—i. e., a value more or less the same as that of the resting potential. By contrast, for Na^+ ions, $\Delta\psi_G$ is much higher than the resting potential, at +0.07 V. Na^+ ions therefore immediately flow into the cell when Na^+ channels open (see p. 350). The disequilibrium between Na^+ and K^+ ions is constantly maintained by the enzyme *Na^+/K^+-ATPase,* which consumes ATP.

B. Proton motive force ◑

Hydronium ions ("H^+ ions") can also develop electrochemical gradients. Such a **proton gradient** plays a decisive part in cellular ATP synthesis (see p. 142). As usual, the energy content of the gradient depends on the concentration gradients—i. e., on the **pH difference** ΔpH between the two sides of the membrane. In addition, the **membrane potential** $\Delta\psi$ also makes a contribution. Together, these two values give the **proton motive force** Δp, a measure for the work that the H^+ gradient can do. The proton gradient across the inner mitochondrial membrane thus delivers approximately 24 kJ per mol H^+.

C. Energy conservation in proton gradients ◑

Proton gradients can be built up in various ways. A very unusual type is represented by **bacteriorhodopsin** (**1**), a *light-driven proton pump* that various bacteria use to produce energy. As with rhodopsin in the eye, the light-sensitive component used here is covalently bound retinal (see p. 358). In photosynthesis (see p. 130), reduced *plastoquinone* (QH_2) transports protons, as well as electrons, through the membrane (**Q cycle**, **2**). The formation of the proton gradient by the **respiratory chain** is also coupled to redox processes (see p. 140). In complex III, a Q cycle is responsible for proton translocation (not shown). In *cytochrome c oxidase* (complex IV, **3**), H^+ transport is coupled to electron flow from cytochrome *c* to O_2.

In each of these cases, the H^+ gradient is utilized by an **ATP synthase** (**4**) to form ATP. ATP synthases consist of two components—a proton channel (F_0) and an inwardly directed protein complex (F_1), which conserves the energy of back-flowing protons through ATP synthesis (see p. 142).

A. Electrochemical gradient

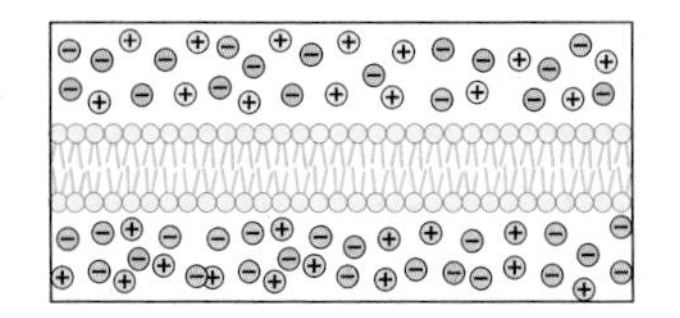

⊕ Na ⊕ K ⊖ Cl ⊖ Organic anions

1. Cause

Ion	Concentrations		Permeability coefficient
	Cytoplasm (mM)	Extracelluar space (mM)	$(cm \cdot s^{-1} \cdot 10^9)$
$K^{\oplus}$	100	5	500
$Na^{\oplus}$	15	150	5
$Ca^{2\oplus}$	0.0002	2	
$Cl^{\ominus}$	13	150	10
Organic anions	138	34	0

2. Concentrations

$$\Delta\Psi_G = \frac{R \cdot T}{F \cdot n} \cdot \ln \frac{C_{outside}}{C_{inside}}$$

R = gas constant n = Ion charge
T = temperature (K) F = Faraday constant

3. Nernst equation

B. Proton motive force

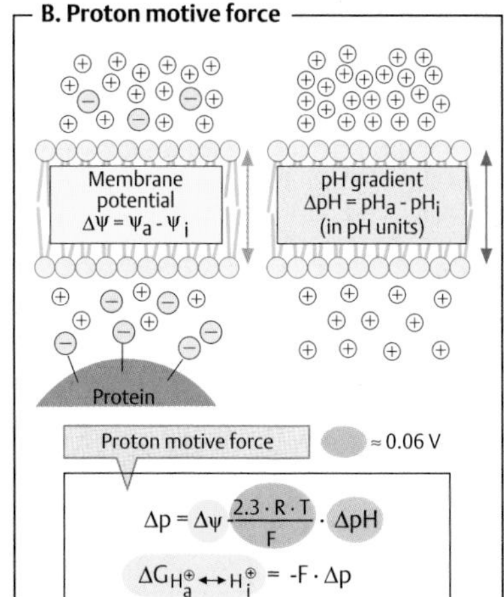

C. Energy conservation in proton gradients

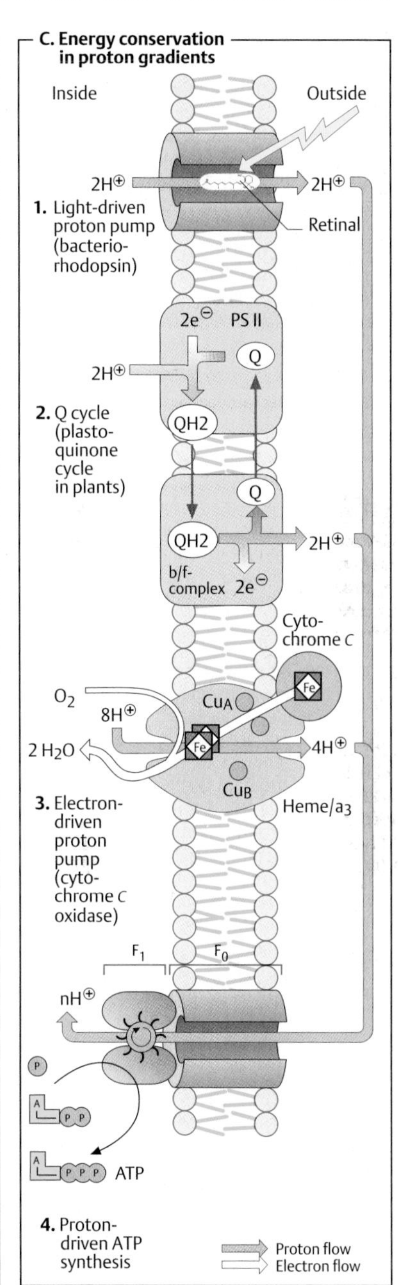

Photosynthesis: light reactions

Sunlight is the most important source of energy for nearly all living organisms. With the help of **photosynthesis,** light energy is used to produce organic substances from CO_2 and water. This property of *phototrophic organisms* (plants, algae, and some bacteria) is exploited by *heterotrophic* organisms (e. g., animals), which are dependent on a supply of organic substances in their diet (see p. 112). The atmospheric *oxygen* that is vital to higher organisms is also derived from photosynthesis.

A. Photosynthesis: overview ◑

The chemical balance of photosynthesis is simple. Six molecules of CO_2 are used to form one hexose molecule (right). The hydrogen required for this reduction process is taken from water, and molecular oxygen is formed as a by-product (left). Light energy is required, since water is a very poor reducing agent and is therefore not capable of reducing CO_2.

In the light-dependent part of photosynthesis—the "**light reactions** "—H_2O molecules are split into protons, electrons, and oxygen atoms. The electrons undergo *excitation* by light energy and are raised to an energy level that is high enough to reduce $NADP^+$. The $NADPH+H^+$ formed in this way, in contrast to H_2O, is capable of "fixing" CO_2 reductively—i. e., of incorporating it into organic bonds. Another product of the light reactions is ATP, which is also required for CO_2 fixation. If $NADPH+H^+$, ATP, and the appropriate enzymes are available, CO_2 fixation can also take place in darkness. This process is therefore known as the "**dark reaction.**"

The excitation of electrons to form NADPH is a complex photochemical process that involves **chlorophyll,** a tetrapyrrole dye containing Mg^{2+} that bears an extra phytol *residue* (see p. 132).

B. Light reactions ○

In green algae and higher plants, photosynthesis occurs in **chloroplasts.** These are organelles, which—like mitochondria—are surrounded by two membranes and contain their own DNA. In their interior, the *stroma, thylakoids* or flattened membrane sacs are stacked on top of each other to form *grana*. The inside of the thylakoid is referred to as the *lumen.* The light reactions are catalyzed by enzymes located in the thylakoid membrane, whereas the dark reactions take place in the stroma.

As in the respiratory chain (see p. 140), the light reactions cause electrons to pass from one redox system to the next in an **electron transport chain.** However, the *direction of transport* is opposite to that found in the respiratory chain. In the respiratory chain, electrons flow from $NADH+H^+$ to O_2, with the production of water and energy.

In photosynthesis, electrons are taken up from water and transferred to $NADP^+$, with an *expenditure of energy.* Photosynthetic electron transport is therefore energetically "uphill work." To make this possible, the transport is stimulated at two points by the *absorption of light energy.* This occurs through two **photosystems**—protein complexes that contain large numbers of chlorophyll molecules and other pigments (see p. 132). Another component of the transport chain is the **cytochrome b/f complex,** an aggregate of integral membrane proteins that includes two cytochromes (b_{563} and f). **Plastoquinone**, which is comparable to ubiquinone, and two soluble proteins, the coppercontaining **plastocyanin** and **ferredoxin**, function as mobile electron carriers. At the end of the chain, there is an enzyme that transfers the electrons to $NADP^+$.

Because photosystem II and the cytochrome b/f complex release protons from reduced plastoquinone into the lumen (via a Q cycle), photosynthetic electron transport establishes an **electrochemical gradient** across the thylakoid membrane (see p. 126), which is used for ATP synthesis by an *ATP synthase.* ATP and $NADPH+H^+$, which are both needed for the dark reactions, are formed in the stroma.

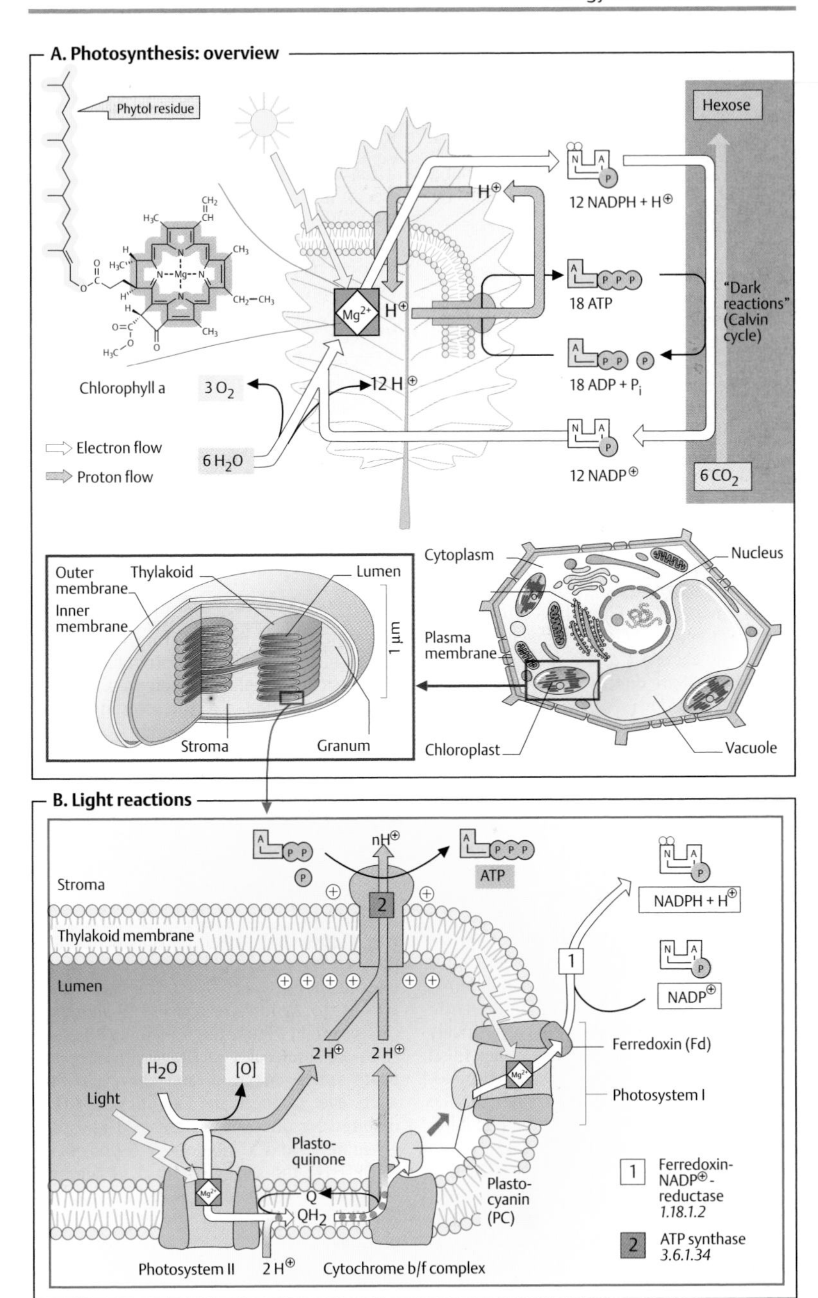
A. Photosynthesis: overview
Phytol residue
Chlorophyll a
Hexose
12 NADPH + H⊕
18 ATP
18 ADP + Pi
12 NADP⊕
"Dark reactions" (Calvin cycle)
6 CO2
3 O2
12 H⊕
6 H2O
Electron flow
Proton flow
Outer membrane
Thylakoid
Lumen
Inner membrane
1 µm
Stroma
Granum
Cytoplasm
Nucleus
Plasma membrane
Chloroplast
Vacuole
B. Light reactions
Stroma
nH⊕
ATP
Thylakoid membrane
Lumen
NADPH + H⊕
NADP⊕
Ferredoxin (Fd)
Photosystem I
H2O
[O]
Light
2 H⊕
2 H⊕
Plasto-quinone
Q
QH2
Plasto-cyanin (PC)
Photosystem II
2 H⊕
Cytochrome b/f complex
1 Ferredoxin-NADP⊕-reductase 1.18.1.2
2 ATP synthase 3.6.1.34

Photosynthesis: dark reactions

The "light reactions" in photosynthesis bring about two strongly endergonic reactions—the reduction of $NADP^+$ to $NADPH+H^+$ and ATP synthesis (see p. 122). The chemical energy needed for this is produced from radiant energy by two photosystems.

A. Photosystem II ○

The photosynthetic electron transport chain in plants starts in **photosystem II** (PS II; see p. 128). PS II consists of numerous protein subunits (brown) that contain bound **pigments**—i. e., dye molecules that are involved in the absorption and transfer of light energy.

The schematic overview of PS II presented here (**1**) only shows the important pigments. These include a special chlorophyll molecule, the *reaction center* P_{680}; a neighboring Mg^{2+} free chlorophyll (*pheophytin*); and two bound *plastoquinones* (Q_A and Q_B). A third quinone (Q_P) is not linked to PS II, but belongs to the plastoquinone pool. The white arrows indicate the direction of electron flow from water to Q_P. Only about 1% of the chlorophyll molecules in PS II are *directly* involved in photochemical excitation (see p. 128). Most of them are found, along with other pigments, in what are known as light-harvesting or antenna complexes (green). The energy of light quanta striking these can be passed on to the reaction center, where it can be utilized.

In Fig. **2**, photosynthetic electron transport in PS II is separated into the individual steps involved. Light energy from the light-harvesting complexes (**a**) raises an electron of the chlorophyll in the reaction center to an excited *"singlet state."* The excited electron is immediately passed on to the neighboring pheophytin. This leaves behind an "electron gap" in the reaction center—i. e., a positively charged P_{680} radical (**b**). This gap is now filled by an electron removed from an H_2O molecule by the *water-splitting enzyme* (b). The excited electron passes on from the pheophytin via Q_A to Q_B, converting the latter into a *semiquinone radical* (**c**). Q_B is then reduced to *hydroquinone* by a second excited electron, and is then exchanged for an oxidized quinone (Q_P) from the plastoquinone pool. Further transport of electrons from the plastoquinone pool takes place as described on the preceding page and shown in **B**.

B. Redox series ○

It can be seen from the *normal potentials* E^0 (see p. 18) of the most important redox systems involved in the light reactions why two excitation processes are needed in order to transfer electrons from H_2O to $NADP^+$. After excitation in PS II, E^0 *rises* from around –1 V back to positive values in plastocyanin (PC)—i. e., the energy of the electrons has to be increased again in PS I. If there is no $NADP^+$ available, photosynthetic electron transport can still be used for ATP synthesis. During *cyclic photophosphorylation,* electrons return from ferredoxin (Fd) via the plastoquinone pool to the b/f complex. This type of electron transport does not produce any NADPH, but does lead to the formation of an H^+ gradient and thus to ATP synthesis.

C. Calvin cycle ○

The synthesis of hexoses from CO_2 is only shown in a very simplified form here; a complete reaction scheme is given on p. 407. The actual **CO_2 fixation**—i. e., the incorporation of CO_2 into an organic compound—is catalyzed by *ribulose bisphosphate carboxylase/oxygenase* ("rubisco"). Rubisco, the most abundant enzyme on Earth, converts ribulose 1,5-bisphosphate, CO_2 and water into *two molecules* of 3-phosphoglycerate. These are then converted, via 1,3-bisphosphoglycerate and 3-phosphoglycerate, into glyceraldehyde 3-phosphate (glyceral 3-phosphate). In this way, 12 glyceraldehyde 3-phosphates are synthesized from six CO_2. Two molecules of this intermediate are used by gluconeogenesis reactions to synthesize *glucose 6-phosphate* (bottom right). From the remaining 10 molecules, six molecules of *ribulose 1,5-bisphosphate* are regenerated, and the cycle then starts over again. In the Calvin cycle, ATP is required for phosphorylation of 3-phosphoglycerate and ribulose 5-phosphate. $NADPH+H^+$, the second product of the light reaction, is consumed in the reduction of 1,3-bisphosphoglycerate to glyceraldehyde 3-phosphate.

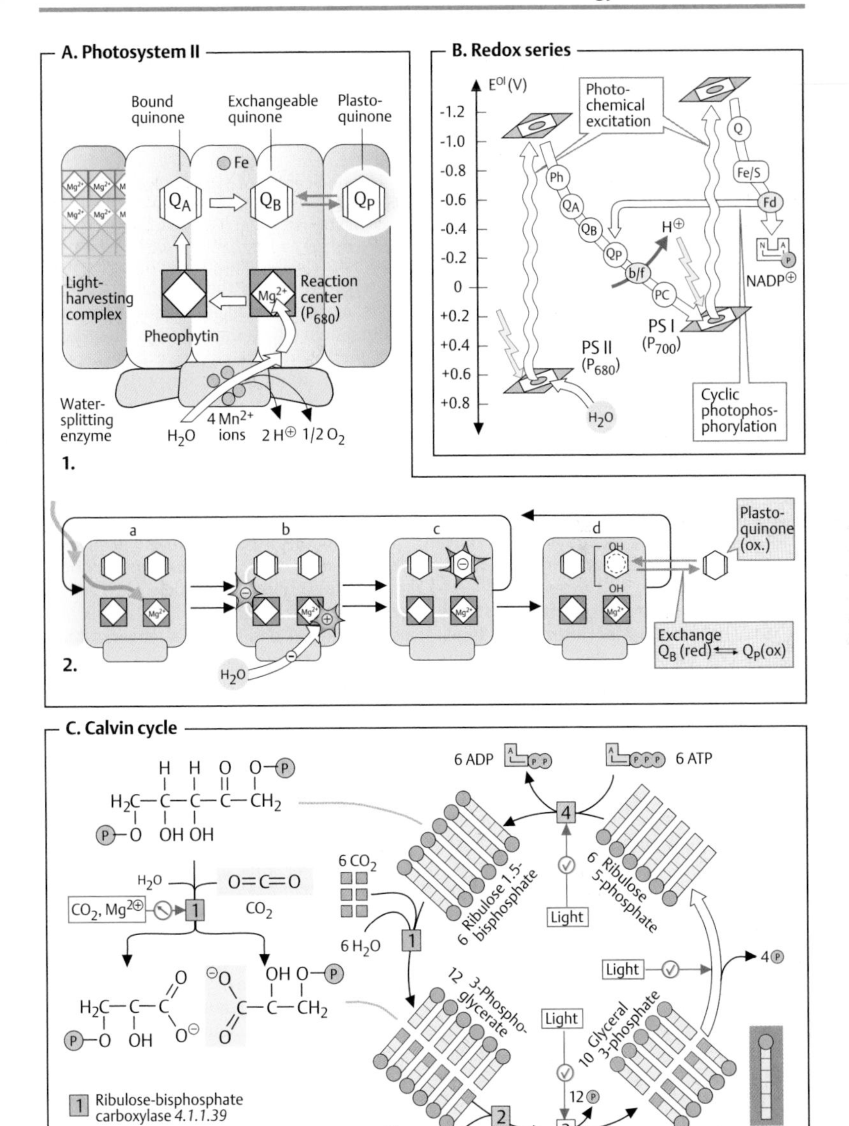
A. Photosystem II
Bound quinone
Exchangeable quinone
Plasto-quinone
Fe
Q_A
Q_B
Q_P
Light-harvesting complex
Reaction center (P_{680})
Pheophytin
Water-splitting enzyme
H_2O
4 Mn^{2+} ions
2 H⊕
1/2 O_2
1.
B. Redox series
$E^{0\prime}$ (V)
-1.2
-1.0
-0.8
-0.6
-0.4
-0.2
0
+0.2
+0.4
+0.6
+0.8
Photo-chemical excitation
Ph
QA
QB
QP
b/f
PC
Q
Fe/S
Fd
H⊕
NADP⊕
PS I (P_{700})
PS II (P_{680})
H_2O
Cyclic photophos-phorylation
a
b
c
d
OH
OH
Plasto-quinone (ox.)
Exchange Q_B (red) ⟷ Q_P(ox)
H_2O
2.
C. Calvin cycle
CO_2, $Mg^{2⊕}$
H_2O
O=C=O
CO_2
6 CO_2
6 H_2O
6 ADP
6 ATP
6 Ribulose 1.5-bisphosphate
6 Ribulose 5-phosphate
Light
12 3-Phospho-glycerate
10 Glyceral 3-phosphate
4 P
12 P
12 ATP
12 ADP
12 NADPH
12 NADP⊕
Glucose 6-phosphate
Gluco-neogenesis
2 Glyceraldehyde 3-phosphate
1 Ribulose-bisphosphate carboxylase 4.1.1.39
2 Phosphoglycerate kinase 2.7.2.3
3 Glyceraldehyde 3-phosphate dehydrogenase (NADP⊕) 1.2.1.13
4 Phosphoribulokinase 2.7.1.19

Molecular models: membrane proteins

The plates show, in simplified form, the structures of **cytochrome *c* oxidase** (**A**; complex IV of the respiratory chain) and of **photosystem I** of a cyanobacterium (**B**). These two molecules are among the few integral membrane proteins for which the structure is known in detail. Both structures were determined by X-ray crystallography.

A. Cytochrome *c* oxidase ○

The enzyme cytochrome *c* oxidase ("COX," *EC 1.9.3.1*) catalyzes the final step of the respiratory chain. It receives electrons from the small heme protein cytochrome *c* and transfers them to molecular oxygen, which is thereby reduced to water (see p. 140). At the same time, 2–4 protons per water molecule formed are pumped from the matrix into the intermembrane space.

Mammalian COX (the illustration shows the enzyme from bovine heart) is a dimer that has two identical subunits with masses of 204 kDa each. Only one subunit is shown in detail here; the other is indicated by gray lines. Each subunit consists of 13 different polypeptides, which all span the inner mitochondrial membrane. Only polypeptides I (light blue) and II (dark blue) and the linked cofactors are involved in electron transport. The other chains, which are differently expressed in the different organs, probably have regulatory functions. The two heme groups, heme a (orange) and heme a_1 (red) are bound in polypeptide 1. The copper center Cu_A consists of two copper ions (green), which are coordinated by amino acid residues in polypeptide II. The second copper (Cu_B) is located in polypeptide I near heme a_3.

To reduce an O_2 molecule to two molecules of H_2O, a total of four electrons are needed, which are supplied by cytochrome *c* (pink, top left) and initially given off to Cu_A. From there, they are passed on via heme a and heme a_3 to the enzyme's reaction center, which is located between heme a_3 and Cu_B. The reduction of the oxygen takes place in several steps, without any intermediate being released. The four protons needed to produce water and the H^+ ions pumped into the intermembrane space are taken up by two channels (D and K, not shown). The mechanism that links proton transport to electron transfer is still being investigated.

B. Reaction center of *Synechococcus elongatus* ○

Photosystem I (PS I) in the cyanobacterium *Synechococcus elongatus* is the first system of this type for which the structure has been solved in atomic detail. Although the bacterial photosystem differs slightly from the systems in higher plants, the structure provides valuable hints about the course of the light reactions in photosynthesis (see p. 128). The functioning of the photosystem is discussed in greater detail on p. 130.

The functional form of PS I in *S. elongatus* consists of a trimer with a mass of more than 10^6 Da that is integrated into the membrane. Only one of the three subunits is shown here. This consists of 12 different polypeptides (gray-blue), 96 chlorophyll molecules (green), 22 carotenoids (orange), several phylloquinones (yellow), and other components. Most of the chlorophyll molecules are so-called **antenna pigments**. These collect light energy and conduct it to the **reaction center**, which is located in the center of the structure and therefore not visible. In the reaction center, an electron is excited and transferred via various intermediate steps to a ferredoxin molecule (see p. 128). The **chlorophylls** (see formula) are heme-like pigments with a highly modified tetrapyrrole ring, a central Mg^{2+} ion, and an apolar phytol side chain. Shown here is chlorophyll a, which is also found in the reaction center of the *S. elongatus* photosystem.

The yellow and orange-colored **carotenoids**—e. g., *β-carotene* (see formula)—are auxiliary pigments that serve to protect the chloroplasts from oxidative damage. Dangerous radicals can be produced during the light reaction—particularly *singlet oxygen*. Carotenoids prevent compounds of this type from arising, or render them inactive. Carotenoids are also responsible for the coloring of leaves seen during fall. They are left behind when plants break down chlorophyll in order to recover the nitrogen it contains.

A. Cytochrome C oxidase

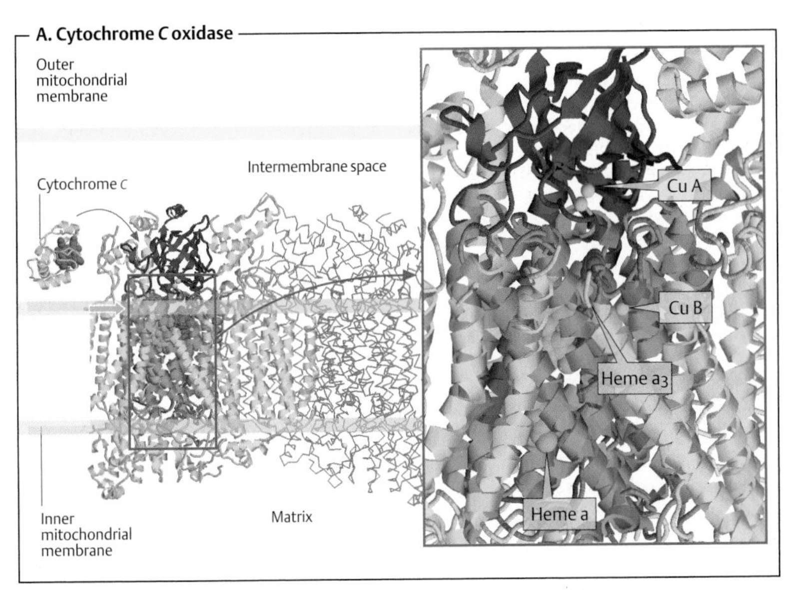

B. Photosystem I

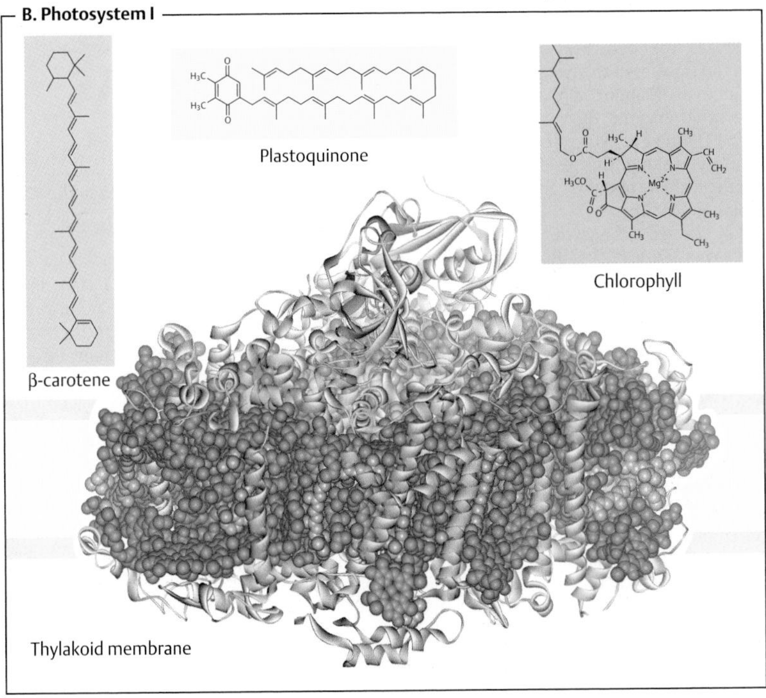

Oxoacid dehydrogenases

The intermediary metabolism has *multienzyme complexes* which, in a complex reaction, catalyze the **oxidative decarboxylation** of 2-oxoacids and the transfer to coenzyme A of the acyl residue produced. NAD^+ acts as the electron acceptor. In addition, thiamine diphosphate, lipoamide, and FAD are also involved in the reaction. The *oxoacid dehydrogenases* include a) the *pyruvate dehydrogenase complex* (PDH, pyruvate → acetyl CoA), b) the *2-oxoglutarate dehydrogenase complex* of the tricarboxylic acid cycle (ODH, 2-oxoglutarate → succinyl CoA), and c) the *branched chain dehydrogenase complex*, which is involved in the catabolism of valine, leucine, and isoleucine (see p. 414).

A. Pyruvate dehydrogenase: reactions ◑

The pyruvate dehydrogenase reaction takes place in the mitochondrial matrix (see p. 210). Three different enzymes [E1–E3] form the PDH multienzyme complex (see **B**).

[1] Initially, *pyruvate dehydrogenase* [E1] catalyzes the decarboxylation of pyruvate and the transfer of the resulting hydroxyethyl residue to **thiamine diphosphate** (TPP, **1a**). The same enzyme then catalyzes oxidation of the TPP-bound hydroxyethyl group to yield an acetyl residue. This residue and the reducing equivalents obtained are then transferred to **lipoamide** (**1b**).

[2] The second enzyme, *dihydrolipoamide acetyltransferase* [E2], shifts the acetyl residue from lipoamide to **coenzyme A** (**2**), with dihydrolipoamide being left over.

[3] The third enzyme, *dihydrolipoamide dehydrogenase* [E3], reoxidizes dihydrolipoamide, with **$NADH+H^+$** being formed. The electrons are first taken over by enzyme-bound **FAD** (**3a**) and then transferred via a catalytically active disulfide bond in the E3 subunit (not shown) to soluble NAD^+ (**3b**).

The five different **coenzymes** involved are associated with the enzyme components in different ways. Thiamine diphosphate is non-covalently bound to E1, whereas lipoamide is covalently bound to a lysine residue of E2 and FAD is bound as a *prosthetic group* to E3. NAD^+ and coenzyme A, being soluble coenzymes, are only temporarily associated with the complex.

An important aspect of PDH catalysis is the spatial relationship between the components of the complex. The covalently bound lipoamide coenzyme is part of a mobile domain of E2, and is therefore highly mobile. This structure is known as the *lipoamide arm*, and swings back and forth between E1 and E3 during catalysis. In this way, lipoamide can interact with the TPP bound at E1, with solute coenzyme A, and also with the FAD that serves as the electron acceptor in E3.

B. PDH complex of *Escherichia coli* ○

The PDH complex of the bacterium *Escherichia coli* has been particularly well studied. It has a molecular mass of $5.3 \cdot 10^6$, and with a diameter of more than 30 nm it is larger than a ribosome. The complex consists of a total of 60 polypeptides (**1**, **2**): 24 molecules of E2 (eight trimers) form the almost cube-shaped core of the complex. Each of the six surfaces of the cube is occupied by a dimer of E3 components, while each of the twelve edges of the cube is occupied by dimers of E1 molecules. Animal oxoacid dehydrogenases have similar structures, but differ in the numbers of subunits and their molecular masses.

Further information

The PDH reaction, which is practically irreversible, occupies a strategic position at the interface between carbohydrate and fatty acid metabolism, and also supplies acetyl residues to the tricarboxylic acid cycle. PDH activity is therefore strictly regulated (see p. 144). **Interconversion** is particularly important in animal cells (see p. 120). Several PDH-specific *protein kinases* inactivate the E1 components through phosphorylation, while equally specific *protein phosphatases* reactivate it again. The binding of the kinases and phosphatases to the complex is in turn regulated by metabolites. For example, high concentrations of acetyl CoA promote binding of kinases and thereby inhibit the reaction, while Ca^{2+} increases the activity of the phosphatase. Insulin activates PDH via inhibition of phosphorylation.

A. Pyruvate dehydrogenase: reactions

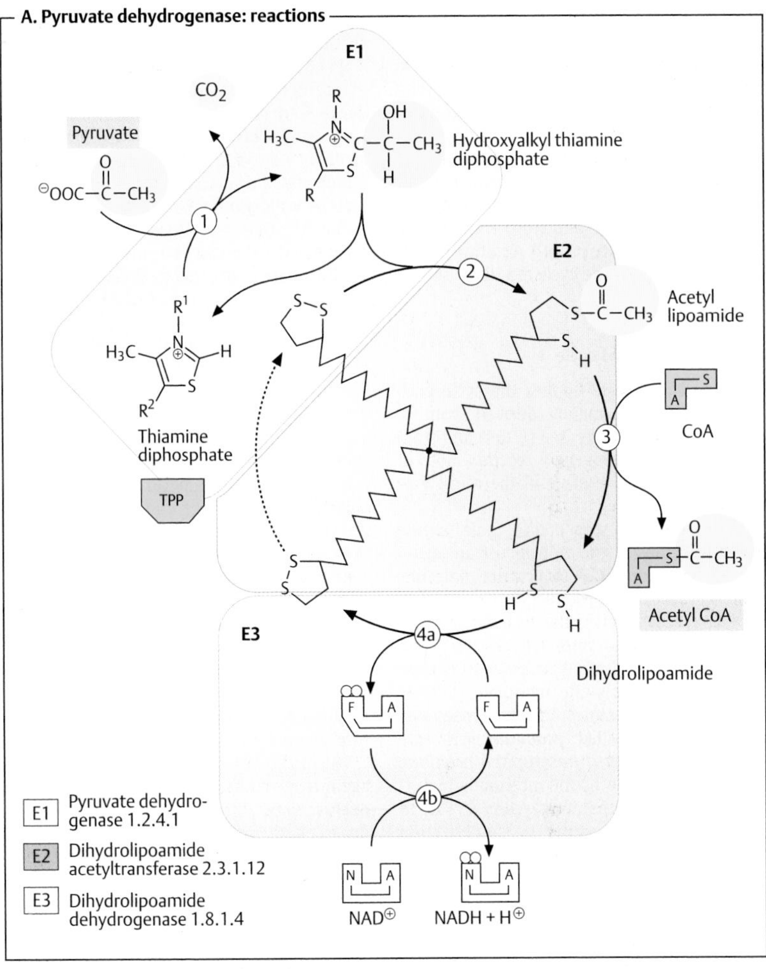

B. PDH complex of *Escherichia coli*

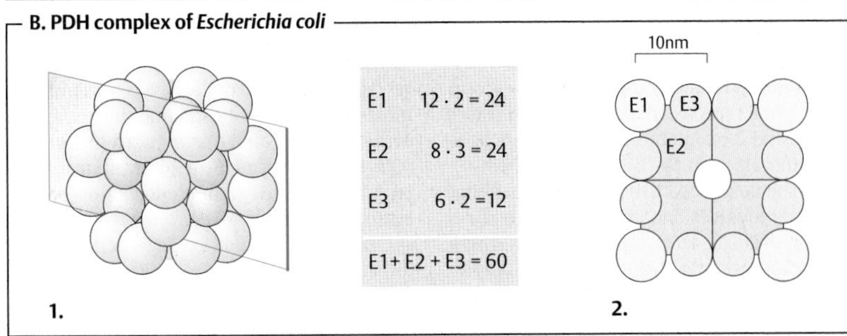

Tricarboxylic acid cycle: reactions

The **tricarboxylic acid cycle** (TCA cycle, also known as the citric acid cycle or Krebs cycle) is a cyclic metabolic pathway in the mitochondrial matrix (see p. 210). In eight steps, it oxidizes acetyl residues (CH_3-CO-) to carbon dioxide (CO_2). The reducing equivalents obtained in this process are transferred to NAD^+ or ubiquinone, and from there to the respiratory chain (see p. 140). Additional metabolic functions of the cycle are discussed on p. 138.

A. Tricarboxylic acid cycle ◑

The acetyl-CoA that supplies the cycle with acetyl residues is mainly derived from *β-oxidation* of fatty acids (see p. 164) and from the *pyruvate dehydrogenase reaction.* Both of these processes take place in the mitochondrial matrix.

[1] In the first step of the cycle, *citrate synthase* catalyzes the transfer of an acetyl residue from **acetyl CoA** to a carrier molecule, oxaloacetic acid. The product of this reaction, **tricarboxylic acid,** gives the cycle its name.

[2] In the next step, tricarboxylic acid undergoes isomerization to yield **isocitrate.** In the process, only the hydroxyl group is shifted within the molecule. The corresponding enzyme is called *aconitate hydratase* ("aconitase"), because unsaturated *aconitate* arises as an enzyme-bound intermediate during the reaction (not shown; see p. 8). Due to the properties of aconitase, the isomerization is absolutely *stereospecific.* Although citrate is not chiral, isocitrate has two chiral centers, so that it could potentially appear in *four* isomeric forms. However, in the tricarboxylic acid cycle, only one of these stereoisomers, (*2R,3S*)-isocitrate, is produced.

[3] The first oxidative step now follows. *Isocitrate dehydrogenase* oxidizes the hydroxyl group of isocitrate into an oxo group. At the same time, a carboxyl group is released as CO_2, and **2-oxoglutarate** (also known as α-ketoglutarate) and NADH+H^+ are formed.

[4] The next step, the formation of **succinyl CoA,** also involves one oxidation and one decarboxylation. It is catalyzed by *2-oxoglutarate dehydrogenase,* a multienzyme complex closely resembling the PDH complex (see p. 134). NADH+H^+ is once again formed in this reaction.

[5] The subsequent cleavage of the thioester succinylCoA into **succinate** and coenzyme A by *succinic acid-CoA ligase* (succinyl CoA synthetase, succinic thiokinase) is strongly *exergonic* and is used to synthesize a phosphoric acid anhydride bond (*"substrate level phosphorylation"*, see p. 124). However, it is not ATP that is produced here as is otherwise usually the case, but instead **guanosine triphosphate (GTP).** However, GTP can be converted into ATP by a *nucleoside diphosphate kinase* (not shown).

[6] Via the reactions described so far, the acetyl residue has been completely oxidized to CO_2. At the same time, however, the carrier molecule oxaloacetate has been reduced to succinate. Three further reactions in the cycle now regenerate oxaloacetate from succinate. Initially, *succinate dehydrogenase* oxidizes succinate to **fumarate.** In contrast to the other enzymes in the cycle, succinate dehydrogenase is an integral protein of the inner mitochondrial membrane. It is therefore also assigned to the respiratory chain as complex II. Although succinate dehydrogenase contains FAD as a prosthetic group, **ubiquinone** is the real electron acceptor of the reaction.

[7] Water is now added to the double bond of fumarate by *fumarate hydratase* (*"fumarase"*), and chiral *(2S)*-**malate** is produced.

[8] In the last step of the cycle, malate is again oxidized by *malate dehydrogenase* into **oxaloacetate**, with NADH+H^+ again being produced. With this reaction, the cycle is complete and can start again from the beginning. As the equilibrium of the reaction lies well on the side of malate, the formation of oxaloacetic acid by reaction [8] depends on the strongly exergonic reaction [1], which immediately removes it from the equilibrium.

The **net outcome** is that each rotation of the tricarboxylic acid cycle converts one acetyl residue and two molecules of H_2O into two molecules of CO_2. At the same time, one GTP, three NADH+H^+ and one reduced ubiquinone (QH_2) are produced. By oxidative phosphorylation (see p. 122), the cell obtains around nine molecules of ATP from these reduced coenzymes (see p. 146). Together with the directly formed GTP, this yields a total of 10 ATP per acetyl group.

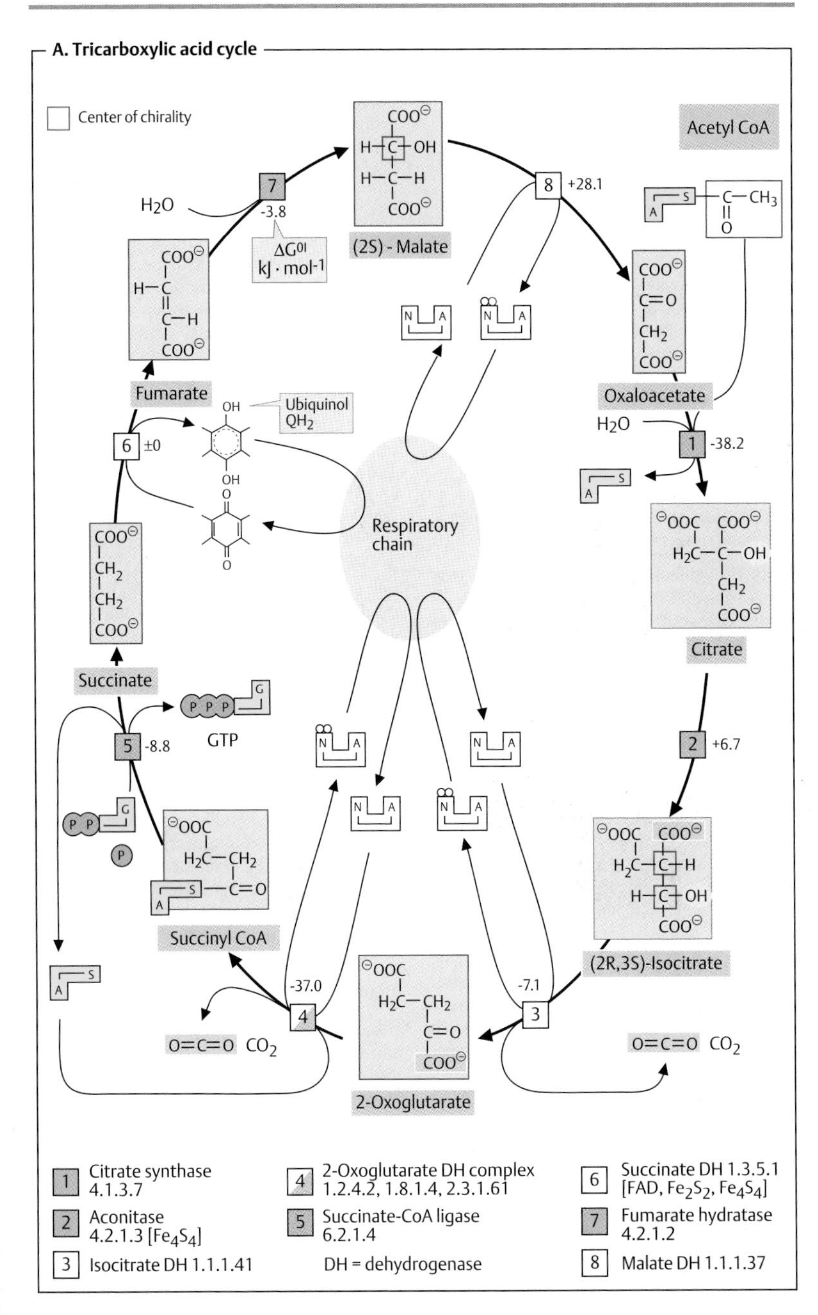
A. Tricarboxylic acid cycle
Center of chirality
Acetyl CoA
(2S) - Malate
Fumarate
Oxaloacetate
Citrate
(2R,3S)-Isocitrate
2-Oxoglutarate
Succinyl CoA
Succinate
Ubiquinol QH_2
Respiratory chain
GTP
$\Delta G^{0\prime}$ $kJ \cdot mol^{-1}$
+28.1
-38.2
+6.7
-7.1
-37.0
-8.8
±0
-3.8
H_2O
CO_2
1 Citrate synthase 4.1.3.7
2 Aconitase 4.2.1.3 [Fe_4S_4]
3 Isocitrate DH 1.1.1.41
4 2-Oxoglutarate DH complex 1.2.4.2, 1.8.1.4, 2.3.1.61
5 Succinate-CoA ligase 6.2.1.4
DH = dehydrogenase
6 Succinate DH 1.3.5.1 [FAD, Fe_2S_2, Fe_4S_4]
7 Fumarate hydratase 4.2.1.2
8 Malate DH 1.1.1.37

Tricarboxylic acid cycle: functions

A. Tricarboxylic acid cycle: functions ◑

The tricarboxylic acid cycle (see p. 136) is often described as the "hub of intermediary metabolism." It has both catabolic and anabolic functions—it is **amphibolic.**

As a **catabolic pathway**, it initiates the **"terminal oxidation"** of energy substrates. Many catabolic pathways lead to intermediates of the tricarboxylic acid cycle, or supply metabolites such as pyruvate and acetyl-CoA that can enter the cycle, where their C atoms are oxidized to CO_2. The reducing equivalents (see p. 14) obtained in this way are then used for *oxidative phosphorylation*—i. e., to aerobically synthesize ATP (see p. 122).

The tricarboxylic acid cycle also supplies important **precursors for anabolic pathways.** Intermediates in the cycle are converted into:

- Glucose (gluconeogenesis; precursors: oxaloacetate and malate—see p. 154)
- Porphyrins (precursor: succinyl-CoA—see p. 192)
- Amino acids (precursors: 2-oxoglutarate, oxaloacetate—see p. 184)
- Fatty acids and isoprenoids (precursor: citrate—see below)

The intermediates of the tricarboxylic acid cycle are present in the mitochondria only in very small quantities. After the oxidation of acetyl-CoA to CO_2, they are constantly regenerated, and their concentrations therefore remain constant, averaged over time. Anabolic pathways, which remove intermediates of the cycle (e. g., gluconeogenesis) would quickly use up the small quantities present in the mitochondria if metabolites did not reenter the cycle at other sites to replace the compounds consumed. Processes that replenish the cycle in this way are called **anaplerotic reactions.**

The degradation of most amino acids is anaplerotic, because it produces either intermediates of the cycle or pyruvate (*glucogenic amino acids;* see p. 180). Gluconeogenesis is in fact largely sustained by the degradation of amino acids. A particularly important anaplerotic step in animal metabolism leads from pyruvate to oxaloacetic acid. This ATP-dependent reaction is catalyzed by *pyruvate carboxylase* [1]. It allows pyruvate yielding amino acids and lactate to be used for gluconeogenesis.

By contrast, *acetyl CoA does not have anaplerotic effects* in animal metabolism. Its carbon skeleton is completely oxidized to CO_2 and is therefore no longer available for biosynthesis. Since fatty acid degradation only supplies acetyl CoA, animals are unable to convert fatty acids into glucose. During periods of hunger, it is therefore not the fat reserves that are initially drawn on, but proteins. In contrast to fatty acids, the amino acids released are able to maintain the blood glucose level (see p. 308).

The tricarboxylic acid cycle not only takes up acetyl CoA from fatty acid degradation, but also supplies the material for the *biosynthesis of fatty acids* and isoprenoids. Acetyl CoA, which is formed in the matrix space of mitochondria by pyruvate dehydrogenase (see p. 134), is not capable of passing through the inner mitochondrial membrane. The acetyl residue is therefore condensed with oxaloacetate by mitochondrial *citrate synthase* to form citrate. This then leaves the mitochondria by antiport with malate (right; see p. 212). In the cytoplasm, it is cleaved again by ATP-dependent *citrate lyase* [4] into acetyl-CoA and oxaloacetate. The oxaloacetate formed is reduced by a cytoplasmic *malate dehydrogenase* to malate [2], which then returns to the mitochondrion via the antiport already mentioned. Alternatively, the malate can be oxidized by *"malic enzyme"* [5], with decarboxylation, to pyruvate. The $NADPH+H^+$ formed in this process is also used for fatty acid biosynthesis.

Additional information

Using the so-called **glyoxylic acid cycle**, plants and bacteria are able to convert acetyl-CoA into succinate, which then enters the tricarboxylic acid cycle. For these organisms, fat degradation therefore functions as an anaplerotic process. In plants, this pathway is located in special organelles, the *glyoxysomes.*

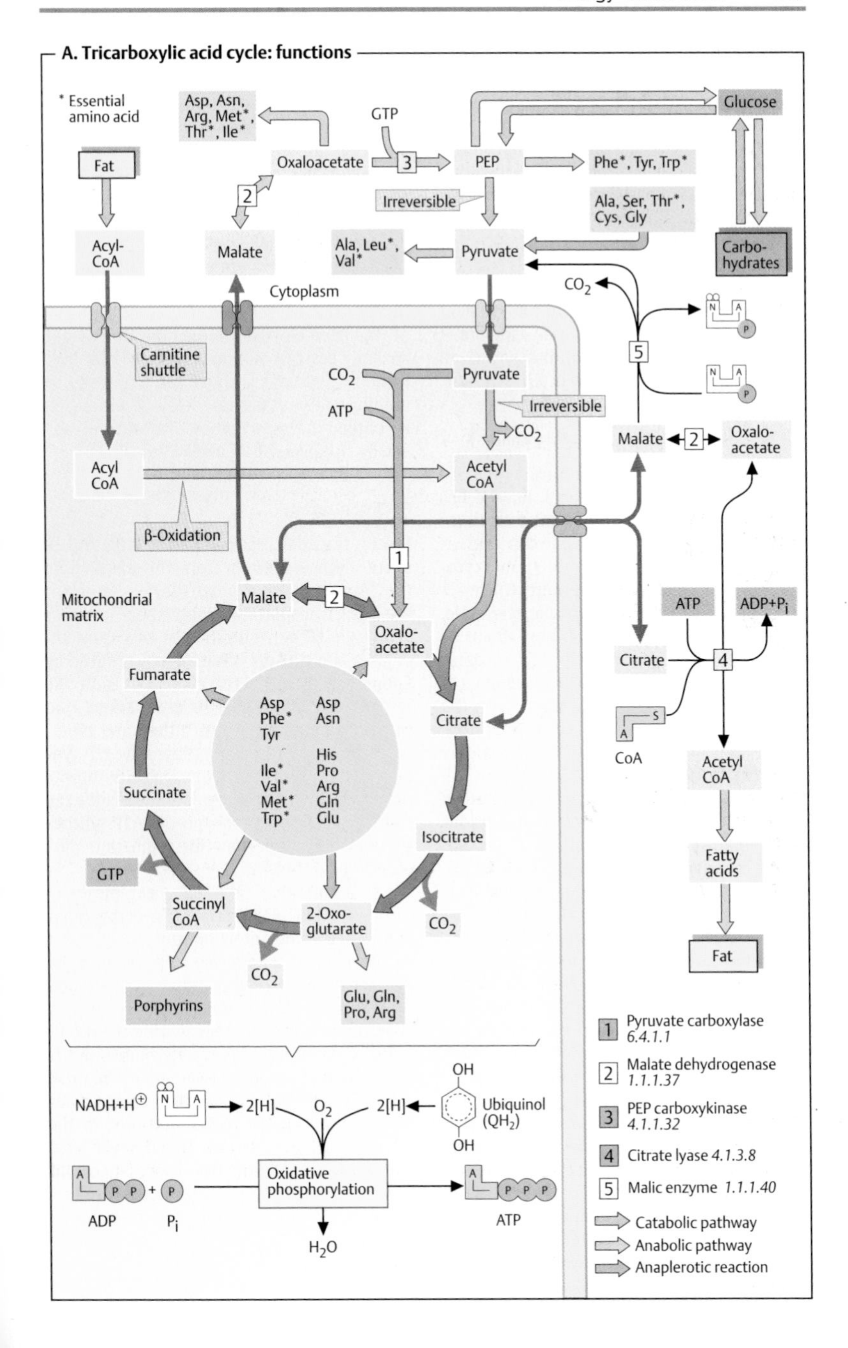
A. Tricarboxylic acid cycle: functions
* Essential amino acid
Asp, Asn, Arg, Met*, Thr*, Ile*
GTP
Glucose
Fat
Oxaloacetate
3
PEP
Phe*, Tyr, Trp*
2
Irreversible
Ala, Ser, Thr*, Cys, Gly
Acyl-CoA
Malate
Ala, Leu*, Val*
Pyruvate
Carbo-hydrates
CO_2
Cytoplasm
Carnitine shuttle
5
CO_2
Pyruvate
ATP
Irreversible
CO_2
Malate
2
Oxalo-acetate
Acyl CoA
Acetyl CoA
β-Oxidation
1
Mitochondrial matrix
Malate
2
ATP
ADP+P_i
Oxalo-acetate
Citrate
4
Fumarate
Asp Phe* Tyr
Asp Asn
Citrate
His
Ile* Val* Met* Trp*
Pro Arg Gln Glu
CoA
Acetyl CoA
Succinate
Isocitrate
GTP
Fatty acids
Succinyl CoA
2-Oxo-glutarate
CO_2
CO_2
Fat
Porphyrins
Glu, Gln, Pro, Arg
1 Pyruvate carboxylase 6.4.1.1
2 Malate dehydrogenase 1.1.1.37
3 PEP carboxykinase 4.1.1.32
4 Citrate lyase 4.1.3.8
5 Malic enzyme 1.1.1.40
Catabolic pathway
Anabolic pathway
Anaplerotic reaction
NADH+H⊕
2[H]
O_2
2[H]
OH
Ubiquinol (QH_2)
OH
Oxidative phosphorylation
ADP
P_i
ATP
H_2O

Respiratory chain

The **respiratory chain** is one of the pathways involved in *oxidative phosphorylation* (see p. 122). It catalyzes the steps by which electrons are transported from NADH+H^+ or reduced ubiquinone (QH_2) to molecular oxygen. Due to the wide difference between the redox potentials of the donor (NADH+H^+ or QH_2) and the acceptor (O_2), this reaction is strongly exergonic (see p. 18). Most of the energy released is used to establish a proton gradient across the inner mitochondrial membrane (see p. 126), which is then ultimately used to synthesize ATP with the help of *ATP synthase.*

A. Components of the respiratory chain ◑

The **electron transport chain** consists of three protein complexes (**complexes I, III, and IV**), which are integrated into the inner mitochondrial membrane, and two mobile carrier molecules—**ubiquinone** (coenzyme Q) and **cytochrome** *c*. *Succinate dehydrogenase,* which actually belongs to the tricarboxylic acid cycle, is also assigned to the respiratory chain as **complex II**. *ATP synthase* (see p. 142) is sometimes referred to as **complex V**, although it is not involved in electron transport. With the exception of complex I, detailed structural information is now available for every complex of the respiratory chain.

All of the complexes in the respiratory chain are made up of numerous polypeptides and contain a series of different protein bound **redox coenzymes** (see pp. 104, 106). These include *flavins* (FMN or FAD in complexes I and II), *iron–sulfur clusters* (in I, II, and III), and *heme groups* (in II, III, and IV). Of the more than 80 polypeptides in the respiratory chain, only 13 are coded by the mitochondrial genome (see p. 210). The remainder are encoded by nuclear genes, and have to be imported into the mitochondria after being synthesized in the cytoplasm (see p. 228).

Electrons enter the respiratory chain in various different ways. In the oxidation of NADH+H^+ by *complex I,* electrons pass via FMN and Fe/S clusters to ubiquinone (Q). Electrons arising during the oxidation of succinate, acyl CoA, and other substrates are passed to ubiquinone by *succinate dehydrogenase* or other *mitochondrial dehydrogenases* via enzyme-bound $FADH_2$ and the electron-transporting flavoprotein (ETF; see p. 164). Ubiquinol passes electrons on to *complex III,* which transfers them via two b-type heme groups, one Fe/S cluster, and heme c_1 to the small heme protein *cytochrome c.* Cytochrome *c* then transports the electrons to complex IV—*cytochrome c oxidase.* Cytochrome *c* oxidase contains redox-active components in the form of two copper centers (Cu_A and Cu_B) and hemes a and a_3, through which the electrons finally reach *oxygen* (see p. 132). As the result of the two-electron reduction of O_2, the strongly basic O^{2-} anion is produced (at least formally), and this is converted into water by binding of two protons. The electron transfer is coupled to the **formation of a proton gradient** by complexes I, III, and IV (see p. 126).

B. Organization ○

Proton transport via complexes I, III, and IV takes place *vectorially* from the matrix into the intermembrane space. When electrons are being transported through the respiratory chain, the H^+ concentration in this space increases—i. e., the pH value there is reduced by about one pH unit. For each H_2O molecule formed, around 10 H^+ ions are pumped into the intermembrane space. If the inner membrane is intact, then generally only *ATP synthase* (see p. 142) can allow protons to flow back into the matrix. This is the basis for the coupling of electron transport to ATP synthesis, which is important for regulation purposes (see p. 144).

As mentioned, although complexes I through V are all integrated into the inner membrane of the mitochondrion, they are not usually in contact with one another, since the electrons are transferred by ubiquinone and cytochrome *c*. With its long apolar side chain, ubiquinone is freely mobile within the membrane. Cytochrome *c* is water-soluble and is located on the *outside* of the inner membrane.

NADH oxidation via complex I takes place on the *inside* of the membrane—i. e., in the matrix space, where the tricarboxylic acid cycle and β-oxidation (the most important sources of NADH) are also located. O_2 reduction and ATP formation also take place in the matrix.

A. Components of the respiratory chain

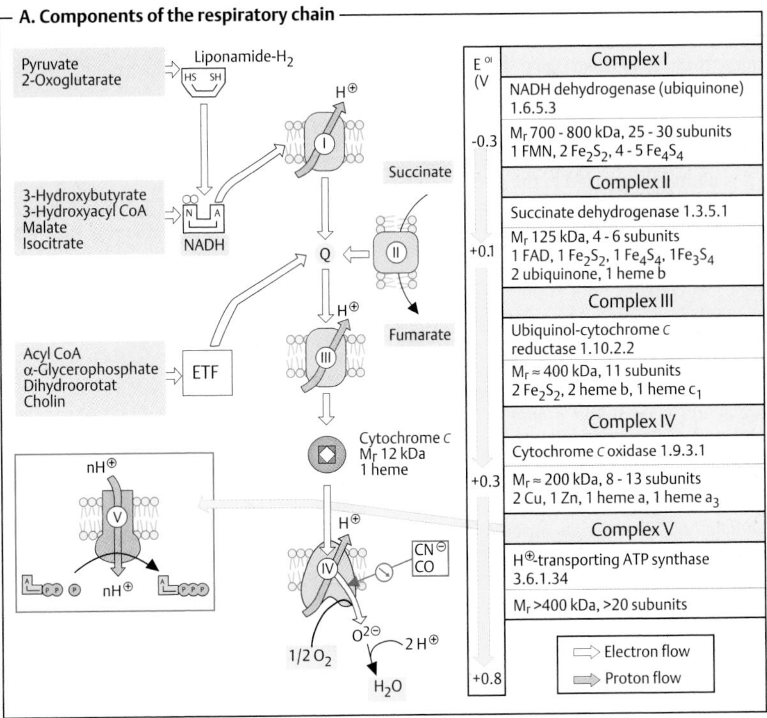

Complex I
NADH dehydrogenase (ubiquinone) 1.6.5.3
M_r 700 - 800 kDa, 25 - 30 subunits 1 FMN, 2 Fe_2S_2, 4 - 5 Fe_4S_4
Complex II
Succinate dehydrogenase 1.3.5.1
M_r 125 kDa, 4 - 6 subunits 1 FAD, 1 Fe_2S_2, 1 Fe_4S_4, 1Fe_3S_4 2 ubiquinone, 1 heme b
Complex III
Ubiquinol-cytochrome c reductase 1.10.2.2
$M_r \approx$ 400 kDa, 11 subunits 2 Fe_2S_2, 2 heme b, 1 heme c_1
Complex IV
Cytochrome c oxidase 1.9.3.1
$M_r \approx$ 200 kDa, 8 - 13 subunits 2 Cu, 1 Zn, 1 heme a, 1 heme a_3
Complex V
$H^{\oplus}$-transporting ATP synthase 3.6.1.34
M_r >400 kDa, >20 subunits

B. Organization

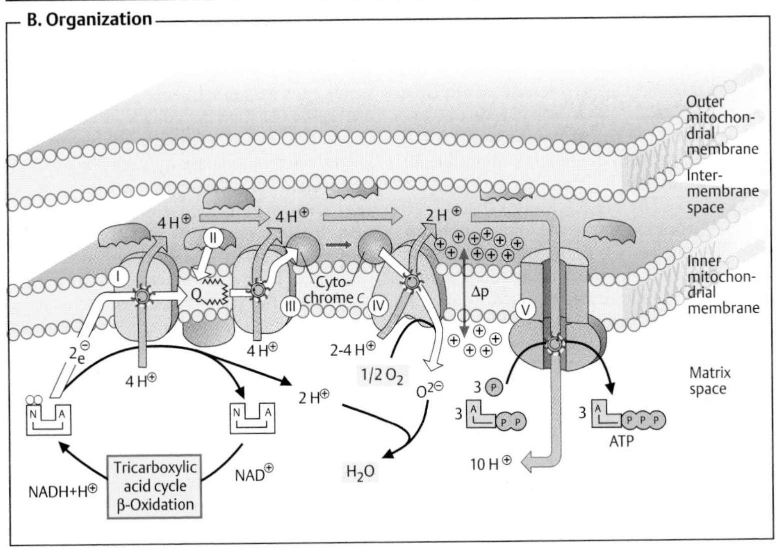

ATP synthesis

In the **respiratory chain** (see p. 140), electrons are transferred from NADH or ubiquinol (QH_2) to O_2. The energy obtained in this process is used to establish a proton gradient across the inner mitochondrial membrane. ATP synthesis is ultimately coupled to the return of protons from the intermembrane space into the matrix.

A. Redox systems of the respiratory chain ◑

The electrons provided by NADH do not reach oxygen directly, but instead are transferred to it in various steps. They pass through at least 10 intermediate redox systems, most of which are bound as **prosthetic groups** in complexes I, III, and IV. The large number of coenzymes involved in electron transport may initially appear surprising. However, as discussed on p. 18, in redox reactions, the *change in free enthalpy* ΔG—i. e., the chemical work that is done—depends only on the difference in redox potentials ΔE between the donor and the acceptor. Introducing additional redox systems does not alter the reaction's overall energy yield. In the case of the respiratory chain, the difference between the normal potential of the donor ($NAD^+/NADH+H^+$, $E^{0'} = -0.32$ V) and that of the acceptor (O_2/H_2O, $E^{0'} = +0.82$ V) corresponds to an energy difference $\Delta G^{0'}$ of more than 200 kJ · mol^{-1}. This large amount is divided into smaller, more manageable "packages," the size of which is determined by the difference in redox potentials between the respective *intermediates*. It is assumed that this division is responsible for the astonishingly high energy yield (about 60%) achieved by the respiratory chain.

The illustration shows the important redox systems involved in mitochondrial electron transport and their approximate redox potentials. These potentials determine the path followed by the electrons, as the members of a **redox series** have to be arranged in order of increasing redox potential if transport is to occur spontaneously (see p. 32).

In complex 1, the electrons are passed from *NADH+H⁺* first to *FMN* (see p. 104) and then on to several *iron–sulfur (Fe/S) clusters*. These redox systems are only stable in the interior of proteins. Depending on the type, Fe/S clusters may contain two to six iron ions, which form complexes with inorganic sulfide and the SH groups of cysteine residues (see p. 286). *Ubiquinone* (coenzyme Q; see p. 104) is a mobile carrier that takes up electrons from complexes I and II and from reduced ETF and passes them on to complex III. *Heme groups* are also involved in electron transport in a variety of ways. Type b hemes correspond to that found in hemoglobin (see p. 280). Heme c in cytochrome *c* is covalently bound to the protein, while the tetrapyrrole ring of heme a is isoprenylated and carries a formyl group. In complex IV, a *copper ion* (Cu_B) and heme a_3 react directly with oxygen.

B. ATP synthase ◑

The ATP synthase (*EC 3.6.1.34*, complex V) that transports H^+ is a complex molecular machine. The enzyme consists of two parts—a *proton channel* (F_o, for "oligomycin-sensitive") that is integrated into the membrane; and a *catalytic unit* (F_1) that protrudes into the matrix. The F_o part consists of 12 membrane-spanning c-peptides and one a-subunit. The "head" of the F_1 part is composed of three α and three β subunits, between which there are three active centers. The "stem" between F_o and F_1 consists of one γ and one ε subunit. Two more polypeptides, b and δ, form a kind of "stator," fixing the α and β subunits relative to the F_o part.

The catalytic cycle can be divided into three phases, through each of which the three active sites pass in sequence. First, ADP and P_i are bound (**1**), then the anhydride bond forms (**2**), and finally the product is released (**3**). Each time protons pass through the F_o channel protein into the matrix, all three active sites change from their current state to the next. It has been shown that the energy for proton transport is initially converted into a rotation of the γ subunit, which in turn cyclically alters the conformation of the α and β subunits, which are stationary relative to the F_o part, and thereby drives ATP synthesis.

A. Redox systems of the respiratory chain

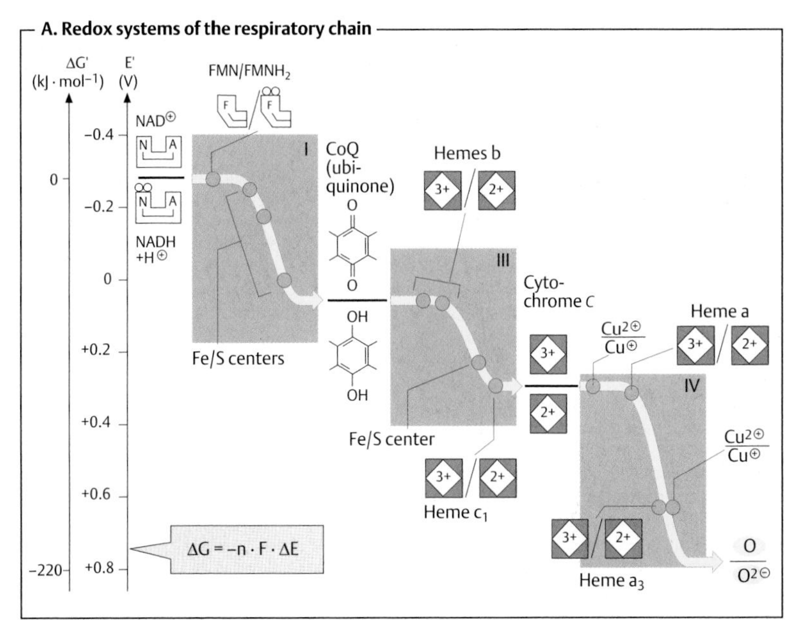

B. ATP synthase

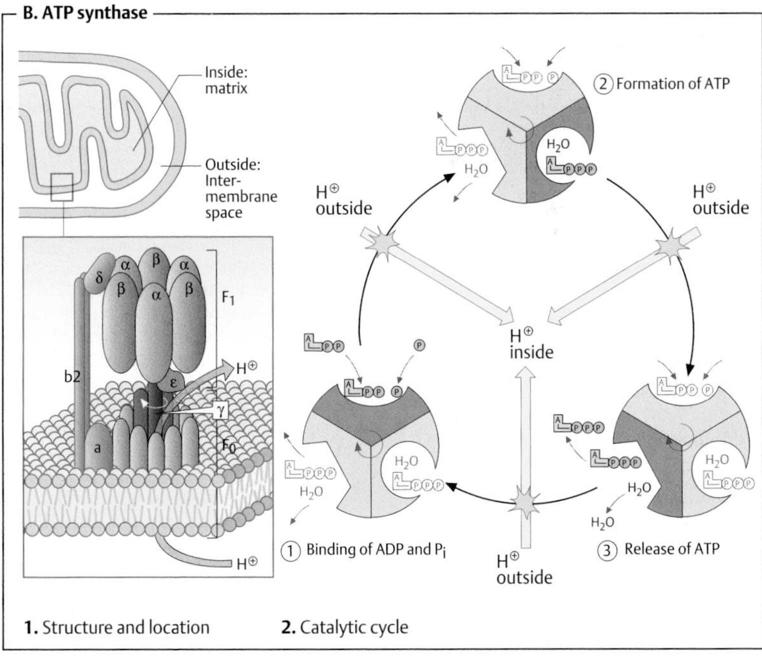

1. Structure and location

2. Catalytic cycle

Regulation

The amount of nutrient degradation and ATP synthesis have to be continually adjusted to the body's changing energy requirements. The need to coordinate the production and consumption of ATP is already evident from the fact that the *total amounts* of coenzymes in the organism are low. The human body forms about 65 kg ATP per day, but only contains 3–4 g of adenine nucleotides (AMP, ADP, and ATP). Each ADP molecule therefore has to be phosphorylated to ATP and dephosphorylated again many thousand times a day.

A. Respiratory control ◑

The simple regulatory mechanism which ensures that ATP synthesis is "automatically" coordinated with ATP consumption is known as **respiratory control.** It is based on the fact that the different parts of the oxidative phosphorylation process are *coupled* via shared coenzymes and other factors (left).

If a cell is not using any ATP, hardly any ADP will be available in the mitochondria. Without ADP, *ATP synthase* (**3**) is unable to break down the proton gradient across the inner mitochondrial membrane. This in turn inhibits electron transport in the respiratory chain (**2**), which means that $NADH+H^+$ can no longer be reoxidized to NAD^+. Finally, the resulting high $NADH/NAD^+$ ratio inhibits the tricarboxylic acid cycle (**C**), and thus slows down the degradation of the substrate SH_2 (**1**). Conversely, high rates of ATP utilization stimulate nutrient degradation and the respiratory chain via the same mechanism.

If the formation of a proton gradient is prevented (right), substrate oxidation (**1**) and electron transport (**2**) proceed much more rapidly. However, instead of ATP, only heat is produced.

B. Uncouplers ◑

Substances that functionally separate oxidation and phosphorylation from one another are referred to as uncouplers. They break down the proton gradient by allowing H^+ ions to pass from the intermembrane space back into the mitochondrial matrix without the involvement of ATP synthase. Uncoupling effects are produced by **mechanical damage** to the inner membrane (**1**) or by lipid-soluble substances that can transport protons through the membrane, such as **2,4-dinitrophenol** (DNP, **2**). **Thermogenin** (uncoupling protein-1, UCP-1, **3**)—an ion channel (see p. 222) in mitochondria of *brown fat* tissue—is a naturally occurring uncoupler. Brown fat is found, for example, in newborns and in hibernating animals, and serves exclusively to generate heat. In cold periods, norepinephrine activates the *hormone-sensitive lipase* (see p. 162). Increased lipolysis leads to the production of large quantities of free fatty acids. Like DNP, these bind H^+ ions in the intermembrane space, pass the UCP in this form, and then release the protons in the matrix again. This makes fatty acid degradation independent of ADP availability—i.e., it takes place at maximum velocity and only produces heat (**A**). It is becoming increasingly clear that there are also UCPs in other cells, which are controlled by hormones such as thyroxine (see p. 374). This regulates the ATP yield and what is known as the basal metabolic rate.

C. Regulation of the tricarboxylic acid cycle ◑

The most important factor in the regulation of the cycle is the **$NADH/NAD^+$ ratio.** In addition to *pyruvate dehydrogenase* (PDH) and *oxoglutarate dehydrogenase* (ODH; see p. 134), *citrate synthase* and *isocitrate dehydrogenase* are also inhibited by NAD^+ deficiency or an excess of $NADH+H^+$. With the exception of isocitrate dehydrogenase, these enzymes are also subject to **product inhibition** by acetyl-CoA, succinyl-CoA, or citrate.

Interconversion processes (see p. 120) also play an important role. They are shown here in detail using the example of the PDH complex (see p. 134). The *inactivating protein kinase* [1a] is inhibited by the substrate pyruvate and is activated by the products acetyl-CoA and $NADH+H^+$. The *protein phosphatase* [1b]—like *isocitrate dehydrogenase* [3] and the *ODH complex* [4]—is activated by Ca^{2+}. This is particularly important during muscle contraction, when large amounts of ATP are needed. *Insulin* also activates the PDH complex (through inhibition of phosphorylation) and thereby promotes the breakdown of glucose and its conversion into fatty acids.

A. Respiratory control

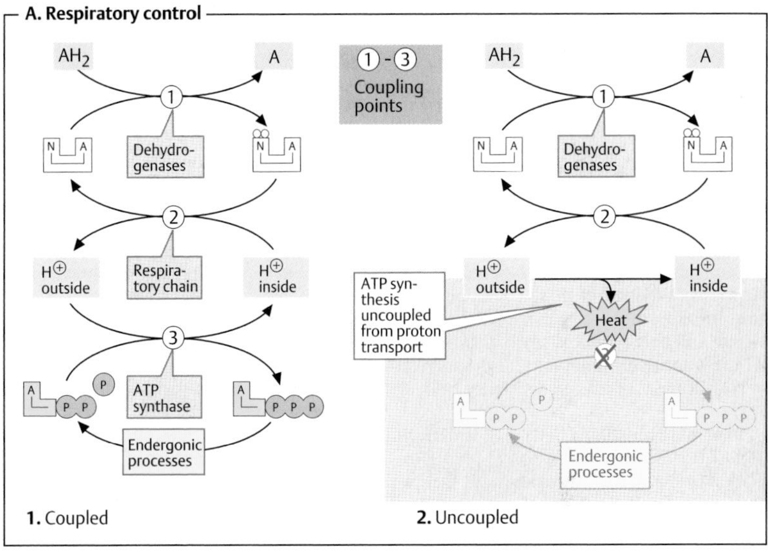

B. Uncouplers

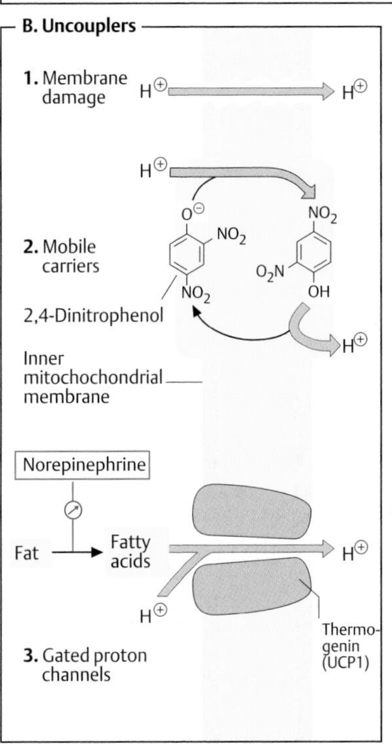

C. Regulation of the tricarboxylic acid cycle

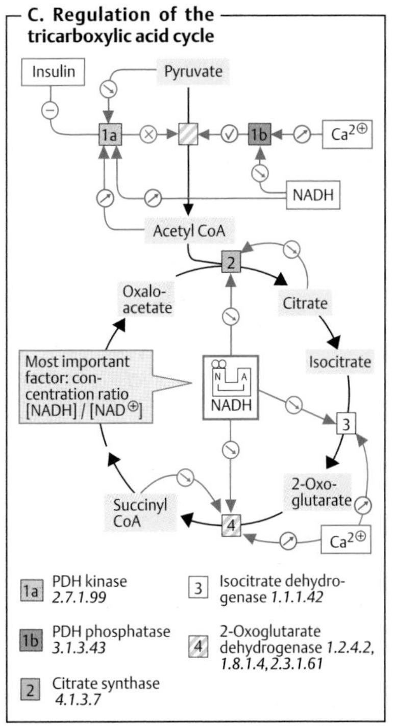

Respiration and fermentation

A. Aerobic and anaerobic oxidation of glucose ◑

In the presence of oxygen (i.e., in *aerobic* conditions), most animal cells are capable of "respiring" various types of nutrient (lipids, amino acids, and carbohydrates)—i.e., using oxidative processes to break them down completely. If oxygen is lacking (i.e., in *anaerobic* conditions), only glucose can be used for ATP synthesis. Although in these conditions glucose breakdown in animals already ends in lactate and only produces small quantities of ATP, it is decisively important for the survival of cells at times of oxygen deficiency.

In **aerobic conditions** (left), ATP is derived almost exclusively from oxidative phosphorylation (see p. 140). **Fatty acids** enter the mitochondria with the help of carnitine (see p. 164), and are broken down there into CoA-bound acetyl residues. **Glucose** is converted into pyruvate by glycolysis (see p. 150) in the cytoplasm. Pyruvate is then also transported into the mitochondrial matrix, where it is oxidatively decarboxylated by the pyruvate dehydrogenase complex (see p. 134) to yield acetyl-CoA. The reducing equivalents (2 $NADH+H^+$ per glucose) that arise in glycolysis enter the mitochondrial matrix via the malate shuttle (see p. 212). The acetyl residues that are formed are oxidized to CO_2 in the tricarboxylic acid cycle (see p. 136). Breakdown of **amino acids** also produces acetyl residues or products that can directly enter the tricarboxylic acid cycle (see p. 180). The reducing equivalents that are obtained are transferred to oxygen via the respiratory chain as required. In the process, chemical energy is released, which is used (via a proton gradient) to synthesize ATP (see p. 140).

In the absence of oxygen—i.e., in **anaerobic conditions**—the picture changes completely. Since O_2 is missing as the electron acceptor for the respiratory chain, $NADH+H^+$ and QH_2 can no longer be reoxidized. Consequently, not only is mitochondrial ATP synthesis halted, but also almost the whole metabolism in the mitochondrial matrix. The main reason for this is the high $NADH+H^+$ concentration and lack of NAD^+, which inhibit the tricarboxylic acid cycle and the pyruvate dehydrogenase reaction (see p. 144). β-Oxidation and the malate shuttle, which are dependent on free NAD^+, also come to a standstill. Since amino acid degradation is also no longer able to contribute to energy production, the cell becomes totally dependent on ATP synthesized via the degradation of glucose by **glycolysis.** For this process to proceed continuously, the $NADH+H^+$ formed in the cytoplasm has to be constantly reoxidized. Since this can no longer occur in the mitochondria, in anaerobic conditions animal cells reduce pyruvate to lactate and pass it into the blood. This type of process is called **fermentation** (see p. 148). The ATP yield is low, with only two ATPs per glucose arising during lactate synthesis.

To estimate the number of ATP molecules formed in an aerobic state, it is necessary to know the **P/O quotient**—i.e., the molar ratio between synthesized ATP ("P") and the water formed ("O"). During transport of two electrons from $NADH+H^+$ to oxygen, about 10 protons are transported into the intermembrane space, while from ubiquinol (QH_2), the number is only six. ATP synthase (see p. 142) probably requires three H^+ to synthesize one ATP, so that maximum P/O quotients of around **3 or 2** are possible. This implies a yield of up to 38 ATP per mol of glucose. However, the actual value is much lower. It needs to be taken into account that the transport of specific metabolites into the mitochondrial matrix and the exchange of ATP^{4-} for ADP^{3-} are also driven by the proton gradient (see p. 212). The P/O quotients for the oxidation of $NADH+H^+$ and QH_2 are therefore more in the range of **2.5 and 1.5**. If the energy balance of aerobic glycolysis is calculated on this basis, the result is a yield of around **32 ATP per glucose**. However, this value is also not constant, and can be adjusted as required by the cell's own uncouplers (UCPs; see p. 144) and other mechanisms.

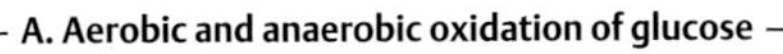

A. Aerobic and anaerobic oxidation of glucose

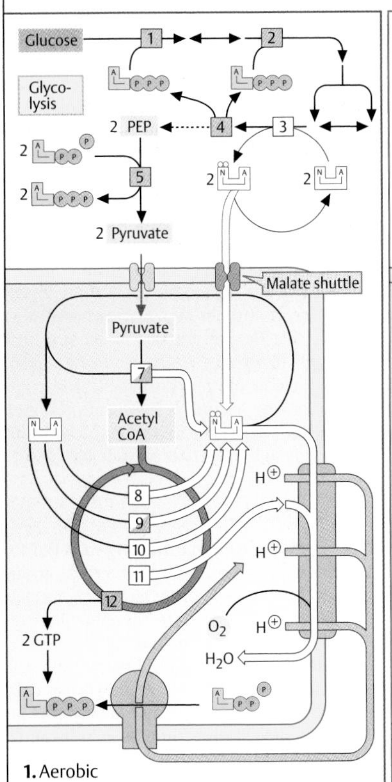

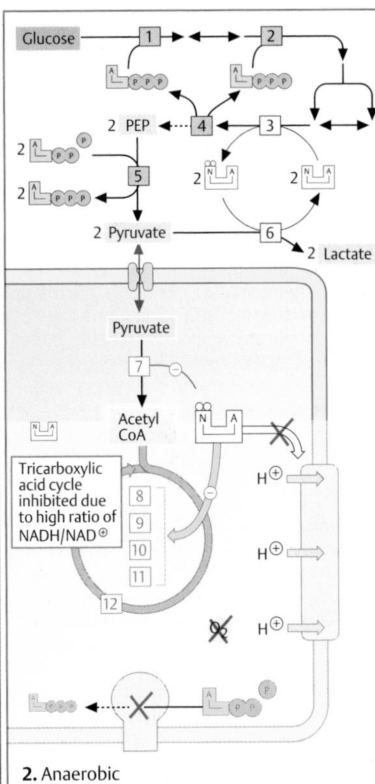

ATP	Coenzymes	Enzymes	Coenzymes	ATP
–1	–1 ATP	1 Hexokinase	–1 ATP	–1
–2	–1 ATP	2 6-Phosphofructokinase	–1 ATP	–2
+3	+5 ATP ← +2 NADH	3 Glyceraldehyde-3(P)DH	+2 NADH	–2
+5	+2 ATP	4 Phosphoglycerate kinase	+2 ATP NAD$^+$	0
+7	+2 ATP	5 Pyruvate kinase	+2 ATP recycled	+2
		6 Lactate dehydrogenase	–2 NADH	
+12	+5 ATP ← +2 NADH	7 Pyruvate dehydrogenase		
+17	+5 ATP ← +2 NADH	8 Isocitrate dehydrogenase		
+22	+5 ATP ← +2 NADH	9 Oxoglutarate dehydrogenase		
+27	+5 ATP ← +2 NADH	10 Malate dehydrogenase		
+30	+3 ATP ← +2 QH_2	11 Succinate dehydrogenase		
+32	+2 ATP ← +2 GTP	12 Succinate-CoA ligase		

Sum: 32 ATP/glucose

DH = dehydrogenase

Sum: 2 ATP/glucose

Fermentations

As discussed on p. 146, degradation of glucose to pyruvate is the only way for most organisms to synthesize ATP in the *absence of oxygen*. The NADH+H$^+$ that is also formed in this process has to be constantly reoxidized to NAD$^+$ in order to maintain glycolysis and thus ATP synthesis. In the animal organism, this is achieved by the reduction of pyruvate to lactate. In microorganisms, there are many other forms of NAD$^+$ regeneration. Processes of this type are referred to as **fermentations.** Microbial fermentation processes are often used to produce foodstuffs and alcoholic beverages, or to preserve food. Features common to all fermentation processes are that they start with pyruvate and only occur under *anaerobic conditions.*

A. Lactic acid and propionic acid fermentation ○

Many milk products, such as sour milk, yogurt, and cheese are made by *bacterial lactic acid fermentation* (**1**). The reaction is the same as in animals. Pyruvate, which is mainly derived from degradation of the disaccharide *lactose* (see p. 38), is reduced to lactate by *lactate dehydrogenase* [1]. Lactic acid fermentation also plays an important role in the production of sauerkraut and silage. These products usually keep for a long time, because the *pH reduction* that occurs during fermentation inhibits the growth of putrefying bacteria.

Bacteria from the genera *Lactobacillus* and *Streptococcus* are involved in the first steps of dairy production (**3**). The raw materials produced by their effects usually only acquire their final properties after additional fermentation processes. For example, the characteristic taste of Swiss cheese develops during a subsequent propionic acid fermentation. In this process, bacteria from the genus *Propionibacterium* convert pyruvate to propionate in a complex series of reactions (**2**).

B. Alcoholic fermentation ○

Alcoholic beverages are produced by the fermentation of plant products that have a high carbohydrate content. Pyruvate, which is formed from glucose, is initially decarboxylated by *pyruvate decarboxylase* [2], which does not occur in animal metabolism, to produce acetaldehyde (ethanal). When this is reduced by alcohol dehydrogenase [3], with NADH being consumed, *ethanol* [3] is formed.

Yeasts, unicellular fungi that belong to the eukaryotes (**3**), rather than bacteria, are responsible for this type of fermentation. Yeasts are also often used in baking. They produce CO_2 and ethanol, which raise the dough. Brewers' and bakers' yeasts *(Saccharomyces cerevisiae)* are usually haploid and reproduce asexually by budding (**3**). They can live both aerobically and anaerobically. Wine is produced by other types of yeast, some of which already live on the grapes. To promote the formation of ethanol, efforts are made to generally exclude oxygen during alcoholic fermentation—for example, by covering dough with a cloth when it is rising and by fermenting liquids in barrels that exclude air.

C. Beer brewing ○

Barley is the traditional starting material for the brewing of beer. Although cereal grains contain starch, they hardly have any *free* sugars. The barley grains are therefore first allowed to germinate so that starch-cleaving *amylases* are formed. Careful warming of the sprouting grain produces **malt.** This is then ground, soaked in water, and kept warm for a certain time. In the process, a substantial proportion of the starch is broken down into the disaccharide *maltose* (see p. 38). The product (the wort) is then boiled, **yeast** and **hops** are added, and the mixture is allowed to ferment for several days. The addition of hops makes the beer less perishable and gives it its slightly bitter taste. Other substances contained in hops act as sedatives and diuretics.

A. Lactic acid and propionic acid fermentation

B. Alcoholic fermentation

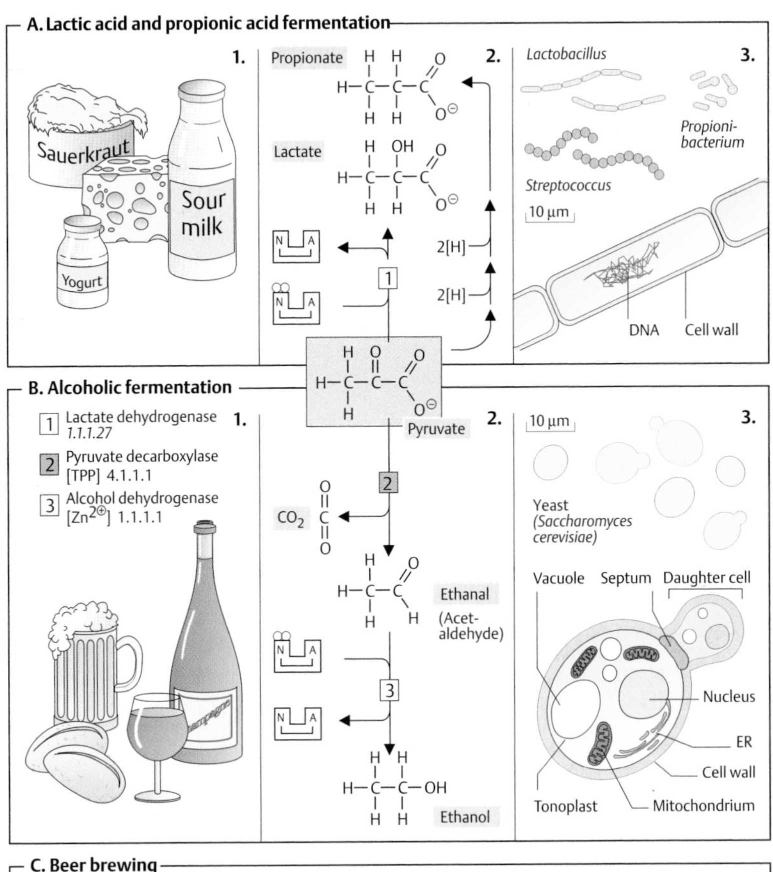

C. Beer brewing

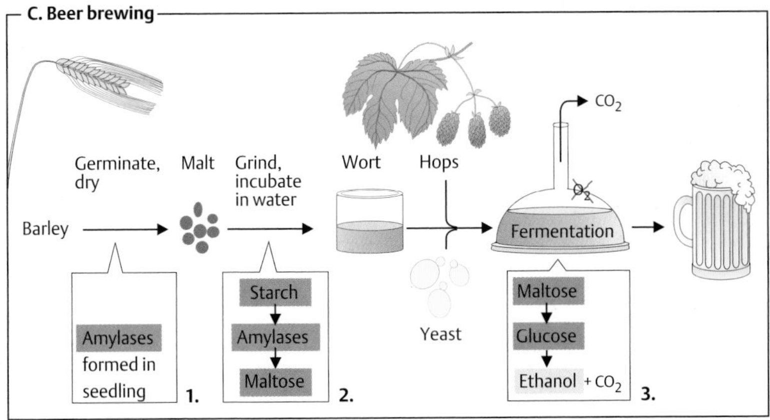

Glycolysis

A. Balance ●

Glycolysis is a catabolic pathway in the cytoplasm that is found in almost all organisms—irrespective of whether they live aerobically or anaerobically. The balance of glycolysis is simple: glucose is broken down into two molecules of pyruvate, and in addition two molecules of ATP and two of NADH+H$^+$ are formed.

In the presence of oxygen, pyruvate and NADH+H$^+$ reach the mitochondria, where they undergo further transformation (**aerobic glycolysis**; see p. 146). In anaerobic conditions, *fermentation products* such as lactate or ethanol have to be formed in the cytoplasm from pyruvate and NADH+H$^+$, in order to regenerate NAD$^+$ so that glycolysis can continue (**anaerobic glycolysis**; see p. 146). In the anaerobic state, glycolysis is the only means of obtaining ATP that animal cells have.

B. Reactions ◑

Glycolysis involves ten individual steps, including three isomerizations and four phosphate transfers. The only redox reaction takes place in step [6].

[1] Glucose, which is taken up by animal cells from the blood and other sources, is first phosphorylated to **glucose 6-phosphate**, with ATP being consumed. The glucose 6-phosphate is not capable of leaving the cell.

[2] In the next step, glucose 6-phosphate is isomerized into **fructose 6-phosphate**.

[3] Using ATP again, another phosphorylation takes place, giving rise to **fructose 1,6-bisphosphate.** *Phosphofructokinase* is the most important key enzyme in glycolysis (see p. 144).

[4] Fructose 1,6-bisphosphate is broken down by *aldolase* into the C_3 compounds **glyceraldehyde 3-phosphate** (also known as glyceral 3-phosphate) and **glycerone 3-phosphate** (dihydroxyacetone 3-phosphate).

[5] The latter two products are placed in fast equilibrium by *triosephosphate isomerase.*

[6] Glyceraldehyde 3-phosphate is now oxidized by *glyceraldehyde-3-phosphate dehydrogenase,* with NADH+H$^+$ being formed. In this reaction, *inorganic phosphate* is taken up into the molecule (*substrate-level phosphorylation*; see p. 124), and **1,3-bisphosphoglycerate** is produced. This intermediate contains a *mixed acid–anhydride bond,* the phosphate part of which is at a high chemical potential.

[7] Catalyzed by *phosphoglycerate kinase,* this phosphate residue is transferred to ADP, producing **3-phosphoglycerate** and ATP. The ATP balance is thus once again in equilibrium.

[8] As a result of shifting of the remaining phosphate residue within the molecule, the isomer **2-phosphoglycerate** is formed.

[9] Elimination of water from 2-phosphoglycerate produces the phosphate ester of the *enol form* of pyruvate—**phospho*enol*pyruvate** (PEP). This reaction also raises the second phosphate residue to a high potential.

[10] In the last step, *pyruvate kinase* transfers this residue to ADP. The remaining enol pyruvate is immediately rearranged into **pyruvate**, which is much more stable. Along with step [7] and the thiokinase reaction in the tricarboxylic acid cycle (see p. 136), the pyruvate kinase reaction is one of the three reactions in animal metabolism that are able to produce ATP independently of the respiratory chain.

In glycolysis, two molecules of ATP are initially used for activation ([1], [3]). Later, two ATPs are formed *per C_3 fragment.* Overall, therefore, there is a small net gain of 2 mol ATP per mol of glucose.

C. Energy profile ○

The energy balance of metabolic pathways depends not only on the standard changes in enthalpy ΔG^0, but also on the concentrations of the metabolites (see p. 18). Fig. **C** shows the *actual* enthalpy changes ΔG for the individual steps of glycolysis in erythrocytes.

As can be seen, only three reactions ([1], [3], and [10]), are associated with large changes in free enthalpy. In these cases, the equilibrium lies well on the side of the products (see p. 18). All of the other steps are freely reversible. The same steps are also followed—in the reverse direction—in gluconeogenesis (see p. 154), with the same enzymes being activated as in glucose degradation. The non-reversible steps [1], [3], and [10] are bypassed in glucose biosynthesis (see p. 154).

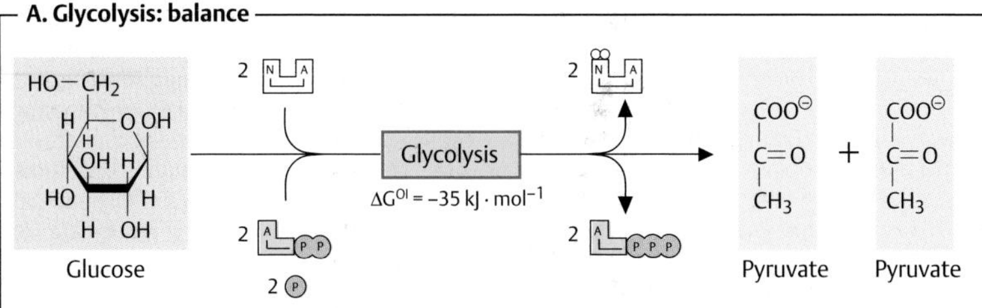
A. Glycolysis: balance
Glucose
Glycolysis
ΔG°' = –35 kJ · mol–1
Pyruvate
Pyruvate

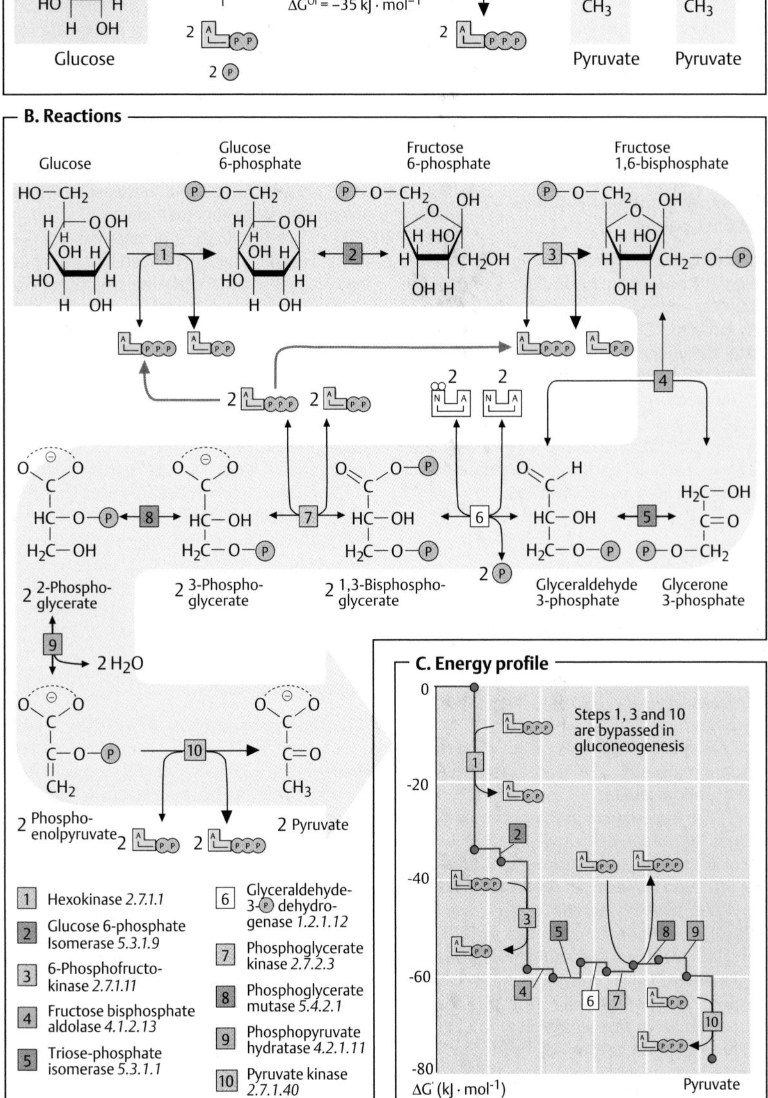
B. Reactions
Glucose
Glucose 6-phosphate
Fructose 6-phosphate
Fructose 1,6-bisphosphate
2 2-Phospho-glycerate
2 3-Phospho-glycerate
2 1,3-Bisphospho-glycerate
Glyceraldehyde 3-phosphate
Glycerone 3-phosphate
2 H2O
2 Phospho-enolpyruvate
2 Pyruvate
1 Hexokinase 2.7.1.1
2 Glucose 6-phosphate Isomerase 5.3.1.9
3 6-Phosphofructo-kinase 2.7.1.11
4 Fructose bisphosphate aldolase 4.1.2.13
5 Triose-phosphate isomerase 5.3.1.1
6 Glyceraldehyde-3-P dehydro-genase 1.2.1.12
7 Phosphoglycerate kinase 2.7.2.3
8 Phosphoglycerate mutase 5.4.2.1
9 Phosphopyruvate hydratase 4.2.1.11
10 Pyruvate kinase 2.7.1.40
C. Energy profile
Steps 1, 3 and 10 are bypassed in gluconeogenesis
0
-20
-40
-60
-80
ΔG' (kJ · mol-1)
Pyruvate

Pentose phosphate pathway

The pentose phosphate pathway (PPP, also known as the *hexose monophosphate pathway)* is an oxidative metabolic pathway located in the cytoplasm, which, like glycolysis, starts from glucose 6-phosphate. It supplies two important precursors for anabolic pathways: **NADPH+H+,** which is required for the biosynthesis of fatty acids and isoprenoids, for example (see p. 168), and **ribose 5-phosphate,** a precursor in nucleotide biosynthesis (see p. 188).

A. Pentose phosphate pathway: oxidative part ◑

The **oxidative segment** of the PPP converts glucose 6-phosphate to ribulose 5-phosphate. One CO_2 and two $NADPH+H^+$ are formed in the process. Depending on the metabolic state, the much more complex **regenerative part** of the pathway (see **B**) can convert some of the pentose phosphates back to hexose phosphates, or it can pass them on to glycolysis for breakdown. In most cells, less than 10% of glucose 6-phosphate is degraded via the pentose phosphate pathway.

B. Reactions ◑

[1] The **oxidative part** starts with the oxidation of **glucose 6-phosphate** by *glucose-6-phosphate dehydrogenase.* This forms $NADPH+H^+$ for the first time. The second product, **6-phosphogluconolactone**, is an intramolecular ester (*lactone*) of 6-phosphogluconate.

[2] A specific hydrolase then cleaves the lactone, exposing the carboxyl group of **6-phosphogluconate.**

[3] The last enzyme in the oxidative part is *phosphogluconate dehydrogenase* [3], which releases the carboxylate group of 6-phosphogluconate as CO_2 and at the same time oxidizes the hydroxyl group at C_3 to an oxo group. In addition to a second $NADPH+H^+$, this also produces the ketopentose **ribulose 5-phosphate.** This is converted by an isomerase to ribose 5-phosphate, the initial compound for nucleotide synthesis (top).

The **regenerative part** of the PPP is only shown here schematically. A complete reaction scheme is given on p. 408. The function of the regenerative branch is to adjust the *net* production of $NADPH+H^+$ and pentose phosphates to the cell's current requirements. Normally, the demand for $NADPH+H^+$ is much higher than that for pentose phosphates. In these conditions, the reaction steps shown first convert six ribulose 5-phosphates to five molecules of fructose 6-phosphate and then, by isomerization, regenerate five glucose 6-phosphates. These can once again supply $NADPH+H^+$ to the oxidative part of the PPP. Repeating these reactions finally results in the oxidation of one glucose 6-phosphate into six CO_2. Twelve $NADPH+H^+$ arise in the same process. In sum, no pentose phosphates are produced via this pathway.

In the recombination of sugar phosphates in the regenerative part of the PPP, there are two enzymes that are particularly important:

[5] *Transaldolase* transfers C_3 units from sedoheptulose 7-phosphate, a ketose with seven C atoms, to the aldehyde group of glyceraldehyde 3-phosphate.

[4] *Transketolase,* which contains thiamine diphosphate, transfers C_2 fragments from one sugar phosphate to another.

The reactions in the regenerative segment of the PPP are freely reversible. It is therefore easily possible to use the regenerative part of the pathway to convert hexose phosphates into pentose phosphates. This can occur when there is a high demand for pentose phosphates—e. g., during DNA replication in the S phase of the cell cycle (see p. 394).

Additional information

When energy in the form of ATP is required in addition to $NADPH+H^+$, the cell is able to channel the products of the regenerative part of the PPP (fructose 6-phosphate and glyceraldehyde 3-phosphate) into glycolysis. Further degradation is carried out via the tricarboxylic acid cycle and the respiratory chain to CO_2 and water. Overall, the cell in this way obtains 12 mol $NADPH+H^+$ and around 150 mol ATP from 6 mol glucose 6-phosphate. PPP activity is stimulated by *insulin* (see p. 388). This not only increases the rate of glucose degradation, but also produces additional $NADPH+H^+$ for fatty acid synthesis (see p. 168).

A. Pentose phosphate pathway: oxidative part

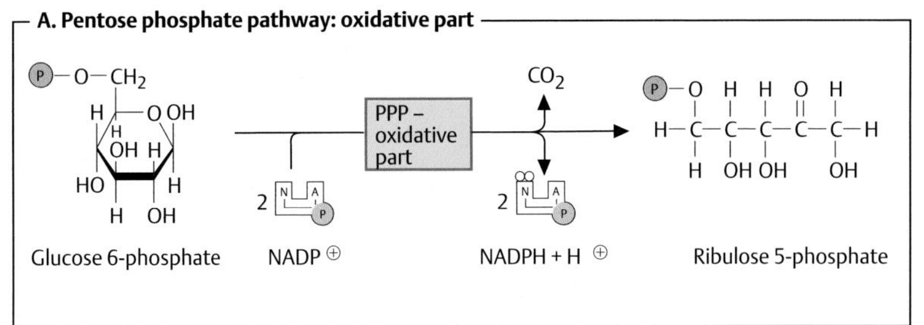

B. Reactions

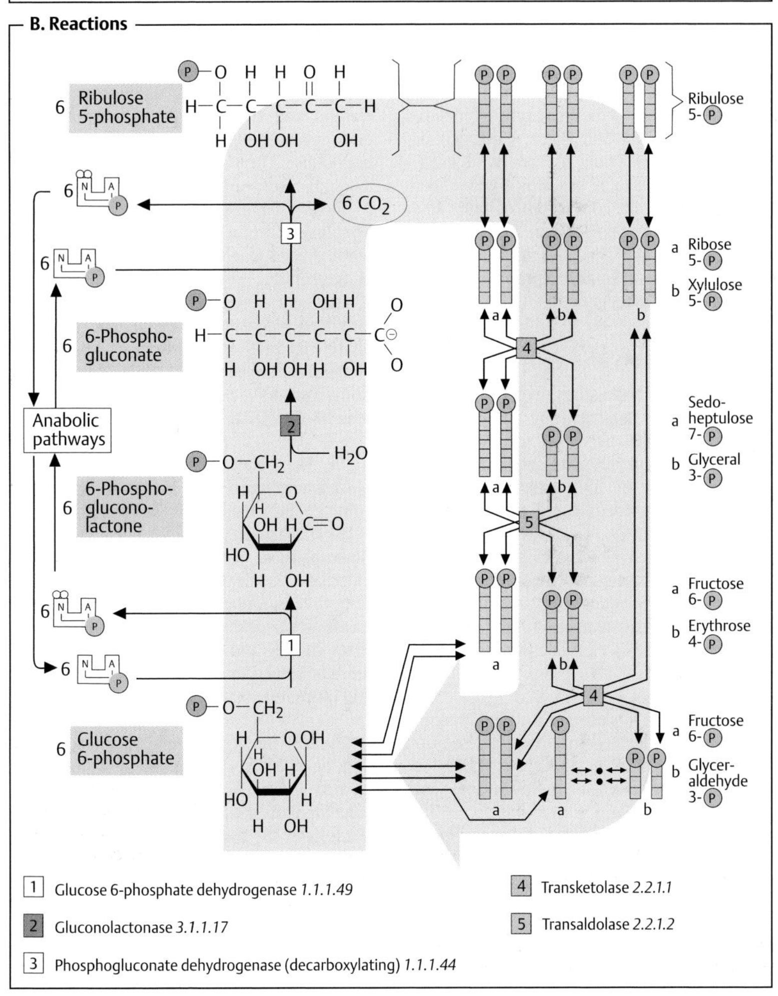

1 Glucose 6-phosphate dehydrogenase *1.1.1.49*

2 Gluconolactonase *3.1.1.17*

3 Phosphogluconate dehydrogenase (decarboxylating) *1.1.1.44*

4 Transketolase *2.2.1.1*

5 Transaldolase *2.2.1.2*

Gluconeogenesis

Some tissues, such as *brain* and *erythrocytes*, depend on a constant supply of glucose. If the amount of carbohydrate taken up in food is not sufficient, the blood sugar level can be maintained for a limited time by *degradation of hepatic glycogen* (see p. 156). If these reserves are also exhausted, de-novo synthesis of glucose (**gluconeogenesis**) begins. The **liver** is also mainly responsible for this (see p. 310), but the tubular cells of the **kidney** also show a high level of gluconeogenetic activity (see p. 328). The main precursors for gluconeogenesis are **amino acids** derived from muscle proteins. Another important precursor is **lactate,** which is formed in erythrocytes and muscle proteins when there is oxygen deficiency. **Glycerol** produced from the degradation of fats can also be used for gluconeogenesis. However, the conversion of fatty acids into glucose is *not* possible in animal metabolism (see p. 138). The human organism can synthesize several hundred grams of glucose per day by gluconeogenesis.

A. Gluconeogenesis ◑

Many of the reaction steps involved in gluconeogenesis are catalyzed by the same enzymes that are used in glycolysis (see p. 150). Other enzymes are specific to gluconeogenesis and are only synthesized, under the influence of *cortisol* and *glucagon* when needed (see p. 158). Glycolysis takes place exclusively when needed in the cytoplasm, but gluconeogenesis also involves the *mitochondria* and the *endoplasmic reticulum* (ER). Gluconeogenesis consumes 4 ATP (3 ATP + 1 GTP) per glucose—i. e., twice as many as glycolysis produces.

[1] **Lactate** as a precursor for gluconeogenesis is mainly derived from muscle (see Cori cycle, p. 338) and erythrocytes. LDH (see p. 98) oxidizes lactate to pyruvate, with $NADH+H^+$ formation.

[2] The first steps of actual gluconeogenesis take place in the *mitochondria*. The reason for this "detour" is the equilibrium state of the pyruvate kinase reaction (see p. 150). Even coupling to ATP hydrolysis would not be sufficient to convert pyruvate *directly* into phospho*enol* pyruvate (PEP). **Pyruvate** derived from lactate or amino acids is therefore initially transported into the mitochondrial matrix, and—in a biotin-dependent reaction catalyzed by *pyruvate carboxylase*—is carboxylated there to **oxaloacetate**. Oxaloacetate is also an intermediate in the tricarboxylic acid cycle. *Amino acids* with breakdown products that enter the cycle or supply pyruvate can therefore be converted into glucose (see p. 180).

[3] The oxaloacetate formed in the mitochondrial matrix is initially reduced to **malate**, which can leave the mitochondria via inner membrane transport systems (see p. 212).

[4] In the cytoplasm, oxaloacetate is reformed and then converted into **phosphoenol pyruvate** by a GTP-dependent *PEP carboxykinase*. The subsequent steps up to fructose 1,6-bisphosphate represent the reverse of the corresponding reactions involved in glycolysis. One additional ATP per C_3 fragment is used for the synthesis of 1,3-bisphosphoglycerate.

Two gluconeogenesis-specific phosphatases then successively cleave off the phosphate residues from **fructose 1,6-bisphosphate**. In between these reactions lies the isomerization of fructose 6-phosphate to **glucose 6-phosphate**—another glycolytic reaction.

[5] The reaction catalyzed by *fructose 1,6-bisphosphatase* is an important regulation point in gluconeogenesis (see p. 158).

[6] The last enzyme in the pathway, *glucose 6-phosphatase*, occurs in the liver, but not in muscle. It is located in the interior of the smooth endoplasmic reticulum. Specific transporters allow glucose 6-phosphate to enter the ER and allow the **glucose** formed there to return to the cytoplasm. From there, it is ultimately released into the blood.

Glycerol initially undergoes phosphorylation at C-3 [7]. The **glycerol 3-phosphate** formed is then oxidized by an NAD^+-dependent dehydrogenase to form **glycerone 3-phosphate** [8] and thereby channeled into gluconeogenesis. An FAD-dependent mitochondrial enzyme is also able to catalyze this reaction (known as the "glycerophosphate shuttle"; see p. 212).

A. Gluconeogenesis

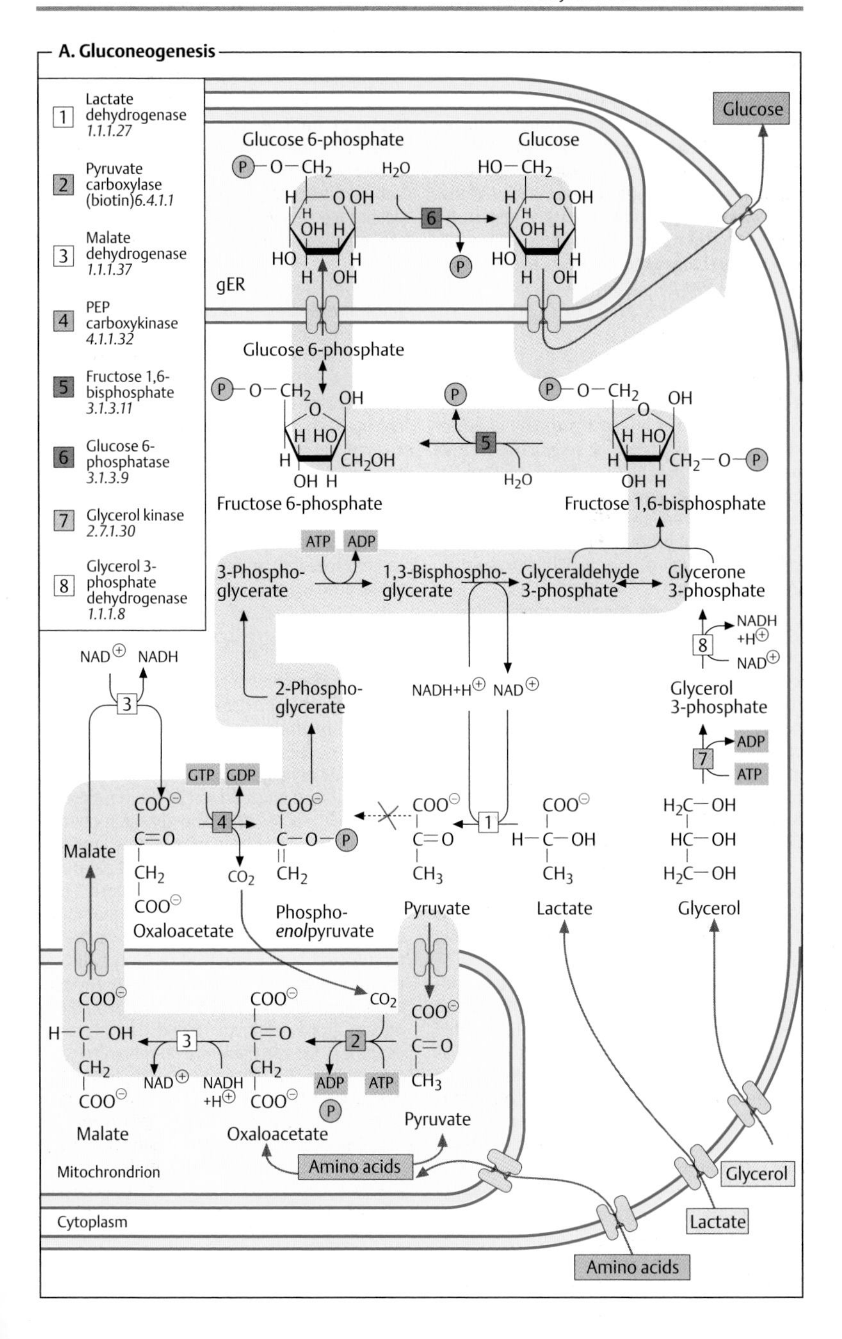

Glycogen metabolism

Glycogen (see p. 40) is used in animals as a **carbohydrate reserve,** from which glucose phosphates and glucose can be released when needed. Glucose storage itself would not be useful, as high concentrations within cells would make them strongly hypertonic and would therefore cause an influx of water. By contrast, insoluble glycogen has only low osmotic activity.

A. Glycogen balance ◑

Animal glycogen, like amylopectin in plants, is a *branched homopolymer of glucose.* The glucose residues are linked by an $\alpha 1 \rightarrow 4$-glycosidic bond. Every tenth or so glucose residue has an additional $\alpha 1 \rightarrow 6$ bond to another glucose. These branches are extended by additional $\alpha 1 \rightarrow 4$-linked glucose residues. This structure produces tree-shaped molecules consisting of up to 50000 residues ($M > 1 \cdot 10^7$ Da).

Hepatic glycogen is never completely degraded. In general, only the nonreducing ends of the "tree" are shortened, or—when glucose is abundant—elongated. The reducing end of the tree is linked to a special protein, **glycogenin**. Glycogenin carries out autocatalytic covalent bonding of the first glucose at one of its tyrosine residues and elongation of this by up to seven additional glucose residues. It is only at this point that *glycogen synthase* becomes active to supply further elongation.

[1] The formation of glycosidic bonds between sugars is *endergonic.* Initially, therefore, the activated form—**UDP-glucose**—is synthesized by reaction of glucose 1-phosphate with UTP (see p. 110).

[2] *Glycogen synthase* now transfers glucose residues one by one from UDP-glucose to the non-reducing ends of the available "branches."

[3] Once the growing chain has reached a specific length (> 11 residues), the *branching enzyme* cleaves an oligosaccharide consisting of 6–7 residues from the end of it, and adds this into the interior of the same chain or a neighboring one with $\alpha 1 \rightarrow 6$ linkage. These **branches** are then further extended by glycogen synthase.

[4] The branched structure of glycogen allows rapid release of sugar residues. The most important degradative enzyme, *glycogen phosphorylase,* cleaves residues from a non-reducing end one after another as **glucose 1-phosphate**. The larger the number of these ends, the more phosphorylase molecules can attack simultaneously. The formation of glucose 1-phosphate instead of glucose has the advantage that no ATP is needed to channel the released residues into glycolysis or the PPP.

[5] [6] Due to the structure of glycogen phosphorylase, degradation comes to a halt four residues away from each branching point. Two more enzymes overcome this blockage. First, a *glucanotransferase* moves a trisaccharide from the side chain to the end of the main chain [5]. A *1,6-glucosidase* [6] then cleaves the single remaining residue as a free glucose and leaves behind an unbranched chain that is once again accessible to phosphorylase.

The **regulation of glycogen metabolism** by interconversion, and the role of hormones in these processes, are discussed on p. 120.

B. Glycogen balance ◑

The human organism can store up to 450 g of glycogen—one-third in the **liver** and almost all of the remainder in **muscle.** The glycogen content of the other organs is low.

Hepatic glycogen is mainly used to maintain the *blood glucose level* in the postresorptive phase (see p. 308). The glycogen content of the liver therefore varies widely, and can decline to almost zero in periods of extended hunger. After this, gluconeogenesis (see p. 154) takes over the glucose supply for the organism. *Muscle glycogen* serves as an *energy reserve* and is not involved in blood glucose regulation. Muscle does not contain any glucose 6-phosphatase and is therefore unable to release glucose into the blood. The glycogen content of muscle therefore does not fluctuate as widely as that of the liver.

A. Glycogen metabolism

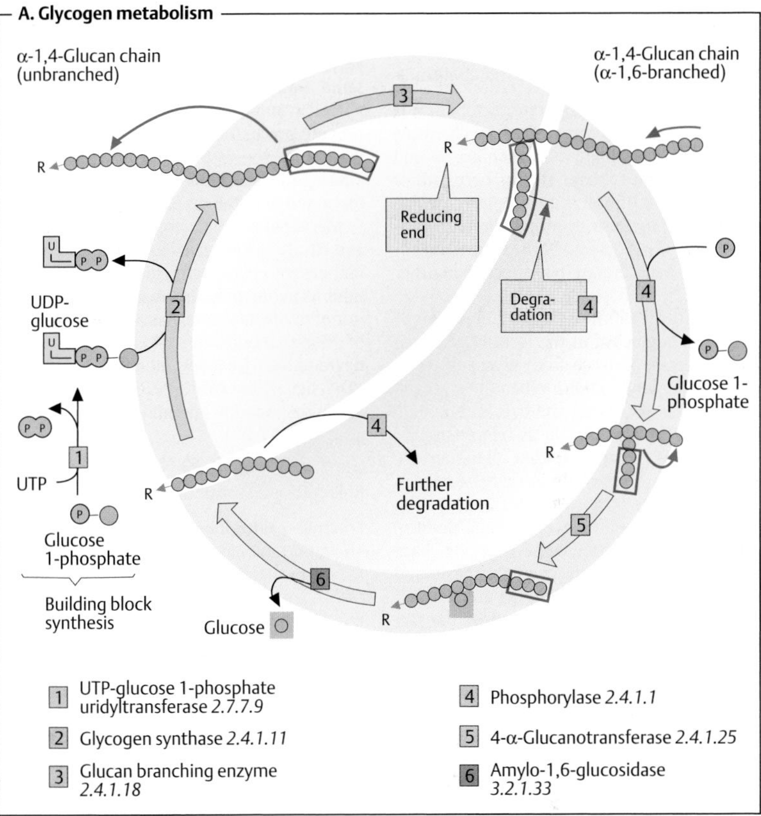

B. Glycogen balance

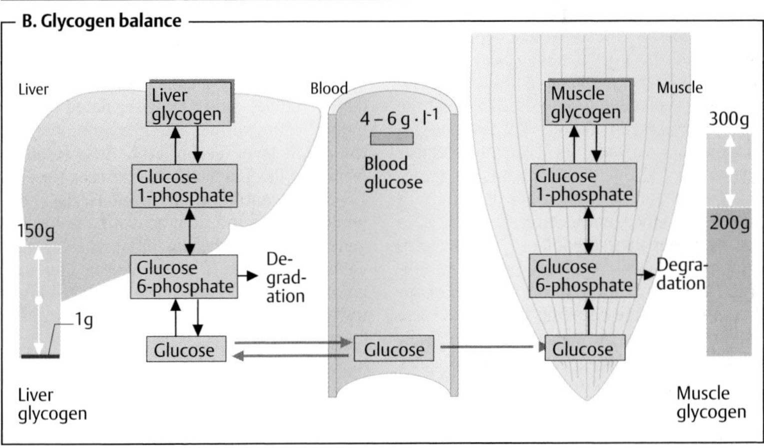

Regulation

A. Regulation of carbohydrate metabolism ◑

In all organisms, carbohydrate metabolism is subject to complex regulatory mechanisms involving *hormones, metabolites,* and *coenzymes.* The scheme shown here (still a simplified one) applies to the liver, which has central functions in carbohydrate metabolism (see p. 306). Some of the control mechanisms shown here are not effective in other tissues.

One of the liver's most important tasks is to store excess glucose in the form of glycogen and to release glucose from glycogen when required *(buffer function).* When the glycogen reserves are exhausted, the liver can provide glucose by de novo synthesis (*gluconeogenesis*; see p. 154). In addition, like all tissues, the liver breaks glucose down via glycolysis. These functions have to be coordinated with each other. For example, there is no point in glycolysis and gluconeogenesis taking place *simultaneously,* and glycogen synthesis and glycogen degradation should not occur simultaneously either. This is ensured by the fact that two *different* enzymes exist for important steps in both pathways, each of which catalyzes only the anabolic or the catabolic reaction. The enzymes are also regulated differently. Only these key enzymes are shown here.

Hormones. The hormones that influence carbohydrate metabolism include the peptides insulin and glucagon; a glucocorticoid, cortisol; and a catecholamine, epinephrine (see p. 380). **Insulin** activates *glycogen synthase* ([1]; see p. 388), and induces several enzymes involved in glycolysis [3, 5, 7]. At the same time, insulin inhibits the synthesis of enzymes involved in gluconeogenesis (*repression;* [4, 6, 8, 9]). **Glucagon,** the antagonist of insulin, has the opposite effect. It induces gluconeogenesis enzymes [4, 6, 8, 9] and represses *pyruvate kinase* [7], a key enzyme of glycolysis. Additional effects of glucagon are based on the *interconversion* of enzymes and are mediated by the second messenger cAMP. This inhibits glycogen synthesis [1] and activates glycogenolysis [2]. Epinephrine acts in a similar fashion. The inhibition of *pyruvate kinase* [7] by glucagon is also due to interconversion.

Glucocorticoids—mainly **cortisol** (see p. 374)—induce all of the key enzymes involved in gluconeogenesis [4, 6, 8, 9]. At the same time, they also induce enzymes involved in amino acid degradation and thereby provide precursors for gluconeogenesis. Regulation of the expression of *PEP carboxykinase,* a key enzyme in gluconeogenesis, is discussed in detail on p. 244.

Metabolites. High concentrations of **ATP** and **citrate** inhibit glycolysis by allosteric regulation of *phosphofructokinase.* ATP also inhibits *pyruvate kinase.* **Acetyl-CoA,** an inhibitor of *pyruvate kinase,* has a similar effect. All of these metabolites arise from glucose degradation *(feedback inhibition).* **AMP** and **ADP,** signals for ATP deficiency, activate glycogen degradation and inhibit gluconeogenesis.

B. Fructose 2,6-bisphosphate ◑

Fructose 2,6-bisphosphate (Fru-2,6-bP) plays an important part in carbohydrate metabolism. This metabolite is formed in small quantities from fructose 6-phosphate and has purely *regulatory functions.* It stimulates glycolysis by allosteric activation of *phosphofructokinase* and inhibits gluconeogenesis by inhibition of *fructose 1,6-bisphosphatase.*

The synthesis and degradation of Fru-2,6-bP are catalyzed by one and the same protein [10a, 10b]. If the enzyme is present in an unphosphorylated form [10a], it acts as a kinase and leads to the formation of Fru-2,6-bP. After phosphorylation by cAMP-dependent protein kinase A (PK-A), it acts as a phosphatase [10b] and now catalyzes the degradation of Fru-2,6-bP to fructose 6-phosphate. The equilibrium between [10a] and [10b] is regulated by hormones. Epinephrine and glucagon increase the cAMP level (see p. 120). As a result of increased PK-A activity, this reduces the Fru-2,6-bP concentration and inhibits glycolysis, while at the same time activating gluconeogenesis. Conversely, via [10a], insulin activates the synthesis of Fru-2,6-bP and thus glycolysis. In addition, insulin also inhibits the action of glucagon by reducing the cAMP level (see p. 120).

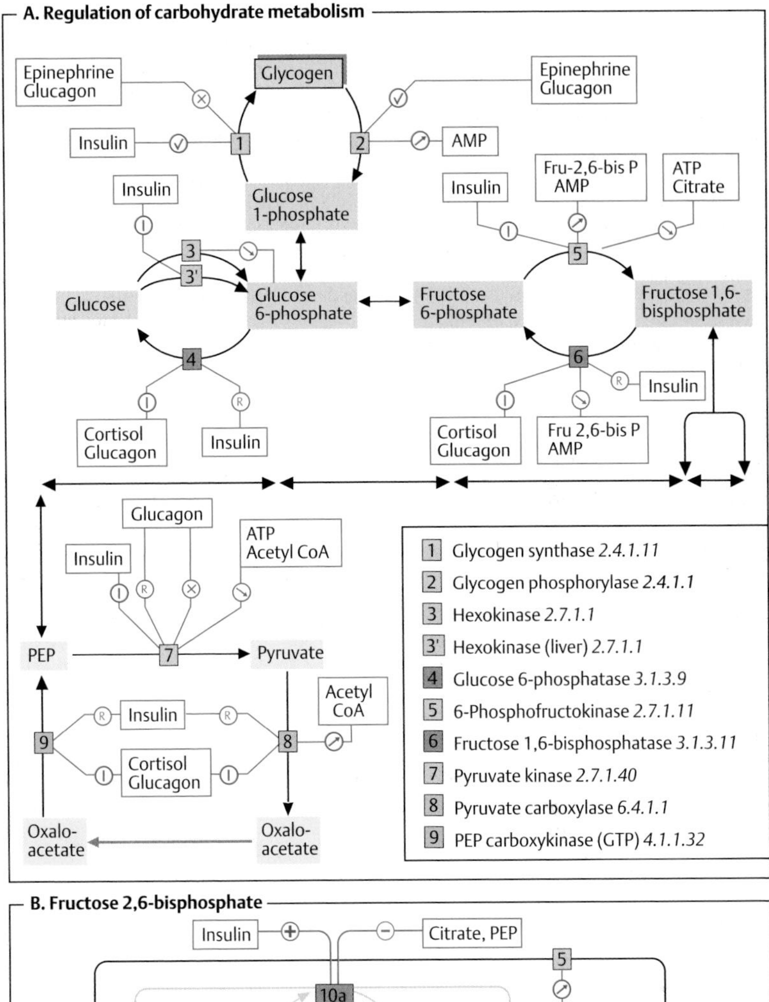
A. Regulation of carbohydrate metabolism
Epinephrine
Glucagon
Glycogen
Epinephrine
Glucagon
Insulin
1
2
AMP
Insulin
Glucose
1-phosphate
Insulin
Fru-2,6-bis P
AMP
ATP
Citrate
3
3'
5
Glucose
Glucose
6-phosphate
Fructose
6-phosphate
Fructose 1,6-
bisphosphate
4
6
Insulin
Cortisol
Glucagon
Insulin
Cortisol
Glucagon
Fru 2,6-bis P
AMP
Glucagon
Insulin
ATP
Acetyl CoA
PEP
7
Pyruvate
Acetyl
CoA
Insulin
9
8
Cortisol
Glucagon
Oxalo-
acetate
Oxalo-
acetate
1 Glycogen synthase 2.4.1.11
2 Glycogen phosphorylase 2.4.1.1
3 Hexokinase 2.7.1.1
3' Hexokinase (liver) 2.7.1.1
4 Glucose 6-phosphatase 3.1.3.9
5 6-Phosphofructokinase 2.7.1.11
6 Fructose 1,6-bisphosphatase 3.1.3.11
7 Pyruvate kinase 2.7.1.40
8 Pyruvate carboxylase 6.4.1.1
9 PEP carboxykinase (GTP) 4.1.1.32

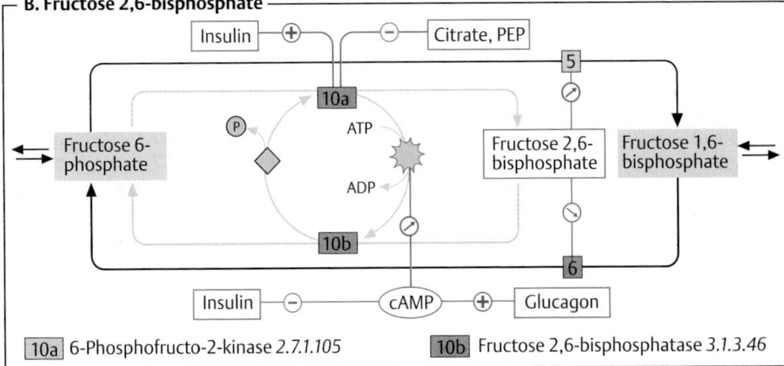
B. Fructose 2,6-bisphosphate
Insulin
Citrate, PEP
5
10a
P
ATP
Fructose 6-
phosphate
Fructose 2,6-
bisphosphate
Fructose 1,6-
bisphosphate
ADP
10b
6
Insulin
cAMP
Glucagon
10a 6-Phosphofructo-2-kinase 2.7.1.105
10b Fructose 2,6-bisphosphatase 3.1.3.46

Diabetes mellitus

Diabetes mellitus is a very common metabolic disease that is caused by absolute or relative insulin deficiency. The lack of this peptide hormone (see p. 76) mainly affects carbohydrate and lipid metabolism. Diabetes mellitus occurs in two forms. In **type 1** diabetes (insulin-dependent diabetes mellitus, IDDM), the insulin-forming cells are destroyed in young individuals by an autoimmune reaction. The less severe **type 2** diabetes (non-insulin-dependent diabetes mellitus, NIDDM) usually has its first onset in elderly individuals. The causes have not yet been explained in detail in this type.

A. Insulin biosynthesis ○

Insulin is produced by the B cells of the *islets of Langerhans* in the pancreas. As is usual with secretory proteins, the hormone's precursor (*preproinsulin*) carries a signal peptide that directs the peptide chain to the interior of the endoplasmic reticulum (see p. 210). *Proinsulin* is produced in the ER by cleavage of the signal peptide and formation of disulfide bonds. Proinsulin passes to the Golgi apparatus, where it is packed into vesicles—the *β-granules.* After cleavage of the *C peptide, mature insulin* is formed in the β-granules and is stored in the form of zinc-containing hexamers until secretion.

B. Effects of insulin deficiency ◑

The effects of insulin on **carbohydrate metabolism** are discussed on p. 158. In simplified terms, they can be described as *stimulation of glucose utilization* and *inhibition of gluconeogenesis.* In addition, the transport of glucose from the blood into most tissues is also insulin-dependent (exceptions to this include the liver, CNS, and erythrocytes).

The **lipid metabolism** of adipose tissue is also influenced by the hormone. In these cells, insulin stimulates the reorganization of glucose into fatty acids. This is mainly based on activation of *acetyl CoA carboxylase* (see p. 162) and increased availability of NADPH+H^+ due to increased PPP activity (see p. 152). On the other hand, insulin also inhibits the degradation of fat by hormone-sensitive lipases (see p. 162) and prevents the breakdown of muscle protein.

The effects of insulin *deficiency* on metabolism are shown by arrows in the illustration. Particularly noticeable is the increase in the glucose concentration in the blood, from 5 mM to 9 mM (90 mg · dL^{-1}) or more (**hyperglycemia,** elevated blood glucose level). In *muscle* and *adipose tissue* – the two most important glucose consumers—glucose uptake and glucose utilization are impaired by insulin deficiency. Glucose utilization in the *liver* is also reduced. At the same time, gluconeogenesis is stimulated, partly due to increased proteolysis in the muscles. This increases the blood sugar level still further. When the capacity of the *kidneys* to resorb glucose is exceeded (at plasma concentrations of 9 mM or more), glucose is excreted in the urine (**glucosuria**).

The increased degradation of fat that occurs in insulin deficiency also has serious effects. Some of the fatty acids that accumulate in large quantities are taken up by the liver and used for lipoprotein synthesis **(hyperlipidemia)**, and the rest are broken down into acetyl CoA. As the tricarboxylic acid cycle is not capable of taking up such large quantities of acetyl CoA, the excess is used to form **ketone bodies** (*acetoacetate* and *β-hydroxybutyrate* see p. 312). As H^+ ions are released in this process, diabetics not receiving adequate treatment can suffer severe **metabolic acidosis** (diabetic coma). The *acetone* that is also formed gives these patients' breath a characteristic odor. In addition, large amounts of ketone body anions appear in the urine (**ketonuria**).

Diabetes mellitus can have serious secondary effects. A constantly raised blood sugar level can lead in the long term to changes in the blood vessels (diabetic angiopathy), kidney damage (nephropathy) and damage to the nervous system (neuropathy), as well as to cataracts in the eyes.

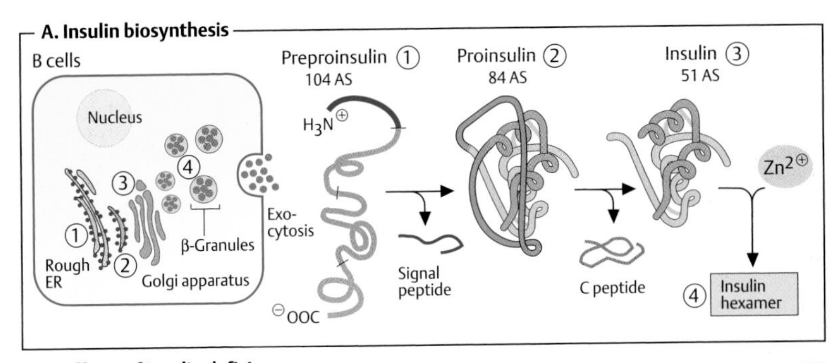
A. Insulin biosynthesis
B cells
Nucleus
Rough ER
Golgi apparatus
β-Granules
Exo-cytosis
Preproinsulin ①
104 AS
$H_3N^{\oplus}$
$^{\ominus}OOC$
Signal peptide
Proinsulin ②
84 AS
C peptide
Insulin ③
51 AS
$Zn^{2\oplus}$
④ Insulin hexamer

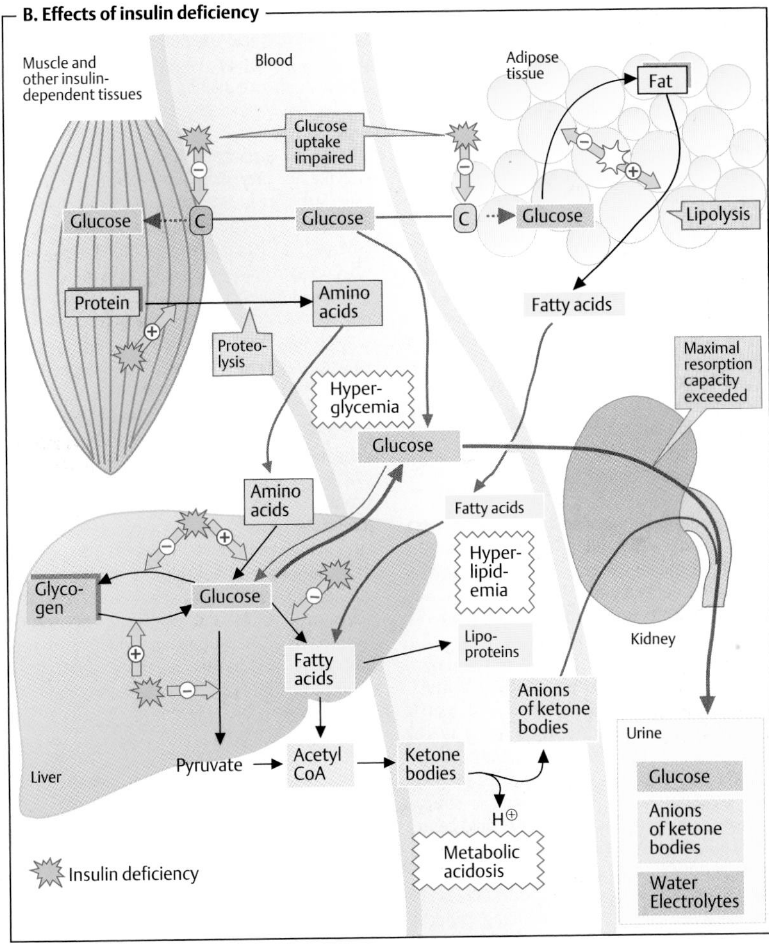
B. Effects of insulin deficiency
Muscle and other insulin-dependent tissues
Blood
Adipose tissue
Fat
Glucose uptake impaired
Glucose
C
Glucose
C
Glucose
Lipolysis
Protein
Amino acids
Proteo-lysis
Fatty acids
Maximal resorption capacity exceeded
Hyper-glycemia
Glucose
Amino acids
Fatty acids
Hyper-lipid-emia
Glyco-gen
Glucose
Fatty acids
Lipo-proteins
Kidney
Anions of ketone bodies
Liver
Pyruvate
Acetyl CoA
Ketone bodies
$H^{\oplus}$
Metabolic acidosis
Urine
Glucose
Anions of ketone bodies
Water Electrolytes
Insulin deficiency

Overview

A. Fat metabolism ◑

Fat metabolism in adipose tissue (top). Fats (triacylglycerols) are the most important energy reserve in the animal organism. They are mostly stored in insoluble form in the cells of adipose tissue—the *adipocytes*—where they are constantly being synthesized and broken down again.

As precursors for the biosynthesis of fats (**lipogenesis**), the adipocytes use triacylglycerols from lipoproteins (VLDLs and chylomicrons; see p. 278), which are formed in the liver and intestines and delivered by the blood. *Lipoprotein lipase* [1], which is located on the inner surface of the blood capillaries, cleaves these triacylglycerols into glycerol and fatty acids, which are taken up by the adipocytes and converted back into fats.

The degradation of fats (**lipolysis**) is catalyzed in adipocytes by *hormone-sensitive lipase* [2]—an enzyme that is regulated by various hormones by cAMP-dependent *interconversion* (see p. 120). The amount of fatty acids released depends on the activity of this lipase; in this way, the enzyme regulates the plasma levels of fatty acids.

In the blood plasma, fatty acids are transported in free form—i. e., non-esterified. Only short-chain fatty acids are soluble in the blood; longer, less water-soluble fatty acids are transported bound to albumin.

Degradation of fatty acids in the liver (left). Many tissues take up fatty acids from the blood plasma in order to synthesize fats or to obtain energy by oxidizing them. The metabolism of fatty acids is particularly intensive in the hepatocytes in the liver.

The most important process in the degradation of fatty acids is **β-oxidation**—a metabolic pathway in the mitochondrial matrix (see p. 164). Initially, the fatty acids in the cytoplasm are activated by binding to coenzyme A into **acyl CoA** [3]. Then, with the help of a transport system (the carnitine shuttle [4]; see p. 164), the activated fatty acids enter the mitochondrial matrix, where they are broken down into **acetyl CoA**. The resulting acetyl residues can be oxidized to CO_2 in the tricarboxylic acid cycle, producing reduced coenzyme and ATP derived from it by oxidative phosphorylation. If acetyl CoA production exceeds the energy requirements of the hepatocytes—as is the case when there is a high level of fatty acids in the blood plasma (typical in hunger and diabetes mellitus)—then the excess is converted into **ketone bodies** (see p. 312). These serve exclusively to supply other tissues with energy.

Fat synthesis in the liver (right). Fatty acids and fats are mainly synthesized in the liver and in adipose tissue, as well as in the kidneys, lungs, and mammary glands. Fatty acid biosynthesis occurs in the cytoplasm—in contrast to fatty acid degradation. The most important precursor is **glucose**, but certain amino acids can also be used.

The first step is carboxylation of **acetyl CoA** to **malonyl CoA**. This reaction is catalyzed by *acetyl-CoA carboxylase* [5], which is the *key enzyme in fatty acid biosynthesis.* Synthesis into fatty acids is carried out by *fatty acid synthase* [6]. This multifunctional enzyme (see p. 168) starts with one molecule of acetyl-CoA and elongates it by adding malonyl groups in seven reaction cycles until palmitate is reached. One CO_2 molecule is released in each reaction cycle. The fatty acid therefore grows by two carbon units each time. $NADPH+H^+$ is used as the reducing agent and is derived either from the *pentose phosphate pathway* (see p. 152) or from *isocitrate dehydrogenase* and *malic enzyme* reactions.

The elongation of the fatty acid by *fatty acid synthase* concludes at C_{16}, and the product, **palmitate** (16:0), is released. Unsaturated fatty acids and long-chain fatty acids can arise from palmitate in subsequent reactions. Fats are finally synthesized from activated fatty acids (acyl CoA) and glycerol 3-phosphate (see p. 170). To supply peripheral tissues, fats are packed by the hepatocytes into lipoprotein complexes of the VLDL type and released into the blood in this form (see p. 278).

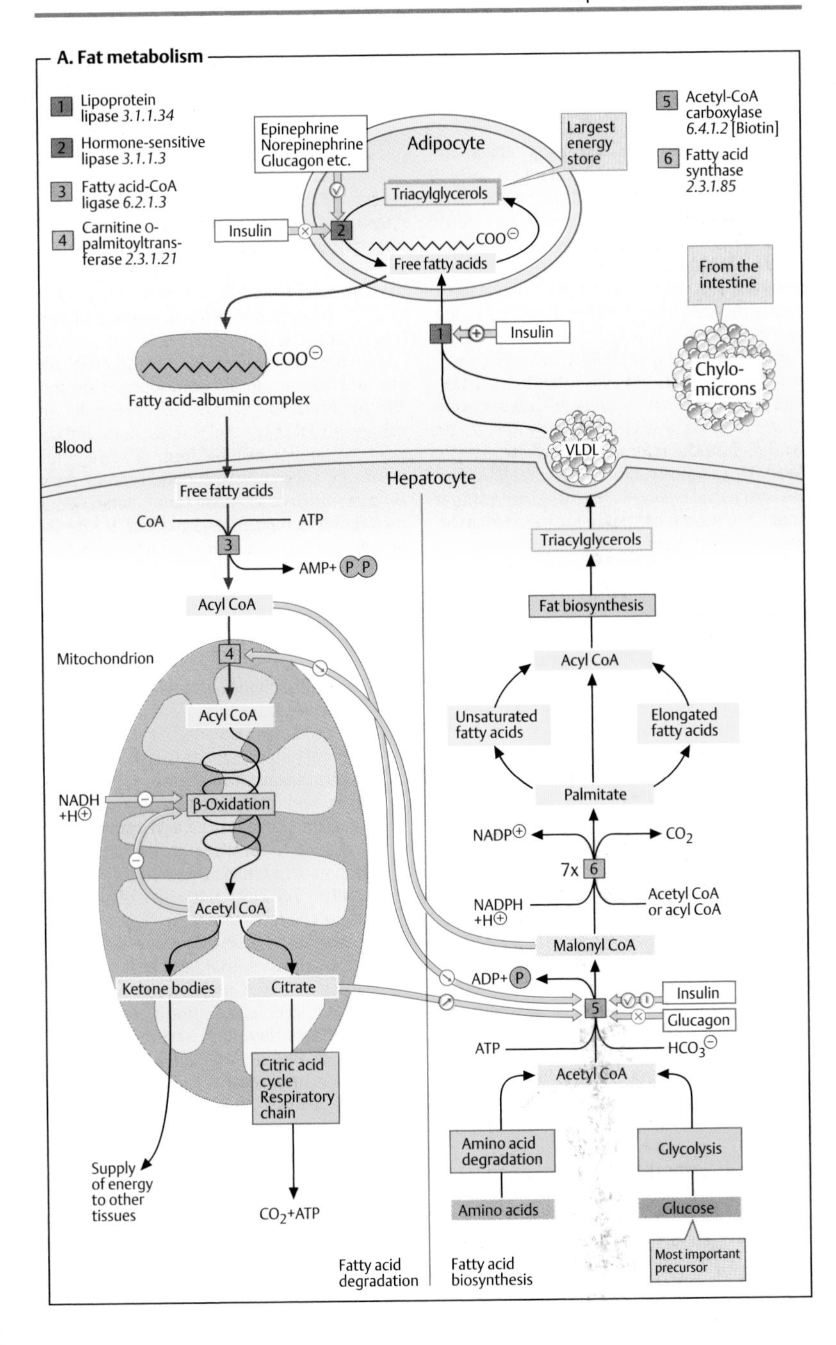
A. Fat metabolism
1 Lipoprotein lipase 3.1.1.34
2 Hormone-sensitive lipase 3.1.1.3
3 Fatty acid-CoA ligase 6.2.1.3
4 Carnitine O-palmitoyltransferase 2.3.1.21
5 Acetyl-CoA carboxylase 6.4.1.2 [Biotin]
6 Fatty acid synthase 2.3.1.85
Epinephrine
Norepinephrine
Glucagon etc.
Adipocyte
Largest energy store
Triacylglycerols
Insulin
COO⊖
Free fatty acids
From the intestine
Insulin
Chylo-microns
Fatty acid-albumin complex
Blood
VLDL
Hepatocyte
Free fatty acids
CoA
ATP
AMP+ P P
Acyl CoA
Triacylglycerols
Fat biosynthesis
Mitochondrion
Acyl CoA
Acyl CoA
Unsaturated fatty acids
Elongated fatty acids
NADH +H⊕
β-Oxidation
Palmitate
NADP⊕
CO2
7x
Acetyl CoA
Acetyl CoA or acyl CoA
NADPH +H⊕
Malonyl CoA
Ketone bodies
Citrate
ADP+ P
Insulin
Glucagon
ATP
HCO3⊖
Acetyl CoA
Citric acid cycle Respiratory chain
Amino acid degradation
Glycolysis
Supply of energy to other tissues
CO2+ATP
Amino acids
Glucose
Most important precursor
Fatty acid degradation
Fatty acid biosynthesis

Fatty acid degradation

A. Fatty acid degradation: β-oxidation

After uptake by the cell, fatty acids are activated by conversion into their CoA derivatives—**acyl CoA** is formed. This uses up two energy-rich anhydride bonds of ATP per fatty acid (see p. 162). For channeling into the mitochondria, the acyl residues are first transferred to *carnitine* and then transported across the inner membrane as **acyl carnitine** (see **B**).

The degradation of the fatty acids occurs in the mitochondrial matrix through an oxidative cycle in which C_2 units are successively cleaved off as **acetyl CoA** (*activated acetic acid*). Before the release of the acetyl groups, each CH_2 group at C-3 of the acyl residue (the β-C atom) is oxidized to the keto group—hence the term **β-oxidation** for this metabolic pathway. Both spatially and functionally, it is closely linked to the tricarboxylic acid cycle (see p. 136) and to the respiratory chain (see p. 140).

[1] The first step is dehydrogenation of **acyl CoA** at C-2 and C-3. This yields an unsaturated Δ^2-enoyl-CoA derivative with a *trans*-configured double bond. The two hydrogen atoms are initially transferred from FAD-containing *acyl CoA dehydrogenase* to the **electron-transferring flavoprotein (ETF)**. *ETF dehydrogenase* [5] passes them on from ETF to ubiquinone (coenzyme Q), a component of the *respiratory chain* (see p. 140). Other FAD-containing mitochondrial dehydrogenases are also able to supply the respiratory chain with electrons in this fashion.

There are three *isoenzymes* (see p. 98) of *acyl CoA dehydrogenase* that are specialized for long-chain fatty acids (12–18 C atoms), medium-chain fatty acids (4–14), and short-chain fatty acids (4–8).

[2] The next step in fatty acid degradation is the addition of a water molecule to the double bond of the enoyl CoA (*hydration*), with formation of **β-hydroxyacyl CoA**.

[3] In the next reaction, the OH group at C-3 is oxidized to a carbonyl group (*dehydrogenation*). This gives rise to **β-ketoacyl CoA**, and the reduction equivalents are transferred to NAD^+, which also passes them on to the *respiratory chain*.

[4] β-Ketoacyl-CoA is now broken down by an *acyl transferase* into **acetyl CoA** and an **acyl CoA shortened by 2 C atoms** (*"thioclastic cleavage"*).

Several cycles are required for complete degradation of long-chain fatty acids—eight cycles in the case of stearyl-CoA (C18:0), for example. The acetyl CoA formed can then undergo further metabolism in the *tricarboxylic acid cycle* (see p. 136), or can be used for biosynthesis. When there is an excess of acetyl CoA, the liver can also form ketone bodies (see p. 312).

When oxidative degradation is complete, one molecule of palmitic acid supplies around 106 molecules of ATP, corresponding to an energy of 3300 $kJ \cdot mol^{-1}$. This high energy yield makes fats an ideal form of storage for metabolic energy. Hibernating animals such as polar bears can meet their own energy requirements for up to 6 months solely by fat degradation, while at the same time producing the vital water they need via the respiratory chain ("respiratory water").

B. Fatty acid transport

The inner mitochondrial membrane has a group-specific transport system for fatty acids. In the cytoplasm, the acyl groups of activated fatty acids are transferred to **carnitine** by *carnitine acyltransferase* [1]. They are then channeled into the matrix by an acylcarnitine/carnitine antiport as **acyl carnitine**, in exchange for free carnitine. In the matrix, the mitochondrial enzyme *carnitine acyltransferase* catalyzes the return transfer of the acyl residue to CoA.

The carnitine shuttle is the rate-determining step in mitochondrial fatty acid degradation. Malonyl CoA, a precursor of fatty acid biosynthesis, inhibits *carnitine acyltransferase* (see p. 162), and therefore also inhibits uptake of fatty acids into the mitochondrial matrix.

The most important regulator of β-oxidation is the $NAD^+/NADH+H^+$ ratio. If the respiratory chain is not using any $NADH+H^+$, then not only the tricarboxylic acid cycle (see p. 136) but also β-oxidation come to a standstill due to the lack of NAD^+.

A. Fatty acid degradation: β-oxidation

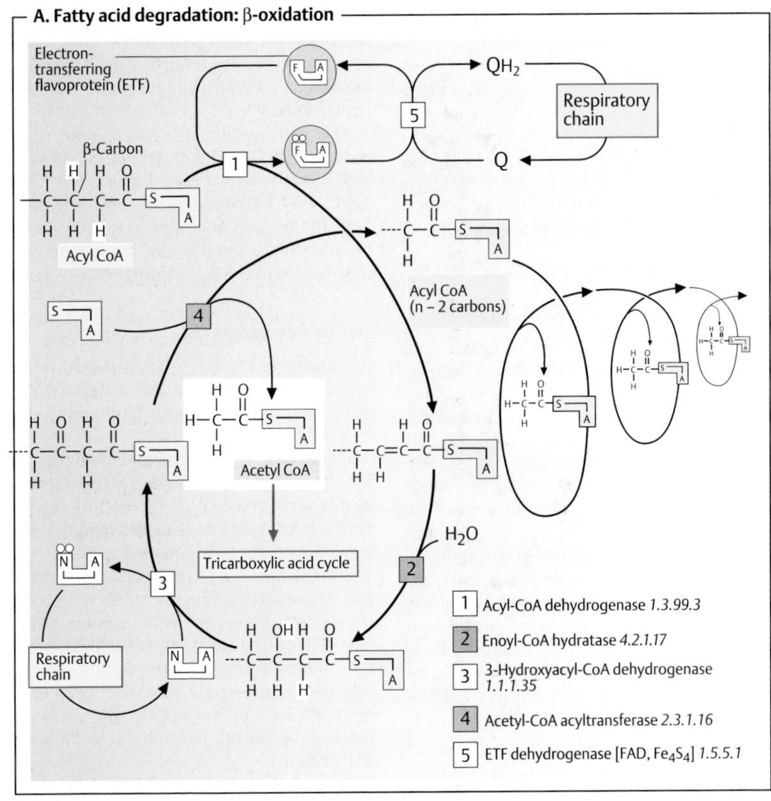

B. Fatty acid transport

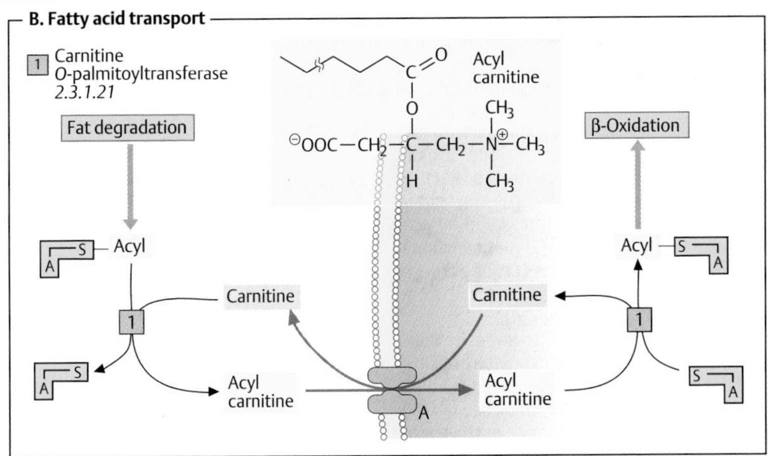

Minor pathways of fatty acid degradation

Most fatty acids are saturated and even-numbered. They are broken down via β-**oxidation** (see p.164). In addition, there are special pathways involving degradation of unsaturated fatty acids (**A**), degradation of fatty acids with an odd number of C atoms (**B**), α and ω oxidation of fatty acids, and degradation in peroxisomes.

A. Degradation of unsaturated fatty acids ○

Unsaturated fatty acids usually contain a *cis* double bond at position 9 or 12—e. g., linoleic acid (18:2; 9,12). As with saturated fatty acids, degradation in this case occurs via β-oxidation until the C-9-*cis* double bond is reached. Since *enoyl-CoA hydratase* only accepts substrates with *trans* double bonds, the corresponding enoyl-CoA is converted by an isomerase from the *cis*-Δ^3, *cis*- Δ^6 isomer into the *trans*-Δ^3,*cis*-Δ^6 isomer [1]. Degradation by β-oxidation can now continue until a shortened *trans*-Δ^2, *cis*-Δ^4 derivative occurs in the next cycle. This cannot be isomerized in the same way as before, and instead is reduced in an NADPH-dependent way to the *trans*-Δ^3 compound [2]. After rearrangement by *enoyl-CoA isomerase* [1], degradation can finally be completed via normal β-oxidation.

B. Degradation of oddnumbered fatty acids ○

Fatty acids with an odd number of C atoms are treated in the same way as "normal" fatty acids—i. e., they are taken up by the cell with ATP-dependent activation to acyl CoA and are transported into the mitochondria with the help of the carnitine shuttle and broken down there by β–oxidation (see p. 164). In the last step, **propionyl CoA** arises instead of acetyl CoA. This is first carboxylated by *propionyl CoA carboxylase* into **(*S*)-methylmalonyl CoA** [3], which—after isomerization into the (*R*) enantiomer (not shown; see p. 411)—is isomerized into **succinyl CoA** [4].

Various coenzymes are involved in these reactions. The carboxylase [3] requires *biotin*, and the mutase [4] is dependent on *coenzyme* B_{12} (5′-deoxyadenosyl cobalamin; see p. 108). Succinyl-CoA is an intermediate in the tricarboxylic acid cycle and is available for *gluconeogenesis* through conversion into oxaloacetate. Odd-numbered fatty acids from propionyl-CoA can therefore be used to synthesize glucose.

This pathway is also important for ruminant animals, which are dependent on symbiotic microorganisms to break down their food. The microorganisms produce large amounts of propionic acid as a degradation product, which the host can channel into the metabolism in the way described.

Further information ○

In addition to the degradation pathways described above, there are also additional special pathways for particular fatty acids found in food.

α **Oxidation** is used to break down methyl-branched fatty acids. It takes place through step-by-step removal of C_1 residues, begins with a hydroxylation, does not require coenzyme A, and does not produce any ATP.

ω **Oxidation**—i. e., oxidation starting at the end of the fatty acid—also starts with a hydroxylation catalyzed by a *monooxygenase* (see p. 316), and leads via subsequent oxidation to fatty acids with two carboxyl groups, which can undergo β-oxidation from both ends until C_8 or C_6 dicarboxylic acids are reached, which can be excreted in the urine in this form.

Degradation of unusually long fatty acids. An alternative form of β-oxidation takes place in *hepatic peroxisomes*, which are specialized for the degradation of particularly long fatty acids (n > 20). The degradation products are acetyl-CoA and hydrogen peroxide (H_2O_2), which is detoxified by the *catalase* (see p. 32) common in peroxisomes.

Enzyme defects are also known to exist in the minor pathways of fatty acid degradation. In **Refsum disease**, the methyl-branched phytanic acid (obtained from vegetable foods) cannot be degraded by α-oxidation. In **Zellweger syndrome**, a peroxisomal defect means that long-chain fatty acids cannot be degraded.

A. Degradation of unsaturated fatty acids

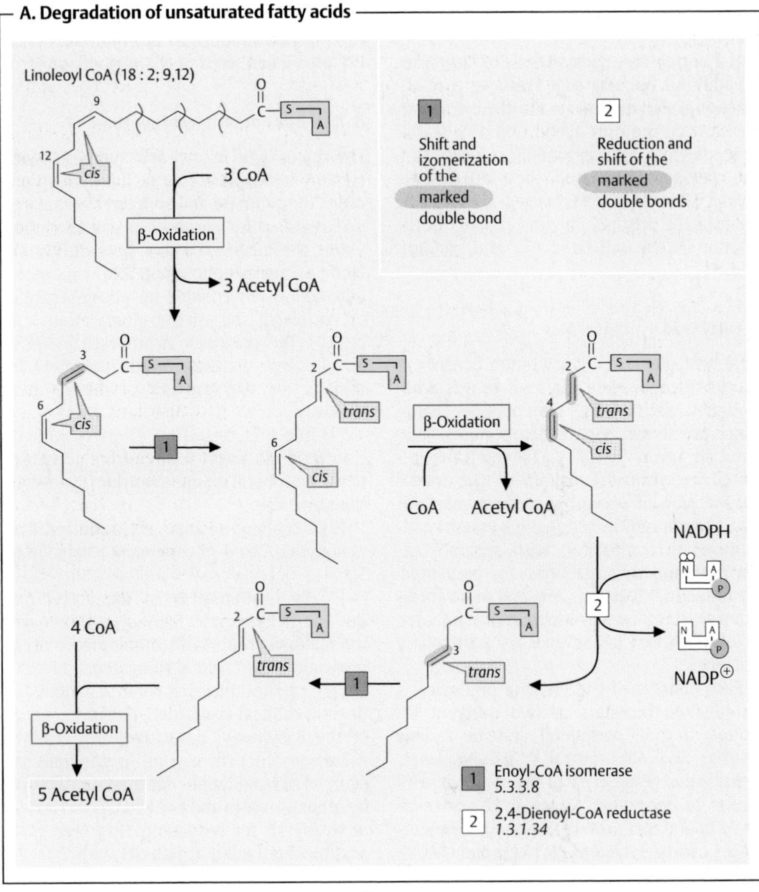

B. Degradation of odd-numbered fatty acids

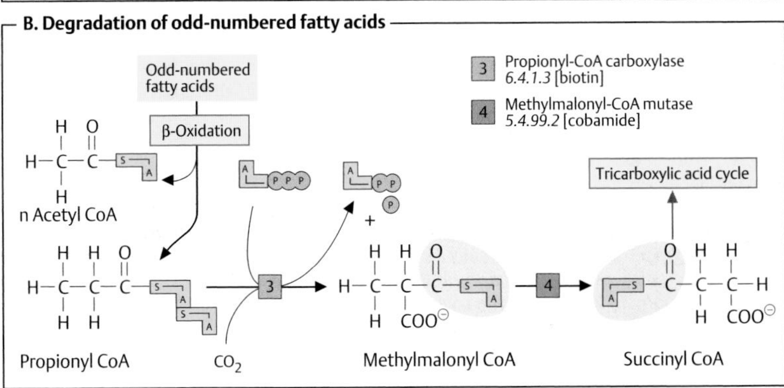

Fatty acid synthesis

In the vertebrates, biosynthesis of fatty acids is catalyzed by *fatty acid synthase,* a multifunctional enzyme. Located in the cytoplasm, the enzyme requires acetyl CoA as a starter molecule. In a cyclic reaction, the acetyl residue is elongated by one C_2 unit at a time for seven cycles. $NADPH+H^+$ is used as a reducing agent in the process. The end product of the reaction is the saturated C_{16} acid, *palmitic acid.*

A. Fatty acid synthase ◑

Fatty acid synthase in vertebrates consists of two identical peptide chains—i. e., it is a homodimer. Each of the two peptide chains, which are shown here as hemispheres, catalyzes all seven of the partial reactions required to synthesize palmitate. The spatial compression of several successive reactions into a single multifunctional enzyme has advantages in comparison with separate enzymes. Competing reactions are prevented, the individual reactions proceed in a coordinated way as if on a production line, and due to low diffusion losses they are particularly efficient.

Each subunit of the enzyme binds acetyl residues as thioesters at two different SH groups: at one peripheral *cysteine residue* (CysSH) and one central *4′-phosphopantetheine group* (Pan-SH). Pan-SH, which is very similar to coenzyme A (see p. 12), is covalently bound to a protein segment of the synthase known as the *acyl-carrier protein* (ACP). This part functions like a long arm that passes the substrate from one reaction center to the next. The two subunits of fatty acid synthase cooperate in this process; the enzyme is therefore only capable of functioning as a dimer.

Spatially, the enzyme activities are arranged into three different domains. **Domain 1** catalyzes the entry of the substrates acetyl CoA and malonyl CoA by *[ACP]-S-acetyltransferase* [1] and *[ACP]-Smalonyl transferase* [2] and subsequent condensation of the two partners by *3-oxoacyl-[ACP]-synthase* [3]. **Domain 2** catalyzes the conversion of the 3-oxo group to a CH_2 group by *3-oxoacyl-[ACP]-reductase* [4], *3-hydroxyacyl-[ACP]-dehydratase* [5], and *enoyl-[ACP]-reductase* [6]. Finally, **domain 3** serves to release the finished product by *acyl-[ACP]-hydrolase* [7] after seven steps of chain elongation.

B. Reactions of fatty acid synthase ◑

The key enzyme in fatty acid synthesis is **acetyl CoA carboxylase** (see p. 162), which precedes the synthase and supplies the malonyl-CoA required for elongation. Like all carboxylases, the enzyme contains covalently bound *biotin* as a prosthetic group and is hormone-dependently *inactivated* by phosphorylation or *activated* by dephosphorylation (see p. 120). The precursor *citrate* (see p. 138) is an allosteric activator, while *palmitoyl-CoA* inhibits the end product of the synthesis pathway.

[1] The first cycle (n = 1) starts with the transfer of an acetyl residue from acetyl CoA to the peripheral cysteine residue (Cys-SH). At the same time,

[2] a malonyl residue is transferred from malonyl CoA to 4-phosphopantetheine (Pan-SH).

[3] By condensation of the acetyl residue—or (in later cycles) the acyl residue—with the malonyl group, with simultaneous decarboxylation, the chain is elongated.

[4]–[6] The following three reactions (reduction of the 3-oxo group, dehydrogenation of the 3-hydroxyl derivative, and renewed reduction of it) correspond in principle to a reversal of β-oxidation, but they are catalyzed by other enzymes and use $NADPH+H^+$ instead of $NADH+H^+$ for reduction. They lead to an acyl residue bound at Pan-SH with 2n + 2 C atoms (n = the number of the cycle). Finally, depending on the length of the product,

[1′] The acyl residue is transferred back to the peripheral cysteine, so that the next cycle can begin again with renewed loading of the ACP with a malonyl residue, or:

[7] After seven cycles, the completed **palmitic acid** is hydrolytically released.

In all, one acetyl-CoA and seven malonyl-CoA are converted with the help of 14 $NADPH+H^+$ into one palmitic acid, 7 CO_2, 6 H_2O, 8 CoA and 14 $NADP^+$. Acetyl CoA carboxylase also uses up seven ATP.

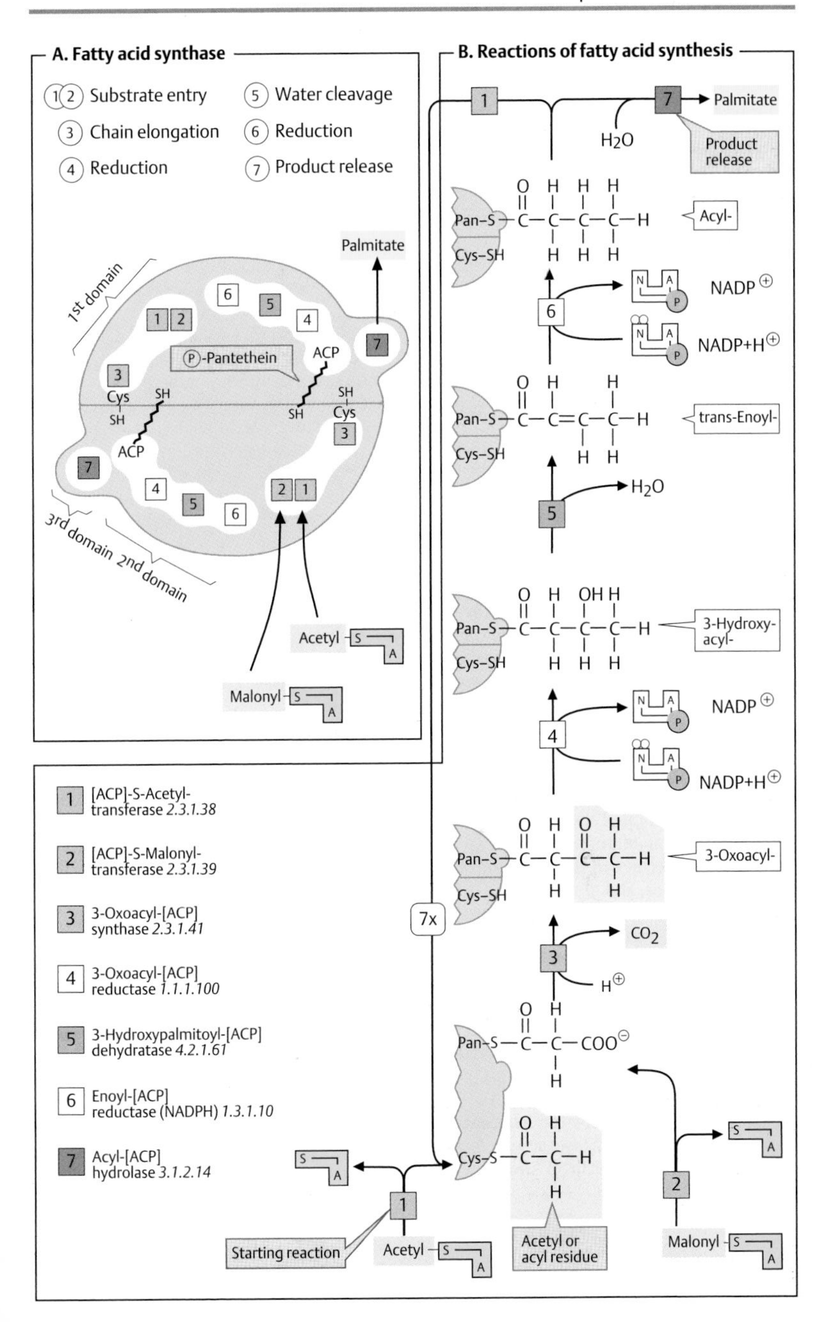

A. Fatty acid synthase
1 2 Substrate entry
3 Chain elongation
4 Reduction
5 Water cleavage
6 Reduction
7 Product release
Palmitate
1st domain
P-Pantethein
ACP
Cys
SH
3rd domain
2nd domain
Acetyl
Malonyl
B. Reactions of fatty acid synthesis
Palmitate
H2O
Product release
Pan-S
Cys-SH
Acyl-
NADP+
NADP+H+
trans-Enoyl-
3-Hydroxy-acyl-
3-Oxoacyl-
CO2
H+
7x
Cys-S
Acetyl or acyl residue
Starting reaction
Acetyl
Malonyl
1 [ACP]-S-Acetyl-transferase 2.3.1.38
2 [ACP]-S-Malonyl-transferase 2.3.1.39
3 3-Oxoacyl-[ACP] synthase 2.3.1.41
4 3-Oxoacyl-[ACP] reductase 1.1.1.100
5 3-Hydroxypalmitoyl-[ACP] dehydratase 4.2.1.61
6 Enoyl-[ACP] reductase (NADPH) 1.3.1.10
7 Acyl-[ACP] hydrolase 3.1.2.14

Biosynthesis of complex lipids

A. Biosynthesis of fats and phospholipids ◑

Complex lipids, such as neutral fats (triacylglycerols), phospholipids, and glycolipids, are synthesized via common reaction pathways. Most of the enzymes involved are associated with the membranes of the smooth endoplasmic reticulum.

The synthesis of fats and phospholipids starts with **glycerol 3-phosphate**. This compound can arise via two pathways:

[1] By reduction from the glycolytic intermediate **glycerone 3-phosphate** (dihydroxyacetone 3-phosphate; enzyme: *glycerol-3-phosphate dehydrogenase (NAD+) 1.1.1.8*), or:

[2] By phosphorylation of **glycerol** deriving from fat degradation (enzyme: *glycerol kinase 2.7.1.30*).

[3] Esterification of glycerol 3-phosphate with a long-chain fatty acid produces a strongly amphipathic **lysophosphatidate** (enzyme: *glycerol-3-phosphate acyltransferase 2.3.1.15*). In this reaction, an acyl residue is transferred from the activated precursor **acyl-CoA** to the hydroxy group at C-1.

[4] A second esterification of this type leads to a **phosphatidate** (enzyme: *1-acylglycerol-3-phosphate acyltransferase 2.3.1.51*). Unsaturated acyl residues, particularly oleic acid, are usually incorporated at C-2 of the glycerol. Phosphatidates (anions of phosphatidic acids) are the key molecules in the biosynthesis of fats, phospholipids, and glycolipids.

[5] To biosynthesize fats (triacylglycerols), the phosphate residue is again removed by hydrolysis (enzyme: *phosphatidate phosphatase 3.1.3.4*). This produces **diacylglycerols (DAG)**.

[6] Transfer of an additional acyl residue to DAG forms **triacylglycerols** (enzyme: *diacylglycerol acyltransferase 2.3.1.20*). This completes the biosynthesis of neutral fats. They are packaged into VLDLs by the liver and released into the blood. Finally, they are stored by adipocytes in the form of insoluble fat droplets.

The biosynthesis of most phospholipids also starts from DAG.

[7] Transfer of a phosphocholine residue to the free OH group gives rise to **phosphatidylcholine** (lecithin; enzyme: *1-alkyl-2-acetylglycerolcholine phosphotransferase 2.7.8.16*). The phosphocholine residue is derived from the precursor CDP-choline (see p. 110). **Phosphatidylethanolamine** is similarly formed from CDP-ethanolamine and DAG. By contrast, **phosphatidylserine** is derived from phosphatidylethanolamine by an exchange of the amino alcohol. Further reactions serve to interconvert the phospholipids—e. g., phosphatidylserine can be converted into phosphatidylethanolamine by decarboxylation, and the latter can then be converted into phosphatidylcholine by methylation with S-adenosyl methionine (not shown; see also p. 409). The biosynthesis of **phosphatidylinositol** starts from phosphatidate rather than DAG.

[8] In the lumen of the intestine, fats from food are mainly broken down into **monoacylglycerols** (see p. 270). The cells of the intestinal mucosa re-synthesize these into neutral fats. This pathway also passes via **DAG** (enzyme: *acylglycerolpalmitoyl transferase 2.3.1.22*).

[9] Transfer of a CMP residue gives rise first to **CDP-diacylglycerol** (enzyme: *phosphatidatecytidyl transferase 2.3.1.22*).

[10] Substitution of the CMP residue by inositol then provides **phosphatidylinositol** (**PtdIns**; enzyme: *CDPdiacylglycerolinositol-3-phosphatidyl transferase 2.7.8.11*).

[12] An additional phosphorylation (enzyme: *phosphatidylinositol-4-phosphate kinase 2.7.1.68*) finally provides **phosphaditylinositol-4,5-bisphosphate** (PIP_2, $PtdIns(4,5)P_2$). PIP_2 is the precursor for the second messengers *2,3-diacylglycerol* (DAG) and *inositol-1,4,5-trisphosphate* ($InsP_3$, IP_3; see p. 367).

The biosynthesis of the **sphingolipids** is shown in schematic form on p. 409.

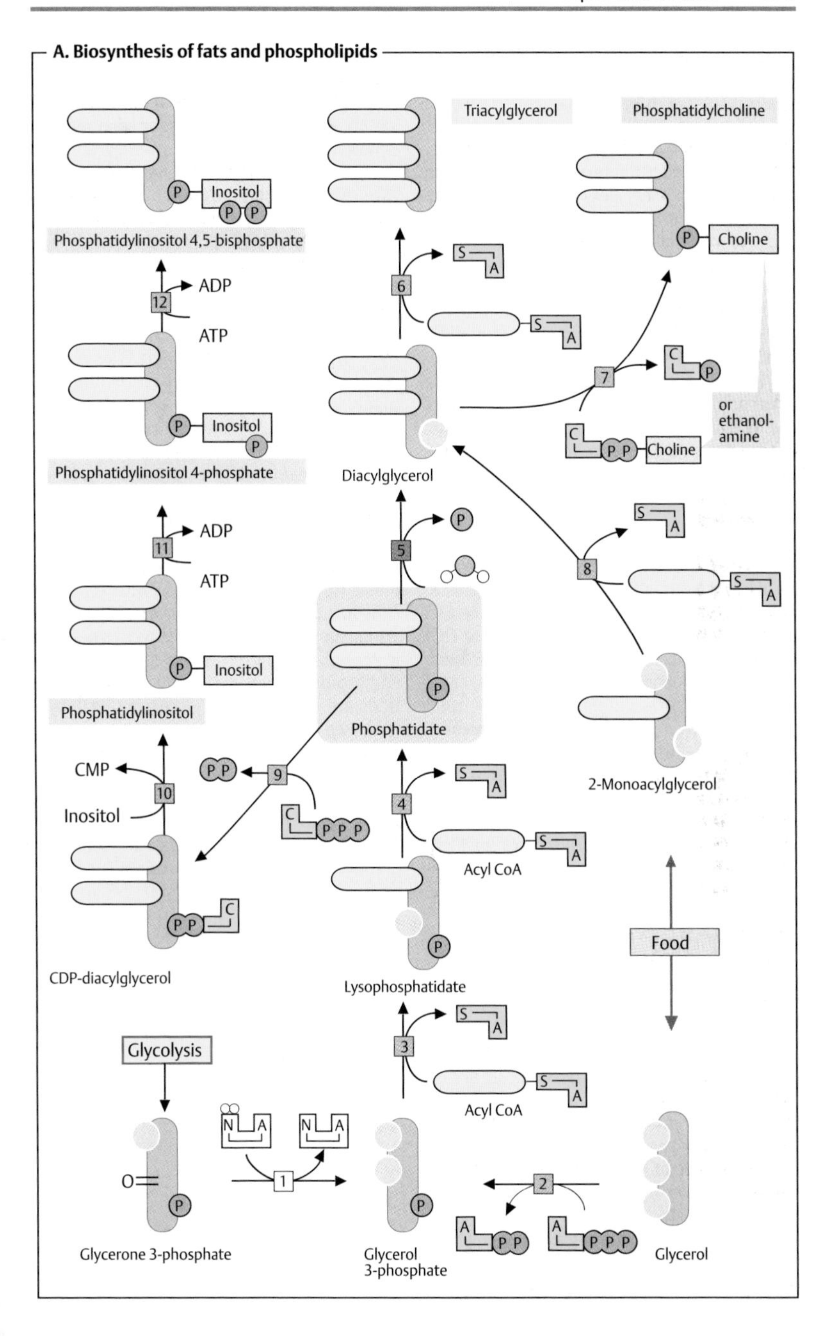
A. Biosynthesis of fats and phospholipids
Inositol
Phosphatidylinositol 4,5-bisphosphate
Triacylglycerol
Phosphatidylcholine
Choline
ADP
ATP
Inositol
Phosphatidylinositol 4-phosphate
Diacylglycerol
or ethanol-amine
Choline
ADP
ATP
Inositol
Phosphatidylinositol
Phosphatidate
2-Monoacylglycerol
CMP
Inositol
Acyl CoA
CDP-diacylglycerol
Lysophosphatidate
Food
Glycolysis
Acyl CoA
Glycerone 3-phosphate
Glycerol 3-phosphate
Glycerol

Biosynthesis of cholesterol

Cholesterol is a major constituent of the *cell membranes* of animal cells (see p. 216). It would be possible for the body to provide its full daily cholesterol requirement (ca. 1 g) by synthesizing it itself. However, with a mixed diet, only about half of the cholesterol is derived from *endogenous biosynthesis*, which takes place in the intestine and skin, and mainly in the liver (about 50%). The rest is taken up from *food*. Most of the cholesterol is incorporated into the lipid layer of plasma membranes, or converted into **bile acids** (see p. 314). A very small amount of cholesterol is used for biosynthesis of the **steroid hormones** (see p. 376). In addition, up to 1 g cholesterol per day is released into the *bile* and thus excreted.

A. Cholesterol biosynthesis ○

Cholesterol is one of the isoprenoids, synthesis of which starts from **acetyl CoA** (see p. 52). In a long and complex reaction chain, the C_{27} sterol is built up from C_2 components. The biosynthesis of cholesterol can be divided into four sections. In the first (**1**), **mevalonate,** a C_6 compound, arises from three molecules of **acetyl CoA.** In the second part (**2**), mevalonate is converted into **isopentenyl diphosphate,** the "active isoprene." In the third part (**3**), six of these C_5 molecules are linked to produce **squalene,** a C_{30} compound. Finally, squalene undergoes cyclization, with three C atoms being removed, to yield cholesterol (**4**). The illustration only shows the most important intermediates in biosynthesis.

(1) Formation of mevalonate. The conversion of acetyl CoA to acetoacetyl CoA and then to *3-hydroxy-3-methylglutaryl CoA* (3-HMG CoA) corresponds to the biosynthetic pathway for *ketone bodies* (details on p. 312). In this case, however, the synthesis occurs not in the mitochondria as in ketone body synthesis, but in the smooth endoplasmic reticulum. In the next step, the 3-HMG group is cleaved from the CoA and at the same time reduced to mevalonate with the help of NADPH+H^+. *3-HMG CoA reductase* is the *key enzyme* in cholesterol biosynthesis. It is regulated by *repression* of transcription (effectors: oxysterols such as cholesterol) and by *interconversion* (effectors: hormones). Insulin and thyroxine stimulate the enzyme and glucagon inhibits it by cAMP-dependent phosphorylation. A large supply of cholesterol from food also inhibits 3-HMG-CoA reductase.

(2) Formation of isopentenyl diphosphate. After phosphorylation, mevalonate is decarboxylated to *isopentenyl diphosphate,* with consumption of ATP. This is the component from which all of the isoprenoids are built (see p. 53).

(3) Formation of squalene. Isopentenyl diphosphate undergoes isomerization to form dimethylallyl diphosphate. The two C_5 molecules condense to yield geranyl diphosphate, and the addition of another isopentenyl diphosphate produces farnesyl diphosphate. This can then undergo dimerization, in a *head-to-head reaction,* to yield squalene. Farnesyl diphosphate is also the starting-point for other polyisoprenoids, such as dolichol (see p. 230) and ubiquinone (see p. 52).

(4) Formation of cholesterol. Squalene, a linear isoprenoid, is cyclized, with O_2 being consumed, to form lanosterol, a C_{30} sterol. Three methyl groups are cleaved from this in the subsequent reaction steps, to yield the end product cholesterol. Some of these reactions are catalyzed by *cytochrome P450 systems* (see p. 318).

The endergonic biosynthetic pathway described above is located entirely in the *smooth endoplasmic reticulum.* The energy needed comes from the CoA derivatives used and from ATP. The reducing agent in the formation of mevalonate and squalene, as well as in the final steps of cholesterol biosynthesis, is NADPH+H^+.

The division of the intermediates of the reaction pathway into three groups is characteristic: CoA compounds, diphosphates, and highly lipophilic, poorly soluble compounds (squalene to cholesterol), which are bound to *sterol carriers* in the cell.

A. Cholesterol biosynthesis

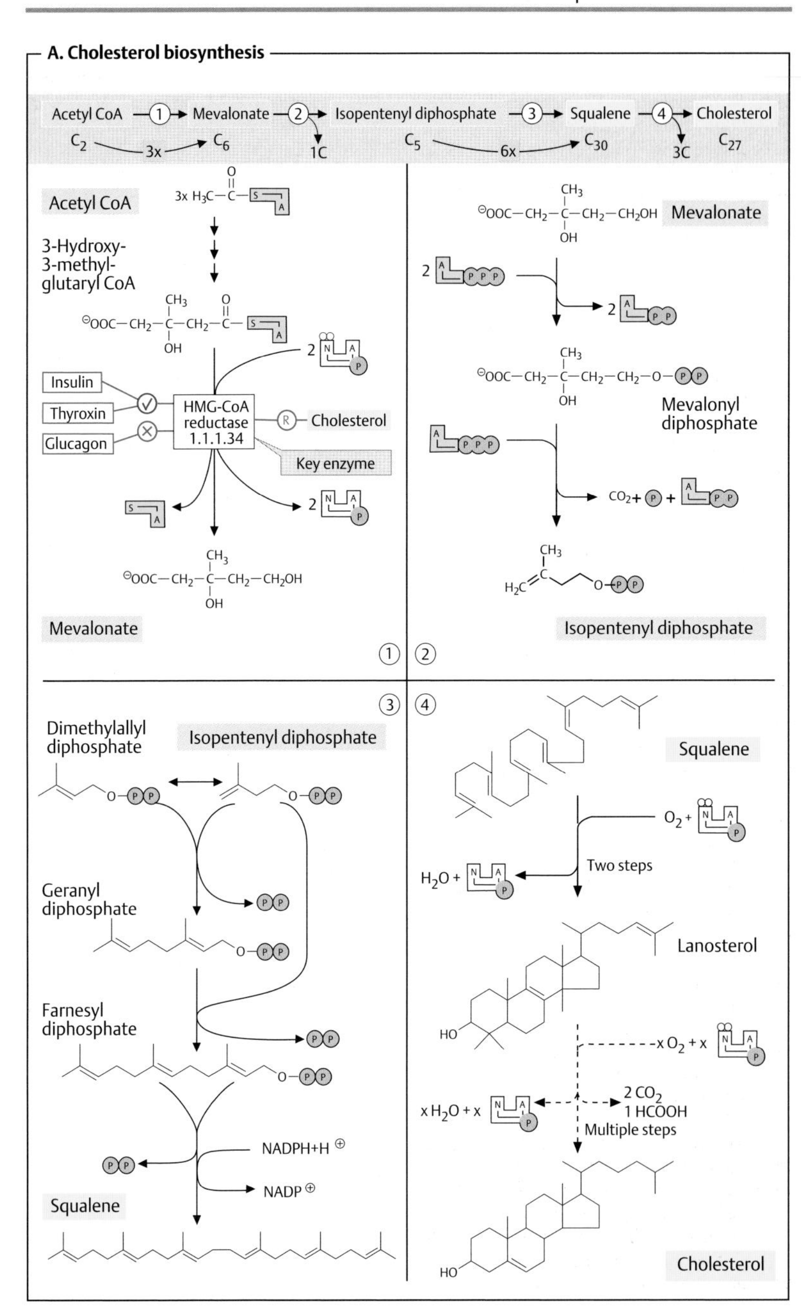

Protein metabolism: overview

Quantitatively, proteins are the most important group of endogenous macromolecules. A person weighing 70 kg contains about 10 kg protein, with most of it located in muscle. By comparison, the proportion made up by other nitrogencontaining compounds is minor. The organism's nitrogen balance is therefore primarily determined by protein metabolism. Several hormones—mainly *testosterone* and *cortisol*—regulate the nitrogen balance (see p. 374).

A. Protein metabolism: overview ●

In adults, the **nitrogen balance** is generally in *equilibrium*—i. e., the quantities of protein nitrogen taken in and excreted per day are approximately equal. If only some of the nitrogen taken in is excreted again, then the balance is *positive*. This is the case during growth, for example. *Negative* balances are rare and usually occur due to disease.

Proteins taken up in food are initially broken down in the gastrointestinal tract into amino acids, which are resorbed and distributed in the organism via the blood (see p. 266). The human body is not capable of synthesizing 8–10 of the 20 proteinogenic amino acids it requires (see p. 60). These amino acids are **essential**, and have to be supplied from food (see p. 184).

Proteins are constantly being lost via the intestine and, to a lesser extent, via the kidneys. To balance these inevitable losses, at least 30 g of protein have to be taken up with food every day. Although this minimum value is barely reached in some countries, in the industrial nations the protein content of food is usually much higher than necessary. As it is not possible to store amino acids, up to 100 g of excess amino acids per day are used for biosynthesis or degraded in the liver in this situation. The nitrogen from this excess is converted into urea (see p. 182) and excreted in the urine in this form. The carbon skeletons are used to synthesize carbohydrates or lipids (see p. 180), or are used to form ATP.

It is thought that adults break down 300–400 g of protein per day into amino acids (**proteolysis**). On the other hand, approximately the same amount of amino acids is reincorporated into proteins (**protein biosynthesis**). The body's high level of protein turnover is due to the fact that many proteins are relatively *short-lived.* On average, their half-lives amount to 2–8 days. The *key enzymes* of the intermediary metabolism have even shorter half-lives. They are sometimes broken down only a few hours after being synthesized, and are replaced by new molecules. This constant process of synthesis and degradation makes it possible for the cells to quickly adjust the quantities, and therefore the activity, of important enzymes in order to meet current requirements. By contrast, structural proteins such as the histones, hemoglobin, and the components of the cytoskeleton are particularly long-lived.

Almost all cells are capable of carrying out **biosynthesis** of proteins (top left). The formation of peptide chains by **translation** at the ribosome is described in greater detail on pp. 250–253. However, the functional forms of most proteins arise only after a series of additional steps. To begin with, supported by auxiliary proteins, the biologically active conformation of the peptide chain has to be formed (**folding**; see pp. 74, 232). During subsequent "post-translational" **maturation**, many proteins remove part of the peptide chain again and attach additional groups—e.g., oligosaccharides or lipids. These processes take place in the endoplasmic reticulum and in the Golgi apparatus (see p. 232). Finally, the proteins have to be transported to their site of action (**sorting**; see p. 228).

Some *intracellular* protein degradation (**proteolysis**) takes place in the lysosomes (see p. 234). In addition, there are protein complexes in the cytoplasm, known as *proteasomes,* in which incorrectly folded or old proteins are degraded. These molecules are recognized by a special **marking** (see p. 176). The proteasome also plays an important part in the presentation of antigens by immune cells (see p. 296).

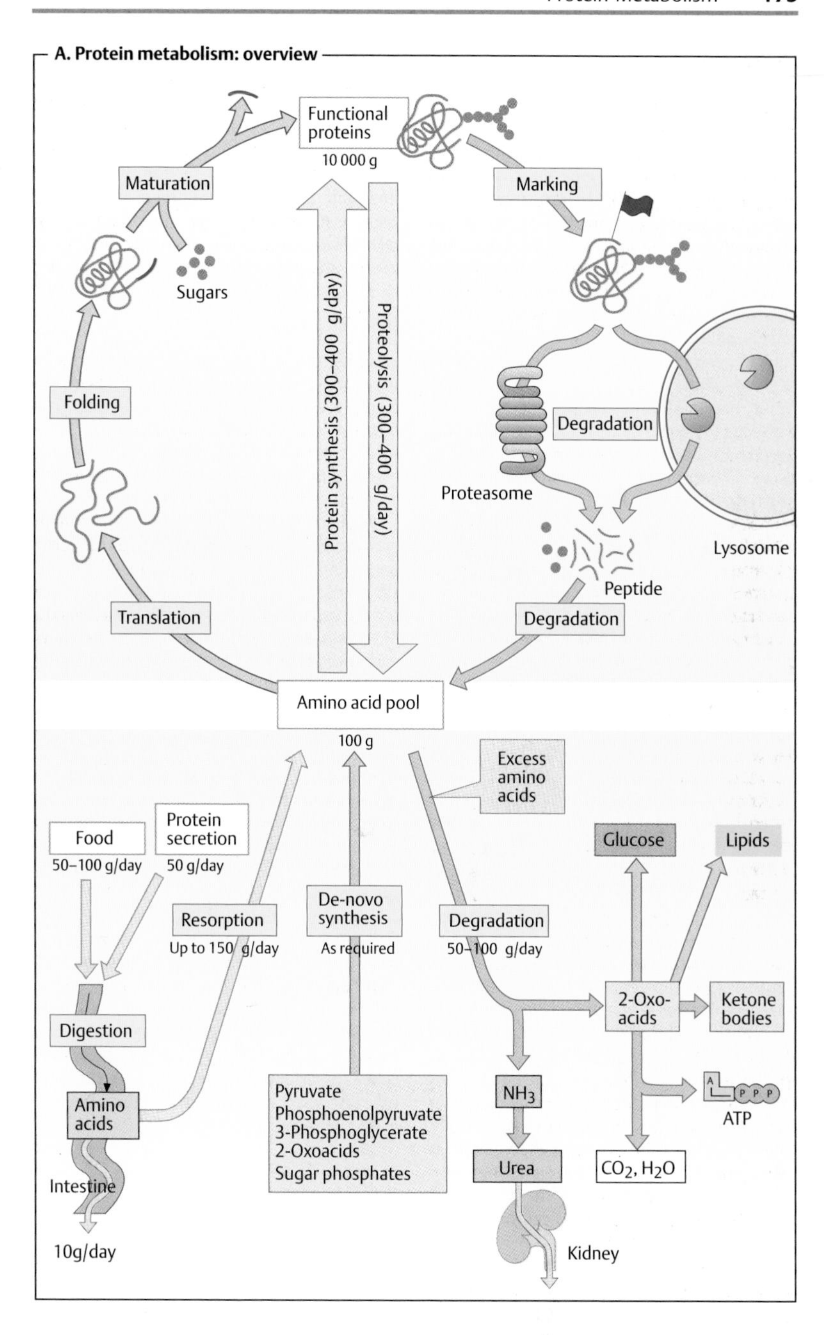
A. Protein metabolism: overview
Functional proteins
10 000 g
Maturation
Marking
Sugars
Folding
Protein synthesis (300–400 g/day)
Proteolysis (300–400 g/day)
Degradation
Proteasome
Lysosome
Peptide
Translation
Degradation
Amino acid pool
100 g
Excess amino acids
Food
50–100 g/day
Protein secretion
50 g/day
Resorption
Up to 150 g/day
De-novo synthesis
As required
Degradation
50–100 g/day
Glucose
Lipids
Digestion
Amino acids
Intestine
10g/day
Pyruvate
Phosphoenolpyruvate
3-Phosphoglycerate
2-Oxoacids
Sugar phosphates
NH_3
Urea
Kidney
2-Oxo-acids
Ketone bodies
ATP
CO_2, H_2O

Proteolysis

A. Proteolytic enzymes ◑

Combinations of several enzymes with different specificities are required for complete degradation of proteins into free amino acids. **Proteinases** and **peptidases** are found not only in the gastrointestinal tract (see p. 268), but also inside the cell (see below).

The proteolytic enzymes are classified into **endopeptidases** and **exopeptidases**, according to their site of attack in the substrate molecule. The *endopeptidases* or *proteinases* cleave peptide bonds *inside* peptide chains. They "recognize" and bind to short sections of the substrate's sequence, and then hydrolyze bonds between particular amino acid residues in a relatively specific way (see p. 94). The **proteinases** are classified according to their reaction mechanism. In *serine proteinases,* for example (see **C**), a serine residue in the enzyme is important for catalysis, while in *cysteine proteinases,* it is a cysteine residue, and so on.

The exopeptidases attack peptides from their termini. Peptidases that act at the N terminus are known as **aminopeptidases**, while those that recognize the C terminus are called **carboxypeptidases**. The **dipeptidases** only hydrolyze dipeptides.

B. Proteasome ○

The functional proteins in the cell have to be protected in order to prevent premature degradation. Some of the intracellularly active proteolytic enzymes are therefore enclosed in lysosomes (see p. 234). The proteinases that act there are also known as **cathepsins**. Another carefully regulated system for protein degradation is located in the cytoplasm. This consists of large protein complexes (mass $2 \cdot 10^6$ Da), the **proteasomes**. Proteasomes contain a barrel-shaped core consisting of 28 subunits that has a sedimentation coefficient (see p. 200) of 20 S. Proteolytic activity (shown here by the scissors) is localized in the interior of the 20-S core and is therefore protected. The openings in the barrel are sealed by 19-S particles with a complex structure that control access to the core.

Proteins destined for degradation in the proteasome (e.g., incorrectly folded or old molecules) are marked by covalent linkage with chains of the small protein **ubiquitin**. The ubiquitin is previously activated by the introduction of reactive thioester groups. Molecules marked with ubiquitin ("ubiquitinated") are recognized by the 19S particle, unfolded using ATP, and then shifted into the interior of the nucleus, where degradation takes place. Ubiquitin is not degraded, but is reused after renewed activation.

C. Serine proteases ○

A large group of proteinases contain serine in their active center. The serine proteases include, for example, the digestive enzymes *trypsin, chymotrypsin,* and *elastase* (see pp. 94 and 268), many *coagulation factors* (see p. 290), and the fibrinolytic enzyme *plasmin* and its *activators* (see p. 292).

As described on p. 270, pancreatic proteinases are secreted as **proenzymes** (zymogens). Activation of these is also based on proteolytic cleavages. This is illustrated here in detail using the example of **trypsinogen**, the precursor of trypsin (**1**). Activation of trypsinogen starts with cleavage of an N-terminal hexapeptide by *enteropeptidase* (enterokinase), a specific serine proteinase that is located in the membrane of the intestinal epithelium. The cleavage product (β-trypsin) is already catalytically active, and it cleaves additional trypsinogen molecules at the sites marked in red in the illustration (autocatalytic cleavage). The precursors of chymotrypsin, elastase, and carboxypeptidase A, among others, are also activated by trypsin.

The active center of trypsin is shown in Fig. **2**. A serine residue in the enzyme (Ser-195), supported by a histidine residue and an aspartate residue (His-57, Asp-102), nucleophilically attacks the bond that is to be cleaved (red arrow). The cleavage site in the substrate peptide is located on the C-terminal side of a lysine residue, the side chain of which is fixed in a special "binding pocket" of the enzyme (left) during catalysis (see p. 94).

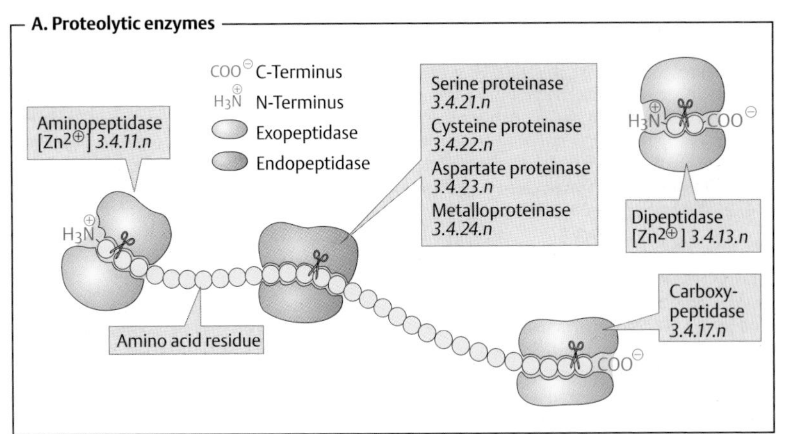
A. Proteolytic enzymes
COO⊖ C-Terminus
H3N⊕ N-Terminus
Exopeptidase
Endopeptidase
Aminopeptidase [Zn2⊕] 3.4.11.n
Serine proteinase 3.4.21.n
Cysteine proteinase 3.4.22.n
Aspartate proteinase 3.4.23.n
Metalloproteinase 3.4.24.n
Dipeptidase [Zn2⊕] 3.4.13.n
Carboxy-peptidase 3.4.17.n
Amino acid residue
H3N
COO

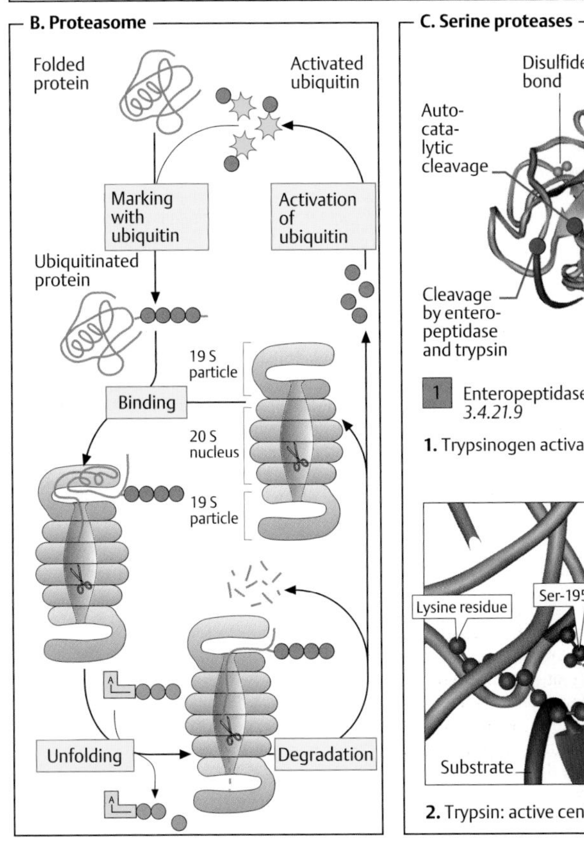
B. Proteasome
Folded protein
Activated ubiquitin
Marking with ubiquitin
Activation of ubiquitin
Ubiquitinated protein
19 S particle
20 S nucleus
19 S particle
Binding
Unfolding
Degradation

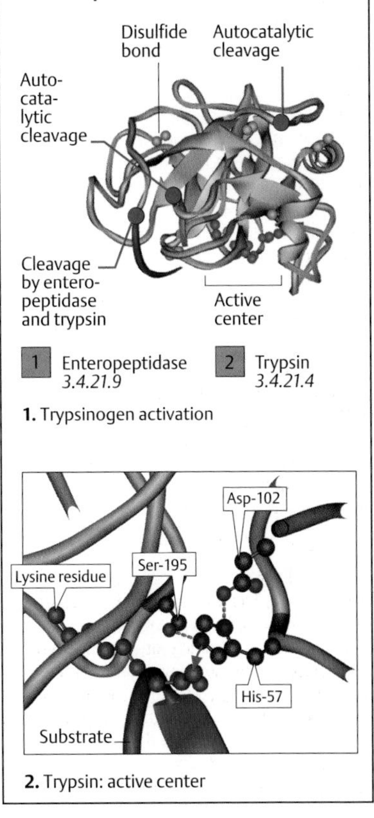
C. Serine proteases
Disulfide bond
Autocatalytic cleavage
Auto-cata-lytic cleavage
Cleavage by entero-peptidase and trypsin
Active center
1 Enteropeptidase 3.4.21.9
2 Trypsin 3.4.21.4
1. Trypsinogen activation
Asp-102
Ser-195
Lysine residue
His-57
Substrate
2. Trypsin: active center

Transamination and deamination

Amino nitrogen accumulates during protein degradation. In contrast to carbon, amino nitrogen is not suitable for oxidative energy production. If they are not being reused for biosynthesis, the amino groups of amino acids are therefore incorporated into urea (see p. 182) and excreted in this form.

A. Transamination and deamination ◑

Among the NH_2 transfer reactions, **transaminations** (**1**) are particularly important. They are catalyzed by *transaminases*, and occur in both catabolic and anabolic amino acid metabolism. During transamination, the amino group of an amino acid (amino acid 1) is transferred to a 2-oxoacid (oxoacid 2). From the amino acid, this produces a 2-oxoacid (a), while from the original oxoacid, an amino acid is formed (b). The NH_2 group is temporarily taken over by enzyme-bound **pyridoxal phosphate** (PLP; see p. 106), which thus becomes pyridoxamine phosphate.

If the NH_2 is released as ammonia, the process is referred to as **deamination**. There are different mechanisms for this (see p. 180). A particularly important one is **oxidative deamination** (**2**). In this reaction, the α-amino group is initially *oxidized* into an imino group (2a), and the reducing equivalents are transferred to NAD^+ or $NADP^+$. In the second step, the imino group is then cleaved by *hydrolysis*. As in transamination, this produces a 2-oxoacid (**C**). Oxidative deamination mainly takes place in the liver, where glutamate is broken down in this way into 2-oxoglutarate and ammonia, catalyzed by *glutamate dehydrogenase*. The reverse reaction initiates biosynthesis of the amino acids in the glutamate family (see p. 184).

B. Mechanism of transamination ○

In the absence of substrates, the aldehyde group of pyridoxal phosphate is covalently bound to a lysine residue of the transaminase (**1**). This type of compound is known as an **aldimine** or "Schiff's base." During the reaction, amino acid 1 (**A, 1a**) displaces the lysine residue, and a new aldimine is formed (**2**). The double bond is then shifted by isomerization. The ketimine (**3**) is hydrolyzed to yield the 2-oxoacid and **pyridoxamine phosphate** (**4**).

In the second part of the reaction (see **A**, 1b), these steps take place *in the opposite direction:* pyridoxamine phosphate and the second 2-oxoacid form a ketimine, which is isomerized into aldimine. Finally, the second amino acid is cleaved and the coenzyme is regenerated.

C. NH_3 metabolism in the liver ◑

In addition to urea synthesis itself (see p. 182), the precursors NH_3 and aspartate are also mainly formed in the liver. Amino nitrogen arising in tissue is transported to the liver by the blood, mainly in the form of **glutamine** (Gln) and **alanine** (Ala; see p. 338). In the liver, Gln is hydrolytically deaminated by *glutaminase* [3] into **glutamate** (Glu) and NH_3. The amino group of the alanine is transferred by *alanine transaminase* [1] to **2-oxoglutarate** (2-OG; formerly known as α-ketoglutarate). This transamination (**A**) produces another glutamate. NH_3 is finally released from glutamate by oxidative deamination (**A**). This reaction is catalyzed by *glutamate dehydrogenase* [4], a typical liver enzyme. **Aspartate** (Asp), the second amino group donor in the urea cycle, also arises from glutamate. The *aspartate transaminase* [2] responsible for this reaction is found with a high level of activity in the liver, as is *alanine transaminase* [1].

Transaminases are also found in other tissues, from which they leak from the cells into the blood when injury occurs. Measurement of serum enzyme activity (**serum enzyme diagnosis**; see also p. 98) is an important method of recognizing and monitoring the course of such injuries. Transaminase activity in the blood is for instance important for diagnosing liver disease (e.g., hepatitis) and myocardial disease (cardiac infarction).

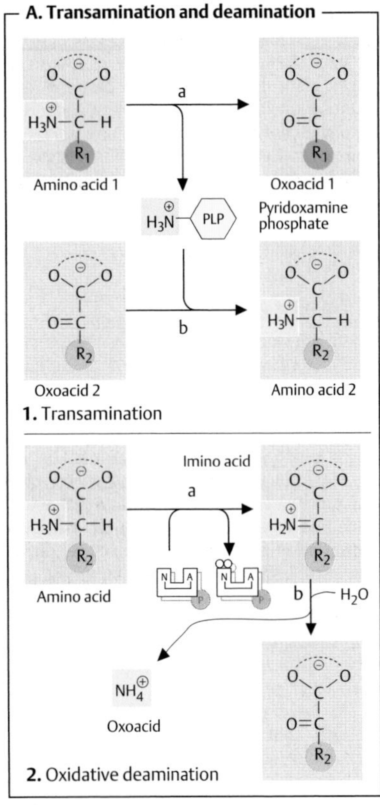
A. Transamination and deamination
a
Amino acid 1
Oxoacid 1
PLP
Pyridoxamine phosphate
b
Oxoacid 2
Amino acid 2
1. Transamination
Imino acid
a
Amino acid
b
H_2O
NH_4^+
Oxoacid
2. Oxidative deamination

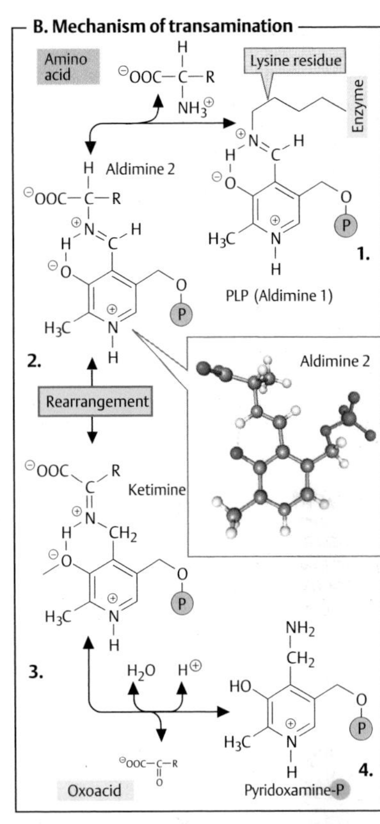
B. Mechanism of transamination
Amino acid
Lysine residue
Enzyme
Aldimine 2
PLP (Aldimine 1)
1.
2.
Rearrangement
Aldimine 2
Ketimine
3.
H_2O
Oxoacid
Pyridoxamine-P
4.

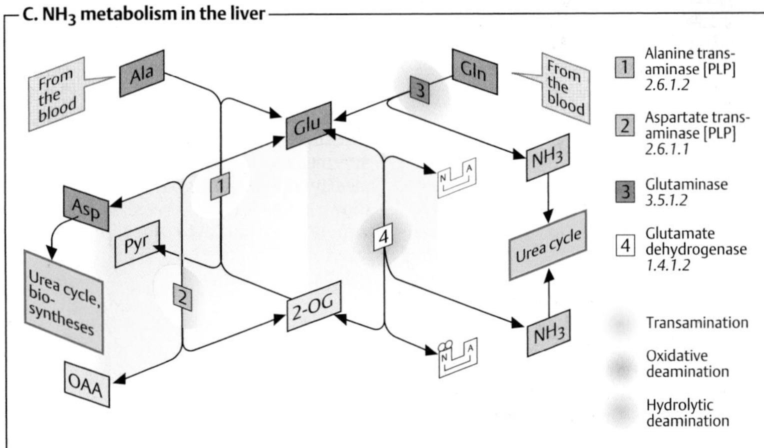
C. NH_3 metabolism in the liver
From the blood
Ala
Gln
From the blood
Glu
NH_3
Asp
Pyr
Urea cycle
Urea cycle, bio-syntheses
2-OG
NH_3
OAA
1 Alanine transaminase [PLP] 2.6.1.2
2 Aspartate transaminase [PLP] 2.6.1.1
3 Glutaminase 3.5.1.2
4 Glutamate dehydrogenase 1.4.1.2
Transamination
Oxidative deamination
Hydrolytic deamination

Amino acid degradation

A large number of metabolic pathways are available for amino acid degradation, and an overview of these is presented here. Further details are given on pp. 414 and 415.

A. Amino acid degradation : overview ◑

During the degradation of most amino acids, the α-amino group is initially removed by **transamination** or **deamination**. Various mechanisms are available for this, and these are discussed in greater detail in **B.** The carbon skeletons that are left over after deamination undergo further degradation in various ways.

During degradation, the 20 proteinogenic amino acids produce only seven different **degradation products** (highlighted in pink and violet). Five of these metabolites (2-oxoglutarate, succinyl CoA, fumarate, oxaloacetate, and pyruvate) are precursors for gluconeogenesis and can therefore be converted into glucose by the liver and kidneys (see p. 154). Amino acids whose degradation supplies one of these five metabolites are therefore referred to as **glucogenic amino acids**. The first four degradation products listed are already intermediates in the tricarboxylic acid cycle, while pyruvate can be converted into oxaloacetate by *pyruvate carboxylase* and thus made available for gluconeogenesis (green arrow).

With two exceptions (lysine and leucine; see below), all of the proteinogenic amino acids are also glucogenic. Quantitatively, they represent the most important precursors for gluconeogenesis. At the same time, they also have an **anaplerotic** effect—i. e., they replenish the tricarboxylic acid cycle in order to feed the anabolic reactions that originate in it (see p. 138).

Two additional degradation products (acetoacetate and acetyl CoA) cannot be channeled into gluconeogenesis in animal metabolism, as there is no means of converting them into precursors of gluconeogenesis. However, they can be used to synthesize ketone bodies, fatty acids, and isoprenoids. Amino acids that supply acetyl CoA or acetoacetate are therefore known as **ketogenic amino acids.** Only leucine and lysine are *purely* ketogenic. Several amino acids yield degradation products that are both *glucogenic and ketogenic*. This group includes phenylalanine, tyrosine, tryptophan, and isoleucine.

Degradation of acetoacetate to acetyl CoA takes place in two steps (not shown). First, acetoacetate and succinyl CoA are converted into acetoacetyl CoA and succinate (enzyme: *3-oxoacid-CoA transferase 2.8.3.5*). Acetoacetyl CoA is then broken down by β-oxidation into two molecules of acetyl CoA (see p. 164), while succinate can be further metabolized via the tricarboxylic acid cycle.

B. Deamination ○

There are various ways of releasing ammonia (NH_3) from amino acids, and these are illustrated here using the example of the amino acids glutamine, glutamate, alanine, and serine.

[1] In the branched-chain amino acids (Val, Leu, Ile) and also tyrosine and ornithine, degradation starts with a **transamination**. For alanine and aspartate, this is actually the only degradation step. The mechanism of transamination is discussed in detail on p. 178.

[2] **Oxidative deamination**, with the formation of $NADH+H^+$, only applies to glutamate in animal metabolism. The reaction mainly takes place in the liver and releases NH_3 for urea formation (see p. 178).

[3] Two amino acids—asparagine and glutamine—contain acid–amide groups in the side chains, from which NH_3 can be released by hydrolysis (**hydrolytic deamination**). In the blood, glutamine is the most important transport molecule for amino nitrogen. Hydrolytic deamination of glutamine in the liver also supplies the urea cycle with NH_3.

[4] **Eliminating deamination** takes place in the degradation of histidine and serine. H_2O is first eliminated here, yielding an unsaturated intermediate. In the case of serine, this intermediate is first rearranged into an imine (not shown), which is hydrolyzed in the second step into NH_3 and pyruvate, with H_2O being taken up. H_2O does not therefore appear in the reaction equation.

A. Amino acid degradation: overview

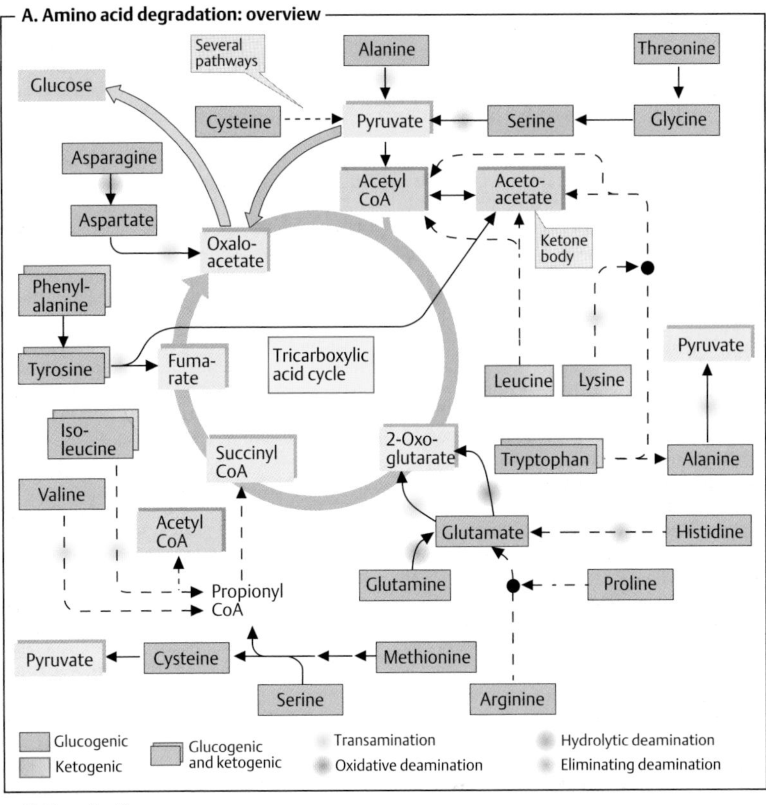

B. Deamination

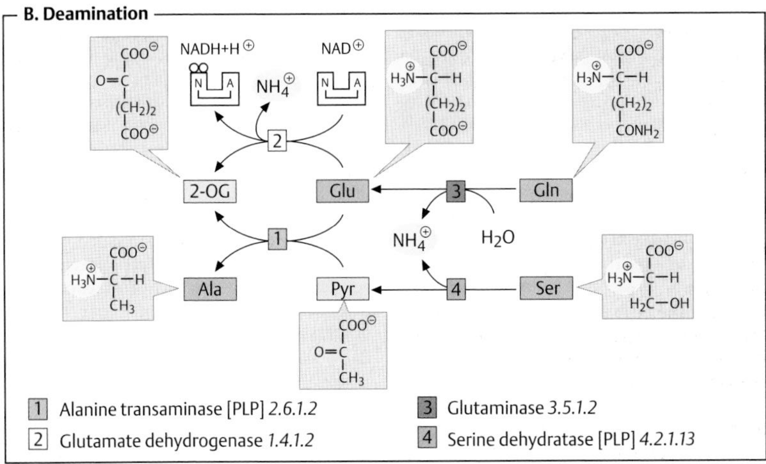

Urea cycle

Amino acids are mainly broken down in the liver. Ammonia is released either directly or indirectly in the process (see p. 178). The degradation of nucleobases also provides significant amounts of ammonia (see p. 186).

Ammonia (NH_3) is a relatively strong **base**, and at physiological pH values it is mainly present in the form of the **ammonium ion** NH_4^+ (see p. 30). NH_3 and NH_4^+ are toxic, and at higher concentrations cause brain damage in particular. Ammonia therefore has to be effectively inactivated and excreted. This can be carried out in various ways. Aquatic animals can excrete NH_4^+ directly. For example, fish excrete NH_4^+ via the gills (*ammonotelic animals*). Terrestrial vertebrates, including humans, hardly excrete any NH_3, and instead, most ammonia is converted into urea before excretion (*ureotelic animals*). Birds and reptiles, by contrast, form *uric acid*, which is mainly excreted as a solid in order to save water (*uricotelic animals*).

The reasons for the neurotoxic effects of ammonia have not yet been explained. It may disturb the metabolism of glutamate and its precursor glutamine in the brain (see p. 356).

A. Urea cycle ◑

Urea ($H_2N–CO–NH_2$) is the diamide of carbonic acid. In contrast to ammonia, it is **neutral** and therefore relatively **non-toxic**. The reason for the lack of basicity is the molecule's mesomeric characteristics. The free electron pairs of the two nitrogen atoms are *delocalized* over the whole structure, and are therefore no longer able to bind protons. As a small, uncharged molecule, urea is able to cross biological membranes easily. In addition, it is easily transported in the blood and excreted in the urine.

Urea is produced **only in the liver**, in a cyclic sequence of reactions (the **urea cycle**) that starts in the mitochondria and continues in the cytoplasm. The two nitrogen atoms are derived from NH_4^+ (the second has previously been incorporated into aspartate; see below). The keto group comes from **hydrogen carbonate** (HCO_3^-), or CO_2 that is in equilibrium with HCO_3^-.

[1] In the first step, **carbamoyl phosphate** is formed in the mitochondria from hydrogen carbonate (HCO_3^-) and NH_4^+, with two ATP molecules being consumed. In this compound, the carbamoyl residue ($–O–CO–NH_2$) is at a high chemical potential. In hepatic mitochondria, enzyme [1] makes up about 20% of the matrix proteins.

[2] In the next step, the carbamoyl residue is transferred to the non-proteinogenic amino acid **ornithine**, converting it into **citrulline**, which is also non-proteinogenic. This is passed into the cytoplasm via a transporter.

[3] The second NH_2 group of the later urea molecule is provided by **aspartate**, which condenses with citrulline into **argininosuccinate**. ATP is cleaved into AMP and diphosphate (PP_i) for this endergonic reaction. To shift the equilibrium of the reaction to the side of the product, diphosphate is removed from the equilibrium by hydrolysis.

[4] Cleavage of fumarate from argininosuccinate leads to the proteinogenic amino acid **arginine**, which is synthesized in this way in animal metabolism.

[5] In the final step, isourea is released from the guanidinium group of the arginine by hydrolysis (not shown), and is immediately rearranged into **urea**. In addition, ornithine is regenerated and returns via the ornithine transporter into the mitochondria, where it becomes available for the cycle once again.

The **fumarate** produced in step [4] is converted via malate to oxaloacetate [6, 7], from which **aspartate** is formed again by transamination [9]. The glutamate required for reaction [9] is derived from the glutamate dehydrogenase reaction [8], which fixes the second NH_4^+ in an organic bond. Reactions [6] and [7] also occur in the tricarboxylic acid cycle. However, in urea formation they take place in the cytoplasm, where the appropriate isoenzymes are available.

The rate of urea formation is mainly controlled by reaction [1]. ***N*-acetyl glutamate**, as an allosteric effector, activates *carbamoyl-phosphate synthase*. In turn, the concentration of acetyl glutamate depends on arginine and ATP levels, as well as other factors.

A. Urea cycle

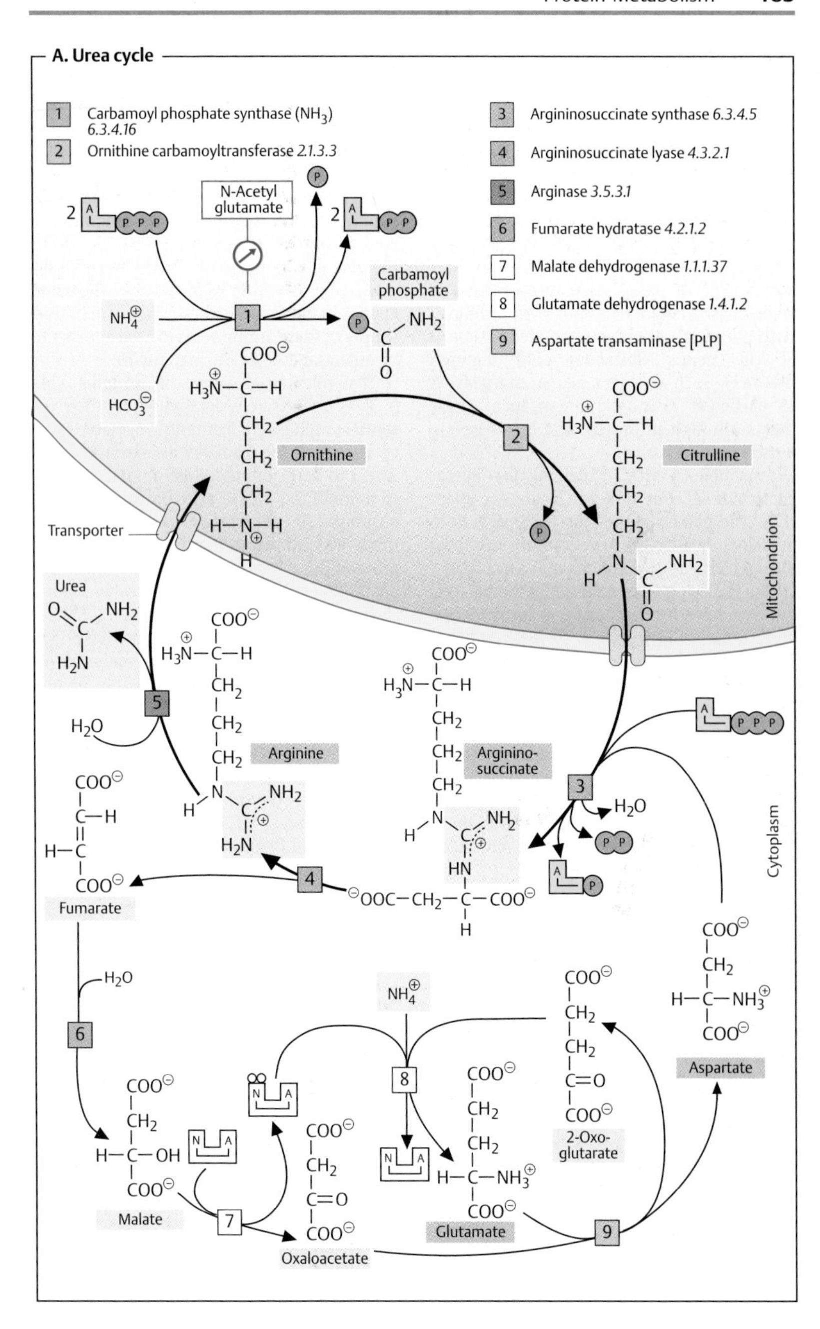
1 Carbamoyl phosphate synthase (NH_3) 6.3.4.16
2 Ornithine carbamoyltransferase 2.1.3.3
3 Argininosuccinate synthase 6.3.4.5
4 Argininosuccinate lyase 4.3.2.1
5 Arginase 3.5.3.1
6 Fumarate hydratase 4.2.1.2
7 Malate dehydrogenase 1.1.1.37
8 Glutamate dehydrogenase 1.4.1.2
9 Aspartate transaminase [PLP]
N-Acetyl glutamate
Carbamoyl phosphate
Ornithine
Citrulline
Transporter
Mitochondrion
Urea
Arginine
Arginino-succinate
Cytoplasm
Fumarate
Malate
Oxaloacetate
Glutamate
2-Oxo-glutarate
Aspartate

Amino acid biosynthesis

A. Symbiotic nitrogen fixation ○

Practically unlimited quantities of elementary nitrogen (N_2) are present in the atmosphere. However, before it can enter the natural nitrogen cycle, it has to be reduced to NH_3 and incorporated into amino acids ("fixed"). Only a few species of bacteria and bluegreen algae are capable of fixing atmospheric nitrogen. These exist freely in the soil, or in **symbiosis** with plants. The symbiosis between bacteria of the genus *Rhizobium* and legumes (*Fabales*)—such as clover, beans, and peas—is of particular economic importance. These plants are high in protein and are therefore nutritionally valuable.

In symbiosis with *Fabales*, bacteria live as *bacteroids* in **root nodules** inside the plant cells. The plant supplies the bacteroids with nutrients, but it also benefits from the fixed nitrogen that the symbionts make available.

The N_2-fixing enzyme used by the bacteria is *nitrogenase*. It consists of two components: an *Fe protein* that contains an $[Fe_4S_4]$ cluster as a redox system (see p. 106), accepts electrons from *ferredoxin*, and donates them to the second component, the *Fe–Mo protein*. This molybdenum-containing protein transfers the electrons to N_2 and thus, via various intermediate steps, produces ammonia (NH_3). Some of the reducing equivalents are transferred in a side-reaction to H^+. In addition to NH_3, hydrogen is therefore always produced as well.

B. Amino acid biosynthesis: overview ◑

The proteinogenic amino acids (see p. 60) can be divided into **five families** in relation to their biosynthesis. The members of each family are derived from common precursors, which are all produced in the tricarboxylic acid cycle or in catabolic carbohydrate metabolism. An overview of the biosynthetic pathways is shown here; further details are given on pp. 412 and 413.

Plants and microorganisms are able to synthesize all of the amino acids from scratch, but during the course of evolution, mammals have lost the ability to synthesize approximately half of the 20 proteinogenic amino acids. These **essential amino acids** therefore have to be supplied in food. For example, animal metabolism is no longer capable of carrying out de-novo synthesis of the **aromatic amino acids** (tyrosine is only non-essential because it can be formed from phenylalanine when there is an adequate supply available). The **branched-chain amino acids** (valine, leucine, isoleucine, and threonine) as well as **methionine** and **lysine**, also belong to the essential amino acids. Histidine and arginine are essential in rats; whether the same applies in humans is still a matter of debate. A supply of these amino acids in food appears to be essential at least during growth.

The nutritional value of proteins (see p. 360) is decisively dependent on their essential amino acid content. Vegetable proteins—e. g., those from cereals—are low in lysine and methionine, while animal proteins contain all the amino acids in balanced proportions. As mentioned earlier, however, there are also plants that provide high-value protein. These include the soy bean, one of the plants that is supplied with NH_3 by symbiotic N_2 fixers (**A**).

Non-essential amino acids are those that arise by transamination from 2-oxoacids in the intermediary metabolism. These belong to the **glutamate family** (Glu, Gln, Pro, Arg, derived from 2-oxoglutarate), the **aspartate family** (only Asp and Asn in this group, derived from oxaloacetate), and **alanine**, which can be formed by transamination from pyruvate. The amino acids in the **serine family** (Ser, Gly, Cys) and **histidine**, which arise from intermediates of glycolysis, can also be synthesized by the human body.

A. Symbiotic nitrogen fixation

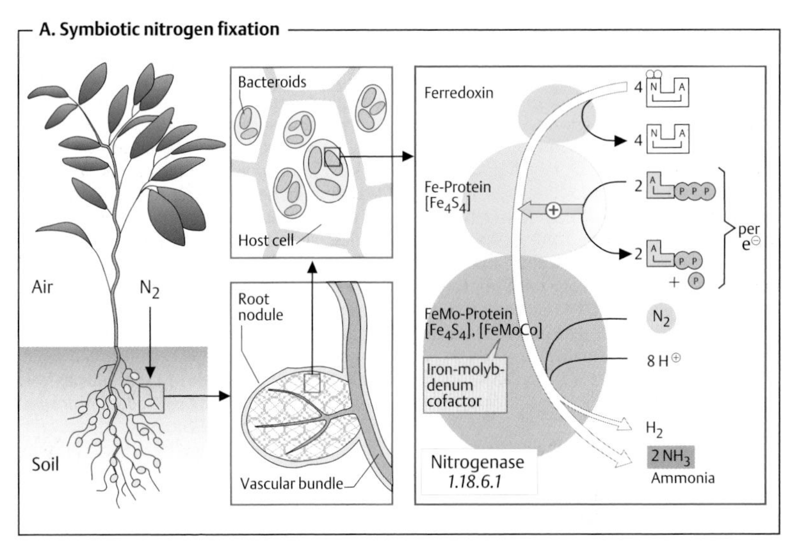

B. Amino acid biosynthesis: overview

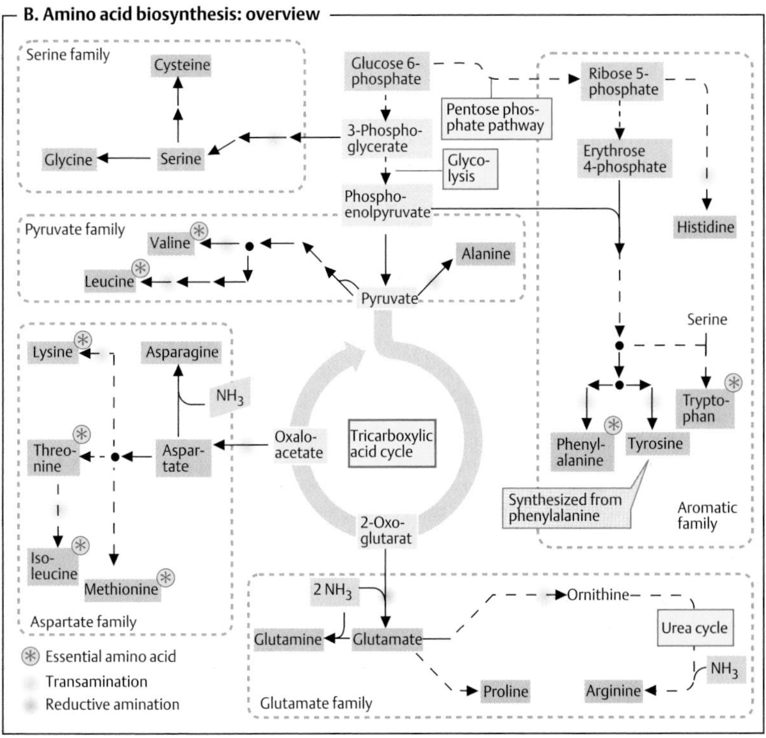

Nucleotide degradation

The nucleotides are among the most complex metabolites. Nucleotide biosynthesis is elaborate and requires a high energy input (see p. 188). Understandably, therefore, bases and nucleotides are not completely degraded, but instead mostly recycled. This is particularly true of the purine bases adenine and guanine. In the animal organism, some 90% of these bases are converted back into nucleoside monophosphates by linkage with phosphoribosyl diphosphate (PRPP) (enzymes [1] and [2]). The proportion of pyrimidine bases that are recycled is much smaller.

A. Degradation of nucleotides ◑

The principles underlying the degradation of purines (**1**) and pyrimidines (**2**) differ. In the human organism, purines are degraded into uric acid and excreted in this form. The purine ring remains intact in this process. In contrast, the ring of the pyrimidine bases (uracil, thymine, and cytosine) is broken down into small fragments, which can be returned to the metabolism or excreted (for further details, see p. 419).

Purine (left). The purine nucleotide **guanosine monophosphate** (**GMP**, **1**) is degraded in two steps—first to the *guanosine* and then to *guanine* (Gua). Guanine is converted by deamination into another purine base, *xanthine*.

In the most important degradative pathway for **adenosine monophosphate** (**AMP**), it is the nucleotide that deaminated, and *inosine monophosphate* (IMP) arises. In the same way as in GMP, the purine base *hypoxanthine* is released from IMP. A single enzyme, *xanthine oxidase* [3], then both converts hypoxanthine into xanthine and xanthine into **uric acid**. An oxo group is introduced into the substrate in each of these reaction steps. The oxo group is derived from *molecular oxygen;* another reaction product is *hydrogen peroxide* (H_2O_2), which is toxic and has to be removed by peroxidases.

Almost all mammals carry out further degradation of uric acid with the help of *uricase*, with further opening of the ring to **allantoin**, which is then excreted. However, the primates, including humans, are not capable of synthesizing allantoin. *Uric acid* is therefore the form of the purines excreted in these species. The same applies to birds and many reptiles. Most other animals continue purine degradation to reach allantoic acid or urea and glyoxylate.

Pyrimidine (right). In the degradation of pyrimidine nucleotides (**2**), the free bases *uracil* (Ura) and *thymine* (Thy) are initially released as important intermediates. Both are further metabolized in similar ways. The pyrimidine ring is first reduced and then hydrolytically cleaved. In the next step, *β-alanine* arises by cleavage of CO_2 and NH_3 as the degradation product of uracil. When there is further degradation, *β-alanine* is broken down to yield acetate, CO_2, and NH_3. Propionate, CO_2, and NH_3 arise in a similar way from *β-aminoisobutyrate,* the degradation product of thymine (see p. 419).

B. Hyperuricemia ○

The fact that purine degradation in humans already stops at the uric acid stage can lead to problems, since—in contrast to allantoin—uric acid is *poorly soluble in water.* When large amounts of uric acid are formed or uric acid processing is disturbed, excessive concentrations of uric acid can develop in the blood (*hyperuricemia*). This can result in the accumulation of uric acid crystals in the body. Deposition of these crystals in the joints can cause very painful attacks of **gout.**

Most cases of hyperuricemia are due to disturbed uric acid excretion via the kidneys (1). A high-purine diet (e. g., meat) may also have unfavorable effects (2). A rare hereditary disease, *Lesch–Nyhan syndrome,* results from a defect in *hypoxanthine phosphoribosyltransferase* (A, enzyme [1]). The impaired recycling of the purine bases caused by this leads to hyperuricemia and severe neurological disorders.

Hyperuricemia can be treated with *allopurinol,* a competitive inhibitor of xanthine oxidase. This substrate analogue differs from the substrate hypoxanthine only in the arrangement of the atoms in the 5-ring.

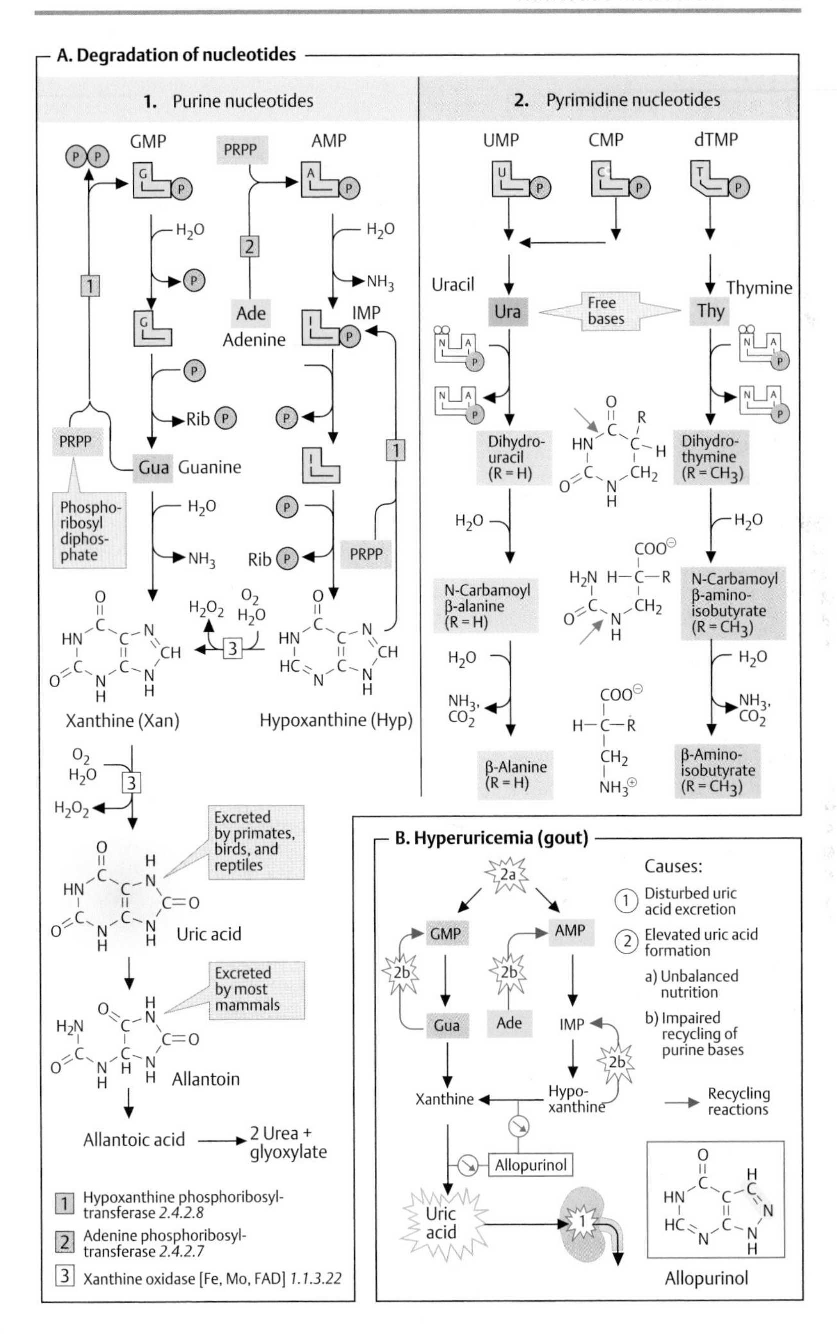
A. Degradation of nucleotides
1. Purine nucleotides
GMP
PRPP
AMP
H2O
NH3
Ade
Adenine
IMP
Rib P
PRPP
Gua Guanine
Phosphoribosyl diphosphate
H2O
NH3
Rib P
PRPP
H2O2
O2
H2O
Xanthine (Xan)
Hypoxanthine (Hyp)
O2
H2O
H2O2
Excreted by primates, birds, and reptiles
Uric acid
Excreted by most mammals
Allantoin
Allantoic acid
2 Urea + glyoxylate
1 Hypoxanthine phosphoribosyltransferase 2.4.2.8
2 Adenine phosphoribosyltransferase 2.4.2.7
3 Xanthine oxidase [Fe, Mo, FAD] 1.1.3.22
2. Pyrimidine nucleotides
UMP
CMP
dTMP
Uracil
Ura
Free bases
Thy
Thymine
Dihydrouracil (R = H)
Dihydrothymine (R = CH3)
H2O
N-Carbamoyl β-alanine (R = H)
N-Carbamoyl β-aminoisobutyrate (R = CH3)
H2O
NH3, CO2
β-Alanine (R = H)
β-Aminoisobutyrate (R = CH3)
B. Hyperuricemia (gout)
Causes:
1 Disturbed uric acid excretion
2 Elevated uric acid formation
a) Unbalanced nutrition
b) Impaired recycling of purine bases
GMP
AMP
Gua
Ade
IMP
Xanthine
Hypoxanthine
Recycling reactions
Allopurinol
Uric acid
Allopurinol

Purine and pyrimidine biosynthesis

The bases occurring in nucleic acids are derivatives of the aromatic heterocyclic compounds *purine* and *pyrimidine* (see p. 80). The biosynthesis of these molecules is complex, but is vital for almost all cells. The synthesis of the nucleobases is illustrated here schematically. Complete reaction schemes are given on pp. 417 and 418.

A. Components of nucleobases ○

The **pyrimidine ring** is made up of three components: the nitrogen atom N-1 and carbons C-4 to C-6 are derived from *aspartate,* carbon C-2 comes from HCO_3^-, and the second nitrogen (N-3) is taken from the amide group of *glutamine*.

The synthesis of the **purine ring** is more complex. The only major component is *glycine*, which donates C-4 and C-5, as well as N-7. All of the other atoms in the ring are incorporated individually. C-6 comes from HCO_3^-. Amide groups from *glutamine* provide the atoms N-3 and N-9. The amino group donor for the inclusion of N-1 is *aspartate,* which is converted into fumarate in the process, in the same way as in the urea cycle (see p. 182). Finally, the carbon atoms C-2 and C-8 are derived from formyl groups in N^{10}-formyl-tetrahydrofolate (see p. 108).

B. Pyrimidine and purine synthesis ○

The major intermediates in the biosynthesis of nucleic acid components are the mononucleotides *uridine monophosphate* (UMP) in the pyrimidine series and *inosine monophosphate* (IMP, base: hypoxanthine) in the purines. The synthetic pathways for pyrimidines and purines are fundamentally different. For the pyrimidines, the pyrimidine ring is first constructed and then linked to ribose 5′-phosphate to form a nucleotide. By contrast, synthesis of the purines starts directly from ribose 5′-phosphate. The ring is then built up step by step on this carrier molecule.

The precursors for the synthesis of the pyrimidine ring are **carbamoyl phosphate**, which arises from glutamate and HCO_3^- (**1a**) and the amino acid **aspartate.** These two components are linked to *N*-**carbamoyl aspartate** (**1b**) and then converted into **dihydroorotate** by closure of the ring (**1c**). In mammals, steps 1a to 1c take place in the cytoplasm, and are catalyzed by a single multifunctional enzyme. In the next step (**1d**), dihydroorotate is oxidized to **orotate** by an FMN-dependent dehydrogenase. Orotate is then linked with **phosphoribosyl diphosphate** (PRPP) to form the nucleotide **orotidine 5′-monophosphate (OMP).** Finally, decarboxylation yields **uridine 5′-monophosphate (UMP).**

Purine biosynthesis starts with PRPP (the names of the individual intermediates are given on p. 417). Formation of the ring starts with transfer of an amino group, from which the later N-9 is derived (**2a**). Glycine and a formyl group from N^{10}-formyl-THF then supply the remaining atoms of the five-membered ring (**2b**, **2c**). Before the five-membered ring is closed (in step **2f**), atoms N-3 and C-6 of the later six-membered ring are attached (**2d**, **2e**). Synthesis of the ring then continues with N-1 and C-2 (**2g**, **2i**). In the final step (**2j**), the six-membered ring is closed, and **inosine 5′-monophosphate** arises. However, the IMP formed does not accumulate, but is rapidly converted into AMP and GMP. These reactions and the synthesis of the other nucleotides are discussed on p. 190.

Further information

The regulation of bacterial *aspartate carbamoyltransferase* by ATP and CTP has been particularly well studied, and is discussed on p. 116. In animals, in contrast to prokaryotes, it is not ACTase but *carbamoyl-phosphate synthase* that is the key enzyme in pyrimidine synthesis. It is activated by ATP and PRPP and inhibited by UTP.

The biosynthesis of the purines is also regulated by *feedback inhibition.* ADP and GDP inhibit the formation of PRRPP from ribose-5′-phosphate. Similarly, step **2a** is inhibited by AMP and GMP.

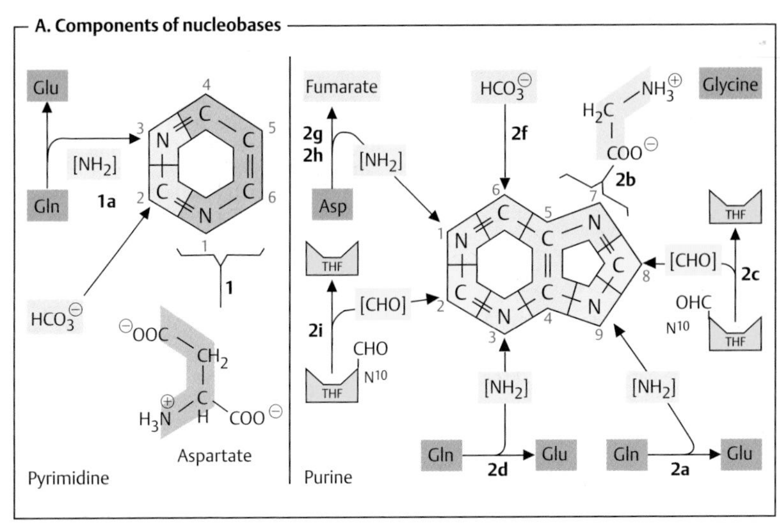
A. Components of nucleobases
Glu
Gln
[NH2]
1a
HCO3⊖
Aspartate
Pyrimidine
Fumarate
2g
2h
[NH2]
Asp
THF
[CHO]
2i
CHO
N10
HCO3⊖
2f
NH3⊕
Glycine
COO⊖
2b
[CHO]
2c
OHC
[NH2]
Gln
2d
Glu
Gln
2a
Glu
Purine

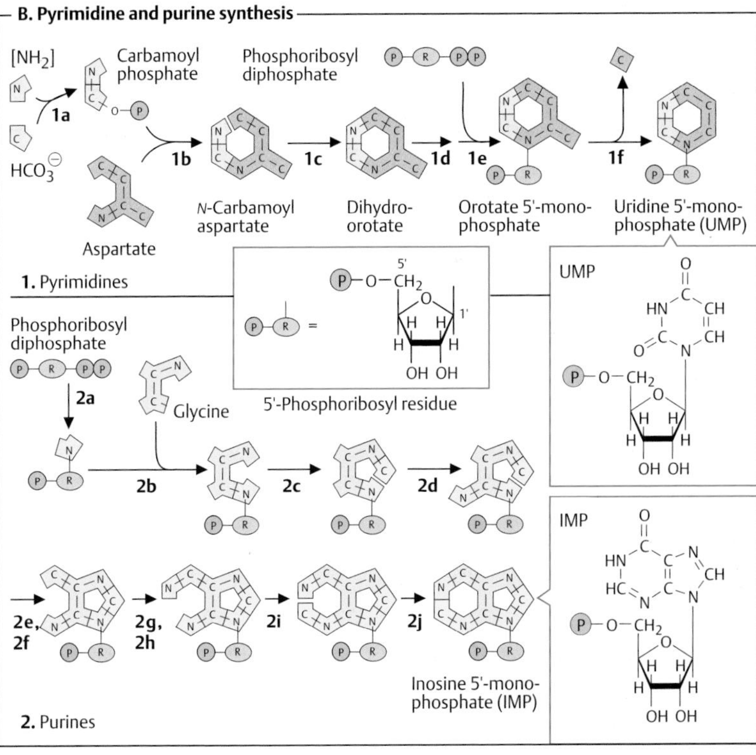
B. Pyrimidine and purine synthesis
[NH2]
Carbamoyl phosphate
Phosphoribosyl diphosphate
1a
HCO3⊖
1b
1c
1d
1e
1f
Aspartate
N-Carbamoyl aspartate
Dihydro-orotate
Orotate 5'-mono-phosphate
Uridine 5'-mono-phosphate (UMP)
1. Pyrimidines
5'-Phosphoribosyl residue
UMP
Phosphoribosyl diphosphate
2a
Glycine
2b
2c
2d
2e, 2f
2g, 2h
2i
2j
Inosine 5'-mono-phosphate (IMP)
IMP
2. Purines

Nucleotide biosynthesis

De novo synthesis of purines and pyrimidines yields the monophosphates IMP and UMP, respectively (see p. 188). All other nucleotides and deoxynucleotides are synthesized from these two precursors. An overview of the pathways involved is presented here; further details are given on p. 417. Nucleotide synthesis by recycling of bases (the salvage pathway) is discussed on p. 186.

A. Nucleotide synthesis: overview ◑

The synthesis of **purine nucleotides** (**1**) starts from **IMP**. The base it contains, *hypoxanthine*, is converted in two steps each into adenine or guanine. The nucleoside monophosphates **AMP** and **GMP** that are formed are then phosphorylated by *nucleoside phosphate kinases* to yield the diphosphates **ADP** and **GDP**, and these are finally phosphorylated into the triphosphates **ATP and GTP**. The nucleoside triphosphates serve as components for RNA, or function as coenzymes (see p. 106). Conversion of the ribonucleotides into deoxyribonucleotides occurs at the level of the *diphosphates* and is catalyzed by *nucleoside diphosphate reductase* (**B**).

The biosynthetic pathways for the **pyrimidine nucleotides** (**2**) are more complicated. The first product, **UMP**, is phosphorylated first to the diphosphate and then to the triphosphate, **UTP**. *CTP synthase* then converts UTP into **CTP**. Since pyrimidine nucleotides are also reduced to deoxyribonucleotides at the diphosphate level, CTP first has to be hydrolyzed by a *phosphatase* to yield **CDP** before **dCDP and dCTP** can be produced.

The DNA component deoxythymidine triphosphate (**dTTP**) is synthesized from UDP in several steps. The base thymine, which only occurs in DNA (see p. 80), is formed by methylation of **dUMP** at the nucleoside monophosphate level. *Thymidylate synthase* and its helper enzyme *dihydrofolate reductase* are important target enzymes for cytostatic drugs (see p. 402).

B. Ribonucleotide reduction ○

2′-Deoxyribose, a component of DNA, is not synthesized as a free sugar, but arises at the diphosphate level by reduction of ribonucleoside diphosphates. This reduction is a complex process in which several proteins are involved. The reducing equivalents needed come from **NADPH+H$^+$**. However, they are not transferred directly from the coenzyme to the substrate, but first pass through a *redox series* that has several steps (**1**).

In the first step, *thioredoxin reductase* reduces a small redox protein, **thioredoxin**, via enzyme-bound FAD. This involves cleavage of a disulfide bond in thioredoxin. The resulting SH groups in turn reduce a catalytically active disulfide bond in *nucleoside diphosphate reductase* ("ribonucleotide reductase"). The free SH groups formed in this way are the actual electron donors for the reduction of ribonucleotide diphosphates.

In eukaryotes, ribonucleotide reductase is a tetramer consisting of two R1 and two R2 subunits. In addition to the **disulfide bond** mentioned, a **tyrosine radical** in the enzyme also participates in the reaction (**2**). It initially produces a substrate radical (**3**). This cleaves a water molecule and thereby becomes radical cation. Finally, the deoxyribose residue is produced by reduction, and the tyrosine radical is regenerated.

The regulation of ribonucleotide reductase is complex. The substrate-specificity and activity of the enzyme are controlled by two allosteric binding sites (a and b) in the R1 subunits. ATP and dATP increase or reduce the activity of the reductase by binding at site a. Other nucleotides interact with site b, and thereby alter the enzyme's specificity.

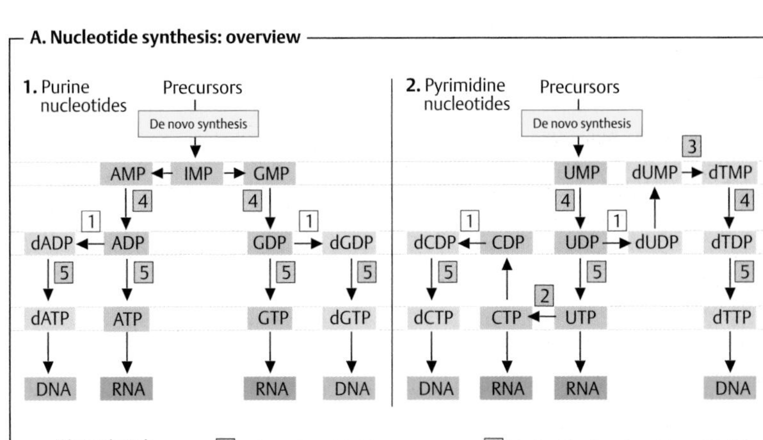
A. Nucleotide synthesis: overview
1. Purine nucleotides
Precursors
De novo synthesis
AMP
IMP
GMP
dADP
ADP
GDP
dGDP
dATP
ATP
GTP
dGTP
DNA
RNA
RNA
DNA
2. Pyrimidine nucleotides
Precursors
De novo synthesis
UMP
dUMP
dTMP
dCDP
CDP
UDP
dUDP
dTDP
dCTP
CTP
UTP
dTTP
DNA
RNA
RNA
DNA
1 Ribonucleoside diphosphate reductase 1.17.4.1
2 CTP synthase 6.3.4.2
3 Thymidylate synthase 2.1.1.45
4 Nucleoside phosphate kinase 2.7.4.4
5 Nucleoside diphosphate kinase 2.7.4.6

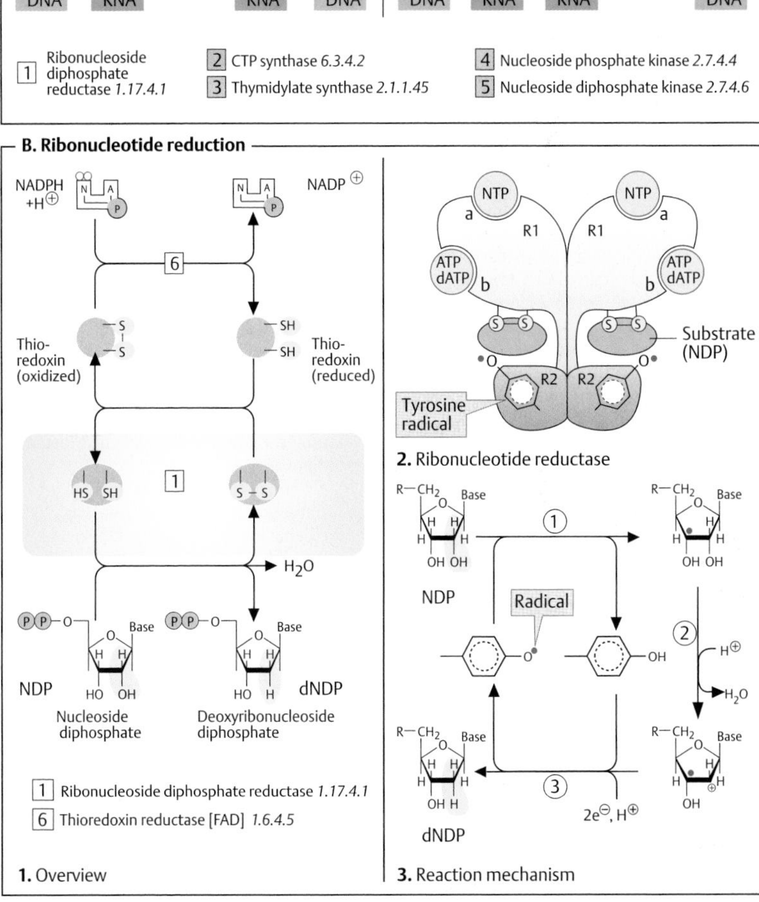
B. Ribonucleotide reduction
NADPH +H⊕
NADP⊕
6
Thio-redoxin (oxidized)
Thio-redoxin (reduced)
HS SH
S – S
1
H2O
NDP
dNDP
Base
Nucleoside diphosphate
Deoxyribonucleoside diphosphate
1 Ribonucleoside diphosphate reductase 1.17.4.1
6 Thioredoxin reductase [FAD] 1.6.4.5
1. Overview
NTP
a
R1
ATP dATP
b
Substrate (NDP)
R2
Tyrosine radical
2. Ribonucleotide reductase
R–CH2
Base
NDP
Radical
H⊕
H2O
2e⊖, H⊕
dNDP
3. Reaction mechanism

Heme biosynthesis

Heme, an iron-containing tetrapyrrole pigment, is a component of O_2-binding proteins (see p. 106) and a coenzyme of various oxidoreductases (see p. 32). Around 85% of heme biosynthesis occurs in the bone marrow, and a much smaller percentage is formed in the liver. Both mitochondria and cytoplasm are involved in heme synthesis.

A. Biosynthesis of heme ○

Synthesis of the tetrapyrrole ring starts in the mitochondria.

[1] **Succinyl CoA** (upper left), an intermediate in the tricarboxylic acid cycle, undergoes condensation with **glycine** and subsequent decarboxylation to yield **5-aminolevulinate** (ALA). The *ALA synthase* responsible for this step is the key enzyme of the whole pathway. Synthesis of ALA synthase is *repressed* and existing enzyme is inhibited by heme, the end product of the pathway. This is a typical example of end-product or *feedback inhibition.*

[2] 5-Aminolevulinate now leaves the mitochondria. In the cytoplasm, two molecules condense to form **porphobilinogen,** a compound that already contains the pyrrole ring. *Porphobilinogen synthase* is inhibited by lead ions. This is why acute lead poisoning is associated with increased concentrations of ALA in the blood and urine.

[3] The tetrapyrrole structure characteristic of the porphyrins is produced in the next steps of the synthetic pathway. *Hydroxymethylbilane synthase* catalyzes the linkage of four porphobilinogen molecules and cleavage of an NH_2 group to yield **uroporphyrinogen III.**

[4] Formation of this intermediate step requires a second enzyme, *uroporphyrinogen III synthase.* If this enzyme is lacking, the "wrong" isomer, uroporphyrinogen I, is formed.

The tetrapyrrole structure of uroporphyrinogen III is still very different from that of heme. For example, the central iron atom is missing, and the ring contains only eight of the 11 double bonds. In addition, the ring system only carries charged R side chains (four acetate and four propionate residues). As heme groups have to act in the apolar interior of proteins, most of the polar side chains have to be converted into less polar groups.

[5] Initially, the four acetate residues (R_1) are decarboxylated into methyl groups. The resulting **coproporphyrinogen III** returns to the mitochondria again. The subsequent steps are catalyzed by enzymes located either on or inside the *inner mitochondrial membrane.*

[6] An *oxidase* first converts two of the propionate groups (R_2) into vinyl residues. The formation of **protoporphyrinogen IX** completes the modification of the side chains.

[7] In the next step, another oxidation produces the conjugated π-electron system of **protoporphyrin IX.**

[8] Finally, a divalent iron is incorporated into the ring. This step also requires a specific enzyme, *ferrochelatase.* The **heme b** or **Fe-protoporphyrin IX** formed in this way is found in hemoglobin and myoglobin, for example (see p. 280), where it is noncovalently bound, and also in various oxidoreductases (see p. 106).

Further information

There are a large number of hereditary or acquired disturbances of porphyrin synthesis, known as **porphyrias**, some of which can cause severe clinical pictures. Several of these diseases lead to the excretion of heme precursors in feces or urine, giving them a dark red color. Accumulation of porphyrins in the skin can also occur, and exposure to light then causes disfiguring, poorly healing blisters. Neurological disturbances are also common in the porphyrias.

It is possible that the medieval legends about human vampires ("Dracula") originated in the behavior of porphyria sufferers (avoidance of light, behavioral disturbances, and drinking of blood in order to obtain heme—which markedly improves some forms of porphyria).

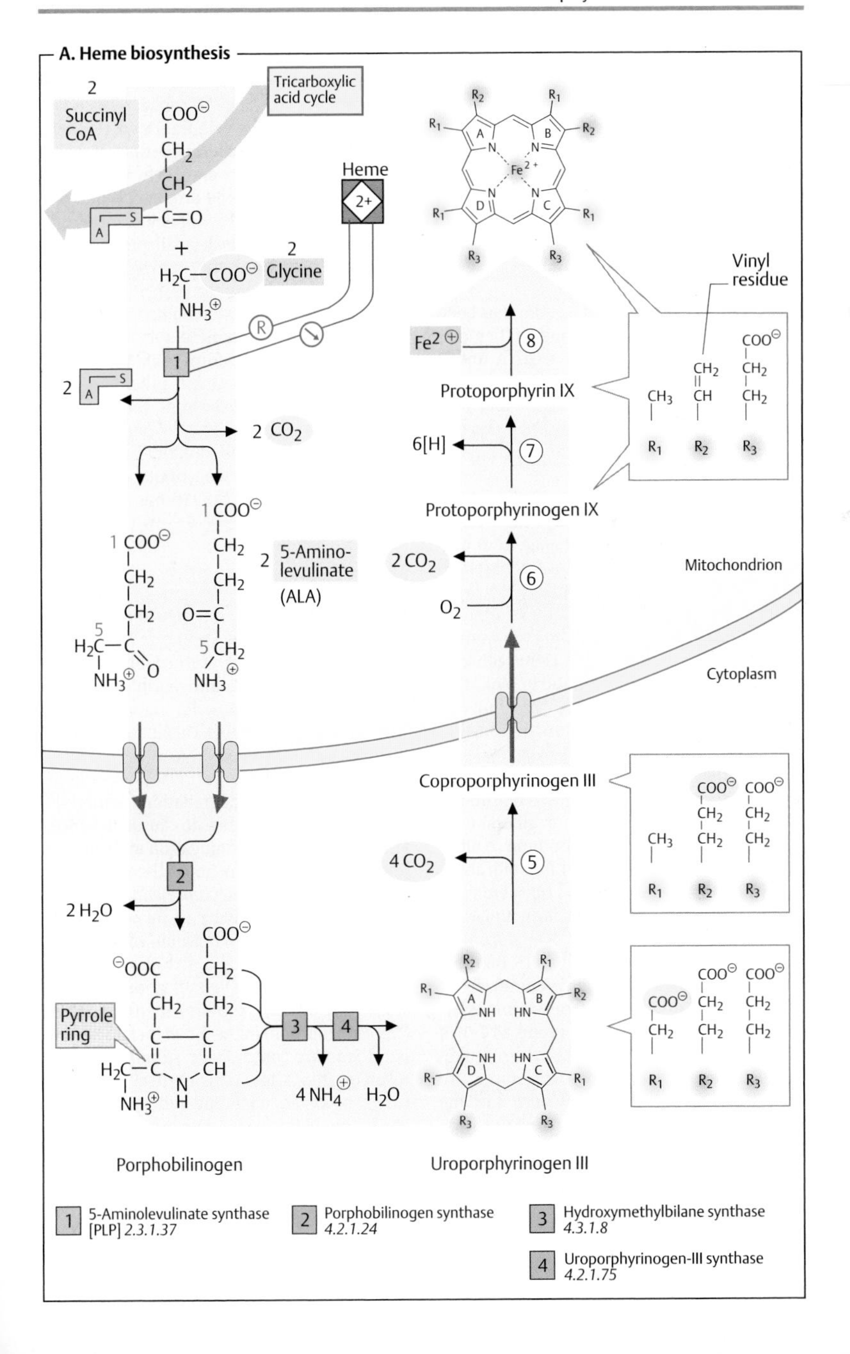
A. Heme biosynthesis
2
Succinyl CoA
Tricarboxylic acid cycle
Heme
2+
2
Glycine
2 CO2
1 COO⊖
5
2
5-Amino-levulinate
(ALA)
2 H2O
Pyrrole ring
Porphobilinogen
4 NH4⊕
H2O
Uroporphyrinogen III
4 CO2
Coproporphyrinogen III
Cytoplasm
Mitochondrion
2 CO2
O2
Protoporphyrinogen IX
6[H]
Protoporphyrin IX
Fe2⊕
Vinyl residue
Fe2+
1 5-Aminolevulinate synthase [PLP] 2.3.1.37
2 Porphobilinogen synthase 4.2.1.24
3 Hydroxymethylbilane synthase 4.3.1.8
4 Uroporphyrinogen-III synthase 4.2.1.75

Heme degradation

A. Degradation of heme groups ○

Heme is mainly found in the human organism as a prosthetic group in erythrocyte hemoglobin. Around 100–200 million aged erythrocytes per hour are broken down in the human organism. The degradation process starts in reticuloendothelial cells in the spleen, liver, and bone marrow.

[1] After the protein part (globin) has been removed, the tetrapyrrole ring of heme is oxidatively cleaved between rings A and B by *heme oxygenase.* This reaction requires molecular oxygen and NADPH+H^+, and produces green **biliverdin**, as well as CO (carbon monoxide) and Fe^{2+}, which remains available for further use (see p. 286).

[2] In another redox reaction, biliverdin is reduced by *biliverdin reductase* to the orange-colored **bilirubin.** The color change from purple to green to yellow can be easily observed in vivo in a bruise or hematoma.

The color of heme and the other *porphyrin systems* (see p. 106) results from their numerous conjugated double bonds. Heme contains a cyclic conjugation (highlighted in pink) that is removed by reaction [1]. Reaction [2] breaks the π system down into two smaller separate systems (highlighted in yellow).

For further degradation, bilirubin is transported to the liver via the blood. As bilirubin is poorly soluble, it is bound to **albumin** for transport. Some drugs that also bind to albumin can lead to an increase in free bilirubin.

[3] The hepatocytes take up bilirubin from the blood and conjugate it in the endoplasmic reticulum with the help of **UDP-glucuronic acid** into the more easily soluble **bilirubin monoglucuronides** and **diglucuronides**. To do this, *UDP-glucuronosyltransferase* forms ester-type bonds between the OH group at C-1 of glucuronic acid and the carboxyl groups in bilirubin (see p. 316). The glucuronides are then excreted by active transport into the **bile**, where they form what are known as the **bile pigments.**

Glucuronide synthesis is the rate-determining step in hepatic bilirubin metabolism. Drugs such as *phenobarbital,* for example, can induce both conjugate formation and the transport process.

Some of the bilirubin conjugates are broken down further in the intestine by bacterial *β-glucuronidases.* The bilirubin released is then reduced further via intermediate steps into colorless **stercobilinogen**, some of which is oxidized again into orange to yellow-colored stercobilin. The end products of bile pigment metabolism in the intestine are mostly excreted in feces, but a small proportion is resorbed *(enterohepatic circulation;* see p. 314). When high levels of heme degradation are taking place, stercobilinogen appears as **urobilinogen** in the urine, where oxidative processes darken it to form **urobilin.**

In addition to hemoglobin, other *heme proteins* (myoglobin, cytochromes, catalases, and peroxidases; see p. 32) also supply heme groups that are degraded via the same pathway. However, these contribute only about 10–15% to a total of ca. 250 mg of bile pigment formed per day.

Further information

Hyperbilirubinemias. An elevated bilirubin level (> 10 mg · L^{-1}) is known as *hyperbilirubinemia.* When this is present, bilirubin diffuses from the blood into peripheral tissue and gives it a yellow color (jaundice). The easiest way of observing this is in the white conjunctiva of the eyes.

Jaundice can have various causes. If increased erythrocyte degradation (hemolysis) produces more bilirubin, it causes *hemolytic jaundice.* If bilirubin conjugation in the liver is impaired—e. g., due to hepatitis or liver cirrhosis—it leads to *hepatocellular jaundice,* which is associated with an increase in unconjugated (*"indirect"*) bilirubin in the blood. By contrast, if there is a disturbance of bile drainage (*obstructive jaundice,* due to gallstones or pancreatic tumors), then conjugated (*"direct"*) bilirubin in the blood increases. *Neonatal jaundice* (physiologic jaundice) usually resolves after a few days by itself. In severe cases, however, unconjugated bilirubin can cross the blood–brain barrier and lead to brain damage (*kernicterus*).

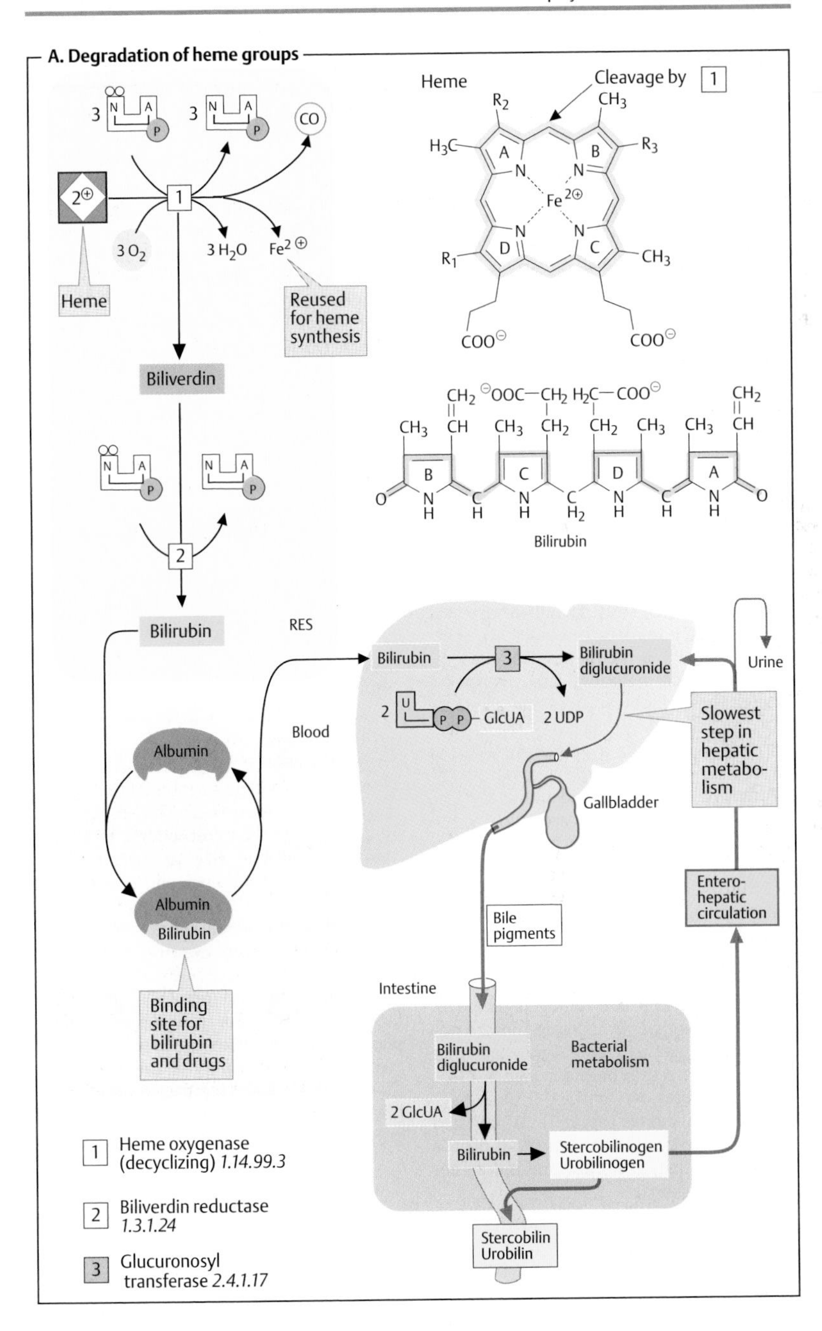
A. Degradation of heme groups
Heme
Cleavage by 1
3
3
CO
1
$2^{\oplus}$
$3\ O_2$
$3\ H_2O$
$Fe^{2\oplus}$
Heme
Reused for heme synthesis
Biliverdin
2
Bilirubin
RES
Bilirubin
3
Bilirubin diglucuronide
2
GlcUA
2 UDP
Urine
Slowest step in hepatic metabolism
Blood
Albumin
Gallbladder
Albumin
Bilirubin
Entero-hepatic circulation
Bile pigments
Binding site for bilirubin and drugs
Intestine
Bilirubin diglucuronide
Bacterial metabolism
2 GlcUA
Bilirubin
Stercobilinogen
Urobilinogen
Stercobilin
Urobilin
1 Heme oxygenase (decyclizing) 1.14.99.3
2 Biliverdin reductase 1.3.1.24
3 Glucuronosyl transferase 2.4.1.17

Structure of cells

A. Comparison of prokaryotes and eukaryotes ●

Present-day living organisms can be divided into two large groups—the prokaryotes and eukaryotes. The **prokaryotes** are represented by bacteria *(eubacteria* and *archaebacteria).* These are almost all small unicellular organisms only a few microns (10^{-6} m) in size. The **eukaryotes** include *fungi, plants,* and *animals* and comprise both unicellular and multicellular organisms. Multicellular eukaryotes are made up of a wide variety of cell types that are specialized for different tasks. Eukaryotic cells are much larger than prokaryotic ones (volume ratio approximately 2000 : 1). The most important distinguishing feature of these cells in comparison with the prokaryotes is the fact that they have a **nucleus** (*karyon* in Greek—hence the term).

In comparison with the prokaryotes, eukaryotic cells have greater specialization and complexity in their structure and functioning. Eukaryotic cells are structured into *compartments* (see below). The metabolism and synthesis of macromolecules are distributed through these reaction spaces and are separately regulated. In prokaryotes, these functions are organized in a simpler fashion and are spatially closely related.

Although the storage and transfer of genetic information function according to the same principle in the prokaryotes and eukaryotes, there are also differences. Eukaryotic DNA consists of very long, linear molecules with a total of 10^7 to more than 10^{10} base pairs (bp), only a small fraction of which are used for genetic information. In eukaryotes, the genes (20000–50000 per genome) are usually interrupted by non-coding regions (*introns*). Eukaryotic DNA is located in the nucleus, where together with histones and other proteins it forms the chromatin (see p. 238).

In prokaryotes, by contrast, DNA is ring-shaped, much shorter (up to $5 \cdot 10^6$ bp), and located in the cytoplasm. Almost all of it is used for information storage, and it does not contain any introns.

B. Structure of an animal cell ●

In the human body alone, there are at least 200 different cell types. The illustration outlines the basic structures of an animal cell in an extremely simplified way. The details given regarding the proportion of the compartments relative to cell volume (highlighted in yellow) and their numbers per cell frequency (blue) refer to mammalian hepatocytes (liver cells). The figures can vary widely from cell type to cell type.

The eukaryotic cell is subdivided by membranes. On the outside, it is enclosed by a **plasma membrane.** Inside the cell, there is a large space containing numerous components in solution—the **cytoplasm**. Additional membranes divide the internal space into *compartments* (confined reaction spaces). Welldefined compartments of this type are known as **organelles.**

The largest organelle is the **nucleus** (see p. 208). It is easily recognized using the light microscope. The **endoplasmic reticulum** (ER), a closed network of shallow sacs and tubules (see pp. 226ff.), is linked with the outer membrane of the nucleus. Another membrane-bound organelle is the **Golgi apparatus** (see p. 228), which resembles a bundle of layered slices. The **endosomes** and **exosomes** are bubble-shaped compartments (vesicles) that are involved in the exchange of substances between the cell and its surroundings. Probably the most important organelles in the cell's metabolism are the **mitochondria,** which are around the same size as bacteria (see pp. 210ff.). The **lysosomes** and **peroxisomes** are small, globular organelles that carry out specific tasks. The whole cell is traversed by a framework of proteins known as the **cytoskeleton** (see pp. 204ff.).

In addition to these organelles, plant cells (see p. 43) also have plastids—eg., **chloroplasts,** in which photosynthesis takes place (see p. 128). In their interior, there is a large, fluid-filled **vacuole**. Like bacteria and fungi, plant cells have a rigid cell wall consisting of polysaccharides and proteins.

A. Comparison of prokaryotes and eukaryotes

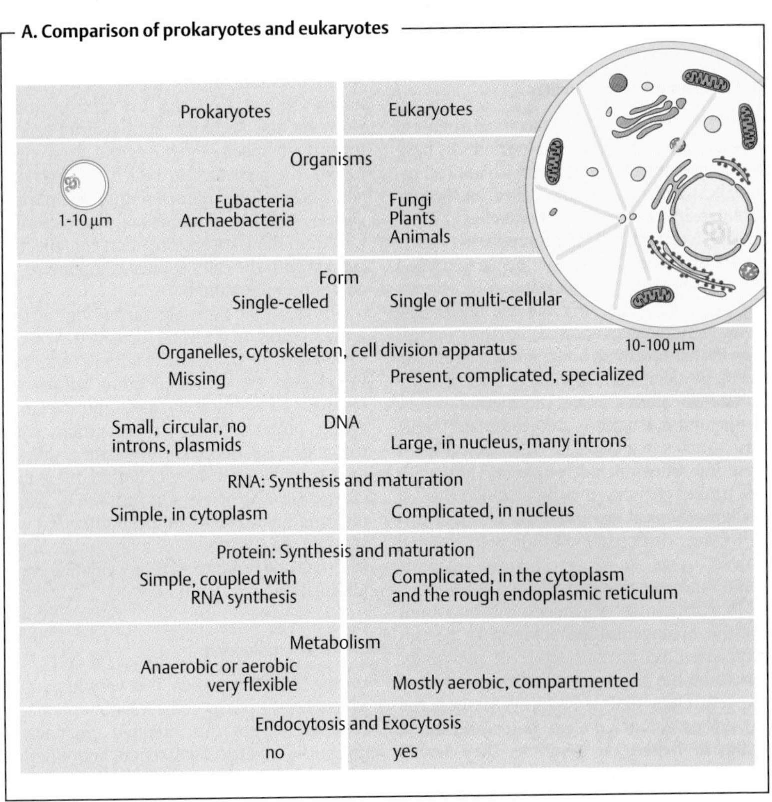

Prokaryotes	Eukaryotes
Organisms	
Eubacteria Archaebacteria	Fungi Plants Animals
Form	
Single-celled	Single or multi-cellular
Organelles, cytoskeleton, cell division apparatus	
Missing	Present, complicated, specialized
DNA	
Small, circular, no introns, plasmids	Large, in nucleus, many introns
RNA: Synthesis and maturation	
Simple, in cytoplasm	Complicated, in nucleus
Protein: Synthesis and maturation	
Simple, coupled with RNA synthesis	Complicated, in the cytoplasm and the rough endoplasmic reticulum
Metabolism	
Anaerobic or aerobic very flexible	Mostly aerobic, compartmented
Endocytosis and Exocytosis	
no	yes

B. Structure of an animal cell

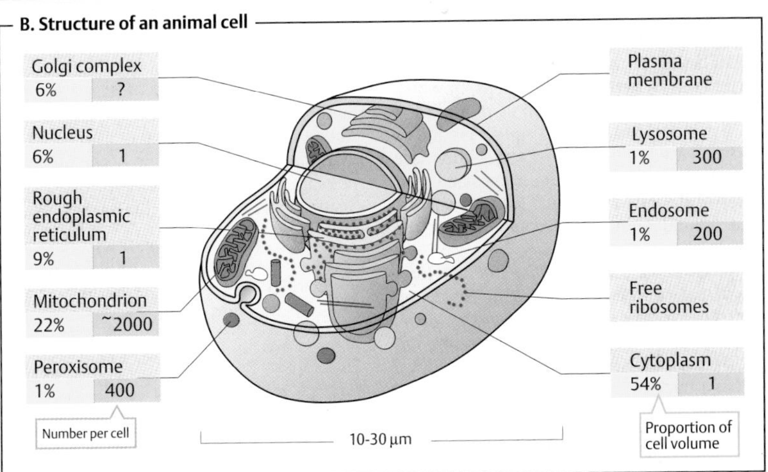

Cell fractionation

A. Isolation of cell organelles ○

To investigate the individual compartments of the cell (see p. 196), various procedures have been developed to enrich and isolate cell organelles. These are mainly based on the size and density of the various organelles.

The isolation of cell components starts with **disruption** of the tissue being examined and subsequent **homogenization** of it (breaking down the cells) in a suitable buffer (see below). Homogenization using the "Potter" (the Potter–Elvehjem homogenizer, a rotating Teflon pestle in a glass cylinder) is particularly suitable for animal tissue. This method is very gentle and is therefore used to isolate fragile structures and molecules. Other cell disruption procedures include **enzymatic lysis** with the help of enzymes that break down the cell wall, **mechanical disruption** by grinding frozen tissue, cutting or smashing with rotating knives, large pressure changes, osmotic shock, and repeated freezing and thawing.

To isolate intact organelles, it is important for the homogenization solution to be *isotonic*—i. e., the osmotic value of the buffer has to be the same as that of the interior of the cell. If hypotonic solutions were used, the organelles would take up water and burst, while in hypertonic solutions they would shrink.

Homogenization is followed by coarse **filtration** through gauze to remove intact cells and connective-tissue fragments. The actual fractionation of cellular components is then carried out by **centrifugation steps**, in which the gravitational force (given as multiples of the earth's gravity, $g = 9.81\ m \cdot s^{-2}$) is gradually increased (*differential centrifugation*; see p. 200). Due to the different shapes and densities of the organelles, this leads to successive sedimentation of each type out of the suspension.

Nuclei already sediment at low accelerations that can be achieved with bench-top centrifuges. Decanting the residue (the "supernatant") and carefully suspending the sediment (or "pellet") in an isotonic medium yields a fraction that is enriched with nuclei. However, this fraction may still contain other cellular components as contaminants—e. g., fragments of the cytoskeleton.

Particles that are smaller and less dense than the nuclei can be obtained by step-by-step acceleration of the gravity on the supernatant left over from the first centrifugation. However, this requires very powerful centrifuges (high-speed centrifuges and ultracentrifuges). The sequence in which the fractions are obtained is: **mitochondria, membrane vesicles**, and **ribosomes**. Finally, the supernatant from the last centrifugation contains the **cytosol** with the cell's soluble components, in addition to the buffer.

The isolation steps are carried out at low temperatures on principle (usually 0–5 °C), to slow down degradation reactions—e. g., due to released enzymes and other influencing factors. The addition of thiols and chelating agents protects functional SH groups from oxidation. Isolated cell organelles quickly lose their biological activity despite these precautions. Nevertheless, it is possible by working carefully to isolate mitochondria that will still take up substrates for a few hours in the test tube and produce ATP via oxidative phosphorylation.

B. Marker molecules ○

During cell fractionation, it is very important to analyze the purity of the fractions obtained. Whether or not the intended organelle is present in a particular fraction, and whether or not the fraction contains other components, can be determined by analyzing characteristic **marker molecules.** These are molecules that occur exclusively or predominantly in one type of organelle. For example, the activity of organelle-specific enzymes **(marker enzymes)** is often assessed. The distribution of marker enzymes in the cell reflects the compartmentation of the processes they catalyze. These reactions are discussed in greater detail here under the specific organelles.

A. Isolation of cell organelles

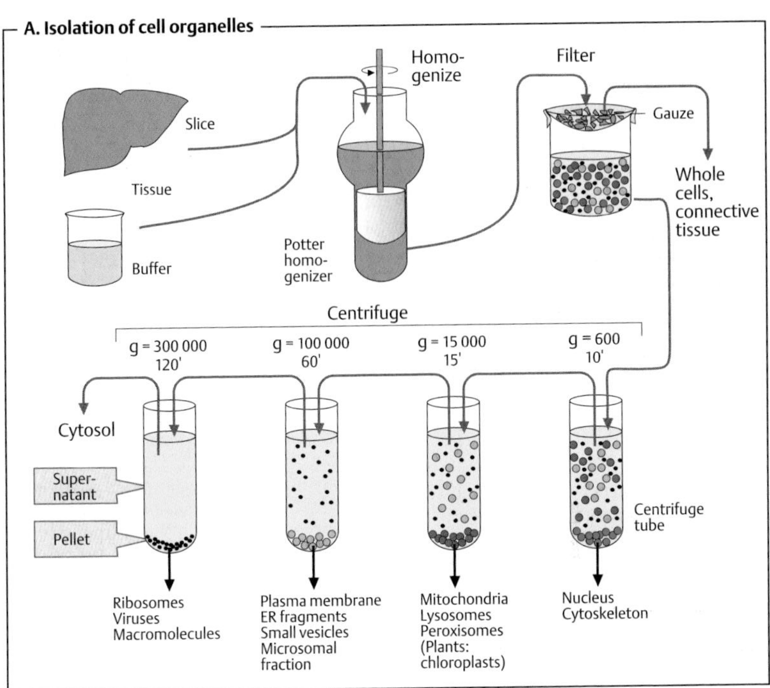

B. Marker molecules

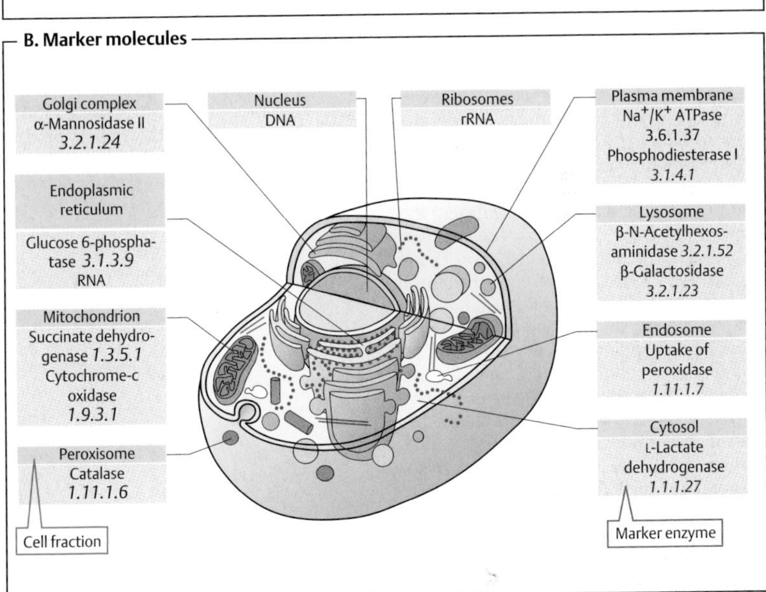

Centrifugation

A. Principles of centrifugation ○

In a solution, particles whose density is higher than that of the solvent sink (*sediment*), and particles that are lighter than it *float* to the top. The greater the difference in density, the faster they move. If there is no difference in density (isopyknic conditions), the particles *hover*. To take advantage of even tiny differences in density to separate various particles in a solution, gravity can be replaced with the much more powerful "centrifugal force" provided by a centrifuge.

Equipment. The acceleration achieved by centrifugation is expressed as a multiple of the earth's gravitational force ($g = 9.81\ m \cdot s^{-2}$). Bench-top centrifuges can reach acceleration values of up to 15000 g, while highspeed refrigerated centrifuges can reach 50000 g and ultracentrifuges, which operate with refrigeration and in a vacuum, can reach 500000 g. Two types of rotor are available in high-powered centrifuges: *fixed angle rotors* and *swingout rotors* that have movable bucket containers. The tubes or buckets used for centrifugation are made of plastic and have to be very precisely adjusted to avoid any imbalances that could lead to accidents.

Theory. The velocity (v) of particle sedimentation during centrifugation depends on the angular velocity ω of the rotor, its effective radius (r_{eff}, the distance from the axis of rotation), and the particle's sedimentation properties. These properties are expressed as the **sedimentation coefficient** S (1 Svedberg, $= 10^{-13}$ s). The sedimentation coefficient depends on the mass M of the particle, its shape (expressed as the coefficient of friction, f), and its density (expressed as the reciprocal density $\bar{v}$, "partial specific volume").

At the top right, the diagram shows the densities and sedimentation coefficients for biomolecules, cell organelles, and viruses. Proteins and protein-rich structures have densities of around $1.3\ g \cdot cm^{-3}$, while nucleic acids show densities of up to $2\ g \cdot cm^{-3}$. Equilibrium sedimentation of nucleic acids therefore requires high-density media—e. g., concentrated solutions of cesium chloride (CsCl). To allow comparison of S values measured in different media, they are usually corrected to values for water at 20 °C ("S_{20W}").

B. Density gradient centrifugation ○

Density gradient centrifugation is used to separate macromolecules that differ only slightly in size or density. Two techniques are commonly used.

In **zonal centrifugation,** the sample being separated (e. g., a cell extract or cells) is placed on top of the centrifugation solution as a thin layer. During centrifugation, the particles move through the solution due to their greater density. The rate of movement basically depends on their molecular mass (see **A**, formulae). Centrifugation stops before the particles reach the bottom of the tube. Drilling a hole into the centrifugation tube and allowing the contents to drip out makes it possible to collect the different particles in separate fractions. During centrifugation, the solution tube is stabilized in the tube by a **density gradient**. This consists of solutions of carbohydrates or colloidal silica gel, the concentration of which increases from the surface of the tube to the bottom. Density gradients prevent the formation of convection currents, which would impair the separation of the particles.

Isopyknic centrifugation, which takes much longer, starts with a CsCl solution in which the sample material (e. g., DNA, RNA, or viruses) is homogeneously distributed. A density gradient only forms *during* centrifugation, as a result of sedimentation and diffusion processes. Each particle moves to the region corresponding to its own *buoyant density*. Centrifugation stops once equilibrium has been reached. The samples are obtained by fractionation, and their concentration is measured using the appropriate methods.

A. Principles of centrifugation

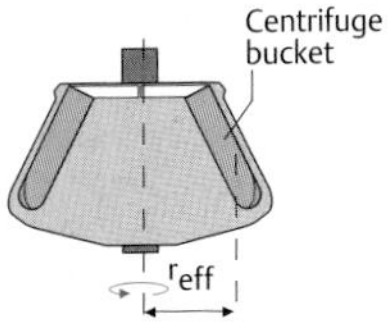

Centrifuge bucket

r_{eff}

Fixed angle rotor

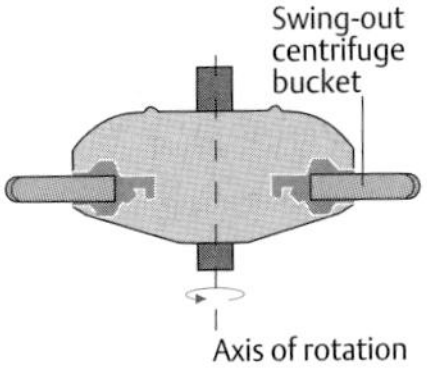

Swing-out centrifuge bucket

Axis of rotation

Swing-out bucket rotor

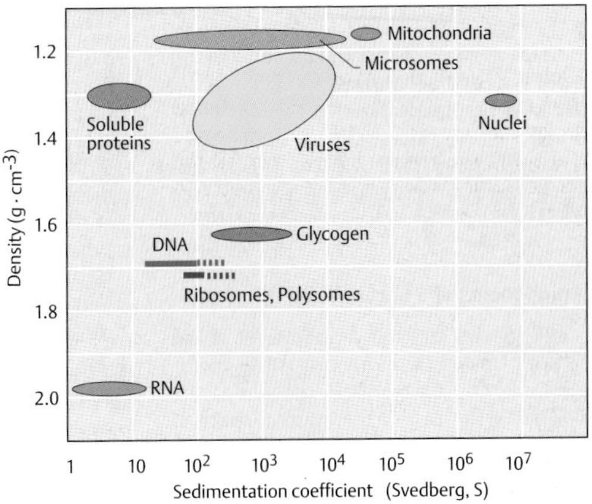

Mitochondria

Microsomes

Soluble proteins

Nuclei

Viruses

Glycogen

DNA

Ribosomes, Polysomes

RNA

Density ($g \cdot cm^{-3}$)

1.2 1.4 1.6 1.8 2.0

1 10 10^2 10^3 10^4 10^5 10^6 10^7

Sedimentation coefficient (Svedberg, S)

g: Gravitational acceleration

v: Sedimentation velocity ($cm \cdot s^{-1}$)

ω: Angular velocity ($rad \cdot s^{-1}$)

r_{eff}: Effective radius (cm)

$$g = \omega^2 \cdot r_{eff}$$

$$v = \omega^2 \cdot r_{eff} \cdot s$$

$$s = \frac{M \cdot (1 - \bar{v} \cdot r)}{f}$$

s: Sedimentation coefficient ($S = 10^{-13}$ s)

M: Molecular mass

$\bar{v}$: Partial specific particle volume ($cm^3 \cdot g^{-1}$)

r: Density of the solution ($g \cdot cm^3$)

f: Coefficient of friction

B. Density gradient centrifugation

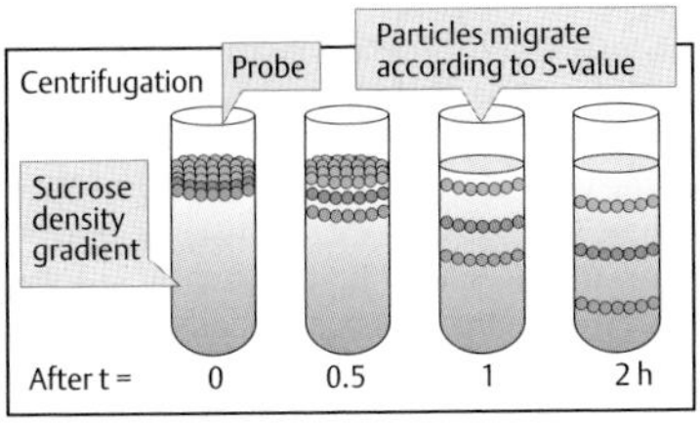

Centrifugation

Probe

Particles migrate according to S-value

Sucrose density gradient

After t = 0 0.5 1 2 h

Zonal centrifugation

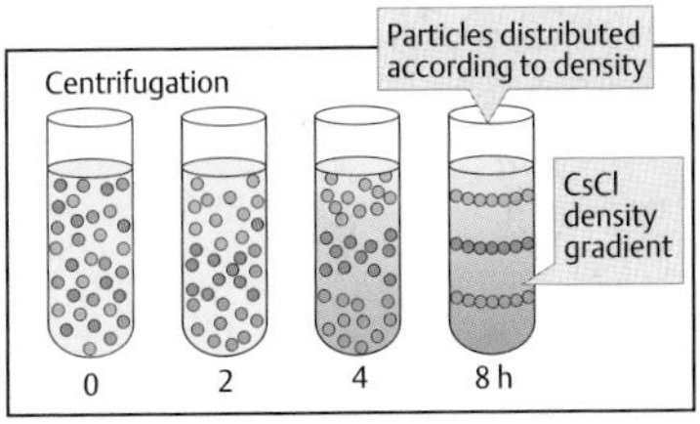

Centrifugation

Particles distributed according to density

CsCl density gradient

0 2 4 8 h

Isopyknic centrifugation

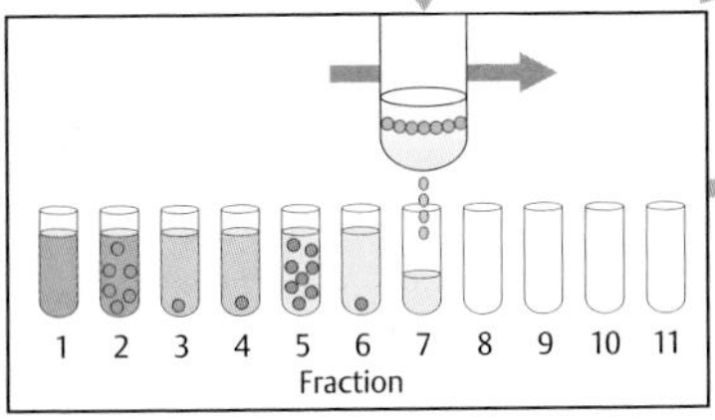

1 2 3 4 5 6 7 8 9 10 11

Fraction

Fractionation

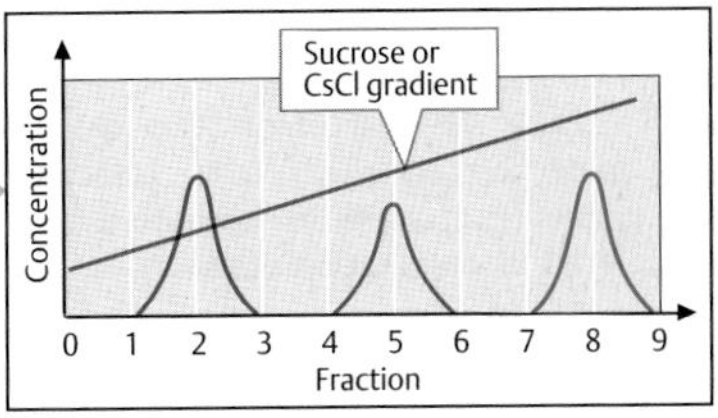

Concentration

Sucrose or CsCl gradient

0 1 2 3 4 5 6 7 8 9

Fraction

Detection

Cell components and cytoplasm

The Gram-negative bacterium *Escherichia coli (E. coli)* is a usually harmless symbiont in the intestine of mammals. The structure and characteristics of this organism have been particularly well characterized. *E. coli* is also frequently used in genetic engineering (see p. 258).

A. Components of a bacterial cell ◑

A single *E. coli* cell has a **volume** of about 0.88 μm^3. One-sixth of this consists of membranes and one-sixth is DNA (known as the "nucleoid"). The rest of the internal space of the cell is known as **cytoplasm** (not "cytosol"; see p. 198).

The main component of *E. coli*—as in all cells—is **water** (70%). The other components are **macromolecules** (proteins, nucleic acids, polysaccharides), **small organic molecules**, and **inorganic ions**. The majority of the macromolecules are proteins, which represent ca. 55% of the dry mass of the cell. When a number of assumptions are made about the distribution and size (average mass 40 kDa) of proteins, it can be estimated that there are approximately 250000 protein molecules in the cytoplasm of an *E. coli* cell. In eukaryotic cells, which are about a thousand times larger, it is estimated that the number of protein molecules is in the order of several billion.

B. Looking inside a bacterial cell ○

The illustration shows a schematic view inside the **cytoplasm** of *E. coli*, magnified approximately one million times. At this magnification, a single carbon atom would be the size of a grain of salt, and an ATP molecule would be as large as a grain of rice. The detail shown is 100 nm long, corresponding to about 1/600th of the volume of a cell in *E. coli*. To make the macromolecules clearer, small molecules such as water, cofactors, and metabolites have all been omitted from the illustration. The section of the cytoplasm shown contains:

- Several hundred **macromolecules**, which are needed for protein biosynthesis—i.e., 30 ribosomes, more than 100 protein factors, 30 aminoacyl–tRNA synthases, 340 tRNA molecules, 2–3 mRNAs (each of which is 10 times the length of the section shown), and six molecules of RNA polymerase.
- About 330 other enzyme molecules, including 130 glycolytic enzymes and 100 enzymes from the tricarboxylic acid cycle.
- 30000 **small organic molecules** with masses of 100–1000 Da—e.g., metabolites of the intermediary metabolism and coenzymes. These are shown at a magnification 10 times higher in the bottom right corner.
- And finally, 50000 **inorganic ions.** The rest consists of water.

The illustration shows that the cytoplasm of cells is a compartment densely packed with macromolecules and smaller organic molecules. The distances between organic molecules are small. They are only separated by a few water molecules.

All of the molecules are in motion. Due to constant collisions, however, they do not advance in a straight path but move in zigzags. Due to their large mass, proteins are particularly slow. However, they do cover an average of 5 nm in 1 ms—a distance approximately equal to their own length. Statistically, a protein is capable of reaching any point in a bacterial cell in less than a second.

C. Biochemical functions of the cytoplasm ◑

In eukaryotes, the cytoplasm, representing slightly more than 50% of the cell volume, is the most important cellular compartment. It is the *central reaction space of the cell.* This is where many important pathways of the intermediary metabolism take place—e.g., glycolysis, the pentose phosphate pathway, the majority of gluconeogenesis, and fatty acid synthesis. Protein biosynthesis (translation; see p. 250) also takes place in the cytoplasm. By contrast, fatty acid degradation, the tricarboxylic acid cycle, and oxidative phosphorylation are located in the mitochondria (see p. 210).

A. Components of a bacterial cell

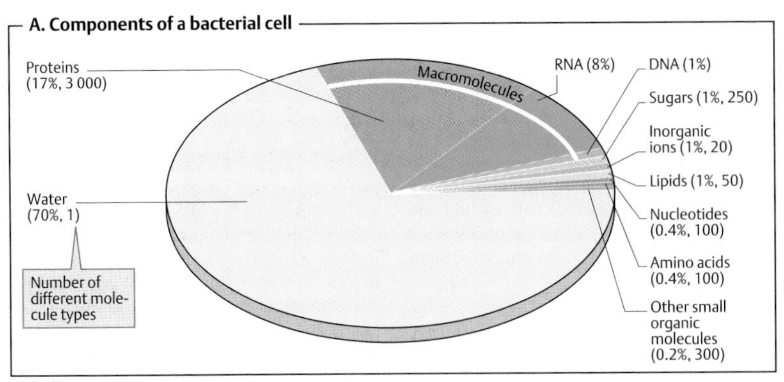

B. View into a bacterial cell

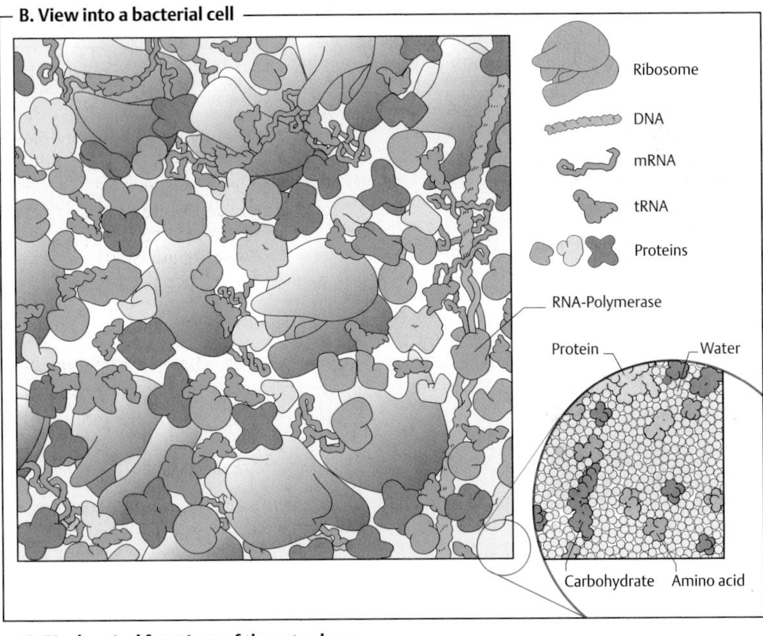

C. Biochemical functions of the cytoplasm

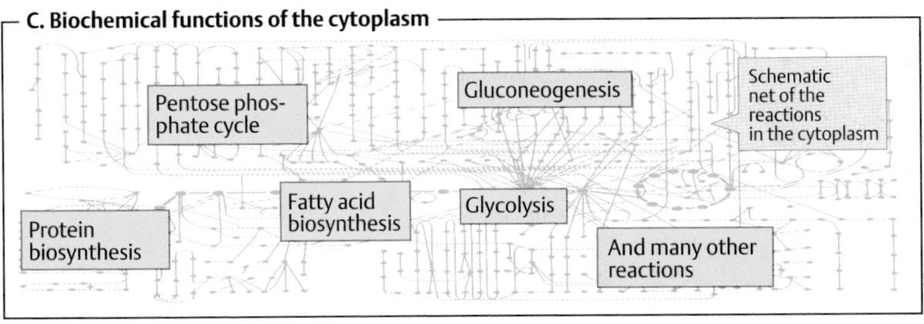

Cytoskeleton: components

The cytoplasm of eukaryotic cells is traversed by three-dimensional scaffolding structures consisting of filaments (long protein fibers), which together form the **cytoskeleton**. These filaments are divided into three groups, based on their *diameters*: **microfilaments** (6–8 nm), **intermediate filaments** (ca. 10 nm), and **microtubules** (ca. 25 nm). All of these filaments are polymers assembled from protein components.

A. Actin ◑

Actin, the most abundant protein in eukaryotic cells, is the protein component of the **microfilaments** (actin filaments). Actin occurs in two forms—a monomolecular form (**G actin**, globular actin) and a polymer (**F actin**, filamentous actin). G actin is an asymmetrical molecule with a mass of 42 kDa, consisting of two domains. As the ionic strength increases, G actin aggregates reversibly to form F actin, a helical homopolymer. G actin carries a firmly bound ATP molecule that is slowly hydrolyzed in F actin to form ADP. Actin therefore also has enzyme properties (*ATPase* activity).

As individual G actin molecules are always oriented in the same direction relative to one another, F actin consequently has *polarity*. It has two different ends, at which polymerization takes place at different rates. If the ends are not stabilized by special proteins (as in muscle cells), then at a critical concentration of G actin the (+) end of F actin will constantly grow, while the (–) end simultaneously decays. These partial processes can be blocked by fungal toxins experimentally. **Phalloidin**, a toxin contained in the *Amanita phalloides* mushroom, inhibits decay by binding to the (–) end. By contrast, **cytochalasins**, mold toxins with cytostatic effects, block polymerization by binding to the (+) end.

Actin-associated proteins. The cytoplasm contains more than 50 different proteins that bind specifically to G actin and F actin. Their actin uptake has various different functions. This type of bonding can serve to regulate the G actin pool (example: *profilin*), influence the polymerization rate of G actin (*villin*), stabilize the chain ends of F actin (*fragin*, *β-actinin*), attach filaments to one another or to other cell components (*villin*, *α-actinin*, *spectrin*), or disrupt the helical structure of F actin (*gelsolin*). The activity of these proteins is regulated by protein kinases via Ca^{2+} and other second messengers (see p. 386).

B. Intermediate filaments ◑

The components of the intermediate filaments belong to five related protein families. They are specific for particular cell types. Typical representatives include the *cytokeratins*, *desmin*, *vimentin*, *glial fibrillary acidic protein* (GFAP), and *neurofilament*. These proteins all have a rod-shaped basic structure in the center, which is known as a *superhelix* ("coiled coil"; see keratin, p. 70). The dimers are arranged in an antiparallel fashion to form tetramers. A staggered head-to-head arrangement produces **protofilaments**. Eight protofilaments ultimately form an intermediary filament.

Free protein monomers of intermediate filaments rarely occur in the cytoplasm, in contrast to microfilaments and microtubules. Their polymerization leads to stable polymers that have no polarity.

C. Tubulins ◑

The basic components of the tube-shaped **microtubules** are α- and **β-tubulin** (53 and 55 kDa). These form α,β-heterodimers, which in turn polymerize to form linear protofilaments. Thirteen protofilaments form a ring-shaped complex, which then grows into a long tube as a result of further polymerization.

Like microfilaments, microtubules are dynamic structures with (+) and (–) ends. The (–) end is usually stabilized by bonding to the centrosome. The (+) end shows *dynamic instability*. It can either grow slowly or shorten rapidly. GTP, which is bound by the microtubules and gradually hydrolyzed into GDP, plays a role in this. Various proteins can also be associated with microtubules.

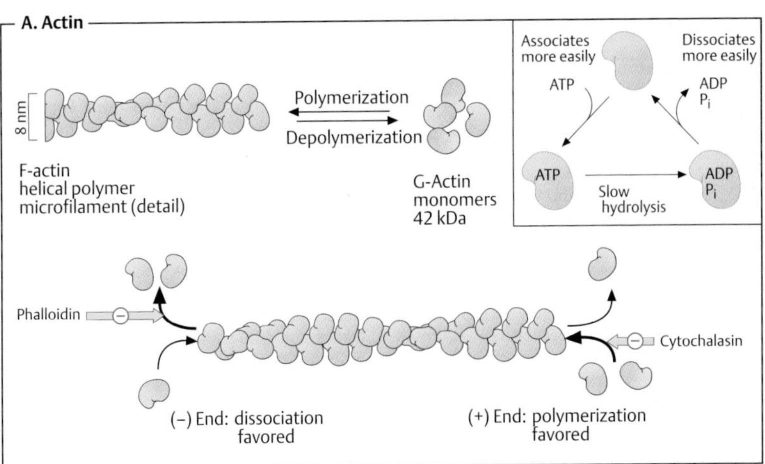
A. Actin
8 nm
Polymerization
Depolymerization
F-actin
helical polymer
microfilament (detail)
G-Actin
monomers
42 kDa
Associates
more easily
Dissociates
more easily
ATP
ADP
Pi
ATP
ADP
Pi
Slow
hydrolysis
Phalloidin
Cytochalasin
(–) End: dissociation
favored
(+) End: polymerization
favored

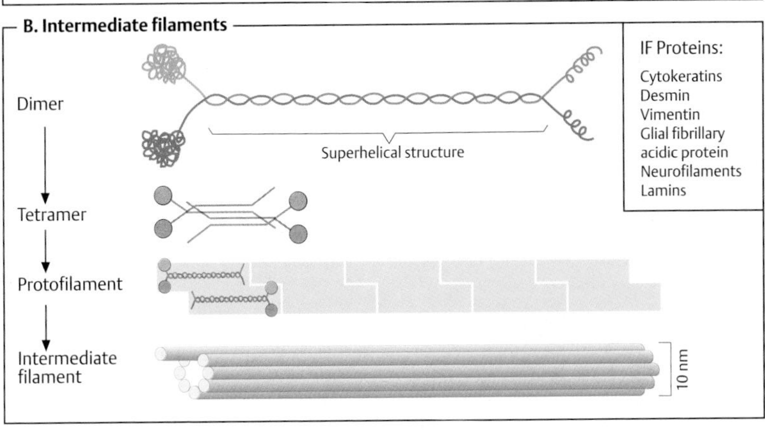
B. Intermediate filaments
Dimer
Superhelical structure
Tetramer
Protofilament
Intermediate
filament
10 nm
IF Proteins:
Cytokeratins
Desmin
Vimentin
Glial fibrillary
acidic protein
Neurofilaments
Lamins

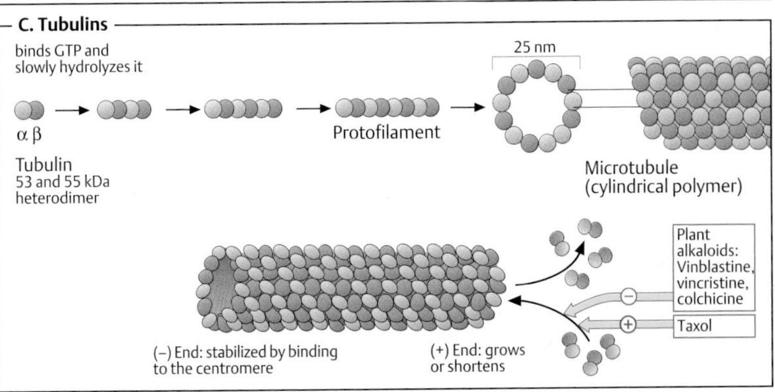
C. Tubulins
binds GTP and
slowly hydrolyzes it
α β
Tubulin
53 and 55 kDa
heterodimer
Protofilament
25 nm
Microtubule
(cylindrical polymer)
Plant
alkaloids:
Vinblastine,
vincristine,
colchicine
Taxol
(–) End: stabilized by binding
to the centromere
(+) End: grows
or shortens

Structure and functions

The cytoskeleton carries out three major tasks:

- It represents the cell's **mechanical scaffolding**, which gives it its typical shape and connects membranes and organelles to each other. This scaffolding has dynamic properties; it is constantly being synthesized and broken down to meet the cell's requirements and changing conditions.
- It acts as the **motor for movement** of animal cells. Not only muscle cells (see p. 332), but also cells of noncontractile tissues contain many different *motor proteins*, which they use to achieve coordinated and directed movement. Cell movement, shape changes during growth, cytoplasmic streaming, and cell division are all made possible by components of the cytoskeleton.
- It serves as a **transport track** within the cell. Organelles and other large protein complexes can move along the filaments with the help of the motor proteins.

A. Microfilaments and intermediate filaments ○

The illustration schematically shows a detail of the **microvilli** of an intestinal epithelial cell as an example of the structure and function of the components of the cytoskeleton (see also **C1**).

Microfilaments of *F actin* traverse the microvilli in ordered bundles. The microfilaments are attached to each other by actin–associated proteins, particularly *fimbrin* and *villin. Calmodulin* and a myosin–like *ATPase* connect the microfilaments laterally to the plasma membrane. *Fodrin,* another microfilament–associated protein, anchors the actin fibers to each other at the base, as well as attaching them to the cytoplasmic membrane and to a network of **intermediate filaments.** In this example, the microfilaments have a mainly static function. In other cases, actin is also involved in dynamic processes. These include muscle contraction (see p. 332), cell movement, phagocytosis by immune cells, the formation of microspikes and lamellipodia (cellular extensions), and the acrosomal process during the fusion of sperm with the egg cell.

B. Microtubules ○

Only the cell's **microtubules** are shown here. They radiate out in all directions from a center near the nucleus, the **centrosome.** The tube-shaped microtubules are constantly being synthesized and broken down at their (+) ends. In the centriole, the (–) end is blocked by associated proteins (see p. 204). The (+) end can also be stabilized by associated proteins—e. g., when the microtubules have reached the cytoplasmic membrane.

The microtubules are involved in defining the shape of the cell and also serve as guiding tracks for the transport of organelles. Together with associated proteins (*dynein, kinesin*), microtubules are able to carry out mechanical work—e. g., during the transport of mitochondria, the movement of cilia (hair-like cell protrusions in the lungs, intestinal epithelium, and oviduct) and the beating of the flagella of sperm. Microtubules also play a special role in the mitotic period of cell division (see p. 394).

C. Architecture ○

The complex structure and net-like density of the cytoskeleton is illustrated here using three examples in which the cytoskeletal components are visualized with the help of antibodies.

1. The border of an *intestinal epithelial cell* is seen here (see also **B**). There are **microfilaments** (**a**) passing from the interior of the cell out into the microvilli. The filaments are firmly held together by **spectrin** (**b**), an associated protein, and they are anchored to **intermediate filaments** (**c**).

2. Only **microtubules** are seen in this *fibroblast cell.* They originate from the microtubule organizing center (centrosome) and radiate out as far as the plasma membrane.

3. **Keratin filaments** are visible here in an *epithelial cell.* Keratin fibers belong to the group of **intermediate filaments** (see pp. 70, 204; **d** = nucleus).

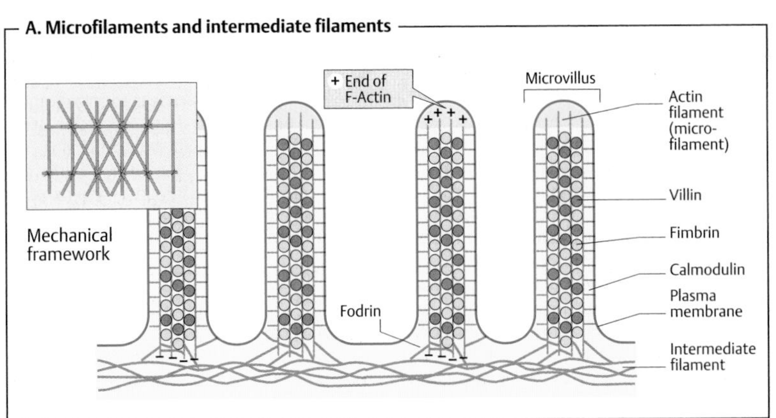

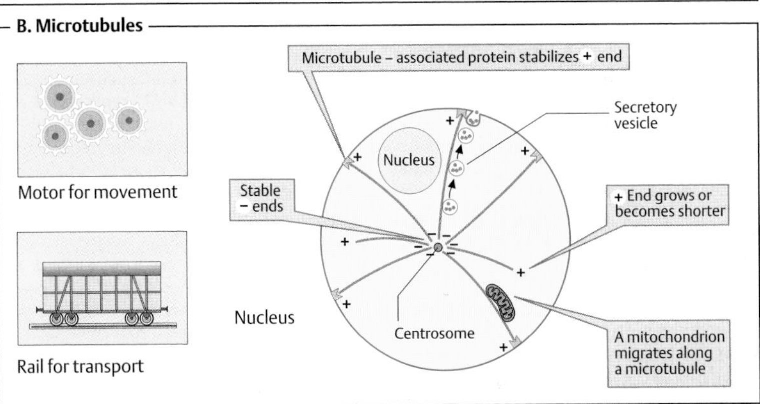

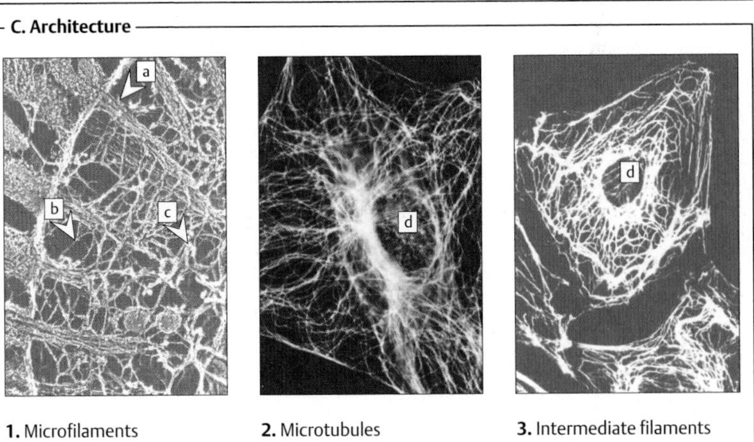

1. Microfilaments **2.** Microtubules **3.** Intermediate filaments

Nucleus

A. Nucleus ●

The nucleus is the largest organelle in the eukaryotic cell. With a diameter of about 10 μm, it is easily recognizable with the light microscope. This is the location for *storage, replication,* and *expression* of genetic information.

The nucleus is separated from the cytoplasm by the **nuclear envelope**, which consists of the **outer** and **inner nuclear membranes**. Each of the two nuclear membranes has two layers, and the membranes are separated from each other by the **perinuclear space**. The outer nuclear membrane is continuous with the rough endoplasmic reticulum and is covered with ribosomes. The inner side of the membrane is covered with a protein layer (the nuclear lamina), in which the nuclear structures are anchored.

The nucleus contains almost all of the cell's **DNA** (around 1% of which is mitochondrial DNA). Together with histones and structural proteins, the nuclear DNA forms the **chromatin** (see p. 238). It is only during cell division that chromatin condenses into *chromosomes,* which are also visible with the light microscope. During this phase, the nuclear membrane temporarily disintegrates.

During the phase between cell divisions, the *interphase,* it is possible to distinguish between the more densely packed **heterochromatin** and loose **euchromatin** using an electron microscope. Active *transcription* of DNA into mRNA takes place in the region of the euchromatin. A particularly electron-dense region is noticeable in many nuclei—the **nucleolus** (several nucleoli are sometimes present). The DNA in the nucleolus contains numerous copies of the genes for rRNAs (see p. 242). They are constantly undergoing transcription, leading to a high local concentration of RNA.

B. Nuclear pores ◑

The exchange of substances between the nucleus and the cytoplasm is mediated by **pore complexes** with complicated structures, which traverse the nuclear membrane. The nuclear pores consist of numerous proteins that form several connected rings of varying diameter. Low-molecular structures and small proteins can enter the nucleus without difficulty. By contrast, larger proteins (over 40 kDa) can only pass through the nuclear pores if they carry a **nuclear localization sequence** consisting of four successive basic amino acids inside their peptide chains (see p. 228). mRNAs and rRNAs formed in the nucleus cross the pores into the cytoplasm as complexes with proteins (see below).

C. Relationships between the nucleus and cytoplasm ◑

Almost all of the RNA in the cell is synthesized in the nucleus. In this process, known as **transcription**, the information stored in DNA is transcribed into RNA (see p. 242). As mentioned above, *ribosomal RNA* (rRNA) is mainly produced in the nucleolus, while *messenger* and *transfer RNA* (mRNA and tRNA) are formed in the region of the euchromatin. Enzymatic duplication of DNA—**replication**—also only takes place in the nucleus (see p. 240).

The nucleotide components required for transcription and replication have to be imported into the nucleus from the cytoplasm. Incorporation of these components into RNA leads to primary products, which are then altered by cleavage, excision of introns, and the addition of extra nucleotides (**RNA maturation**; see p. 242). It is only once these process have been completed that the RNA molecules formed in the nucleus can be exported into the cytoplasm for protein synthesis (**translation**; see p. 250).

The nucleus is not capable of synthesizing proteins. All of the nuclear proteins therefore have to be imported—the *histones* with which DNA is associated in chromatin, and also the so-called *non-histone proteins* (DNA polymerases and RNA polymerases, auxiliary and structural proteins, transcription factors, and ribosomal proteins). Ribosomal RNA (rRNA) already associates with proteins in the nucleolus to form ribosome precursors.

A special metabolic task carried out by the nucleus is **biosynthesis of NAD$^+$**. The immediate precursor of this coenzyme, *nicotinamide mononucleotide* (NMN$^+$), arises in the cytoplasm and is then transported into the nucleolus, where it is enzymatically converted into the dinucleotide NAD$^+$. Finally, NAD$^+$ then returns to the cytoplasm.

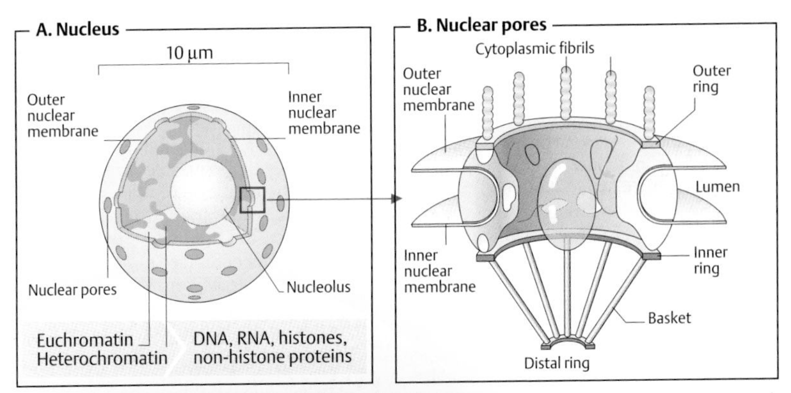
A. Nucleus
10 µm
Outer nuclear membrane
Inner nuclear membrane
Nuclear pores
Nucleolus
Euchromatin
Heterochromatin
DNA, RNA, histones, non-histone proteins
B. Nuclear pores
Cytoplasmic fibrils
Outer nuclear membrane
Outer ring
Lumen
Inner nuclear membrane
Inner ring
Basket
Distal ring

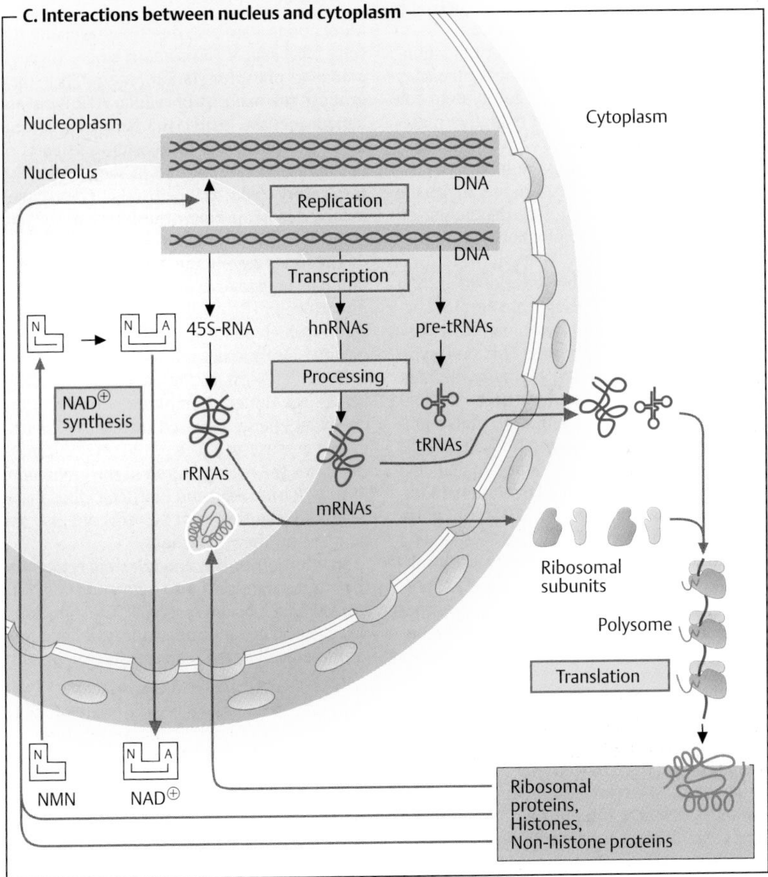
C. Interactions between nucleus and cytoplasm
Nucleoplasm
Nucleolus
Cytoplasm
DNA
Replication
DNA
Transcription
N
N A
45S-RNA
hnRNAs
pre-tRNAs
NAD⊕ synthesis
Processing
rRNAs
tRNAs
mRNAs
Ribosomal subunits
Polysome
Translation
N
N A
NMN
NAD⊕
Ribosomal proteins,
Histones,
Non-histone proteins

Structure and functions

A. Mitochondrial structure

Mitochondria are bacteria-sized organelles (about 1 × 2 μm in size), which are found in large numbers in almost all eukaryotic cells. Typically, there are about 2000 mitochondria per cell, representing around 25% of the cell volume. Mitochondria are enclosed by two membranes—a smooth **outer** membrane and a markedly folded or tubular **inner mitochondrial membrane**, which has a large surface and encloses the **matrix space**. The folds of the inner membrane are known as **cristae**, and tube-like protrusions are called **tubules**. The **intermembrane space** is located between the inner and the outer membranes.

The number and shape of the mitochondria, as well as the numbers of cristae they have, can differ widely from cell type to cell type. Tissues with intensive oxidative metabolism—e. g., heart muscle—have mitochondria with particularly large numbers of cristae. Even within one type of tissue, the shape of the mitochondria can vary depending on their functional status. Mitochondria are mobile, plastic organelles.

Mitochondria probably developed during an early phase of evolution from aerobic bacteria that entered into symbiosis with primeval anaerobic eukaryotes. This **endosymbiont theory** is supported by many findings. For example, mitochondria have a ring-shaped DNA (four molecules per mitochondrion) and have their own ribosomes. The mitochondrial genome became smaller and smaller during the course of evolution. In humans, it still contains 16 569 base pairs, which code for two rRNAs, 22 tRNAs, and 13 proteins. Only these 13 proteins (mostly subunits of respiratory chain complexes) are produced in the mitochondrion. All of the other mitochondrial proteins are coded by the nuclear genome and have to be imported into the mitochondria after translation in the cytoplasm (see p. 228). The mitochondrial envelope consisting of two membranes also supports the endosymbiont theory. The inner membrane, derived from the former symbiont, has a structure reminiscent of prokaryotes. It contains the unusual lipid cardiolipin (see p. 50), but hardly any cholesterol (see p. 216).

Both mitochondrial membranes are very rich in proteins. **Porins** (see p. 214) in the outer membrane allow small molecules (< 10 kDa) to be exchanged between the cytoplasm and the intermembrane space. By contrast, the inner mitochondrial membrane is completely impermeable even to small molecules (with the exception of O_2, CO_2, and H_2O). Numerous **transporters** in the inner membrane ensure the import and export of important metabolites (see p. 212). The inner membrane also transports **respiratory chain** complexes, **ATP synthase**, and other enzymes. The matrix is also rich in enzymes (see **B**).

B. Metabolic functions

Mitochondria are also described as being the cell's *biochemical powerhouse*, since—through **oxidative phosphorylation** (see p. 112)—they produce the majority of cellular ATP. **Pyruvate dehydrogenase** (PDH), the **tricarboxylic acid cycle**, **β-oxidation** of fatty acids, and parts of the **urea cycle** are located in the matrix. The **respiratory chain**, **ATP synthesis**, and enzymes involved in **heme biosynthesis** (see p. 192) are associated with the inner membrane.

The inner membrane itself plays an important part in oxidative phosphorylation. As it is impermeable to protons, the respiratory chain—which pumps protons from the matrix into the intermembrane space via complexes I, III, and IV—establishes a **proton gradient** across the inner membrane, in which the chemical energy released during NADH oxidation is conserved (see p. 126). ATP synthase then uses the energy stored in the gradient to form ATP from ADP and inorganic phosphate. Several of the transport systems are also dependent on the H^+ gradient.

In addition to the endoplasmic reticulum, the mitochondria also function as an intracellular **calcium reservoir**. The mitochondria also play an important role in "programmed cell death"—**apoptosis** (see p. 396).

A. Mitochondrial structure

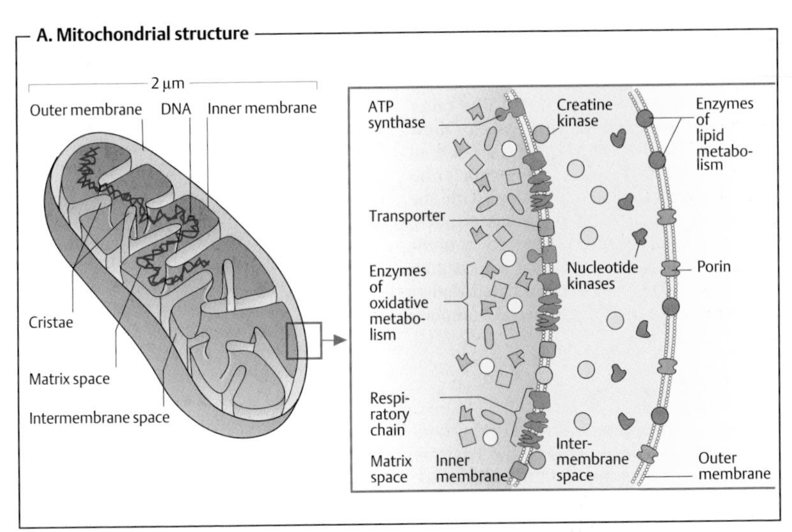

B. Metabolic functions

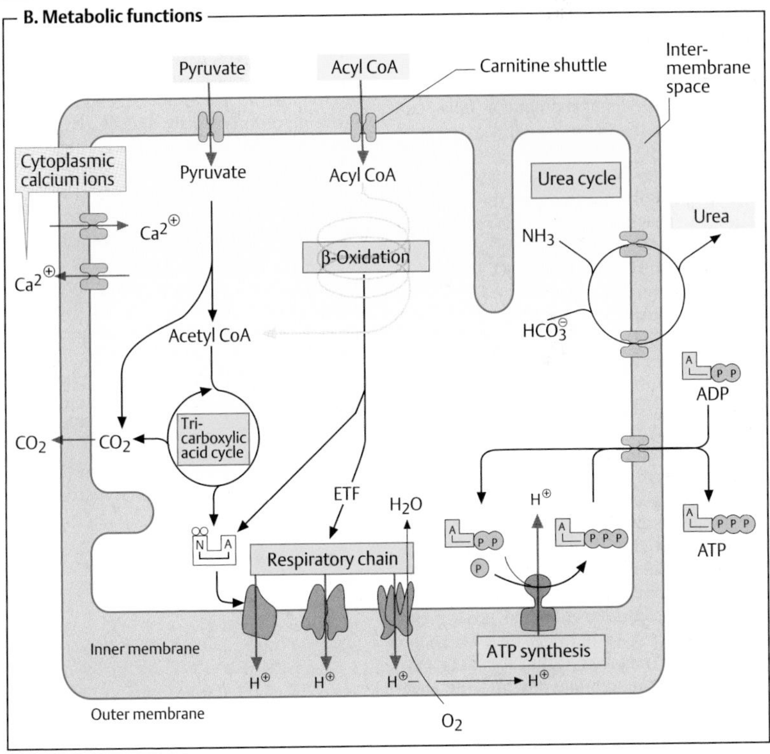

Transport systems

Mitochondria are surrounded by an inner and an outer membrane (see p. 210). The **outer membrane** contains porins, which allow smaller molecules up to 10 kDa in size to pass. By contrast, the **inner membrane** is also impermeable to small molecules (with the exception of water and the gases O_2, CO_2, and NH_3). All of the other substrates of mitochondrial metabolism, as well as its products, therefore have to be moved through the inner membrane with the help of special **transporters**.

A. Transport systems ◑

The transport systems of the inner mitochondrial membrane use various mechanisms. Metabolites or ions can be transported alone (uniport, **U**), together with a second substance (symport, **S**), or in exchange for another molecule (antiport, **A**). Active transport—i. e., transport coupled to ATP hydrolysis—does not play an important role in mitochondria. The driving force is usually the **proton gradient** across the inner membrane (blue star) or the general membrane potential (red star; see p. 126).

The **pyruvate** (left) formed by glycolysis in the cytoplasm is imported into the matrix in antiport with OH^-. The OH^- ions react in the intermembrane space with the H^+ ions abundantly present there to form H_2O. This maintains a concentration gradient of OH^-. The import of **phosphate** ($H_2PO_4^-$) is driven in a similar way. The exchange of the **ATP** formed in the mitochondrion for **ADP** via an *adenine nucleotide translocase* (center) is also dependent on the H^+ gradient. ATP with a quadruple negative charge is exchanged for ADP with a triple negative charge, so that overall one negative charge is transported into the H^+-rich intermembrane space. The import of malate by the *tricarboxylate transporter*, which is important for the malate shuttle (see **B**) is coupled to the export of citrate, with a net export of one negative charge to the exterior again. In the opposite direction, malate can leave the matrix in antiport for phosphate. When Ca^{2+} is imported (right), the metal cation follows the membrane potential. An electroneutral antiport for two H^+ or two Na^+ serves for Ca^{2+} export.

B. Malate and glycerophosphate shuttles ◑

Two systems known as "shuttles" are available to allow the **import of reducing equivalents** that arise from glycolysis in the cytoplasm in the form of NADH+H$^+$. There is no transporter in the inner membrane for NADH+H$^+$ itself.

In the **malate shuttle** (left)—which operates in the heart, liver, and kidneys, for example—oxaloacetic acid is reduced to malate by malate dehydrogenase (MDH, [2a]) with the help of NADH+H$^+$. In antiport for 2-oxoglutarate, malate is transferred to the matrix, where the mitochondrial isoenzyme for MDH [2b] regenerates oxaloacetic acid and NADH+H$^+$. The latter is reoxidized by complex I of the respiratory chain, while oxaloacetic acid, for which a transporter is not available in the inner membrane, is first transaminated to aspartate by aspartate aminotransferase (AST, [3a]). Aspartate leaves the matrix again, and in the cytoplasm once again supplies oxaloacetate for step [2a] and glutamate for return transport into the matrix [3b]. On balance, only NADH+H$^+$ is moved from the cytoplasm into the matrix; ATP is not needed for this.

The **glycerophosphate shuttle** (right) was discovered in insect muscle, but is also active in the skeletal musculature and brain in higher animals. In this shuttle, NADH+H$^+$ formed in the cytoplasm is used to reduce glycerone 3-phosphate, an intermediate of glycolysis (see p. 150) to glycerol 3-phosphate. Via porins, this enters the intermembrane space and is oxidized again there on the exterior side of the inner membrane by the flavin enzyme *glycerol 3-phosphate dehydrogenase* back into glycerone 3-phosphate. The reducing equivalents are passed on to the respiratory chain via **ubiquinone (coenzyme Q)**.

The **carnitine shuttle** for transporting acyl residues into the mitochondrial matrix is discussed on p. 164.

A. Transport systems

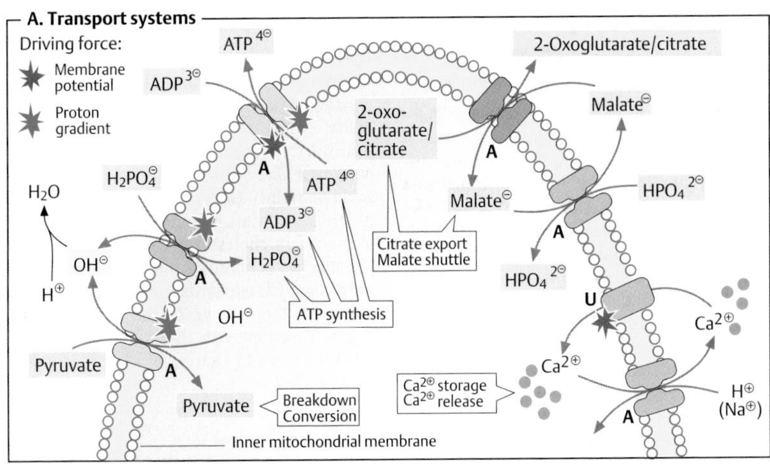

B. Malate and glycerophosphate shuttle

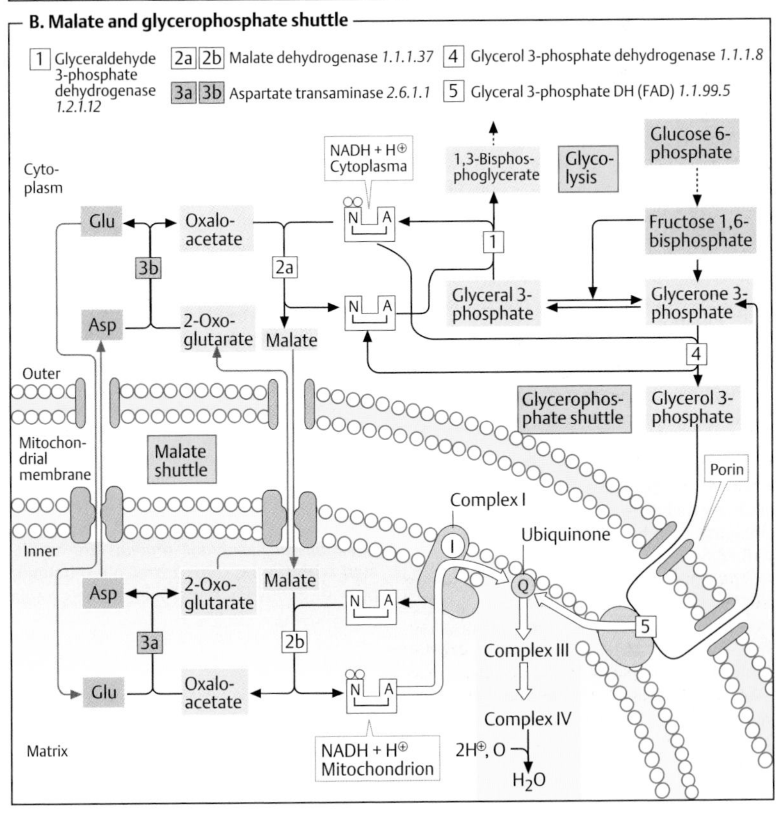

Structure and components

A. Structure of the plasma membrane ◑

All biological membranes are constructed according to a standard pattern. They consist of a continuous **bilayer of amphipathic lipids** approximately 5 nm thick, into which **proteins** are embedded. In addition, some membranes also carry **carbohydrates** (mono- and oligosaccharides) on their exterior, which are bound to lipids and proteins. The proportions of lipids, proteins, and carbohydrates differ markedly depending on the type of cell and membrane (see p. 216).

Membrane lipids are strongly *amphipathic molecules* with a polar hydrophilic "head group" and an apolar hydrophobic "tail." In membranes, they are primarily held together by the hydrophobic effect (see p. 28) and weak Van der Waals forces, and are therefore mobile relative to each other. This gives membranes a more or less fluid quality.

The **fluidity** of membranes primarily depends on their lipid composition and on temperature. At a specific **transition temperature**, membranes pass from a semicrystalline state to a more fluid state. The double bonds in the alkyl chains of unsaturated acyl residues in the membrane lipids disturb the semicrystalline state. The *higher* the proportion of unsaturated lipids present, therefore, the *lower* the transition temperature. The cholesterol content also influences membrane fluidity. While cholesterol increases the fluidity of semicrystalline, closely-packed membranes, it *stabilizes* fluid membranes that contain a high proportion of unsaturated lipids.

Like lipids, proteins are also mobile within the membrane. If they are not fixed in place by special mechanisms, they float within the lipid layer as if in a two-dimensional liquid; biological membranes are therefore also described as being a "fluid mosaic."

Lipids and proteins can shift easily *within* one layer of a membrane, but switching between the two layers (*"flip/flop"*) is not possible for proteins and is only possible with difficulty for lipids (with the exception of cholesterol). To move to the other side, phospholipids require special auxiliary proteins (translocators, "flipases").

B. Membrane lipids ◑

The illustration shows a model of a small section of a membrane. The **phospholipids** are the most important group of membrane lipids. They include *phosphatidylcholine* (lecithin), *phosphatidylethanolamine*, *phosphatidylserine*, *phosphatidylinositol*, and *sphingomyelin* (for their structures, see p. 50). In addition, membranes in animal cells also contain **cholesterol** (with the exception of inner mitochondrial membranes). **Glycolipids** (a *ganglioside* is shown here) are mainly found on the outside of the plasma membrane. Together with the *glycoproteins*, they form the exterior coating of the cell (the *glycocalyx*).

C. Membrane proteins ◑

Proteins can be anchored in or on membranes in various ways. **Integral membrane proteins** cross right through the lipid bilayer. The sections of the peptide chains that lie within the bilayer usually consist of 20 to 25 mainly hydrophobic amino acid residues that form a right-handed α-helix.

Type I and II membrane proteins only contain *one* **transmembrane helix** of this type, while type III proteins contain several. Rarely, type I and II polypeptides can aggregate to form a type IV transmembrane protein. Several groups of integral membrane proteins—e.g., the porins (see p. 212)—penetrate the membrane with antiparallel β-sheet structures. Due to its shape, this tertiary structure is known as a "β-barrel."

Type V and VI proteins carry **lipid anchors.** These are fatty acids (palmitic acid, myristic acid), isoprenoids (e.g., farnesol), or glycolipids such as glycosyl phosphatidylinositol (GPI) that are covalently bound to the peptide chain.

Peripheral membrane proteins are associated with the head groups of phospholipids or with another integral membrane protein (not shown).

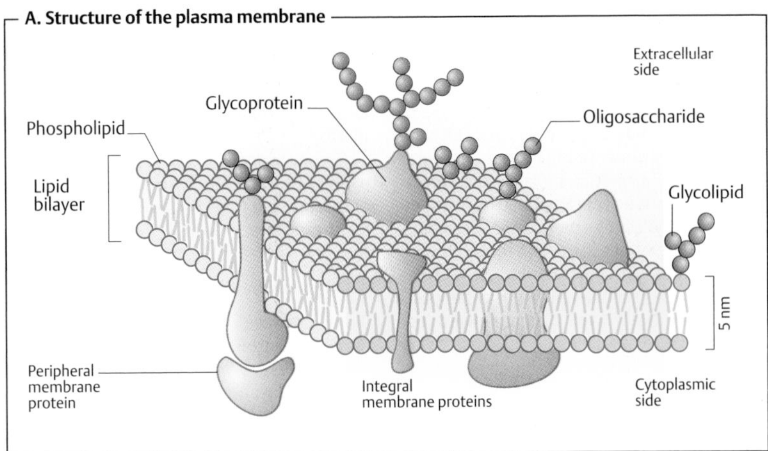
A. Structure of the plasma membrane
Extracellular side
Glycoprotein
Oligosaccharide
Phospholipid
Lipid bilayer
Glycolipid
5 nm
Peripheral membrane protein
Integral membrane proteins
Cytoplasmic side

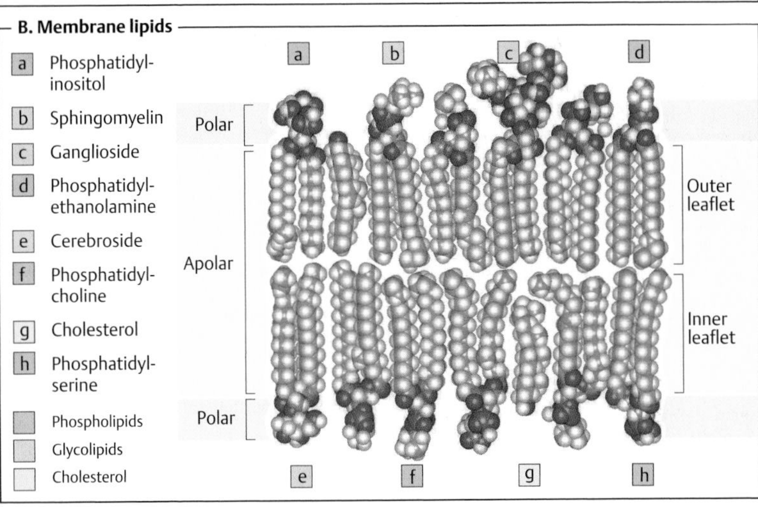
B. Membrane lipids
a Phosphatidyl-inositol
b Sphingomyelin
c Ganglioside
d Phosphatidyl-ethanolamine
e Cerebroside
f Phosphatidyl-choline
g Cholesterol
h Phosphatidyl-serine
Phospholipids
Glycolipids
Cholesterol
Polar
Apolar
Polar
Outer leaflet
Inner leaflet
a
b
c
d
e
f
g
h

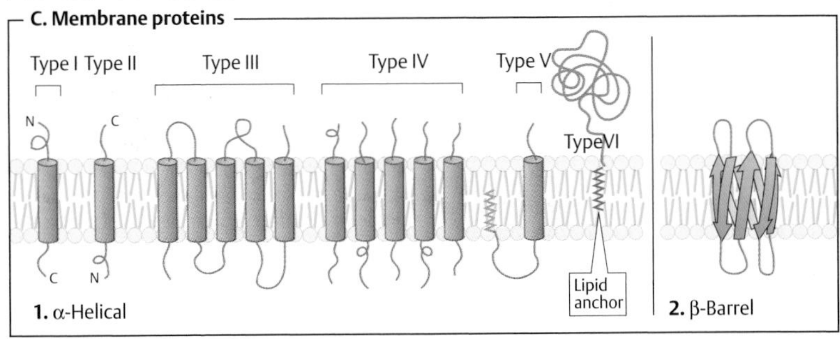
C. Membrane proteins
Type I Type II
Type III
Type IV
Type V
Type VI
N
C
C
N
Lipid anchor
1. α-Helical
2. β-Barrel

Functions and composition

The most important membranes in animal cells are the *plasma membrane*, the inner and outer *nuclear membranes*, the membranes of the *endoplasmic reticulum* (ER) and the *Golgi apparatus*, and the inner and outer *mitochondrial membranes*. *Lysosomes*, *peroxisomes*, and various *vesicles* are also separated from the cytoplasm by membranes. In plants, additional membranes are seen in the plastids and vacuoles. All membranes show *polarity*—i. e., there is a difference in the composition of the inner layer (facing toward the cytoplasm) and the outer layer (facing away from it).

A. Functions of membranes ●

Membranes and their components have the following functions:

1. **Enclosure and insulation** of cells and organelles. The *enclosure* provided by the plasma membrane protects cells from their environment both mechanically and chemically. The plasma membrane is essential for maintaining differences in the concentration of many substances between the intracellular and extracellular compartments.

2. **Regulated transport of substances**, which determines the *internal milieu* and is a precondition for *homeostasis*—i. e., the maintenance of constant concentrations of substances and physiological parameters. Regulated and selective transport of substances through pores, channels, and transporters (see p. 218) is necessary because the cells and organelles are enclosed by membrane systems.

3. **Reception of extracellular signals** and transfer of these signals to the inside of the cell (see pp. 384ff.), as well as the production of signals.

4. **Enzymatic catalysis** of reactions. Important enzymes are located in membranes at the interface between the lipid and aqueous phases. This is where reactions with apolar substrates occur. Examples include *lipid biosynthesis* (see p. 170) and the *metabolism of apolar xenobiotics* (see p. 316). The most important reactions in energy conversion—i. e., *oxidative phosphorylation* (see p. 140) and *photosynthesis* (see p. 128)—also occur in membranes.

5. **Interactions with other cells** for the purposes of cell fusion and tissue formation, as well as communication with the extracellular matrix.

6. **Anchoring of the cytoskeleton** (see p. 204) to maintain the shape of cells and organelles and to provide the basis for movement processes.

B. Composition of membranes ◑

Biological membranes consist of **lipids**, **proteins**, and **carbohydrates** (see p. 214). These components occur in varying proportions (left). Proteins usually account for the largest proportion, at around half. By contrast, carbohydrates, which are only found on the side facing away from the cytoplasm, make up only a few percent. An extreme composition is seen in *myelin*, the insulating material in nerve cells, three-quarters of which consists of lipids. By contrast, the inner *mitochondrial membrane* is characterized by a very low proportion of lipids and a particularly high proportion of proteins.

When the individual proportions of **lipids** in membranes are examined more closely (right part of the illustration), typical patterns for particular cells and tissues are also found. The illustration shows the diversity of the membrane lipids and their approximate quantitative composition. *Phospholipids* are predominant in membrane lipids in comparison with *glycolipids* and *cholesterol*. Triacylglycerols (neutral fats) are not found in membranes.

Cholesterol is found almost exclusively in eukaryotic cells. Animal membranes contain substantially more cholesterol than plant membranes, in which cholesterol is usually replaced by other sterols. There is no cholesterol at all in prokaryotes (with a few exceptions). The inner mitochondrial membrane of eukaryotes is also low in cholesterol, while it is the only membrane that contains large amounts of cardiolipin. These facts both support the endosymbiotic theory of the development of mitochondria (see p. 210).

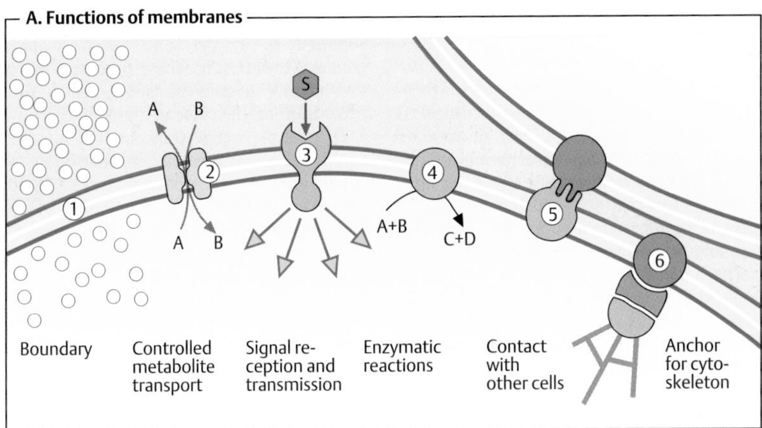
A. Functions of membranes
S
A
B
1
2
3
4
5
6
A
B
A+B
C+D
Boundary
Controlled metabolite transport
Signal reception and transmission
Enzymatic reactions
Contact with other cells
Anchor for cytoskeleton

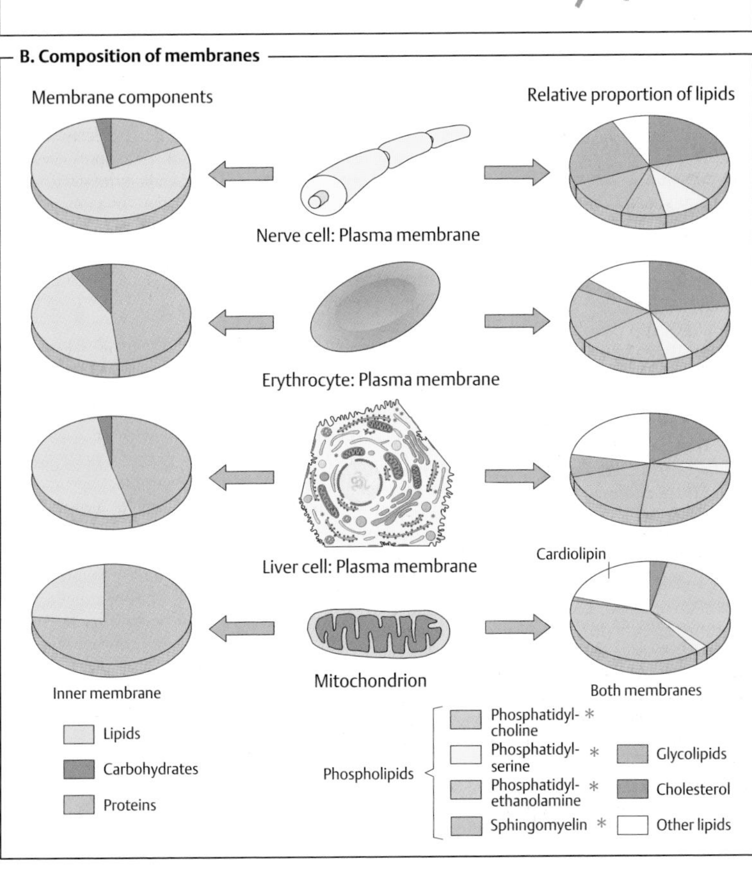
B. Composition of membranes
Membrane components
Relative proportion of lipids
Nerve cell: Plasma membrane
Erythrocyte: Plasma membrane
Liver cell: Plasma membrane
Cardiolipin
Mitochondrion
Inner membrane
Both membranes
Lipids
Carbohydrates
Proteins
Phospholipids
Phosphatidylcholine *
Phosphatidylserine *
Phosphatidylethanolamine *
Sphingomyelin *
Glycolipids
Cholesterol
Other lipids

Transport processes

A. Permeability ●

Only small, uncharged molecules such as gases, water, ammonia, glycerol, or urea are able to pass through biological membranes by *free diffusion*. With increasing **size**, even compounds of this type are no longer able to pass through. Membranes are impermeable to glucose and other sugars, for example.

The **polarity** of a molecule is also important. Apolar substances, such as benzene, ethanol, diethyl ether, and many narcotic agents are able to enter biological membranes easily. By contrast, membranes are impermeable to strongly polar compounds, particularly those that are electrically charged. To be able to take up or release molecules of this type, cells have specialized *channels* and *transporters* in their membranes (see below).

B. Passive and active transport ●

Free diffusion is the simplest form of membrane transport. When it is supported by integral membrane proteins, it is known as **facilitated diffusion** (or facilitated transport).

1. **Channel proteins** have a polar pore through which ions and other hydrophilic compounds can pass. For example, there are channels that allow selected ions to pass (**ion channels**; see p. 222) and **porins** that allow molecules below a specific size to pass in a more or less nonspecific fashion (see p. 212).

2. **Transporters** recognize and bind the molecules to be transported and help them to pass through the membrane as a result of a conformational change. These proteins (permeases) are thus comparable with enzymes—although with the difference that they "catalyze" vectorial transport rather than an enzymatic reaction. Like enzymes, they show a certain *affinity* for each molecule transported (expressed as the dissociation constant, K_d in $mol \cdot L^{-1}$) and a *maximum transport capacity* (V).

Free diffusion and transport processes facilitated by ion channels and transport proteins always follow a *concentration gradient*—i.e., the direction of transport is from the site of higher concentration to the site of lower concentration. In ions, the *membrane potential* also plays a role; the processes are summed up by the term *"electrochemical gradient"* (see p. 126). These processes therefore involve **passive transport**, which runs "downhill" on the slope of a gradient.

By contrast, **active transport** can also run "uphill"—i.e., against a concentration or charge gradient. It therefore requires an input of *energy*, which is usually supplied by the hydrolysis of ATP (see p. 124). The transporter first binds its "cargo" on one side of the membrane. ATP-dependent phosphorylation then causes a conformation change that releases the cargo on the other side of the membrane (see p. 220). A non-spontaneous transport process can also take place through coupling to another active transport process (known as *secondary active transport;* see p. 220).

Using the transport systems in the membranes, cells regulate their volume, internal pH value, and ionic environment. They concentrate metabolites that are important for energy metabolism and biosynthesis, and exclude toxic substances. Transport systems also serve to establish *ion gradients*, which are required for oxidative phosphorylation and stimulation of muscle and nerve cells, for example (see p. 350).

C. Transport processes ◑

Another classification of transport processes is based on the number of particles transported and the direction in which they move. When a *single* molecule or ion passes through the membrane with the help of a channel or transporter, the process is described as a **uniport** (example: the transport of glucose into liver cells). Simultaneous transport of *two different* particles can take place either as a **symport** (example: the transport of amino acids or glucose together with Na^+ ions into intestinal epithelial cells) or as an **antiport**. Ions are often transported in an antiport in exchange for another similarly charged ion. This process is **electroneutral** and therefore more energetically favorable (example: the exchange of HCO_3^- for Cl^- at the erythrocyte membrane).

A. Permeability of membranes

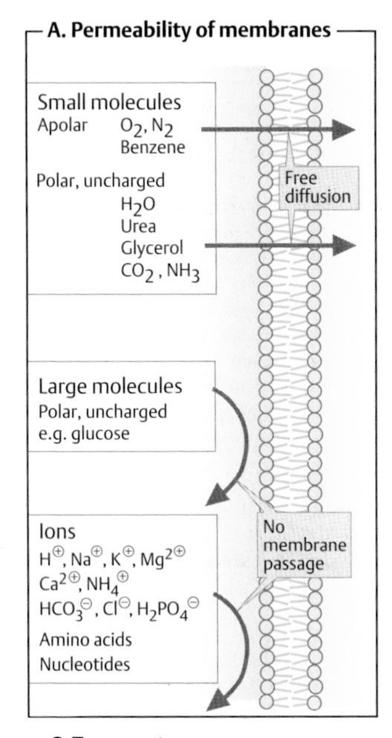

B. Passive and active transport

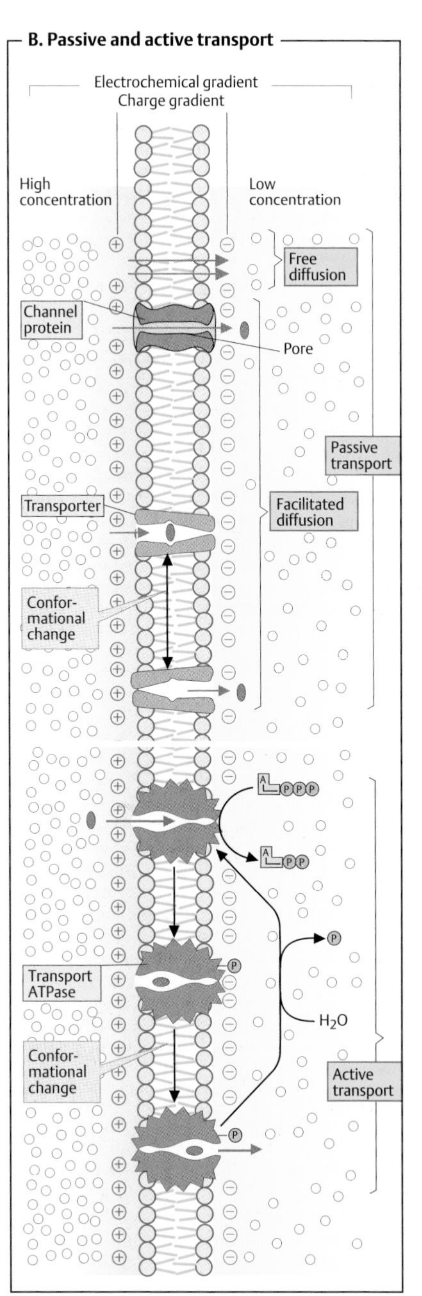

C. Transport processes

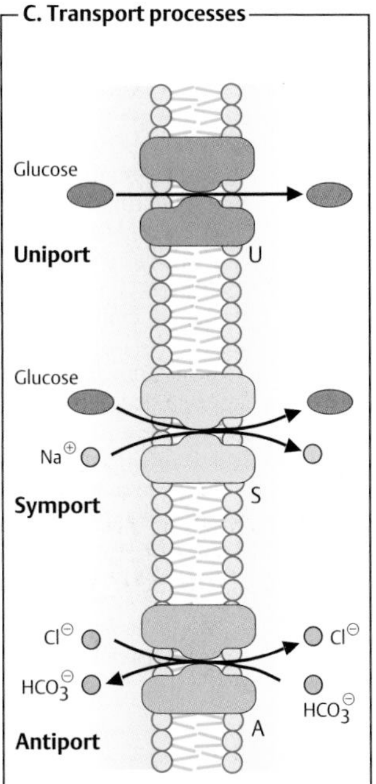

Transport proteins

Illustrations **B–D** show transporters whose structure has been determined experimentally or established on analogy with other known structures. They all belong to group III of the α-helical transmembrane proteins (see p. 214).

A. Transport mechanisms ◑

Some cells couple the "pure" transport forms discussed on p. 218—i. e., passive transport (**1**) and active transport (**2**)—and use this mechanism to take up metabolites. In **secondary active transport** (**3**), which is used for example by epithelial cells in the small intestine and kidney to take up glucose and amino acids, there is a **symport** (S) located on the luminal side of the membrane, which takes up the metabolite M together with an Na^+ ion. An ATP-dependent Na^+ transporter (Na^+/K^+ ATPase; see p. 350) on the other side keeps the intracellular Na+ concentration low and thus indirectly drives the uptake of M. Finally, a uniport (U) releases M into the blood.

B. Glucose transporter Glut-1 ○

The glucose transporters (Glut) are a family of related membrane proteins with varying distribution in the organs. Glut-1 and Glut-3 have a relatively high affinity for glucose (K_d = 1 mM). They occur in nearly all cells, and ensure continuous glucose uptake. Glut-2 is found in the liver and pancreas. This form has a lower affinity (K_d = 15–20 mM). The rate of glucose uptake by Glut-2 is therefore strongly dependent on the blood glucose level (normally 4–8 mM). Transport by Glut-4 (K_d = 5 mM), which is mainly expressed in muscle and fat cells, is controlled by insulin, which increases the number of transporters on the cell surface (see p. 388). Glut-5 mediates secondary active resorption of glucose in the intestines and kidney (see **A**).

Glut-1 consists of a single peptide chain that spans the membrane with 12 α-helices of different lengths. The glucose is bound by the peptide loops that project on each side of the membrane.

C. Aquaporin-1 ○

Aquaporins help water to pass through biological membranes. They form hydrophilic pores that allow H_2O molecules, but not hydrated ions or larger molecules, to pass through. Aquaporins are particularly important in the kidney, where they promote the reuptake of water (see p. 328). Aquaporin-2 in the renal collecting ducts is regulated by **antidiuretic hormone** (ADH, vasopressin), which via cAMP leads to shifting of the channels from the ER into the plasma membrane.

Aquaporin-1, shown here, occurs in the proximal tubule and in Henle's loop. It contains eight transmembrane helices with different lengths and orientations. The yellow-colored residues form a narrowing that only H_2O molecules can overcome.

D. Sarcoplasmic Ca^{2+} pump ○

Transport ATPases transport cations—they are "ion pumps." ATPases of the **F type**—e. g., mitochondrial ATP synthase (see p. 142)—use H^+ transport for *ATP synthesis*. Enzymes of the **V type**, using up ATP, "pump" protons into lysosomes and other acidic cell compartments (see p. 234). **P type** transport ATPases are particularly numerous. These are ATP-driven cation transporters that undergo covalent phosphorylation during the transport cycle.

The **Ca^{2+} ATPase** shown also belongs to the P type. In muscle, its task is to pump the Ca^{2+} released into the cytoplasm to trigger muscle contraction back into the sarcoplasmic reticulum (SR; see p. 334). The molecule (**1**) consists of a single peptide chain that is folded into various domains. In the transmembrane part, which is formed by numerous α-helices, there are binding sites for two Ca^{2+} ions (blue) ATP is bound to the cytoplasmic N domain (green).

Four different stages can be distinguished in the enzyme's catalytic cycle (**2**). First, binding of ATP to the N domain leads to the uptake of two Ca^{2+} into the transmembrane part (**a**). Phosphorylation of an aspartate residue in the P domain (**b**) and dissociation of ADP then causes a conformation change that releases the Ca^{2+} ions into the SR (**c**). Finally, dephosphorylation of the aspartate residue restores the initial conditions (**d**).

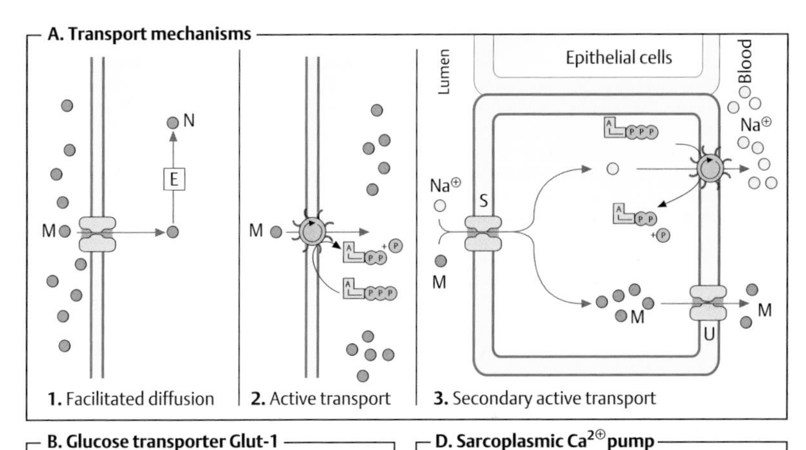
A. Transport mechanisms
N
E
M
M
Lumen
Epithelial cells
Blood
Na⊕
Na⊕
S
M
M
M
U
1. Facilitated diffusion
2. Active transport
3. Secondary active transport

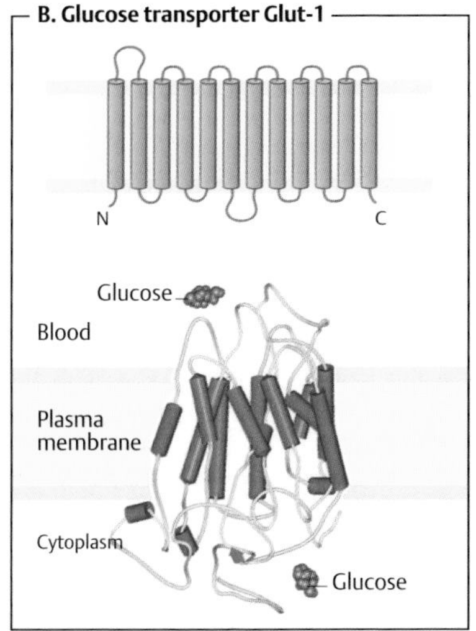
B. Glucose transporter Glut-1
N
C
Glucose
Blood
Plasma membrane
Cytoplasm
Glucose

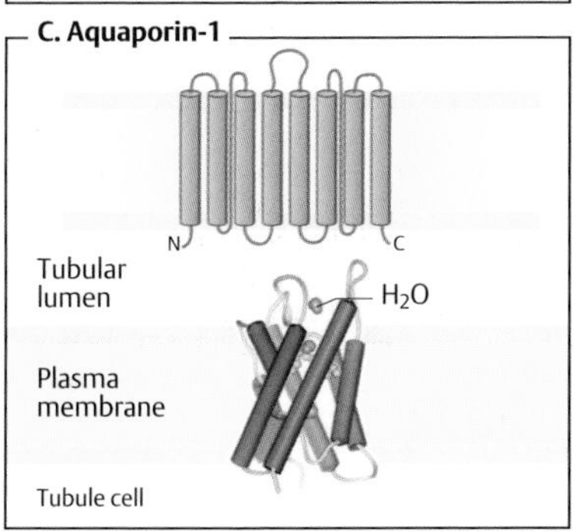
C. Aquaporin-1
N
C
Tubular lumen
H2O
Plasma membrane
Tubule cell

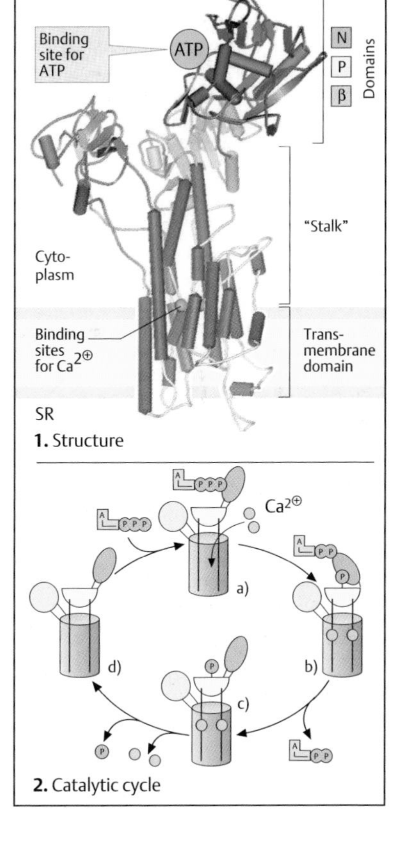
D. Sarcoplasmic Ca2⊕ pump
Binding site for ATP
ATP
N
P
β
Domains
"Stalk"
Cyto-plasm
Binding sites for Ca2⊕
Trans-membrane domain
SR
1. Structure
Ca2⊕
a)
b)
c)
d)
2. Catalytic cycle

Ion channels

Ion channels facilitate the diffusion of ions through biological membranes. Some ion channels open and close depending on the membrane potential (**voltage-gated channels, A**) or in response to specific ligands (**ligand-gated channels, B**). Other channels operate passively. In these cases, transport depends only on the concentration gradient (**C**).

A. Voltage-gated Na^+ channel ○

Voltage-gated Na^+ channels play a decisive part in the conduction of electrical impulses in the nervous system (see p. 350). These channels open when the membrane potential in their environment reverses. Due to the high equilibrium potential for Na^+ (see p. 126), an inflow of Na^+ ions takes place, resulting in local **depolarization** of the membrane, which propagates by activation of neighboring voltage-dependent Na^+ channels. A spreading depolarization wave of this type is known as an **action potential** (see p. 350). Externally directed K^+ channels are involved in the repolarization of the membrane. In their functioning, these resemble the much more simply structured K^+ channels shown in **C**. The Ca^{2+} channels that trigger exocytosis of vesicles (see p. 228) are also controlled by the action potential.

The voltage-gated Na^+ channels in higher animals are large complexes made up of several subunits. The α-subunit shown here mediates Na^+ transport. It consists of a very long peptide chain (around 2000 amino acid residues), which is folded into four domains, each with six transmembrane helices (left). The S6 helices of all the domains (blue) together form a centrally located hydrophilic pore which can be made narrow or wide depending on the channel's functional status. The six S4 helices (green) function as voltage sensors.

The current conception of the way in which the opening and closing mechanism functions is shown in a highly simplified form on the right. For the sake of clarity, only one of the four domains (domain IV) is shown. The S4 helices contain several positively charged residues. When the membrane is polarized (**a**), the surplus negative charges on the inner side keep the helix in the membrane. If this attraction is removed as a result of local depolarization, the S4 helices are thought to snap upwards like springs and thus open the central pore (**b**).

B. Nicotinic acetylcholine receptor ○

Many receptors for neurotransmitters function as ligand-gated channels for Na^+, K^+ and/or Ca^{2+} ions (see p. 354). The ones that have been studied in the greatest detail are the nicotinic receptors for **acetylcholine** (see p. 352). These consist of five separate but structurally closely related subunits. Each forms four transmembrane helices, the second of which is involved in the central pore in each case. The type of monomer and its arrangement in the complex is not identical in all receptors of this type. In the neuromuscular junction (see p. 334), the arrangement αβγαδ is found (**1**).

In the interior of the structure, acetylcholine binds to the two α-subunits and thus opens the pore for a short time (1–2 ms). Negatively charged residues are arranged in three groups in a ring shape inside it. They are responsible for the receptor's ion specificity. It is thought that binding of the neurotransmitter changes the position of the subunits in such a way that the pore expands (**3**). The bound acetylcholine dissociates again and is hydrolytically inactivated (see p. 356). The receptor is thus able to function again.

C. K^+ channel in *Streptomyces lividans* ○

The only detailed structures of ion channels established so far are those of potassium channels like that of an outwardly directed K^+ channel in the bacterium *Streptomyces lividans*. It consists of four identical subunits (blue, yellow, green, and red), each of which contains two long α-helices and one shorter one. In the interior of the cell (bottom), the K^+ ions (violet) enter the structure's central channel. Before they are released to the outside, they have to pass through what is known as a "selectivity filter." In this part of the channel, several C=O groups in the peptide chain form a precisely defined opening that is only permeable to non-hydrated K^+ ions.

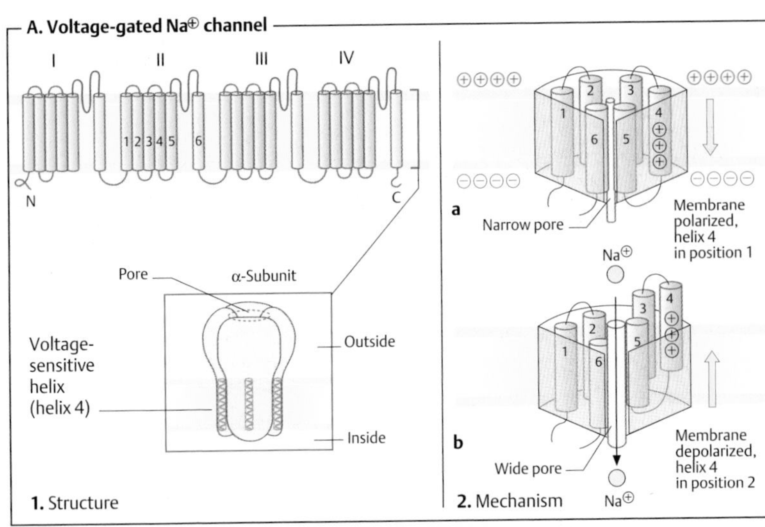
A. Voltage-gated Na⊕ channel
I
II
III
IV
1 2 3 4 5
6
N
C
Pore
α-Subunit
Outside
Voltage-sensitive helix (helix 4)
Inside
1. Structure
1
2
3
4
5
6
Narrow pore
Membrane polarized, helix 4 in position 1
a
Na⊕
Wide pore
Membrane depolarized, helix 4 in position 2
b
2. Mechanism
Na⊕

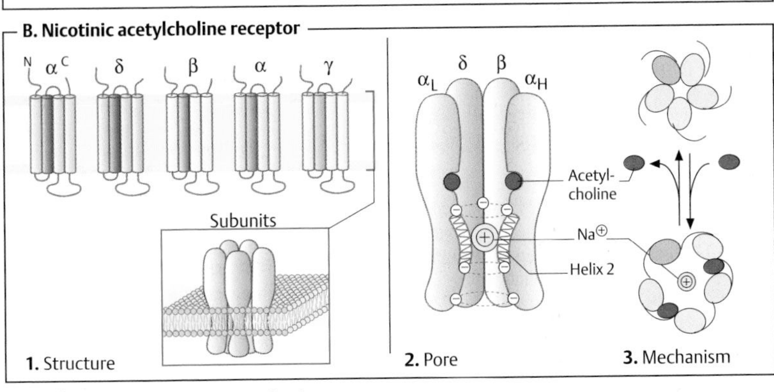
B. Nicotinic acetylcholine receptor
N
α
C
δ
β
α
γ
Subunits
1. Structure
$α_L$
δ
β
$α_H$
Acetyl-choline
Na⊕
Helix 2
2. Pore
3. Mechanism

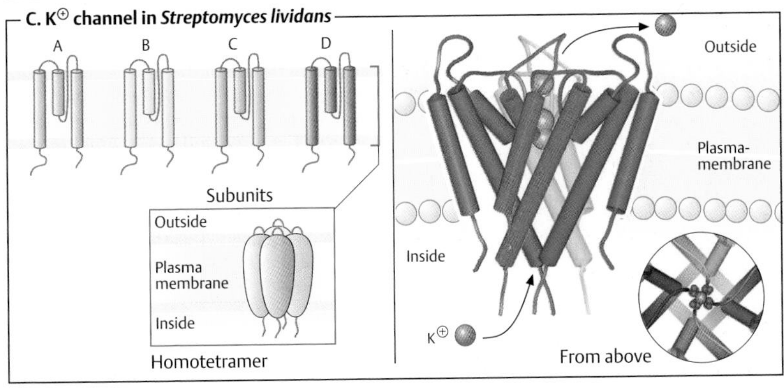
C. K⊕ channel in Streptomyces lividans
A
B
C
D
Subunits
Outside
Plasma membrane
Inside
Homotetramer
Outside
Plasma-membrane
Inside
K⊕
From above

Membrane receptors

To receive and pass on chemical or physical signals, cells are equipped with **receptor proteins**. Many of these are integral membrane proteins in the plasma membrane, where they receive signals from their surroundings. Other receptor proteins are located in intercellular membranes. The receptors for lipophilic hormones are among the few that function in a soluble form. They regulate gene transcription in the nucleus (see p. 378).

A. Principle of receptor action ●

Membrane-located receptors can be divided into three parts, which have different tasks. The **receptor domain** reacts specifically to a given signal. Signals of this type can be of a purely physical nature. For example, many organisms react to light. In this way, plants adapt growth and photosynthesis to light conditions, while animals need light receptors for visual processing (**C**; see p. 358). Mechanoreceptors are involved in hearing and in pressure regulation, among other things. Channels that react to action potentials (see p. 350) can be regarded as receptors for electrical impulses.

However, most receptors do not react to physical stimuli, but rather to signal molecules. Receptors for these chemical signals contain binding sites in the receptor domain that are complementary to each ligand. In this respect, they resemble enzymes (see p. 94). As the **effector domain** of the receptor is usually separated by a membrane, a mechanism for **signal transfer** between the domains is needed. Little is yet known regarding this. It is thought that conformation changes in the receptor protein play a decisive part. Some receptors dimerize after binding of the ligand, thereby bringing the effector domains of two molecules into contact (see p. 392).

The way in which the effector works differs from case to case. By binding or interconversion, many receptors activate special **mediator proteins**, which then trigger a signal cascade (signal transduction; see p. 384). Other receptors function as **ion channels**. This is particularly widespread in receptors for neurotransmitters (see p. 354).

B. Insulin receptor ◑

The receptor for the hormone insulin (see p. 76) belongs to the family of **1-helix receptors**.

These molecules span the membrane with only one α-helix. The subunits of the dimeric receptor (red and blue) each consist of two polypeptides (α and β) bound by disulfide bonds. The α-chains together bind the insulin, while the β-chains contain the transmembrane helix and, at the C-terminus, domains with **tyrosine kinase** activity. In the activated state, the kinase domains phosphorylate themselves and also mediator proteins (receptor substrates) that set in motion cascades of further phosphorylations (see pp. 120 and 388).

C. 7-helix receptors ◑

A large group of receptors span the membrane with α-helices seven times. These are known as **7-helix receptors**. Via their effector domains, they bind and activate trimeric proteins, which in turn bind and hydrolyze GTP and are therefore called G proteins. Most G proteins, in turn, activate or inhibit enzymes that create secondary signaling molecules (**second messengers**; see p. 386). Other G proteins regulate ion channels. The illustration shows the complex of the light receptor **rhodopsin**, with the associated G protein **transducin** (see p. 358). The GTP-binding α-subunit (green) and the γ-subunit (violet) of transducin are anchored in the membrane via lipids (see p. 214). The β-subunit is shown in detail on p. 72.

D. T-cell receptor ◑

The cells of the immune system communicate with each other particularly intensively. The **T-cell receptor** plays a central role in the activation of T lymphocytes (see p. 296). The cell at the top has been infected with a virus, and it indicates this by presenting a viral peptide (violet) with the help of a class I **MHC protein** (yellow and green). The combination of the two molecules is recognized by the dimeric T-cell receptor (blue) and converted into a signal that activates the T cell (bottom) and thereby enhances the immune response to the virus.

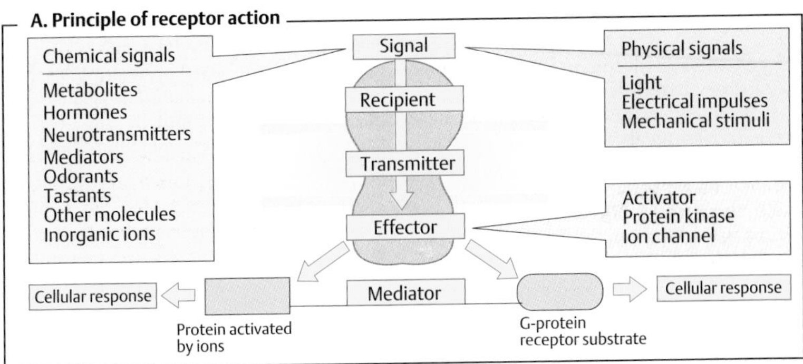
A. Principle of receptor action
Chemical signals
Metabolites
Hormones
Neurotransmitters
Mediators
Odorants
Tastants
Other molecules
Inorganic ions
Signal
Recipient
Transmitter
Effector
Physical signals
Light
Electrical impulses
Mechanical stimuli
Activator
Protein kinase
Ion channel
Cellular response
Mediator
Cellular response
Protein activated by ions
G-protein receptor substrate

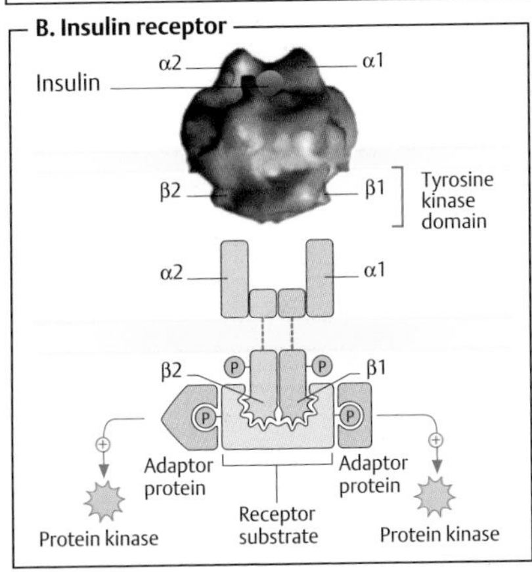
B. Insulin receptor
Insulin
α2
α1
β2
β1
Tyrosine kinase domain
α2
α1
β2
β1
Adaptor protein
Adaptor protein
Receptor substrate
Protein kinase
Protein kinase

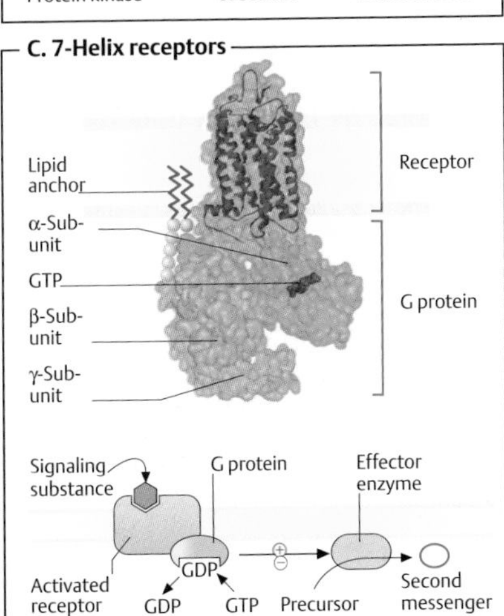
C. 7-Helix receptors
Lipid anchor
α-Sub-unit
GTP
β-Sub-unit
γ-Sub-unit
Receptor
G protein
Signaling substance
G protein
Effector enzyme
Activated receptor
GDP
GDP
GTP
Precursor
Second messenger

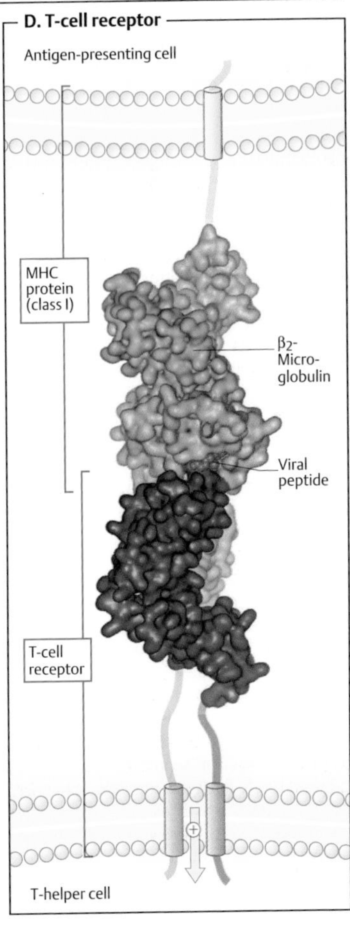
D. T-cell receptor
Antigen-presenting cell
MHC protein (class I)
β2-Micro-globulin
Viral peptide
T-cell receptor
T-helper cell

ER: structure and function

The endoplasmic reticulum (ER) is an extensive closed membrane system consisting of tubular and saccular structures. In the area of the nucleus, the ER turns into the external nuclear membrane. Morphologically, a distinction is made between the **rough ER** (rER) and the **smooth ER** (sER). Large numbers of ribosomes are found on the membranes of the rER, which are lacking on the sER. On the other hand, the sER is rich in membrane-bound **enzymes**, which catalyze partial reactions in the lipid metabolism as well as biotransformations.

A. Rough endoplasmic reticulum and the Golgi apparatus ◑

The **rER** (**1**) is a site of active *protein biosynthesis*. This is where proteins destined for membranes, lysosomes, and export from the cell are synthesized. The remaining proteins are produced in the cytoplasm on ribosomes that are not bound to membranes.

Proteins synthesized at the rER (**1**) are folded and modified after translation (protein maturation; see p. 230). They remain either in the rER as membrane proteins, or pass with the help of transport vesicles (**2**) to the Golgi apparatus (**3**). Transport vesicles are formed by budding from existing membranes, and they disappear again by fusing with them (see p. 228).

The **Golgi apparatus** (**3**) is a complex network, also enclosed, consisting of flattened membrane saccules ("cisterns"), which are stacked on top of each other in layers. Proteins mature here and are sorted and packed. A distinction is made between the *cis, medial,* and *trans* Golgi regions, as well as a *trans* Golgi network (tGN). The *post-translational modification of proteins,* which starts in the ER, continues in these sections.

From the Golgi apparatus, the proteins are transported by vesicles to various targets in the cells—e. g., to lysosomes (**4**), the plasma membrane (**6**), and secretory vesicles (**5**) that release their contents into the extracellular space by fusion with the plasma membrane (**exocytosis**; see p. 228). Protein transport can either proceed continuously (*constitutive*), or it can be *regulated* by chemical signals. The decision regarding which pathway a protein will take and whether its transport will be constitutive or regulated depends on the signal sequences or signal structures that proteins carry with them like address labels (see p. 228). In addition to proteins, the Golgi apparatus also transports membrane lipids to their targets.

B. Smooth endoplasmic reticulum ◑

Regions of the ER that have no bound ribosomes are known as the **smooth endoplasmic reticulum** (sER). In most cells, the proportion represented by the sER is small. A marked sER is seen in cells that have an active lipid metabolism, such as hepatocytes and Leydig cells. The sER is usually made up of branching, closed tubules.

Membrane-located enzymes in the sER catalyze **lipid synthesis**. Phospholipid synthesis (see p. 170) is located in the sER, for example, and several steps in cholesterol biosynthesis (see p. 172) also take place there. In endocrine cells that form *steroid hormones,* a large proportion of the reaction steps involved also take place in the sER (see p. 376).

In the liver's hepatocytes, the proportion represented by the sER is particularly high. It contains enzymes that catalyze so-called **biotransformations**. These are reactions in which apolar foreign substances, as well as endogenous substances—e. g., steroid hormones—are chemically altered in order to inactivate them and/or prepare them for conjugation with polar substances (phase I reactions; see p. 316). Numerous *cytochrome P450 enzymes* are involved in these conversions (see p. 318) and can therefore be regarded as the major molecules of the sER.

The sER also functions as an intracellular **calcium store**, which normally keeps the Ca^{2+} level in the cytoplasm low. This function is particularly marked in the *sarcoplasmic reticulum,* a specialized form of the sER in muscle cells (see p. 334). For release and uptake of Ca^{2+}, the membranes of the sER contain signal-controlled Ca^{2+} channels and energy-dependent Ca^{2+} ATPases (see p. 220). In the lumen of the sER, the high Ca^{2+} concentration is buffered by Ca^{2+}-binding proteins.

A. Rough endoplasmic reticulum and Golgi apparatus

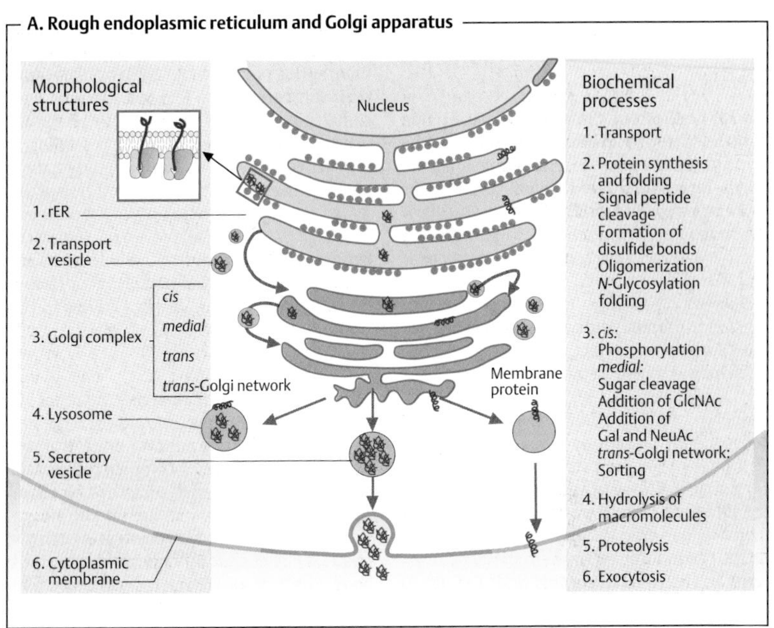

B. Smooth endoplasmic reticulum

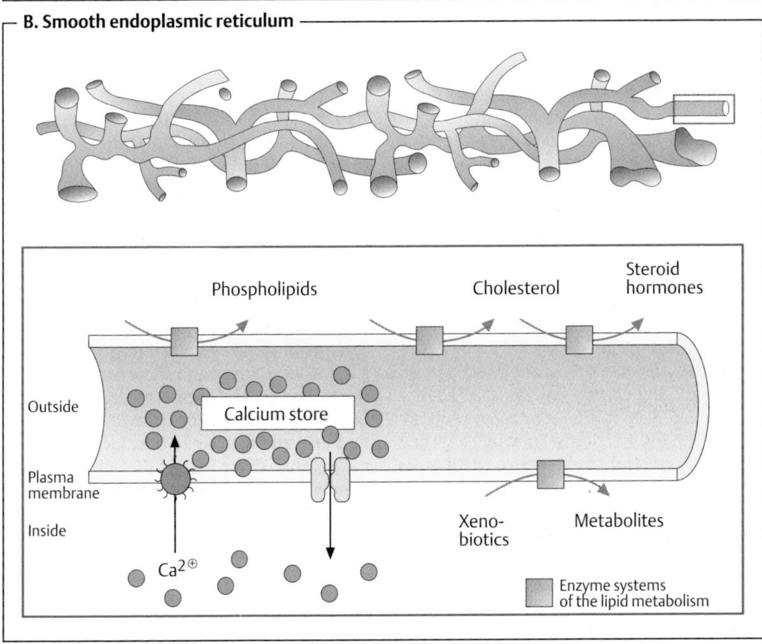

Protein sorting

A. Protein sorting ◑

The biosynthesis of all proteins starts on free ribosomes (top). However, the paths that the proteins follow soon diverge, depending on which target they are destined for. Proteins that carry a *signal peptide for the ER* (**1**) follow the *secretory pathway* (right). Proteins that do not have this signal follow the *cytoplasmic pathway* (left).

Secretory pathway. Ribosomes that synthesize a protein with a signal peptide for the ER settle on the ER (see p. 228). The peptide chain is transferred into the lumen of the rER. The presence or absence of other signal sequences and signal regions determines the subsequent transport pathway.

Proteins that have *stop-transfer sequences* (**4**) remain as integral membrane proteins in the ER membrane. They then pass into other membranes via vesicular transport (see p. 226). From the rER, their pathway then leads to the Golgi apparatus and then on to the plasma membrane. Proteins destined to remain in the rER—e.g., enzymes—find their way back from the Golgi apparatus to the rER with the help of a *retention signal* (**2**). Other proteins move from the Golgi apparatus to the lysosomes (**3**; see p. 234), to the cell membrane (integral membrane proteins or constitutive exocytosis), or are transported out of the cell (**9;** signal-regulated exocytosis) by secretory vesicles (**8**).

Cytoplasmic pathway. Proteins that do not have a signal peptide for the ER are synthesized in the cytoplasm on free ribosomes, and remain in that compartment. Special signals mediate further transport into the mitochondria (**5**; see p. 232), the nucleus (**6**; see p. 208) or peroxisomes (**7**).

B. Translocation signals ○

Signal peptides are short sections at the N or C terminus, or within the peptide chain. Areas on the protein surface that are formed by various sections of the chain or by various chains are known as **signal regions**. Signal peptides and signal regions are *structural signals* that are usually recognized by *receptors* on organelles (see **A**). They move the proteins, with the help of additional proteins, into the organelles (*selective protein transfer*). Structural signals can also activate *enzymes* that modify the proteins and thereby determine their subsequent fate. Examples include lysosomal proteins (see p. 234) and membrane proteins with lipid anchors (see p. 214).

After they have been used, signal peptides at the N terminus are cleaved off by specific hydrolases (symbol: scissors). In proteins that contain several successive signal sequences, this process can expose the subsequent signals. By contrast, signal peptides that have to be read several times are not cleaved.

C. Exocytosis ○

Exocytosis is a term referring to processes that allow cells to expel substances (e.g., hormones or neurotransmitters) quickly and in large quantities. Using a complex protein machinery, secretory vesicles fuse completely or partially with the plasma membrane and release their contents. Exocytosis is usually *regulated* by chemical or electrical signals. As an example, the mechanism by which neurotransmitters are released from synapses (see p. 348) is shown here, although only the most important proteins are indicated.

The decisive element in exocytosis is the interaction between proteins known as SNAREs that are located on the vesicular membrane (v-SNAREs) and on the plasma membrane (t-SNAREs). In the resting state (**1**), the v-SNARE *synaptobrevin* is blocked by the vesicular protein *synaptotagmin*. When an action potential reaches the presynaptic membrane, voltage-gated Ca^{2+} channels open (see p. 348). Ca^{2+} flows in and triggers the machinery by conformational changes in proteins. Contact takes place between synaptobrevin and the t-SNARE *synaptotaxin* (**2**). Additional proteins known as SNAPs bind to the SNARE complex and allow fusion between the vesicle and the plasma membrane (**3**). The process is supported by the hydrolysis of GTP by the auxiliary protein *Rab*.

The toxin of the bacterium *Clostridium botulinum*, one of the most poisonous substances there is, destroys components of the exocytosis machinery in synapses through enzymatic hydrolysis, and in this way blocks neurotransmission.

A. Protein sorting

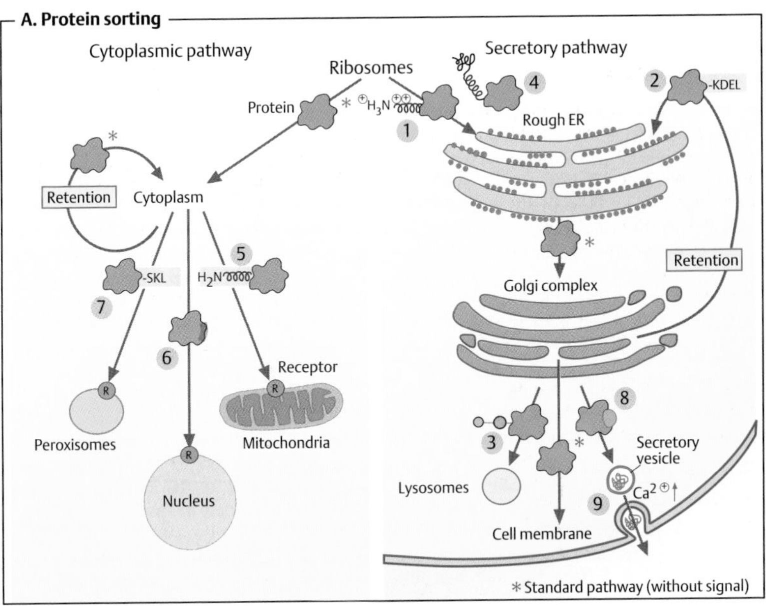

B. Translocation signals

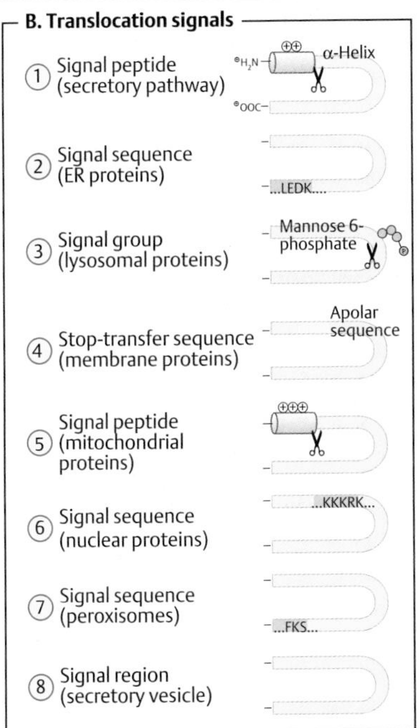

C. Exocytosis

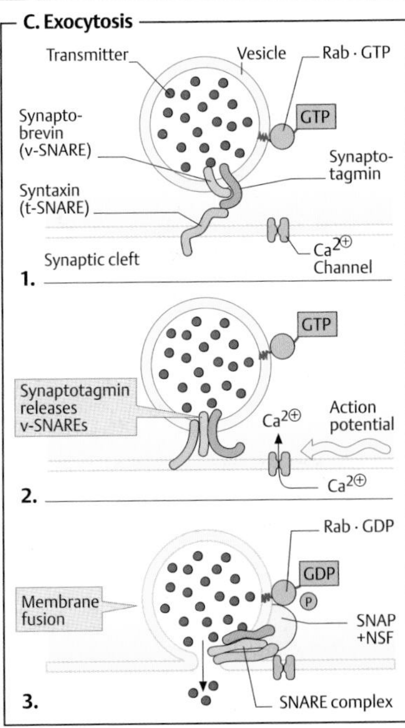

Protein synthesis and maturation

A. Protein synthesis in the rER ◑

With all proteins, protein biosynthesis (**Translation**; for details, see p. 250) starts on free ribosomes in the cytoplasm (**1**). Proteins that are exported out of the cell or into lysosomes, and membrane proteins of the ER and the plasma membrane, carry a **signal peptide for the ER** at their N-terminus. This is a section of 15–60 amino acids in which one or two strongly basic residues (Lys, Arg) near the N-terminus are followed by a strongly hydrophobic sequence of 10–15 residues (see p. 228).

As soon as the signal peptide (red) appears on the surface of the ribosome (**2**), an RNA-containing *signal recognition particle* (**SRP**, green) binds to the sequence and initially interrupts translation (**3**). The SRP then binds to an **SRP receptor** in the rER membrane, and in this way attaches the ribosome to the ER (**4**). After this, the SRP dissociates from the signal peptide and from the SRP receptor and is available again for step 3. This endergonic process is driven by GTP hydrolysis (**5**). Translation now resumes. The remainder of the protein, still unfolded, is gradually introduced into a channel (the **translocon**) in the lumen of the rER (**6**), where a *signal peptidase* located in the inner ER membrane cleaves the signal peptide while translation is still taking place (**7**). This converts the **preprotein** into a **proprotein**, from which the mature protein finally arises after additional post-translational modifications (**8**) in the ER and in the Golgi apparatus.

If the growing polypeptide contains a **stop-transfer signal** (see p. 228), then this hydrophobic section of the chain remains stuck in the membrane outside the translocon, and an *integral membrane protein* arises. In the course of translation, an additional signal sequence can re-start the transfer of the chain through the translocon. Several repetitions of this process produce integral membrane proteins with several transmembrane helices (see p. 214).

B. Protein glycosylation ○

Most extracellular proteins contain covalently bound oligosaccharide residues. For example, all plasma proteins with the exception of albumin are glycosylated. Together with glycolipids, numerous **glycoproteins** on the cell surface form the **glycocalyx**. Inside the ER, the carbohydrate parts of the glycoproteins are cotranslationally transferred to the growing chain, and are then converted into their final form while passing through the ER and Golgi apparatus.

N-bound oligosaccharides (see p. 44) are always bound to the acid-amide group of asparagine residues. If a **glycosylation sequence** (–Asn–X–Ser(Thr)–, where X can be any amino acid) appears in the growing peptide chain, then a *transglycosylase* in the ER membrane [1] transfers a previously produced **core oligosaccharide** consisting of 14 hexose residues *en-bloc* from the carrier molecule **dolichol diphosphate** to the peptide.

Dolichol is a long-chain isoprenoid (see p. 52) consisting of 10–20 isoprene units, which is embedded in the ER membrane. A hydroxyl group at the end of the molecule is bound to diphosphate, on which the nuclear oligosaccharide is built up in an extended reaction sequence (not shown here in detail). The core structure consists of two residues of N-acetylglucosamine (GlcNAc), a branched group of nine mannose residues (Man) and three terminal glucose resides (Glc).

As the proprotein passes through the ER, *glycosidases* [2] remove the glucose residues completely and the mannoses partially ("trimming"), thereby producing the **mannose-rich type** of oligosaccharide residues. Subsequently, various *glycosyltransferases* [3] transfer additional monosaccharides (e.g., GlcNAc, galactose, fucose, and N-acetylneuraminic acid; see p. 38) to the mannose-rich intermediate and thereby produce the **complex type** of oligosaccharide. The structure of the final oligosaccharide depends on the type and activity of the glycosyltransferases present in the ER of the cell concerned, and is therefore genetically determined (although indirectly).

A. Protein synthesis in the rough endoplasmic reticulum

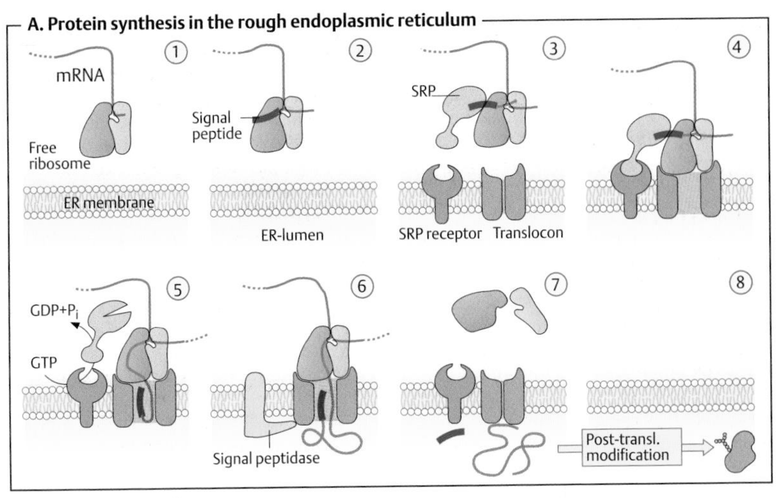

B. Protein glycosylation

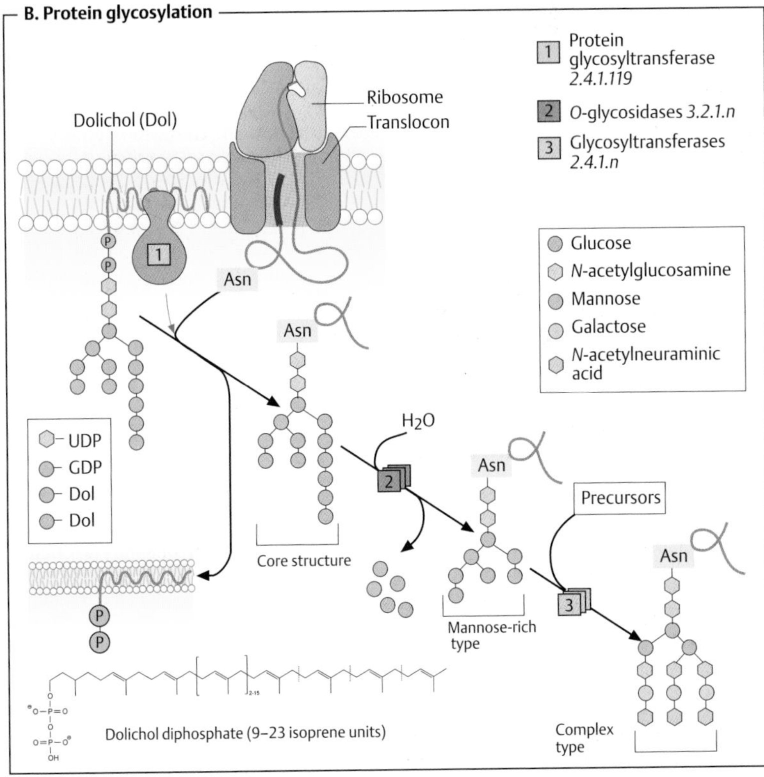

Protein maturation

After translation, proteins destined for the secretory pathway (see p. 228) first have to fold into their native conformation within the rER (see p. 230). During this process they are supported by various *auxiliary proteins.*

A. Protein folding in the rER ●

To prevent incorrect folding of the growing protein during protein biosynthesis, **chaperones** (see **B**) in the lumen of the rER bind to the peptide chain and stabilize it until translation has been completed. Binding protein (**BiP**) is an important chaperone in the ER.

Many secretory proteins—e.g., pancreatic ribonuclease (RNAse; see p. 74)—contain several disulfide bonds that are only formed oxidatively from SH groups after translation. The eight cysteine residues of the RNAse can in principle form 105 different pairings, but only the combination of the four disulfide bonds shown on p. 75 provides active enzyme. Incorrect pairings can block further folding or lead to unstable or insoluble conformations. The enzyme *protein disulfide isomerase* [1] accelerates the equilibration between paired and unpaired cysteine residues, so that incorrect pairs can be quickly split before the protein finds its final conformation.

Most peptide bonds in proteins take on the *trans* conformation (see p. 66). Only bonds with proline residues (–X–Pro–) can be present in both *cis* and *trans* forms.

In the protein's native conformation, every X–Pro bond has to have the correct conformation (*cis* or *trans*). As the uncatalyzed transition between the two forms is very slow, there is a *proline* cis–trans *isomerase* [2] in the ER that accelerates the conversion.

B. Chaperones and chaperonins ●

Most proteins fold spontaneously into their native conformation, even in the test tube. In the cell, where there are very high concentrations of proteins (around 350 g · L^{-1}), this is more difficult. In the unfolded state, the apolar regions of the peptide chain (yellow) tend to aggregate—due to the hydrophobic effect (see p. 28)—with other proteins or with each other to form insoluble products (2). In addition, unfolded proteins are susceptible to proteinases. To protect partly folded proteins, there are auxiliary proteins called **chaperones** because they guard immature proteins against damaging contacts. Chaperones are formed increasingly during temperature stress and are therefore also known as **heat-shock proteins** (hsp). Several classes of hsp are distinguished. Chaperones of the hsp70 type (Dna K in bacteria) are common, as are type hsp60 **chaperonins** (GroEL/ES in bacteria). Class hsp90 chaperones have special tasks (see p. 378).

While small proteins can often reach their native conformation without any help (1), larger molecules require hsp70 proteins for protection against aggregation which bind as monomers and can dissociate again, dependent on ATP (3). By contrast, type hsp60 chaperonins form large, barrel-shaped complexes with 14 subunits in which proteins can fold independently while shielded from their environment (4). The function of hsp60 has been investigated in detail in the bacterial chaperonin **GroEL** (right). The barrel has two chambers, which are closed with a lid (**GroES**) during folding of the guest protein. Driven by ATP hydrolysis, the chambers open and close alternately—i. e., the release of the fully folded protein from one chamber is coupled to the uptake of an unfolded peptide in the second chamber.

C. Protein import in mitochondria ●

Class hsp70 chaperones are also needed for translocation of nuclear-coded proteins from the cytoplasm into the mitochondria (see p. 228). As two membranes have to be crossed to reach the matrix, there are two translocator complexes: **TOM** ("transport outer membrane") and **TIM** ("transport inner membrane"). For transport, proteins are unfolded in the cytoplasm and protected by hsp70. TOM recognizes the positively charged signal sequence at the protein's N terminus (see p. 228) and with the help of the membrane potential threads the chains through the central pores of the two complexes. Inside TIM, further hsp70 molecules bind and pull the chain completely into the matrix. As with import into the ER, the signal peptide is proteolytically removed by a signal peptidase during translocation.

A. Protein folding in the rER

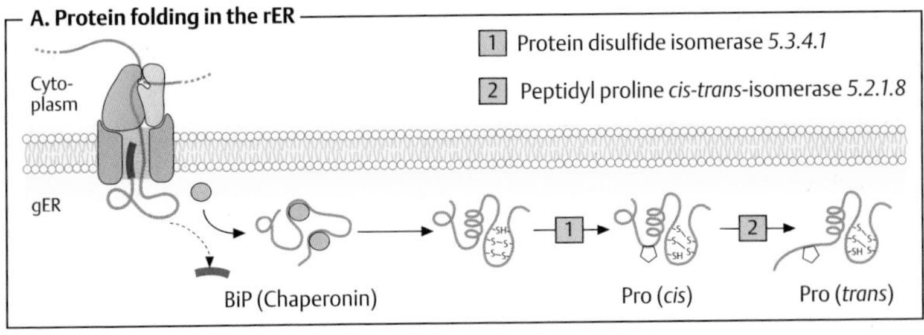

B. Chaperones and chaperonins

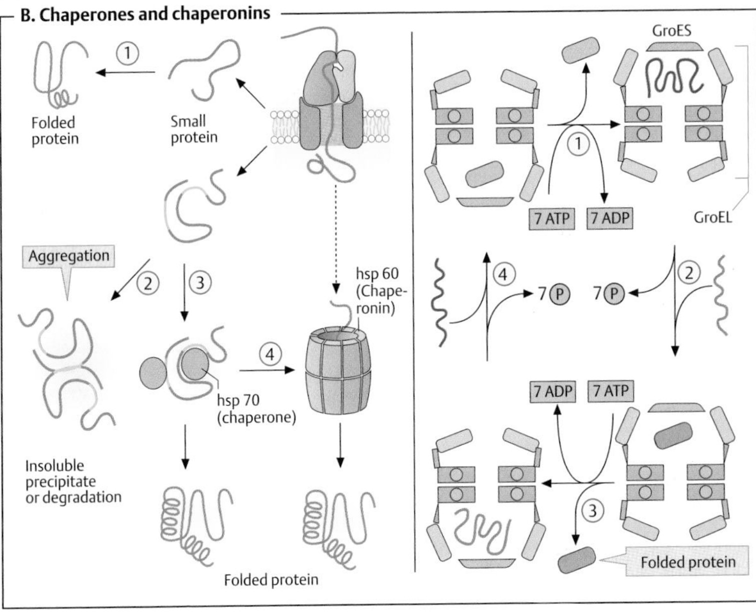

C. Protein import in mitochondria

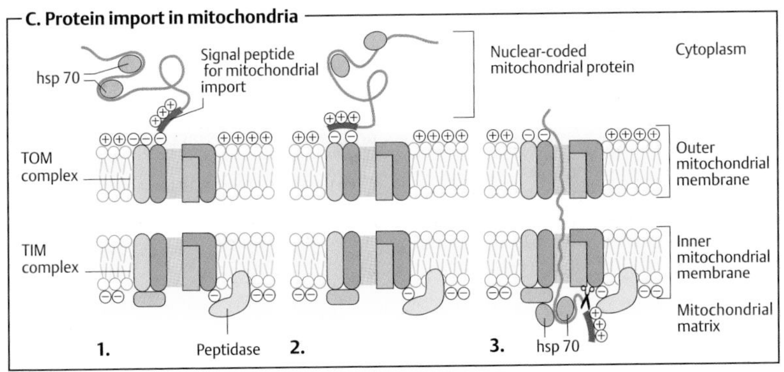

Lysosomes

A. Structure and contents ◑

Animal lysosomes are organelles with a diameter of 0.2–2.0 μm with various shapes that are surrounded by a single membrane. There are usually several hundred lysosomes per cell. ATP-driven V-type proton pumps are active in their membranes (see p. 220). As these accumulate H^+ in the lysosomes, the content of lysosomes with pH values of 4.5–5 is much more acidic than the cytoplasm (pH 7–7.3).

The lysosomes are the cell's "stomach," serving to break down various cell components. For this purpose, they contain some 40 different types of **hydrolases**, which are capable of breaking down every type of macromolecule. The marker enzyme of lysosomes is *acid phosphatase.* The pH optimum of lysosomal enzymes is adjusted to the acid pH value and is also in the range of pH 5. At neutral pH, as in the cytoplasm, lysosomal enzymes only have low levels of activity. This appears to be a mechanism for protecting the cells from digesting themselves in case lysosomal enzymes enter the cytoplasm at any time. In plants and fungi, the **cell vacuoles** (see p. 43) have the function of lysosomes.

B. Functions ◑

Lysosomes serve for enzymatic degradation of macromolecules and cell organelles, which are supplied in various ways. The example shows the degradation of an overaged mitochondrion by *autophagy.* To accomplish this, the lysosome encloses the organelle (**1**). During this process, the **primary lysosome** converts into a **secondary lysosome**, in which the hydrolytic degradation takes place (**2**). Finally, **residual bodies** contain the indigestible residues of the lysosomal degradation process. Lysosomes are also responsible for the degradation of macromolecules and particles taken up by cells via *endocytosis* and *phagocytosis*—e. g., lipoproteins, proteohormones, and bacteria (*heterophagy*). In the process, lysosomes fuse with the **endosomes** (**3**) in which the endocytosed substances are supplied.

C. Synthesis and transport of lysosomal proteins ◑

Primary lysosomes arise in the region of the Golgi apparatus. **Lysosomal proteins** are synthesized in the rER and are glycosylated there as usual (**1**; see p. 228). The next steps are specific for lysosomal proteins (right part of the illustration). In a two-step reaction, terminal mannose residues (Man) are phosphorylated at the C-6 position of the mannose. First, *N*-acetylglucosamine 1-phosphate is transferred to the OH group at C-6 in a terminal mannose residue, and *N*-acetylglucosamine is then cleaved again. Lysosomal proteins therefore carry a terminal **mannose 6-phosphate** (Man6–P; **2**).

The membranes of the Golgi apparatus contain receptor molecules that bind Man 6–P. They recognize lysosomal proproteins by this residue and bind them (**3**). With the help of *clathrin,* the receptors are concentrated locally. This allows the appropriate membrane sections to be pinched off and transported to the endolysosomes with the help of transport vesicles (**4**), from which primary lysosomes arise through maturation (**5**). Finally, the phosphate groups are removed from Man 6–P (**6**).

The *Man 6–P receptors* are reused. The fall in the pH value in the endolysosomes releases the receptors from the bound proteins (**7**) which are then transported back to the Golgi apparatus with the help of transport vesicles.

Further information

Many hereditary diseases are due to genetic defects in lysosomal enzymes. The metabolism of glycogen (→ *glycogenoses*), lipids (→ *lipidoses*), and proteoglycans (→ *mucopolysaccharidoses*) is particularly affected. As the lysosomal enzymes are indispensable for the intracellular breakdown of macromolecules, unmetabolized macromolecules or degradation products accumulate in the lysosomes in these diseases and lead to irreversible cell damage over time. In the longer term, enlargement takes place, and in severe cases there may be failure of the organ affected—e. g., the liver.

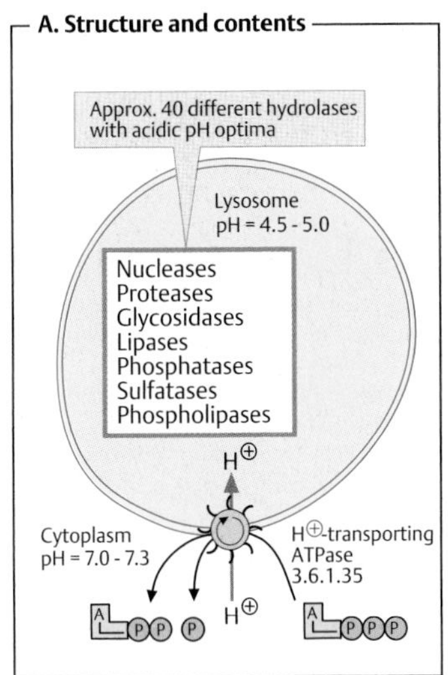
A. Structure and contents
Approx. 40 different hydrolases with acidic pH optima
Lysosome
pH = 4.5 - 5.0
Nucleases
Proteases
Glycosidases
Lipases
Phosphatases
Sulfatases
Phospholipases
H⊕
Cytoplasm
pH = 7.0 - 7.3
H⊕-transporting ATPase
3.6.1.35
H⊕

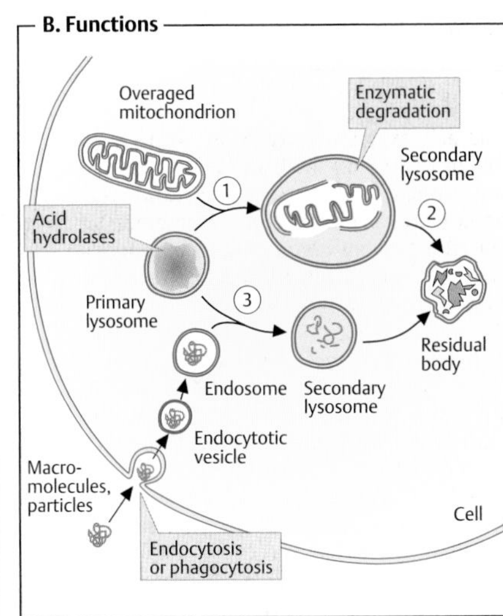
B. Functions
Overaged mitochondrion
Enzymatic degradation
Secondary lysosome
Acid hydrolases
Primary lysosome
Endosome
Secondary lysosome
Residual body
Endocytotic vesicle
Macro-molecules, particles
Cell
Endocytosis or phagocytosis

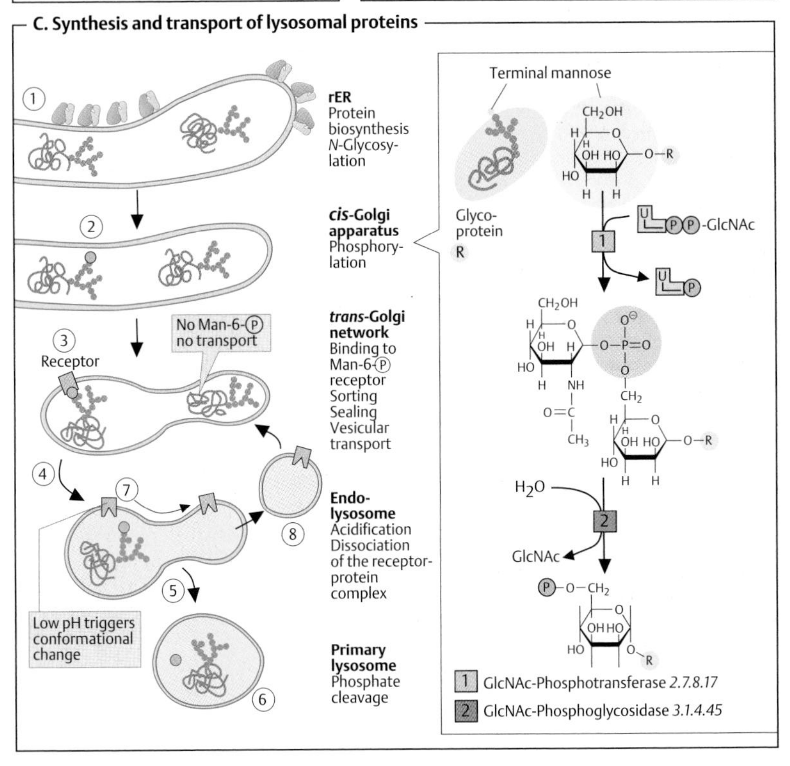
C. Synthesis and transport of lysosomal proteins
rER
Protein biosynthesis
N-Glycosylation
cis-Golgi apparatus
Phosphorylation
trans-Golgi network
Binding to Man-6-P receptor
Sorting
Sealing
Vesicular transport
No Man-6-P no transport
Receptor
Endo-lysosome
Acidification
Dissociation of the receptor-protein complex
Low pH triggers conformational change
Primary lysosome
Phosphate cleavage
Terminal mannose
Glyco-protein
R
-GlcNAc
H2O
GlcNAc
1 GlcNAc-Phosphotransferase 2.7.8.17
2 GlcNAc-Phosphoglycosidase 3.1.4.45

Molecular genetics: overview

Nucleic acids (**DNA** and various **RNAs**) are of central importance in the storage, transmission, and expression of genetic information. The decisive factor involved is their ability to enter into specific base pairings with each other (see p. 84). The individual processes involved, which are summed up in an overview here, are discussed in more detail on the following pages.

A. Expression and transmission of genetic information ●

Storage. The genetic information of all cells is stored in the base sequence of their DNA (RNA only occurs as a genetic material in viruses; see p. 404). Functional sections of DNA that code for inheritable structures or functions are referred to as **genes**. The 30 000–40 000 human genes represent only a few percent of the **genome**, which consists of approximately $5 \cdot 10^9$ base pairs (bp). Most genes code for proteins—i. e., they contain the information for the sequence of amino acid residues of a protein (its sequence). Every amino acid residue is represented in DNA by a code word (a **codon**) consisting of a sequence of three base pairs (a triplet). At the level of DNA, codons are defined as sequences of the sense strand read in the 5′→3′ direction (see p. 84). A DNA codon for the amino acid phenylalanine, for example, is thus *TTC* (**2**).

Replication. During cell division, all of the genetic information has to be passed on to the daughter cells. To achieve this, the whole of the DNA is copied during the S phase of the cell cycle (see p. 394). In this process, each strand serves as a matrix for the synthesis of a complementary second strand (**1**; see p. 240).

Transcription. For *expression* of a gene—i. e., synthesis of the coded protein—the DNA sequence information has to be converted into a protein sequence. As DNA itself is not involved in protein synthesis, the information is transferred from the nucleus to the site of synthesis in the cytoplasm. To achieve this, the *template strand* in the relevant part of the gene is transcribed into an **RNA** (hnRNA). The sequence of this RNA is thus complementary to that of the *template strand* (**3**), but—with the exception of the exchange of thymine for uracil—it is identical to that of the sense strand. In this way, the DNA triplet *TTC* gives rise in hnRNA to the RNA codon *UUC*.

RNA maturation. In eukaryotes, the hnRNA initially formed is modified several times before it can leave the nucleus as **messenger RNA** (mRNA, **4**). During RNA maturation, superfluous ("intervening") sequences (introns) are removed from the molecule, and both ends of the transcript are protected by the addition of further nucleotides (see p. 246).

Translation. Mature mRNA enters the cytoplasm, where it binds to **ribosomes**, which convert the RNA information into a peptide sequence. The ribosomes (see p. 250) consist of more than 100 proteins and several RNA molecules (rRNA; see p. 82). rRNA plays a role as a ribosomal structural element and is also involved in the binding of mRNA to the ribosome and the formation of the peptide bond.

The actual information transfer is based on the interaction between the mRNA codons and another type of RNA, **transfer RNA** (tRNA; see p. 82). tRNAs, of which there are numerous types, always provide the correct amino acid to the ribosome according to the sequence information in the mRNA. tRNAs are loaded with an amino acid residue at the 3′ end. Approximately in the middle, they present the triplet that is complementary to each mRNA codon, known as the *anticodon* (*GAA* in the example shown). If the codon *UUC* appears on the mRNA, the anticodon binds a molecule of Phe-t-RNA^{Phe} to the mRNA (**5**) and thus brings the phenylalanine residue at the other end of the molecule into a position in which it can take over the growing polypeptide chain from the neighboring tRNA (**6**).

Amino acid activation. Before binding to the ribosomes, tRNAs are loaded with the correct amino acids by specific ligases (**7**; see p. 248). It is the amino acid tRNA ligases that carry out the transfer (translation) of the genetic information from the nucleic acid level to the protein level.

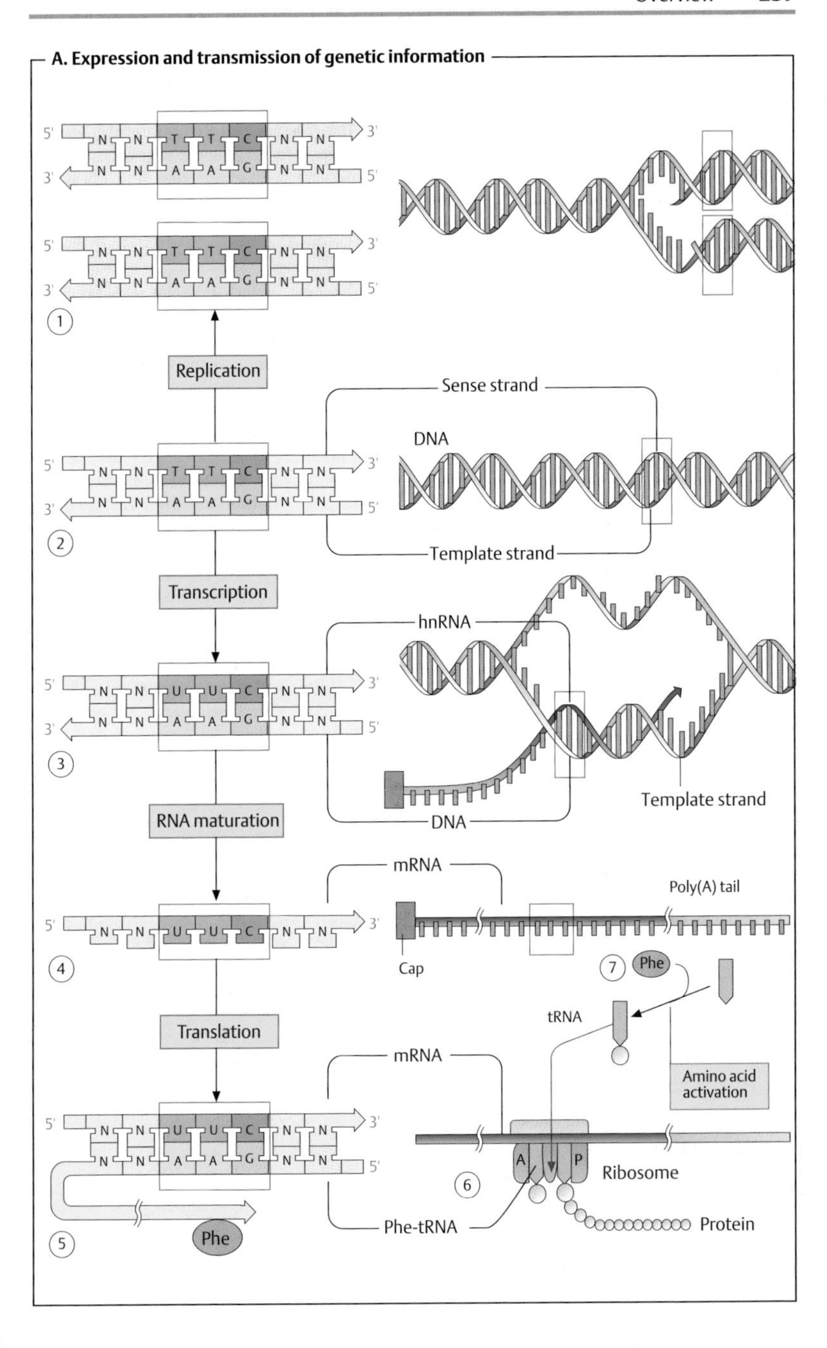
A. Expression and transmission of genetic information
5'
3'
N
T
C
A
G
1
Replication
Sense strand
DNA
2
Template strand
Transcription
hnRNA
U
3
Template strand
DNA
RNA maturation
mRNA
Poly(A) tail
4
Cap
7
Phe
tRNA
Translation
mRNA
Amino acid activation
A
P
Ribosome
6
Phe-tRNA
Protein
5
Phe

Genome

A. Chromatin ◑

In the nuclei of eukaryotes (see p. 196), DNA is closely associated with proteins and RNA. These nucleoprotein complexes, with a DNA proportion of approximately one-third, are known as **chromatin**. It is only during cell division (see p. 394) that chromatin condenses into **chromosomes** that are visible under light microscopy. During interphase, most of the chromatin is loose, and in these conditions a morphological distinction can be made between tightly packed **heterochromatin** and the less dense **euchromatin**. Euchromatin is the site of active transcription.

The proteins contained in chromatin are classified as either histone or non-histone proteins. **Histones** (**B**) are small, strongly basic proteins that are directly associated with DNA. They contribute to the structural organization of chromatin, and their basic amino acids also neutralize the negatively charged phosphate groups, allowing the dense packing of DNA in the nucleus. This makes it possible for the 46 DNA molecules of the diploid human genome, with their $5 \cdot 10^9$ base pairs (bp) and a total length of about 2 m, to be accommodated in a nucleus with a diameter of only 10 μm. Histones also play a central role in regulating transcription (see p. 244).

Two histone molecules each of types **H2A** (blue), **H2B** (green), **H3** (yellow), and **H4** (red) form an octameric complex, around which 146 bp of DNA are wound in 1.8 turns. These particles, with a diameter of 7 nm, are referred to as **nucleosomes**. Another histone (**H1**) binds to DNA segments that are not directly in contact with the histone octamers ("linker" DNA). It covers about 20 bp and supports the formation of spirally wound superstructures with diameters of 30 nm, known as **solenoids**. When chromatin condenses into chromosomes, the solenoids form **loops** about 200 nm long, which already contain about 80 000 bp. The loops are bound to a protein framework (the **nuclear scaffolding**), which in turn organizes some 20 loops to form **minibands**. A large number of stacked minibands finally produces a chromosome. In the chromosome, the DNA is so densely packed that the smallest human chromosome already contains more than 50 million bp.

The **non-histone proteins** are very heterogeneous. This group includes *structural proteins* of the nucleus, as well as many *enzymes* and *transcription factors* (see p. 118), which selectively bind to specific segments of DNA and regulate gene expression and other processes.

B. Histones ○

The histones are remarkable in several ways. With their high proportions of lysine and arginine (blue shading), they are strongly basic, as mentioned above. In addition, their amino acid sequence has hardly changed at all in the course of evolution. This becomes clear when one compares the histone sequences in mammals, plants, and fungi (yeasts are single-celled fungi; see p. 148). For example, the H4 histones in humans and wheat differ only in a single amino acid residue, and there are only a few changes between humans and yeast. In addition, all of these changes are "conservative"—i. e., the size and polarity barely differ. It can be concluded from this that the histones were already "optimized" when the last common predecessor of animals, plants, and fungi was alive on Earth (more than 700 million years ago). Although countless mutations in histone genes have taken place since, almost all of these evidently led to the extinction of the organisms concerned.

The histones in the octamer carry N-terminal mobile "tails" consisting of some 20 amino acid residues that project out of the nucleosomes and are important in the regulation of chromatin structure and in controlling gene expression (see **A2**; only two of the eight tails are shown in full length). For example, the condensation of chromatin into chromosomes is associated with *phosphorylation* (P) of the histones, while the transcription of genes is initiated by *acetylation* (A) of lysine residues in the N-terminal region (see p. 244).

A. Chromatin

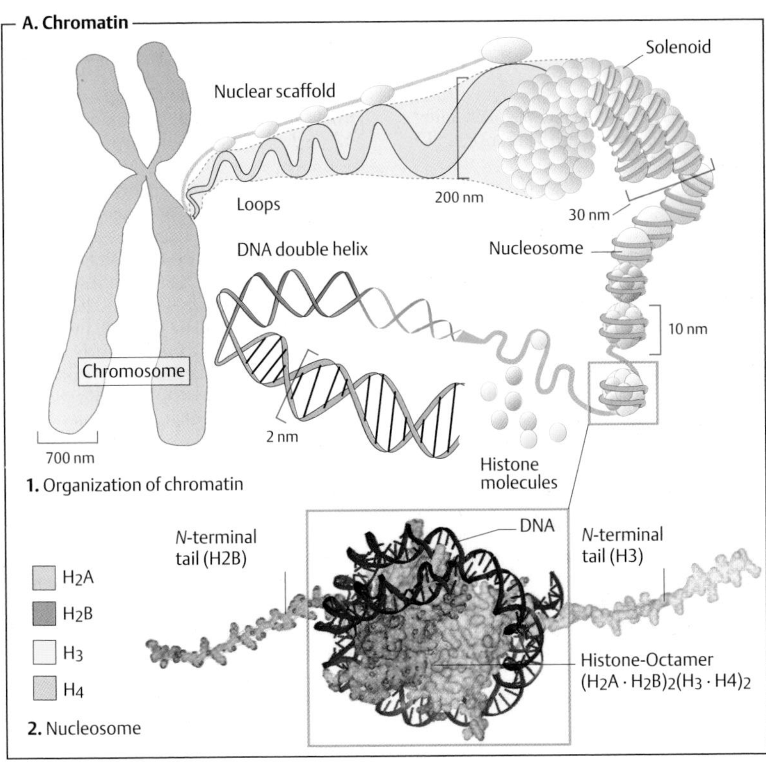

B. Histone

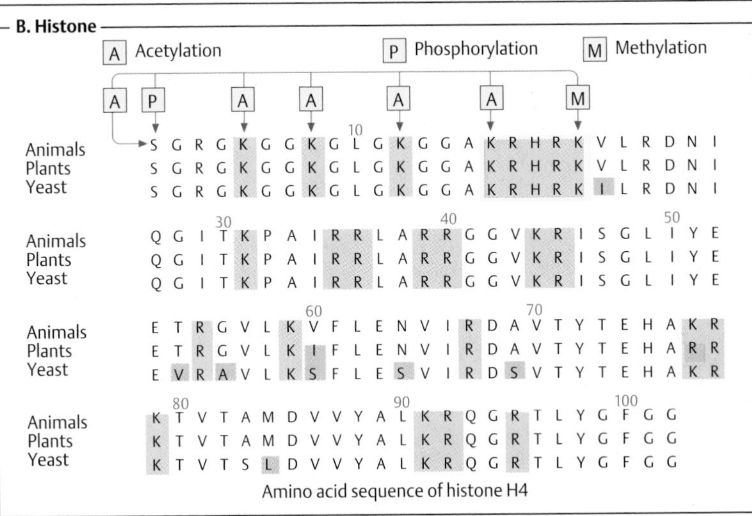

Replication

For genetic information to be passed on during cell division, a complete copy of the genome has to be produced before each mitosis. This process is known as DNA **replication**.

A. Mechanism of DNA polymerases ◑

Replication is catalyzed by *DNA-dependent DNA polymerases.* These enzymes require a single strand of DNA, known as the **template**. Beginning at a short starting sequence of RNA (the **primer**), they synthesize a second complementary strand on the basis of this template, and thus create a complete DNA double helix again. The substrates of the DNA polymerases are the four deoxynucleoside triphosphates **dATP**, **dGTP**, **dCTP**, and **dTTP**. In each step, base pairing first binds the nucleotide that is complementary to the current base in the template strand. The α-phosphate residue of the newly bound nucleoside triphosphate is then subjected to nucleophilic attack by the 3′-OH group of the nucleotide incorporated immediately previously. This is followed by the elimination of diphosphate and the formation of a new phosphoric acid diester bond. These steps are repeated again for each nucleotide. The mechanism described means that the matrix can only be read in the 3′→5′ direction. In other words, the newly synthesized strand always grows in the **5′→3′ direction**. The same mechanism is also used in transcription by *DNA-dependent RNA polymerases* (see p. 242). Most DNA and RNA polymerases consist of more than 10 subunits, the role of which is still unclear to some extent.

B. Replication in *E. coli* ◑

Although replication in prokaryotes is now well understood, many details in eukaryotes are still unclear. However, it is certain that the process is in principle similar. A simplified scheme of replication in the bacterium *Escherichia coli* is shown here.

In bacteria, replication starts at a specific point in the circular DNA—the **origin of replication**—and proceeds in both directions. This results in two diverging **replication forks**, in which the two strands are replicated simultaneously. Numerous proteins are involved in the processes taking place in this type of fork, and only the most important are shown here. The two strands of the initial DNA (**1**) are shown in blue and violet, while the newly formed strands are pink and orange.

Each fork (**2**) contains two molecules of **DNA polymerase III** and a number of helper proteins. The latter include **DNA topoisomerases** and **single-strand-binding proteins.** Topoisomerases are enzymes that unwind the superhelical DNA double strand (*gyrase, topoisomerase II*) and then separate it into the two individual strands (*helicase, topoisomerase I*). Since the template strand is always read from 3′ to 5′ (see above), only one of the strands (known as the **leading** strand; violet/pink) can undergo *continuous* replication. For the **lagging strand** (light blue), the reading direction is the *opposite* of the direction of movement of the fork. In this matrix, the new strand is first synthesized in individual pieces, which are known as **Okazaki fragments** after their discoverer (green/orange).

Each fragment starts with a short RNA primer (green), which is necessary for the functioning of the DNA polymerase and is synthesized by a special *RNA polymerase* (*"primase,"* not shown). The primer is then extended by *DNA polymerase III* (orange). After 1000–2000 nucleotides have been included, synthesis of the fragment is interrupted and a new one is begun, starting with another RNA primer that has been synthesized in the interim. The individual Okazaki fragments are initially not bound to one another and still have RNA at the 5′ end (**3**). At some distance from the fork, *DNA polymerase I* therefore starts to remove the RNA primer and replace it with DNA components. Finally, the gaps still remaining are closed by a *DNA ligase.* In DNA double helices formed in this way, only *one* of the strands has been newly synthesized—i. e., replication is *semiconservative.*

In bacteria, some 1000 nucleotides are replicated per second. In eukaryotes, replication takes place more slowly (about 50 nucleotides $\cdot$ s^{-1}) and the genome is larger. Thousands of replication forks are therefore active simultaneously in eukaryotes.

A. Mechanism of DNA polymerases

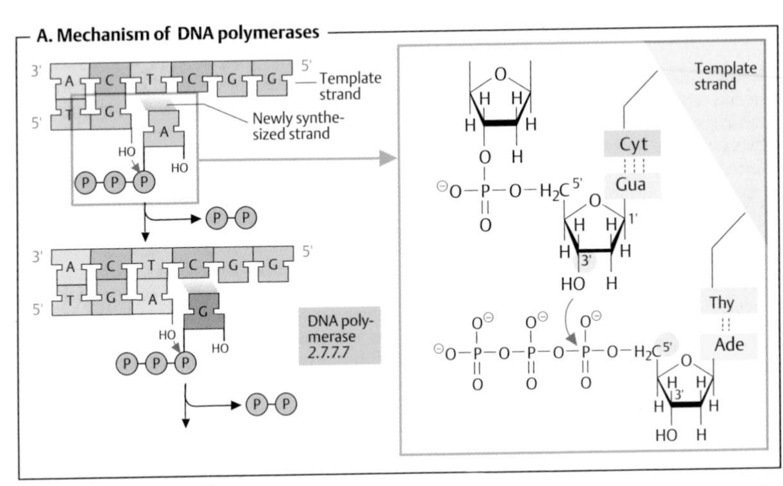

B. Replication in *E. coli*

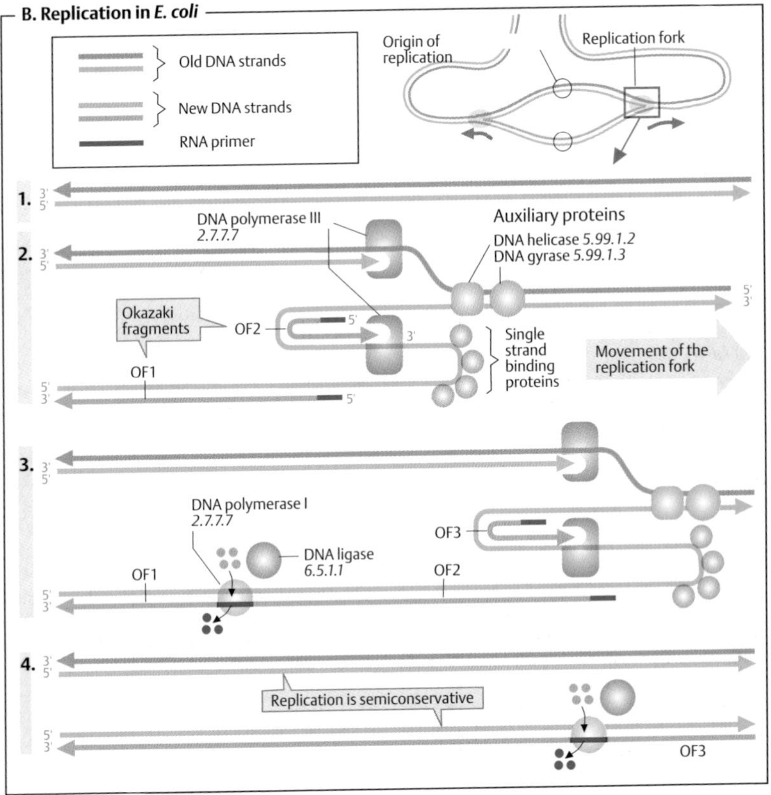

Transcription

For the genetic information stored in DNA to become effective, it has to be rewritten (transcribed) into RNA. DNA only serves as a template and is not altered in any way by the **transcription** process. Transcribable segments of DNA that code for a defined product are called **genes.** It is estimated that the mammalian genome contains 30 000–40 000 genes, which together account for less than 5% of the DNA.

A. Transcription and maturation of RNA: overview ◑

Transcription is catalyzed by *DNA-dependent RNA polymerases.* These act in a similar way to DNA polymerases (see p. 240), except that they incorporate *ribonucleotides* instead of deoxyribonucleotides into the newly synthesized strand; also, they do not require a primer. Eukaryotic cells contain at least three different types of RNA polymerase. *RNA polymerase I* synthesizes an RNA with a sedimentation coefficient (see p. 200) of 45 S, which serves as precursor for three ribosomal RNAs. The products of *RNA polymerase II* are hnRNAs, from which mRNAs later develop, as well as precursors for snRNAs. Finally, *RNA polymerase III* transcribes genes that code for tRNAs, 5S rRNA, and certain snRNAs. These precursors give rise to functional RNA molecules by a process called **RNA maturation** (see p. 246). Polymerases II and III are inhibited by *α-amanitin,* a toxin in the *Amanita phalloides* mushroom.

B. Organization of the PEP-CK gene ○

The way in which a typical eukaryotic gene is organized is illustrated here using a gene that codes for a key enzyme in gluconeogenesis (see p. 154)—the *phosphoenolpyruvate carboxykinase* (PEP-CK).

In the rat, the PEP-CK gene is nearly 7 kbp (kilobase pairs) long. Only 1863 bp, distributed over 10 coding segments (**exons**, dark blue) carry the information for the protein's 621 amino acids. The remainder is allotted to the promoter (pink) and intervening sequences (**introns**, light blue). The gene's promoter region (approximately 1 kbp) serves for regulation (see p. 188). Transcription starts at the 3′ end of the promoter ("transcription start") and continues until the polyadenylation sequence (see below) is reached. The primary transcript (**hnRNA**) still has a length of about 6.2 kbp. During RNA maturation, the noncoding sequences corresponding to the introns are removed, and the two ends of the hnRNA are modified. The translatable mRNA still has half the length of the hnRNA and is modified at both ends (see p. 246).

In many eukaryotic genes, the proportion of introns is even higher. For example, the gene for *dihydrofolate reductase* (see p. 402) is over 30 kbp long. The information is distributed over six exons, which together have a length of only about 6 kbp.

C. Transcription process ◑

As mentioned above, *RNA polymerase II* (green) binds to the 3′ end of the promoter region. A sequence that is important for this binding is known as the **TATA box**—a short A- and T-rich sequence that varies slightly from gene to gene. A typical base sequence *("consensus sequence")* is ...TATAAA... Numerous proteins known as *basal transcription factors* are necessary for the interaction of the polymerase with this region. Additional factors can promote or inhibit the process (transcriptional control; see p. 244). Together with the polymerase, they form the **basal transcription complex**.

At the end of **initiation** (**2**), the polymerase is repeatedly phosphorylated, frees itself from the basal complex, and starts moving along the DNA in the 3′ direction. The enzyme separates a short stretch of the DNA double helix into two single strands. The complementary nucleoside triphosphates are bound by base pairing in the *template strand* and are linked step by step to the hnRNA as it grows in the 5′→3′ direction (**3**). Shortly after the beginning of **elongation**, the 5′ end of the transcript is protected by a "cap" (see p. 246). Once the polyadenylation sequence has been reached (typical sequence: ...AATAA...), the transcript is released (**4**). Shortly after this, the RNA polymerase stops transcribing and dissociates from the DNA.

A. Transcription and maturation of RNA: overview

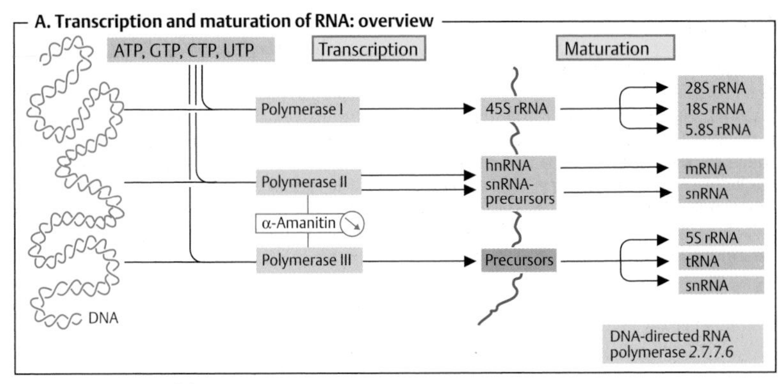

B. Organization of the PEP-CK gene

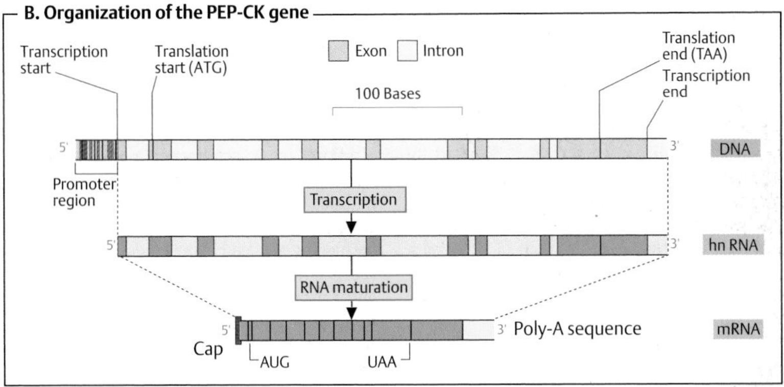

C. Process of transcription

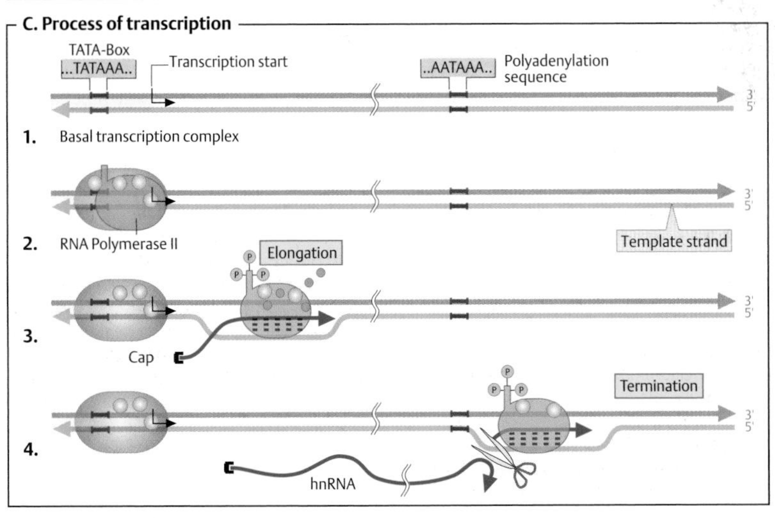

Transcriptional control

Although all cells contain the complete genome, they only use a fraction of the information in it. The genes known as "housekeeping genes," which code for structural molecules and enzymes of intermediate metabolism, are the only ones that undergo constant transcription. The majority of genes are only active in certain cell types, in specific metabolic conditions, or during differentiation. Which genes are transcribed and which are not is regulated by **transcriptional control** (see also p. 118). This involves *control elements* (*cis*-active elements) in the gene's promoter region and gene-specific regulatory proteins (transcription factors, *trans* -active factors), which bind to the control elements and thereby activate or inhibit transcription.

A. Initiation of transcription ○

In the higher organisms, DNA is blocked by histones (see p. 238) and is therefore not capable of being transcribed without special positive regulation. In eukaryotes, it is therefore histones that play the role of *repressors* (see p. 118). For transcription to be set in motion at all, the chromatin first has to be restructured.

In the resting state, the lysine residues in the N-terminal "tail" of the histones (see p. 238) are not acetylated. In this state, which can be produced by *histone deacetylases* [1], the nucleosomes are stable. It is only the interaction of activator and regulator proteins with their control elements that allows the binding of coactivator complexes that have *histone acetylase* activity [2]. They acetylate the histone tails and thereby loosen the nucleosome structure sufficiently for the basal transcription complex to form.

This consists of *DNA-dependent RNA polymerase II* and basal transcription factors (TFIIX, X = A – H). First, the basal factor TFIID binds to the promoter. TFIID, a large complex of numerous proteins, contains *TATA box-binding protein* (TBP) and so-called TAFs (TBP-associated factors). The polymerase is attached to this core with the help of TFIIB. Before transcription starts, additional TFs have to bind, including TFIIH, which has *helicase* activity and separates the two strands of DNA during elongation. In all, some 35 different proteins are involved in the basal complex. This alone, however, is still not sufficient for transcription to start. In addition, positive signals have to be emitted by more distant *trans*-active factors, integrated by the coactivator/mediator complex, and passed on to the basal complex (see **B**).

The actual signal for starting elongation consists of the multiple phosphorylation of a domain in the C-terminal region of the polymerase. In phosphorylated form, it releases itself from the basal complex along with a few TFs and starts to synthesize hnRNA.

B. Regulation of PEP-CK transcription ○

Phosphoenolpyruvate carboxykinase (PEP-CK), a key enzyme in gluconeogenesis, is regulated by several hormones, all of which affect the transcription of the PEP-CK gene. Cortisol, glucagon, and thyroxin induce PEP-CK, while insulin inhibits its induction (see p. 158).

More than ten **control elements** (dark red), distributed over approximately 1 kbp, have so far been identified in the promoter of the PEP-CK gene (top). These include response elements for the glucocorticoid receptor (GRE; see p. 378), for the thyroxin receptor (TRE), and for the steroid-like retinoic acid (AF-1). Additional control elements (CRE, cAMP-responsive element) bind the transcription factor C/EBP, which is activated by cAMP-dependent protein kinase A through phosphorylation. This is the way in which glucagon, which raises the cAMP level (see p. 158), works. Control element P1 binds the hormone-independent factor NF-1 (nuclear factor-1). All proteins that bind to the control elements mentioned above are in contact with a **coactivator/mediator complex** (CBP/p300), which integrates their input like a computer and transmits the result in the form of stronger or weaker signals to the basal transcription complex. Inhibition of PEP-CK transcription by insulin is mediated by an insulin-responsive element (IRE) in the vicinity of the GRE. Binding of an as yet unknown factor takes place here, inhibiting the binding of the glucocorticoid receptor to the GRE.

A. Initiation of transcription

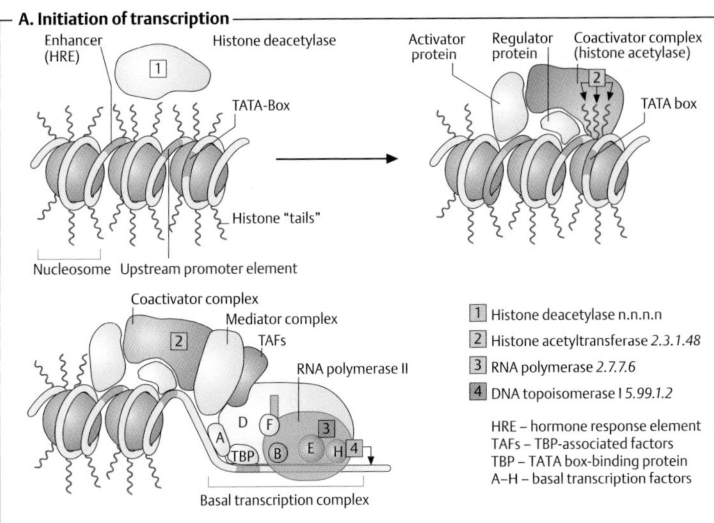

B. Regulation of PEP-CK transcription

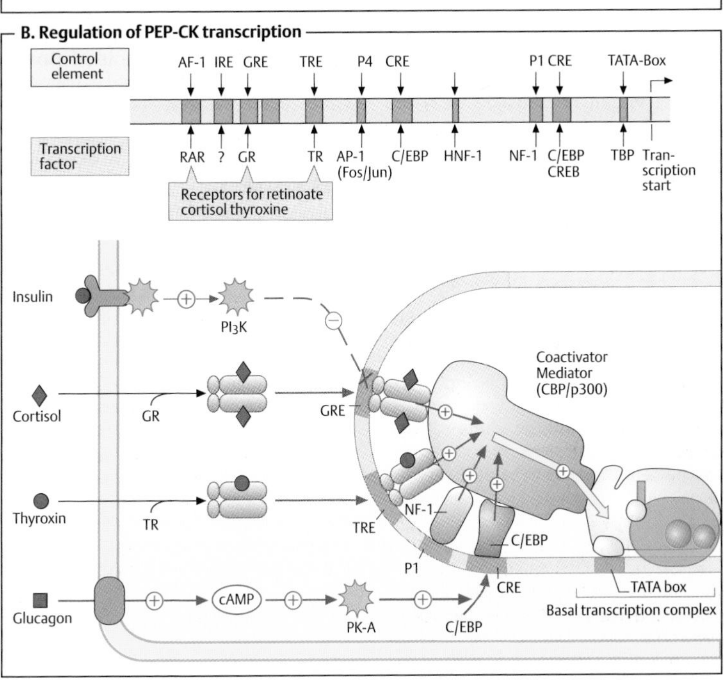

RNA maturation

Before the hnRNA produced by RNA polymerase II (see p. 242) can leave the nucleus in order to serve as a template for protein synthesis in the cytoplasm, it has to undergo several modifications first. Even during transcription, the two ends of the transcript have additional nucleotides added (**A**). The sections that correspond to the intervening gene sequences in the DNA (introns) are then cut out (splicing; see **B**). Other transcripts—e. g., the 45 S precursor of rRNA formed by polymerase I (see p. 242)—are broken down into smaller fragments by nucleases before export into the cytoplasm.

A. 5′ and 3′ modification of mRNA ◑

Shortly after transcription begins in eukaryotes, the end of the growing RNA is blocked in several reaction steps by a structure known as a "**cap**." In hnRNAs, this consists of a GTP residue that is methylated at N-7 of the guanine ring. The β-phosphate residue of the cap is linked to the free 5′-OH group of the terminal ribose via an ester bond. After the "polyadenylation signal" has been reached (typical sequence: ...AAUAAA...; see p. 242), a **polyadenylate "tail"** consisting of up to 200 AMP nucleotides is also added at the free 3′ end of the transcript. This reaction is catalyzed by a special *polyadenylate polymerase.* It is only at this point that the mRNA leaves the nucleus as a complex with RNA-binding proteins.

Both the cap and the poly-A tail play a vital part in initiating eukaryotic translation (see p. 250). They help position the ribosome correctly on the mRNA near to the starting codon. The protection which the additional nucleotides provide against premature enzymatic degradation appears to be of lesser importance.

B. Splicing of hnRNA ◑

Immediately after transcription, the hnRNA introns are removed and the exons are linked to form a continuous coding sequence. This process, known as **splicing**, is supported by complicated RNA–protein complexes in the nucleus, the so-called **spliceosomes**. The components of these macromolecular machines are called **snRNPs** (*small nuclear ribonucleoprotein particles,* pronounced "snurps"). SnRNPs occur in five different forms (U1, U2, U4, U5, and U6). They consist of numerous *proteins* and one molecule of *snRNA* each (see p. 82).

To ensure that the RNA message is not destroyed, splicing has to take place in a very precise fashion. The start and end of the hnRNA introns are recognized by a characteristic sequence (...AGGT... at the 5′ end or ...[C,U]AGG... at the 3′ end). Another important structure is the so-called *branching point* inside the intron. Its sequence is less conserved than the terminal splicing sites, but it always contains one adenosine residue (A). During splicing, the 2′-OH group of this residue—supported by the spliceosome (see **C**)—attacks the phosphoric acid diester bond at the 5′ end of the intron and cleaves it (**b**). Simultaneously, an unusual 2′→5′ bond is formed inside the intron, which thereby takes on a *lasso shape* (**c**; see formula). In the second step of the splicing process, the free 3′-OH group at the end of the 5′ terminal exon attacks the A–G bond at the 3′ end of the intron. As a result, the two exons are linked and the intron is released, still in a lasso shape (d).

C. Spliceosome ○

As described above, it is residues of the hnRNA that carry out bond cleavage and bond formation during the splicing process. It is therefore not a protein enzyme that acts as a catalyst here, but rather an RNA. Catalytic RNAs of this type are called *ribozymes* (see also p. 88). The task of the spliceosomes is to fix and orientate the reacting groups by establishing base pairings between snRNAs and segments of the hnRNA. The probable situation before the adenosine attack at the branching point on the 5′ splicing site (see **B**, Fig. **b**) is shown schematically on the right side of the illustration. In this phase, the U1 snRNA fixes the 5′ splicing site, U2 fixes the branching site, and U5 fixes the ends of the two exons.

A. 5' and 3' modification of mRNA

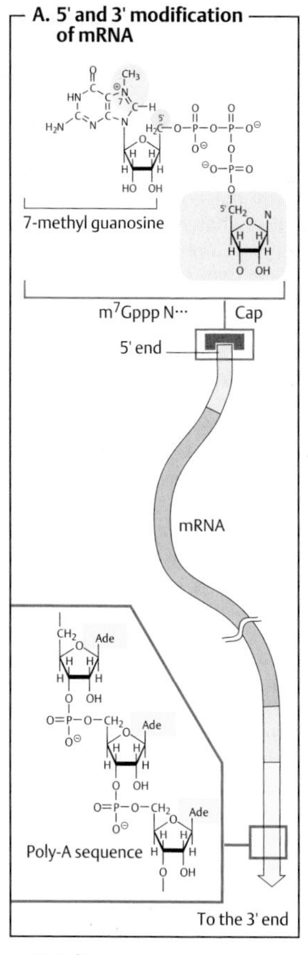

B. Splicing of hnRNA: mechanism

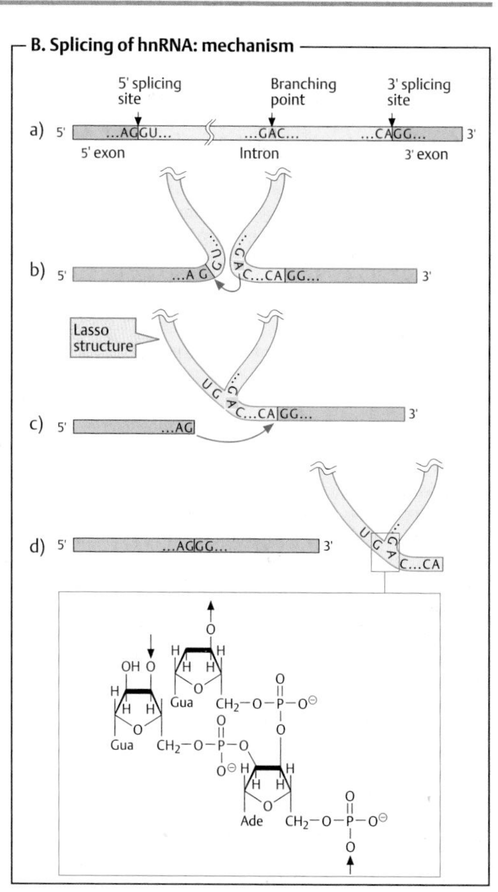

C. Spliceosome

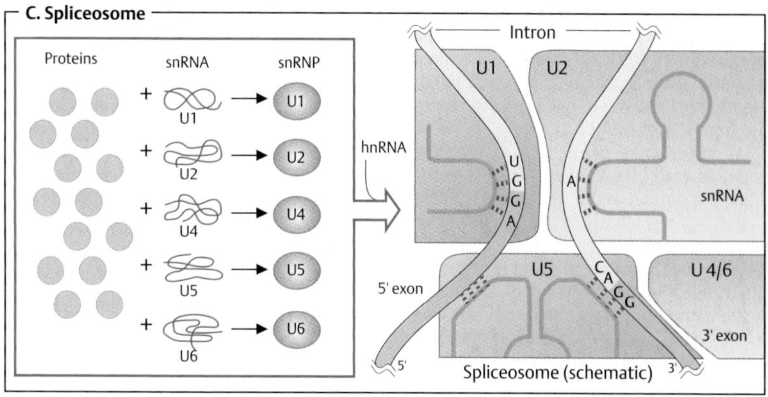

Amino acid activation

A. The genetic code ◑

Most of the genetic information stored in the genome codes for the amino acid sequences of proteins. For these proteins to be expressed, a text in "nucleic acid language" therefore has to be translated into "protein language." This is the origin of the use of the term **translation** to describe protein biosynthesis. The dictionary used for the translation is the **genetic code.**

As there are 20 proteinogenic amino acids (see p. 60), the nucleic acid language has to contain at least as many words **(codons).** However, there are only four letters in the nucleic acid alphabet (A, G, C, and U or T). To obtain 20 different words from these, each word has to be at least three letters long (with two letters, there would only be $4^2 = 16$ possibilities). And in fact the codons do consist of three sequential bases **(triplets).**

Figure 1 shows the standard code in "DNA language" (i. e., as a sequence of triplets in the *sense strand* of DNA, read in the 5'→3' direction; see p. 84), represented as a circular diagram. The scheme is read from the inside to the outside. For example, the triplet CAT codes for the amino acid histidine. With the exception of the exchange of U for T, the DNA codons are identical to those of mRNA.

As the genetic code provides $4^3 = 64$ codons for the 20 amino acids, there are several synonymous codons for most amino acids—the code is **degenerate**. Three triplets do not code for amino acids, but instead signal the end of translation **(stop codons)**. Another special codon, the **start codon**, marks the start of translation. The code shown here is almost universally applicable; only the mitochondria (see p. 210) and a few microorganisms deviate from it slightly.

As an example of the way in which the code is read, Fig. 2 shows small sections from the normal and a mutated form of the β-globin gene (see p. 280), as well as the corresponding mRNA and protein sequences. The **point mutation** shown, which is relatively frequent, leads to replacement of a glutamate residue in position 6 of the β-chain by valine (GAG → GTG). As a consequence, the mutated hemoglobin tends to aggregate in the deoxygenated form. This leads to sickle-shaped distortions of the erythrocytes and disturbances of O_2 transport (sickle-cell anemia).

B. Amino acid activation ◑

Some 20 different *amino acid tRNA ligases* in the cytoplasm each bind one type of tRNA (see p. 82) with the corresponding amino acid. This reaction, known as **amino acid activation**, is endergonic and is therefore coupled to ATP cleavage in two steps.

First, the amino acid is bound by the enzyme and reacts there with ATP to form diphosphate and an "energy–rich" mixed acid anhydride (**aminoacyl adenylate**). In the second step, the 3'-OH group (in other ligases it is the 2'-OH group) of the terminal ribose residue of the tRNA takes over the amino acid residue from the aminoacyl adenylate. In **aminoacyl tRNAs**, the carboxyl group of the amino acid is therefore esterified with the ribose residue of the terminal adenosine of the sequence ...CCA-3'.

The accuracy of translation primarily depends on the specificity of the amino acid tRNA ligases, as incorrectly incorporated amino acid residues are not recognized by the ribosome later. A *"proofreading mechanism"* in the active center of the ligase therefore ensures that incorrectly incorporated amino acid residues are immediately removed again. On average, an error only occurs once every 1300 amino acid residues. This is a surprisingly low rate considering how similar some amino acids are—e. g., leucine and isoleucine.

C. Asp–tRNA ligase (dimer) ○

The illustration shows the ligase responsible for the activation of aspartate. Each subunit of the dimeric enzyme (protein parts shown in orange) binds one molecule of $tRNA^{Asp}$ (blue). The active centers can be located by the bound ATP (green). They are associated with the 3' end of the tRNA. Another domain in the protein (upper left) is responsible for "recognition" of the tRNA anticodon.

A. The genetic code

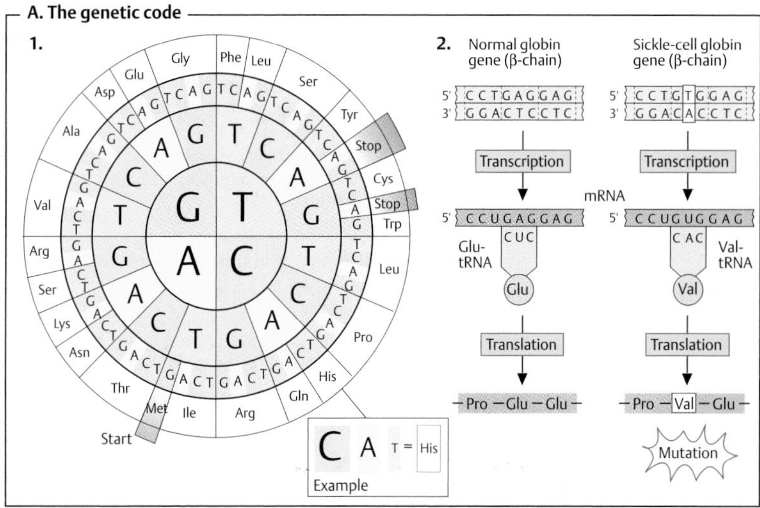

B. Amino acid activation

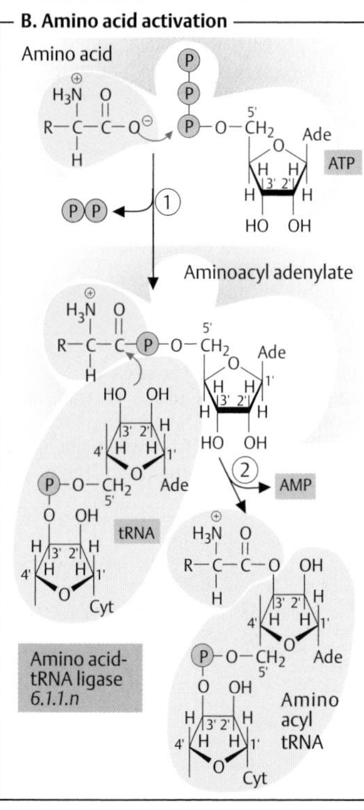

C. Asp-tRNA-Ligase (Dimer)

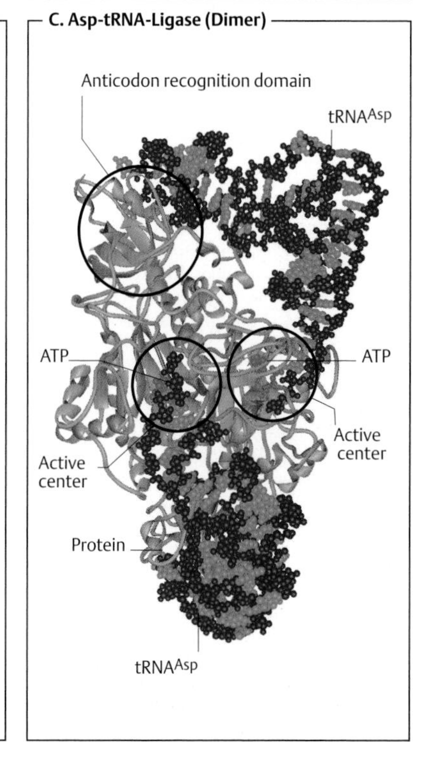

Translation I: initiation

Like amino acid activation (see p. 248), protein biosynthesis (**translation**) takes place in the cytoplasm. It is catalyzed by complex nucleoprotein particles, the **ribosomes**, and mainly requires GTP to cover its energy requirements.

A. Structure of the eukaryotic ribosome ◑

Ribosomes consist of two subunits of different size, made up of **ribosomal RNA (rRNA)** and nearly 80 **proteins** (the number of proteins applies to rat liver ribosomes). It is customary to give the sedimentation coefficients (see p. 200) of ribosomes and their components instead of their masses. For example, the eukaryotic ribosome has a sedimentation coefficient of 80 Svedberg units (80 S), while the sedimentation coefficients of its subunits are 40 S and 60 S (S values are not additive).

The smaller 40 S subunit consists of one molecule of 18 S rRNA and 33 protein molecules. The larger 60 S subunit contains three types of rRNA with sedimentation coefficients of 5 S, 5.8 S, and 28 S and 47 proteins. In the presence of mRNA, the subunits assemble to form the complete ribosome, with a mass about 650 times larger than that of a hemoglobin molecule.

The arrangement of the individual components of a ribosome has now been determined for prokaryotic ribosomes. It is known that filamentous mRNA passes through a cleft between the two subunits near the characteristic "horn" on the small subunit. tRNAs also bind near this site. The illustration shows the size of a tRNA molecule for comparison.

Prokaryotic ribosomes have a similar structure, but are somewhat smaller than those of eukaryotes (sedimentation coefficient 70 S for the complete ribosome, 30 S and 50 S for the subunits). Mitochondrial and chloroplast ribosomes are comparable to prokaryotic ones.

B. Polysomes ◑

In cells that are carrying out intensive protein synthesis, ribosomes are often found in a linear arrangement like a string of pearls; these are known as **polysomes**. This arrangement arises because several ribosomes are translating a single mRNA molecule simultaneously. The ribosome first binds near the **start codon** (AUG; see p. 248) at the 5′ end of the mRNA (top). During translation, the ribosome moves in the direction of the 3′ end until it reaches a **stop codon** (UAA, UAG, or UGA). At this point, the newly synthesized chain is released, and the ribosome dissociates again into its two subunits.

C. Initiation of translation in E. coli ◑

Protein synthesis in prokaryotes is in principle the same as in eukaryotes. However, as the process is simpler and has been better studied in prokaryotes, the details involved in translation are discussed here and on p. 252 using the example of the bacterium *Escherichia coli.*

The first phase of translation, **initiation**, involves several steps. First, two proteins, **initiation factors** IF–1 and IF–3, bind to the 30 S subunit (**1**). Another factor, IF–2, binds as a complex with GTP (**2**). This allows the subunit to associate with the mRNA and makes it possible for a special tRNA to bind to the start codon (**3**). In prokaryotes, this starter tRNA carries the substituted amino acid *N-formylmethionine* (fMet). In eukaryotes, it carries an unsubstituted *methionine.* Finally, the 50 S subunit binds to the above complex (**4**). During steps 3 and 4, the initiation factors are released again, and the GTP bound to IF–2 is hydrolyzed to GDP and P_i.

In the **70 S initiation complex**, formylmethionine tRNA is initially located at a binding site known as the **peptidyl site (P)**. A second binding site, the **acceptor site (A)**, is not yet occupied during this phase of translation. Sometimes, a third tRNA binding site is defined as an *exit site (E)*, from which uncharged tRNAs leave the ribosome again (see p. 252; not shown).

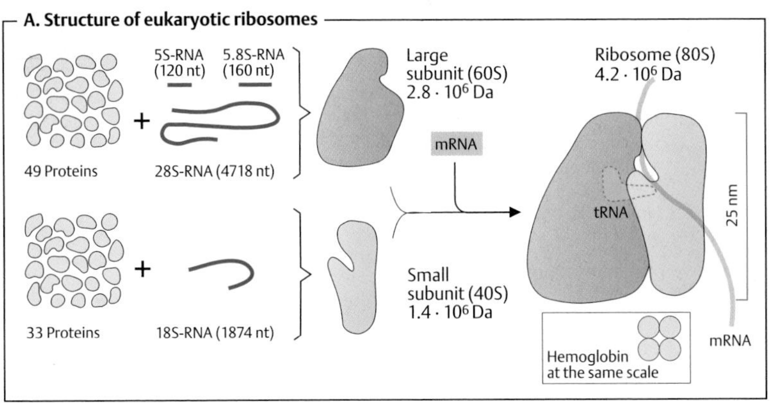
A. Structure of eukaryotic ribosomes
5S-RNA (120 nt)
5.8S-RNA (160 nt)
49 Proteins
28S-RNA (4718 nt)
Large subunit (60S) 2.8 · 10⁶ Da
mRNA
Ribosome (80S) 4.2 · 10⁶ Da
tRNA
25 nm
33 Proteins
18S-RNA (1874 nt)
Small subunit (40S) 1.4 · 10⁶ Da
Hemoglobin at the same scale
mRNA

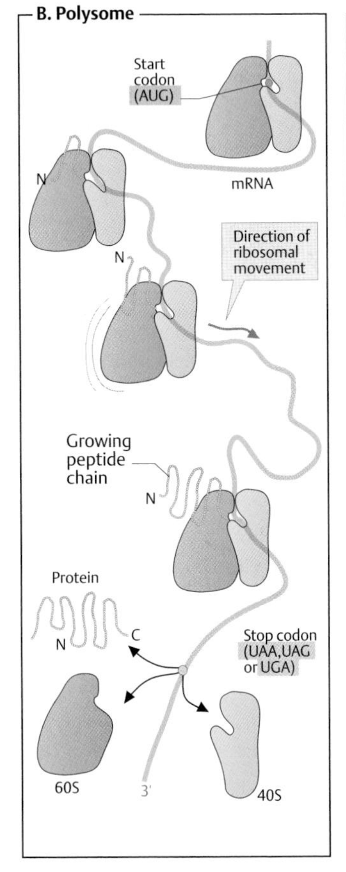
B. Polysome
Start codon (AUG)
N
mRNA
N
Direction of ribosomal movement
Growing peptide chain
N
Protein
N
C
Stop codon (UAA, UAG or UGA)
60S
3'
40S

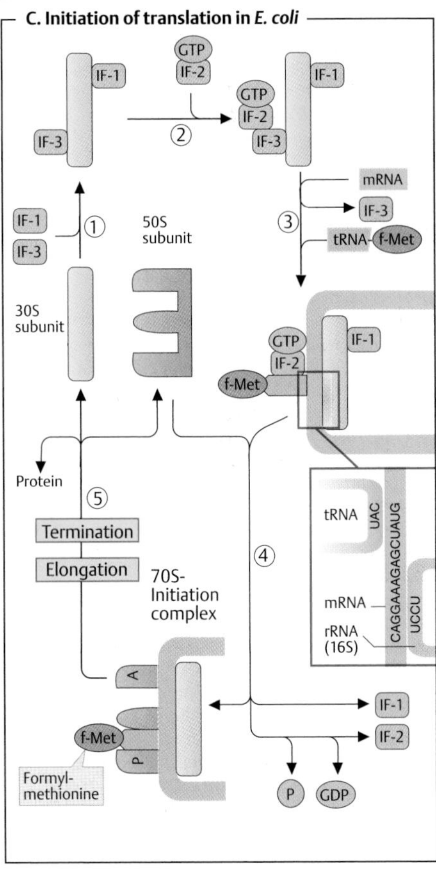
C. Initiation of translation in E. coli
GTP
IF-2
IF-1
IF-3
2
GTP
IF-2
IF-3
IF-1
mRNA
IF-3
3
tRNA
f-Met
IF-1
IF-3
1
50S subunit
30S subunit
GTP
IF-2
IF-1
f-Met
Protein
5
tRNA
UAC
CAGGAAAGAGCUAUG
UCCU
mRNA
rRNA (16S)
Termination
Elongation
4
70S-Initiation complex
A
P
f-Met
Formyl-methionine
IF-1
IF-2
P
GDP

Translation II: elongation and termination

After translation has been initiated (see p. 250), the peptide chain is extended by the addition of further amino acid residues (**elongation**) until the ribosome reaches a stop codon on the mRNA and the process is interrupted (**termination**).

A. Elongation and termination of protein biosynthesis in E. coli ◑

Elongation can be divided into four phases:

[1] **Binding of aminoacyl tRNA.** First, the peptidyl site (P) of the ribosome is occupied by a tRNA that carries at its 3′ end the complete peptide chain formed up to this point (top left). A second tRNA, loaded with the next amino acid (Val–$tRNA^{Val}$ in the example shown), then binds via its complementary anticodon (see p. 82) to the mRNA codon exposed at the acceptor site (in this case GUG). The tRNA binds as a complex with a GTP-containing protein, the *elongation factor Tu* (EF–Tu) (**1a**). It is only after the bound GTP has been hydrolyzed to GDP and phosphate that EF–Tu dissociates again (**1b**). As the binding of the tRNA to the mRNA is still loose before this, GTP hydrolysis acts as a delaying factor, making it possible to check whether the correct tRNA has been bound. A further protein, the *elongation factor Ts* (EF-Ts), later catalyzes the exchange of GDP for GTP and in this way regenerates the EF–Tu · GTP complex. EF-Tu is related to the G proteins involved in signal transduction (see p. 384).

[2] **Synthesis of the peptide bond** takes place in the next step. Ribosomal *peptidyltransferase* catalyzes (without consumption of ATP or GTP) the transfer of the peptide chain from the tRNA at the P site to the NH_2 group of the amino acid residue of the tRNA at the A site. The ribosome's peptidyltransferase activity is not located in one of the ribosomal proteins, but in the 28 S rRNA. Catalytically active RNAs of this type are known as *ribozymes* (cf. p. 246). It is thought that the few surviving ribozymes are remnants of the *"RNA world"*—an early phase of evolution in which proteins were not as important as they are today.

[3] After the transfer of the growing peptide to the A site, the free tRNA at the P site dissociates and another GTP–containing elongation factor (EF-G · GTP) binds to the ribosome. Hydrolysis of the GTP in this factor provides the energy for **translocation** of the ribosome. During this process, the ribosome moves three bases along the mRNA in the direction of the 3′ end. The tRNA carrying the peptide chain is stationary relative to the mRNA and reaches the ribosome's P site during translocation, while the next mRNA codon (in this case GUG) appears at the A site.

[4] The uncharged Val-tRNA then dissociates from the E site. The ribosome is now ready for the next elongation cycle.

When one of the three stop codons (UAA, UAG, or UGA) appears at the A site, **termination** starts.

[5] There are no complementary tRNAs for the stop codons. Instead, two *releasing factors* bind to the ribosome. One of these factors (RF–1) catalyzes hydrolytic cleavage of the ester bond between the tRNA and the C–terminus of the peptide chain, thereby releasing the protein.

[6] Hydrolysis of GTP by factor RF–3 supplies the energy for the dissociation of the whole complex into its component parts.

Energy requirements in protein synthesis are high. Four energy-rich phosphoric acid anhydride bonds are hydrolyzed for each amino acid residue. Amino acid activation uses up two energy-rich bonds per amino acid (ATP → AMP + PP; see p. 248), and two GTPs are consumed per elongation cycle. In addition, initiation and termination each require one GTP per chain.

Further information

In eukaryotic cells, the number of initiation factors is larger and initiation is therefore more complex than in prokaryotes. The cap at the 5′ end of mRNA and the polyA tail (see p. 246) play important parts in initiation. However, the elongation and termination processes are similar in all organisms. The individual steps of bacterial translation can be inhibited by antibiotics (see p. 254).

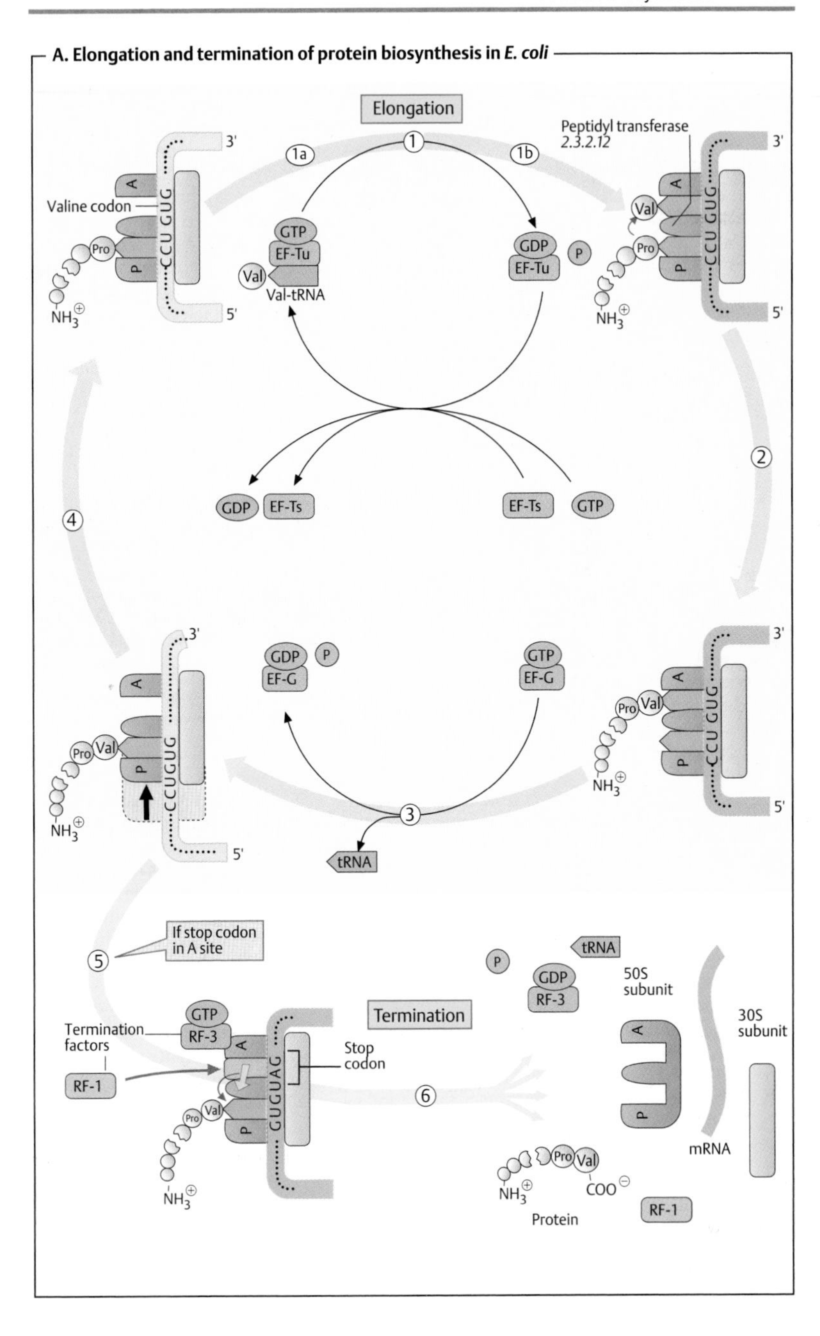
A. Elongation and termination of protein biosynthesis in E. coli
Elongation
Peptidyl transferase
2.3.2.12
1
1a
1b
3'
5'
Valine codon
CCU GUG
A
P
Pro
Val
NH3
GTP
EF-Tu
Val-tRNA
GDP
EF-Tu
P
2
GDP
EF-Ts
EF-Ts
GTP
4
GDP
EF-G
GTP
EF-G
3
tRNA
5
If stop codon in A site
Termination
Termination factors
GTP
RF-3
RF-1
Stop codon
GUGUAG
6
tRNA
GDP
RF-3
50S subunit
30S subunit
mRNA
COO
Protein
RF-1

Antibiotics

A. Antibiotics: overview ◑

Antibiotics are substances which, even at low concentrations, inhibit the growth and reproduction of bacteria and fungi. The treatment of infectious diseases would be inconceivable today without antibiotics. Substances that only restrict the reproduction of bacteria are described as having *bacteriostatic* effects (or *fungistatic* for fungi). If the target cells are killed, then the term *bactericidal* (or *fungicidal*) is used. Almost all antibiotics are produced by microorganisms—mainly bacteria of the genus *Streptomyces* and certain fungi. However, there are also synthetic antibacterial substances, such as sulfonamides and gyrase inhibitors.

A constantly increasing problem in antibiotic treatment is the development of resistant pathogens that no longer respond to the drugs available. The illustration shows a few of the therapeutically important antibiotics and their sites of action in the bacterial metabolism.

Substances known as **intercalators**, such as *rifamycin* and *actinomycin D* (bottom) are deposited in the DNA double helix and thereby interfere with replication and transcription (**B**). As DNA is the same in all cells, intercalating antibiotics are also toxic for eukaryotes, however. They are therefore only used as cytostatic agents (see p. 402). Synthetic inhibitors of DNA topoisomerase II (see p. 240), known as **gyrase inhibitors** (center), restrict replication and thus bacterial reproduction.

A large group of antibiotics attack bacterial ribosomes. These **inhibitors of translation** (left) include the *tetracyclines*—broad-spectrum antibiotics that are effective against a large number of different pathogens. The *aminoglycosides*, of which *streptomycin* is the best-known, affect all phases of translation. *Erythromycin* impairs the normal functioning of the large ribosomal subunit. *Chloramphenicol*, one of the few natural nitro compounds, inhibits ribosomal peptidyltransferase. Finally, *puromycin* mimics an aminoacyl tRNA and therefore leads to premature interruption of elongation.

The β–**lactam antibiotics** (bottom right) are also frequently used. The members of this group, the *penicillins* and *cephalosporins*, are synthesized by fungi and have a reactive β-lactam ring. They are mainly used against Gram-positive pathogens, in which they inhibit cell wall synthesis (**C**).

The first synthetic antibiotics were the **sulfonamides** (right). As analogues of p-aminobenzoic acid, these affect the synthesis of *folic acid*, an essential precursor of the coenzyme THF (see p. 108). **Transport antibiotics** (top center) have the properties of ion channels (see p. 222). When they are deposited in the plasma membrane, it leads to a loss of ions that damages the bacterial cells.

B. Intercalators ○

The effects of intercalators (see also p. 262) are illustrated here using the example of the **daunomycin–DNA complex**, in which two daunomycin molecules (red) are inserted in the double helix (blue). The antibiotic's ring system inserts itself between G/C base pairs (bottom), while the sugar moiety occupies the minor groove in the DNA (above). This leads to a localized change of the DNA conformation that prevents replication and transcription.

C. Penicillin as a "suicide substrate" ○

The site of action in the β–lactam antibiotics is *muramoylpentapeptide carboxypeptidase*, an enzyme that is essential for cross–linking of bacterial cell walls. The antibiotic resembles the substrate of this enzyme (a peptide with the C-terminal sequence D-Ala–D–Ala) and is therefore reversibly bound in the active center. This brings the β–lactam ring into proximity with an essential serine residue of the enzyme. Nucleophilic substitution then results in the formation of a stable covalent bond between the enzyme and the inhibitor, blocking the active center (see p. 96). In dividing bacteria, the loss of activity of the enzyme leads to the formation of unstable cell walls and eventually death.

A. Antibiotics: overview

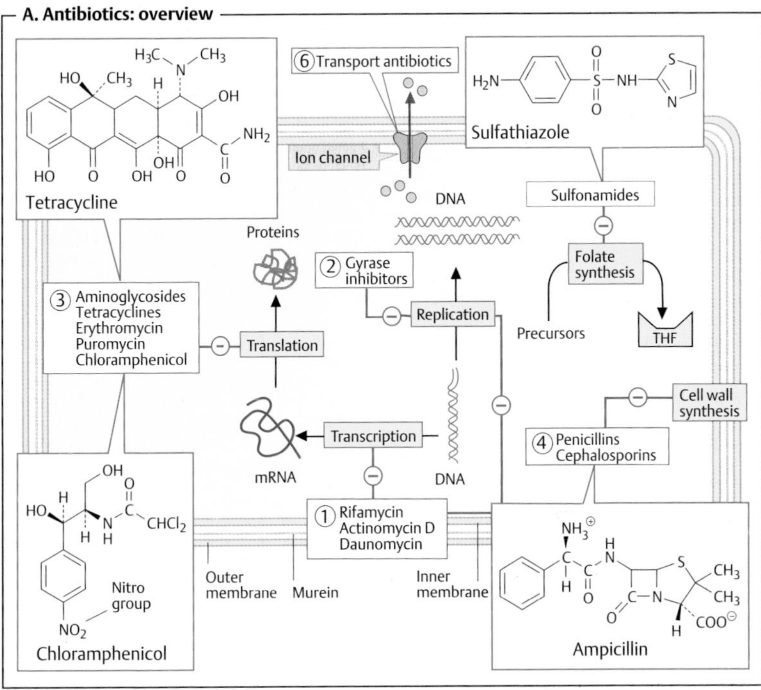

B. Intercalators

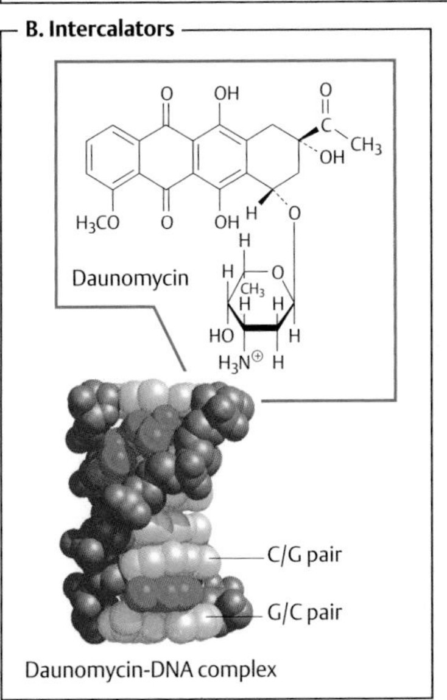

C. Penicillin as "suicide substrate"

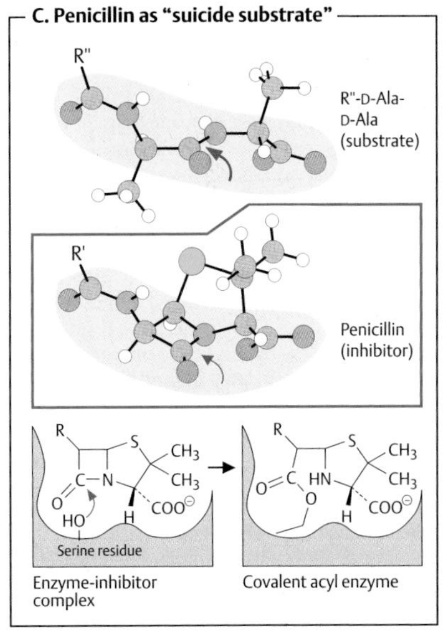

Mutation and repair

Genetic information is set down in the base sequence of DNA. Changes in the DNA bases or their sequence therefore have *mutagenic* effects. Mutagens often also damage growth regulation in cells, and they are then also *carcinogenic* (see p. 400). Gene alterations (**mutations**) are one of the decisive positive factors in biological evolution. On the other hand, an excessive mutation frequency would threaten the survival of individual organisms or entire species. For this reason, every cell has **repair mechanisms** that eliminate most of the DNA changes arising from mutations (**C**).

A. Mutagenic agents ◑

Mutations can arise as a result of physical or chemical effects, or they can be due to accidental errors in DNA replication and recombination.

The principal physical mutagen is **ionizing radiation** (α, β, and γ radiation, X-rays). In cells, it produces **free radicals** (molecules with unpaired electrons), which are extremely reactive and can damage DNA. Short-wavelength **ultraviolet light** (UV light) also has mutagenic effects, mainly in skin cells (sunburn). The most common chemical change due to UV exposure is the formation of **thymine dimers**, in which two neighboring thymine bases become covalently linked to one another (**2**). This results in errors when the DNA is read during replication and transcription.

Only a few examples of the group of **chemical mutagens** are shown here. *Nitrous acid* (HNO_2; salt: nitrite) and *hydroxylamine* (NH_2OH) both deaminate bases; they convert cytosine to uracil and adenine to inosine.

Alkylating compounds carry reactive groups that can form covalent bonds with DNA bases (see also p. 402). *Methylnitrosamines* (**3**) release the reactive methyl cation (CH_3^+), which methylates OH and NH_2 groups in DNA. The dangerous carcinogen *benzo [a]pyrene* is an aromatic hydrocarbon that is only converted into the active form in the organism (**4**; see p. 316). Multiple hydroxylation of one of the rings produces a reactive epoxide that can react with NH_2 groups in guanine residues, for example. Free radicals of benzo[*a*]pyrene also contribute to its toxicity.

B. Effects ◑

Nitrous acid causes **point mutations** (**1**). For example, C is converted to U, which in the next replication pairs with A instead of G. The alteration thus becomes permanent. Mutations in which a number of nucleotides not divisible by three are inserted or removed lead to reading errors in whole segments of DNA, as they move the reading frame (**frameshift mutations**). This is shown in Fig. **2** using a simple example. From the inserted C onwards, the resulting mRNA is interpreted differently during translation, producing a completely new protein sequence.

C. Repair mechanisms ○

An important mechanism for the removal of DNA damage is **excision repair** (**1**). In this process, a specific *excision endonuclease* removes a complete segment of DNA on both sides of the error site. Using the sequence of the opposite strand, the missing segment is then replaced by a *DNA polymerase*. Finally, a *DNA ligase* closes the gaps again.

Thymine dimers can be removed by **photoreactivation** (**2**). A specific *photolyase* binds at the defect and, when illuminated, cleaves the dimer to yield two single bases again.

A third mechanism is **recombination repair** (**3**, shown in simplified form). In this process, the defect is omitted during replication. The gap is closed by shifting the corresponding sequence from the correctly replicated second strand. The new gap that results is then filled by polymerases and ligases. Finally, the original defect is corrected by excision repair as in Fig. **1** (not shown).

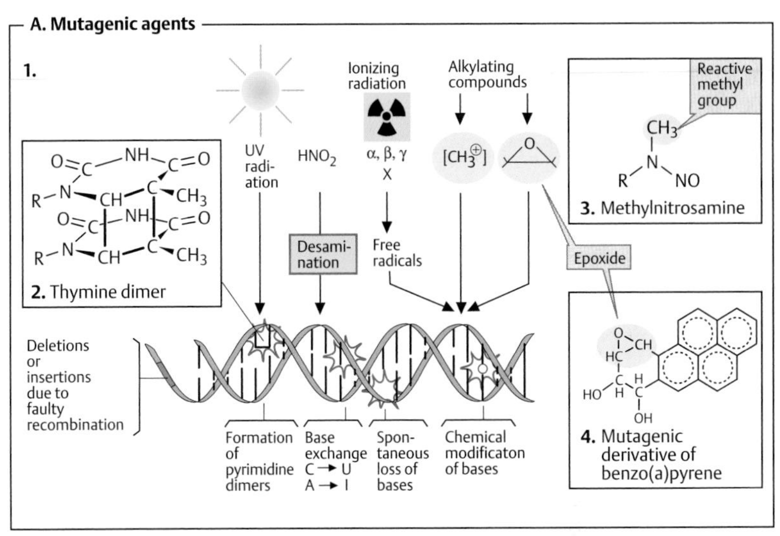
A. Mutagenic agents
1.
UV radiation
HNO2
Ionizing radiation
α, β, γ
X
Alkylating compounds
[CH3⊕]
Reactive methyl group
CH3
R N NO
3. Methylnitrosamine
2. Thymine dimer
Desamination
Free radicals
Epoxide
Deletions or insertions due to faulty recombination
Formation of pyrimidine dimers
Base exchange C → U A → I
Spontaneous loss of bases
Chemical modificaton of bases
4. Mutagenic derivative of benzo(a)pyrene

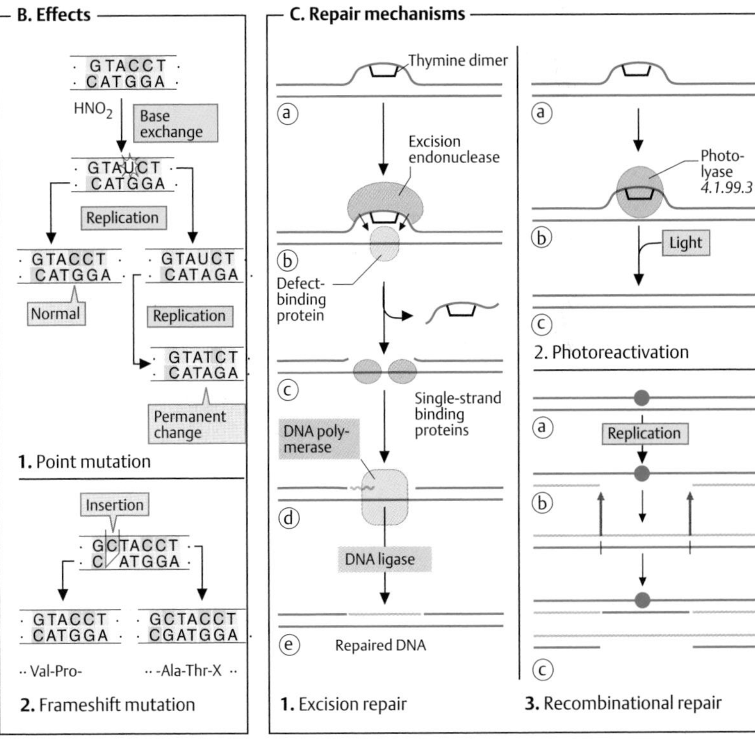
B. Effects
GTACCT
CATGGA
HNO2
Base exchange
GTAUCT
CATGGA
Replication
GTACCT
CATGGA
Normal
GTAUCT
CATAGA
Replication
GTATCT
CATAGA
Permanent change
1. Point mutation
Insertion
GCTACCT
C ATGGA
GTACCT
CATGGA
GCTACCT
CGATGGA
·· Val-Pro-
·· -Ala-Thr-X ··
2. Frameshift mutation
C. Repair mechanisms
Thymine dimer
a
Excision endonuclease
b
Defect-binding protein
c
Single-strand binding proteins
DNA polymerase
d
DNA ligase
e
Repaired DNA
1. Excision repair
a
Photo-lyase
4.1.99.3
b
Light
c
2. Photoreactivation
a
Replication
b
c
3. Recombinational repair

DNA cloning

The growth of molecular genetics since 1970 has mainly been based on the development and refinement of methods of analyzing and manipulating DNA. **Genetic engineering** has practical applications in many fields. For example, it has provided new methods of diagnosing and treating diseases, and it is now also possible to create targeted changes in specific characteristics of organisms. Since biological risks cannot be completely ruled out with these procedures, it is particularly important to act responsibly when dealing with genetic engineering. A short overview of important methods involved in genetic engineering is provided here and on the following pages.

A. Restriction endonucleases ◑

In many genetic engineering procedures, defined DNA fragments have to be isolated and then newly combined with other DNA segments. For this purpose, enzymes are used that can cut DNA and join it together again inside the cell. Of particular importance are *restriction endonucleases*—a group of bacterial enzymes that cleave the DNA double strand in a sequence-specific way. The numerous restriction enzymes known are named using abbreviations based on the organism from which they originate. The example used here is *Eco*RI, a nuclease isolated from the bacterium *Escherichia coli.*

Like many other restriction endonucleases, EcoRI cleaves DNA at the site of a *palindrome*—i. e., a short segment of DNA in which both the strand and counter-strand have the same sequence (each read in the 5′→3′ direction). In this case, the sequence is 5′-GAATTC-3′. *Eco*RI, a homodimer, cleaves the phosphoric acid diester bonds in both strands between G and A. This results in the formation of complementary *overhanging* or *"sticky" ends* (AATT), which are held together by base pairing. However, they are easily separated—e. g. by heating. When the fragments are cooled, the overhanging ends hybridize again in the correct arrangement. The cleavage sites can then be sealed again by a *DNA ligase*.

B. DNA cloning ○

Most DNA segments—e. g., genes—occur in very small quantities in the cell. To be able to work with them experimentally, a large number of identical copies (**"clones"**) first have to be produced. The classic procedure for cloning DNA takes advantage of the ability of bacteria to take up and replicate short, circular DNA fragments known as **plasmids**.

The segment to be cloned is first cut out of the original DNA using restriction endonucleases (see above; for the sake of simplicity, cleavage using *Eco*RI alone is shown here, but in practice two different enzymes are usually used). As a vehicle (*"vector"*), a plasmid is needed that has only *one Eco*RI cleavage site. The plasmid rings are first opened by cleavage with *Eco*RI and then mixed with the isolated DNA fragments. Since the fragment and the vector have the same overhanging ends, some of the molecules will hybridize in such a way that the fragment is incorporated into the vector DNA. When the cleavage sites are now closed again using *DNA ligase*, a newly combined *("recombinant") plasmid* arises.

By pretreating a large number of host cells, one can cause some of them to take up the plasmid (a process known as **transformation**) and replicate it along with their own genome when reproducing. To ensure that only host bacteria that contain the plasmid replicate, plasmids are used that give the host *resistance to a particular antibiotic. When the bacteria are incubated in the presence of this antibiotic, only the cells containing the plasmid will replicate. The plasmid is then isolated from these cells, cleaved with Eco*RI again, and the fragments are separated using agarose gel electrophoresis (see p. 262). The desired fragment can be identified using its size and then extracted from the gel and used for further experiments.

A. Restriction endonucleases

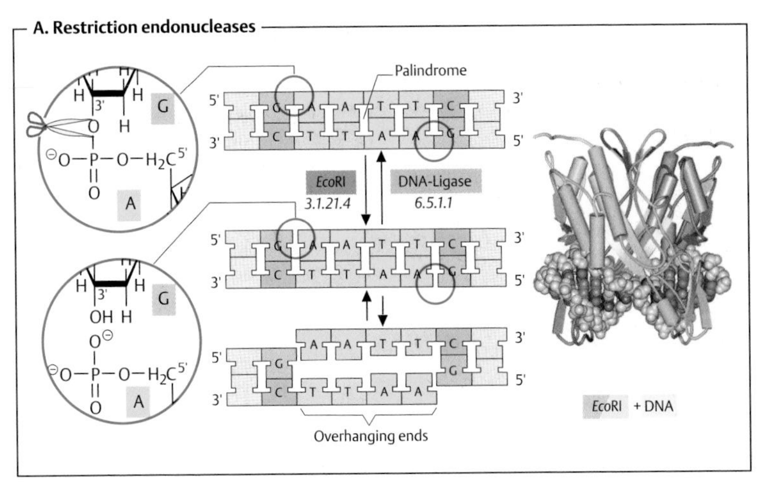

B. DNA cloning

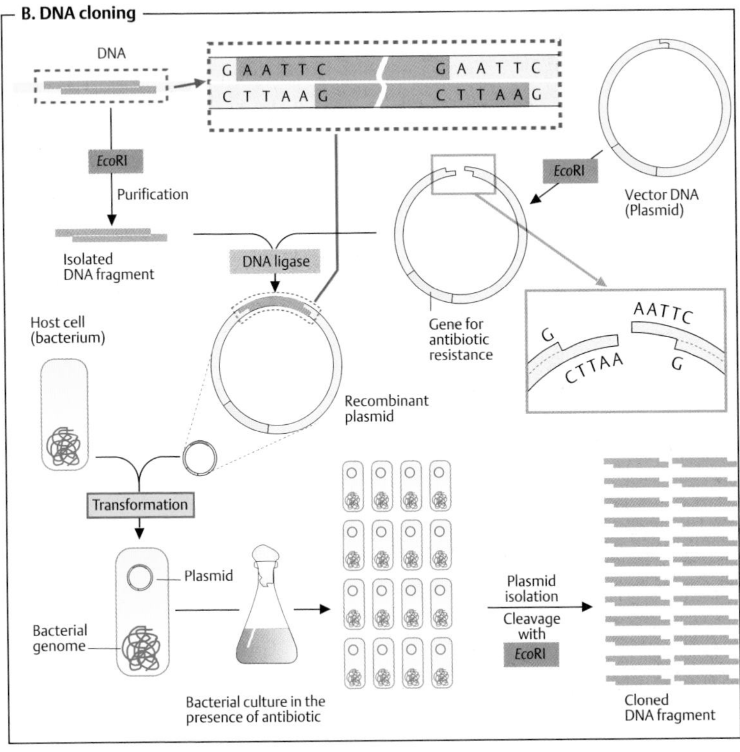

DNA sequencing

A. Gene libraries ○

It is often necessary in genetic engineering to isolate a DNA segment when its details are not fully known—e.g., in order to determine its nucleotide sequence. In this case, one can use what are known as **DNA libraries**. A DNA library consists of a large number of *vector DNA molecules* containing different fragments of *foreign DNA*. For example, it is possible to take all of the mRNA molecules present in a cell and transcribe them into DNA. These DNA fragments (known as copy DNA or **cDNA**) are then randomly introduced into vector molecules.

A library of genomic DNA can be established by cleaving the total DNA from a cell into small fragments using restriction endonucleases (see p. 258), and then incorporating these into vector DNA. Suitable vectors for gene libraries include **bacteriophages**, for example ("phages" for short). Phages are viruses that only infect bacteria and are replicated by them (see p. 404). Gene libraries have the advantage that they can be searched for specific DNA segments, using hybridization with oligonucleotides.

The first step is to strongly dilute a small part of the library (10^5–10^6 phages in a small volume), mix it with host bacteria, and plate out the mixture onto nutrient medium. The bacteria grow and form a continuous cloudy layer of cells. Bacteria infected by phages grow more slowly. In their surroundings, the bacterial "lawn" is less dense, and a clearer circular zone known as a **plaque** forms. The bacteria in this type of plaque exclusively contain the offspring of a single phage from the library.

The next step is to make an impression of the plate on a plastic foil, which is then heated. This causes the phage DNA to adhere to the foil. When the foil is incubated with a DNA fragment that hybridizes to the DNA segment of interest (a **gene probe**), the probe binds to the sites on the imprint at which the desired DNA is attached. Binding of the gene probe can be detected by prior radioactive or other labeling of the probe. Phages from the positive plaques in the original plate are then isolated and replicated. Restriction cleavage finally provides large amounts of the desired DNA.

B. Sequencing of DNA ○

The nucleotide sequence of DNA is nowadays usually determined using the so-called **chain termination method**. In single-strand sequencing, the DNA fragment (**a**) is cloned into the DNA of **phage M13** (see p. 404), from which the coded single strand can be easily isolated. This is *hybridized* with a **primer**—a short, synthetically produced DNA fragment that binds to 3′ end of the introduced DNA segment (**b**).

Based on this hybrid, the missing second strand can now be generated in the test tube by adding the four *deoxyribonucleoside triphosphates* (*dNTP*) and a suitable *DNA polymerase* (**c**). The trick lies in also adding small amounts of *dideoxynucleoside triphosphates (ddNTP)*. Incorporating a ddNTP leads to the *termination of second-strand synthesis*. This can occur whenever the corresponding dNTP ought to be incorporated. The illustration shows this in detail using the example of ddGTP. In this case, fragments are obtained that each include the primer plus three, six, eight, 13, or 14 additional nucleotides. *Four separate reactions*, each with a different ddNTP, are carried out (**c**), and the products are placed side by side on a supporting material. The fragments are then separated by *gel electrophoresis* (see p. 76), in which they move in relation to their length.

Following *visualization* (**d**), the sequence of the fragments in the individual lanes is simply read from bottom to top (**e**) to directly obtain the nucleotide sequence. A detail from such a sequencing gel and the corresponding protein sequence are shown in Fig. **2**.

In a more modern procedure, the four ddNTPs are covalently marked with fluorescent dyes, which produce a different color for each ddNTP on laser illumination. This allows the sequence in which the individual fragments appear at the lower end of the gel to be continuously recorded and directly stored in digital form.

A. Gene libraries

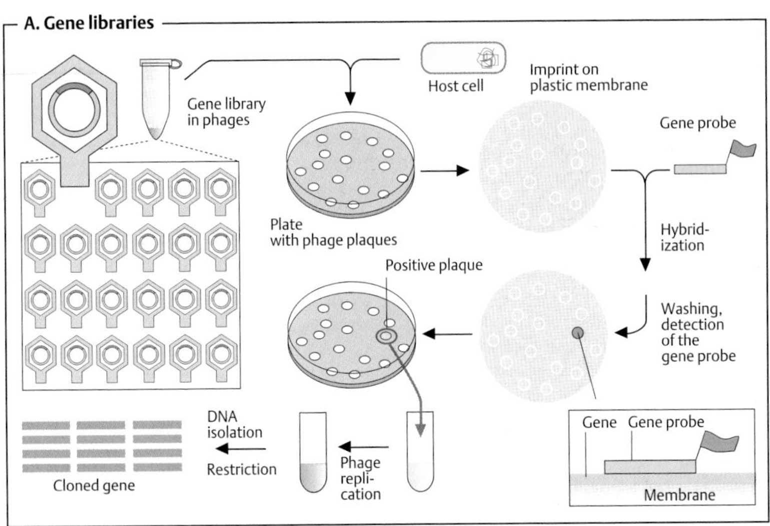

B. Sequencing of DNA

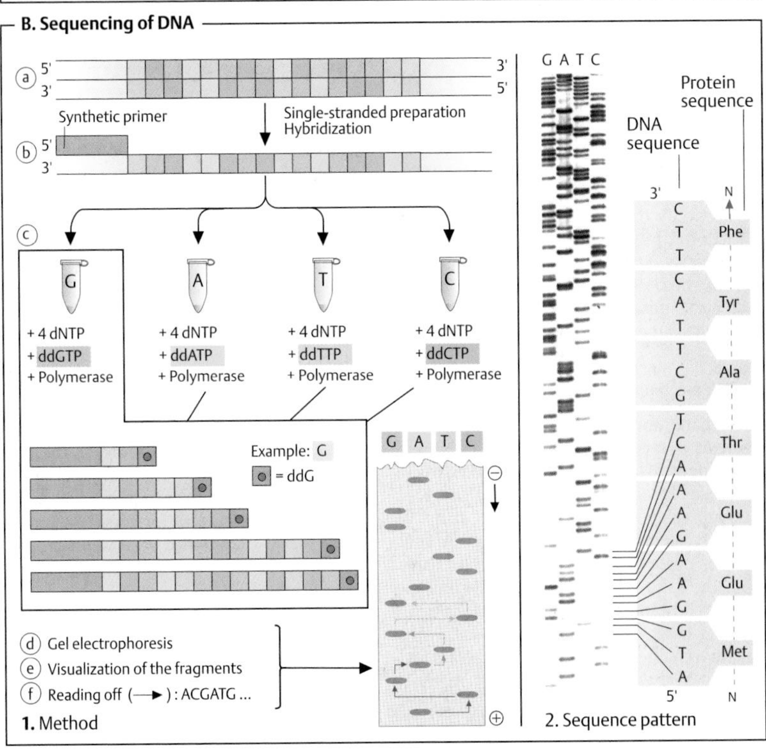

PCR and protein expression

A. Polymerase chain reaction (PCR) ◑

The polymerase chain reaction (PCR) is an important procedure in genetic engineering that allows any DNA segment to be replicated (**amplified**) without the need for restriction enzymes, vectors, or host cells (see p. 258). However, the nucleotide sequence of the segment has to be known. Two oligonucleotides (**primers**) are needed, which each hybridize with one of the strands at each end of the DNA segment to be amplified; also needed are sufficient quantities of the four **deoxyribonucleoside triphosphates** and a special heat-tolerant DNA polymerase. The primers are produced by chemical synthesis, and the polymerase is obtained from thermostable bacteria.

First, the starter is heated to around 90 °C to separate the DNA double helix into single strands (**a**; cf. p. 84). The mixture is then cooled to allow hybridization of the primers (**b**). Starting from the primers, complementary DNA strands are now synthesized in both directions by the polymerase (**c**). This *cycle* (cycle 1) is *repeated 20–30 times* with the same reaction mixture (cycle 2 and subsequent cycles). The cyclic heating and cooling are carried out by *computer-controlled thermostats*.

After only the third cycle, double strands start to form with a length equal to the distance between the two primers. The proportion of these approximately doubles during each cycle, until almost all of the newly synthesized segments have the correct length.

B. DNA electrophoresis ○

The separation of DNA fragments by electrophoresis is technically simpler than protein electrophoresis (see p. 78). The mobility of molecules in an electrical field of a given strength depends on the size and shape of the molecules, as well as their charge. In contrast to proteins, in which all three factors vary, the ratio of mass to charge in nucleic acids is constant, as all of the nucleotide components have similar masses and carry one negative charge. When electrophoresis is carried out in a wide-meshed support material that does not separate according to size and shape, the mobility of the molecules depends on their mass alone. The supporting material generally used in genetic engineering is a gel of the polysaccharide **agarose** (see p. 40). Agarose gels are not very stable and are therefore poured horizontally into a plastic chamber in which they are used for separation (top).

To make the separated fragments visible, after running the procedure the gels are placed in solutions of **ethidium bromide**. This is an intercalator (see p. 254) that shows strong fluorescence in UV light after binding to DNA, although it barely fluoresces in an aqueous solution. The result of separating two PCR amplificates (lanes 1 and 2) is shown in the lower part of the illustration. Comparing their distances with those of polynucleotides of known lengths (lane 3; bp = base pairs) yields lengths of approximately 800 bp for fragment 1 and 1800 bp for fragment 2. After staining, the bands can be cut out of the gel and the DNA can be extracted from them and used for further experiments.

C. Overexpression of proteins ◑

To treat some diseases, proteins are needed that occur in such small quantities in the organism that isolating them on a large scale would not be economically feasible. Proteins of this type can be obtained by *overexpression* in bacteria or eukaryotic cells. To do this, the corresponding *gene* is isolated from human DNA and cloned into an **expression plasmid** as described on p. 258. In addition to the gene itself, the plasmid also has to contain DNA segments that allow replication by the host cell and transcription of the gene. After *transformation* and *replication* of suitable host cells, **induction** is used in a targeted fashion to trigger efficient *transcription* of the gene. *Translation* of the mRNA formed in the host cell then gives rise to large amounts of the desired protein. Human insulin (see p. 76), plasminogen activators for dissolving blood clots (see p. 292), and the growth hormone somatotropin are among the proteins produced in this way.

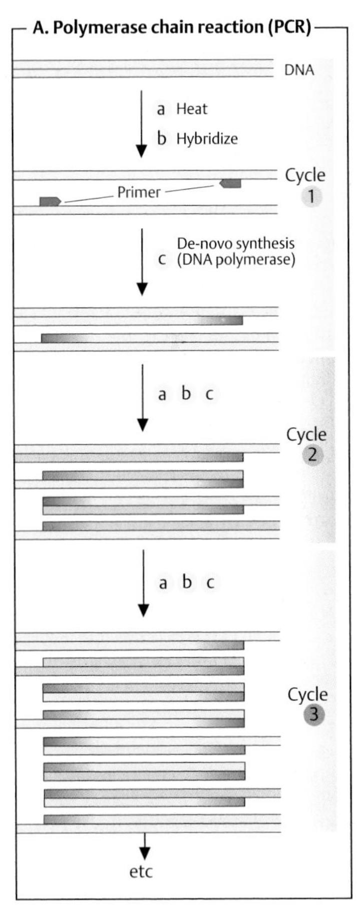
A. Polymerase chain reaction (PCR)
DNA
a Heat
b Hybridize
Cycle 1
Primer
De-novo synthesis
c (DNA polymerase)
a b c
Cycle 2
a b c
Cycle 3
etc

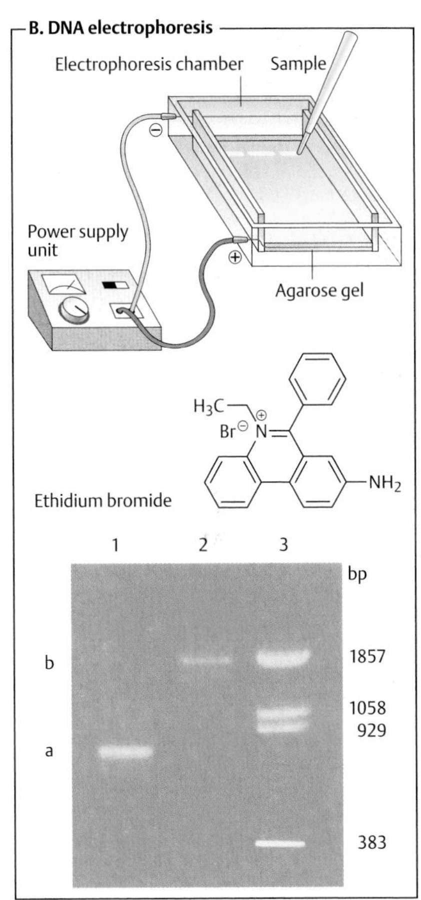
B. DNA electrophoresis
Electrophoresis chamber
Sample
Power supply unit
Agarose gel
Ethidium bromide
H3C
Br
N
NH2
1
2
3
bp
b
a
1857
1058
929
383

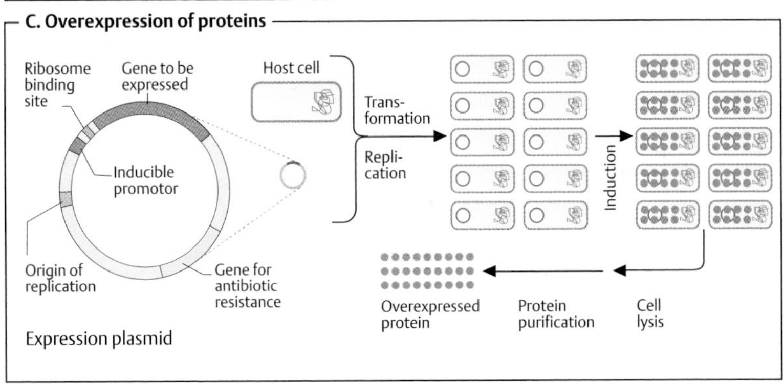
C. Overexpression of proteins
Ribosome binding site
Gene to be expressed
Host cell
Inducible promotor
Origin of replication
Gene for antibiotic resistance
Expression plasmid
Trans-formation
Repli-cation
Induction
Overexpressed protein
Protein purification
Cell lysis

Genetic engineering in medicine

Genetic engineering procedures are becoming more and more important in medicine for diagnostic purposes (**A–C**). New genetic approaches to the treatment of severe diseases are still in the developmental stage ("gene therapy," **D**).

A. DNA fingerprinting ○

DNA fingerprinting is used to link small amounts of biological material—e. g., traces from the site of a crime—to a specific person. The procedure now used is based on the fact that the human genome contains non-coding repetitive DNA sequences, the length of which varies from individual to individual. **Short tandem repeats** (**STRs**) thus exist in which dinucleotides (e. g., -T-X-) are frequently repeated. Each STR can occur in five to 15 different lengths (alleles), of which one individual possesses only one or two. When the various allele combinations for several STRs are determined after PCR amplification of the DNA being investigated, a "genetic fingerprint" of the individual from whom the DNA originates is obtained. Using comparative material—e. g., saliva samples—definite identification is then possible.

B. Diagnosis of sickle-cell anemia using RFLP ○

This example illustrates a procedure for diagnosing a point mutation in the β-globin gene that leads to sickle-cell anemia (see p. 248). The mutation in the first exon of the gene destroys a cleavage site for the restriction endonuclease *Mst*II (see p. 258). When the DNA of healthy and diseased individuals is cleaved with *Mst*II, different fragments are produced in the region of the β-globin gene, which can be separated by electrophoresis and then demonstrated using specific probes (see p. 260). In addition, heterozygotic carriers of the sickle-cell gene can be distinguished from homozygotic ones.

C. Identification of viral DNA using RT-PCR ○

In viral infections, it is often difficult to determine the species of the pathogen precisely. **RT-PCR** can be used to identify RNA viruses. In this procedure, reverse transcriptase (see p. 404) is used to transcribe the viral RNA into dsDNA, and then PCR is employed to amplify a segment of this DNA with virus-specific primers. In this way, an amplificate with a characteristic length can be obtained for each pathogen and identified using gel electrophoresis as described above.

D. Gene therapy ◑

Many diseases, such as hereditary metabolic defects and tumors, can still not be adequately treated. About 10 years ago, projects were therefore initiated that aimed to treat diseases of this type by transferring genes into the affected cells (gene therapy). The illustration combines conceivable and already implemented approaches to gene therapy for metabolic defects (left) and tumors (right). None of these procedures has yet become established in clinical practice.

If a mutation leads to failure of an enzyme E1 (left), its substrate B will no longer be converted into C and will accumulate. This can lead to cell damage by B itself or by a toxic product formed from it. Treatment with intact E1 is not possible, as the proteins are not capable of passing through the cell membrane. By contrast, it is in principle possible to introduce **foreign genes** into the cell using viruses as vectors (adenoviruses or retroviruses are mainly used). Their gene products could replace the defective E1 or convert B into a harmless product. Another approach uses the so-called **antisense DNA** (bottom right). This consists of polynucleotides that hybridize with the mRNA for specific cellular proteins and thereby prevent their translation. In the case shown, the synthesis of E2 could be blocked, for example.

The main problem in chemotherapy for **tumors** is the lack of tumor-specificity in the highly toxic cytostatic agents used (see p. 402). Attempts are therefore being made to introduce into tumor cells genes with products that are only released from a precursor to form active cytostatics once they have reached their target (left). Other gene products are meant to force the cells into apoptosis (see p. 396) or make them more susceptible to attack by the immune system. To steer the viral vectors to the tumor (targeting), attempts are being made to express proteins on the virus surface that are bound by tumor-specific receptors. Fusion with a tumor-specific promoter could also help limit the effect of the foreign gene to the tumor cells.

A. DNA fingerprinting

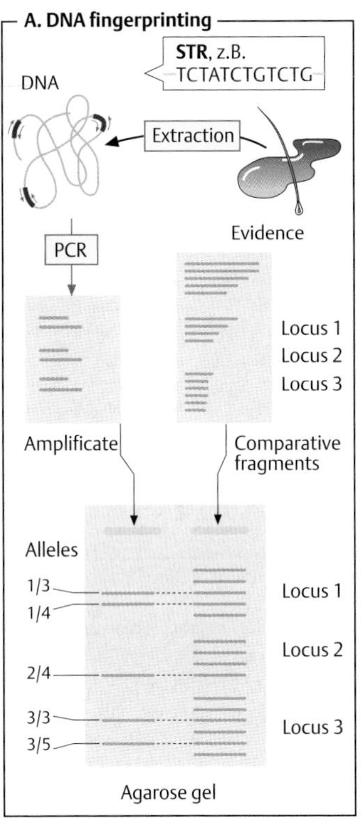

B. Diagnosis of sickle-cell anemia using RFLP

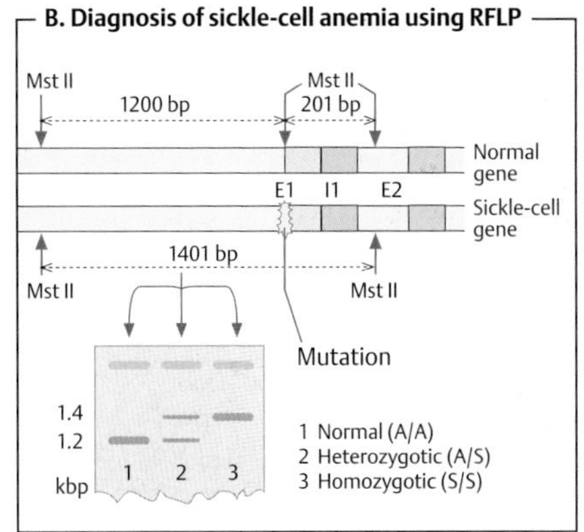

C. Evidence of viral DNA using RT-PCR

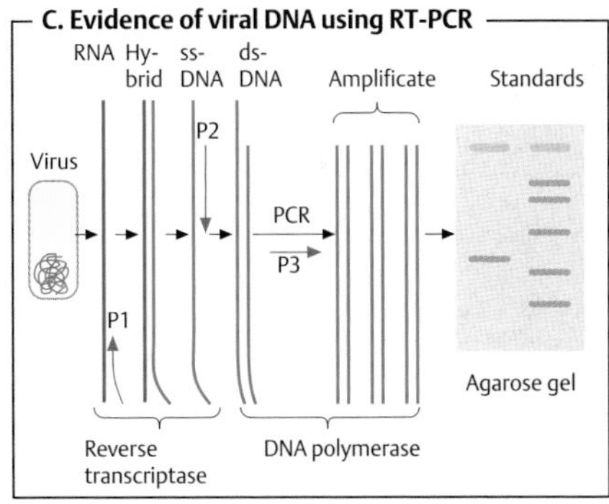

D. Gene therapy

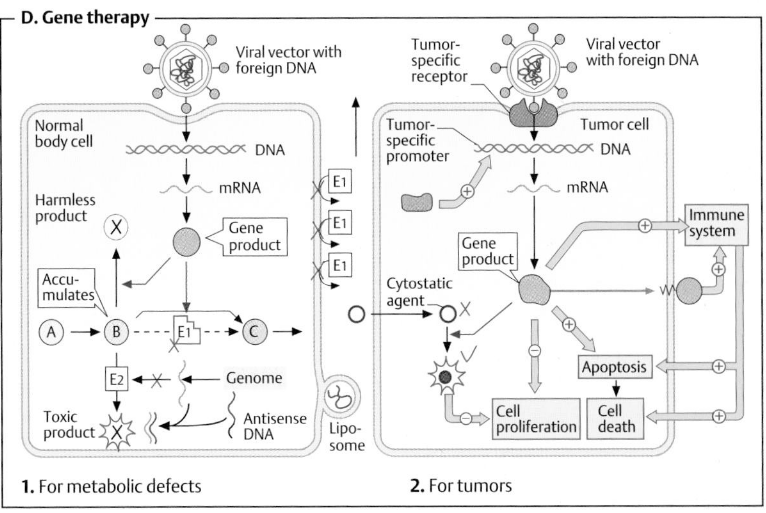

1. For metabolic defects

2. For tumors

Digestion: overview

Most components of food (see p. 360) cannot be resorbed directly by the organism. It is only after they have been broken down into smaller molecules that the organism can take up the essential nutrients. *Digestion* refers to the mechanical and enzymatic breakdown of food and the resorption of the resulting products.

A. Hydrolysis and resorption of food components ●

Following mechanical fragmentation of food during chewing in the mouth, the process of enzymatic degradation starts in the stomach. For this purpose, the chyme is mixed with *digestive enzymes* that occur in the various digestive secretions or in membrane-bound form on the surface of the intestinal epithelium (see p. 268). Almost all digestive enzymes are *hydrolases* (class 3 enzymes; see p. 88); they catalyze the cleavage of composite bonds with the uptake of water.

Proteins are first denatured by the stomach's *hydrochloric acid* (see p. 270), making them more susceptible to attack by the *endopeptidases* (proteinases) present in gastric and pancreatic juice. The peptides released by endopeptidases are further degraded into amino acids by *exopeptidases*. Finally, the amino acids are resorbed by the intestinal mucosa in cotransport with Na^+ ions (see p. 220). There are separate transport systems for each of the various groups of amino acids.

Carbohydrates mainly occur in food in the form of polymers (starches and glycogen). They are cleaved by *pancreatic amylase* into oligosaccharides and are then hydrolyzed by *glycosidases*, which are located on the surface of the intestinal epithelium, to yield monosaccharides. Glucose and galactose are taken up into the enterocytes by secondary active cotransport with Na^+ ions (see p. 220). In addition, monosaccharides also have passive transport systems in the intestine.

Nucleic acids are broken down into their components by *nucleases* from the pancreas and small intestine (ribonucleases and deoxyribonucleases). Further breakdown yields the nucleobases (purine and pyrimidine derivatives), pentoses (ribose and deoxyribose), phosphate, and nucleosides (nucleobase pentose). These cleavage products are resorbed by the intestinal wall in the region of the jejunum.

Lipids are a special problem for digestion, as they are not soluble in water. Before enzymatic breakdown, they have to be emulsified by *bile salts* and *phospholipids* in the bile (see p. 314). At the water–lipid interface, *pancreatic lipase* then attacks triacylglycerols with the help of *colipase* (see p. 270). The cleavage products include *fatty acids, 2-monoacylglycerols, glycerol*, and *phosphate* from phospholipid breakdown. After resorption into the epithelial cells, fats are resynthesized from fatty acids, glycerol and 2-monoacylglycerols and passed into the lymphatic system (see p. 272). The lipids in milk are more easily digested, as they are already present in emulsion; on cleavage, they mostly provide short-chain fatty acids.

Inorganic components such as water, electrolytes, and *vitamins* are directly absorbed by the intestine.

High-molecular-weight indigestible components, such as the fibrous components of plant cell walls, which mainly consist of cellulose and lignin, pass through the bowel unchanged and form the main component of feces, in addition to cells shed from the intestinal mucosa. Dietary fiber makes a positive contribution to digestion as a *ballast material* by binding water and promoting intestinal peristalsis.

The food components resorbed by the epithelial cells of the intestinal wall in the region of the jejunum and ileum are transported directly to the liver via the *portal vein*. Fats, cholesterol, and lipid-soluble vitamins are exceptions. These are first released by the enterocytes in the form of *chylomicrons* (see p. 278) into the *lymph system*, and only reach the blood via the thoracic duct.

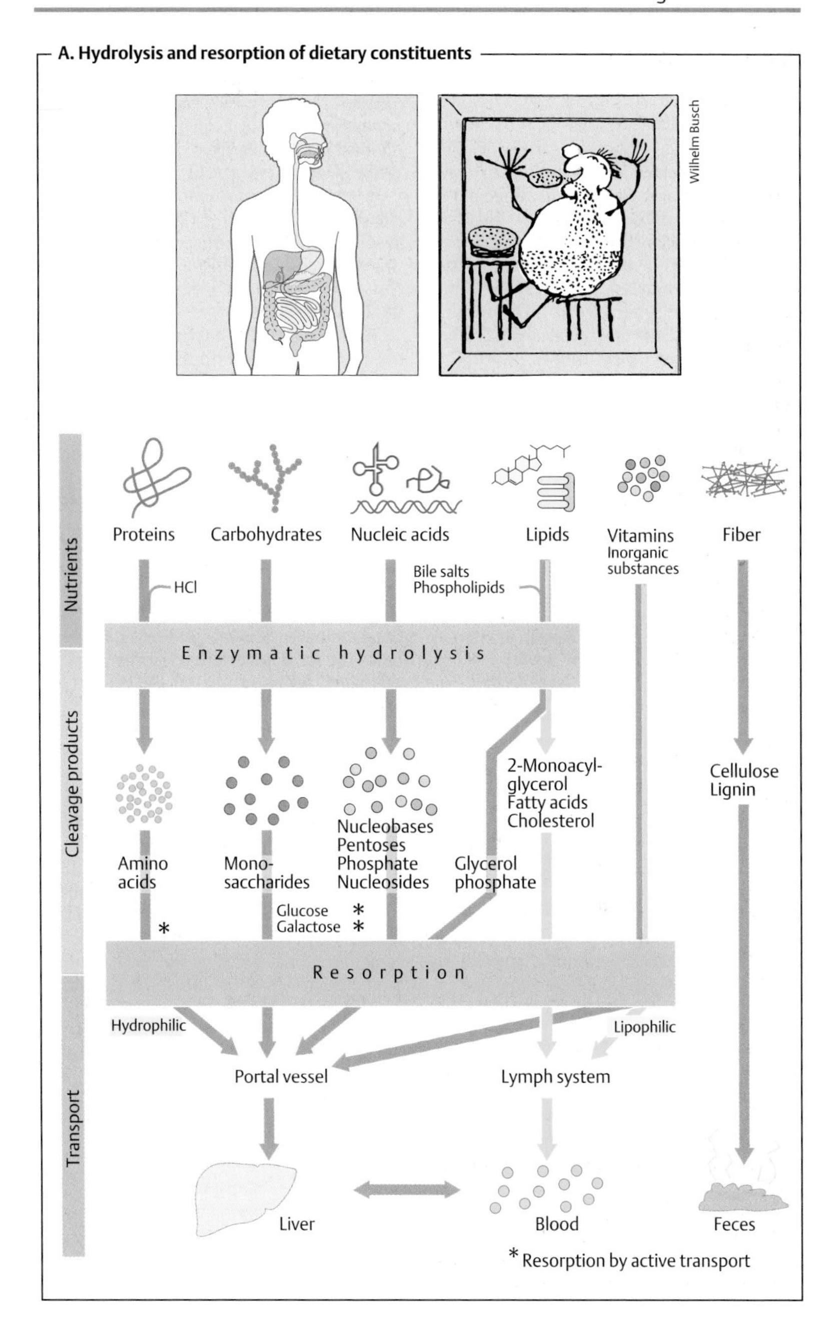
A. Hydrolysis and resorption of dietary constituents
Wilhelm Busch
Nutrients
Cleavage products
Transport
Proteins
Carbohydrates
Nucleic acids
Lipids
Vitamins
Inorganic substances
Fiber
HCl
Bile salts
Phospholipids
Enzymatic hydrolysis
Amino acids
Mono-saccharides
Nucleobases
Pentoses
Phosphate
Nucleosides
Glycerol phosphate
2-Monoacyl-glycerol
Fatty acids
Cholesterol
Cellulose
Lignin
Glucose *
Galactose *
*
Resorption
Hydrophilic
Lipophilic
Portal vessel
Lymph system
Liver
Blood
Feces
* Resorption by active transport

Digestive secretions

A. Digestive juices ◑

Saliva. The salivary glands produce a slightly alkaline secretion which—in addition to water and salts—contains *glycoproteins* (mucins) as lubricants, *antibodies,* and *enzymes.* *α-Amylase* attacks polysaccharides, and a lipase hydrolyzes a small proportion of the neutral fats. α-Amylase and *lysozyme,* a murein-cleaving enzyme (see p. 40), probably serve to regulate the oral bacterial flora rather than for digestion (see p. 340).

Gastric juice. In the stomach, the chyme is mixed with gastric juice. Due to its hydrochloric acid content, this secretion of the gastric mucosa is strongly acidic (pH 1–3; see p. 270). It also contains *mucus* (mainly glycoproteins known as mucins), which protects the mucosa from the hydrochloric acid, *salts,* and *pepsinogen*—the proenzyme ("zymogen") of the aspartate proteinase *pepsin* (see pp. 176, 270). In addition, the gastric mucosa secretes what is known as *"intrinsic factor"*—a glycoprotein needed for resorption of vitamin B_{12} ("extrinsic factor") in the bowel.

In the stomach, pepsin and related enzymes initiate the enzymatic digestion of proteins, which takes 1–3 hours. The acidic gastric contents are then released into the duodenum in batches, where they are neutralized by alkaline pancreatic secretions and mixed with cystic bile.

Pancreatic secretions. In the acinar cells, the pancreas forms a secretion that is alkaline due to its HCO_3^- content, the buffer capacity of which is sufficient to neutralize the stomach's hydrochloric acid. The pancreatic secretion also contains many *enzymes* that catalyze the hydrolysis of high-molecular-weight food components. All of these enzymes are hydrolases with pH optimums in the neutral or weakly alkaline range. Many of them are formed and secreted as proenzymes and are only activated in the bowel lumen (see p. 270).

Trypsin, chymotrypsin, and *elastase* are endopeptidases that belong to the group of serine proteinases (see p. 176). Trypsin hydrolyzes specific peptide bonds on the C side of the basic amino acids Arg and Lys, while chymotrypsin prefers peptide bonds of the apolar amino acids Tyr, Trp, Phe, and Leu (see p. 94). Elastase mainly cleaves on the C side of the aliphatic amino acids Gly, Ala, Val, and Ile. Smaller peptides are attacked by *carboxypeptidases,* which as exopeptidases cleave individual amino acids from the C-terminal end of the peptides (see p. 176).

α-Amylase, the most important endoglycosidase in the pancreas, catalyzes the hydrolysis of α1→4 bonds in the polymeric carbohydrates starch and glycogen. This releases maltose, maltotriose, and a mixture of other oligosaccharides.

Various pancreatic enzymes hydrolyze lipids, including *lipase* with its auxiliary protein *colipase* (see p. 270), *phospholipase* A_2, and *sterol esterase.* Bile salts activate the lipid-cleaving enzymes through micelle formation (see below).

Several hydrolases—particularly *ribonuclease* (RNAse) and *deoxyribonuclease* (DNAse)—break down the nucleic acids contained in food.

Bile. The liver forms a thin secretion (bile) that is stored in the gallbladder after water and salts have been extracted from it. From the gallbladder, it is released into the duodenum. The most important constituents of bile are *water* and inorganic *salts, bile acids* and *bile salts* (see p. 314), *phospholipids, bile pigments,* and *cholesterol.* Bile salts, together with phospholipids, emulsify insoluble food lipids and activate the lipases. Without bile, fats would be inadequately cleaved, if at all, resulting in "fatty stool" (steatorrhea). Resorption of fat-soluble vitamins would also be affected.

Small-intestinal secretions. The glands of the small intestine (the Lieberkühn and Brunner glands) secrete additional digestive enzymes into the bowel. Together with enzymes on the microvilli of the intestinal epithelium (peptidases, glycosidases, etc.), these enzymes ensure almost complete hydrolysis of the food components previously broken down by the endoenzymes.

A. Digestive juices

⤓ Endoenzyme
--◂ Exoenzyme

Components	Function or substrate

Saliva

Daily secretion 1.0–1.5 l
pH 7

Component	Function or substrate
Water	Moistens food
Salts	
Mucus	Lubricant
Antibodies	Bind to bacteria
α-Amylases (3.2.1.1)	Cleave starch ⤓
Lysozyme (3.2.1.17)	Attacks bacterial cell walls ⤓

Gastric juice

Daily secretion 2–3 l
pH 1

Component	Function or substrate
Water	
Salts	
HCl	Denatures proteins, kills bacteria
Mucus	Protects stomach lining
Pepsins (3.4.23.1-3)	Cleave proteins ⤓
Chymosin (3.4.23.4)	Precipitates casein ⤓
Triacylglycerol lipase (3.1.1.3)	Cleaves fats
Intrinsic factor	Protects vitamin B_{12}

Bile

Daily secretion 0.6 l
pH 6.9–7.7

Component	Function or substrate
Water	
$HCO_3^{\ominus}$	Neutralizes gastric juice
Bile salts	Facilitate lipid digestion
Phospholipids	Facilitate lipid digestion
Bile pigments	Waste products
Cholesterol	Waste product

Pancreatic secretions

Daily secretion 0.7–2.5 l
pH 7.7 (7.5–8.8)

Component	Function or substrate
Water	
$HCO_3^{\ominus}$	Neutralizes gastric juice
Trypsin (3.4.21.4)	Proteins ⤓
Chymotrypsin (3.4.21.1)	Proteins ⤓
Elastase (3.4.21.36)	Proteins ⤓
Carboxypeptidases (3.4.n.n)	Peptides --◂
α-Amylase (3.2.1.1)	Starch and glycogen ⤓
Triacylglycerol lipase (3.1.1.3)	Fats
Co-Lipase	Cofactor for lipase
Phospholipase A_2 (3.1.1.4)	Phospholipids
Sterol esterase (3.1.1.13)	Cholesterol esters
Ribonuclease (3.1.27.5)	RNA
Deoxyribonuclease I (3.1.21.1)	DNA

Secretions of the small intestine

Daily secretion unknown
pH 6.5–7.8

Component	Function or substrate
Aminopeptidases (3.4.11.n)	Peptides --◂
Dipeptidases (3.4.13.n)	Dipeptides
α-Glucosidase (3.2.1.20)	Oligosaccharides ⤓
Oligo-1,6-glucosidase (3.2.1.10)	Oligosaccharides ⤓
β-Galactosidase (3.2.1.23)	Lactose
Sucrose α-glucosidase (3.2.1.48)	Sucrose
α, α-Trehalase (3.2.1.28)	Trehalose
Alkaline phosphatase (3.1.3.1)	Phosphoric acid esters
Polynucleotidases (3.1.3.n)	Nucleic acids, nucleotides ⤓
Nucleosidases (3.2.2.n)	Nucleosides --◂
Phospholipases (3.1.n.n)	Phospholipids

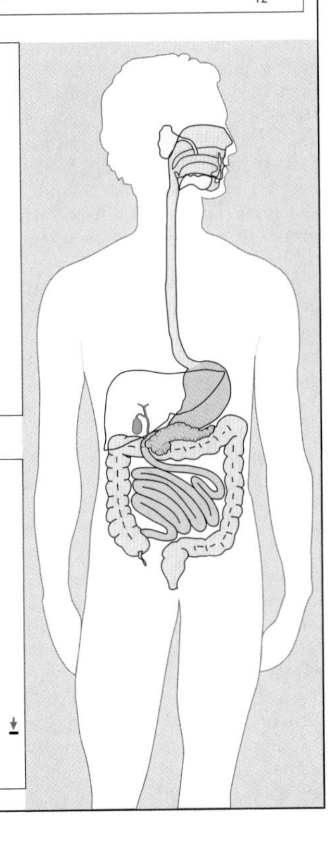

Digestive processes

Gastric juice is the product of several cell types. The *parietal cells* produce hydrochloric acid, *chief cells* release pepsinogen, and *accessory cells* form a mucin-containing mucus.

A. Formation of hydrochloric acid ◑

The secretion of **hydrochloric acid** (H^+ and Cl^-) by the parietal cells is an active process that uses up ATP and takes place against a concentration gradient (in the gastric lumen, with a pH of 1, the H^+ concentration is some 10^6 times higher than in the parietal cells, which have a pH of 7).

The precursors of the exported H^+ ions are carbon dioxide (CO_2) and water (H_2O). CO_2 diffuses from the blood into the parietal cells, and in a reaction catalyzed by *carbonate dehydratase* (carbonic anhydrase [2]), it reacts with H_2O to form H^+ and hydrogen carbonate (HCO_3^-). The H^+ ions are transported into the gastric lumen in exchange for K^+ by a membrane-bound *H^+/K^+-exchanging ATPase* [1] (a transport ATPase of the P type; see p. 220). The remaining hydrogen carbonate is released into the interstitium in electroneutral antiport in exchange for chloride ions (Cl^-), and from there into the blood. The Cl^- ions follow the secreted protons through a channel into the gastric lumen.

The hydrochloric acid in gastric juice is important for digestion. It activates pepsinogen to form pepsin (see below) and creates an optimal pH level for it to take effect. It also denatures food proteins so that they are more easily attacked by proteinases, and it kills micro-organisms.

Regulation. HCl secretion is stimulated by the peptide hormone *gastrin*, the mediator *histamine* (see p. 380), and—via the neurotransmitter *acetylcholine*—by the autonomous nervous system. The peptide *somatostatin* and certain *prostaglandins* (see p. 390) have inhibitory effects. Together with cholecystokinin, secretin, and other peptides, gastrin belongs to the group of **gastrointestinal hormones** (see p. 370). All of these are formed in the gastrointestinal tract and mainly act in the vicinity of the site where they are formed—i. e., they are paracrine hormones (see p. 372). While gastrin primarily enhances HCl secretion, *cholecystokinin* and *secretin* mainly stimulate pancreatic secretion and bile release.

B. Zymogen activation ◑

To prevent self-digestion, the pancreas releases most proteolytic enzymes into the duodenum in an inactive form as *proenzymes* (zymogens). Additional protection from the effects of premature activation of pancreatic proteinases is provided by *proteinase inhibitors* in the pancreatic tissue, which inactivate active enzymes by complex formation (right).

Trypsinogen plays a key role among the proenzymes released by the pancreas. In the bowel, it is proteolytically converted into active trypsin (see p. 176) by **enteropeptidase**, a membrane enzyme on the surface of the enterocytes. Trypsin then autocatalytically activates additional trypsinogen molecules and the other proenzymes (left).

C. Fat digestion ◑

Due to the "hydrophobic effect" (see p. 28), water-insoluble neutral fats in the aqueous environment of the bowel lumen would aggregate into drops of fat in which most of the molecules would not be accessible to pancreatic lipase. The amphipathic substances in bile (bile acids, bile salts, phospholipids) create an **emulsion** in which they occupy the surface of the droplets and thereby prevent them from coalescing into large drops. In addition, the bile salts, together with the auxiliary protein colipase, mediate binding of *triacylglycerol lipase* [1] to the emulsified fat droplets. Activation of the lipase is triggered by a conformation change in the C-terminal domain of the enzyme, which uncovers the active center.

During passage through the intestines, the active lipase breaks down the triacylglycerols in the interior of the droplets into free fatty acids and amphipathic monoacylglycerols. Over time, smaller **micelles** develop (see p. 28), in the envelope of which monoacylglycerols are present in addition to bile salts and phospholipids. Finally, the components of the micelles are resorbed by the enterocytes in ways that have not yet been explained.

Monoacylglycerols and fatty acids are reassembled into fats again (see p. 272), while the bile acids return to the liver (enterohepatic circulation; see p. 314).

A. Formation of hydrochloric acid

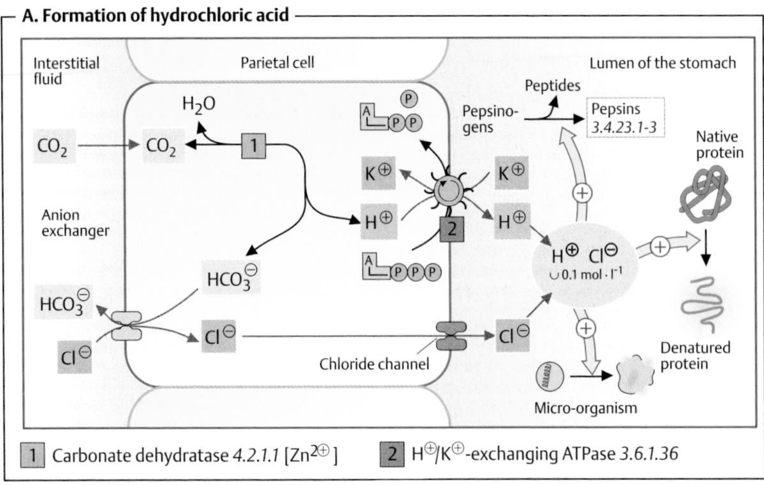

1 Carbonate dehydratase *4.2.1.1* [$Zn^{2\oplus}$]   2 $H^{\oplus}/K^{\oplus}$-exchanging ATPase *3.6.1.36*

B. Zymogen activation

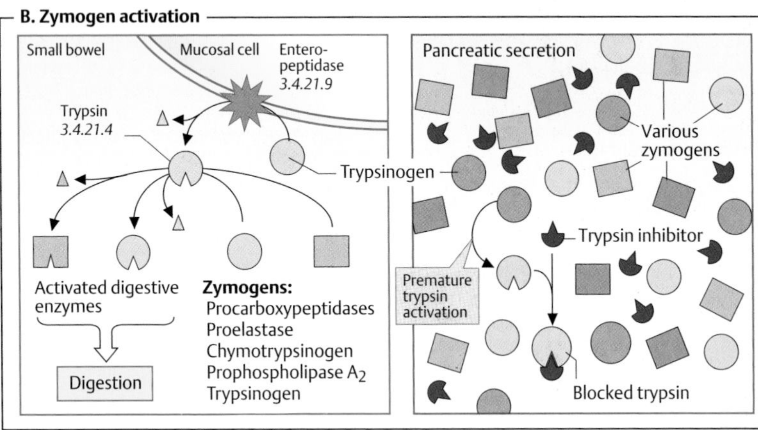

C. Fat digestion

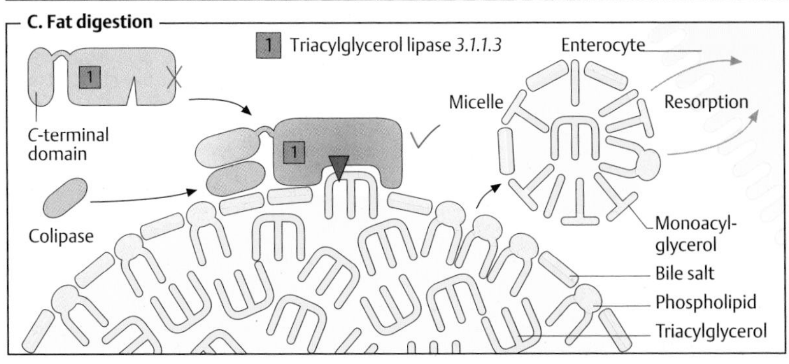

Resorption

Enzymatic hydrolysis in the digestive tract breaks down foodstuffs into their resorbable components. *Resorption* of the cleavage products takes place primarily in the small intestine. Only ethanol and short-chain fatty acids are already resorbed to some extent in the stomach.

The resorption process is facilitated by the large inner surface of the intestine, with its brush-border cells. Lipophilic molecules penetrate the plasma membrane of the mucosal cells by simple diffusion, whereas polar molecules require *transporters* (facilitated diffusion; see p. 218). In many cases, carrier-mediated cotransport with Na^+ ions can be observed. In this case, the difference in the concentration of the sodium ions (high in the intestinal lumen and low in the mucosal cells) drives the import of nutrients against a concentration gradient (secondary active transport; see p. 220). Failure of carrier systems in the gastrointestinal tract can result in diseases.

A. Monosaccharides ◑

The cleavage of polymeric carbohydrates by *α-amylase* [1] leads to **oligosaccharides**, which are broken down further by *exoglycosidases* (oligosaccharidases and disaccharidases [2]) on the membrane surface of the brush border. The monosaccharides released in this way then pass with the help of various *sugar-specific transporters* into the cells of intestinal epithelium. *Secondary active transport* serves for the uptake of **glucose** and **galactose**, which are transported against a concentration gradient in cotransport with Na^+. The Na^+ gradient is maintained on the basal side of the cells by *Na^+/K^+-ATPase* [3]. Another passive transporter then releases glucose and galactose into the blood. **Fructose** is taken up by a special type of transporter using facilitated diffusion.

Amino acids (not illustrated)

Protein degradation is initiated by *proteinases*—by pepsins in the stomach and by trypsin, chymotrypsin, and elastase in the small intestine. The resulting peptides are then further hydrolyzed by various *peptidases* into amino acids. Individual amino acid groups have *group-specific amino acid transporters*, some of which transport the amino acids into the enterocytes in cotransport with Na^+ ions (secondary active transport), while others transport them in an Na^+-independent manner through facilitated diffusion. Small peptides can also be taken up.

B. Lipids ◑

Fats and other lipids are poorly soluble in water. The larger the accessible surface is—i. e., the better the fat is emulsified—the easier it is for enzymes to hydrolyze it (see p. 270). Due to the special properties of milk, milk fats already reach the gastrointestinal tract in emulsified form. Digestion of them therefore already starts in the oral cavity and stomach, where lipases in the saliva and gastric juice are available. Lipids that are less accessible—e. g., from roast pork—are emulsified in the small intestine by *bile salts* and *bile phospholipids*. Only then are they capable of being attacked by *pancreatic lipase* [4] (see p. 270).

Fats (triacylglycerols) are mainly attacked by pancreatic lipase at positions 1 and 3 of the glycerol moiety. Cleavage of two fatty acid residues gives rise to **fatty acids** and **2-monoacylglycerols**, which are quantitatively the most important products. However, a certain amount of **glycerol** is also formed by complete hydrolysis. These cleavage products are resorbed by a non-ATP-dependent process that has not yet been explained in detail.

In the mucosal cells, *long-chain fatty acids* are resynthesized by an ATP-dependent ligase [5] to form acyl-CoA and then triacylglycerols (fats; see p. 170). The fats are released into the lymph in the form of **chylomicrons** (see p. 278) and, bypassing the liver, are deposited in the thoracic duct—i. e., the blood system. *Cholesterol* also follows this route.

By contrast, *short-chain fatty acids* (with chain lengths of less than 12 C atoms) pass directly into the blood and reach the liver via the portal vein. Resorbed glycerol can also take this path.

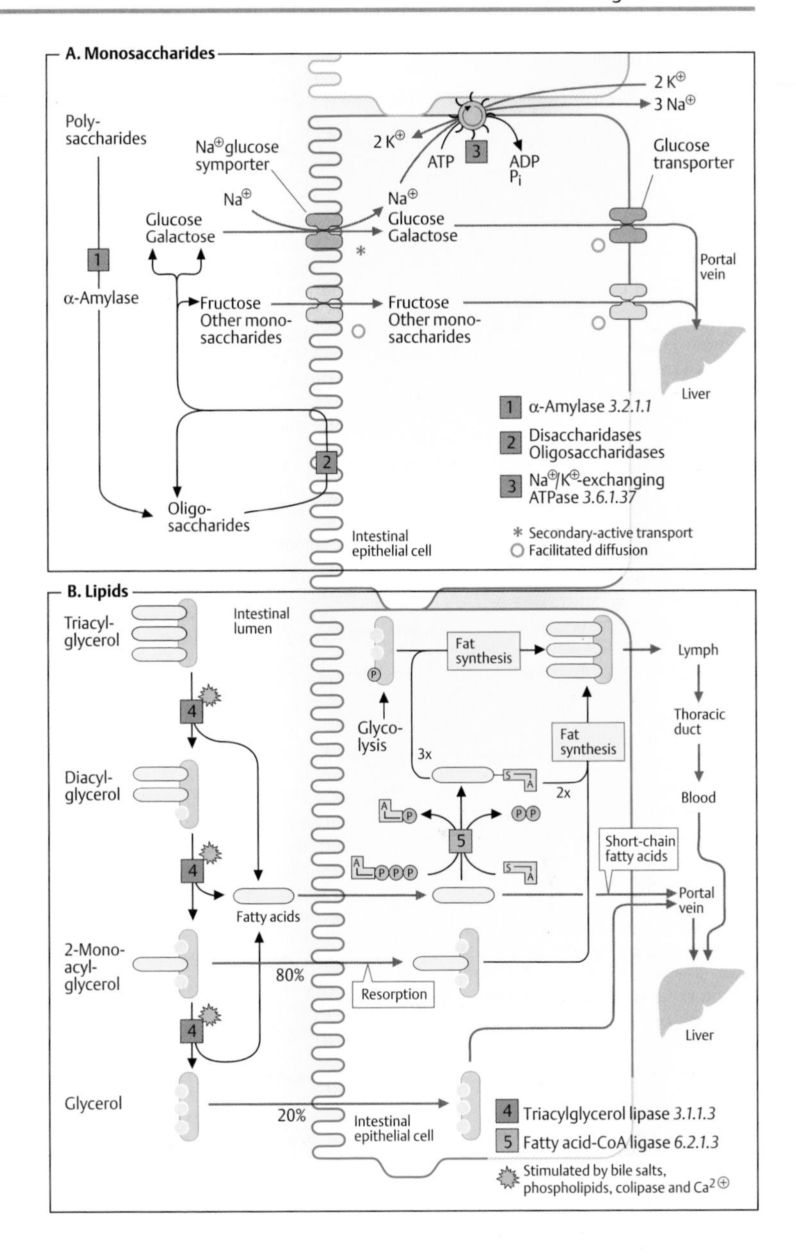
A. Monosaccharides
Poly-
saccharides
1
α-Amylase
Na⊕glucose
symporter
Na⊕
Glucose
Galactose
Fructose
Other mono-
saccharides
2K⊕
3 Na⊕
2 K⊕
ATP
3
ADP
Pi
Na⊕
Glucose
Galactose
*
Fructose
Other mono-
saccharides
Glucose
transporter
Portal
vein
Liver
2
Oligo-
saccharides
Intestinal
epithelial cell
1 α-Amylase 3.2.1.1
2 Disaccharidases
Oligosaccharidases
3 Na⊕/K⊕-exchanging
ATPase 3.6.1.37
* Secondary-active transport
○ Facilitated diffusion
B. Lipids
Triacyl-
glycerol
Intestinal
lumen
4
Diacyl-
glycerol
4
Fatty acids
2-Mono-
acyl-
glycerol
80%
4
Glycerol
20%
Intestinal
epithelial cell
Fat
synthesis
Glyco-
lysis
3x
Fat
synthesis
2x
5
Resorption
Lymph
Thoracic
duct
Blood
Short-chain
fatty acids
Portal
vein
Liver
4 Triacylglycerol lipase 3.1.1.3
5 Fatty acid-CoA ligase 6.2.1.3
Stimulated by bile salts,
phospholipids, colipase and Ca2⊕

Blood: composition and functions

Human blood constitutes about 8% of the body's weight. It consists of **cells** and cell fragments in an aqueous medium, the **blood plasma.** The proportion of cellular elements, known as *hematocrit,* in the total volume is approximately 45%.

A. Functions of the blood ●

The blood is the most important transport medium in the body. It serves to keep the "internal milieu" constant (homeostasis) and it plays a decisive role in defending the body against pathogens.

Transport. The *gases* oxygen and carbon dioxide are transported in the blood. The blood mediates *the exchange of substances between organs* and takes up *metabolic end products* from tissues in order to transport them to the lungs, liver, and kidney for excretion. The blood also distributes *hormones* throughout the organism (see p. 370).

Homeostasis. The blood ensures that a balanced distribution of water is maintained between the vascular system, the cells (intracellular space), and the extracellular space. The *acid–base balance* is regulated by the blood in combination with the lungs, liver, and kidneys (see p. 288). The regulation of *body temperature* also depends on the controlled transport of heat by the blood.

Defense. The body uses both non-specific and specific mechanisms to defend itself against pathogens. The defense system includes the *cells of the immune system* and certain *plasma proteins* (see p. 294).

Self-protection. To prevent blood loss when a vessel is injured, the blood has systems for stanching blood flow and coagulating the blood (hemostasis; see p. 290). The dissolution of blood clots (fibrinolysis) is also managed by the blood itself (see p. 292).

B. Cellular elements ◑

The solid elements in the blood are the erythrocytes (red blood cells), leukocytes (white blood cells), and thrombocytes (platelets).

The **erythrocytes** provide for gas transport in the blood. They are discussed in greater detail on pp. 280–285.

The **leukocytes** include various types of granulocyte, monocyte, and lymphocyte. All of these have immune defense functions (see p. 294). The *neutrophil granulocytes, monocytes,* and the *macrophages* derived from monocytes are phagocytes. They can ingest and degrade invading pathogens. The *lymphocytes* are divided into two groups, B lymphocytes and T lymphocytes. B lymphocytes produce *antibodies,* while T lymphocytes regulate the immune response and destroy virus-infected cells and tumor cells. *Eosinophilic* and *basophilic granulocytes* have special tasks for defense against animal parasites.

Thrombocytes are cell fragments that arise in the bone marrow from large precursor cells, the megakaryocytes. Their task is to promote hemostasis (see p. 290).

C. Blood plasma: composition ◑

The **blood plasma** is an aqueous solution of electrolytes, nutrients, metabolites, proteins, vitamins, trace elements, and signaling substances. The fluid phase of coagulated blood is known as **blood serum**. It differs from the plasma in that it lacks fibrin and other coagulation proteins (see p. 290).

Laboratory assessment of the composition of the blood plasma is often carried out in clinical chemistry. Among the electrolytes, there is a relatively high concentration of Na^+, Ca^{2+}, and Cl^- ions in the blood in comparison with the cytoplasm. By contrast, the concentrations of K^+, Mg^{2+}, and phosphate ions are higher in the cells. Proteins also have a higher intracellular concentration. The electrolyte composition of blood plasma is similar to that of seawater, due to the evolution of early forms of life in the sea. The solution known as *"physiological saline"* (NaCl at a concentration of 0.15 mol · L^{-1}) is almost isotonic with blood plasma.

A list of particularly important **metabolites** in the blood plasma is given on the right.

A. Functions of the blood

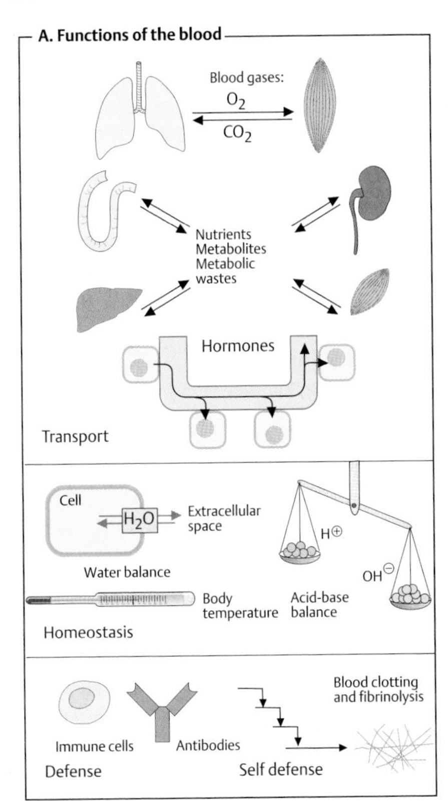

B. Cellular elements

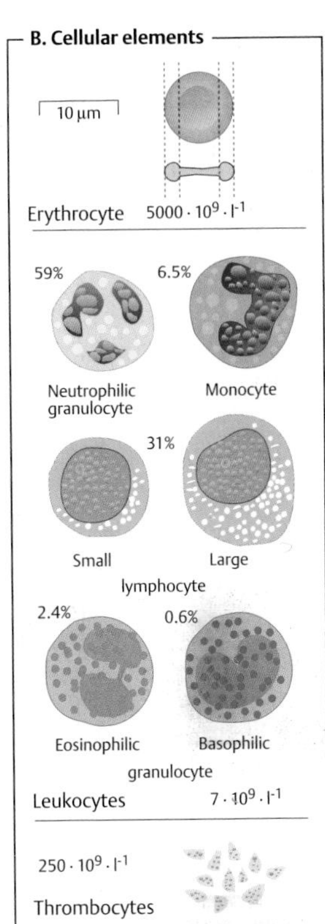

C. Blood plasma: composition

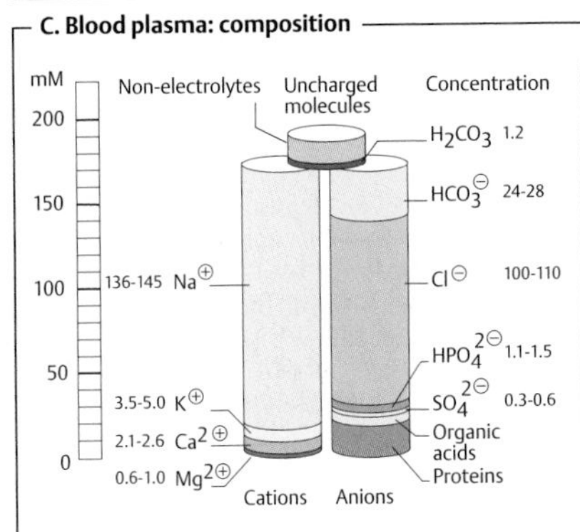

Metabolite	Concentration (mM)
Glucose	3.6 – 6.1
Lactate	0.4 – 1.8
Pyruvate	0.07 – 0.11
Urea	3.5 – 9.0
Uric acid	0.18 – 0.54
Creatinin	0.06 – 0.13
Amino acids	2.3 – 4.0
Ammonia	0.02 – 0.06
Lipids (total)	5.5 – 6.0 $g \cdot l^{-1}$
Triacylglycerols	1.0 – 1.3 $g \cdot l^{-1}$
Cholesterol	1.7 – 2.1 $g \cdot l^{-1}$

Plasma proteins

Quantitatively, proteins are the most important part of the soluble components of the blood plasma. With concentrations of between 60 and 80 g · L^{-1}, they constitute approximately 4% of the body's total protein. Their tasks include transport, regulation of the water balance, hemostasis, and defense against pathogens.

A. Plasma proteins ◑

Some 100 different proteins occur in human blood plasma. Based on their behavior during electrophoresis (see below), they are broadly divided into *five fractions:* **albumins** and α_1–, α_2–, β– and γ-**globulins**. Historically, the distinction between the albumins and globulins was based on differences in the proteins' solubility –albumins are soluble in pure water, whereas globulins only dissolve in the presence of salts.

The most frequent protein in the plasma, at around 45 g · L^{-1}, is **albumin**. Due to its high concentration, it plays a crucial role in maintaining the blood's colloid osmotic pressure and represents an important amino acid reserve for the body. Albumin has binding sites for apolar substances and therefore functions as a transport protein for long-chain fatty acids, bilirubin, drugs, and some steroid hormones and vitamins. In addition, serum albumin binds Ca^{2+} and Mg^{2+} ions. It is the only important plasma protein that is not glycosylated.

The albumin fraction also includes *transthyretin* (prealbumin), which together with other proteins transports the hormone thyroxine and its metabolites.

The table also lists important **globulins** in blood plasma, with their mass and function. The α- and β-globulins are involved in the transport of lipids (lipoproteins; see p. 278), hormones, vitamins, and metal ions. In addition, they provide coagulation factors, protease inhibitors, and the proteins of the complement system (see p. 298). Soluble antibodies (immunoglobulins; see p. 300) make up the γ-globulin fraction.

Synthesis and degradation. Most plasma proteins are synthesized by the liver. Exceptions to this include the immunoglobulins, which are secreted by B lymphocytes known as plasma cells (see p. 302) and peptide hormones, which derive from endocrine gland cells.

With the exception of albumin, almost all plasma proteins are *glycoproteins.* They carry oligosaccharides in *N*-and *O*-glycosidic bonds (see p. 44). *N*-acetylneuraminic acid (sialic acid; see p. 38) often occurs as a terminal carbohydrate among sugar residues. *Neuraminidases* (sialidases) on the surface of the vascular endothelia gradually cleave the sialic acid residues and thereby release galactose units on the surfaces of the proteins. These *asialoglycoproteins* ("asialo-" = without sialic acid) are recognized and bound by galactose receptors on hepatocytes. In this way, the liver takes up aged plasma proteins by endocytosis and breaks them down. The oligosaccharides on the protein surfaces thus determine the half-life of plasma proteins, which is a period of days to weeks.

In healthy individuals, the concentration of plasma proteins is constant. Diseases in organs that are involved in protein synthesis and breakdown can shift the protein pattern. For example, via cytokines (see p. 392), severe injuries trigger increased synthesis of *acute-phase proteins,* which include C-reactive protein, haptoglobin, fibrinogen, complement factor C-3, and others. The concentrations of individual proteins are altered in some diseases (known as *dysproteinemias*).

B. Carrier electrophoresis ◑

Proteins and other electrically charged macromolecules can be separated using electrophoresis (see also pp. 78, 262). Among the various procedures used, **carrier electrophoresis** on cellulose acetate foil (CAF) is particularly simple. Using this method, serum proteins—which at slightly alkaline pH values all move towards the anode, due to their excess of negative charges—can be separated into the five fractions mentioned. After the proteins have been stained with dyes, the resulting bands can be quantitatively assessed using densitometry.

A. Plasma proteins

Group	Protein	M_r in kDa	Function
Albumins:	Transthyretin	50-66	Transport of thyroxin and triiodothyronin
	Albumin: 45 g · l^{-1}	67	Maintenance of osmotic pressure; transport of fatty acids, bilirubin, bile acids, steroid hormones, pharmaceuticals and inorganic ions.
α_1-Globulins:	Antitrypsin	51	Inhibition of trypsin and other proteases
	Antichymotrypsin	58-68	Inhibition of chymotrypsin
	Lipoprotein (HDL)	200-400	Transport of lipids
	Prothrombin	72	Coagulation factor II, thrombin precursor (*3.4.21.5*)
	Transcortin	51	Transport of cortisol, corticosterone and progesterone
	Acid glycoprotein	44	Transport of progesterone
	Thyroxin-binding globulin	54	Transport of iodothyronins
α_2-Globulins:	Ceruloplasmin	135	Transport of copper ions
	Antithrombin III	58	Inhibition of blood clotting
	Haptoglobin	100	Binding of hemoglobin
	Cholinesterase (3.1.1.8)	ca. 350	Cleavage of choline esters
	Plasminogen	90	Precursor of plasmin (*3.4.21.7*), breakdown of blood clots
	Macroglobulin	725	Binding of proteases, transport of zinc ions
	Retinol-binding protein	21	Transport of vitamin A
	Vitamin D-binding protein	52	Transport of calciols
β-Globulins:	Lipoprotein (LDL)	2.000-4.500	Transport of lipids
	Transferrin	80	Transport of iron ions
	Fibrinogen	340	Coagulation factor I
	Sex hormone-binding globulin	65	Transport of testosterone and estradiol
	Transcobalamin	38	Transport of vitamin B_{12}
	C-reactive protein	110	Complement activation
γ-Globulins:	IgG	150	Late antibodies
	IgA	162	Mucosa-protecting antibodies
	IgM	900	Early antibodies
	IgD	172	B-lymphocyte receptors
	IgE	196	Reagins

B. Electrophoresis

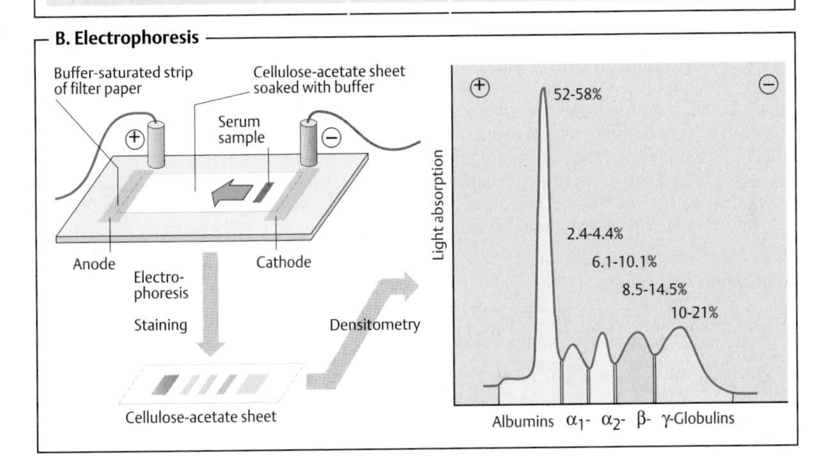

Lipoproteins

Most lipids are barely soluble in water, and many have amphipathic properties. In the blood, free triacylglycerols would coalesce into drops that could cause fat embolisms. By contrast, amphipathic lipids would be deposited in the blood cells' membranes and would dissolve them. Special precautions are therefore needed for lipid transport in the blood. While long-chain fatty acids are bound to albumin and short-chain ones are dissolved in the plasma (see p. 276), other lipids are transported in **lipoprotein complexes**, of which there several types in the blood plasma, with different sizes and composition.

A. Composition of lipoprotein complexes ◑

Lipoproteins are spherical or discoid aggregates of **lipids** and **apoproteins**. They consist of a nucleus of apolar lipids (triacylglycerols and cholesterol esters) surrounded by a single-layered shell approximately 2 nm thick of amphipathic lipids (phospholipids and cholesterol; the example shown here is LDL). The shell, in which the apoproteins are also deposited, gives the surfaces of the particles polar properties and thereby prevents them from aggregating into large particles. The larger the lipid nucleus of a lipoprotein is—i. e., the larger the number of apolar lipids it contains—the lower its density is.

Lipoproteins are classified into five groups. In order of decreasing size and increasing density, these are: *chylomicrons, VLDLs* (very-low-density lipoproteins), *IDLs* (intermediate-density lipoproteins), *LDLs* (low-density lipoproteins), and *HDLs* (high-density lipoproteins). The proportions of apoproteins range from 1% in chylomicrons to over 50% in HDLs. These proteins serve less for solubility purposes, but rather function as recognition molecules for the membrane receptors and enzymes that are involved in lipid exchange.

B. Transport functions

The classes of lipoproteins differ not only in their composition, but also in the ways in which they originate and function.

The **chylomicrons** take care of the transport of triacylglycerols from the intestine to the tissues. They are formed in the intestinal mucosa and reach the blood via the lymphatic system (see p. 266). In the peripheral vessels—particularly in muscle and adipose tissue—*lipoprotein lipase* [1] on the surface of the vascular endothelia hydrolyzes most of the triacylglycerols. Chylomicron breakdown is activated by the transfer of apoproteins E and C from HDL. While the fatty acids released and the glycerol are taken up by the cells, the chylomicrons gradually become converted into **chylomicron remnants**, which are ultimately removed from the blood by the liver.

VLDLs, **IDLs**, and **LDLs** are closely related to one another. VLDLs formed in the liver (see p. 312) transport triacylglycerols, cholesterol, and phospholipids to other tissues. Like chylomicrons, they are gradually converted into IDL and LDL under the influence of *lipoprotein lipase* [1]. This process is also stimulated by HDL. Cells that have a demand for cholesterol bind LDL through an interaction between their **LDL receptor** and ApoB-100, and then take up the complete particle through **receptor-mediated endocytosis**. This type of transport is mediated by depressions in the membrane (*"coated pits"*), the interior of which is lined with the protein *clathrin*. After LDL binding, clathrin promotes invagination of the pits and pinching off of vesicles (*"coated vesicles"*). The clathrin then dissociates off and is reused. After fusion of the vesicle with lysosomes, the LDL particles are broken down (see p. 234), and cholesterol and other lipids are used by the cells.

The **HDLs** also originate in the liver. They return the excess cholesterol formed in the tissues to the liver. While it is being transported, cholesterol is acylated by *lecithin cholesterol acyltransferase* (LCAT). The cholesterol esters formed are no longer amphipathic and can be transported in the core of the lipoproteins. In addition, HDLs promote chylomicron and VLDL turnover by exchanging lipids and apoproteins with them (see above).

A. Composition of lipoprotein complexes

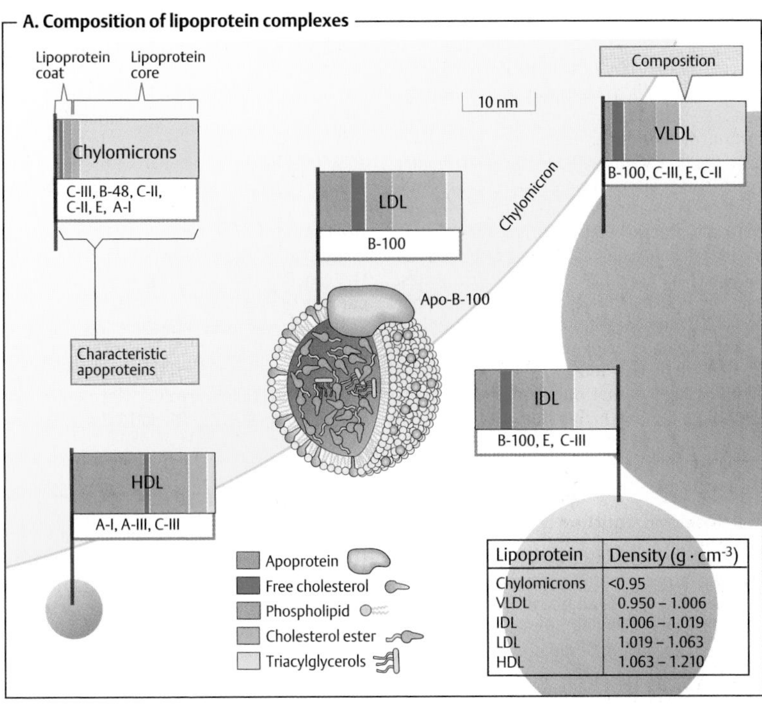

Lipoprotein	Density (g · cm^{-3})
Chylomicrons	<0.95
VLDL	0.950 – 1.006
IDL	1.006 – 1.019
LDL	1.019 – 1.063
HDL	1.063 – 1.210

B. Transport functions

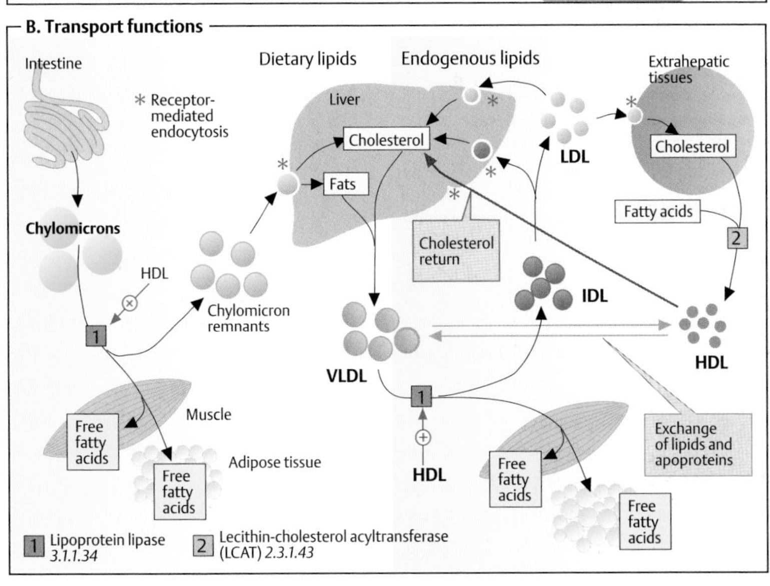

Hemoglobin

The most important task of the red blood cells (erythrocytes) is to **transport** molecular oxygen (**O_2**) from the lungs into the tissues, and carbon dioxide (**CO_2**) from the tissues back into the lungs. To achieve this, the higher organisms require a special transport system, since O_2 is *poorly soluble in water.* For example, only around 3.2 mL O_2 is soluble in 1 L blood plasma. By contrast, the protein **hemoglobin** (Hb), contained in the erythrocytes, can bind a maximum of 220 mL O_2 per liter—70 times the physically soluble amount.

The Hb content of blood, at 140–180 $g \cdot L^{-1}$ in men and 120–160 $g \cdot L^{-1}$ in women, is twice as high as that of the plasma proteins (50–80 $g \cdot L^{-1}$). Hb is therefore also responsible for the majority of the blood proteins' pH buffer capacity (see p. 288).

A. Hemoglobin: structure ◑

In adults, hemoglobin (**HbA**; see below) is a *heterotetramer* consisting of two α-chains and two β-chains, each with masses of 16 kDa. The α- and β-chains have different sequences, but are similarly folded. Some 80% of the amino acid residues form *α-helices,* which are identified using the letters A–H.

Each subunit carries a **heme group** (formula on p. 106), with a central bivalent **iron ion**. When O_2 binds to the heme iron (**Oxygenation** of Hb) and when O_2 is released (**Deoxygenation**), the oxidation stage of the iron does *not* change. Oxidation of Fe^{2+} to Fe^{3+} only occurs occasionally. The oxidized form, *methemoglobin,* is then no longer able to bind O_2. The proportion of Met-Hb is kept low by reduction (see p. 284) and usually amounts to only 1–2%.

Four of the six coordination sites of the iron in hemoglobin are occupied by the nitrogen atoms of the pyrrol rings, and another is occupied by a histidine residue of the globin (the *proximal histidine*). The iron's sixth site is coordinated with oxygen in oxyhemoglobin and with H_2O in deoxyhemoglobin.

B. Hemoglobin: allosteric effects ◑

Like aspartate carbamoyltransferase (see p. 116), Hb can exist in two different states (*conformations*), known as the T form and the R form. The **T form** (for *tense;* left) and has a much *lower O_2* affinity than the R form (for *relaxed;* right).

Binding of O_2 to one of the subunits of the T form leads to a local conformational change that weakens the association between the subunits. Increasing O_2 partial pressure thus means that more and more molecules convert to the higher–affinity R form. This **cooperative interaction** between the subunits increases the O_2 affinity of Hb with increasing O_2 concentrations—i.e., the O_2 saturation curve is **sigmoidal** (see p. 282).

Various **allosteric effectors** influence the equilibrium between the T and R forms and thereby regulate the O_2 binding behavior of hemoglobin (yellow arrows). The most important effectors are *CO_2, H^+,* and *2,3-bisphosphoglycerate* (see p. 282).

Further information

As mentioned above, hemoglobin in adults consists of two α- and two β-chains. In addition to this main form (**HbA_1**, $\alpha_2\beta_2$), adult blood also contains small amounts of a second form with a higher O_2 affinity in which the β-chains are replaced by δ-chains (**HbA_2**, $\alpha_2\delta_2$). Two other forms occur during embryonic and fetal development. In the first three months, **embryonic hemoglobins** are formed, with the structure $\zeta_2\varepsilon_2$ and $\alpha_2\varepsilon_2$. Up to the time of birth, **fetal hemoglobin** then predominates (HbF, $\alpha_2\gamma_2$), and it is gradually replaced by HbA during the first few months of life. Embryonic and fetal hemoglobins have higher O_2 affinities than HbA, as they have to take up oxygen from the maternal circulation.

A. Hemoglobin: structure

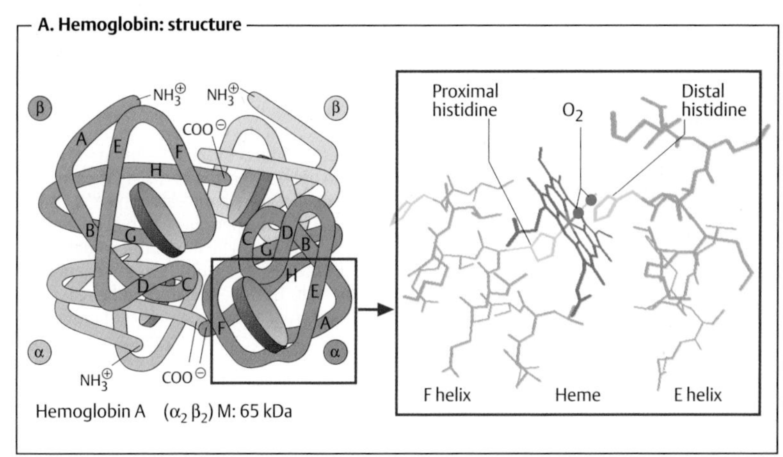

B. Hemoglobin: allosteric effects

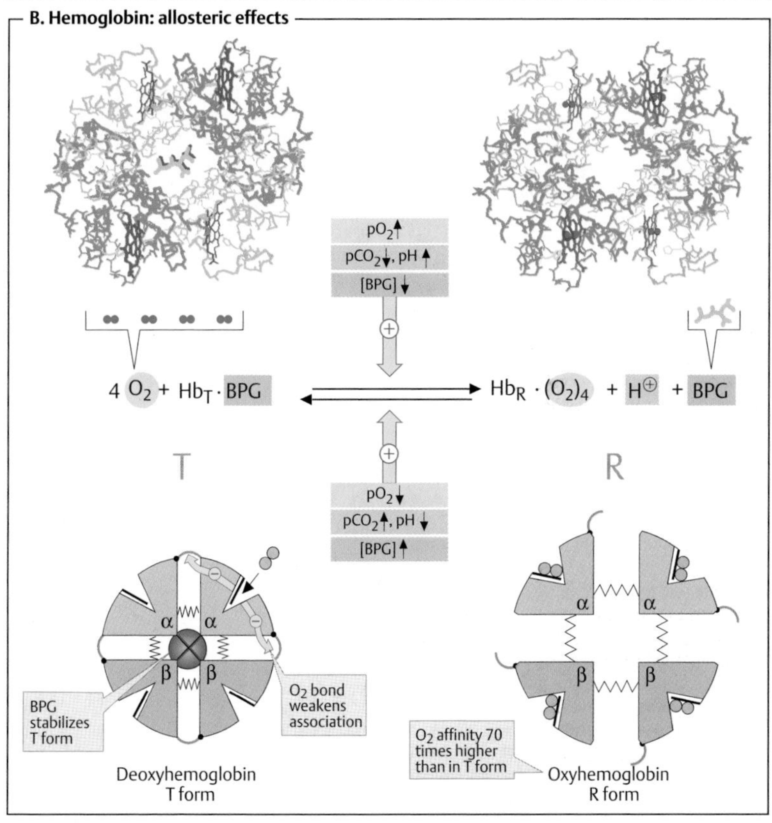

Gas transport

Most tissues are constantly dependent on a supply of molecular oxygen (O_2) to maintain their oxidative metabolism. Due to its poor solubility, O_2 is bound to hemoglobin for transport in the blood (see p. 280). This not only increases the oxygen transport capacity, but also allows regulation of O_2 uptake in the lungs and O_2 release into tissues.

A. Regulation of O_2 transport ◑

When an enzyme reacts to effectors (substrates, activators, or inhibitors) with conformational changes that increase or reduce its activity, it is said to show *allosteric behavior* (see p. 116). Allosteric enzymes are usually oligomers with several subunits that mutually influence each other.

Although **hemoglobin** is not an enzyme (it releases the bound oxygen without changing it), it has all the characteristics of an allosteric protein. Its effectors include oxygen, which as a *positive homotropic effector* promotes its own binding. The **O_2 saturation curve** of hemoglobin is therefore markedly sigmoidal in shape (**2**, curve 2). The non-sigmoidal saturation curve of the muscular protein **myoglobin** is shown for comparison (curve 1). The structure of myoglobin (see p. 336) is similar to that of a subunit of hemoglobin, but as a monomer it does not exhibit any allosteric behavior.

CO_2, H^+, and a special metabolite of erythrocytes—**2,3-bisphosphoglycerate (BPG)**—act as *heterotropic effectors* of hemoglobin. BPG is synthesized from *1,3-bisphosphoglycerate,* an intermediate of glycolysis (see p. 150), and it can be returned to glycolysis again by breakdown into 2-phosphoglycerate (**1**), with loss of an ATP.

BPG binds selectively to *deoxy-Hb,* thereby increasing its amount of equilibrium. The result is *increased O_2 release* at constant pO_2. In the diagram, this corresponds to a **right shift** of the saturation curve (**2**, curve 3). CO_2 and H^+ act in the same direction as BPG. Their influence on the position of the curve has long been known as the **Bohr effect**.

The effects of CO_2 and BPG are *additive.* In the presence of both effectors, the saturation curve of isolated Hb is similar to that of whole blood (curve 4).

B. Hemoglobin and CO_2 transport ◑

Hemoglobin is also decisively involved in the transport of carbon dioxide (CO_2) from the tissues to the lungs.

Some 5% of the CO_2 arising in the tissues is covalently bound to the N terminus of hemoglobin and transported as *carbaminohemoglobin* (not shown). About 90% of the CO_2 is first converted in the periphery into *hydrogen carbonate* (HCO_3^-), which is more soluble (bottom). In the lungs (top), CO_2 is regenerated again from HCO_3^- and can then be exhaled.

These two processes are coupled to the oxygenation and deoxygenation of Hb. *Deoxy-Hb* is a stronger base than oxy-Hb. It therefore binds additional protons (about 0.7 H^+ per tetramer), which promotes the formation of HCO_3^- from CO_2 in the peripheral tissues. The resulting HCO_3^- is released into the plasma via an antiporter in the erythrocyte membrane in exchange for Cl^-, and passes from the plasma to the lungs. In the lungs, the reactions described above then proceed in reverse order: deoxy-Hb is oxygenated and releases protons. The protons shift the HCO_3/CO_2 equilibrium to the left and thereby promote CO_2 release.

O_2 binding to Hb is regulated by H^+ ions (i. e., by the pH value) via the same mechanism. High concentrations of CO_2 such as those in tissues with intensive metabolism locally increase the H^+ concentration and thereby reduce hemoglobin's O_2 affinity (Bohr effect; see above). This leads to increased O_2 release and thus to an improved oxygen supply.

The adjustment of the equilibrium between CO_2 and HCO_3^- is relatively slow in the uncatalyzed state. It is therefore accelerated in the erythrocytes by *carbonate dehydratase* (carbonic anhydrase) [1])—an enzyme that occurs in high concentrations in the erythrocytes.

A. Regulation of O_2 transport

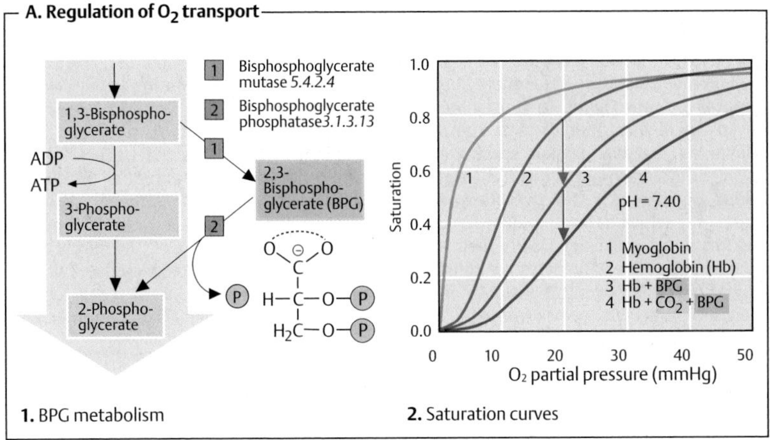

1. BPG metabolism

2. Saturation curves

D. Hemoglobin and CO_2 transport

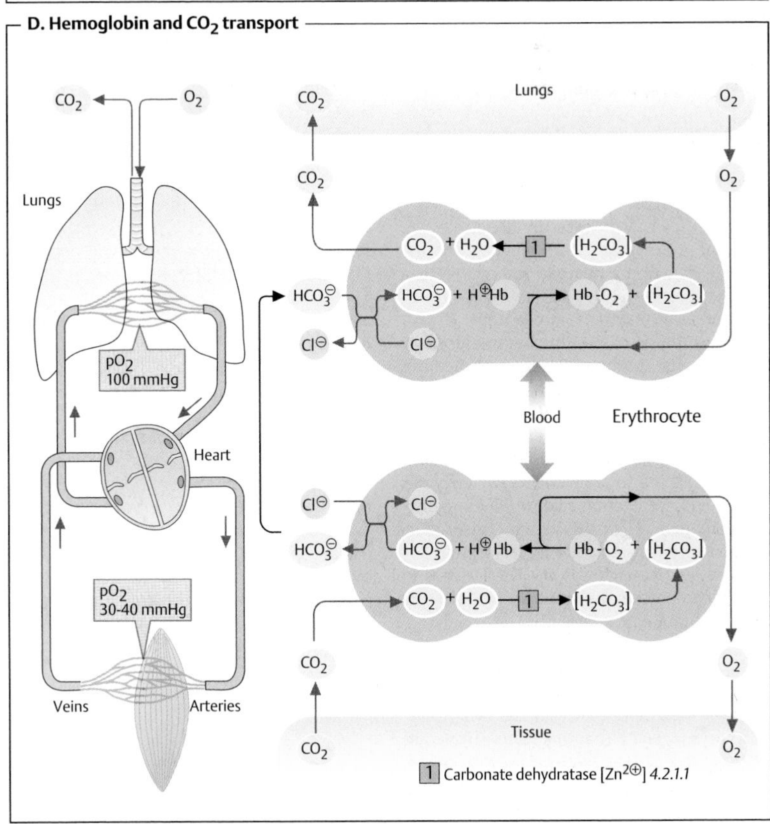

Erythrocyte metabolism

Cells living in aerobic conditions are dependent on molecular oxygen for energy production. On the other hand, O_2 constantly gives rise to small quantities of toxic substances known as **reactive oxygen species** (ROS). These substances are powerful *oxidation agents* or extremely reactive *free radicals* (see p. 32), which damage cellular structures and functional molecules. Due to their role in O_2 transport, the erythrocytes are constantly exposed to high concentrations of O_2 and are therefore particularly at risk from ROS.

A. Reactive oxygen species ◑

The dioxygen molecule (O_2) contains two unpaired electrons—i. e., it is a *diradical.* Despite this, O_2 is relatively stable due to its special electron arrangement. However, if the molecule takes up an extra electron (**a**), the highly reactive **superoxide radical** ($\cdot O_2^-$) arises. Another reduction step (**b**) leads to the **peroxide anion** (O_2^{2-}), which easily binds protons and thus becomes **hydrogen peroxide** (H_2O_2). Inclusion of a third electron (**c**) leads to cleavage of the molecule into the ions O^{2-} and O^-. While O^{2-} can form **water** by taking up two protons, protonation of O^- provides the extremely dangerous **hydroxy radical** ($\cdot OH$). A fourth electron transfer and subsequent protonation also convert O^- into water.

The synthesis of ROS can be catalyzed by iron ions, for example. Reaction of O_2 with FMN or FAD (see p. 32) also constantly produces ROS. By contrast, reduction of O_2 by *cytochrome c-oxidase* (see p. 140) is "clean," as the enzyme does not release the intermediates. In addition to antioxidants (**B**), **enzymes** also provide protection against ROS: *superoxide dismutase* [1] breaks down ("disproportionates") two superoxide molecules into O_2 and the less damaging H_2O_2. The latter is in turn disproportionated into O_2 and H_2O by heme-containing *catalase* [2].

B. Biological antioxidants ◑

To protect them against ROS and other radicals, all cells contain **antioxidants**. These are *reducing agents* that react easily with oxidative substances and thus protect more important molecules from oxidation. Biological antioxidants include vitamins C and E (see pp. 364, 368), coenzyme Q (see p. 104), and several carotenoids (see pp. 132, 364). Bilirubin, which is formed during heme degradation (see p. 194), also serves for protection against oxidation.

Glutathione, a tripeptide that occurs in high concentrations in almost all cells, is particularly important. Glutathione (sequence: Glu–Cys–Gly) contains an atypical γ-peptide bond between Glu and Cys. The thiol group of the cysteine residue is redox-active. Two molecules of the reduced form (GSH, top) are bound to the disulfide (GSSG, bottom) during oxidation.

C. Erythrocyte metabolism ◑

Erythrocytes also have systems that can inactivate ROS (superoxide dismutase, catalase, GSH). They are also able to repair damage caused by ROS. This requires products that are supplied by the erythrocytes' **maintenance metabolism**, which basically only involves anaerobic glycolysis (see p. 150) and the pentose phosphate pathway (PPP; see p. 152).

The **ATP** formed during glycolysis serves mainly to supply *Na^+/K^+-ATPase,* which maintains the erythrocytes' membrane potential. The allosteric effector **2,3-BPG** (see p. 282) is also derived from glycolysis. The PPP supplies **NADPH+H$^+$**, which is needed to regenerate **glutathione** (GSH) from GSSG with the help of *glutathione reductase* [3]. GSH, the most important antioxidant in the erythrocytes, serves as a coenzyme for *glutathione peroxidase* [5]. This selenium-containing enzyme detoxifies H_2O_2 and hydroperoxides, which arise during the reaction of ROS with unsaturated fatty acids in the erythrocyte membrane. The reduction of methemoglobin ($Hb \cdot Fe^{3+}$) to Hb ($Hb \cdot Fe^{2+}$, [4]) is carried out by GSH or ascorbate by a non-enzymatic pathway; however, there are also NAD(P)H-dependent Met-Hb reductases.

A. Reactive oxygen species

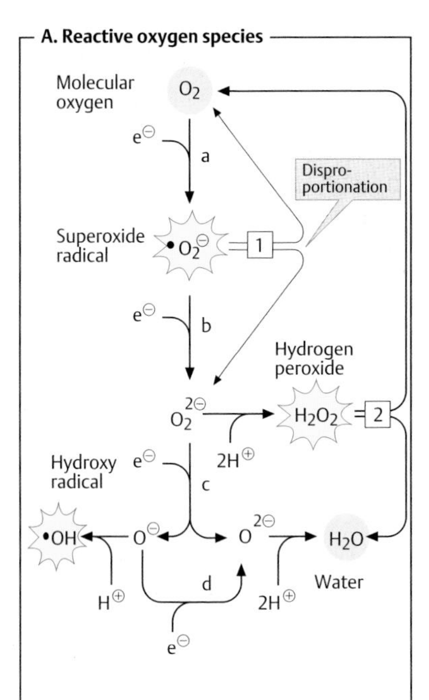

B. Biological antioxidants

Quinols and enols	α-Tocopherol (vitamin E) Ubiquinol (coenzyme Q) Ascorbic acid (vitamin C)
Carotenoids	β-Carotin Lycopin
Others	Glutathione Bilirubin

1. Examples

Glu Cys Gly

$^{\ominus}OOC-CH(NH_3^{\oplus})-(CH_2)_2-CO-NH-CH(CH_2SH)-CO-NH-CH_2-COO^{\ominus}$

2 Glutathione
2 (GSH)

Oxidation ↓↑ Reduction

$^{\ominus}OOC-CH(NH_3^{\oplus})-(CH_2)_2-CO-NH-CH(CH_2-S-S-CH_2)-CO-NH-CH_2-COO^{\ominus}$
$^{\ominus}OOC-CH(NH_3^{\oplus})-(CH_2)_2-CO-NH-CH-CO-NH-CH_2-COO^{\ominus}$

Glutathione disulfide (GSSG)

2. Glutathione

C. Erythrocyte metabolism

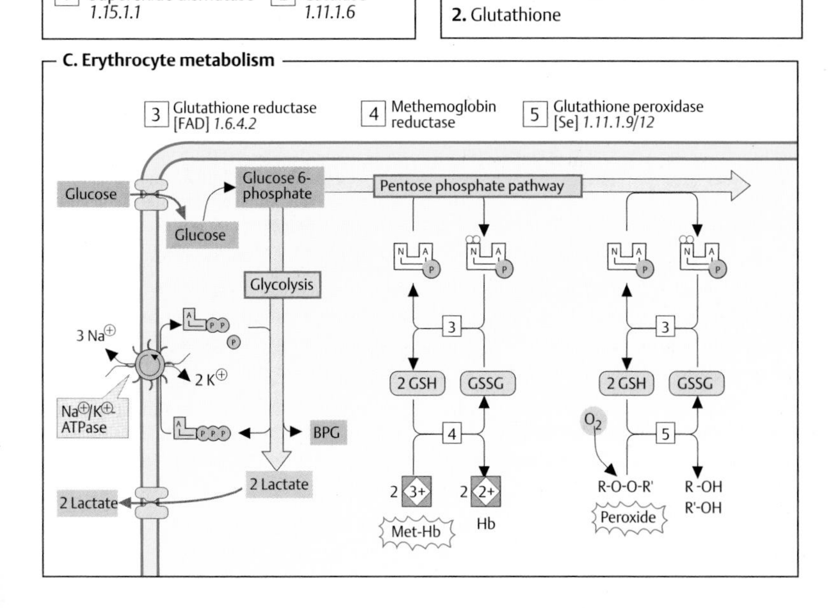

Iron metabolism

A. Distribution of iron ◑

Iron (Fe) is quantitatively the most important trace element (see p. 362). The human body contains 4–5 g iron, which is almost exclusively present in protein-bound form. Approximately three-quarters of the total amount is found in **heme proteins** (see pp. 106, 192), mainly hemoglobin and myoglobin. About 1% of the iron is bound in **iron–sulfur clusters** (see p. 106), which function as cofactors in the respiratory chain, in photosynthesis, and in other redox chains. The remainder consists of iron in transport and storage proteins (transferrin, ferritin; see **B**).

B. Iron metabolism ◑

Iron can only be resorbed by the bowel in bivalent form (i. e., as Fe^{2+}). For this reason, reducing agents in food such as ascorbate (vitamin C; see p. 368) promote **iron uptake**. Via transporters on the luminal and basal side of the enterocytes, Fe^{2+} enters the blood, where it is bound by *transferrin*. Part of the iron that is taken up is stored in the bowel in the form of *ferritin* (see below). Heme groups can also be resorbed by the small intestine.

Most of the resorbed iron serves for the formation of red blood cells in the bone marrow (**erythropoiesis**, top). As discussed on p. 192, it is only in the final step of hem biosynthesis that Fe^{2+} is incorporated by *ferrochelatase* into the previously prepared tetrapyrrol framework.

In the blood, 2.5–3.0 g of hemoglobin iron circulates as a component of the erythrocytes (top right). Over the course of several months, the flexibility of the red blood cells constantly declines due to damage to the membrane and cytoskeleton. Old erythrocytes of this type are taken up by macrophages in the spleen and other organs and broken down. The organic part of the heme is oxidized into bilirubin (see p. 194), while the iron returns to the plasma pool. The quantity of heme iron recycled per day is much larger than the amount resorbed by the intestines.

Transferrin, a β-globulin with a mass of 80 kDa, serves to transport iron in the blood. This monomeric protein consists of two similar domains, each of which binds an Fe^{2+} ion very tightly. Similar iron transport proteins are found in secretions such as saliva, tears, and milk; these are known as lactoferrins (bottom right). Transferrin and the lactoferrins maintain the concentration of *free* iron in body fluids at values below 10^{-10} mol · L^{-1}. This low level prevents bacteria that require free iron as an essential growth factor from proliferating in the body. Like LDLs (see p. 278), transferrin and the lactoferrins are taken up into cells by *receptor-mediated endocytosis.*

Excess iron is incorporated into **ferritin** and stored in this form in the liver and other organs. The ferritin molecule consists of 24 subunits and has the shape of a hollow sphere (bottom left). It takes up Fe^{2+} ions, which in the process are oxidized to Fe^{3+} and then deposited in the interior of the sphere as *ferrihydrate.* Each ferritin molecule is capable of storing several thousand iron ions in this way. In addition to ferritin, there is another storage form, **hemosiderin**, the function of which is not yet clear.

Further information

Disturbances of the iron metabolism are frequent and can lead to severe disease pictures.

Iron deficiency is usually due to blood loss, or more rarely to inadequate iron uptake. During pregnancy, increased demand can also cause iron deficiency states. In severe cases, reduced hemoglobin synthesis can lead to **anemia** ("iron-deficiency anemia"). In these patients, the erythrocytes are smaller and have less hemoglobin. As their membrane is also altered, they are prematurely eliminated in the spleen.

Disturbances resulting from raised iron concentrations are less frequent. Known as **hemochromatoses**, these conditions can have genetic causes, or may be due to repeated administration of blood transfusions. As the body has practically no means of excreting iron, more and more stored iron is deposited in the organs over time in patients with untreated hemochromatosis, ultimately leading to severe disturbances of organ function.

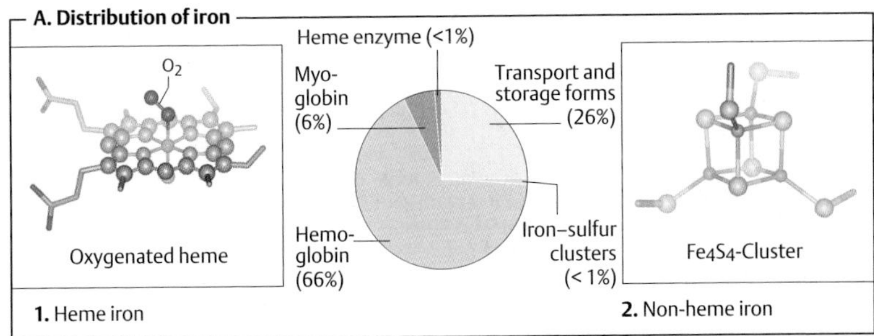
A. Distribution of iron
Heme enzyme (<1%)
Myo-
globin
(6%)
Transport and
storage forms
(26%)
O2
Oxygenated heme
Hemo-
globin
(66%)
Iron–sulfur
clusters
(< 1%)
Fe4S4-Cluster
1. Heme iron
2. Non-heme iron

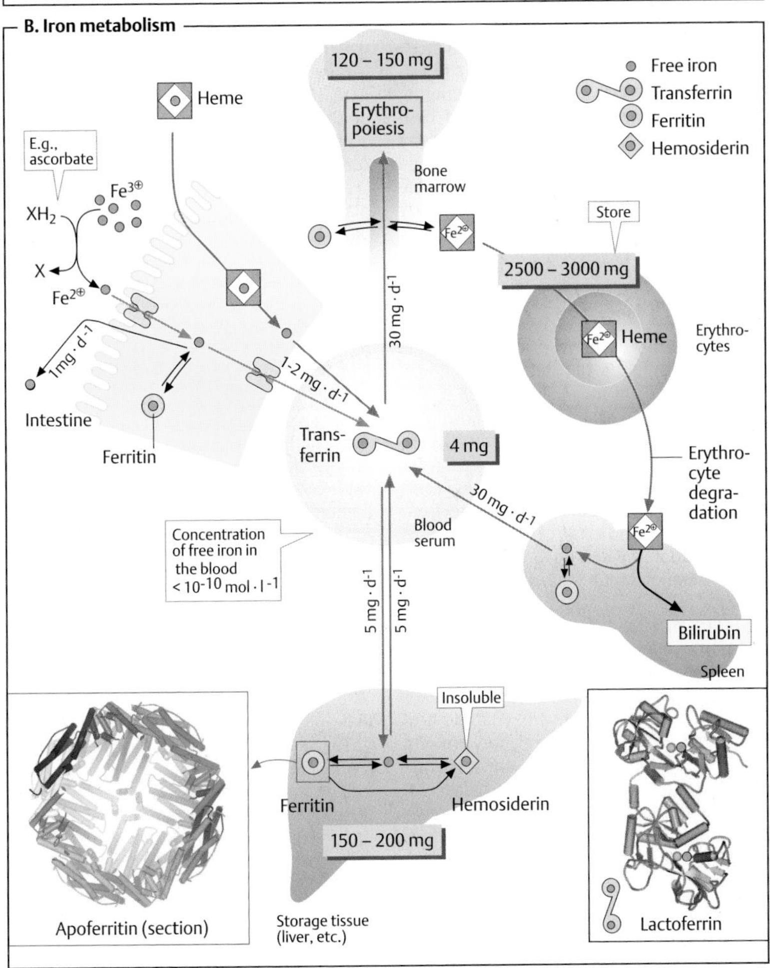
B. Iron metabolism
120 – 150 mg
Erythro-
poiesis
Bone
marrow
Heme
Free iron
Transferrin
Ferritin
Hemosiderin
E.g.,
ascorbate
XH2
X
Fe3⊕
Fe2⊕
Store
2500 – 3000 mg
Heme
Erythro-
cytes
30 mg · d-1
1mg · d-1
1-2 mg · d-1
Intestine
Ferritin
Trans-
ferrin
4 mg
Erythro-
cyte
degra-
dation
30 mg · d-1
Blood
serum
Concentration
of free iron in
the blood
< 10-10 mol · l -1
5 mg · d-1
5 mg · d-1
Bilirubin
Spleen
Insoluble
Ferritin
Hemosiderin
150 – 200 mg
Apoferritin (section)
Storage tissue
(liver, etc.)
Lactoferrin

Acid–base balance

A. Hydrogen ion concentration in the blood plasma

The H^+ concentration in the blood and extracellular space is approximately 40 nM ($4 \cdot 10^{-8}$ mol $\cdot$ L^{-1}). This corresponds to a pH of 7.40. The body tries to keep this value constant, as large shifts in pH are incompatible with life.

The pH value is kept constant by **buffer systems** that cushion minor disturbances in the *acid–base balance* (**C**). In the longer term, the decisive aspect is maintaining a balanced equilibrium between H^+ production and uptake and H^+ release. If the blood's buffering capacity is not sufficient, or if the acid–base balance is not in equilibrium—e. g., in kidney disease or during *hypoventilation* or *hyperventilation*—shifts in the plasma pH value can occur. A reduction by more than 0.03 units is known as **acidosis,** and an increase is called **alkalosis.**

B. Acid–base balance

Protons are mainly derived from two sources—free acids in the diet and sulfur-containing amino acids. *Acids* taken up with food—e. g., citric acid, ascorbic acid, and phosphoric acid—already release protons in the alkaline pH of the intestinal tract. More important for proton balance, however, are the amino acids **methionine** and **cysteine**, which arise from protein degradation in the cells. Their S atoms are oxidized in the liver to form sulfuric acid, which supplies protons by dissociation into sulfate.

During anaerobic glycolysis in the muscles and erythrocytes, glucose is converted into **lactate**, releasing protons in the process (see p. 338). The synthesis of the **ketone bodies** acetoacetic acid and 3-hydroxybutyric acid in the liver (see p. 312) also releases protons. Normally, the amounts formed are small and of little influence on the proton balance. If acids are formed in large amounts, however (e. g., during *starvation* or in *diabetes mellitus;* see p. 160), they strain the buffer systems and can lead to a reduction in pH (**metabolic acidoses**; *lactacidosis* or *ketoacidosis*).

Only the **kidney** is capable of excreting protons in exchange for Na^+ ions (see p. 326). In the urine, the H^+ ions are buffered by NH_3 and phosphate.

C. Buffer systems in the plasma

The **buffering capacity** of a buffer system depends on its concentration and its pK_a value. The strongest effect is achieved if the pH value corresponds to the buffer system's pK_a value (see p. 30). For this reason, weak acids with pK_a values of around 7 are best suited for buffering purposes in the blood.

The most important buffer in the blood is the **CO_2/bicarbonate buffer**. This consists of water, carbon dioxide (CO_2, the anhydride of carbonic acid H_2CO_3), and hydrogen carbonate (HCO_3^-, bicarbonate). The adjustment of the balance between CO_2 and HCO_3^- is accelerated by the zinc-containing enzyme *carbonate dehydratase* (carbonic anhydrase [1]; see also p. 282). At the pH value of the plasma, HCO_3^- and CO_2 are present in a ratio of about 20 : 1. However, the CO_2 in solution in the blood is in equilibrium with the gaseous CO_2 in the pulmonary alveoli. The CO_2/HCO_3^- system is therefore a powerful *open buffer system,* despite having a not entirely optimal pK_a value of 6.1. Faster or slower respiration increases or reduces CO_2 release in the lungs. This shifts the CO_2/HCO_3^- ratio and thus the plasma pH value (respiratory acidosis or alkalosis). In this way, respiration can compensate to a certain extent for changes in plasma pH values. However, it does *not* lead to the excretion of protons.

Due to their high concentration, **plasma proteins**—and **hemoglobin** in the erythrocytes in particular—provide about one-quarter of the blood's buffering capacity. The buffering effect of proteins involves contributions from all of the ionizable side chains. At the pH value of blood, the acidic amino acids (Asp, Glu) and histidine are particularly effective.

The second dissociation step in **phosphate** (H_2PO_4/HPO_4^{2-}) also contributes to the buffering capacity of the blood plasma. Although the pK_a value of this system is nearly optimal, its contribution remains small due to the low total concentration of phosphate in the blood (around 1 mM).

A. Hydrogen ion concentration in the blood plasma

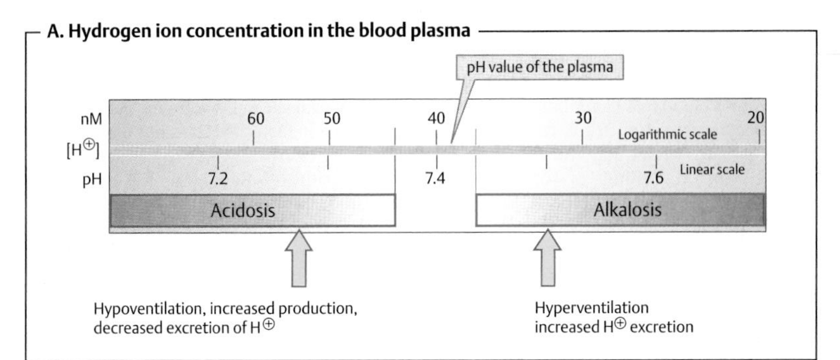

B. Acid–base balance

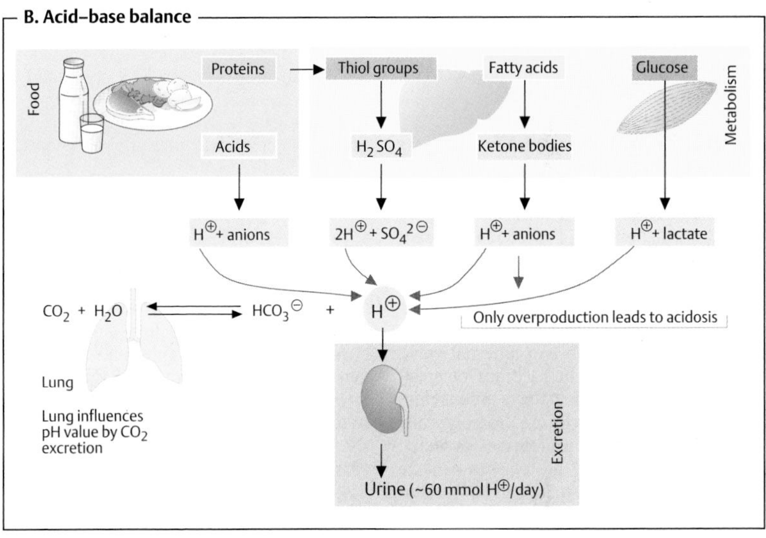

C. Buffer systems in the plasma

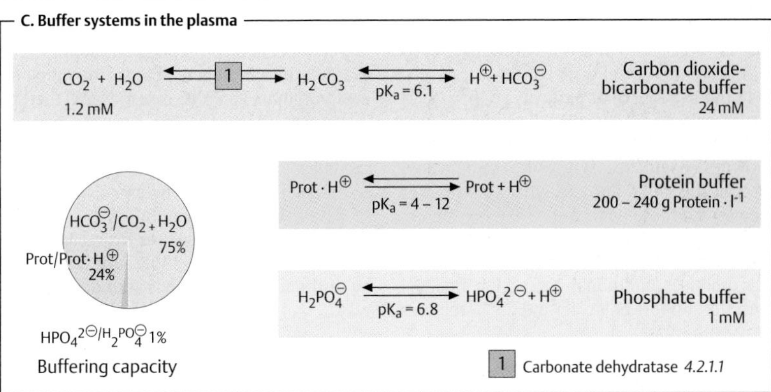

Blood clotting

Following injury to blood vessels, **hemostasis** ensures that blood loss is minimized. Initially, thrombocyte activation leads to contraction of the injured vessel and the formation of a loose clot consisting of thrombocytes (**hemostasis**). Slightly later, the action of the enzyme *thrombin* leads to the formation and deposition in the thrombus of polymeric fibrin (**coagulation, blood clotting**). The coagulation process is discussed here in detail.

A. Blood clotting ◑

The most important reaction in blood clotting is the conversion, catalyzed by *thrombin*, of the soluble plasma protein *fibrinogen* (factor I) into polymeric *fibrin*, which is deposited as a fibrous network in the primary thrombus. Thrombin (factor IIa) is a serine proteinase (see p. 176) that cleaves small peptides from fibrinogen. This exposes binding sites that spontaneously allow the fibrin molecules to aggregate into polymers. Subsequent covalent cross-linking of fibrin by a *transglutaminase* (factor XIII) further stabilizes the thrombus.

Normally, thrombin is present in the blood as an inactive proenzyme (see p. 270). Prothrombin is activated in two different ways, both of which represent cascades of enzymatic reactions in which inactive proenzymes (zymogens, symbol: circle) are proteolytically converted into active **proteinases** (symbol: sector of a circle). The proteinases activate the next proenzyme in turn, and so on. Several steps in the cascade require additional **protein factors** (factors III, Va and VIIIa) as well as anionic **phospholipids** (PL; see below) and $\mathbf{Ca^{2+}}$ ions. Both pathways are activated by injuries to the vessel wall.

In the **extravascular pathway** (right), *tissue thromboplastin* (factor III), a membrane protein in the deeper layers of the vascular wall, activates coagulation factor VII. The activated form of this (VIIa) autocatalytically promotes its own synthesis and also generates the active factors IXa and Xa from their precursors. With the aid of factor VIIIa, PL, and Ca^{2+}, factor IXa produces additional Xa, which finally—with the support of Va, PL, and Ca^{2+}—releases active thrombin.

The **intravascular pathway** (left) is probably also triggered by vascular injuries. It leads in five steps via factors XIIa, XIa, IXa, and Xa to the activation of prothrombin. The significance of this pathway in vivo has been controversial since it was found that a genetic deficiency in factor XII does not lead to coagulation disturbances.

Both pathways depend on the presence of activated **thrombocytes**, on the surface of which several reactions take place. For example, the *prothrombinase complex* (left) forms when factors Xa and II, with the help of Va, bind via Ca^{2+} ions to anionic phospholipids in the thrombocyte membrane. For this to happen, factors II and X have to contain the non-proteinogenic amino acid **γ-carboxyglutamate** (Gla; see p. 62), which is formed in the liver by post-translational carboxylation of the factors. The Gla residues are found in groups in special domains that create contacts to the Ca^{2+} ions. Factors VII and IX are also linked to membrane phospholipids via Gla residues.

Substances that bind Ca^{2+} ions (e.g., *citrate*) prevent Gla-containing factors from attaching to the membrane and therefore inhibit coagulation. Antagonists of *vitamin K*, which is needed for synthesis of the Gla residues (see p. 364) also have anticoagulatory effects. These include dicumarol, for example.

Active **thrombin** not only converts fibrinogen into fibrin, but also indirectly promotes its own synthesis by catalyzing the activation of factors V and VIII. In addition, it catalyzes the activation of factor XIII and thereby triggers the cross-linking of the fibrin.

Regulation of blood clotting (not shown). To prevent the coagulation reaction from becoming excessive, the blood contains a number of anticoagulant substances, including highly effective proteinase inhibitors. For example, *antithrombin III* binds to various serine proteinases in the cascade and thereby inactivates them. *Heparin*, an anticoagulant glycosaminoglycan (see p. 346), potentiates the effect of antithrombin III. *Thrombomodulin*, which is located on the vascular endothelia, also inactivates thrombin. A glycoprotein known as *Protein C* ensures proteolytic degradation of factors V and VIII. As it is activated by thrombin, coagulation is shut down in this way.

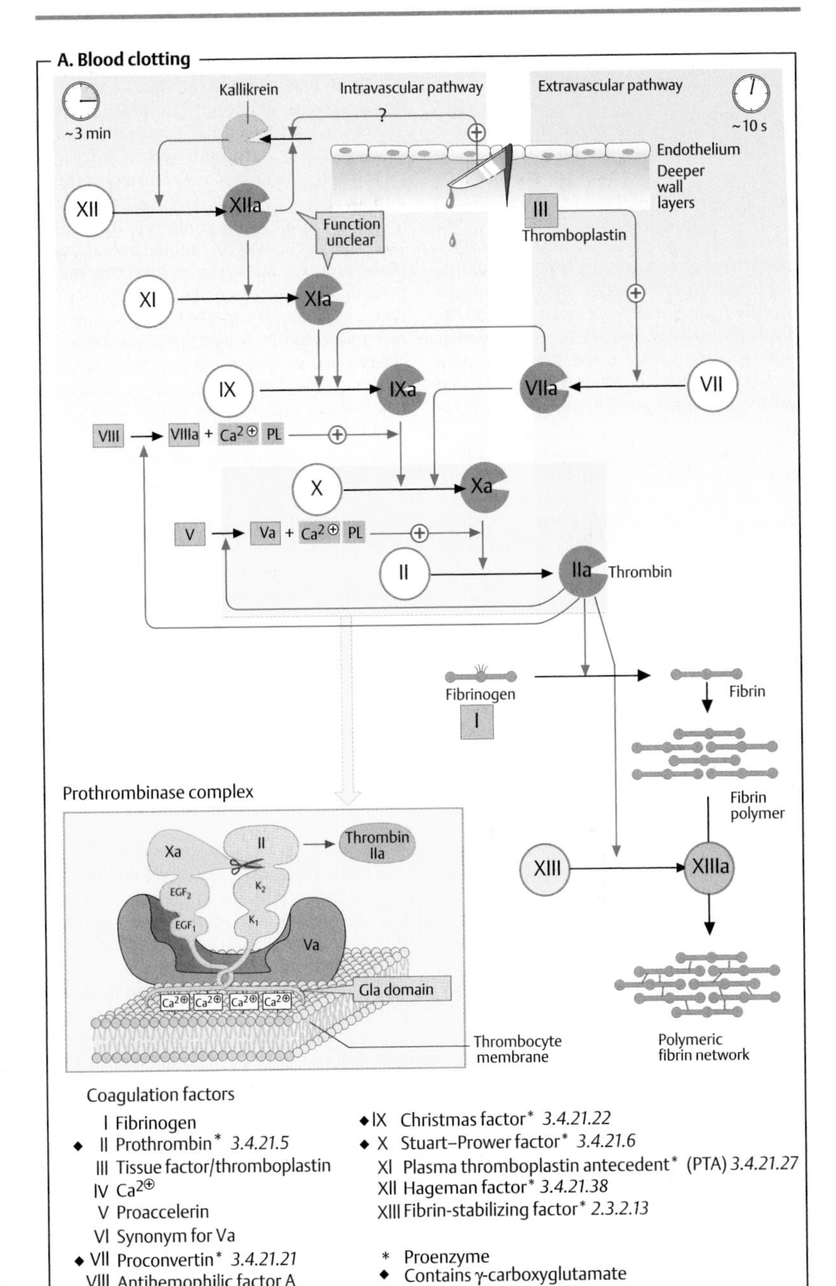
A. Blood clotting
~3 min
Kallikrein
Intravascular pathway
Extravascular pathway
~10 s
?
Endothelium
Deeper wall layers
XII
XIIa
Function unclear
III
Thromboplastin
XI
XIa
IX
IXa
VIIa
VII
VIII
VIIIa + Ca2⊕ PL
X
Xa
V
Va + Ca2⊕ PL
II
IIa
Thrombin
Fibrinogen
I
Fibrin
Fibrin polymer
Prothrombinase complex
Xa
II
Thrombin IIa
EGF2
K2
EGF1
K1
Va
Gla domain
Ca2⊕ Ca2⊕ Ca2⊕ Ca2⊕
Thrombocyte membrane
XIII
XIIIa
Polymeric fibrin network
Coagulation factors
I Fibrinogen
◆ II Prothrombin* 3.4.21.5
III Tissue factor/thromboplastin
IV Ca2⊕
V Proaccelerin
VI Synonym for Va
◆ VII Proconvertin* 3.4.21.21
VIII Antihemophilic factor A
◆ IX Christmas factor* 3.4.21.22
◆ X Stuart–Prower factor* 3.4.21.6
XI Plasma thromboplastin antecedent* (PTA) 3.4.21.27
XII Hageman factor* 3.4.21.38
XIII Fibrin-stabilizing factor* 2.3.2.13
* Proenzyme
◆ Contains γ-carboxyglutamate

Fibrinolysis, blood groups

A. Fibrinolysis ◑

The fibrin thrombus resulting from blood clotting (see p. 290) is dissolved again by *plasmin*, a serine proteinase found in the blood plasma. For this purpose, the precursor *plasminogen* first has to be proteolytically activated by enzymes from various tissues. This group includes the *plasminogen activator* from the kidney (*urokinase*) and *tissue plasminogen activator* (t-PA) from vascular endothelia. By contrast, the plasma protein α_2-antiplasmin, which binds to active plasmin and thereby inactivates it, inhibits fibrinolysis.

Urokinase, t-PA, and streptokinase, a bacterial proteinase with similar activity, are used clinically to dissolve thrombi following *heart attacks*. All of these proteins are expressed recombinantly in bacteria (see p. 262).

B. Blood groups: the ABO system ◑

During blood transfusions, immune reactions can occur that destroy the erythrocytes transfused from the donor. These reactions result from the formation of antibodies (see p. 300) directed to certain surface structures on the erythrocytes. Known as **blood group antigens**, these are *proteins* or *oligosaccharides* that can differ from individual to individual. More than 20 different blood group systems are now known. The ABO system and the Rh system are of particular clinical importance.

In the **ABO system**, the carbohydrate parts of glycoproteins or glycolipids act as antigens. In this relatively simple system, there are four *blood groups* (A, B, AB, and 0). In individuals with blood groups A and B, the antigens consist of tetrasaccharides that only differ in their terminal sugar (galactose or *N*-acetylgalactosamine). Carriers of the AB blood group have both antigens (A *and* B). Blood group 0 arises from an oligosaccharide (the H antigen) that lacks the terminal residue of antigens A and B. The molecular causes for the differences between blood groups are mutations in the *glycosyl transferases* that transfer the terminal sugar to the core oligosaccharide.

Antibodies are only formed against antigens that the individual concerned does *not* possess. For example, carriers of blood group A form antibodies against antigen B ("anti-B"), while carriers of group B form antibodies against antigen A ("anti-A"). Individuals with blood group 0 form both types, and those with blood group AB do not form any of these antibodies.

If blood from blood group A is transfused into the circulation of an individual with blood group B, for example, then the anti-A present there binds to the A antigens. The donor erythrocytes marked in this way are recognized and destroyed by the complement system (see p. 298). In the test tube, *agglutination* of the erythrocytes can be observed when donor and recipient blood are incompatible.

The recipient's serum should not contain any antibodies against the donor erythrocytes, and the donor serum should not contain any antibodies against the recipient's erythrocytes. Donor blood from blood group 0 is unproblematic, as its erythrocytes do not possess any antibodies and therefore do not react with anti-A or anti-B in the recipient's blood. Conversely, blood from the AB group can only be administered to recipients with the AB group, as these are the only ones without antibodies.

In the **Rh system** (not shown), proteins on the surface of the erythrocytes act as antigens. These are known as "rhesus factors," as the system was first discovered in rhesus monkeys.

The rhesus D antigen occurs in 84% of all white individuals, who are therefore *"Rh-positive."* If an Rh-positive child is born to an Rh-negative mother, fetal erythrocytes can enter the mother's circulation during birth and lead to the formation of antibodies (IgG) against the D antigen. This initially has no acute effects on the mother or child. Complications only arise when there is a second pregnancy with an Rh-positive child, as maternal anti-D antibodies cross the placenta to the fetus even before birth and can trigger destruction of the child's Rh-positive erythrocytes (*fetal erythroblastosis*).

A. Fibrinolysis

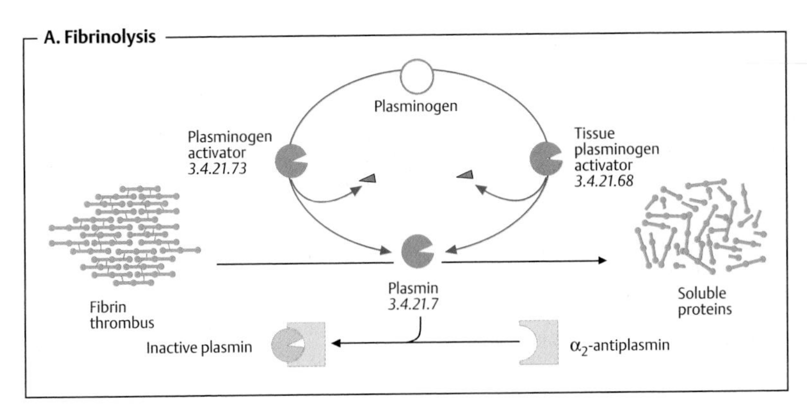

B. Blood groups: the AB0 system

Blood group	A	B	AB	0
Genotypes	AA and A0	BB and B0	AB	00
Antigens	A	B	A, B	H
Antibodies in blood	anti-B	anti-A	—	anti-A anti-B
Frequency in central Europe	40%	16%	4%	40%

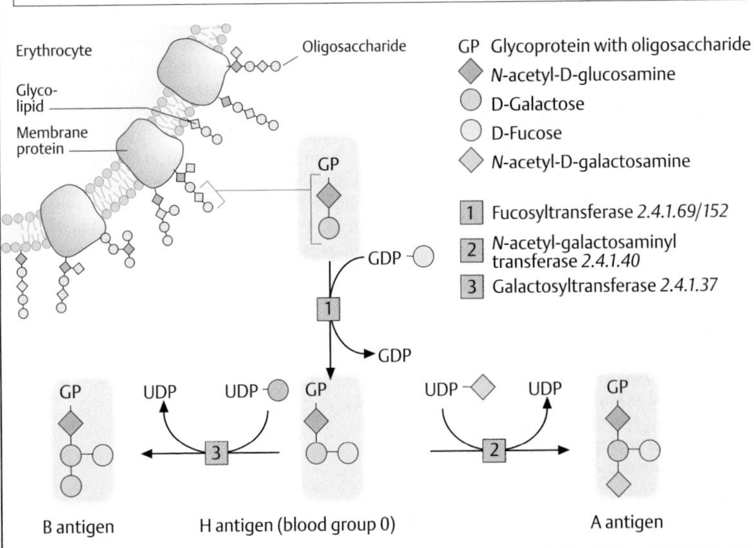

Immune response

Viruses, bacteria, fungi, and parasites that enter the body of vertebrates of are recognized and attacked by the **immune system.** Endogenous cells that have undergone alterations—e. g., tumor cells—are also usually recognized as foreign and destroyed. The immune system is supported by physiological changes in infected tissue, known as **inflammation.** This reaction makes it easier for the immune cells to reach the site of infection.

Two different systems are involved in the immune response. The **innate immune system** is based on receptors that can distinguish between bacterial and viral surface structures or foreign proteins (known as *antigens*) and those that are endogenous. With the help of these receptors, *phagocytes* bind to the pathogens, absorb them by endocytosis, and break them down. The complement system (see p. 298) is also part of the innate system.

The **acquired** (adaptive) **immune system** is based on the ability of the *lymphocytes* to form highly specific antigen receptors "on suspicion," without ever having met the corresponding antigen. In humans, there are several billion different lymphocytes, each of which carries a different antigen receptor. If this type of receptor recognizes "its" cognate antigen, the lymphocyte carrying it is activated and then plays its special role in the immune response.

In addition, a distinction is made between cellular and humoral immune responses. The *T lymphocytes* (T cells) are responsible for **cellular immunity.** They are named after the thymus, in which the decisive steps in their differentiation take place. Depending on their function, another distinction is made between *cytotoxic T cells* (green) and *helper T cells* (blue). **Humoral immunity** is based on the activity of the *B lymphocytes* (B cells, light brown), which mature in the bone marrow. After activation by T cells, B cells are able to release soluble forms of their specific antigen receptors, known as *antibodies* (see p. 300), into the blood plasma. The immune system's "memory" is represented by memory cells. These are particularly long-lived cells that can arise from any of the lymphocyte types described.

A. Simplified diagram of the immune response ◑

Pathogens that have entered the body—e. g., viruses (top)—are taken up by **antigen-presenting cells** (APCs) and proteolytically degraded (**1**). The viral fragments produced in this way are then presented on the surfaces of these cells with the help of special membrane proteins (MHC proteins; see p. 296) (**2**). The APCs include B lymphocytes, macrophages, and dendritic cells such as the skin's Langerhans cells.

The complexes of MHC proteins and viral fragments displayed on the APCs are recognized by T cells that carry a receptor that matches the antigen ("T-cell receptors"; see p. 296) (**3**). Binding leads to activation of the T cell concerned and selective replication of it (**4**, *"clonal selection"*). The proliferation of immune cells is stimulated by *interleukins* (IL). These are a group of more than 20 signaling substances belonging to the cytokine family (see p. 392), with the help of which immune cells communicate with each other. For example, activated macrophages release IL-1 (**5**), while T cells stimulate their own replication and that of other immune cells by releasing IL-2 (**6**).

Depending on their type, activated T cells have different functions. **Cytotoxic T cells** (green) are able to recognize and bind virus-infected body cells or tumor cells (**7**). They then drive the infected cells into apoptosis (see p. 396) or kill them with *perforin*, a protein that perforates the target cell's plasma membrane (**8**).

B lymphocytes, which as APCs present viral fragments on their surfaces, are recognized by **helper T cells** (blue) or their T cell receptors (**9**). Stimulated by interleukins, selective clonal replication then takes place of B cells that carry antigen receptors matching those of the pathogen (**10**). These mature into **plasma cells** (**11**) and finally secrete large amounts of soluble **antibodies** (**12**).

A. Simplified scheme of the immune response

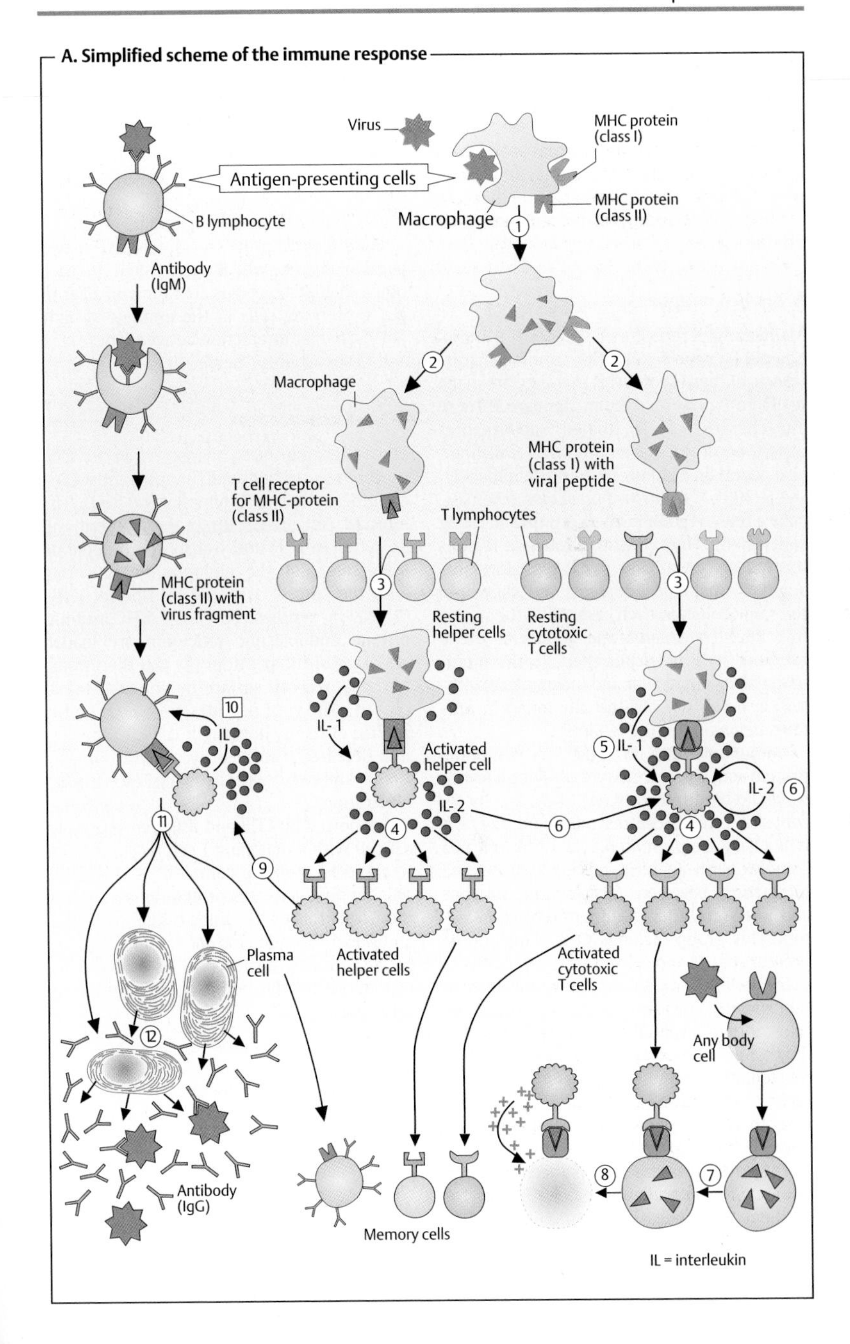

T-cell activation

For the selectivity of the immune response (see p. 294), the cells involved must be able to recognize foreign antigens and proteins on other immune cells safely and reliably. To do this, they have antigen receptors on their cell surfaces and co-receptors that support recognition.

A. Antigen receptors ○

Many antigen receptors belong to the **immunoglobulin superfamily**. The common characteristic of these proteins is that they are made up from "immunoglobulin domains." These are characteristically folded substructures consisting of 70–110 amino acids, which are also found in soluble immunoglobulins (Ig; see p. 300). The illustration shows schematically a few of the important proteins in the Ig superfamily. They consist of constant regions (brown or green) and variable regions (orange). Homologous domains are shown in the same colors in each case. All of the receptors have transmembrane helices at the C terminus, which anchor them to the membranes. Intramolecular and intermolecular disulfide bonds are also usually found in proteins belonging to the Ig family.

Immunoglobulin M (IgM), a membrane protein on the surface of B lymphocytes, serves to bind free antigens to the B cells. By contrast, **T cell receptors** only bind antigens when they are presented by another cell as a complex with an MHC protein (see below). Interaction between MHC-bound antigens and T cell receptors is supported by **co-receptors**. This group includes **CD8**, a membrane protein that is typical in cytotoxic T cells. T helper cells use **CD4** as a co-receptor instead (not shown). The abbreviation "CD" stands for "cluster of differentiation." It is the term for a large group of proteins that are all located on the cell surface and can therefore be identified by antibodies. In addition to CD4 and CD8, there are many other co-receptors on immune cells (not shown).

The **MHC proteins** are named after the *"major histocompatibility complex"*—the DNA segment that codes for them. Human MHC proteins are also known as HLA antigens ("human leukocyte-associated" antigens). Their polymorphism is so large that it is unlikely that any two individuals carry the same set of MHC proteins—except for monozygotic twins.

Class I MHC proteins occur in almost all nucleated cells. They mainly interact with cytotoxic T cells and are the reason for the rejection of transplanted organs. Class I MHC proteins are heterodimers ($\alpha\beta$). The β subunit is also known as β_2-microglobulin.

Class II MHC proteins also consist of two peptide chains, which are related to each other. MHC II molecules are found on all antigen-presenting cells in the immune system. They serve for interaction between these cells and CD4-carrying T helper cells.

B. T-cell activation ◑

The illustration shows an interaction between a virus-infected body cell (bottom) and a CD8-carrying cytotoxic T lymphocyte (top). The infected cell breaks down viral proteins in its cytoplasm (**1**) and transports the peptide fragments into the endoplasmic reticulum with the help of a special transporter (*TAP*) (**2**). Newly synthesized class I MHC proteins on the endoplasmic reticulum are loaded with one of the peptides (**3**) and then transferred to the cell surface by vesicular transport (**4**). The viral peptides are bound on the surface of the α_2 domain of the MHC protein in a depression formed by an insertion as a "floor" and two helices as "walls" (see smaller illustration).

Supported by CD8 and other co-receptors, a T cell with a matching T cell receptor binds to the MHC peptide complex (**5**; cf. p. 224). This binding activates protein kinases in the interior of the T cell, which trigger a chain of additional reactions (*signal transduction;* see p. 388). Finally, destruction of the virus-infected cell by the cytotoxic T lymphocytes takes place.

A. Antigen receptors

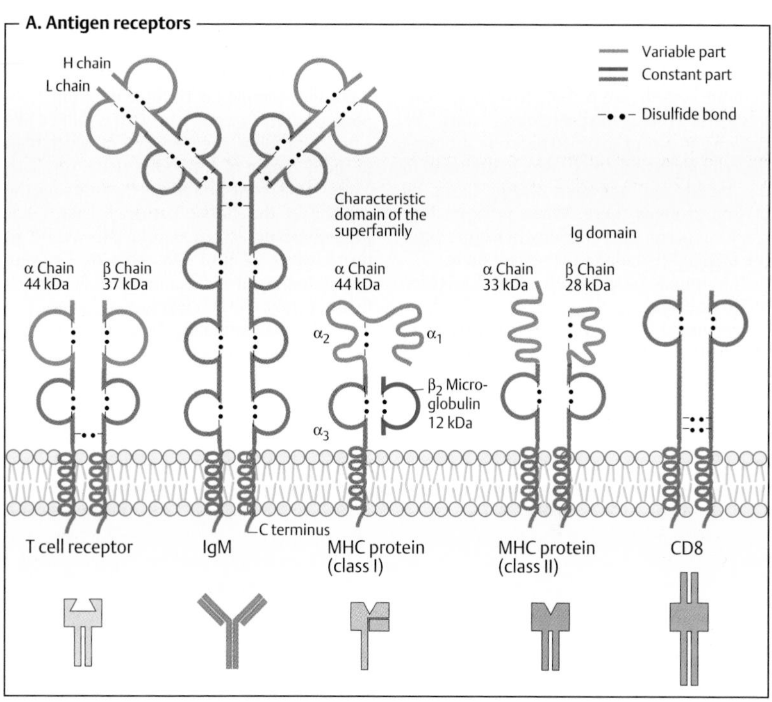

B. T cell activation

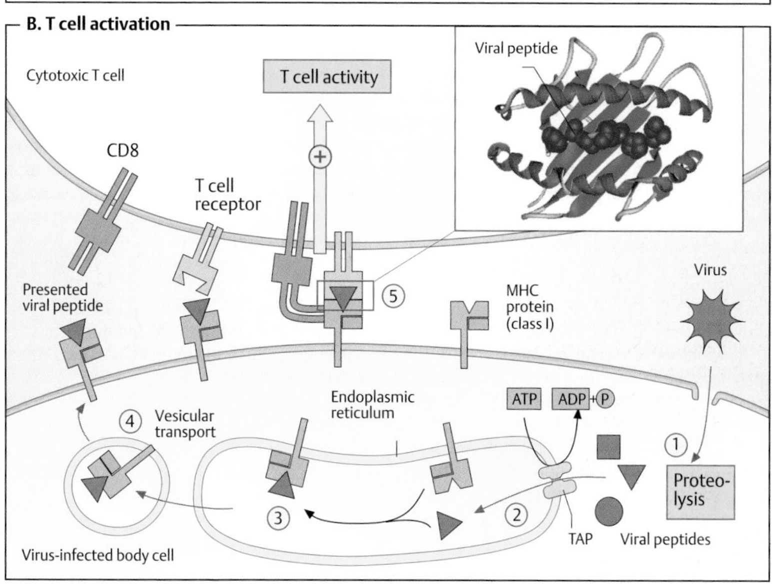

Complement system

The complement system is part of the innate immune system (see p. 294). It supports *non-specific defense* against microorganisms. The system consists of some 30 different proteins, the *"complement factors,"* which are found in the blood and represent about 4% of all plasma proteins there. When inflammatory reactions occur, the complement factors enter the infected tissue and take effect there.

The complement system works in three different ways:

Chemotaxis. Various complement factors attract immune cells that can attack and phagocytose pathogens.

Opsonization. Certain complement factors ("opsonins") bind to the pathogens and thereby mark them as targets for phagocytosing cells (e.g., macrophages).

Membrane attack. Other complement factors are deposited in the bacterial membrane, where they create pores that lyse the pathogen (see below).

A. Complement activation ◑

The reactions that take place in the complement system can be initiated in several ways. During the early phase of infection, lipopolysaccharides and other structures on the surface of the pathogens trigger the *alternative pathway* (right). If antibodies against the pathogens become available later, the antigen–antibody complexes formed activate the *classic pathway* (left). Acute-phase proteins (see p. 276) are also able to start the complement cascade (*lectin pathway,* not shown).

Factors C1 to **C4** (for "complement") belong to the classic pathway, while **factors B** and **D** form the reactive components of the alternative pathway. Factors **C5** to **C9** are responsible for membrane attack. Other components not shown here regulate the system.

As in blood coagulation (see p. 290), the *early components* in the complement system are *serine proteinases,* which mutually activate each other through limited proteolysis. They create a self-reinforcing **enzyme cascade**. Factor **C3**, the products of which are involved in several functions, is central to the complement system.

The **classic pathway** is triggered by the formation of factor C1 at IgG or IgM on the surface of microorganisms (left). C1 is an 18-part molecular complex with three different components (C1q, C1r, and C1s). C1q is shaped like a bunch of tulips, the "flowers" of which bind to the F_c region of antibodies (left). This activates C1r, a *serine proteinase* that initiates the cascade of the classic pathway. First, C4 is proteolytically activated into C4b, which in turn cleaves C2 into C2a and C2b. C4B and C2a together form *C3 convertase* [1], which finally catalyzes the cleavage of C3 into C3a and C3b. Small amounts of C3b also arise from non-enzymatic hydrolysis of C3.

The **alternative pathway** starts with the binding of factors C3b and B to bacterial lipopolysaccharides (endotoxins). The formation of this complex allows cleavage of B by factor D, giving rise to a second form of *C3 convertase (C3bBb).*

Proteolytic cleavage of factor **C3** provides two components with different effects. The reaction exposes a highly *reactive thioester group* in C3b, which reacts with hydroxyl or amino groups. This allows C3b to bind covalently to molecules on the bacterial surface (*opsonization,* right). In addition, C3b initiates a chain of reactions leading to the formation of the *membrane attack complex* (see below). Together with C4a and C5a (see below), the smaller product C3a promotes the inflammatory reaction and has chemotactic effects.

The "late" factors C5 to C9 are responsible for the development of the **membrane attack complex** (bottom). They create an ion-permeable pore in the bacterial membrane, which leads to lysis of the pathogen. This reaction is triggered by *C5 convertase* [2]. Depending on the type of complement activation, this enzyme has the structure *C4b2a3b* or *C3bBb3b,* and it cleaves C5 into C5a and C5b. The complex of C5b and C6 allows deposition of C7 in the bacterial membrane. C8 and numerous C9 molecules—which form the actual pore—then bind to this core.

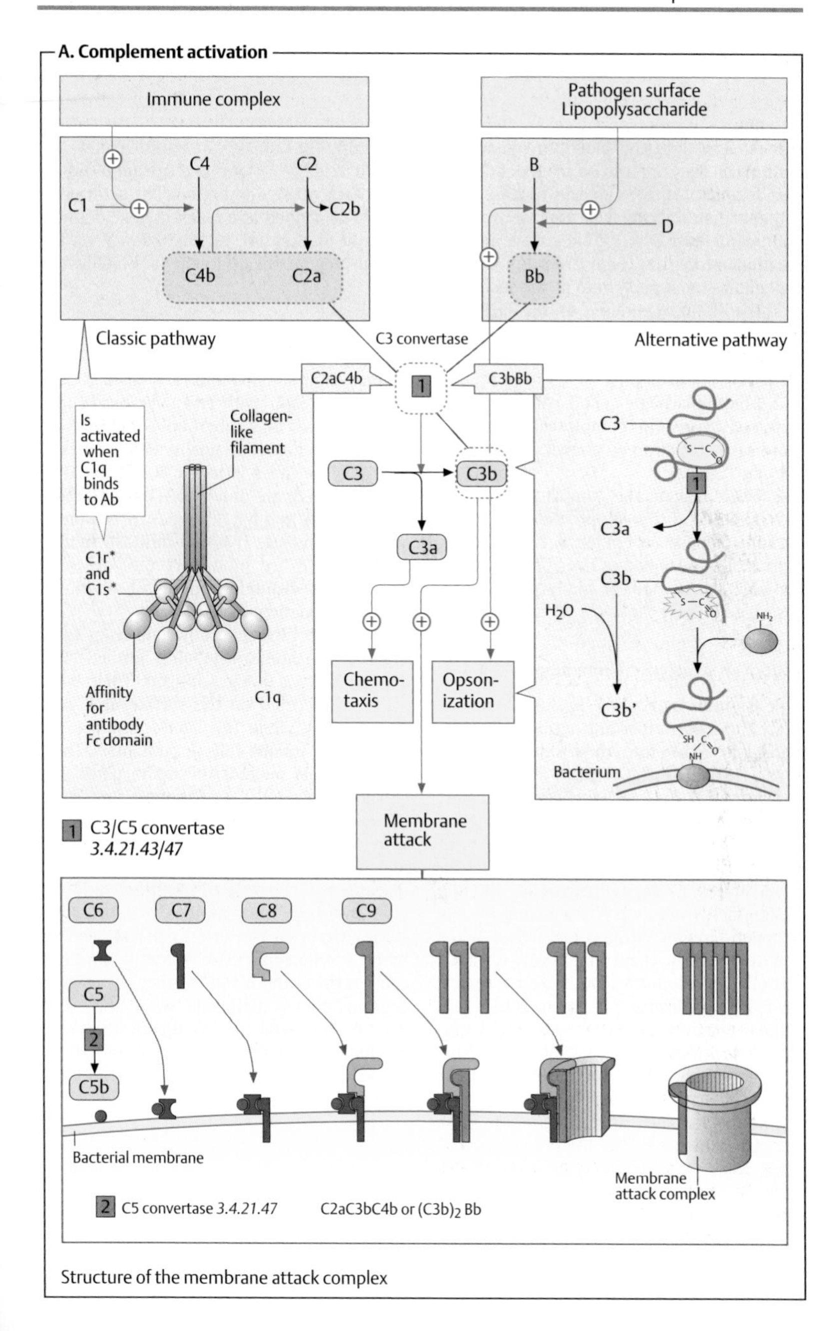
A. Complement activation
Immune complex
Pathogen surface
Lipopolysaccharide
C1
C4
C2
C2b
C4b
C2a
B
D
Bb
Classic pathway
C3 convertase
Alternative pathway
C2aC4b
1
C3bBb
Is activated when C1q binds to Ab
Collagen-like filament
C1r* and C1s*
Affinity for antibody Fc domain
C1q
C3
C3b
C3a
Chemo-taxis
Opson-ization
C3
C3a
C3b
H_2O
C3b'
Bacterium
Membrane attack
1 C3/C5 convertase 3.4.21.43/47
C6
C7
C8
C9
C5
2
C5b
Bacterial membrane
Membrane attack complex
2 C5 convertase 3.4.21.47
C2aC3bC4b or $(C3b)_2$ Bb
Structure of the membrane attack complex

Antibodies

Soluble antigen receptors, which are formed by activated B cells (plasma cells; see p. 294) and released into the blood, are known as **antibodies**. They are also members of the immunoglobulin family (Ig; see p. 296). Antibodies are an important part of the humoral immune defense system. They have no antimicrobial properties themselves, but support the cellular immune system in various ways:

1. They bind to antigens on the surface of pathogens and thereby prevent them from interacting with body cells (*neutralization;* see p. 404, for example).
2. They link single-celled pathogens into aggregates (immune complexes), which are more easily taken up by phagocytes (*agglutination*).
3. They activate the complement system (see p. 298) and thereby promote the innate immune defense system (*opsonization*).

In addition, antibodies have become indispensable aids in medical and biological diagnosis (see p. 304).

A. Domain structure of immunoglobulin G ◑

Type G immunoglobulins (**IgG**) are quantitatively the most important antibodies in the blood, where they form the fraction of γ-globulins (see p. 276). IgGs (mass 150 kDa) are tetramers with two **heavy chains** (H chains; red or orange) and two **light chains** (L chains; yellow). Both H chains are glycosylated (violet; see also p. 43).

The proteinase *papain* cleaves IgG into two F_{ab} fragments and one F_c fragment. The $\mathbf{F_{ab}}$ ("antigen-binding") fragments, which each consist of one L chain and the N-terminal part of an H chain, are able to bind antigens. The $\mathbf{F_c}$ ("crystallizable") fragment is made up of the C-terminal halves of the two H chains. This segment serves to bind IgG to cell surfaces, for interaction with the complement system and antibody transport.

Immunoglobulins are constructed in a modular fashion from several **immunoglobulin domains** (shown in the diagram on the right in Ω form). The H chains of IgG contain four of these domains ($\mathbf{V}_H$, $\mathbf{C}_H1$, $\mathbf{C}_H2$, and $\mathbf{C}_H3$) and the L chains contain two ($\mathbf{C}_L$ and $\mathbf{V}_L$). The letters C and V designate constant or variable regions.

Disulfide bonds link the two heavy chains to each other and also link the heavy chains to the light chains. Inside the domains, there are also disulfide bonds that stabilize the tertiary structure. The domains are approximately 110 amino acids (AA) long and are homologous with each other. The antibody structure evidently developed as a result of gene duplication. In its central region, known as the "hinge" region, the antibodies are highly mobile.

B. Classes of immunoglobulins ◑

Human immunoglobulins are divided into five classes. **IgA** (with two subgroups), **IgD, IgE, IgG** (with four subgroups), and **IgM** are defined by their H chains, which are designated by the Greek letters α, δ, ε, γ, and μ. By contrast, there are only two types of **L chain** (κ and λ). IgD and IgE (like IgG) are tetramers with the structure H_2L_2. By contrast, soluble IgA and IgM are multimers that are held together by disulfide bonds and additional **J peptides** (joining peptides).

The antibodies have different tasks. **IgMs** are the first immunoglobulins formed after contact with a foreign antigen. Their early forms are located on the surface of B cells (see p. 296), while the later forms are secreted from plasma cells as pentamers. Their action targets microorganisms in particular. Quantitatively, **IgGs** are the most important immunoglobulins (see the table showing serum concentrations). They occur in the blood and interstitial fluid. As they can pass the placenta with the help of receptors, they can be transferred from mother to fetus. **IgAs** mainly occur in the intestinal tract and in body secretions. **IgEs** are found in low concentrations in the blood. As they can trigger degranulation of mast cells (see p. 380), they play an important role in allergic reactions. The function of **IgDs** is still unexplained. Their plasma concentration is also very low.

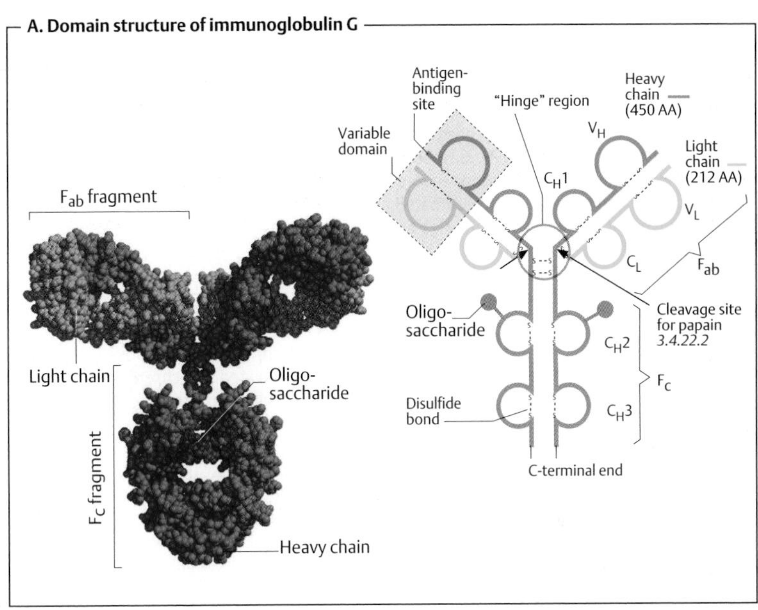
A. Domain structure of immunoglobulin G
Fab fragment
Light chain
Oligo-saccharide
Fc fragment
Heavy chain
Antigen-binding site
"Hinge" region
Heavy chain (450 AA)
Variable domain
VH
Light chain (212 AA)
CH1
VL
CL
Fab
Oligo-saccharide
Cleavage site for papain 3.4.22.2
CH2
Fc
Disulfide bond
CH3
C-terminal end

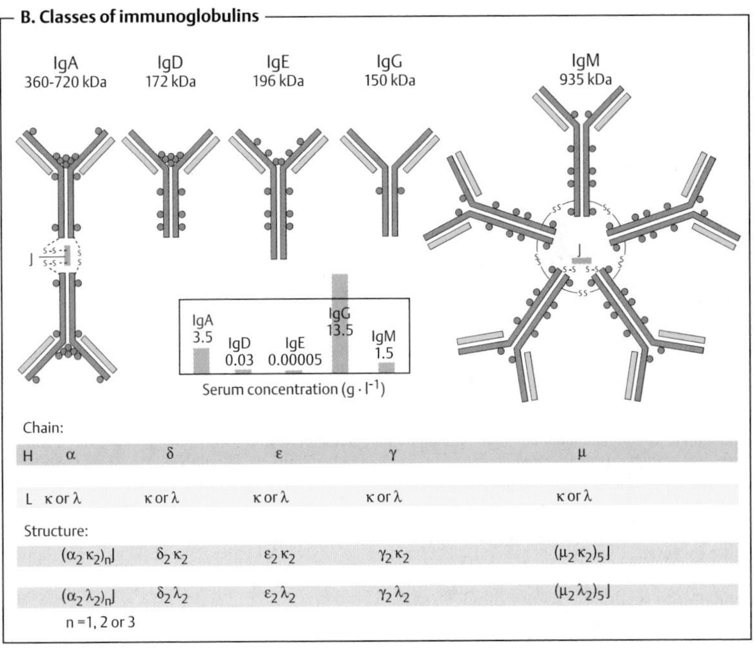
B. Classes of immunoglobulins
IgA 360-720 kDa
IgD 172 kDa
IgE 196 kDa
IgG 150 kDa
IgM 935 kDa
J
IgA 3.5
IgD 0.03
IgE 0.00005
IgG 13.5
IgM 1.5
Serum concentration (g · l-1)
Chain:
H α δ ε γ μ
L κ or λ κ or λ κ or λ κ or λ κ or λ
Structure:
(α2 κ2)nJ δ2 κ2 ε2 κ2 γ2 κ2 (μ2 κ2)5 J
(α2 λ2)nJ δ2 λ2 ε2 λ2 γ2 λ2 (μ2 λ2)5 J
n =1, 2 or 3

Antibody biosynthesis

The acquired (adaptive) immune system (see p. 294) is based on the ability of the lymphocytes to keep an extremely large repertoire of antigen receptors and soluble antibodies ready for use, so that even infections involving new types of pathogen can be combated. The wide range of immunoglobulins (Ig) are produced by genetic recombination and additional mutations during the development and maturation of the individual lymphocytes.

A. Variability of immunoglobulins ◑

It is estimated that more than 10^8 different antibody variants occur in every human being. This **variability** affects both the heavy and the light chains of immunoglobulins.

There are five different types of heavy (H) chain, according to which the antibody classes are defined (α, δ, ε, γ, μ), and two types of light (L) chain (κ and λ; see p. 300). The various Ig types that arise from combinations of these chains are known as **isotypes**. During immunoglobulin biosynthesis, plasma cells can switch from one isotype to another ("gene switch"). **Allotypic variation** is based on the existence of various alleles of the same gene—i. e., genetic differences between individuals. The term **idiotypic variation** refers to the fact that the antigen binding sites in the F_{ab} fragments can be highly variable. Idiotypic variation affects the *variable domains* (shown here in pink) of the light and heavy chains. At certain sites—known as the *hypervariable regions* (shown here in red)—variation is particularly wide; these sequences are directly involved in the binding of the antigen.

B. Causes of antibody variety ◑

There are three reasons for the extremely wide variability of antibodies:

1. **Multiple genes.** Various genes are available to code for the variable protein domains. Only one gene from among these is selected and expressed.

2. **Somatic recombination.** The genes are divided into several segments, of which there are various versions. Various ("untidy") combinations of the segments during lymphocyte maturation give rise to randomly combined new genes ("mosaic genes").

3. **Somatic mutation.** During differentiation of B cells into plasma cells, the coding genes mutate. In this way, the "primordial" *germ-line genes* can become different *somatic genes* in the individual B cell clones.

C. Biosynthesis of a light chain ○

We can look at the basic features of the genetic organization and synthesis of immunoglobulins using the biosynthesis of a mouse κ chain as an example. The gene segments for this light chain are designated L, V, J, and C. They are located on chromosome 6 in the **germ-line DNA** (on chromosome 2 in humans) and are separated from one another by introns (see p. 242) of different lengths.

Some 150 identical **L segments** code for the signal peptide ("leader sequence," 17–20 amino acids) for secretion of the product (see p. 230). The **V segments**, of which there are 150 different variants, code for most of the variable domains (95 of the 108 amino acids). L and V segments always occur in pairs—in tandem, so to speak. By contrast, there are only five variants of the **J segments** (joining segments) at most. These code for a peptide with 13 amino acids that links the variable part of the κ chains to the constant part. A single **C segment** codes for the constant part of the light chain (84 amino acids).

During the differentiation of B lymphocytes, individual *V/J combinations* arise in each B cell. One of the 150 L/V tandem segments is selected and linked to one of the five J segments. This gives rise to a *somatic gene* that is much smaller than the germline gene. Transcription of this gene leads to the formation of the **hnRNA** for the κ chain, from which introns and surplus J segments are removed by splicing (see p. 246). Finally, the completed **mRNA** still contains one each of the L–V–J–C segments and after being transported into the cytoplasm is available for translation. The subsequent steps in Ig biosynthesis follow the rules for the synthesis of membrane-bound or secretory proteins (see p. 230).

A. Variability of immunoglobulins

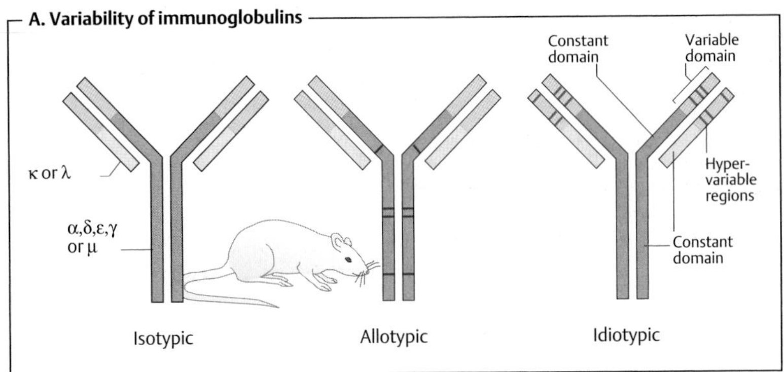

B. Origins of antibody variety

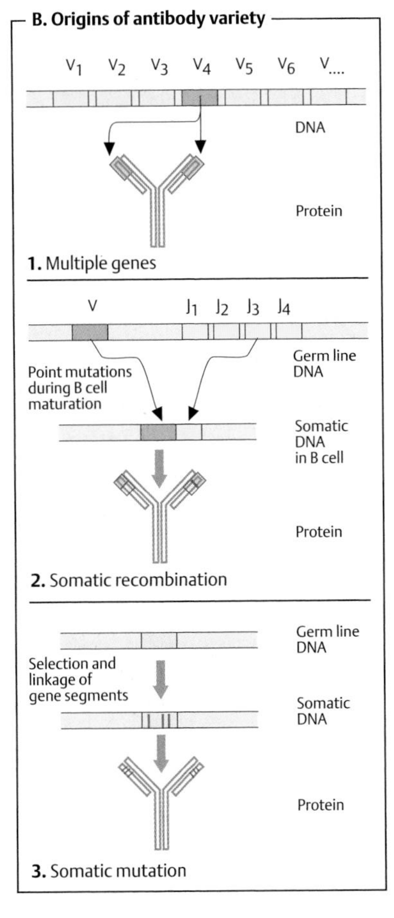

C. Biosynthesis of a light chain

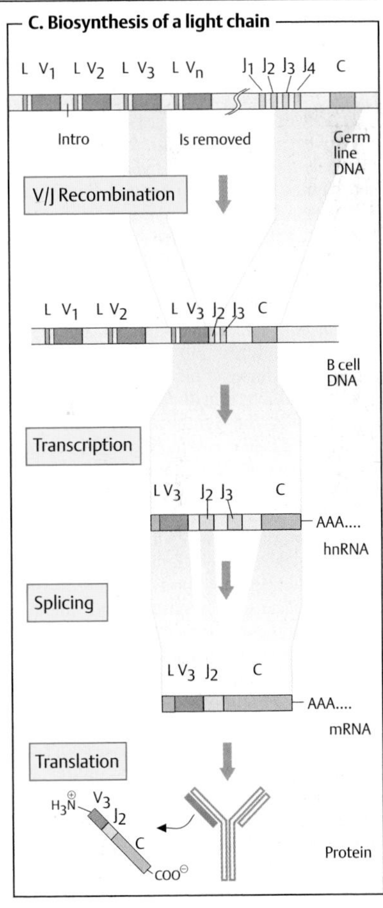

Monoclonal antibodies, immunoassay

A. Monoclonal antibodies ○

Monoclonal antibodies (MABs) are secreted by immune cells that derive from a single antibody-forming cell (from a single cell *clone*). This is why each MAB is directed against only one specific *epitope* of an immunogenic substance, known as an *"antigenic determinant."* Large molecules contain several epitopes, against which various antibodies are formed by various B cells. An antiserum containing a mixture of all of these antibodies is described as being *polyclonal.*

To obtain MABs, **lymphocytes** isolated from the spleen of immunized mice (**1**) are fused with mouse tumor cells (**myeloma cells, 2**). This is necessary because antibody-secreting lymphocytes in culture have a lifespan of only a few weeks. Fusion of lymphocytes with tumor cells gives rise to cell hybrids, known as **hybridomas,** which are potentially immortal.

Successful *fusion* (**2**) is a rare event, but the frequency can be improved by adding polyethylene glycol (PEG). To obtain only successfully fused cells, incubation is required for an extended period in a **primary culture** with *HAT medium* (**3**), which contains hypoxanthine, aminopterin, and thymidine. *Aminopterin,* an analogue of dihydrofolic acid, competitively inhibits *dihydrofolate reductase* and thus inhibits the synthesis of dTMP (see p. 402). As dTMP is essential for DNA synthesis, myeloma cells cannot survive in the presence of aminopterin. Although spleen cells are able to circumvent the inhibitory effect of aminopterin by using *hypoxanthine* and *thymidine,* they have a limited lifespan and die. Only hybridomas survive culture in HAT medium, because they possess both the immortality of the myeloma cells and the spleen cells' metabolic side pathway.

Only a few fused cells actually produce antibodies. To identify these cells, the hybridomas have to be isolated and replicated by **cloning** (**4**). After the clones have been tested for antibody formation, positive cultures are picked out and selected by further cloning (**5**). This results in hybridomas that synthesize *monoclonal antibodies.* Finally, MAB production is carried out in vitro using a bioreactor, or in vivo by producing ascites fluid in mice (**6**).

B. Immunoassay ○

Immunoassays are semiquantitative procedures for assessing substances with low concentrations. In principle, immunoassays can be used to assess any compound against which antibodies are formed.

The basis for this procedure is the *antigen–antibody "reaction"*—i. e., specific binding of an antibody to the molecule being assayed. Among the many different immunoassay techniques that have been developed—e. g., *radioimmunoassay* (RIA), and *chemoluminescence immunoassay* (CIA)—a version of the **enzyme-linked immunoassay** (EIA) is shown here.

The substance to be assayed—e. g., the hormone thyroxine in a serum sample—is pipetted into a microtiter plate (**1**), the walls of which are coated with *antibodies* that specifically bind the hormone. At the same time, a small amount of thyroxine is added to the incubation to which an enzyme known as the *"tracer"* (**1**) has been chemically coupled. The tracer and the hormone being assayed compete for the small number of antibody binding sites available. After binding has taken place (**2**), all of the unbound molecules are rinsed out. The addition of a substrate solution for the enzyme (a *chromogenic solution*) then triggers an indicator reaction (**3**), the products of which can be assessed using photometry (**4**).

The larger the amount of enzyme that can bind to the antibodies on the container's walls, the larger the amount of dye that is produced. Conversely, the larger the amount of the substance being assayed that is present in the sample, the *smaller* the amount of tracer that can be bound by the antibodies. Quantitative analysis can be carried out through parallel measurement using standards with a known concentration.

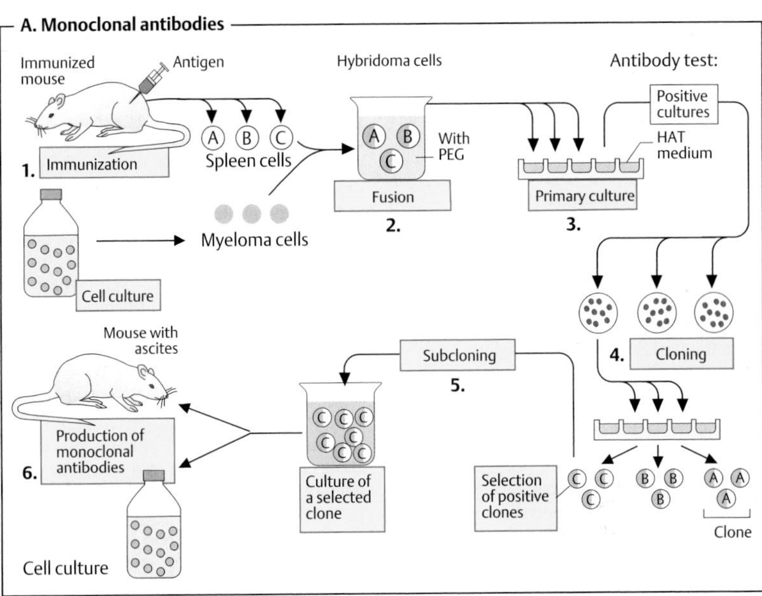
A. Monoclonal antibodies
Immunized mouse
Antigen
Hybridoma cells
Antibody test:
Positive cultures
HAT medium
Spleen cells
With PEG
1. Immunization
Fusion
2.
Primary culture
3.
Myeloma cells
Cell culture
Mouse with ascites
Subcloning
5.
4. Cloning
Production of monoclonal antibodies
6.
Culture of a selected clone
Selection of positive clones
Clone
Cell culture

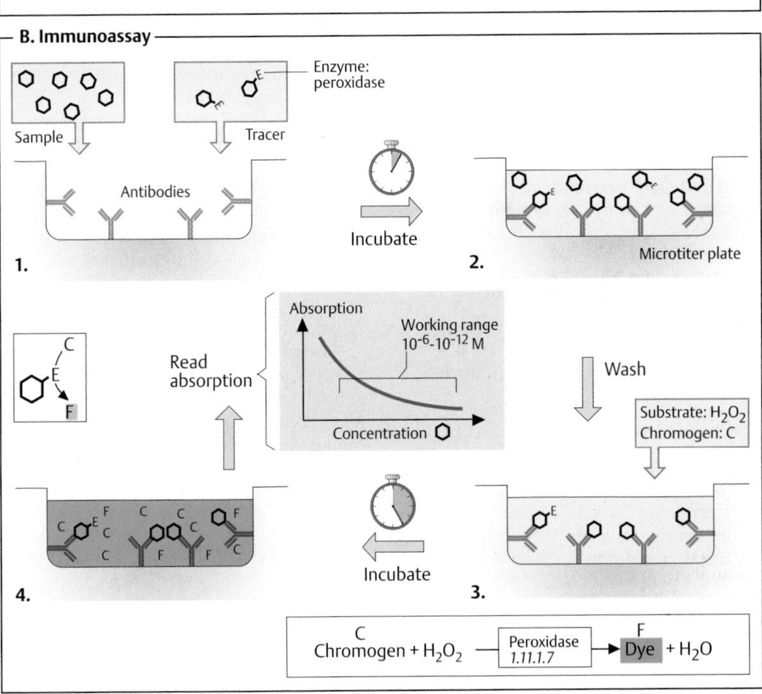
B. Immunoassay
Enzyme: peroxidase
Sample
Tracer
Antibodies
Incubate
1.
2.
Microtiter plate
Absorption
Working range 10^{-6}-10^{-12} M
Read absorption
Concentration
Wash
Substrate: H_2O_2
Chromogen: C
Incubate
4.
3.
C Chromogen + H_2O_2
Peroxidase 1.11.1.7
F Dye + H_2O

Liver: functions

Weighing 1.5 kg, the liver is one of the largest organs in the human body. Although it only represents 2–3% of the body's mass, it accounts for 25–30% of oxygen consumption.

A. Diagram of a hepatocyte ◑

The $3 \cdot 10^{11}$ cells in the liver—particularly the **hepatocytes**, which make up 90% of the cell mass—are the central location for the body's *intermediary metabolism*. They are in close contact with the blood, which enters the liver from the portal vein and the hepatic arteries, flows through capillary vessels known as sinusoids, and is collected again in the central veins of the hepatic lobes. Hepatocytes are particularly rich in endoplasmic reticulum, as they carry out intensive protein and lipid synthesis. The cytoplasm contains granules of insoluble glycogen. Between the hepatocytes, there are bile capillaries through which bile components are excreted.

B. Functions of the liver ●

The most important functions of the liver are:

1. **Uptake** of nutrients supplied by the intestines via the portal vein.
2. Biosynthesis of endogenous compounds and storage, conversion, and degradation of them into excretable molecules (**metabolism**). In particular, the liver is responsible for the biosynthesis and degradation of almost all plasma proteins.
3. **Supply** of the body with metabolites and nutrients.
4. **Detoxification** of toxic compounds by biotransformation.
5. **Excretion** of substances with the bile.

C. Hepatic metabolism ●

The liver is involved in the metabolism of practically all groups of metabolites. Its functions primarily serve to cushion fluctuations in the concentration of these substances in the blood, in order to ensure a constant supply to the peripheral tissues (*homeostasis*).

Carbohydrate metabolism. The liver takes up glucose and other monosaccharides from the plasma. Glucose is then either stored in the form of the polysaccharide *glycogen* or converted into fatty acids. When there is a drop in the blood glucose level, the liver releases glucose again by breaking down glycogen. If the glycogen store is exhausted, glucose can also be synthesized by *gluconeogenesis* from lactate, glycerol, or the carbon skeleton of amino acids (see p. 310).

Lipid metabolism. The liver synthesizes fatty acids from acetate units. The fatty acids formed are then used to synthesize fats and phospholipids, which are released into the blood in the form of *lipoproteins*. The liver's special ability to convert fatty acids into *ketone bodies* and to release these again is also important (see p. 312).

Like other organs, the liver also synthesizes cholesterol, which is transported to other tissues as a component of lipoproteins. Excess cholesterol is converted into bile acids in the liver or directly excreted with the bile (see p. 314).

Amino acid and protein metabolism. The liver controls the plasma levels of the amino acids. Excess amino acids are broken down. With the help of the urea cycle (see p. 182), the nitrogen from the amino acids is converted into urea and excreted via the kidneys. The carbon skeleton of the amino acids enters the intermediary metabolism and serves for glucose synthesis or energy production. In addition, most of the plasma proteins are synthesized or broken down in the liver (see p. 276).

Biotransformation. Steroid hormones and bilirubin, as well as drugs, ethanol, and other xenobiotics are taken up by the liver and inactivated and converted into highly polar metabolites by conversion reactions (see p. 316).

Storage. The liver not only stores energy reserves and nutrients for the body, but also certain mineral substances, trace elements, and vitamins, including iron, retinol, and vitamins A, D, K, folic acid, and B_{12}.

A. Diagram of a hepatocyte

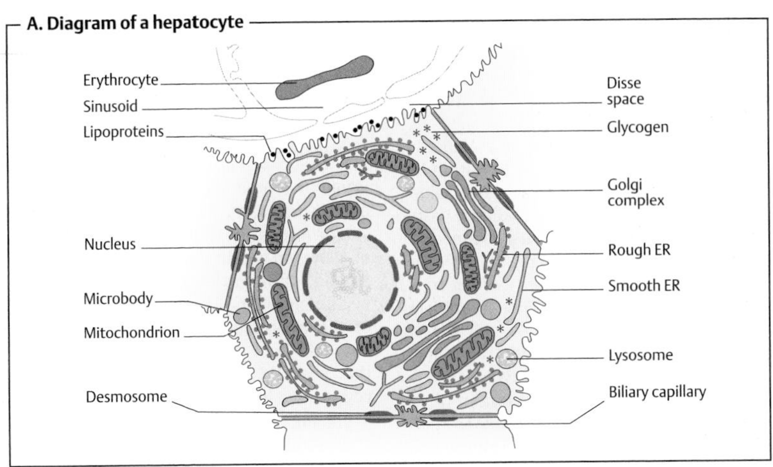

B. Functions of the liver

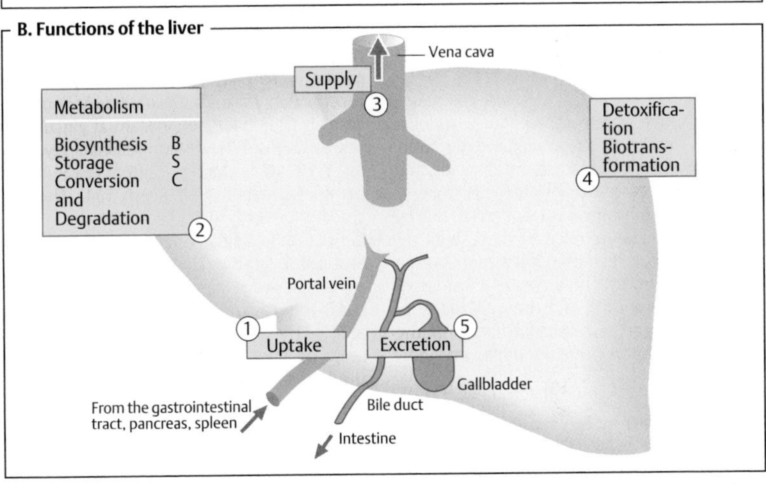

C. Liver metabolism

Carbohydrate metabolism	
Glucose	BSC
Galactose	C
Fructose	C
Mannose	C
Pentoses	BC
Lactate	C
Glycerol	BC
Glycogen	BSC

Lipid metabolism	
Fatty acids	BC
Fats	BC
Ketone bodies	B
Cholesterol	BEC
Bile acids	BE
Vitamins	SC

Amino acid metabolism	
Amino acids	C
Urea	B

Plasma proteins	
Lipoproteins	BC
Albumin	BC
Coagulation factors	BC
Hormones	C
Enzymes	BC

Biotransformation	
Steroid hormones	EC
Bile pigments	EC
Ethanol	C
Drugs	EC

B	Biosynthesis
C	Conversion and degradation
E	Excretion
S	Storage

Buffer function in organ metabolism

All of the body's tissues have a constant requirement for energy substrates and nutrients. The body receives these metabolites with food, but the supply is irregular and in varying amounts. The liver acts here along with other organs, particularly adipose tissue, as a balancing *buffer* and *storage organ.*

In the metabolism, a distinction is made between the *absorptive state (well-fed state)* immediately after a meal and the *postabsorbtive state (state of starvation)*, which starts later and can merge into hunger. The switching of the organ metabolism between the two phases depends on the concentration of energy-bearing metabolites in the blood (plasma level). This is regulated jointly by hormones and by the autonomic nervous system.

A. Absorptive state ◑

The absorptive state continues for 2–4 hours after food intake. As a result of food digestion, the plasma levels of glucose, amino acids, and fats (triacylglycerols) temporarily increase.

The endocrine **pancreas** responds to this by altering its hormone release—there is an increase in *insulin* secretion and a reduction in *glucagon* secretion. The increase in the insulin/glucagon quotient and the availability of substrates trigger an *anabolic phase* in the tissues—particularly liver, muscle, and adipose tissues.

The **liver** forms increased amounts of glycogen and fats from the substrates supplied. Glycogen is stored, and the fat is released into the blood in very low density lipoproteins (VLDLs).

Muscle also refills its glycogen store and synthesizes proteins from the amino acids supplied.

Adipose tissue removes free fatty acids from the lipoproteins, synthesizes triacylglycerols from them again, and stores these in the form of insoluble droplets.

During the absorptive state, the **heart** and **neural tissue** mainly use glucose as an energy source, but they are unable to establish any substantial energy stores. Heart muscle cells are in a sense "omnivorous," as they can also use other substances to produce energy (fatty acids, ketone bodies). By contrast, the central nervous system (CNS) is dependent on glucose. It is only able to utilize ketone bodies after a prolonged phase of hunger (**B**).

B. Postabsorptive state ◑

When the food supply is interrupted, the postabsorbtive state quickly sets in. The pancreatic A cells now release increased amounts of *glucagon*, while the B cells reduce the amount of *insulin* they secrete. The reduced insulin/glucagon quotient leads to switching of the intermediary metabolism. The body now falls back on its energy reserves. To do this, it breaks down *storage substances* (glycogen, fats, and proteins) and shifts energy-supplying metabolites between the organs.

The **liver** first empties its glycogen store (*glycogenolysis;* see p. 156). It does not use the released glucose itself, however, but supplies the other tissues with it. In particular, the brain, adrenal gland medulla, and erythrocytes depend on a constant supply of glucose, as they have no substantial glucose reserves themselves. When the liver's glycogen reserves are exhausted after 12–24 hours, *gluconeogenesis* begins (see p. 154). The precursors for this are derived from the musculature (amino acids) and adipose tissue (glycerol from fat degradation). From the fatty acids that are released (see below), the liver starts to form ketone bodies (*ketogenesis;* see p. 312). These are released into the blood and serve as important energy suppliers during the hunger phase. After 1–2 weeks, the CNS also starts to use ketone bodies to supply part of its energy requirements, in order to save glucose.

In **muscle**, the extensive glycogen reserves are exclusively used for the muscles' own requirements (see p. 320). The slowly initiated protein breakdown in muscle supplies amino acids for gluconeogenesis in the liver.

In **adipose tissue**, glucagon triggers *lipolysis,* releasing fatty acids and glycerol. The fatty acids are used as energy suppliers by many types of tissue (with the exception of brain and erythrocytes). An important recipient of the fatty acids is the liver, which uses them for ketogenesis.

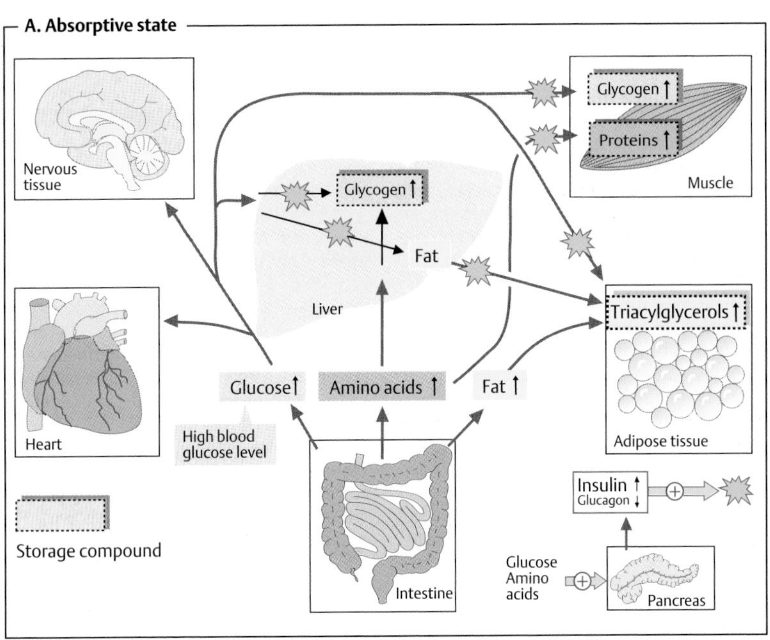
A. Absorptive state
Nervous tissue
Glycogen ↑
Proteins ↑
Muscle
Glycogen ↑
Fat
Liver
Triacylglycerols ↑
Heart
Glucose ↑
Amino acids ↑
Fat ↑
High blood glucose level
Adipose tissue
Insulin ↑
Glucagon ↓
Storage compound
Glucose
Amino acids
Intestine
Pancreas

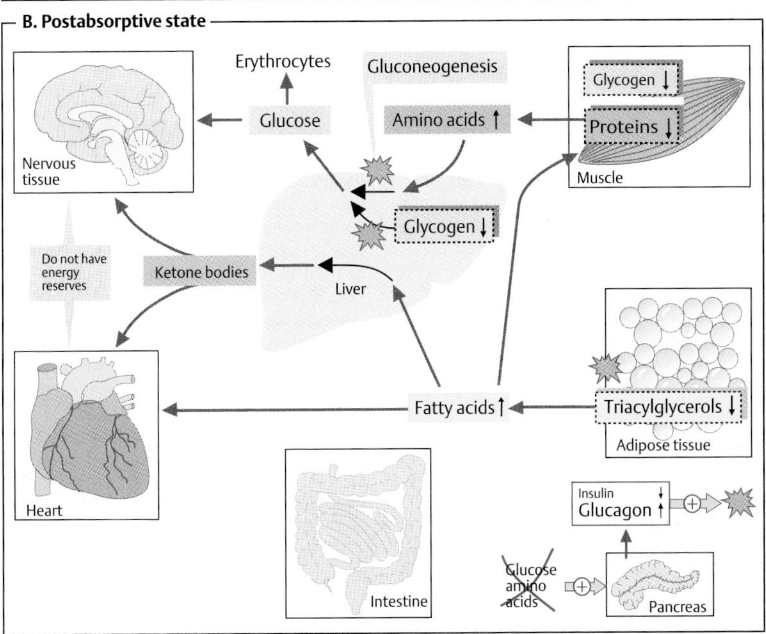
B. Postabsorptive state
Erythrocytes
Gluconeogenesis
Glycogen ↓
Nervous tissue
Glucose
Amino acids ↑
Proteins ↓
Muscle
Glycogen ↓
Do not have energy reserves
Ketone bodies
Liver
Heart
Fatty acids ↑
Triacylglycerols ↓
Adipose tissue
Insulin ↓
Glucagon ↑
Glucose amino acids
Intestine
Pancreas

Carbohydrate metabolism

Besides fatty acids and ketone bodies, glucose is the body's most important energy supplier. The concentration of glucose in the blood (the *"blood glucose level"*) is maintained at 4–6 mM (0.8–1.0 g · L^{-1}) by precise regulation of glucosesupplying and glucose-utilizing processes. Glucose suppliers include the intestines (glucose from food), liver, and kidneys. The liver plays the role of a "glucostat" (see p. 308).

The liver is also capable of forming glucose by converting other sugars—e. g., *fructose* and *galactose*—or by synthesizing from other metabolites. The conversion of lactate to glucose in the *Cori cycle* (see p. 338) and the conversion of alanine to glucose with the help of the *alanine cycle* (see p. 338) are particularly important for the supply of erythrocytes and muscle cells.

Transporters in the plasma membrane of hepatocytes allow insulin-independent transport of glucose and other sugars in both directions. In contrast to muscle, the liver possesses the enzyme *glucose-6-phosphatase*, which can release glucose from glucose-6-phosphate.

A. Gluconeogenesis: overview ◑

Regeneration of glucose (up to 250 g per day) mainly takes place in the liver. The tubule cells of the kidney are also capable of carrying out gluconeogenesis, but due to their much smaller mass, their contribution only represents around 10% of total glucose formation. Gluconeogenesis is regulated by hormones. *Cortisol, glucagon,* and *epinephrine* promote gluconeogenesis, while *insulin* inhibits it (see pp. 158, 244).

The main precursors of gluconeogenesis in the liver are *lactate* from anaerobically working muscle cells and from erythrocytes, glucogenic *amino acids* from the digestive tract and muscles (mainly alanine), and *glycerol* from adipose tissue. The kidney mainly uses amino acids for gluconeogenesis (Glu, Gln; see p. 328).

In mammals, fatty acids and other suppliers of acetyl CoA are not capable of being used for gluconeogenesis, as the acetyl residues formed during β-oxidation in the tricarboxylic acid cycle (see p. 132) are oxidized to CO_2 and therefore cannot be converted into oxaloacetic acid, the precursor for gluconeogenesis.

B. Fructose and galactose metabolism ◑

Fructose is mainly metabolized by the liver, which channels it into glycolysis (left half of the illustration).

A special *ketohexokinase* [1] initially phosphorylates fructose into **fructose 1-phosphate.** This is then cleaved by an *aldolase* [2], which is also fructose-specific, to yield **glycerone 3-phosphate** (dihydroxyacetone phosphate) and **glyceraldehyde.** Glycerone 3-phosphate is already an intermediate of glycolysis (center), while glyceraldehyde can be phosphorylated into glyceraldehyde 3-phosphate by *triokinase* [3].

To a smaller extent, glyceraldehyde is also reduced to glycerol [4] or oxidized to glycerate, which can be channeled into glycolysis following phosphorylation (not shown). The reduction of glyceraldehyde [4] uses up NADH. As the rate of degradation of alcohol in the hepatocytes is limited by the supply of NAD^+, fructose degradation accelerates alcohol degradation (see p. 320).

Outside of the liver, fructose is channeled into the sugar metabolism by reduction at C-2 to yield sorbitol and subsequent dehydration at C-1 to yield glucose (the *polyol pathway;* not shown).

Galactose is also broken down in the liver (right side of the illustration). As is usual with sugars, the metabolism of galactose starts with a phosphorylation to yield **galactose 1-phosphate** [5]. The connection to the glucose metabolism is established by C-4 epimerization to form **glucose 1-phosphate.** However, this does not take place directly. Instead, a *transferase* [6] transfers a uridine 5′-monophosphate (UMP) residue from uridine diphosphoglucose (UDPglucose) to galactose 1-phosphate. This releases glucose 1-phosphate, while galactose 1-phosphate is converted into uridine diphosphogalactose (UDP-galactose). This then is isomerized into UDP-glucose. The *biosynthesis* of galactose also follows this reaction pathway, which is freely reversible up to reaction [5]. Genetic defects of enzymes [5] or [6] can lead to the clinical picture of *galactosemia.*

A. Gluconeogenesis: overview

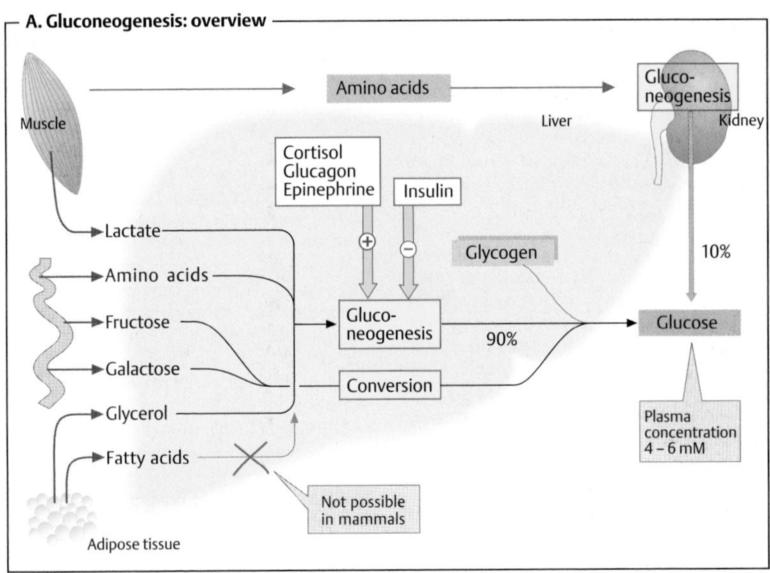

B. Fructose and galactose metabolism

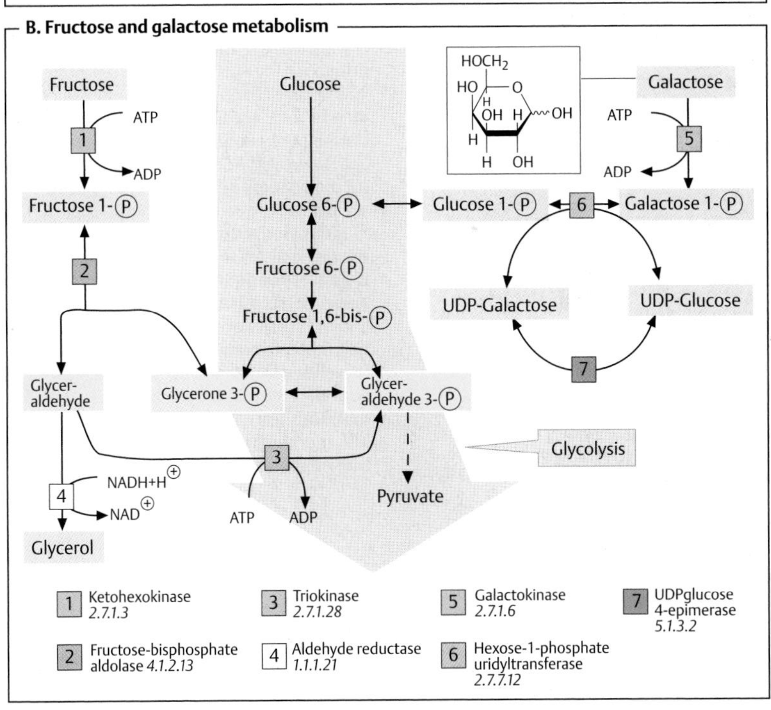

Lipid metabolism

The liver is the most important site for the formation of *fatty acids, fats (triacylglycerols), ketone bodies,* and *cholesterol.* Most of these products are released into the blood. In contrast, the triacylglycerols synthesized in adipose tissue are also stored there.

A. Lipid metabolism ◑

Lipid metabolism in the liver is closely linked to the carbohydrate and amino acid metabolism. When there is a good supply of nutrients in the *resorptive (wellfed) state* (see p. 308), the liver converts glucose via acetyl CoA into **fatty acids**. The liver can also take up fatty acids from *chylomicrons,* which are supplied by the intestine, or from fatty acid–albumin complexes (see p. 162). Fatty acids from both sources are converted into **fats** and **phospholipids**. Together with apoproteins, they are packed into very-low-density lipoproteins (**VLDLs**; see p. 278) and then released into the blood by exocytosis. The VLDLs supply extrahepatic tissue, particularly adipose tissue and muscle.

In the *postresorptive state* (see p. 292)—particularly during fasting and starvation—the lipid metabolism is readjusted and the organism falls back on its own reserves. In these conditions, adipose tissue releases fatty acids. They are taken up by the liver and are mainly converted into **ketone bodies** (**B**).

Cholesterol can be derived from two sources—food or endogenous synthesis from acetyl-CoA. A substantial percentage of endogenous cholesterol synthesis takes place in the liver. Some cholesterol is required for the synthesis of **bile acids** (see p. 314). In addition, it serves as a building block for cell membranes (see p. 216), or can be esterified with fatty acids and stored in lipid droplets. The rest is released together into the blood in the form of lipoprotein complexes (VLDLs) and supplies other tissues. The liver also contributes to the cholesterol metabolism by taking up from the blood and breaking down lipoproteins that contain cholesterol and cholesterol esters (HDLs, IDLs, LDLs; see p. 278).

B. Biosynthesis of ketone bodies ◑

At high concentrations of acetyl-CoA in the liver mitochondria, two molecules condense to form **acetoacetyl CoA** [1]. The transfer of another acetyl group [2] gives rise to 3-hydroxy-3-methylglutaryl-CoA (**HMG CoA**), which after release of acetyl CoA [3] yields free **acetoacetate** (*Lynen cycle*). Acetoacetate can be converted to **3-hydroxybutyrate** by reduction [4], or can pass into **acetone** by nonenzymatic decarboxylation [5]. These three compounds are together referred to as *"ketone bodies,"* although in fact 3-hydroxybutyrate is not actually a ketone. As reaction [3] releases an H^+ ion, metabolic acidosis can occur as a result of increased ketone body synthesis (see p. 288).

The ketone bodies are released by the liver into the blood, in which they are easily soluble. Blood levels of ketone bodies therefore rise during periods of hunger. Together with free fatty acids, 3-hydroxybutyrate and acetoacetate are then the most important energy suppliers in many tissues (including heart muscle). Acetone cannot be metabolized and is exhaled via the lungs or excreted with urine.

To channel ketone bodies into the energy metabolism, acetoacetate is converted with the help of succinyl CoA into succinic acid and acetoacetyl CoA, which is broken down by β-oxidation into acetyl CoA (not shown; see p. 180).

If the production of ketone bodies exceeds the demand for them outside the liver, there is an increase in the concentration of ketone bodies in the plasma (*ketonemia*) and they are also eventually excreted in the urine (*ketonuria*). Both phenomena are observed after prolonged starvation and in inadequately treated *diabetes mellitus.* Severe ketonuria with ketoacidosis can cause electrolyte shifts and loss of consciousness, and is therefore life-threatening (*ketoacidotic coma*).

A. Lipid metabolism

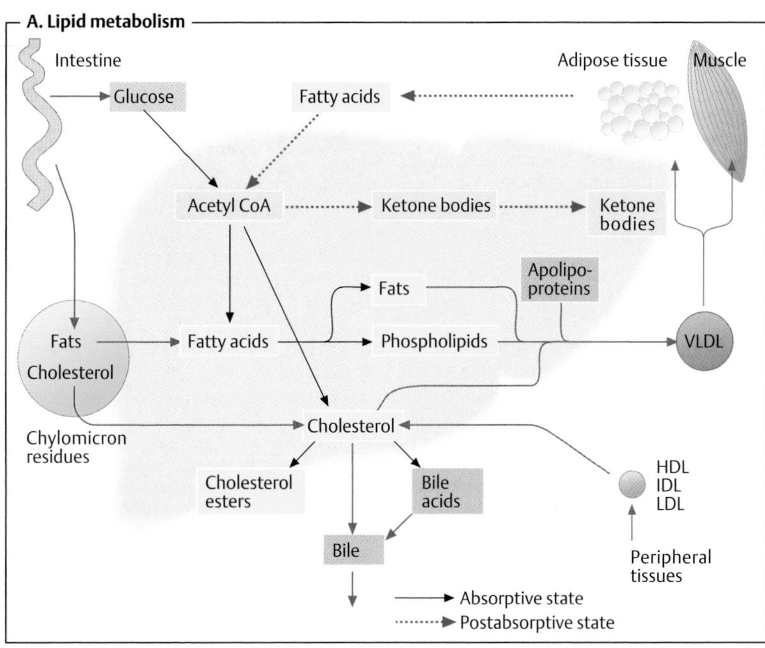

A. Biosynthesis of ketone bodies

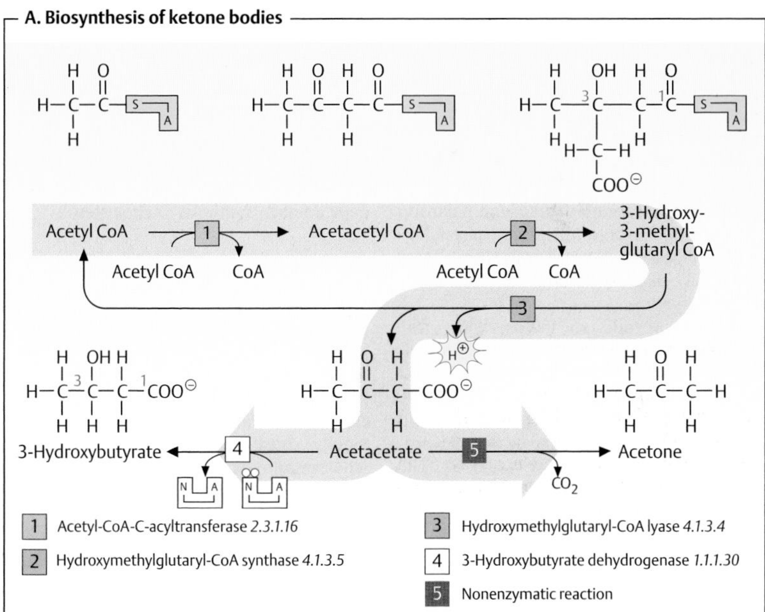

Bile acids

Bile is an important product released by the hepatocytes. It promotes the digestion of fats from food by *emulsifying* them in the small intestine (see p. 2770). The emulsifying components of bile, apart from phospholipids, mainly consist of **bile acids** and **bile salts** (see below). The bile also contains free cholesterol, which is excreted in this way (see p. 312).

A. Bile acids and bile salts ◑

Bile acids are steroids consisting of 24 C atoms carrying one carboxylate group and several hydroxyl groups. They are formed from cholesterol in the liver via an extensive reaction pathway (top). Cytochrome P450 enzymes in the sER of hepatocytes are involved in many of the steps (seep. 318). Initially, the cholesterol double bond is removed. Monooxygenases then introduce one or two additional OH groups into the sterane framework. Finally, the side chain is shortened by three C atoms, and the terminal C atom is oxidized to a carboxylate group.

It is important that the arrangement of the A and B rings is altered from *trans* to *cis* during bile acid synthesis (see p. 54). The result of this is that all of the hydrophilic groups in the bile acids lie on one side of the molecule. Cholesterol, which is weakly amphipathic (top), has a small polar "head" and an extended apolar "tail." By contrast, the much more strongly amphipathic bile acid molecules (bottom) resemble disks with polar top sides and apolar bottom sides. At physiological pH values, the carboxyl groups are almost completely dissociated and therefore negatively charged.

Cholic acid and *chenodeoxycholic acid,* known as the **primary bile acids**, are quantitatively the most important metabolites of cholesterol. After being biosynthesized, they are mostly activated with coenzyme A and then conjugated with *glycine* or the non-proteinogenic amino acid *taurine* (see p. 62). The acid amides formed in this way are known as **conjugated bile acids** or **bile salts**. They are even more amphipathic than the primary products.

Deoxycholic acid and *lithocholic acid* are only formed in the intestine by enzymatic cleavage of the OH group at C-7 (see **B**). They are therefore referred to as **secondary bile acids**.

B. Metabolism of bile salts ◑

Bile salts are exclusively synthesized in the liver (see **A**). The slowest step in their biosynthesis is hydroxylation at position 7 by a *7-α-hydroxylase.* Cholic acid and other bile acids inhibit this reaction (*end-product inhibition*). In this way, the bile acids present in the liver regulate the rate of cholesterol utilization.

Before leaving the liver, a large proportion of the bile acids are activated with CoA and then conjugated with the amino acids *glycine* or *taurine* (**2**; cf. **A**). In this way, cholic acid gives rise to *glycocholic acid* and *taurocholic acid.* The *liver bile* secreted by the liver becomes denser in the gallbladder as a result of the removal of water (*bladder bile*; **3**).

Intestinal bacteria produce enzymes that can chemically alter the bile salts (**4**). The acid amide bond in the bile salts is cleaved, and dehydroxylation at C-7 yields the corresponding secondary bile acids from the primary bile acids (**5**). Most of the intestinal bile acids are resorbed again in the ileum (**6**) and returned to the liver via the portal vein (*enterohepatic circulation*). In the liver, the secondary bile acids give rise to primary bile acids again, from which bile salts are again produced. Of the 15–30g bile salts that are released with the bile per day, only around 0.5g therefore appears in the feces. This approximately corresponds to the amount of daily de novo synthesis of cholesterol.

Further information

The cholesterol excreted with the bile is poorly water-soluble. Together with phospholipids and bile acids, it forms micelles (see p. 270), which keep it in solution. If the proportions of phospholipids, bile acids and cholesterol shift, *gallstones* can arise. These mainly consist of precipitated cholesterol (cholesterol stones), but can also contain Ca^{2+} salts of bile acids and bile pigments (pigment stones).

A. Bile acids and bile salts

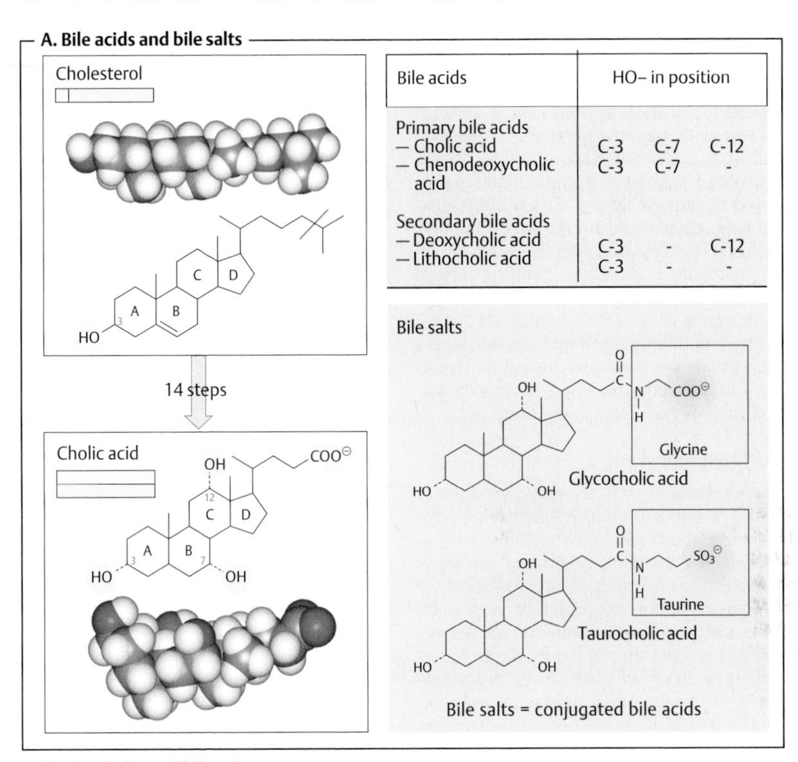

Bile acids	HO– in position		
Primary bile acids			
– Cholic acid	C-3	C-7	C-12
– Chenodeoxycholic acid	C-3	C-7	-
Secondary bile acids			
– Deoxycholic acid	C-3	-	C-12
– Lithocholic acid	C-3	-	-

B. Metabolism of bile salts

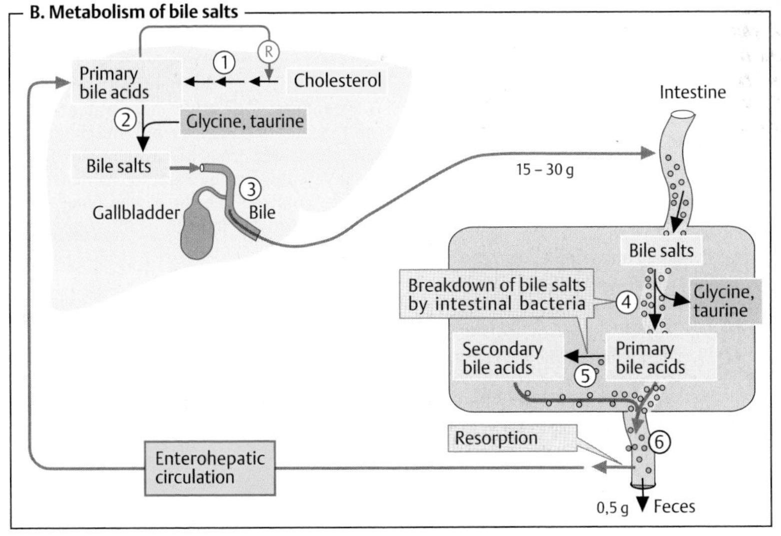

Biotransformations

The body is constantly taking up **foreign substances** (= *xenobiotics*) from food or through contact with the environment, via the skin and lungs. These substances can be natural in origin, or may have been synthetically produced by humans. Many of these substances are toxic, particularly at high concentrations. However, the body has effective mechanisms for inactivating and then excreting foreign substances through *biotransformations*. The mechanisms of biotransformation are similar to those with which **endogenous substances** such as bile pigments and steroid hormones are enzymatically converted. Biotransformations mainly take place in the *liver*.

A. Biotransformations ◑

Phase I reactions (interconversion reactions). Type I reactions introduce functional groups into inert, apolar molecules or alter functional groups that are already present. In many cases, this is what first makes it possible for foreign substances to conjugate with polar molecules via phase II reactions (see below). Phase I reactions usually reduce the *biological activity* or *toxicity* of a substance ("detoxification"). However, some substances only become biologically active as a result of the interconversion reaction (see, for example, benzo[*a*]pyrene, p. 256) or become more toxic after interconversion than the initial substance ("toxification").

Important phase I biotransformation reactions include:

- **Hydrolytic cleavages** of ether, ester, and peptide bonds. Example (**1**) shows hydrolysis of the painkiller *acetylsalicylic acid*.
- **Oxidations.** Hydroxylations, epoxide formation, sulfoxide formation, dealkylation, deamination. For example, benzene is oxidized into phenol, and toluene (methylbenzene) is oxidized into benzoic acid.
- **Reductions.** Reduction of carbonyl, azo-, or nitro- compounds, dehalogenation.
- **Methylations.** Example (**2**) illustrates the inactivation of the catecholamine *norepinephrine* by methylation of a phenolic OH group (see p. 334).
- **Desulfurations.** The reactions take place in the hepatocytes on the smooth endoplasmic reticulum.

Most oxidation reactions are catalyzed by **cytochrome P450 systems** (see p. 318). These monooxygenases are induced by their substrates and show wide specificity. The substrate-specific enzymes of the steroid metabolism (see p. 376) are exceptions to this.

Phase II reactions (conjugate formation). Type II reactions couple their substrates (bilirubin, steroid hormones, drugs, and products of phase I reactions) via ester or amide bonds to highly polar negatively charged molecules. The enzymes involved are transferases, and their products are known as **conjugates.**

The most common type of conjugate formation is coupling with *glucuronate* (GlcUA) as an *O*-or *N*-glucuronide. The coenzyme for the reaction is uridine diphosphate glucuronate, the *"active glucuronate"* (see p. 110). Coupling with the polar glucuronate makes an apolar (hydrophobic) molecule more strongly polar, and it becomes sufficiently water-soluble and capable of being excreted. Example (**3**) shows the glucuronidation of *tetrahydrocortisol*, a metabolite of the glucocorticoid cortisol (see p. 374).

The biosynthesis of sulfate esters with the help of *phosphoadenosine phosphosulfate* (PAPS), the *"active sulfate"*, (see p. 110) and amide formation with *glycine* and *glutamine* also play a role in conjugation. For example, benzoic acid is conjugated with glycine to form the more soluble and less toxic *hippuric acid* (*N*-benzoylglycine; see p. 324).

In contrast with unconjugated compounds, the conjugates are much more water-soluble and capable of being excreted. The conjugates are eliminated from the liver either by the *biliary* route—i. e., by receptor-mediated excretion into the bile—or by the *renal* route, via the blood and kidneys by filtration.

Further information

To detoxify **heavy metals**, the liver contains *metallothioneins*, a group of cysteine-rich proteins with a high affinity for divalent metal ions such as Cd^{2+}, Cu^{2+}, Hg^{2+}, and Zn^{2+}. These metal ions also induce the formation of metallothioneins via a special metal-regulating element (MRE) in the gene's promoter (see p. 244).

A. Biotransformations

Foreign substances:

Drugs
Preservatives
Plasticizers
Pigments
Pesticides etc.

Endogenous substances

Steroid hormones and other low molecular weight substances
Bile pigments

Poorly soluble, biologically active, some toxic

Substrate induction

Phase I reactions

Transformation reactions:

Hydrolytic cleavage,
Epoxide formation
Dealkylation
Deamination
Reduction
Methylation
Desulfuration

Transformation product

Substrate induction

Phase II reactions

Conjugate formation:

Glucuronidation
Esterification with sulfate
Amidation with Gly and Glu

Conjugate

Water soluble, inactive, non-toxic

Bile

Urine

Acetic acid

H_2O CH_3-COOH

$COO^{\ominus}$ CH_3 OH $COO^{\ominus}$

Acetylsalicylic acid → Salicylate

1. Hydrolysis of a drug

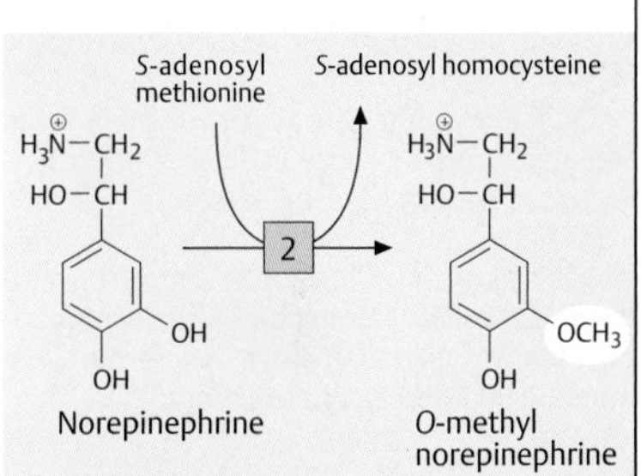

2. Methylation of a hormone/neurotransmitter

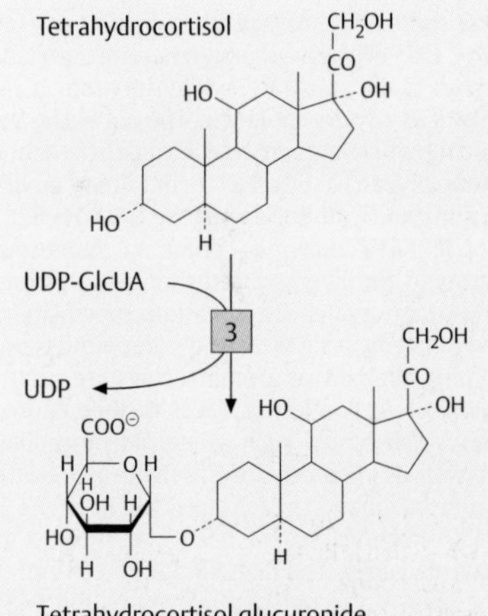

3. Glucuronidation of a hormone

1 Arylesterase *3.1.1.2*

2 Catechol *O*-methyltransferase *2.1.1.6*

3 Glucuronosyltransferase *2.4.1.17*

Cytochrome P450 systems

During the first phase of biotransformation in the liver, compounds that are weakly chemically reactive are enzymatically hydroxylated (see p. 316). This makes it possible for them to be conjugated with polar substances. The hydroxylating enzymes are generally *monooxygenases* that contain a **heme** as the redox-active coenzyme (see p. 106). In the reduced form, the heme can bind carbon monoxide (CO), and it then shows characteristic light absorption at 450 nm. This was what led to this enzyme group being termed **cytochrome P450** (Cyt P450).

Cyt P450 systems are also involved in many other metabolic processes—e. g., the biosynthesis of steroid hormones (see p. 172), bile acids (see p. 314), and eicosanoids (see p. 390), as well as the formation of unsaturated fatty acids (see p. 409). The liver's reddish-brown color is mainly due to the large amounts of P450 enzymes it contains.

A. Cytochrome P450-dependent mono oxygenases: reactions ◑

Cyt P450-dependent monooxygenases catalyze reductive cleavage of molecular oxygen (O_2). One of the two oxygen atoms is transferred to the substrate, while the other is released as a water molecule. The necessary reducing equivalents are transferred to the actual monooxygenase by an FAD-containing auxiliary enzyme from the coenzyme $NADPH+H^+$.

Cyt P450 enzymes occur in numerous forms in the liver, steroid-producing glands, and other organs. The substrate specificity of liver enzymes is low. Apolar compounds containing aliphatic or aromatic rings are particularly easily converted. These include endogenous substances such as steroid hormones, as well as medical drugs, which are inactivated by phase I reactions. This is why Cyt P450 enzymes are of particular interest in pharmacology. The degradation of ethanol in the liver is also partly catalyzed by Cyt P450 enzymes (the "microsomal ethanol-oxidizing system"; see p. 304). As alcohol and drugs are broken down by the same enzyme system, the effects of alcoholic drinks and medical drugs can sometimes be mutually enhancing—even sometimes to the extent of becoming life-threatening.

Only a few examples of the numerous Cyt P450-dependent reactions are shown here. *Hydroxylation* of aromatic rings (**a**) plays a central part in the metabolism of medicines and steroids. Aliphatic methyl groups can also be oxidized to hydroxyl groups (**b**). *Epoxidation* of aromatics (**c**) by Cyt P450 yields products that are highly reactive and often toxic. For example, the mutagenic effect of benzo[*a*]pyrene (see p. 244) is based on this type of interconversion in the liver. In Cyt P450 dependent *dealkylations* (**d**), alkyl substituents of *O*, *N*, or *S* atoms are released as aldehydes.

B. Reaction mechanism ○

The course of Cyt P450 catalysis is in principle well understood. The most important function of the *heme group* consists of converting molecular oxygen into an especially reactive atomic form, which is responsible for all of the reactions described above.

[1] In the resting state, the heme iron is trivalent. Initially, the substrate binds near the heme group.

[2] Transfer of an electron from $FADH_2$ reduces the iron to the divalent form that is able to bind an O_2 molecule (2).

[3] Transfer of a second electron and a change in the valence of the iron reduce the bound O_2 to the peroxide.

[4] A hydroxyl ion is now cleaved from this intermediate. Uptake of a proton gives rise to H_2O and the reactive form of oxygen mentioned above. In this ferryl radical, the iron is formally tetravalent.

[5] The activated oxygen atom inserts itself into a C–H bond in the substrate, thereby forming an OH group.

[6] Dissociation of the product returns the enzyme to its initial state.

A. Cytochrome P450-dependent monooxygenases: reactions

Apolar substrate

$O=O$

$2H^+$

Fe

H_2O

(H–) O

Oxygenated product

1 Monooxygenase *1.14.n.n* [heme P450]

a) Hydroxylation aromatic

b) Hydroxylation alipathic

c) Epoxidation

d) Dealkylation

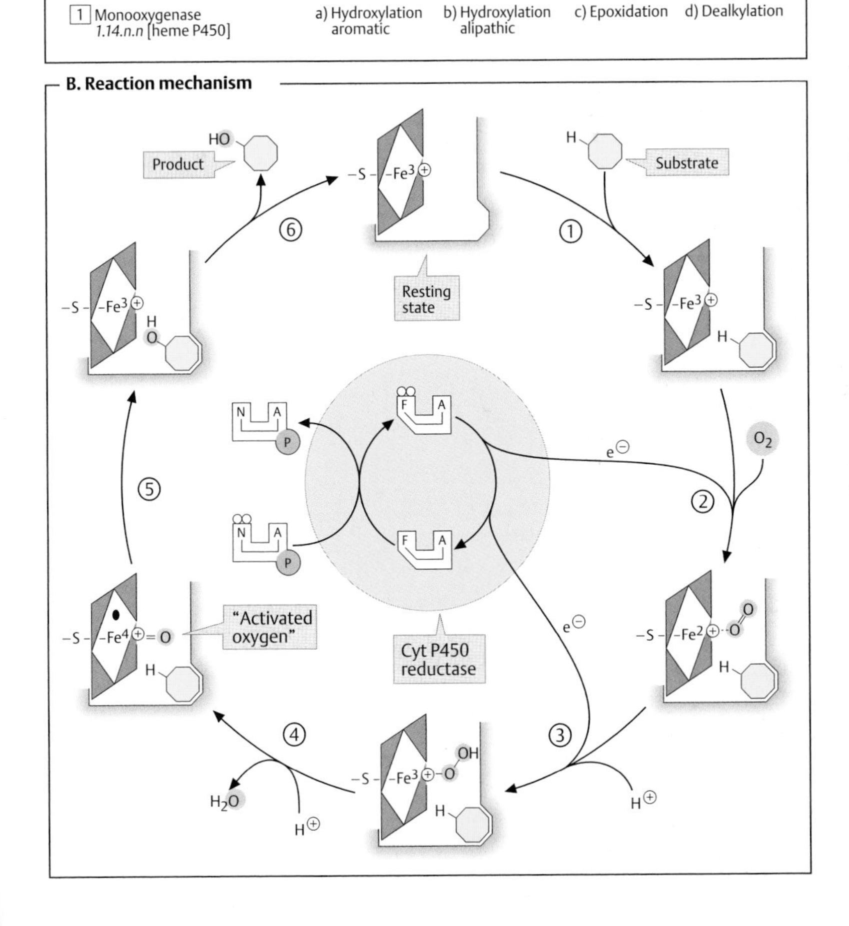

Ethanol metabolism

A. Blood ethanol level

Ethanol (EtOH, "alcohol") naturally occurs in fruit in small quantities. Alcoholic drinks contain much higher concentrations. Their alcohol content is usually given as percent by volume. To estimate alcohol uptake and the blood alcohol level, it is useful to convert the amount to grams of ethanol (density 0.79 kg · L^{-1}). For example, a bottle of beer (0.5 L at 4% v/v alcohol) contains 20 mL = 16 g of ethanol, while a bottle of wine (0.7 L at 12% v/v alcohol) contains 84 mL = 66 g ethanol.

Ethanol is membrane-permeable and is quickly resorbed. The maximum blood level is already reached within 60–90 min after drinking. The *resorption rate* depends on various conditions, however. An empty stomach, a warm drink (e.g., mulled wine), and the presence of sugar and carbonic acid (e.g., in champagne) promote ethanol resorption, whereas a heavy meal reduces it. Ethanol is rapidly distributed throughout the body. A large amount is taken up by the muscles and brain, but comparatively little by adipose tissue and bones. Roughly 70% of the body is accessible to alcohol. Complete resorption of the ethanol contained in one bottle of beer (16 g) by a person weighing 70 kg (distribution in 70 kg · 70/100 = 49 kg) leads to a blood alcohol level of 0.33 per thousand (7.2 mM). The lethal concentration of alcohol is approximately 3.5 per thousand (76 mM).

B. Ethanol metabolism

The major site of ethanol degradation is the liver, although the stomach is also able to metabolize ethanol. Most **ethanol** is initially oxidized by *alcohol dehydrogenase* to form **ethanal** (acetaldehyde). A further oxidization, catalyzed by *aldehyde dehydrogenase*, leads to acetate. Acetate is then converted with the help of *acetate-CoA ligase* to form **acetyl CoA**, using ATP and providing a link to the intermediary metabolism. In addition to cytoplasmic alcohol dehydrogenase, *catalase* and inducible *microsomal alcohol oxidase* ("MEOS"; see p. 318) also contribute to a lesser extent to ethanol degradation. Many of the enzymes mentioned above are induced by ethanol.

The rate of ethanol degradation in the liver is limited by alcohol dehydrogenase activity. The amount of NAD^+ available is the limiting factor. As the maximum degradation rate is already reached at low concentrations of ethanol, the ethanol level therefore declines at a constant rate (zero-order kinetics). The *calorific value* of ethanol is 29.4 kJ · g^{-1}. Alcoholic drinks—particularly in alcoholics—can therefore represent a substantial proportion of dietary energy intake.

C. Liver damage due to alcohol

Alcohol is a socially accepted drug of abuse in Western countries. Due to the high potential for addiction to develop, however, it is actually a "hard" drug and has a much larger number of victims than the opiate drugs, for example. In the brain, ethanol is deposited in membranes due to its amphipathic properties, and it influences receptors for neurotransmitters (see p. 352). The effect of GABA is enhanced, while that of glutamate declines.

High ethanol consumption over many years leads to liver damage. For a healthy man, the limit is about 60 g per day, and for a woman about 50 g. However, these values are strongly dependent on body weight, health status, and other factors.

Ethanol-related high levels of $NADH+H^+$ and acetyl-CoA in the liver lead to increased synthesis of neutral fats and cholesterol. However, since the export of these in the form of VLDLs (see p. 278) is reduced due to alcohol, storage of lipids occurs (**fatty liver**). This increase in the fat content of the liver (from less than 5% to more than 50% of the dry weight) is initially reversible. However, in chronic alcoholism the hepatocytes are increasingly replaced by connective tissue. When **liver cirrhosis** occurs, the damage to the liver finally reaches an irreversible stage, characterized by progressive loss of liver functions.

A. Blood ethanol level

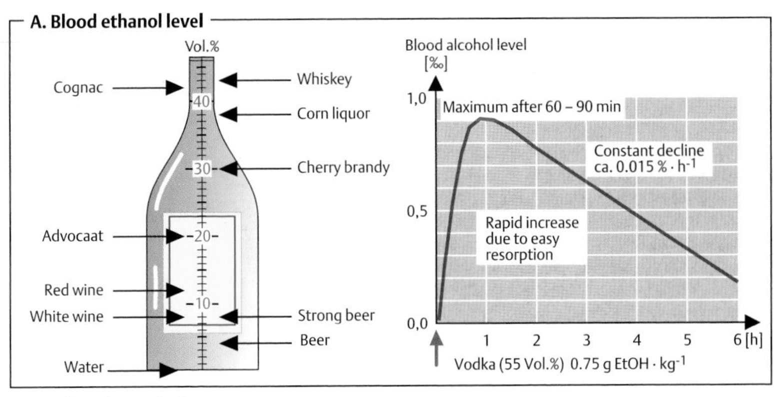

B. Ethanol metabolism

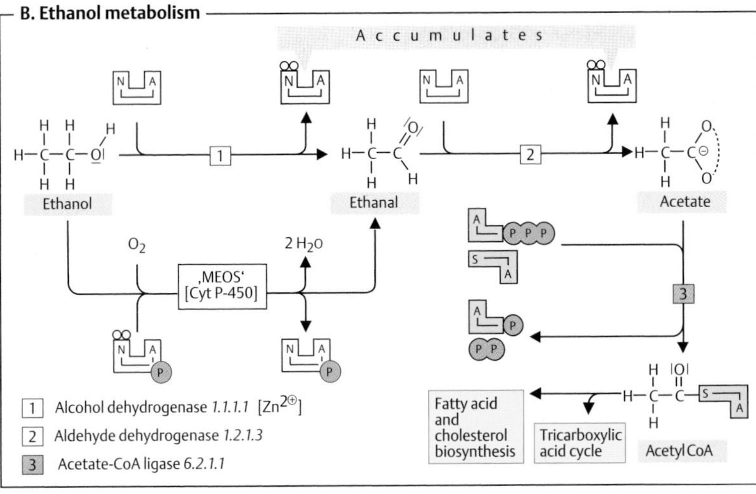

C. Liver damage due to alcohol

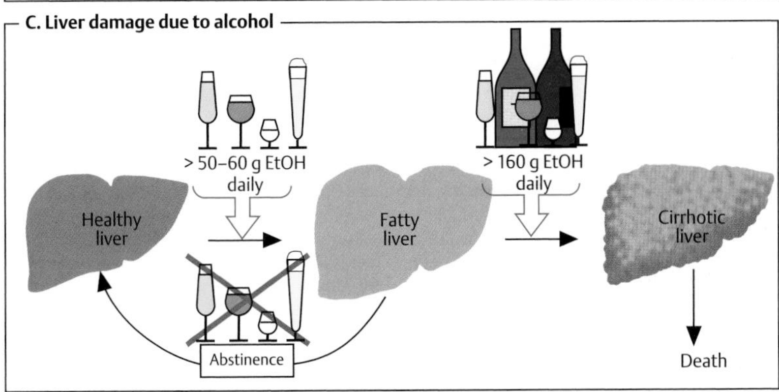

Kidney: functions

A. Functions of the kidneys ●

The kidneys' main function is **excretion** of water and water-soluble substances (**1**). This is closely associated with their role in regulating the body's electrolyte and acid–base balance (**homeostasis, 2**; see pp. 326 and 328). Both excretion and homeostasis are subject to hormonal control. The kidneys are also involved in synthesizing several **hormones** (**3**; see p. 315). Finally, the kidneys also play a role in the **intermediary metabolism** (**4**), particularly in amino acid degradation and gluconeogenesis (see p. 154).

The kidneys are extremely well-perfused organs, with about 1500 L of blood flowing through them every day. Approximately 180 L of primary urine is filtered out of this. Removal of water leads to extreme concentration of the primary urine (to approximately one-hundredth of the initial volume). As a result, only a volume of 0.5–2.0 L of **final urine** is excreted per day.

B. Urine formation ◑

The functional unit of the kidney is the **nephron**. It is made up of the Malpighian bodies or renal corpuscles (consisting of Bowman's capsules and the glomerulus), the proximal tubule, Henle's loop, and the distal tubule, which passes into a collecting duct. The human kidney contains around one million nephrons. The nephrons form urine in the following three phases.

Ultrafiltration. Ultrafiltration of the blood plasma in the glomerulus gives rise to *primary urine*, which is *isotonic* with plasma. The pores in the glomerular basal membrane, which are made up of type IV collagen (see p. 344), have an effective mean diameter of 2.9 nm. This allows all plasma components with a molecular mass of up to about 15 kDa to pass through unhindered. At increasing masses, molecules are progressively held back; at masses greater than 65 kDa, they are completely unable to enter the primary urine. This applies to almost all plasma proteins—which in addition, being anions, are repelled by the negative charge in the basal membrane.

Resorption. All low-molecular weight plasma components enter the primary urine via glomerular filtration. Most of these are transported back into the blood by resorption, to prevent losses of valuable metabolites and electrolytes. In the proximal tubule, organic metabolites (e. g., glucose and other sugars, amino acids, lactate, and ketone bodies) are recovered by secondary active transport (see p. 220). There are several group-specific transport systems for resorbing amino acids, with which hereditary diseases can be associated (e. g., *cystinuria, glycinuria,* and *Hartnup's disease*). HCO_3^-, Na^+, phophate, and sulfate are also resorbed by ATP-dependent (active) mechanisms in the proximal tubule. The later sections of the nephron mainly serve for additional water recovery and regulated resorption of Na^+ and Cl^- (see pp. 326, 328). These processes are controlled by hormones (aldosterone, vasopressin).

Secretion. Some excretable substances are released into the urine by *active transport* in the renal tubules. These substances include H^+ and K^+ ions, urea, and creatinine, as well as drugs such as penicillin.

Clearance. Renal clearance is used as a quantitative measure of renal function. It is defined as the plasma volume cleared of a given substance per unit of time. *Inulin*, a fructose polysaccharide with a mass of ca. 6 kDa (see p. 40) that is neither actively excreted nor resorbed but is freely filtered, has a clearance of $120 mL \cdot min^{-1}$ in healthy individuals.

Further information

Concentrating urine and transporting it through membranes are processes that require large amounts of energy. The kidneys therefore have very high energy demands. In the proximal tubule, the ATP needed is obtained from oxidative metabolism of *fatty acids, ketone bodies,* and several *amino acids.* To a lesser extent, lactate, glycerol, and citric acid are also used. In the distal tubule and Henle's loop, *glucose* is the main substrate for the energy metabolism. The endothelial cells in the proximal tubule are also capable of *gluconeogenesis.* The substrates for this are mainly the carbohydrate skeletons of amino acids. Their amino groups are used as ammonia for buffering urine (see p. 311). Enzymes for peptide degradation and the amino acid metabolism occur in the kidneys at high levels of activity (e. g., amino acid oxidases, amine oxidases, glutaminase).

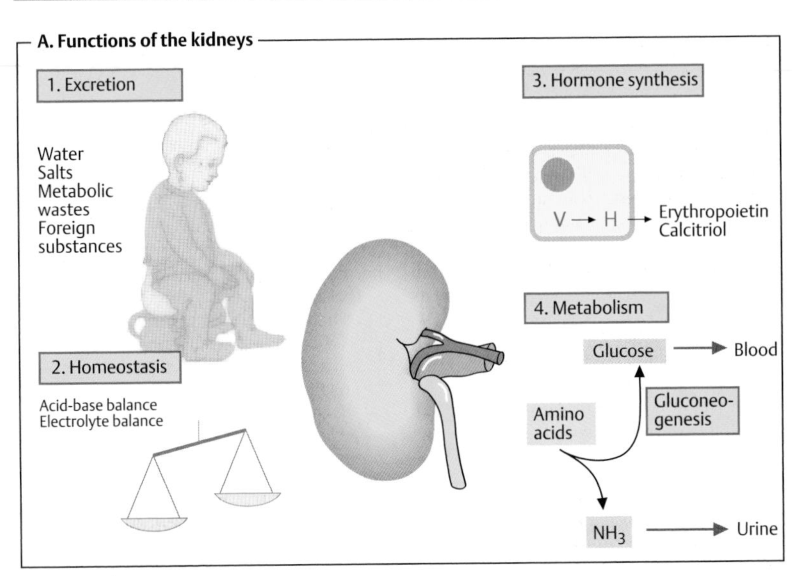
A. Functions of the kidneys
1. Excretion
Water
Salts
Metabolic
wastes
Foreign
substances
2. Homeostasis
Acid-base balance
Electrolyte balance
3. Hormone synthesis
V → H → Erythropoietin
Calcitriol
4. Metabolism
Glucose → Blood
Gluconeo-
genesis
Amino
acids
NH_3 → Urine

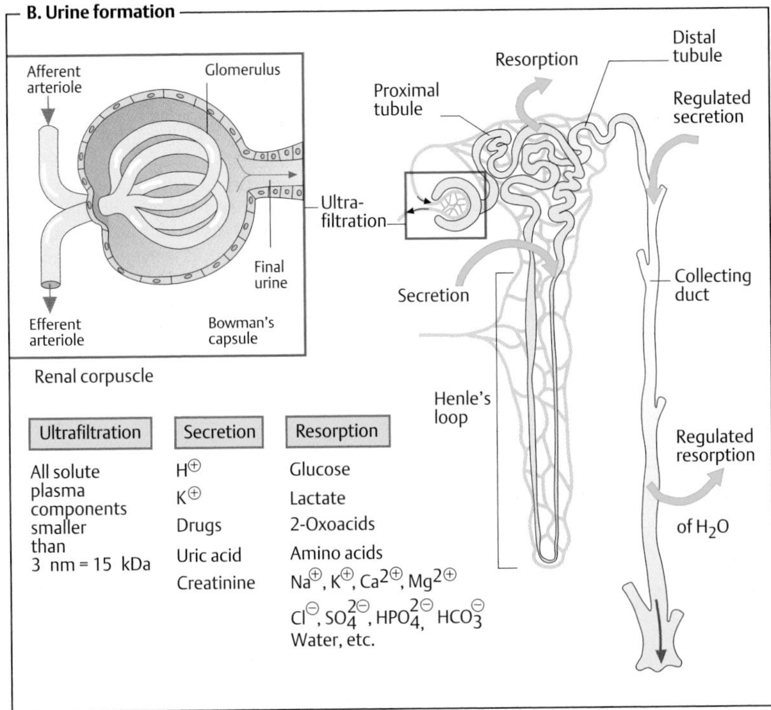
B. Urine formation
Afferent
arteriole
Glomerulus
Final
urine
Efferent
arteriole
Bowman's
capsule
Renal corpuscle
Ultra-
filtration
Proximal
tubule
Resorption
Distal
tubule
Regulated
secretion
Secretion
Collecting
duct
Henle's
loop
Regulated
resorption
of H_2O
Ultrafiltration
All solute
plasma
components
smaller
than
3 nm = 15 kDa
Secretion
$H^{\oplus}$
$K^{\oplus}$
Drugs
Uric acid
Creatinine
Resorption
Glucose
Lactate
2-Oxoacids
Amino acids
$Na^{\oplus}$, $K^{\oplus}$, $Ca^{2\oplus}$, $Mg^{2\oplus}$
$Cl^{\ominus}$, $SO_4^{2\ominus}$, $HPO_4^{2\ominus}$, $HCO_3^{\ominus}$
Water, etc.

Urine

A. Urine

Water and water-soluble compounds are excreted with the urine. The volume and composition of urine are subject to wide variation and depend on food intake, body weight, age, sex, and living conditions such as temperature, humidity, physical activity, and health status. As there is a marked circadian rhythm in urine excretion, the amount of urine and its composition are usually given relative to a 24-hour period.

A human adult produces 0.5–2.0 L urine per day, around 95% of which consists of water. The urine usually has a slightly acidic pH value (around 5.8). However, the pH value of urine is strongly affected by metabolic status. After ingestion of large amounts of plant food, it can increase to over 7.

B. Organic components

Nitrogen-containing compounds are among the most important organic components of urine. **Urea,** which is mainly synthesized in the liver (urea cycle; see p. 182), is the form in which nitrogen atoms from amino acids are excreted. Breakdown of pyrimidine bases also produces a certain amount of urea (see p. 190). When the nitrogen balance is constant, as much nitrogen is excreted as is taken up (see p. 174), and the amount of urea in the urine therefore reflects protein degradation: 70 g protein in food yields approximately 30 g urea in the urine.

Uric acid is the end product of the purine metabolism. When uric acid excretion via the kidneys is disturbed, gout can develop (see p. 190). **Creatinine** is derived from the muscle metabolism, where it arises spontaneously and irreversibly by cyclization of creatine and creatine phosphate (see p. 336). Since the amount of creatinine an individual excretes per day is constant (it is directly proportional to muscle mass), creatinine as an endogenous substance can be used to measure the glomerular filtration rate. The amount of **amino acids** excreted in free form is strongly dependent on the diet and on the efficiency of liver function. Amino acid derivatives are also found in the urine (e. g., **hippurate**, a detoxification product of benzoic acid). Modified amino acids, which occur in special proteins such as *hydroxyproline* in collagen and *3-methylhistidine* in actin and myosin, can be used as indicators of the degradation of these proteins.

Other components of the urine are conjugates with sulfuric acid, glucuronic acid, glycine, and other polar compounds that are synthesized in the liver by biotransformation (see p. 316). In addition, metabolites of many hormones (catecholamines, steroids, serotonin) also appear in the urine and can provide information about hormone production. The proteohormone *chorionic gonadotropin* (hCG, mass ca. 36 kDa), which is formed at the onset of pregnancy, appears in the urine due to its relatively small size. Evidence of hCG in the urine provides the basis for an immunological *pregnancy test.*

The yellow color of urine is due to *urochromes,* which are related to the bile pigments produced by hemoglobin degradation (see p. 194). If urine is left to stand long enough, oxidation of the urochromes may lead to a darkening in color.

C. Inorganic components

The main inorganic components of the urine are the *cations* Na^+, K^+, Ca^{2+}, Mg^{2+}, and NH_4^+ and the *anions* Cl^-, SO_4^{2-}, and HPO_4^{2-}, as well as traces of other ions. In total, Na^+ and Cl^- represent about two-thirds of all the electrolytes in the final urine. Calcium and magnesium occur in the feces in even larger quantities. The amounts of the various inorganic components of the urine also depend on the composition of the diet. For example, in acidosis there can be a marked increase in the excretion of ammonia (see p. 326). Excretion of Na^+, K^+, Ca^{2+}, and phosphate via the kidneys is subject to hormonal regulation (see p. 330).

Further information

Shifts in the concentrations of the *physiological components* of the urine and the appearance of *pathological urine components* can be used to diagnose diseases. Important examples are glucose and ketone bodies, which are excreted to a greater extent in *diabetes mellitus* (see p. 160).

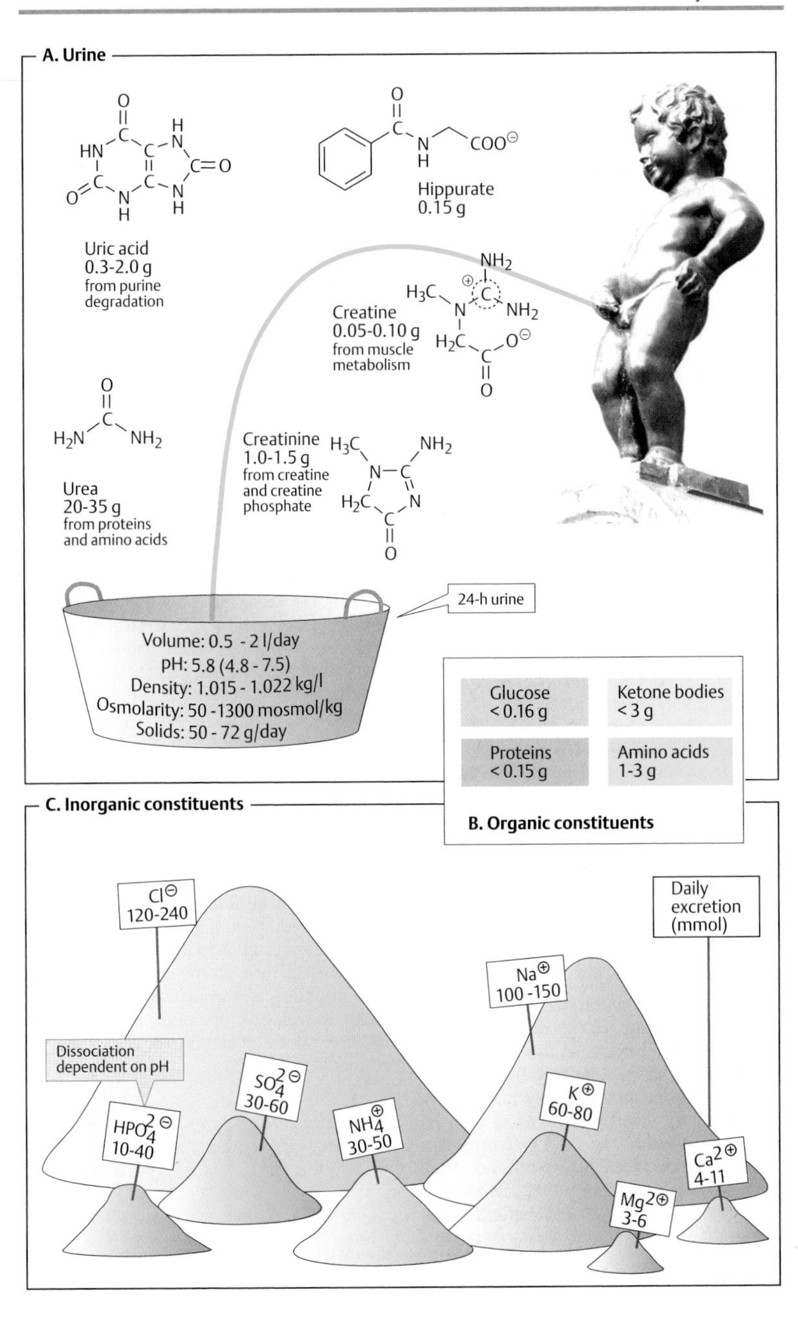
A. Urine
Uric acid
0.3-2.0 g
from purine degradation
Hippurate
0.15 g
Creatine
0.05-0.10 g
from muscle metabolism
Urea
20-35 g
from proteins and amino acids
Creatinine
1.0-1.5 g
from creatine and creatine phosphate
24-h urine
Volume: 0.5 - 2 l/day
pH: 5.8 (4.8 - 7.5)
Density: 1.015 - 1.022 kg/l
Osmolarity: 50 -1300 mosmol/kg
Solids: 50 - 72 g/day
Glucose < 0.16 g
Ketone bodies < 3 g
Proteins < 0.15 g
Amino acids 1-3 g
B. Organic constituents
C. Inorganic constituents
Cl^{-} 120-240
Daily excretion (mmol)
Na^{+} 100 -150
Dissociation dependent on pH
SO_4^{2-} 30-60
K^{+} 60-80
HPO_4^{2-} 10-40
NH_4^{+} 30-50
Ca^{2+} 4-11
Mg^{2+} 3-6

Functions in the acid–base balance

Along with the lungs, the kidneys are particularly involved in keeping the pH value of the extracellular fluid constant (see p. 288). The contribution made by the kidneys particularly involves resorbing HCO_3^- and actively excreting protons.

A. Proton excretion ◑

The renal tubule cells are capable of secreting protons (H^+) from the blood into the urine against a concentration gradient, despite the fact that the H^+ concentration in the urine is up to a thousand times higher than in the blood. To achieve this, carbon dioxide (CO_2) is taken up from the blood and—together with water (H_2O) and with the help of *carbonate dehydratase* (carbonic anhydrase, [1])—converted into hydrogen carbonate ("bicarbonate," HCO_3^-) and one H^+. Formally, this yields carbonic acid H_2CO_3 as an intermediate, but it is not released during the reaction.

The hydrogen carbonate formed in carbonic anhydrase returns to the plasma, where it contributes to the blood's base reserve. The proton is exported into the urine by *secondary active transport* in antiport for Na^+ (bottom right). The driving force for **proton excretion**, as in other secondary active processes, is the Na^+ gradient established by the ATPase involved in the *Na^+/K^+* exchange ("Na^+/K^+ ATPase", see p. 220). This integral membrane protein on the basal side (towards the blood) of tubule cells keeps the Na^+ concentration in the tubule cell low, thereby maintaining Na^+ inflow. In addition to this secondary active H^+ transport mechanism, there is a V-type H^+-transporting ATPase in the distal tubule and collecting duct (see p. 220).

An important function of the secreted H^+ ions is to promote **HCO_3^- resorption** (top right). Hydrogen carbonate, the most important buffering base in the blood, passes into the primary urine quantitatively, like all ions. In the primary urine, HCO_3^- reacts with H^+ ions to form water and CO_2, which returns by free diffusion to the tubule cells and from there into the blood. In this way, the kidneys also influence the CO_2/HCO_3^- buffering balance in the plasma.

B. Ammonia excretion ◑

Approximately 60 mmol of protons are excreted with the urine every day. Buffering systems in the urine catch a large proportion of the H^+ ions, so that the urine only becomes weakly acidic (down to about pH 4.8).

An important buffer in the urine is the *hydrogen phosphate/dihydrogen phosphate system* ($HPO_4^{2-}/H_2PO_4^-$). In addition, *ammonia* also makes a vital contribution to buffering the secreted protons.

Since plasma concentrations of *free* ammonia are low, the kidneys release NH_3 from **glutamine** and other amino acids. At 0.5–0.7 mM, glutamine is the most important amino acid in the plasma and is the preferred form for ammonia transport in the blood. The kidneys take up glutamine, and with the help of *glutaminase* [4], initially release NH_3 from the amide bond hydrolytically. From the **glutamate** formed, a second molecule of NH_3 can be obtained by oxidative deamination with the help of *glutamate dehydrogenase* [5] (see p. 178). The resulting **2-oxoglutarate** is further metabolized in the tricarboxylic acid cycle. Several other amino acids—*alanine* in particular, as well as *serine, glycine,* and *aspartate*—can also serve as suppliers of ammonia.

Ammonia can diffuse freely into the urine through the tubule membrane, while the **ammonium ions** that are formed in the urine are charged and can no longer return to the cell. Acidic urine therefore promotes ammonia excretion, which is normally 30–50 mmol per day. In metabolic **acidosis** (e.g., during fasting or in *diabetes mellitus*), after a certain time increased *induction of glutaminase* occurs in the kidneys, resulting in increased NH_3 excretion. This in turn promotes H^+ release and thus counteracts the acidosis. By contrast, when the plasma pH value shifts towards alkaline values (*alkalosis*), renal excretion of ammonia is reduced.

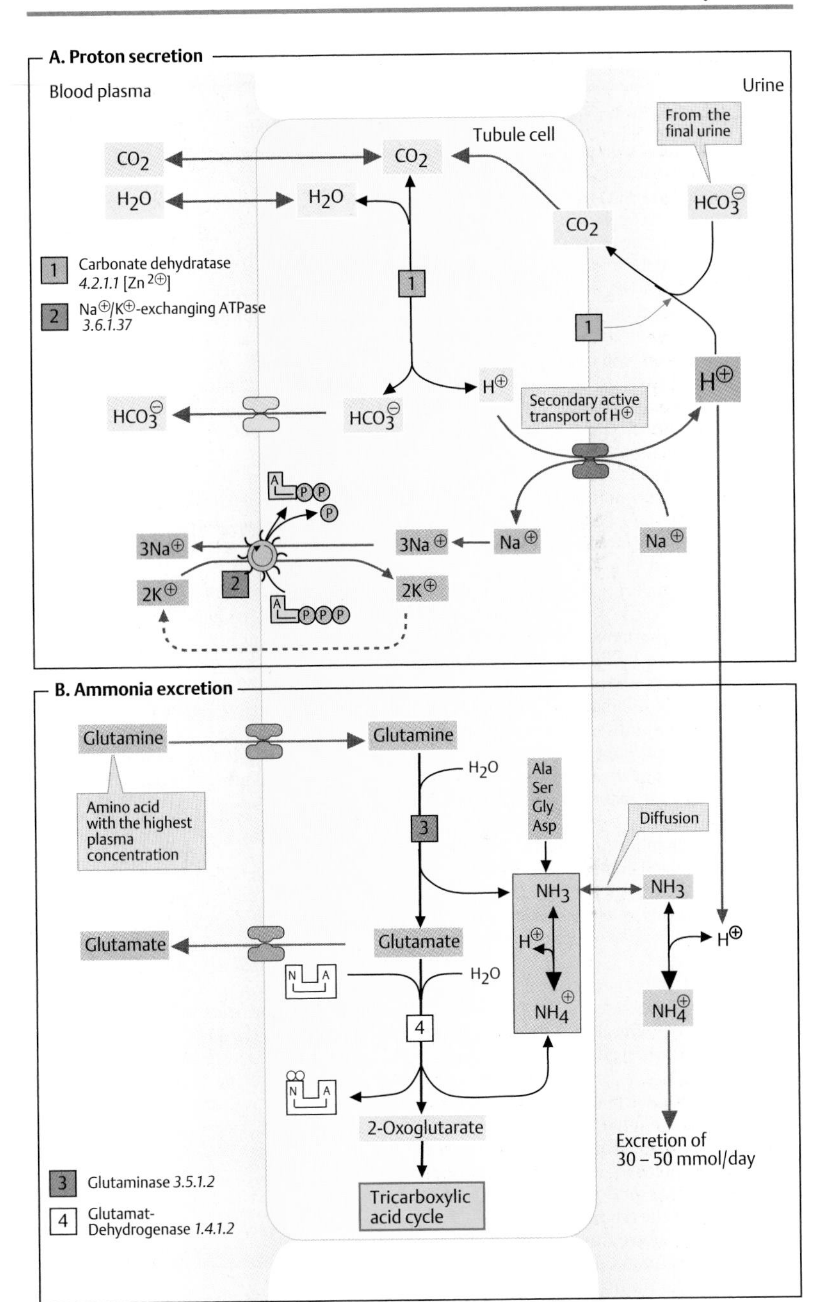
A. Proton secretion
Blood plasma
Urine
Tubule cell
From the final urine
CO_2
CO_2
H_2O
H_2O
CO_2
HCO_3^-
1 Carbonate dehydratase 4.2.1.1 [Zn^{2+}]
2 Na^+/K^+-exchanging ATPase 3.6.1.37
1
1
H^+
H^+
HCO_3^-
HCO_3^-
Secondary active transport of H^+
$3Na^+$
$3Na^+$
Na^+
Na^+
$2K^+$
$2K^+$
2
B. Ammonia excretion
Glutamine
Glutamine
Amino acid with the highest plasma concentration
H_2O
Ala
Ser
Gly
Asp
3
Diffusion
NH_3
NH_3
Glutamate
Glutamate
H^+
H^+
H_2O
NH_4^+
NH_4^+
4
2-Oxoglutarate
Tricarboxylic acid cycle
Excretion of 30 – 50 mmol/day
3 Glutaminase 3.5.1.2
4 Glutamat-Dehydrogenase 1.4.1.2

Electrolyte and water recycling

A. Electrolyte and water recycling

Electrolytes and other plasma components with low molecular weights enter the *primary urine* by ultrafiltration (right). Most of these substances are recovered by energy-dependent resorption (see p. 322). The extent of the resorption determines the amount that ultimately reaches the final urine and is excreted. The illustration does not take into account the *zoning* of transport processes in the kidney (physiology textbooks may be referred to for further details).

Calcium and phosphate ions. Calcium (Ca^{2+}) and phosphate ions are almost completely resorbed from the primary urine by active transport (i.e., in an ATP-dependent fashion). The proportion of Ca^{2+} resorbed is over 99%, while for phosphate the figure is 80–90%. The extent to which these two electrolytes are resorbed is regulated by the three hormones parathyrin, calcitonin, and calcitriol.

The peptide hormone *parathyrin* (PTH), which is produced by the parathyroid gland, stimulates Ca^{2+} resorption in the kidneys and at the same time inhibits the resorption of phosphate. In conjunction with the effects of this hormone in the bones and intestines (see p. 344), this leads to an *increase in the plasma level of* Ca^{2+} and a reduction in the level of phosphate ions.

Calcitonin, a peptide produced in the C cells of the thyroid gland, inhibits the resorption of both calcium and phosphate ions. The result is an overall *reduction in the plasma level* of both ions. Calcitonin is thus a parathyrin antagonist relative to Ca^{2+}.

The steroid hormone *calcitriol*, which is formed in the kidneys (see p. 304), stimulates the resorption of both calcium and phosphate ions and thus increases the plasma level of both ions.

Sodium ions. Controlled resorption of Na^+ from the primary urine is one of the most important functions of the kidney. Na^+ resorption is highly effective, with more than 97% being resorbed. Several mechanisms are involved: some of the Na^+ is taken up passively in the proximal tubule through the junctions between the cells (paracellularly). In addition, there is secondary active transport together with glucose and amino acids (see p. 322). These two pathways are responsible for 60–70% of total Na^+ resorption. In the ascending part of Henle's loop, there is another transporter (shown at the bottom right), which functions electroneutrally and takes up one Na^+ ion and one K^+ ion together with two Cl^- ions. This symport is also dependent on the activity of Na^+/K^+ ATPase [2], which pumps the Na^+ resorbed from the primary urine back into the plasma in exchange for K^+.

The steroid hormone *aldosterone* (see p. 55) increases Na^+ reuptake, particularly in the distal tubule, while *atrial natriuretic peptide* (ANP) originating from the cardiac atrium reduces it. Among other effects, aldosterone induces Na^+/K^+ ATPase and various Na^+ transporters on the luminal side of the cells.

Water. Water resorption in the proximal tubule is a *passive process* in which water follows the osmotically active particles, particularly the Na^+ ions. Fine regulation of water excretion (**diuresis**) takes place in the collecting ducts, where the peptide hormone *vasopressin* (antidiuretic hormone, ADH) operates. This promotes recovery of water by stimulating the transfer of aquaporins (see p. 220) into the plasma membrane of the tubule cells via V_2 receptors. A lack of ADH leads to the disease picture of *diabetes insipidus*, in which up to 30 L of final urine is produced per day.

B. Gluconeogenesis

Apart from the liver, the kidneys are the only organs capable of producing glucose by neosynthesis *(gluconeogenesis;* see p. 154). The main substrate for gluconeogenesis in the cells of the proximal tubule is **glutamine**. In addition, other amino acids and also **lactate**, **glycerol**, and **fructose** can be used as precursors. As in the liver, the key enzymes for gluconeogenesis are induced by *cortisol* (see p. 374). Since the kidneys also have a high level of glucose consumption, they only release very little glucose into the blood.

A. Electrolyte and water recycling

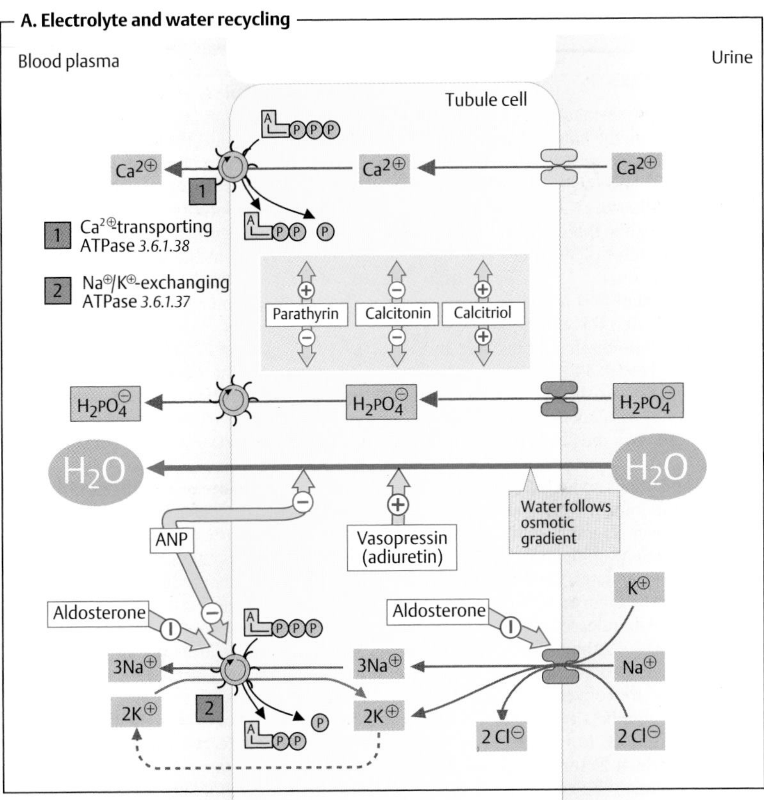

B. Gluconeogenesis

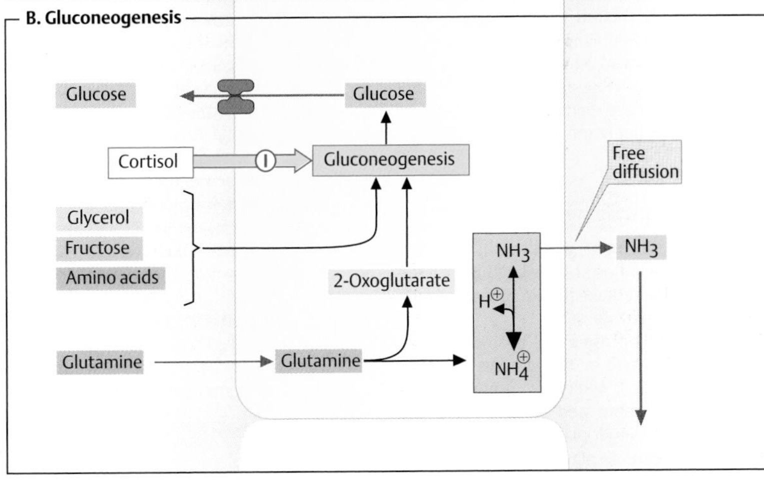

Renal hormones

A. Renal hormones ◑

In addition to their involvement in excretion and metabolism, the kidneys also have endocrine functions. They produce the hormones **erythropoietin** and **calcitriol** and play a decisive part in producing the hormone **angiotensin II** by releasing the enzyme *renin.* Renal prostaglandins (see p. 390) have a local effect on Na^+ resorption.

Calcitriol (vitamin D hormone, 1α,25-dihydroxycholecalciferol) is a hormone closely related to the steroids that is involved in Ca^{2+} homeostasis (see p. 342). In the kidney, it is formed from calcidiol by hydroxylation at C-1. The activity of *calcidiol-1-monooxygenase* [1] is enhanced by the hormone *parathyrin* (PTH).

Erythropoietin is a peptide hormone that is formed predominantly by the kidneys, but also by the liver. Together with other factors known as *"colony-stimulating factors"* (CSF; see p. 392), it regulates the differentiation of stem cells in the bone marrow.

Erythropoietin release is stimulated by hypoxia (low pO_2). Within hours, the hormone ensures that erythrocyte precursor cells in the bone marrow are converted to erythrocytes, so that their numbers in the blood increase. Renal damage leads to reduced erythropoietin release, which in turn results in *anemia.* Forms of anemia with renal causes can now be successfully treated using erythropoietin produced by genetic engineering techniques. The hormone is also administered to dialysis patients. Among athletes and sports professionals, there have been repeated cases of erythropoietin being misused for doping purposes.

B. Renin–angiotensin system ◑

The peptide hormone angiotensin II is not synthesized in a hormonal gland, but in the blood. The kidneys take part in this process by releasing the enzyme renin.

Renin [2] is an aspartate proteinase (see p. 176). It is formed by the kidneys as a precursor (prorenin), which is proteolytically activated into renin and released into the blood. In the blood plasma, renin acts on **angiotensinogen,** a plasma glycoprotein in the α_2-globulin group (see p. 276), which like almost all plasma proteins is synthesized in the liver. The decapeptide cleaved off by renin is called **angiotensin I.** Further cleavage by *peptidyl dipeptidase A* (*angiotensin-converting enzyme,* ACE), a membrane enzyme located on the vascular endothelium in the lungs and other tissues, gives rise to the octapeptide **angiotensin II** [3], which acts as a hormone and neurotransmitter. The lifespan of angiotensin II in the plasma is only a few minutes, as it is rapidly broken down by other peptidases (angiotensinases [4]), which occur in many different tissues.

The plasma level of angiotensin II is mainly determined by the rate at which renin is released by the kidneys. Renin is synthesized by juxtaglomerular cells, which release it when sodium levels decline or there is a fall in blood pressure.

Effects of angiotensin II. Angiotensin II has effects on the kidneys, brain stem, pituitary gland, adrenal cortex, blood vessel walls, and heart via membrane-located receptors. It increases blood pressure by triggering *vasoconstriction* (narrowing of the blood vessels). In the kidneys, it promotes the *retention of* Na^+ and water and reduces potassium secretion. In the brain stem and at nerve endings in the sympathetic nervous system, the effects of angiotensin II lead to increased tonicity (neurotransmitter effect). In addition, it triggers the *sensation of thirst.* In the pituitary gland, angiotensin II stimulates *vasopressin release* (antidiuretic hormone) and corticotropin (ACTH) release. In the adrenal cortex, it increases the *biosynthesis and release of aldosterone,* which promotes sodium and water retention in the kidneys. All of the effects of angiotensin II lead directly or indirectly to *increased blood pressure,* as well as increased *sodium and water retention.* This important hormonal system for blood pressure regulation can be pharmacologically influenced by *inhibitors* at various points:

- Using angiotensinogen analogs that inhibit renin.
- Using angiotensin I analogs that competitively inhibit the enzyme ACE [3].
- Using hormone antagonists that block the binding of angiotensin II to its receptors.

A. Renal hormones

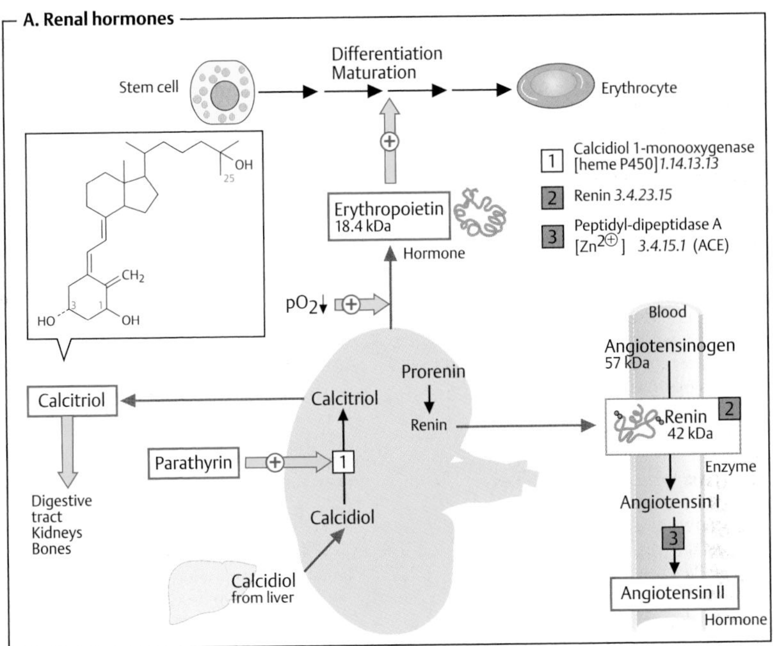

B. Renin angiotensin system

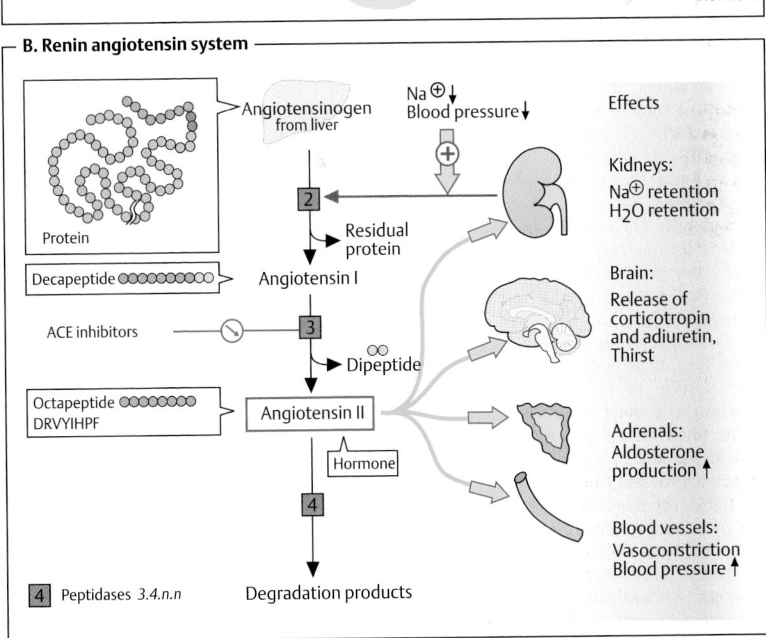

Muscle contraction

The musculature is what makes movements possible. In addition to the **skeletal muscles**, which can be contracted voluntarily, there are also the autonomically activated **heart muscle** and **smooth muscle**, which is also involuntary. In all types of muscle, contraction is based on an interplay between the proteins actin and myosin.

A. Organization of skeletal muscle ◑

Striated muscle consists of parallel bundles of **muscle fibers.** Each fiber is a single large multinucleate cell. The cytoplasm in these cells contains **myofibrils** 2–3 μm thick that can extend over the full length of the muscle fiber.

The *striation* of the muscle fibers is characteristic of skeletal muscle. It results from the regular arrangement of molecules of differing density. The repeating contractile units, the **sarcomeres**, are bounded by Z lines from which thin filaments of **F-actin** (see p. 204) extend on each side. In the A bands, there are also thick parallel filaments of **myosin.** The H bands in the middle of the A bands only contain myosin, while only actin is found on each size of the Z lines.

Myosin is quantitatively the most important protein in the myofibrils, representing 65% of the total. It is shaped like a golf club (bottom right). The molecule is a hexamer consisting of two identical *heavy chains* (2 × 223 kDa) and four *light chains* (each about 20 kDa). Each of the two heavy chains has a globular "head" at its amino end, which extends into a "tail" about 150 nm long in which the two chains are intertwined to form a superhelix. The small subunits are attached in the head area. Myosin is present as a bundle of several hundred stacked molecules in the form of a *"thick myosin filament."* The head portion of the molecule acts as an *ATPase*, the activity of which is modulated by the small subunits.

Actin (42 kDa) is the most important component of the *"thin filaments."* It represents ca. 20–25% of the muscle proteins. **F-actin** is also an important component of the cytoskeleton (see p. 204). This filamentous polymer is held in equilibrium with its monomer, **G-actin.** The other protein components of muscle include tropomyosin and troponin. **Tropomyosin** (64 kDa) attaches to F-actin as a rod-like dimer and connects approximately seven actin units with each other. The heterotrimer **troponin** (78 kDa) is bound to one end of tropomyosin.

In addition to the above proteins, a number of other proteins are also typical of muscle—including *titin* (the largest known protein), *α-* and *β-actinin, desmin,* and *vimentin.*

B. Mechanism of muscle contraction ◑

The *sliding filament model* describes the mechanism involved in muscle contraction. In this model, sarcomeres become shorter when the thin and thick filaments slide alongside each other and telescope together, with ATP being consumed. During contraction, the following reaction cycle is repeated several times:

[**1**] In the initial state, the myosin heads are attached to actin. When ATP is bound, the heads detach themselves from the actin (the "plasticizing" effect of ATP).

[**2**] The myosin head hydrolyzes the bound ATP to ADP and P_i, but initially withholds the two reaction products. ATP cleavage leads to allosteric tension in the myosin head.

[**3**] The myosin head now forms a new bond with a neighboring actin molecule.

[**4**] The actin causes the release of the P_i, and shortly afterwards release of the ADP as well. This converts the allosteric tension in the myosin head into a conformational change that acts like a rowing stroke.

The cycle can be repeated for as long as ATP is available, so that the thick filaments are constantly moving along the thin filaments in the direction of the Z disk. Each rowing stroke of the 500 or so myosin heads in a thick filament produces a contraction of about 10 nm. During strong contraction, the process is repeated about five times per second. This leads to the whole complex of thin filaments moving together; the H band becomes shorter and the Z lines slide closer together.

A. Organization of striated muscle

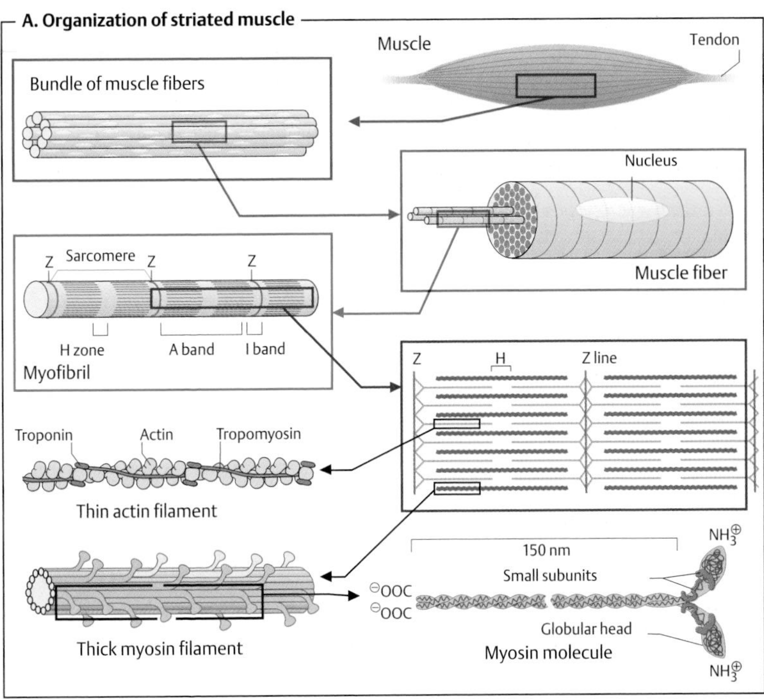

B. Mechanism of muscle contraction

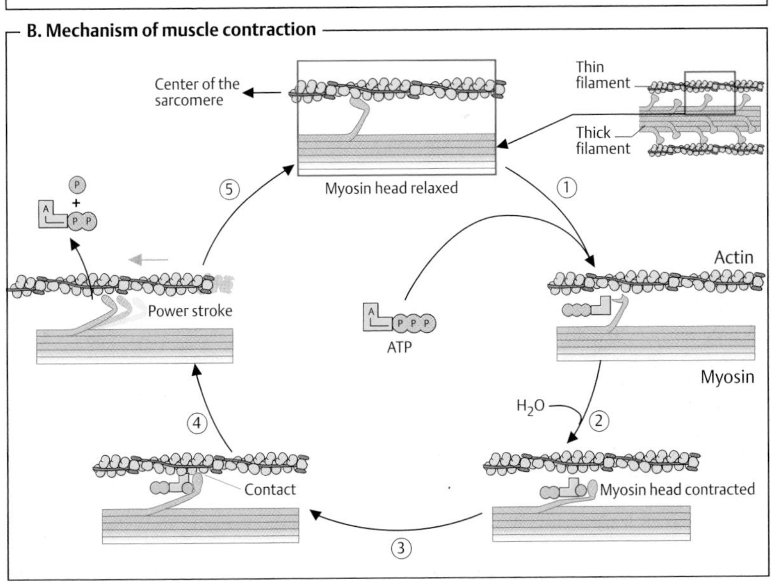

Control of muscle contraction

A. Neuromuscular junction ◑

Muscle contraction is triggered by *motor neurons* that release the neurotransmitter *acetylcholine* (see p. 352). The transmitter diffuses through the narrow synaptic cleft and binds to nicotinic *acetylcholine receptors* on the plasma membrane of the muscle cell (the *sarcolemma*), thereby opening the ion channels integrated into the receptors (see p. 222). This leads to an inflow of Na^+, which triggers an *action potential* (see p. 350) in the sarcolemma. The action potential propagates from the end plate in all directions and constantly stimulates the muscle fiber. With a delay of a few milliseconds, the contractile mechanism responds to this by contracting the muscle fiber.

B. Sarcoplasmic reticulum (SR) ◑

The action potential (**A**) produced at the neuromuscular junction is transferred in the muscle cell into a transient increase in the Ca^{2+} concentration in the cytoplasm of the muscle fiber (the *sarcoplasm*).

In the resting state, the Ca^{2+} level in the sarcoplasm is very low (less than 10^{-7} M). By contrast, the **sarcoplasmic reticulum** (SR), which corresponds to the ER, contains Ca^{2+} ions at a concentration of about 10^{-3} M. The SR is a branched organelle that surrounds the myofibrils like a net stocking inside the muscle fibers (illustrated at the top using the example of a heart muscle cell). The high Ca^{2+} level in the SR is maintained by Ca^{2+}-transporting ATPases (see p. 220). In addition, the SR also contains *calsequestrin*, a protein (55 kDa) that is able to bind numerous Ca^{2+}-ions via acidic amino acid residues.

The transfer of the action potential to the SR is made possible by **transverse tubules** (T tubules), which are open to the extracellular space and establish a close connection with the SR. There is a structure involved in the contact between the T tubule and the SR that was formerly known as the "SR foot" (it involves parts of the *ryanodine receptor;* see p. 386).

At the point of contact with the SR, the action potential triggers the opening of the *Ca^{2+} channels* on the surface of the sarcolemma. Calcium ions then leave the SR and enter the sarcoplasm, where they lead to a rapid increase in Ca^{2+} concentrations. This in turn causes the myofibrils to contract (**C**).

C. Regulation by calcium ions ◑

In relaxed skeletal muscle, the complex consisting of **troponin** and **tropomyosin** blocks the access of the myosin heads to actin (see p. 332). Troponin consists of three different subunits (**T**, **C**, and **I**). The rapid increase in cytoplasmic Ca^{2+} concentrations caused by opening of the calcium channels in the SR leads to binding of Ca^{2+} to the C subunit of troponin, which closely resembles calmodulin (see p. 386). This produces a conformational change in troponin that causes the whole troponin–tropomyosin complex to slip slightly and expose a binding site for myosin (red). This initiates the contraction cycle. After contraction, the sarcoplasmic Ca^{2+} concentration is quickly reduced again by active transport back into the SR. This results in troponin losing the bound Ca^{2+} ions and returning to the initial state, in which the binding site for myosin on actin is blocked. It is not yet clear whether the mechanism described above is the only one that triggers binding of myosin to actin.

When **triggering of contraction** in striated muscle occurs, the following sequence of processes thus takes place:

1. The sarcolemma is depolarized.
2. The action potential is signaled to Ca^{2+} channels in the SR.
3. The Ca^{2+} channels open and the Ca^{2+} level in the sarcoplasm increases.
4. Ca^{2+} binds to troponin C and triggers a conformational change.
5. Troponin causes tropomyosin to slip, and the myosin heads bind to actin.
6. The actin–myosin cycle takes place and the muscle fibers contract.

Conversely, at the **end of contraction**, the following processes take place:

1. The Ca^{2+} level in the sarcoplasm declines due to transport of Ca^{2+} back into the SR.
2. Troponin C loses Ca^{2+} and tropomyosin returns to its original position on the actin molecule.
3. The actin–myosin cycle stops and the muscle relaxes.

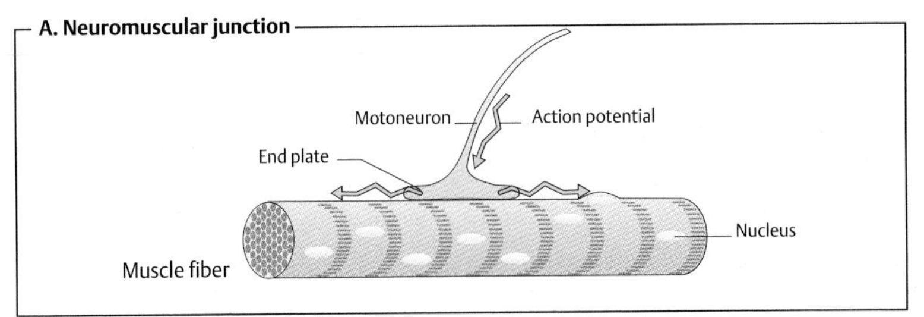
A. Neuromuscular junction
Motoneuron
Action potential
End plate
Nucleus
Muscle fiber

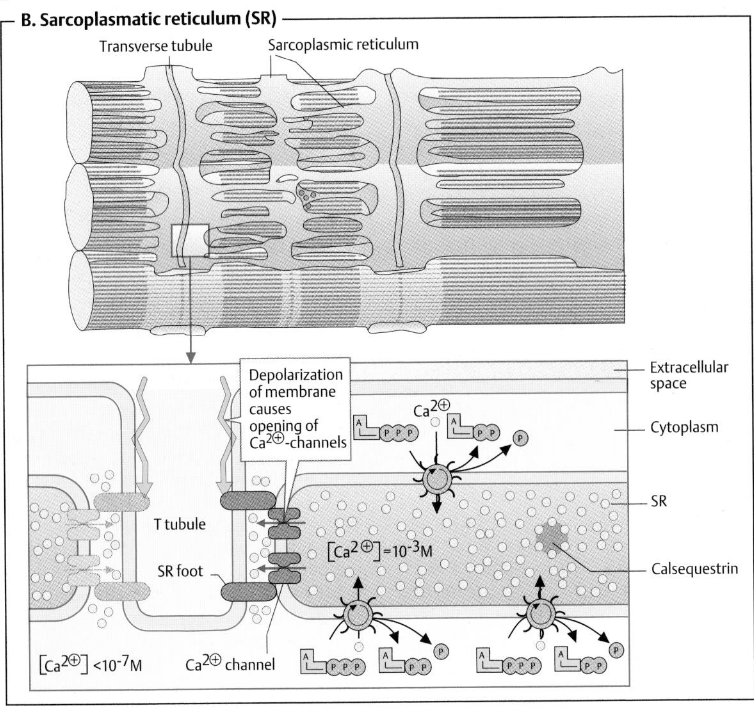
B. Sarcoplasmatic reticulum (SR)
Transverse tubule
Sarcoplasmic reticulum
Depolarization of membrane causes opening of Ca2⊕-channels
Extracellular space
Ca2⊕
Cytoplasm
SR
T tubule
[Ca2⊕]=10-3M
SR foot
Calsequestrin
[Ca2⊕] <10-7M
Ca2⊕ channel

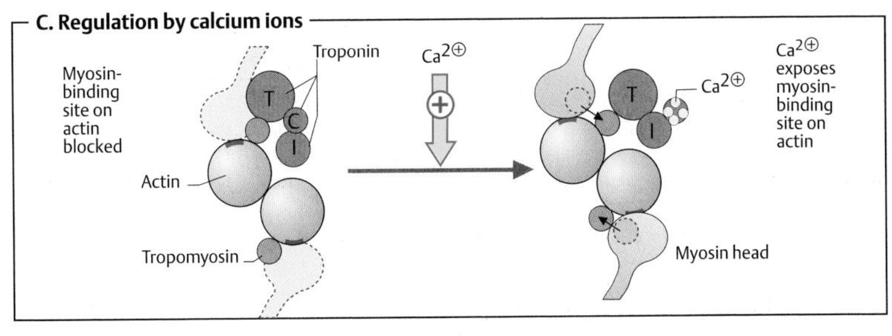
C. Regulation by calcium ions
Myosin-binding site on actin blocked
Troponin
Ca2⊕
Ca2⊕ exposes myosin-binding site on actin
Ca2⊕
Actin
Tropomyosin
Myosin head

Muscle metabolism I

Muscle contraction is associated with a high level of ATP consumption (see p. 332). Without constant resynthesis, the amount of ATP available in the resting state would be used up in less than 1 s of contraction.

A. Energy metabolism in the white and red muscle fibers ◑

Muscles contain two types of fibers, the proportions of which vary from one type of muscle to another. **Red fibers** (type I fibers) are suitable for prolonged effort. Their metabolism is mainly aerobic and therefore depends on an adequate supply of O_2. **White fibers** (type II fibers) are better suited for fast, strong contractions. These fibers are able to form sufficient ATP even when there is little O_2 available. With appropriate training, athletes and sports participants are able to change the proportions of the two fiber types in the musculature and thereby prepare themselves for the physiological demands of their disciplines in a targeted fashion. The expression of functional muscle proteins can also change during the course of training.

Red fibers provide for their ATP requirements mainly (but not exclusively) from **fatty acids**, which are broken down via β-oxidation, the tricarboxylic acid cycle, and the respiratory chain (right part of the illustration). The red color in these fibers is due to the monomeric heme protein **myoglobin**, which they use as an O_2 reserve. Myoglobin has a much higher affinity for O_2 than hemoglobin and therefore only releases its O_2 when there is a severe drop in O_2 partial pressure (cf. p. 282).

At a high level of muscular effort—e. g., during weightlifting or in very fast contractions such as those carried out by the eye muscles—the O_2 supply from the blood quickly becomes inadequate to maintain the aerobic metabolism. White fibers (left part of the illustration) therefore mainly obtain ATP from **anaerobic glycolysis**. They have supplies of **glycogen** from which they can quickly release glucose-1-phosphate when needed (see p. 156). By isomerization, this gives rise to glucose-6-phosphate, the substrate for glycolysis. The NADH+H$^+$ formed during glycolysis has to be reoxidized into NAD$^+$ in order to maintain glucose degradation and thus ATP formation. If there is a lack of O_2, this is achieved by the formation of **lactate**, which is released into the blood and is resynthesized into glucose in the liver (Cori cycle; see p. 338).

Muscle-specific auxiliary reactions for ATP synthesis exist in order to provide additional ATP in case of emergency. **Creatine phosphate** (see **B**) acts as a buffer for the ATP level. Another ATP-supplying reaction is catalyzed by *adenylate kinase* [1] (see also p. 72). This disproportionates two molecules of ADP into ATP and AMP. The AMP is deaminated into IMP in a subsequent reaction [2] in order to shift the balance of the reversible reaction [1] in the direction of ATP formation.

B. Creatine metabolism ◑

Creatine (*N*-methylguanidoacetic acid) and its phosphorylated form **creatine phosphate** (a guanidophosphate) serve as an ATP buffer in muscle metabolism. In creatine phosphate, the phosphate residue is at a similarly high chemical potential as in ATP and is therefore easily transferred to ADP. Conversely, when there is an excess of ATP, creatine phosphate can arise from ATP and creatine. Both processes are catalyzed by *creatine kinase* [5].

In resting muscle, creatine phosphate forms due to the high level of ATP. If there is a risk of a severe drop in the ATP level during contraction, the level can be maintained for a short time by synthesis of ATP from creatine phosphate and ADP. In a nonenzymatic reaction [6], small amounts of creatine and creatine phosphate cyclize constantly to form **creatinine**, which can no longer be phosphorylated and is therefore excreted with the urine (see p. 324).

Creatine does not derive from the muscles themselves, but is synthesized in two steps in the kidneys and liver (left part of the illustration). Initially, the guanidino group of arginine is transferred to glycine in the kidneys, yielding **guanidino acetate** [3]. In the liver, *N*-methylation of guanidino acetate leads to the formation of creatine from this [4]. The coenzyme in this reaction is *S-adenosyl methionine* (SAM; see p. 110).

A. Energy metabolism in the white and red muscle fibers

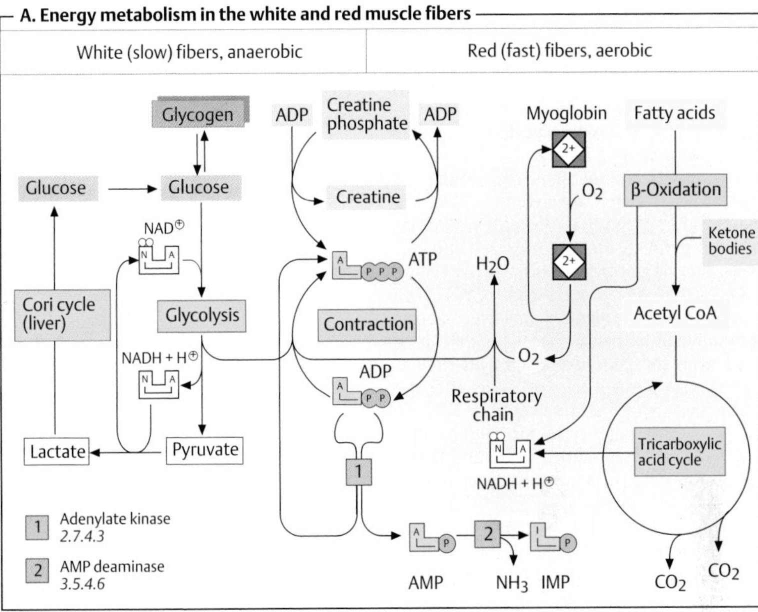

B. Creatine metabolism

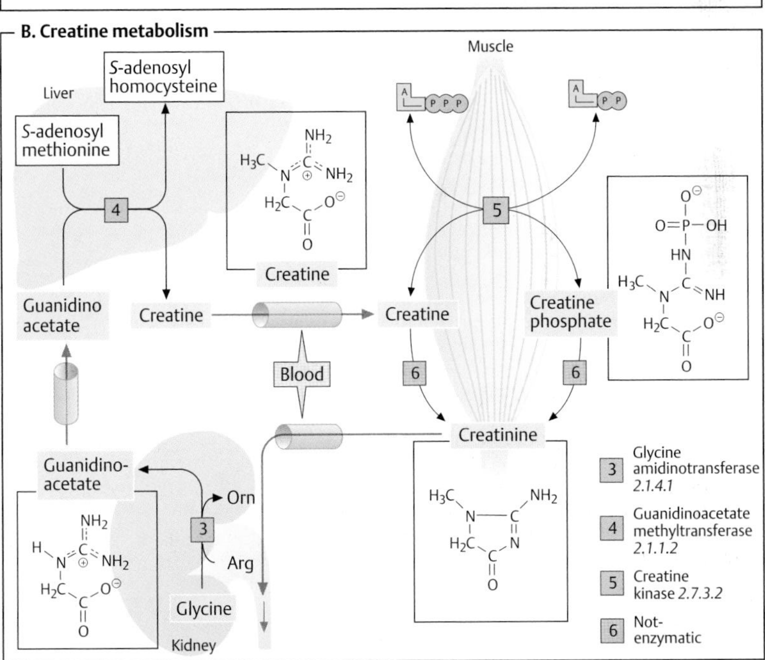

Muscle metabolism II

A. Cori and alanine cycle ◑

White muscle fibers (see p. 336) mainly obtain ATP from **anaerobic glycolysis**—i. e., they convert glucose into lactate. The **lactate** arising in muscle and, in smaller quantities, its precursor **pyruvate** are released into the blood and transported to the liver, where lactate and pyruvate are resynthesized into glucose again via *gluconeogenesis,* with ATP being consumed in the process (see p. 154). The glucose newly formed by the liver returns via the blood to the muscles, where it can be used as an energy source again. This circulation system is called the **Cori cycle**, after the researchers who first discovered it. There is also a very similar cycle for erythrocytes, which do not have mitochondria and therefore produce ATP by anaerobic glycolysis (see p. 284).

The muscles themselves are not capable of gluconeogenesis. Nor would this be useful, as gluconeogenesis requires much more ATP than is supplied by glycolysis. As O_2 deficiencies do not arise in the liver even during intensive muscle work, there is always sufficient energy there available for gluconeogenesis.

There is also a corresponding circulation system for the amino acid alanine. The **alanine cycle** in the liver not only provides alanine as a precursor for gluconeogenesis, but also transports to the liver the amino nitrogen arising in muscles during protein degradation. In the liver, it is incorporated into urea for excretion.

Most of the amino acids that arise in muscle during proteolysis are converted into glutamate and 2-oxo acids by *transamination* (not shown; cf. p. 180). Again by transamination, glutamate and pyruvate give rise to alanine, which after glutamine is the second important form of transport for amino nitrogen in the blood. In the liver, alanine and 2-oxoglutarate are resynthesized into pyruvate and glutamate (see p. 178). Glutamate supplies the urea cycle (see p. 182), while pyruvate is available for gluconeogenesis.

B. Protein and amino acid metabolism ◑

The skeletal muscle is the most important site for degradation of the *branched-chain amino acids* (Val, Leu, Ile; see p. 414), but other amino acids are also broken down in the muscles. **Alanine** and **glutamine** are resynthesized from the components and released into the blood. They transport the nitrogen that arises during amino acid breakdown to the liver (*alanine cycle;* see above) and to the kidneys (see p. 328).

During periods of hunger, **muscle proteins** serve as an energy reserve for the body. They are broken down into amino acids, which are transported to the liver. In the liver, the carbon skeletons of the amino acids are converted into intermediates in the tricarboxylic acid cycle or into acetoacetyl-CoA (see p. 175). These amphibolic metabolites are then available to the energy metabolism and for gluconeogenesis. After prolonged starvation, the brain switches to using ketone bodies in order to save muscle protein (see p. 356).

The synthesis and degradation of muscle proteins are regulated by hormones. **Cortisol** leads to muscle degradation, while **testosterone** stimulates protein formation. Synthetic **anabolics** with a testosterone-like effect have repeatedly been used for doping purposes or for intensive muscle-building.

Further information

Smooth muscle differs from skeletal muscle in various ways. Smooth muscles—which are found, for example, in blood vessel walls and in the walls of the intestines—do not contain any muscle fibers. In smooth-muscle cells, which are usually spindle-shaped, the contractile proteins are arranged in a less regular pattern than in striated muscle. Contraction in this type of muscle is usually not stimulated by nerve impulses, but occurs in a largely spontaneous way. Ca^{2+} (in the form of Ca^{2+}-calmodulin; see p. 386) also activates contraction in smooth muscle; in this case, however, it does not affect troponin, but activates a protein kinase that phosphorylates the light chains in myosin and thereby increases myosin's ATPase activity. Hormones such as epinephrine and angiotensin II (see p. 330) are able to influence vascular tonicity in this way, for example.

A. Cori and alanine cycle

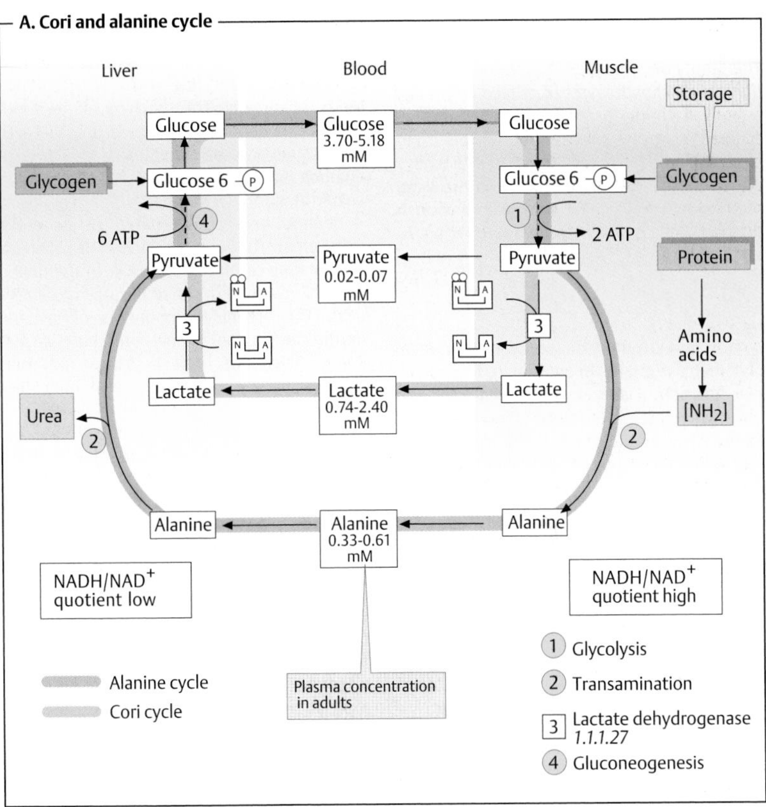

B. Protein and amino acid metabolism

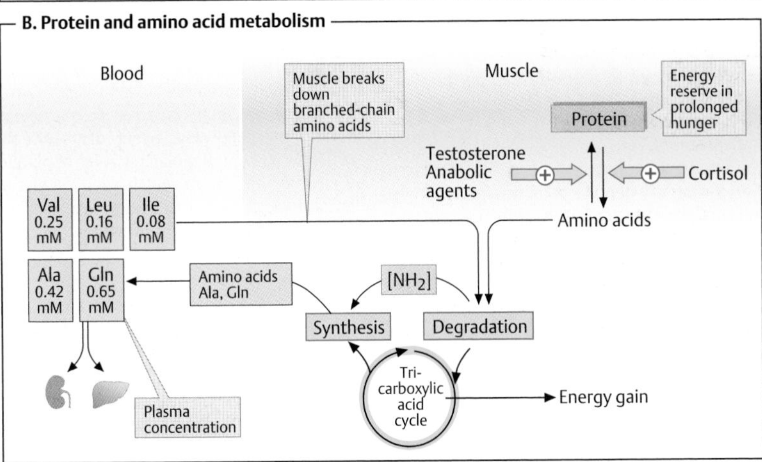

Bone and teeth

The family of connective-tissue cells includes *fibroblasts, chondrocytes* (cartilage cells), and *osteoblasts* (bone-forming cells). They are specialized to secrete extracellular proteins, particularly collagens, and mineral substances, which they use to build up the *extracellular matrix* (see p. 346). By contrast, *osteoclasts* dissolve bone matter again by secreting H^+ and collagenases (see p. 342).

A. Bone ◑

Bone is an extremely dense, specialized form of connective tissue. In addition to its supportive function, it serves to store calcium and phosphate ions. In addition, blood cells are formed in the bone marrow. The most important mineral component of bone is **apatite**, a form of crystalline *calcium phosphate.*

Apatites are complexes of cationic Ca^{2+} matched by HPO_4^{2-}, CO_3^{2-}, OH^-, or F^- as anions. Depending on the counter-ion, apatite can occur in the forms *carbonate apatite* $Ca_{10}(PO_4)_6CO_3$, as *hydroxyapatite* $Ca_{10}(PO_4)_6(OH)_2$, or *fluoroapatite* $Ca_{10}(PO_4)_6F_2$. In addition, alkaline earth carbonates also occur in bone. In adults, more than 1 kg calcium is stored in bone.

Osteoblast and *osteoclast* activity is constantly incorporating Ca^{2+} into bone and removing it again. There are various hormones that regulate these processes: *calcitonin* increases deposition of Ca^{2+} in the bone matrix, while *parathyroid hormone* (PTH) promotes the mobilization of Ca^{2+}, and *calcitriol* improves mineralization (for details, see p. 342).

The most important *organic components* of bone are **collagens** (mainly type I; see p. 344) and **proteoglycans** (see p. 346). These form the extracellular matrix into which the apatite crystals are deposited (*biomineralization*). Various proteins are involved in this not yet fully understood process of bone formation, including collagens and phosphatases. *Alkaline phosphatase* is found in osteoblasts and *acid phosphatase* in osteoclasts. Both of these enzymes serve as *marker enzymes* for bone cells.

B. Teeth ◑

The illustration shows a longitudinal section through an incisor, one of the 32 permanent teeth in humans. The majority of the tooth consists of **dentine**. The crown of the tooth extends beyond the gums, and it is covered in **enamel**. By contrast, the root of the tooth is coated in dental **cement**.

Cement, dentin, and enamel are bone-like substances. The high proportion of inorganic matter they contain (about 97% in the dental enamel) gives them their characteristic hardness. The organic components of cement, dentin, and enamel mainly consist of *collagens and proteoglycans;* their most important mineral component is *apatite,* as in bone (see above).

A widespread form of dental disease, **caries**, is caused by acids that dissolve the mineral part of the teeth by neutralizing the negatively charged counter-ions in apatite (see **A**). Acids occur in food, or are produced by microorganisms that live on the surfaces of the teeth (e.g., *Streptococcus mutans*).

The main product of anaerobic degradation of sugars by these organisms is lactic acid. Other products of bacterial carbohydrate metabolism include extracellular dextrans (see p. 40)—insoluble polymers of glucose that help bacteria to protect themselves from their environment. Bacteria and dextrans are components of *dental plaque,* which forms on inadequately cleaned teeth. When Ca^{2+} salts and other minerals are deposited in plaque as well, *tartar* is formed.

The most important form of protection against caries involves avoiding sweet substances (foods containing saccharose, glucose, and fructose). Small children in particular should not have very sweet drinks freely available to them. Regular removal of plaque by cleaning the teeth and hardening of the dental enamel by fluoridization are also important. Fluoride has a protective effect because fluoroapatite (see **A**) is particularly resistant to acids.

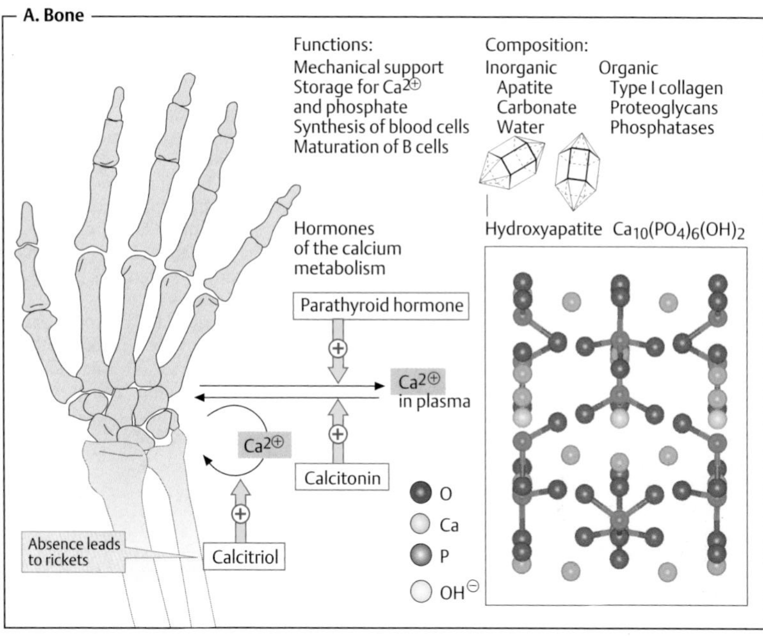
A. Bone
Functions:
Mechanical support
Storage for Ca^{2+} and phosphate
Synthesis of blood cells
Maturation of B cells
Composition:
Inorganic
Apatite
Carbonate
Water
Organic
Type I collagen
Proteoglycans
Phosphatases
Hydroxyapatite $Ca_{10}(PO_4)_6(OH)_2$
Hormones of the calcium metabolism
Parathyroid hormone
Ca^{2+} in plasma
Ca^{2+}
Calcitonin
Calcitriol
Absence leads to rickets
O
Ca
P
OH^{-}

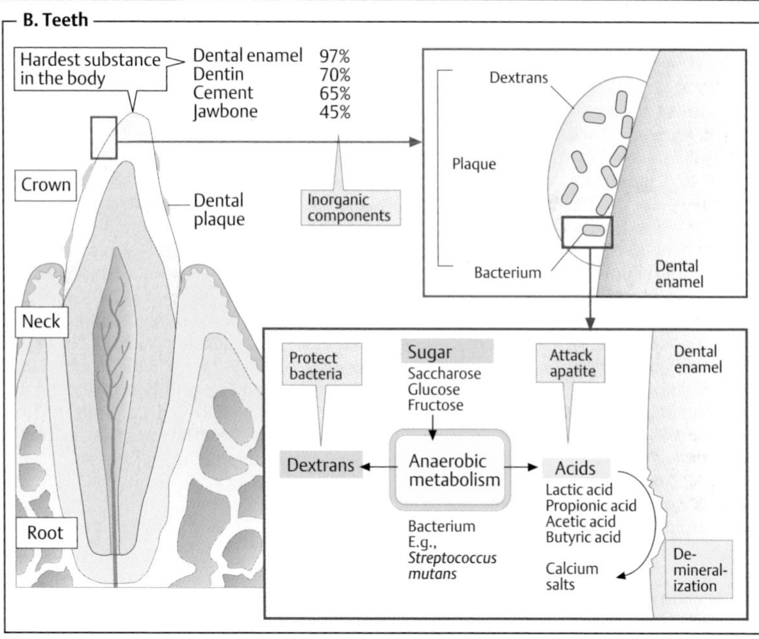
B. Teeth
Hardest substance in the body
Dental enamel 97%
Dentin 70%
Cement 65%
Jawbone 45%
Crown
Dental plaque
Inorganic components
Neck
Root
Dextrans
Plaque
Bacterium
Dental enamel
Protect bacteria
Sugar
Saccharose
Glucose
Fructose
Attack apatite
Dental enamel
Dextrans
Anaerobic metabolism
Acids
Lactic acid
Propionic acid
Acetic acid
Butyric acid
Bacterium
E.g., *Streptococcus mutans*
Calcium salts
De-mineral-ization

Calcium metabolism

A. Functions of calcium ●

The human body contains 1–1.5 kg Ca^{2+}, most of which (about 98%) is located in the mineral substance of bone (see p. 362).

In addition to its role as a **bone component**, calcium functions as a **signaling substance**. Ca^{2+} ions act as *second messengers* in signal transduction pathways (see p. 386), they trigger *exocytosis* (see p. 228) and *muscle contraction* (see p. 334), and they are indispensable as cofactors in *blood coagulation* (see p. 290). Many *enzymes* also require Ca^{2+} for their activity. The intracellular and extracellular concentrations of Ca^{2+} are strictly regulated in order to make these functions possible (see **B**, **C**, and p. 388).

Proteins bind Ca^{2+} via oxygen ligands, particularly carboxylate groups and carbonyl groups of peptide bonds. This also applies to the structure illustrated here, in which a Ca^{2+} ion is coordinated by the oxygen atoms of carboxylate and acid amide groups.

B. Bone remodeling ◑

Deposition of Ca^{2+} in bone (*mineralization*) and Ca^{2+} mobilization from bone are regulated by at least 15 hormones and hormone-like signaling substances. These mainly influence the maturation and activity of bone cells.

Osteoblasts (top) deposit collagen, as well as Ca^{2+} and phosphate, and thereby create new bone matter, while **osteoclasts** (bottom) secrete H^+ ions and collagenases that locally dissolve bone (*bone remodeling*). Osteoblasts and osteoclasts mutually activate each other by releasing **cytokines** (see p. 392) and **growth factors**. This helps keep bone formation and bone breakdown in balance.

The Ca^{2+}-selective hormones calcitriol, parathyroid hormone, and calcitonin influence this interaction in the bone cells. **Parathyroid hormone** promotes Ca^{2+} release by promoting the release of cytokines by osteoblasts. In turn, the cytokines stimulate the development of mature osteoclasts from precursor cells (bottom). **Calcitonin** inhibits this process. At the same time, it promotes the development of osteoblasts (top). *Osteoporosis*, which mainly occurs in women following the menopause, is based (at least in part) on a reduction in **estrogen** levels. Estrogens normally inhibit the stimulation of osteoclast differentiation by osteoblasts. If the effects of estrogen decline, the osteoclasts predominate and excess bone removal occurs.

The effects of the steroid hormone **calcitriol** (see p. 330) in bone are complex. On the one hand, it promotes bone formation by stimulating osteoblast differentiation (top). This is particularly important in small children, in whom calcitriol deficiency can lead to mineralization disturbances (*rickets;* see p. 364). On the other hand, calcitriol increases blood Ca^{2+} levels through increased Ca^{2+} mobilization from bone. An overdose of vitamin D (cholecalciferol), the precursor of calcitriol, can therefore have unfavorable effects on the skeleton similar to those of vitamin deficiency (*hypervitaminosis;* see p. 364).

C. Calcium homeostasis ◑

Ca^{2+} metabolism is balanced in healthy adults. Approximately 1 g Ca^{2+} is taken up per day, about 300 mg of which is resorbed. The same amount is also excreted again. The amounts of Ca^{2+} released from bone and deposited in it per day are much smaller. Milk and milk products, especially cheese, are particularly rich in calcium.

Calcitriol and parathyroid hormone, on the one hand, and calcitonin on the other, ensure a more or less constant level of Ca^{2+} in the blood plasma and in the extracellular space (80–110 mg · 2.0–2.6 mM). The peptide **parathyroid hormone** (PTH; 84 AA) and the steroid **calcitriol** (see p. 374) promote direct or indirect processes that raise the Ca^{2+} level in blood. Calcitriol increases Ca^{2+} resorption in the intestines and kidneys by inducing transporters. Parathyroid hormone supports these processes by stimulating calcitriol biosynthesis in the kidneys (see p. 330). In addition, it directly promotes resorption of Ca^{2+} in the kidneys (see p. 328) and Ca^{2+} release from bone (see **B**). The PTH antagonist **calcitonin** (32 AA) counteracts these processes.

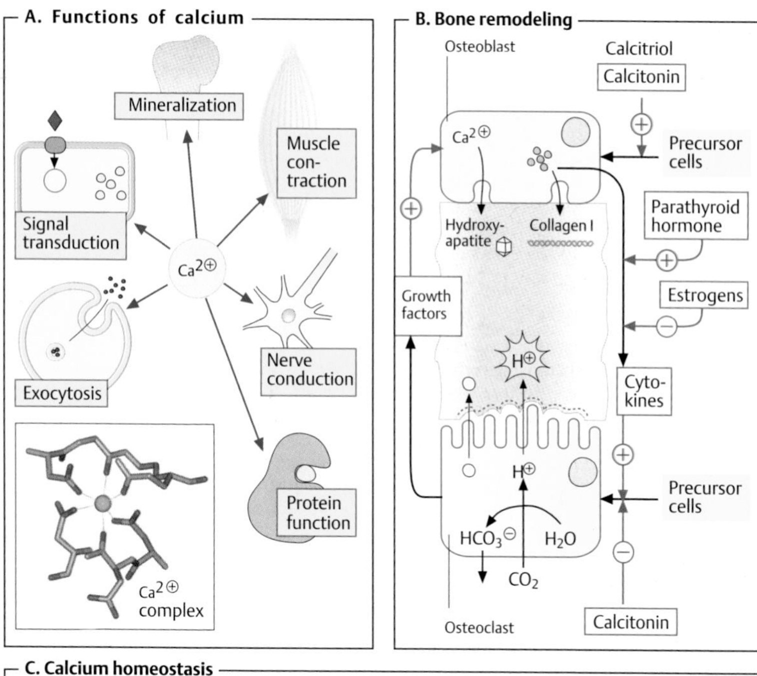
A. Functions of calcium
Mineralization
Muscle con-traction
Signal transduction
Ca2⊕
Exocytosis
Nerve conduction
Protein function
Ca2⊕ complex
B. Bone remodeling
Osteoblast
Calcitriol
Calcitonin
Ca2⊕
Precursor cells
Hydroxy-apatite
Collagen I
Parathyroid hormone
Growth factors
Estrogens
H⊕
Cyto-kines
H⊕
Precursor cells
HCO3⊖
H2O
CO2
Osteoclast
Calcitonin

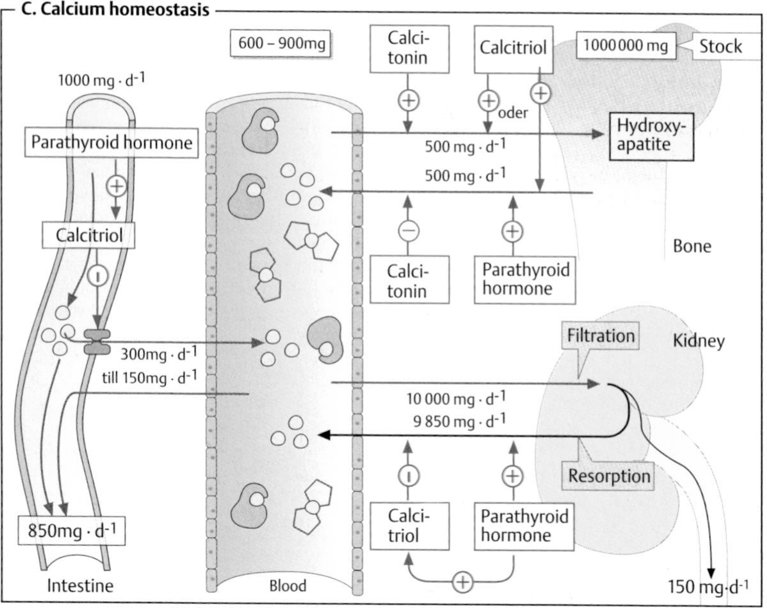
C. Calcium homeostasis
600 – 900mg
Calci-tonin
Calcitriol
1000000 mg
Stock
1000 mg · d-1
Parathyroid hormone
oder
Hydroxy-apatite
500 mg · d-1
500 mg · d-1
Calcitriol
Bone
Calci-tonin
Parathyroid hormone
Filtration
Kidney
300mg · d-1
till 150mg · d-1
10 000 mg · d-1
9 850 mg · d-1
Resorption
850mg · d-1
Calci-triol
Parathyroid hormone
Intestine
Blood
150 mg·d-1

Collagens

Collagens are quantitatively the most abundant of animal proteins, representing 25% of the total. They form insoluble tensile fibers that occur as structural elements of the extracellular matrix and connective tissue throughout the body. Their name (which literally means "glue-producers") is derived from the gelatins that appear as a decomposition product when collagen is boiled.

A. Structure of collagens ◑

Nineteen different collagens are now known, and they are distinguished using roman numerals. They mostly consist of a dextrorotatory **triple helix** made up of three polypeptides (α-chains) (see p. 70).

The triplet **Gly-X-Y** is constantly repeated in the sequence of the triple-helical regions—i. e., every third amino acid in such sequences is a *glycine. Proline* (Pro) is frequently found in positions X or Y; the Y position is often occupied by *4-hydroxyproline* (4Hyp), although *3-hydroxyproline* (3Hyp) and *5-hydroxylysine* (5Hyl) also occur. These hydroxylated amino acids are characteristic components of collagen. They are only produced after protein biosynthesis by **hydroxylation** of the amino acids in the peptide chain (see p. 62).

The formation of Hyp and Hyl residues in procollagen is catalyzed by iron-containing *oxygenases* ("proline and lysine hydroxylase," *EC 1.14.11.1/2*). *Ascorbate* is required to maintain their function. Most of the symptoms of the vitamin C deficiency disease *scurvy* (see p. 368) are explained by disturbed collagen biosynthesis.

The hydroxyproline residues stabilize the triple helix by forming hydrogen bonds between the α-chains, while the hydroxyl groups of hydroxylysine are partly **glycosylated** with a disaccharide (–Glc–Gal).

The various types of collagen consist of different combinations of α-chains (α1 to α3 and other subtypes). Types I, II, and III represent 90% of collagens. The **type I** collagen shown here has the structure $[\alpha 1(I)]_2\alpha 2(1)$.

Numerous **tropocollagen** molecules (mass 285 kDa, length 400 nm) aggregate extracellularly into a defined arrangement, forming cylindrical **fibrils** (20–500 nm in diameter). Under the electron microscope, these fibrils are seen to have a characteristic banding pattern of elements that are repeated every 64–67 nm.

Tropocollagen molecules are firmly linked together, particularly at their ends, by covalent networks of altered lysine side chains. The number of these links increases with age. **Type IV** collagens form networks with a defined mesh size. The size-selective filtering effect of the basal membranes in the renal glomeruli is based on this type of structure (see p. 322).

B. Biosynthesis ◑

The precursor molecule of collagen (*preprocollagen*), formed in the rER, is subject to extensive **post-translational modifications** (see p. 232) in the ER and Golgi apparatus.

Cleavage of the signal peptide gives rise to **procollagen**, which still carries large *propeptides* at each end [1]. During this phase, most proline residues and some lysine residues of procollagen are hydroxylated [2]. The procollagen is then glycosylated at hydroxylysine residues [3]. Intramolecular and intermolecular disulfide bonds form in the propeptides [4], allowing correct positioning of the peptide strands to form a triple helix [5]. It is only after these steps have been completed that procollagen is secreted into the extracellular space by exocytosis. This is where the *N*- and *C*-terminal propeptides are removed proteolytically [6], allowing the staggered aggregation of the **tropocollagen** molecules to form fibrils [7]. Finally, several ε-amino groups in lysine residues are oxidatively converted into aldehyde groups [8]. Covalent links between the molecules then form as a result of condensation [9]. In this way, the fibrils reach their final structure, which is characterized by its high *tensile strength* and *proteinase resistance.*

A. Structure of collagens

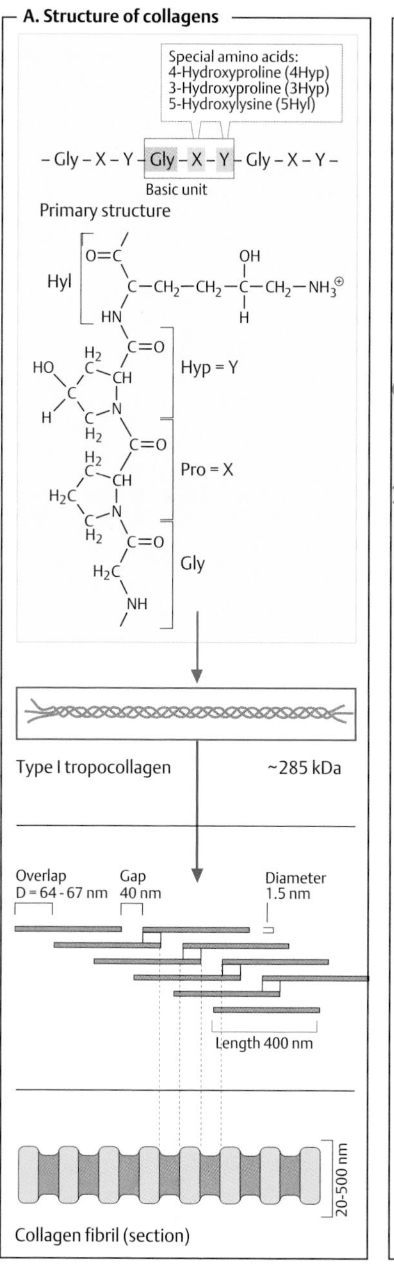

B. Biosynthesis

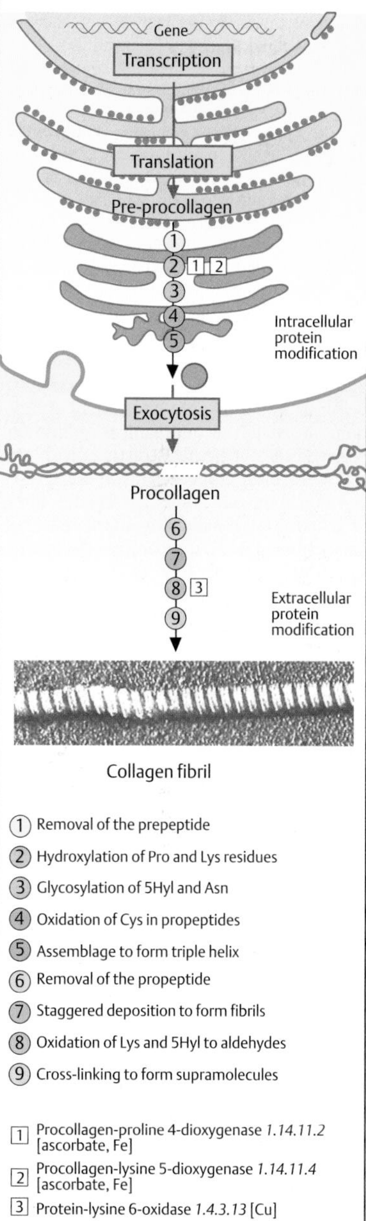

Extracellular matrix

A. Extracellular matrix ◑

The space between the cells (the interstitium) is occupied by a substance with a complex composition known as the **extracellular matrix** (ECM). In many types of tissue—e. g., muscle and liver—the ECM is only a narrow border between the cells, while in others it forms a larger space. In *connective tissue, cartilage,* and *bone,* the ECM is particularly strongly marked and is actually the functional part of the tissue (see p. 340). The illustration shows the three main constituents of the extracellular matrix in a highly schematic way: collagen fibers, network-forming adhesive proteins, and space-filling proteoglycans.

The ECM has a very wide variety of functions: it establishes mechanical connections between cells; it creates structures with special mechanical properties (as in bone, cartilage, tendons, and joints); it creates filters (e. g., in the basal membrane in the renal corpuscles; see p. 322); it separates cells and tissues from each other (e. g., to allow the joints to move freely); and it provides pathways to guide migratory cells (important for embryonic development). The chemical composition of the ECM is just as diverse as its functions.

Collagens (see p. 344), of which there are at least 19 different varieties, form fibers, fibrils, networks, and ligaments. Their characteristic properties are *tensile strength* and *flexibility.* **Elastin** is a fiber protein with a high degree of elasticity.

Adhesive proteins provide the connections between the various components of the extracellular matrix. Important representatives include **laminin** and **fibronectin** (see **B**). These multifunctional proteins simultaneously bind to several other types of matrix component. Cells attach to the cell surface receptors in the ECM with the help of the adhesive proteins.

Due to their polarity and negative charge, **proteoglycans** (see **C**) bind water molecules and cations. As a homogeneous "cement," they fill the gaps between the ECM fibers.

B. Fibronectins ○

Fibronectins are typical representatives of adhesive proteins. They are filamentous dimers consisting of two related peptide chains (each with a mass of 250 kDa) linked to each other by disulfide bonds. The fibronectin molecules are divided into different *domains,* which bind to cell-surface receptors, collagens, fibrin, and various proteoglycans. This is what gives fibronectins their *"molecular glue"* characteristics.

The domain structure in fibronectins is made up of a few types of *peptide module* that are repeated numerous times. Each of the more than 50 modules is coded for by one exon in the fibronectin gene. *Alternative splicing* (see p. 246) of the hnRNA transcript of the fibronectin gene leads to fibronectins with different compositions. The module that causes adhesion to cells contains the characteristic amino acid sequence –Arg–Gly–Asp–Ser–. It is these residues that enable fibronectin to bind to cell-surface receptors, known as **integrins**.

C. Proteoglycans ◑

Proteoglycans are giant molecule complexes consisting of carbohydrates (95%) and proteins (5%), with masses of up to $2 \cdot 10^6$ Da. Their bottlebrush-shaped structure is produced by an axis consisting of **hyaluronate**. This thread-like polysaccharide (see p. 44) has **proteins** attached to it, from which in turn long polysaccharide chains emerge. Like the central hyaluronate, these terminal polysaccharides belong to the glycosaminoglycan group (see p. 44).

The **glycosaminoglycans** are made up of repeating disaccharide units, each of which consists of one *uronic acid* (glucuronic acid or iduronic acid) and one *amino sugar* (*N*-acetylglucosamine or *N*-acetylgalactosamine) (see p. 38). Many of the amino sugars are also esterified with sulfuric acid (sulfated), further increasing their polarity. The proteoglycans bind large amounts of water and fill the gaps between the fibrillar components of the ECM in the form of a hydrated gel. This inhibits the spread of pathogens in the ECM, for example.

A. Extracellular matrix

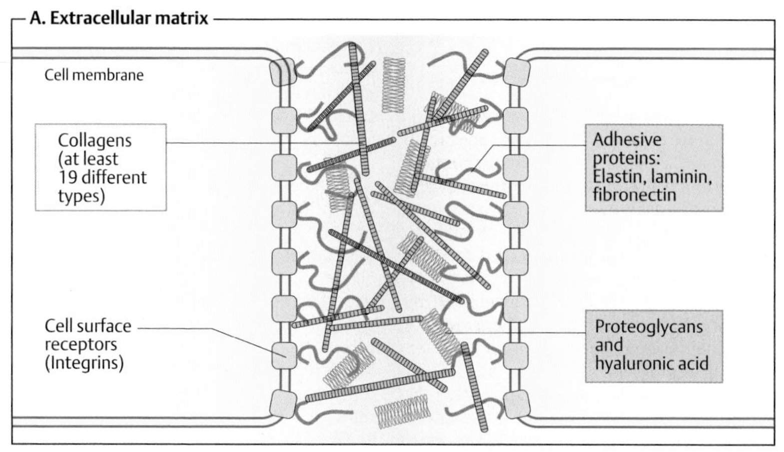

B. Fibronectins

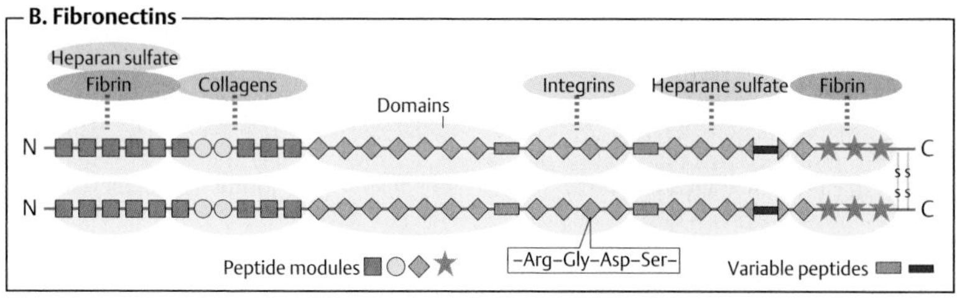

C. Proteoglycans

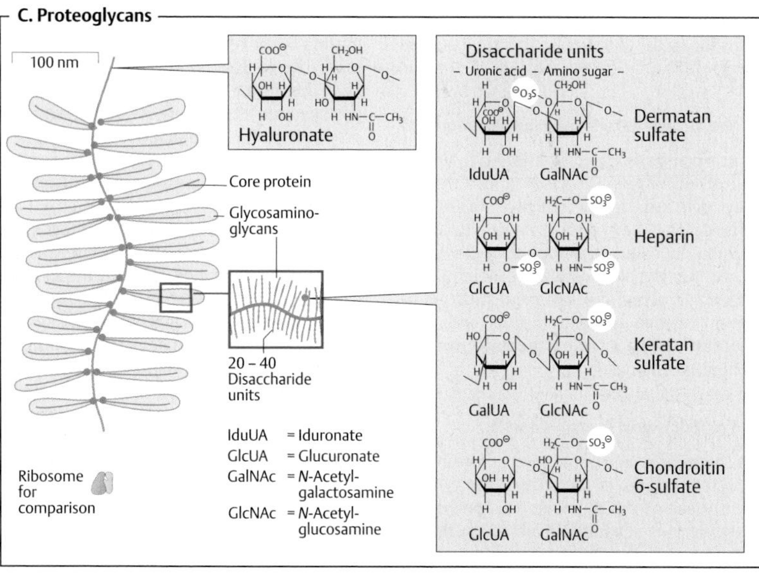

Signal transmission in the CNS

A. Structure of nerve cells ●

Nerve cells (neurons) are easily excitable cells that produce electrical signals and can react to such signals as well. Their structure is markedly different from that of other types of cell. Numerous branching processes project from their **cell body** (soma). Neurons are able to receive signals via **dendrites** and to pass them on via **axons**. The axons, which can be up to 1 m long, are usually surrounded by *Schwann cells*, which cover them with a lipid-rich myelin sheath to improve their electrical insulation.

The transfer of stimuli occurs at the **synapses,** which link the individual neurons to each other as well as linking neurons functionally to muscle fibers. *Neurotransmitters* (see p. 352) are stored in the axonal nerve endings. These signaling substances are released in response to electrical signals in order to excite neighboring neurons (or muscle cells). It is estimated that each neuron in the brain is in contact via synapses with approximately 10 000 other neurons.

There is a noticeably high proportion of lipids in the composition of nerve cells, representing about 50% of their dry weight. In particular, there is a very wide variety of phospholipids, glycolipids, and sphingolipids (see p. 216).

B. Neurotransmitters and neurohormones ◑

Neurosecretions are classed into two groups: **neurotransmitters** are released into the *synaptic cleft* in order to influence neighboring cells (**C**). They have a short range and a short lifespan. By contrast, **neurohormones** are released into the blood, allowing them to cover larger distances. However, the distinction between the two groups is a fluid one; some neurotransmitters simultaneously function as neurohormones.

C. Synaptic signal transmission ◑

All chemical synapses function according to a similar principle. In the area of the synapse, the surface of the signaling cell (*presynaptic membrane*) is separated from the surface of the receiving cell (*postsynaptic membrane*) only by a narrow *synaptic cleft*. When an **action potential** (see p. 350) reaches the presynaptic membrane, *voltage-gated Ca^{2+} channels* integrated into the membrane open and trigger **exocytosis** of the neurotransmitter stored in the presynaptic cell (for details, see p. 228).

Each neuron usually releases *only one type* of **neurotransmitter**. Neurons that release dopamine are referred to as "dopaminergic," for example, while those that release acetylcholine are "cholinergic," etc. The transmitters that are released diffuse through the synaptic cleft and bind on the other side to **receptors** on the postsynaptic membrane. These receptors are integral membrane proteins that have binding sites for neurotransmitters on their exterior (see p. 224).

The receptors for neurotransmitters are divided into two large groups according to the effect produced by binding of the transmitter (for details, see p. 354).

Ionotropic receptors (bottom left) are *ligand-gated ion channels*. When they open as a result of the transmitter's influence, ions flow in due to the membrane potential (see p. 126). If the inflowing ions are cations (Na^+, K^+, Ca^{2+}), **depolarization** of the membrane occurs and an action potential is triggered on the surface of the postsynaptic cell. This is the way in which stimulatory transmitters work (e.g., acetylcholine and glutamate). By contrast, if anions flow in (mainly Cl^-), the result is **hyperpolarization** of the postsynaptic membrane, which makes the production of a postsynaptic action potential more difficult. The action of inhibitory transmitters such as glycine and GABA is based on this effect.

A completely different type of effect is observed in **metabotropic receptors** (bottom right). After binding of the transmitter, these interact on the inside of the postsynaptic membrane with *G proteins* (see p. 384), which in turn activate or inhibit the synthesis of **second messengers**. Finally, second messengers activate or inhibit **protein kinases**, which phosphorylate cellular proteins and thereby alter the behavior of the postsynaptic cells (*signal transduction;* see p. 386).

A. Structure of nerve cells

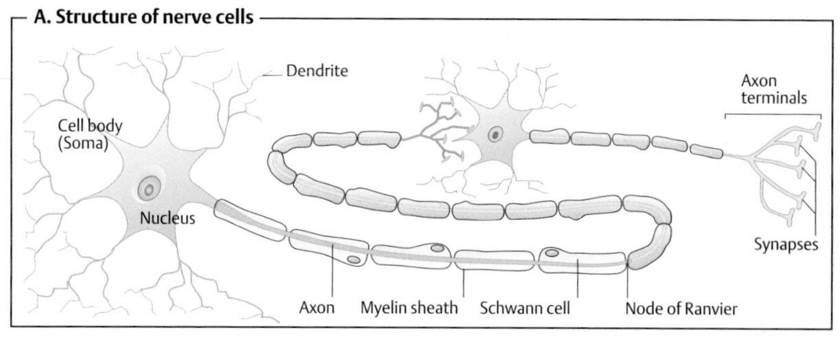

A. Neurotransmitters and neurohormones

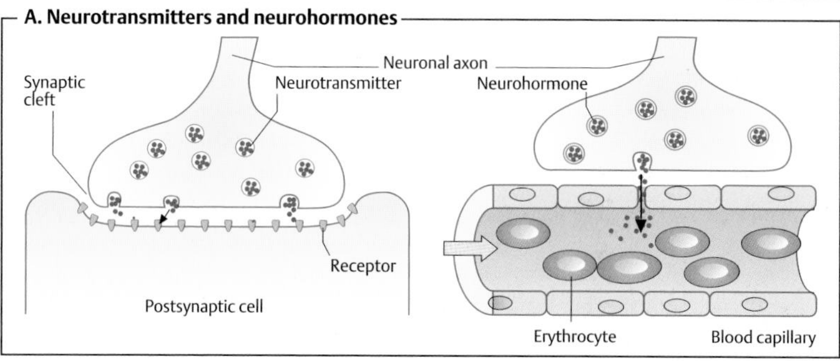

C. Synaptic signal transmission

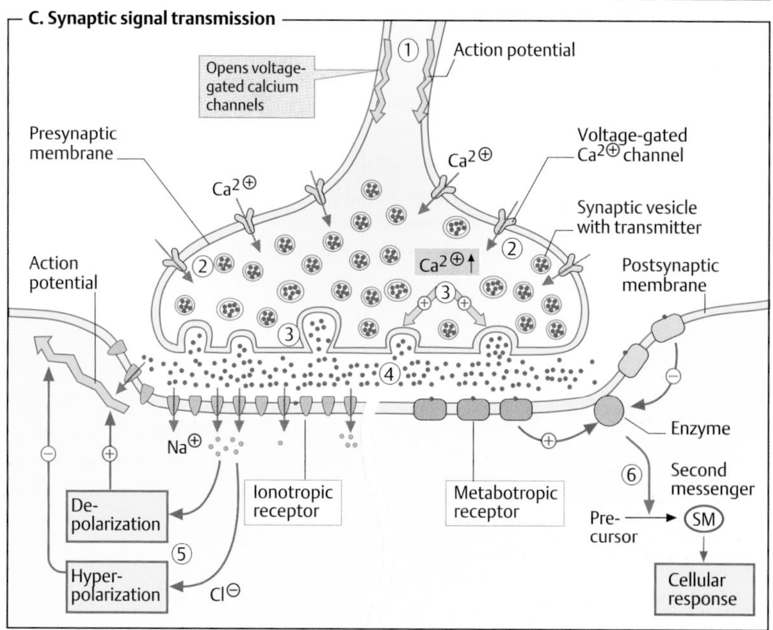

Resting potential and action potential

A. Resting potential

A characteristic property of living cells is the uneven distribution of positively and negatively charged ions on the inside and outside of the plasma membrane. This gives rise to a **membrane potential** (see p. 126)—i. e., there is electrical voltage between the two sides of the membrane, which can only balance out when *ion channels* allow the unevenly distributed ions to move.

At rest, the membrane potential in most cells is –60 to –90 mV. It mainly arises from the activity of *Na^+/K^+ transporting ATPase* ("Na^+/K^+ ATPase"), which occurs on practically all animal cells. Using up ATP, this P-type enzyme (see p. 220) "pumps" three Na^+ ions out of the cell in exchange for two K^+ ions. Some of the K^+ ions, following the concentration gradient, leave the cell again through *potassium channels.* As the protein anions that predominate inside the cell cannot follow them, and inflow of Cl^- ions from the outside is not possible, the result is an excess of positive charges outside the cell, while anions predominate inside it.

An **equilibrium potential** exists for each of the ions involved. This is the value of the membrane potential at which there is no net inflow or outflow of the ions concerned. For K^+ ions, the resting potential lies in the range of the membrane potential, while for Na^+ ions it is much higher at +70 mV. At the first opportunity, Na^+ ions will therefore spontaneously flow into the cell. The occurrence of action potentials is based on this (see **B**).

Nerve cell membranes contain **ion channels** for Na^+, K^+, Cl^-, and Ca^{2+}. These channels are usually closed and open only briefly to let ions pass through. They can be divided into channels that are regulated by membrane potentials (*"voltage-gated"*—e. g., fast Na^+ channels; see p. 222) and those regulated by ligands *("ligand-gated"*—e. g., nicotinic acetylcholine receptors; see p. 222).

B. Action potential

Action potentials are special signals that are used to transmit information in the nervous system. They are triggered by chemical stimuli (or more rarely electrical stimuli). Binding of a neurotransmitter to an ionotropic receptor results in a brief local increase in the membrane potential from –60 mV to about +30 mV. Although the membrane potential quickly returns to the initial value within a few milliseconds (ms) at its site of origin, the depolarization is propagated because neighboring membrane areas are activated during this time period.

[1] The process starts with the opening of voltage-gated Na^+ channels (see p. 222). Due to their high equilibrium potential (see **A**), Na^+ ions flow into the cell and reverse the local membrane potential (**depolarization**).

[2] The Na^+ channels immediately close again, so that the inflow of positive charges is only very brief.

[3] Due to the increase in the membrane potential, voltage-dependent K^+ channels open and K^+ ions flow out. In addition, Na^+/K^+ ATPase (see **A**) pumps the Na+ ions that have entered back out again. This leads to **repolarization** of the membrane.

[4] The two processes briefly lead to the charge even falling below the resting potential (**hyperpolarization**). The K^+ channels also close after a few milliseconds. The nerve cell is then ready for re-stimulation.

Generally, it is always only a very small part of the membrane that is depolarized during an action potential. The process can therefore be repeated again after a short refractory period, when the nerve cell is stimulated again. Conduction of the action potential on the surface of the nerve cell is based on the fact that the local increase in the membrane potential causes neighboring voltage-gated ion channels to open, so that the membrane stimulation spreads over the whole cell in the form of a *depolarization wave.*

A. Resting potential

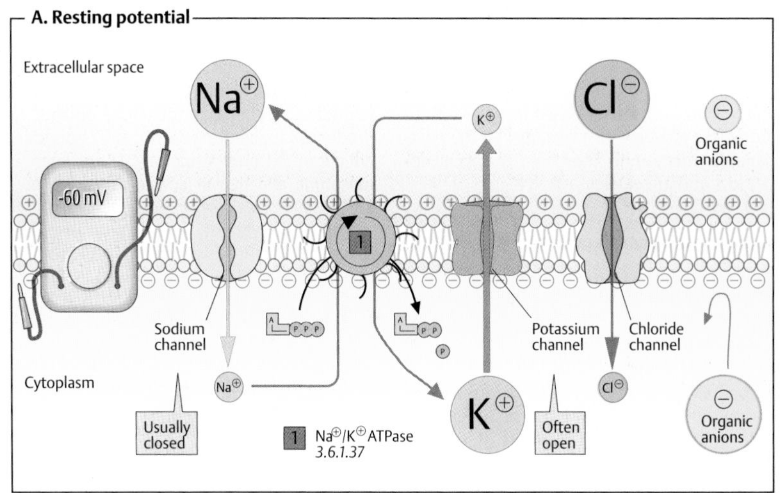

B. Action potential

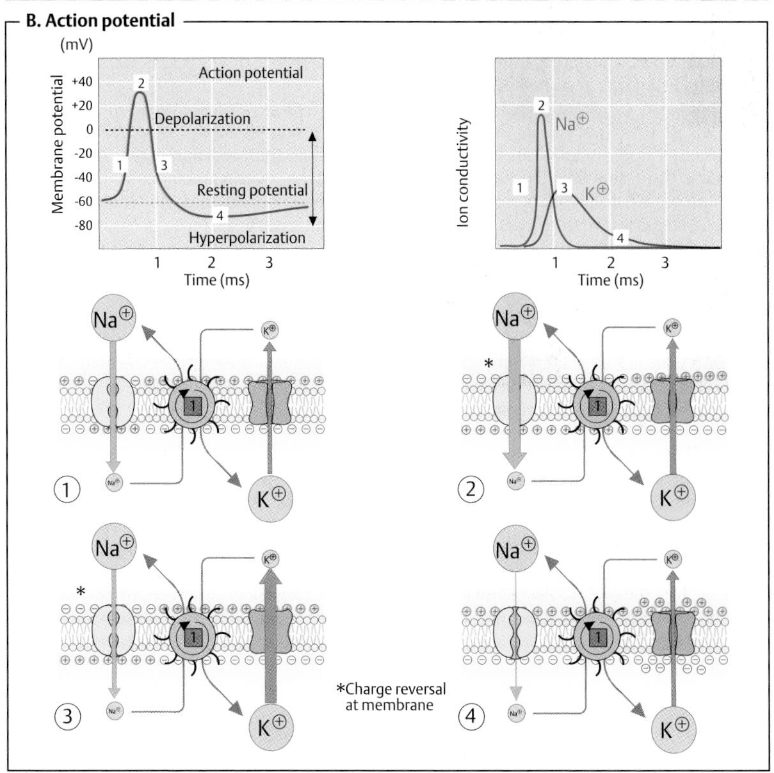

Neurotransmitters

Neurotransmitters in the strict sense are substances that are produced by neurons, stored in the synapses, and released into the synaptic cleft in response to a stimulus. At the postsynaptic membrane, they bind to special receptors and affect their activity.

A. Important neurotransmitters ◑

Neurotransmitters can be classified into several groups according to their chemical structure. The table lists the most important representatives of this family, which has more than 100 members.

Acetylcholine, the acetic acid ester of the cationic alcohol choline (see p. 50) acts at neuromuscular junctions, where it triggers muscle contraction (see p. 334), and in certain parts of the brain and in the autonomous nervous system.

Several proteinogenic **amino acids** (see p. 60) have neurotransmitter effects. A particularly important one is *glutamate*, which acts as a stimulatory transmitter in the CNS. More than half of the synapses in the brain are glutaminergic. The metabolism of glutamate and that of the amine GABA synthesized from it (see below) are discussed in more detail on p. 356. *Glycine* is an inhibitory neurotransmitter with effects in the spinal cord and in parts of the brain.

Biogenic amines arise from amino acids by decarboxylation (see p. 62). This group includes *4-aminobutyrate* (γ-aminobutyric acid, GABA), which is formed from glutamate and is the most important inhibitory transmitter in the CNS. The *catecholamines* norepinephrine and epinephrine (see **B**), *serotonin*, which is derived from tryptophan, and *histamine* also belong to the biogenic amine group. All of them additionally act as hormones or mediators (see p. 380).

Peptides make up the largest group among the neurosecretions. Many peptide hormones—e. g., thyroliberin (TRH) and angiotensin II—simultaneously act as transmitters. Most neuropeptides are small (3–15 AA). At their N-terminus, many of them have a glutamate residue that has been cyclized to form *pyroglutamate* (5-oxoproline, <G), while the C-terminus is often an acid amide ($-NH_2$). This provides better protection against breakdown by peptidases.

Endorphins, dynorphins, and *enkephalins* are a particularly interesting group of neuropeptides. They act as "endogenous opiates" by producing analgetic, sedative, and euphoriant effects in extreme situations. Drugs such as morphine and heroin activate the receptors for these peptides (see p. 354).

Purine derivatives with neurotransmitter function are all derived from adenine-containing nucleotides or nucleosides. ATP is released along with acetylcholine and other transmitters, and among other functions it regulates the emission of transmitters from its synapse of origin. The stimulatory effect of *caffeine* is mainly based on the fact that it binds to adenosine receptors.

B. Biosynthesis of catecholamines ◑

The catecholamines are *biogenic amines* that have a catechol group. Their biosynthesis in the adrenal cortex and CNS starts from **tyrosine**.

[1] Hydroxylation of the aromatic ring initially produces **dopa** (3,4-dihydroxyphenylalanine). This reaction uses the unusual coenzyme *tetrahydrobiopterin* (THB). Dopa (cf. p. 6) is also used in the treatment of Parkinson's disease.

[2] Decarboxylation of dopa yields **dopamine**, an important transmitter in the CNS. In dopaminergic neurons, catecholamine synthesis stops at this point.

[3] The adrenal gland and adrenergic neurons continue the synthesis by hydroxylating dopamine into **norepinephrine** (noradrenaline). *Ascorbic acid* (vitamin C; see p. 368) acts as a hydrogen-transferring coenzyme here.

[4] Finally, *N*-methylation of norepinephrine yields **epinephrine** (adrenaline). The coenzyme for this reaction is *S*-adenosylmethionine (SAM; see p. 110).

The physiological effects of the catecholamines are mediated by a large number of different receptors that are of particular interest in pharmacology. Norepinephrine acts in the autonomic nervous system and certain areas of the brain. Epinephrine is also used as a transmitter by some neurons.

A. Important neurotransmitters

Acetylcholine

$H_3C-C(=O)-O-CH_2-CH_2-N^{\oplus}(CH_3)_3$

Amino acids: Glutamate, Glycine, Dopa

Glutamate: $^{\oplus}H_3N-CH(COO^{\ominus})-(CH_2)_2-COO^{\ominus}$

Glycine: $^{\oplus}H_3N-CH_2-COO^{\ominus}$

Biogenic amines: γ-Aminobutyrate (GABA), Dopa, Dopamine, Norepinephrine, Epinephrine, Serotonine, Histamine

GABA: $^{\oplus}H_3N-CH_2-CH_2-CH_2-COO^{\ominus}$

Serotonin

Histamine

Peptides:

β-Endorphin	YGGFMTSEKSQTPLVTLFKNAITKNAYKKGE
Met- and Leu-enkephalin	YGGFM und YGGFL
Thyroliberin (TRH)	<GHP-NH2
Gonadoliberin (GnRH)	<GHWSYGLRPG-NH2
Substance P	RPKPQQFFGLM
Somatostatin	AGCKNFFWKTFTSC
Angiotensin II	DRVYIHPF
Cholecystokinin (CCK-4)	WMDF-NH2
and many others	

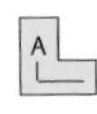

Pyroglutamate (<G) Thyroliberin

Purine derivatives: ATP, ADP, AMP, Adenosine

B. Biosynthesis of the catecholamines

1. Tyrosine 3-monooxygenase [$Fe^{2\oplus}$,THB] *1.14.16.2*
2. Aromatic-L-amino-acid decarboxylase (Dopa decarboxylase) [PLP] *4.1.1.28*
3. Dopamine β-monooxygenase[Cu] *1.14.17.1*
4. Phenylethanolamine *N*-methyltransferase *2.1.1.28*

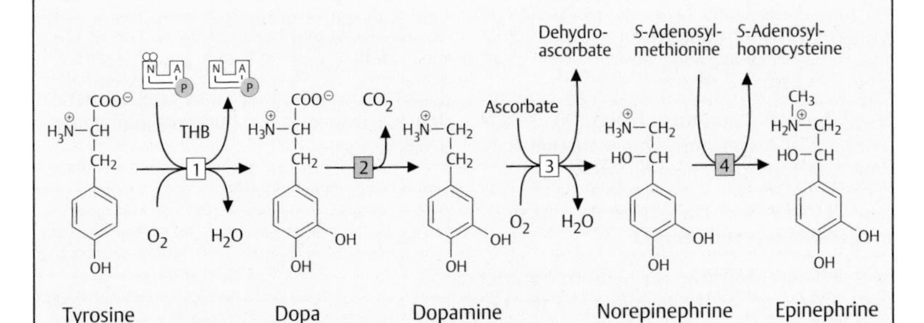

Receptors for neurotransmitters

Like all signaling substances, neurotransmitters (see p. 352) act via receptor proteins. The receptors for neurotransmitters are integrated into the membrane of the postsynaptic cell, where they trigger ion inflow or signal transduction processes (see p. 348).

A. Receptors for neurotransmitters ◑

A considerable number of receptors for neurotransmitters are already known and new ones are continuing to be discovered. The table only lists the most important examples. They are classified into two large groups according to their mode of action.

Ionotropic receptors are *ligand-gated ion channels* (left half of the table). The receptors for stimulatory transmitters (indicated in the table by a ⊕) mediate the inflow of cations (mainly Na^+). When these open after binding of the transmitter, local *depolarization* of the postsynaptic membrane occurs. By contrast, inhibitory neurotransmitters (GABA and glycine) allow Cl^- to flow in. This increases the membrane's negative resting potential and hinders the action of stimulatory transmitters (*hyperpolarization*, ⊖).

Metabotropic receptors (right half of the table) are coupled to G proteins (see p. 386), through which they influence the *synthesis of second messengers*. Receptors that work with type G_s proteins (see p. 386) increase the cAMP level in the postsynaptic cell ([cAMP] ↑), while those that activate G_i proteins reduce it ([cAMP] ↓). Via type G_q proteins, other receptors increase the intracellular Ca^{2+} concentration ($[Ca^{2+}]$ ↑).

There are several **receptor subtypes** for most neurotransmitters. These are distinguished numerically (e.g., D_1 to D_5) or are named after their agonists—i.e., after molecules experimentally found to activate the receptor. For example, one specific subtype of glutamate receptors reacts to NMDA (*N*-methyl-D-aspartate), while another subtype reacts to the compound AMPA, etc.

B. Acetylcholine receptors ◑

Acetylcholine (ACh) was the neurotransmitter first discovered, at the beginning of the last century. It binds to two types of receptor.

The **nicotinic ACh receptor** responds to the alkaloid *nicotine* contained in tobacco (many of the physiological effects of nicotine are based on this). The nicotinic receptor is ionotropic. Its properties are discussed in greater detail on p. 222.

The **muscarinic ACh receptors** (of which there are at least five subtypes) are metabotropic. Their name is derived from the alkaloid *muscarine*, which is found in the fly agaric mushroom (*Amanita muscaria*), for example. Like ACh, muscarine is bound at the receptor, but in contrast to ACh (see **C**), it is not broken down and therefore causes permanent stimulation of muscle.

The muscarinic ACh receptors influence the cAMP level in the postsynaptic cells (M_1, M_3 and M_5 increase it, while subtypes M_2 and M_4 reduce it).

C. Metabolism of acetylcholine ◑

Acetylcholine is synthesized from acetyl-CoA and choline in the cytoplasm of the presynaptic axon [1] and is stored in **synaptic vesicles**, each of which contains around 1000–10 000 ACh molecules. After it is released by exocytosis (see p. 228), the transmitter travels by diffusion to the receptors on the postsynaptic membrane. Catalyzed by *acetylcholinesterase*, hydrolysis of ACh to acetate and choline immediately starts in the synaptic cleft [2], and within a few milliseconds, the ACh released has been eliminated again. The cleavage products **choline** and **acetate** are taken up again by the presynaptic neuron and reused for acetylcholine synthesis [3].

Substances that block the serine residue in the active center of acetylcholinesterase [2]—e.g., the neurotoxin E605 and other *organophosphates*—prevent ACh degradation and thus cause prolonged stimulation of the postsynaptic cell. This impairs nerve conduction and muscle contraction. *Curare*, a paralyzing arrow-poison used by South American Indians, competitively inhibits binding of ACh to its receptor.

A. Receptors for neurotransmitters

Ionotropic

Receptor	Transmitter	Ion(s)	Effect
Acetylcholine (nicotinic)	Acetylcholine	$Na^{\oplus}$	$\oplus$
5HT3	Serotonin	$Na^{\oplus}$	$\oplus$
GABAa	GABA	$Cl^{\ominus}$	$\ominus$
Glycine	Glycine	$Cl^{\ominus}$	$\ominus$
AMPA NMDA Kainate	Glutamate Glutamate Glutamate	$Na^{\oplus}$ $K^{\oplus}$ $Na^{\oplus}$ $K^{\oplus}$ $Ca^{2\oplus}$ $Na^{\oplus}$ $K^{\oplus}$	$\oplus$ $\oplus$ $\oplus$

Metabotropic

Receptor	Transmitter	Effect
Acetylcholine (muscarinic) M1, M3, M5, M2, M4	Acetylcholine	$[Ca^{2\oplus}]\uparrow$ $[cAMP]\downarrow$
$5HT_1$ $5HT_2$ $5HT_4$	Serotonin " "	$[Ca^{2\oplus}]\uparrow$ $[cAMP]\uparrow$ $[cAMP]\downarrow$
α_1 α_2 $\beta_1, \beta_2, \beta_3$	Norepinephrine " "	$[Ca^{2\oplus}]\uparrow$ $[cAMP]\uparrow$ $[cAMP]\downarrow$
D_1, D_5 D_2, D_3, D_4	Dopamine "	$[cAMP]\uparrow$ $[cAMP]\downarrow$
δ, κ, μ	Opioids	$[cAMP]\downarrow$

B. Acetylcholine receptors

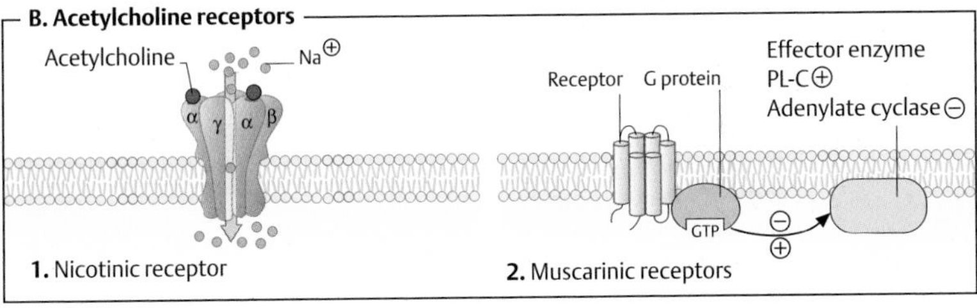

C. Metabolism of acetylcholine

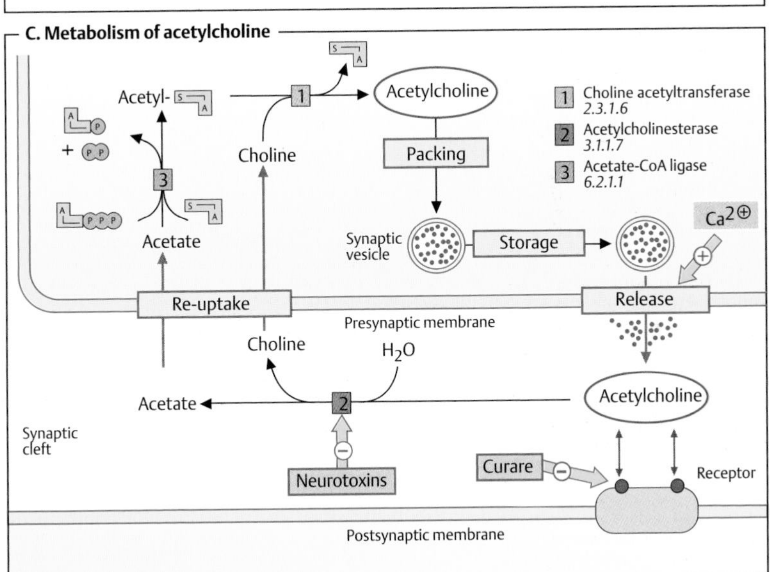

Metabolism

The brain and other areas of the central nervous system (CNS) have high ATP requirements. Although the brain only represents about 2% of the body's mass, it consumes around 20% of the metabolized oxygen and ca. 60% of the glucose. The neurons' high energy requirements are mainly due to ATP-dependent ion pumps (particularly Na^+/K^+ ATPase) and other active transport processes that are needed for nerve conduction (see p. 350).

A. Energy metabolism of the brain ●

Glucose is normally the only metabolite from which the brain is able to obtain adequate amounts of ATP through aerobic glycolysis and subsequent terminal oxidation to CO_2 and H_2O. Lipids are unable to pass the **blood–brain barrier**, and amino acids are also only available in the brain in limited quantities (see **B**). As neurons only have minor glycogen reserves, they are dependent on a constant supply of glucose from the blood. A severe drop in the blood glucose level—as can occur after insulin overdosage in diabetics, for example—rapidly leads to a drop in the ATP level in the brain. This results in loss of consciousness and neurological deficits that can lead to death. Oxygen deficiency (hypoxia) also fint affects the brain. The effects of a brief period of hypoxia are still reversible, but as time progresses irreversible damage increasingly occurs and finally complete loss of function ("brain death").

During periods of starvation, the brain after a certain time acquires the ability to use **ketone bodies** (see p. 312) in addition to glucose to form ATP. In the first weeks of a starvation period, there is a strong increase in the activities of the enzymes required for this in the brain. The degradation of ketone bodies in the CNS saves glucose and thereby reduces the breakdown of muscle protein that maintains gluconeogenesis in the liver during starvation. After a few weeks, the extent of muscle breakdown therefore declines to one-third of the initial value.

B. Glutamate, glutamine, and GABA ◑

The proteinogenic amino acid **glutamate** (Glu) and the biogenic amine **4-aminobutyrate** derived from it are among the most important neurotransmitters in the brain (see p. 352). They are both synthesized in the brain itself. In addition to the neurons, which use Glu or GABA as transmitters, *neuroglia* are also involved in the metabolism of these substances.

Since glutamate and GABA as transmitters must not appear in the extracellular space in an unregulated way, the cells of the neuroglia (center) supply "glutaminergic" and "GABAergic" neurons with the precursor **glutamine** (Gln), which they produce from glutamate with the help of *glutamine synthetase* [1].

GABA neurons (left) and glutamate neurons (right) initially hydrolyze glutamine with the help of *glutaminase* [1] to form glutamate again. The glutamate neurons store this in vesicles and release it when stimulated. The GABA neurons continue the degradation process by using *glutamate decarboxylase* [3] to convert glutamate into the transmitter GABA.

Both types of neuron take up their transmitter again. Some of it also returns to the neuroglia, where glutamate is amidated back into glutamine.

Glutamate can also be produced again from GABA. The reaction sequence needed for this, known as the **GABA shunt**, is characteristic of the CNS. A *transaminase* [4] first converts GABA and 2-oxoglutarate into glutamate and succinate semialdehyde ($^-OOC–CH_2–CH_2–CHO$). In an NAD^+-dependent reaction, the aldehyde is oxidized to succinic acid [5], from which 2-oxoglutarate can be regenerated again via tricarboxylic acid cycle reactions.

The function of glutamate as a stimulatory transmitter in the brain is the cause of what is known as the "Chinese restaurant syndrome." In sensitive individuals, the monosodium glutamate used as a flavor enhancer in Chinese cooking can raise the glutamate level in the brain to such an extent that transient mild neurological disturbances can occur (dizziness, etc.).

A. Energy metabolism of the brain

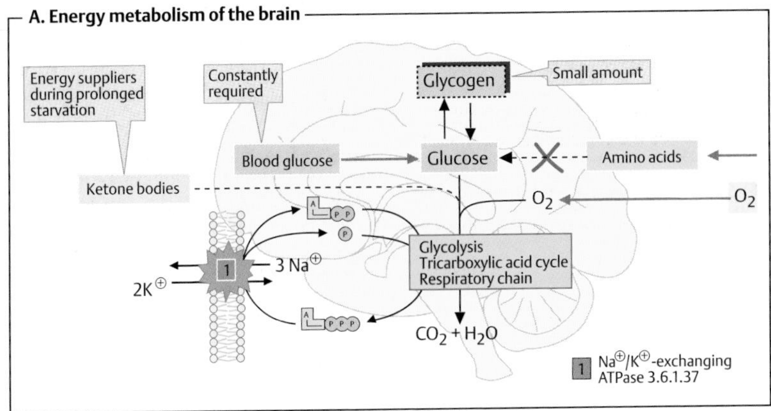

B. Glutamate, glutamine, and GABA

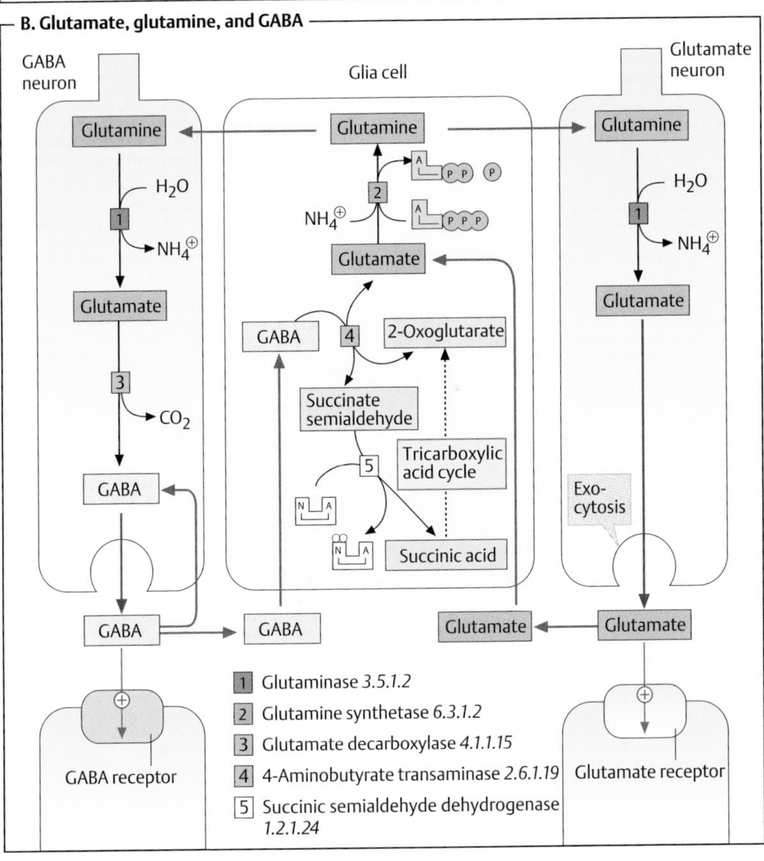

Sight

Two types of photoreceptor cell are found in the human retina—*rods* and *cones.* Rods are sensitive to low levels of light, while the cones are responsible for color vision at higher light intensities.

Signaling substances and many proteins are involved in visual processes. Initially, a **light-induced** *cis–trans* **isomerization** of the pigment retinal triggers a conformational change in the membrane protein *rhodopsin.* Via the G protein *transducin,* which is associated with rhodopsin, an enzyme is activated that breaks down the second messenger *cGMP.* Finally, the cGMP deficiency leads to *hyperpolarization* of the light-sensitive cell, which is registered by subsequent neurons as *reduced neurotransmitter release.*

A. Photoreceptor ◑

The cell illustrated opposite, a **rod**, has a structure divided by membrane discs into which the 7-helix receptor **rhodopsin** is integrated (see p. 224). In contrast to other receptors in the 7-helix class (see p. 384), rhodopsin is a light-sensitive *chromoprotein.* Its protein part, **opsin**, contains the aldehyde **retinal** (see p. 364)—an isoprenoid which is bound to the ε-amino group of a lysine residue as an *aldimine.*

The light absorption of rhodopsin is in the visible range, with a maximum at about 500 nm. The absorption properties of the visual pigment are thus optimally adjusted to the spectral distribution of sunlight.

Absorption of a photon triggers isomerization from the 11-*cis* form of retinal to all-*trans*-retinal (top right). Within milliseconds, this *photochemical process* leads to an allosteric conformational change in rhodopsin. The active conformation (**rhodopsin***) binds and activates the G protein **transducin**. The *signal cascade* (**B**) that now follows causes the rod cells to release less neurotransmitter (glutamate) at their synapses. The adjoining bipolar neurons register this change and transmit it to the brain as a signal for light.

There are several different rhodopsins in the **cones**. All of them contain retinal molecules as light-sensitive components, the absorption properties of which are modulated by the different proportions of opsin they contain in such a way that colors can also be perceived.

B. Signal cascade ◑

Dark (bottom left). Rod cells that are not exposed to light contain relatively high concentrations (70 μM) of the cyclic nucleotide **cGMP** (3′,5′-cycloGMP; cf. cAMP, p. 386), which is synthesized by a *guanylate cyclase* ([2], see p. 388). The cGMP binds to an ion channel in the rod membrane (bottom left) and thus keeps it open. The inflow of cations (Na^+, Ca^{2+}) depolarizes the membrane and leads to release of the neurotransmitter glutamate at the synapse (see p. 356).

Light (bottom right). When the G protein transducin binds to light-activated rhodopsin* (see **A**, on the structure of the complex; see p. 224), it leads to the GDP that is bound to the transducin being exchanged for GTP. In transducin* that has been activated in this way, the GTP-containing α-subunit breaks off from the rest of the molecule and in turn activates a membrane *cGMP phosphodiesterase* [1]. This hydrolyzes cGMP to GMP and thus reduces the level of free cGMP within milliseconds. As a consequence, the cGMP bound at the ion channel dissociates off and the channel closes. As cations are constantly being pumped out of the cell, the membrane potential falls and **hyperpolarization** of the cell occurs, which interrupts glutamate release.

Regeneration. After exposure to light, several processes restore the initial conditions:

1. The α-subunit of transducin* inactivates itself by GTP hydrolysis and thus terminates the activation of cGMP esterase.
2. The reduced Ca^{2+} concentration causes activation of guanylate cyclase, which increases the cGMP level until the cation channels reopen.
3. An isomerase [3] transfers all-*trans* -retinal to the 11-*cis* -form, in which it is available for the next cycle. A dehydrogenase [4] can also allow retinal to be supplied from vitamin A (retinol).

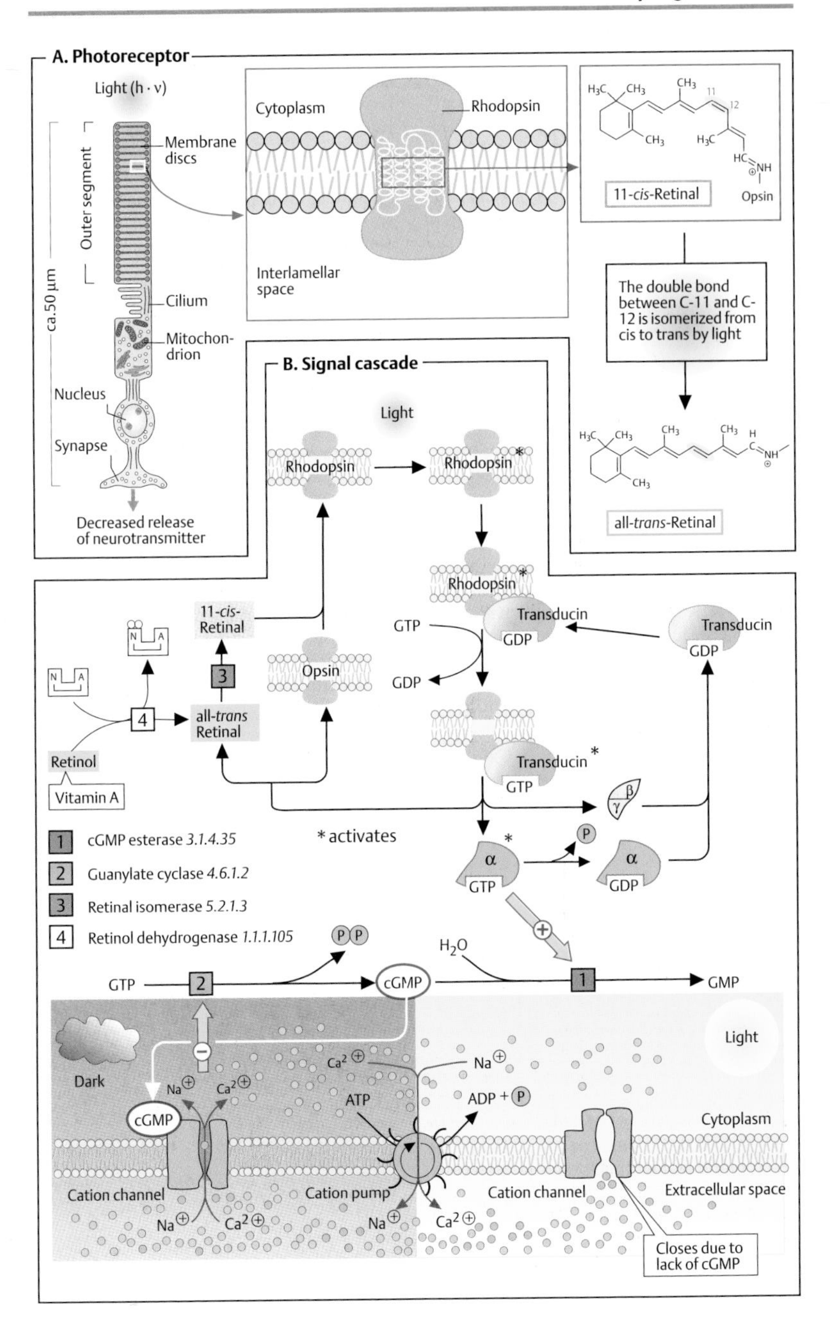
A. Photoreceptor
Light (h · ν)
Membrane discs
Outer segment
ca.50 µm
Cilium
Mitochon-drion
Nucleus
Synapse
Decreased release of neurotransmitter
Cytoplasm
Rhodopsin
Interlamellar space
11-cis-Retinal
Opsin
The double bond between C-11 and C-12 is isomerized from cis to trans by light
all-trans-Retinal
B. Signal cascade
Light
Rhodopsin
Rhodopsin*
Transducin
GDP
GTP
Transducin*
11-cis-Retinal
Opsin
all-trans Retinal
Retinol
Vitamin A
* activates
1 cGMP esterase 3.1.4.35
2 Guanylate cyclase 4.6.1.2
3 Retinal isomerase 5.2.1.3
4 Retinol dehydrogenase 1.1.1.105
cGMP
GMP
Dark
Light
Cation channel
Cation pump
ATP
ADP + P
Cytoplasm
Extracellular space
Closes due to lack of cGMP

Organic substances

A balanced human diet needs to contain a large number of different components. These include *proteins, carbohydrates, fats, minerals* (including water), and *vitamins.* These substances can occur in widely varying amounts and proportions, depending on the type of diet. As several components of the diet are *essential for life,* they have to be regularly ingested with food. Recommended daily minimums for nutrients have been published by the World Health Organization (WHO) and a number of national expert committees.

A. Energy requirement ●

The amount of energy required by a human is expressed in $kJ \cdot d^{-1}$ (kilojoule per day). An older unit is the kilocalorie (kcal; 1 kcal = 4.187 kJ). The figures given are recommended values for adults with a normal body weight. However, actual requirements are based on age, sex, body weight, and in particular on physical activity. In those involved in competitive sports, for example, requirements can increase from 12 000 to 17 000 $kJ \cdot d^{-1}$.

It is recommended that about half of the energy intake should be in the form of carbohydrates, a third at most in the form of fat, and the rest as protein. The fact that *alcoholic beverages* can make a major contribution to daily energy intake is often overlooked. Ethanol has a caloric value of about 30 $kJ \cdot g^{-1}$ (see p. 320).

B. Nutrients ●

Proteins provide the body with amino acids, which are used for endogenous protein biosynthesis. Excess amino acids are broken down to provide energy (see p. 174). Most amino acids are *glucogenic*—i. e., they can be converted into glucose (see p. 180).

Proteins are essential components of the diet, as they provide **essential amino acids** that the human body is not capable of producing on its own (see the table). Some amino acids, including cysteine and histidine, are not absolutely essential, but promote growth in children. Some amino acids are able to substitute for each other in the diet. For example, humans can form tyrosine, which is actually essential, by hydroxylation from phenylalanine, and cysteine from methionine.

The *minimum daily requirement* of protein is 37 g for men and 29 g for women, but the recommended amounts are about twice these values. Requirements in pregnant and breastfeeding women are even higher. Not only the quantity, but also the quality of protein is important. Proteins that lack several essential amino acids or only contain small quantities of them are considered to be of low value, and larger quantities of them are therefore needed. For example, pulses only contain small amounts of methionine, while wheat and corn proteins are poor in lysine. In contrast to vegetable proteins, most animal proteins are high-value (with exceptions such as collagen and gelatin).

Carbohydrates serve as a general and easily available energy source. In the diet, they are present as *monosaccharides* in honey and fruit, or as *disaccharides* in milk and in all foods sweetened with sugar (sucrose). Metabolically usable *polysaccharides* are found in vegetable products (starch) and animal products (glycogen). Carbohydrates represent a substantial proportion of the body's energy supply, but they are not essential.

Fats are primarily important energy suppliers in the diet. Per gram, they provide more than twice as much energy as proteins and carbohydrates. Fats are essential as suppliers of *fat-soluble vitamins* (see p. 364) and as sources of *polyunsaturated fatty acids,* which are needed to biosynthesize eicosanoids (see pp. 48, 390).

Mineral substances and trace elements, a very heterogeneous group of essential nutrients, are discussed in more detail on p. 362. They are usually divided into macrominerals and microminerals.

Vitamins are also indispensable components of the diet. The animal body requires them in very small quantities in order to synthesize coenzymes and signaling substances (see pp. 364–369).

A. Energy requirement

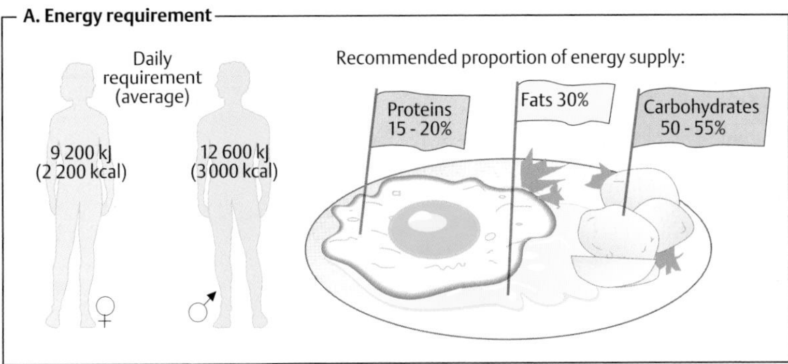

B. Nutrients

	Quantity in body (kg)	Energy content $kJ \cdot g^{-1}$ ($kcal \cdot g^{-1}$)	Daily requirement (g) a b c	General function in metabolism	Essential constituents
Proteins	10	17 (4.1)	♂ 37 55 92 ♀ 29 45 75	Supplier of amino acids Energy source	Essential amino acids:: Val (14) Leu (16) Ile (12) Lys (12) Phe (16) Trp (3) Met (10) Thr (8) (Daily requirement in mg per kg body weight) Cys and His stimulate growth
Carbohydrates	1	17 (4.1)	0 390 240-310	General source of energy (glucose) Energy reserve (glycogen) Roughage (cellulose) Supporting substances (bones, cartilage, mucus)	Non-essential nutritional constituent
Fats	10-15	39 (9.3)	10 80 130	General energy source Most important energy reserve Solvent for vitamins Supplier of essential fatty acids	Polyunsaturated fatty acids: Linoleic acid Linolenic acid Arachidonic acid (together 10 g/day)
Water	35-40	0	2 400 – –	Solvent Cellular building block Dielectric Reaction partner Temperature regulator	
Minerals	3	0		Building blocks Electrolytes Cofactors of enzymes	Macrominerals Microminerals (trace elements)
Vitamins	–	–		Often precursors of coenzymes	Lipid-soluble vitamins Water-soluble vitamins

a: Minimum daily requirement   b: Recommended daily intake   c: Actual daily intake in industrialized nations

Minerals and trace elements

A. Minerals ◑

Water is the most important essential inorganic nutrient in the diet. In adults, the body has a daily requirement of 2–3 L of water, which is supplied from drinks, water contained in solid foods, and from the *oxidation water* produced in the respiratory chain (see p. 140). The special role of water for living processes is discussed in more detail elsewhere (see p. 26).

The elements essential for life can be divided into **macroelements** (daily requirement > 100 mg) and **microelements** (daily requirement < 100 mg). The macroelements include the **electrolytes** sodium (Na), potassium (K), calcium (Ca), and magnesium (Mg), and the nonmetals chlorine (Cl), phosphorus (P), sulfur (S), and iodine (I).

The essential microelements are only required in trace amounts (see also p. 2). This group includes iron (Fe), zinc (Zn), manganese (Mn), copper (Cu), cobalt (Co), chromium (Cr), selenium (Se), and molybdenum (Mo). Fluorine (F) is not essential for life, but does promote healthy bones and teeth. It is still a matter of controversy whether vanadium, nickel, tin, boron, and silicon also belong to the essential trace elements.

The second column in the table lists the average **amounts** of mineral substances in the body of an adult weighing 65 kg. The **daily requirements** listed in the fourth column also apply to an adult, and are *average values.* Children, pregnant and breast-feeding women, and those who are ill generally have higher mineral requirements relative to body weight than men.

As the human body is able to store many minerals, deviations from the daily ration are balanced out over a given period of time. Minerals stored in the body include water, which is distributed throughout the whole body; calcium, stored in the form of apatite in the bones (see p. 340); iodine, stored as thyroglobulin in the thyroid; and iron, stored in the form of ferritin and hemosiderin in the bone marrow, spleen, and liver (see p. 286). The storage site for many trace elements is the liver. In many cases, the metabolism of minerals is regulated by *hormones*—for example, the uptake and excretion of H_2O, Na^+, Ca^{2+}, and phosphate (see p. 328), and storage of Fe^{2+} and I^-.

Resorption of the required mineral substances from food usually depends on the body's requirements, and in several cases also on the composition of the diet. One example of dietary influence is calcium (see p. 342). Its resorption as Ca^{2+} is promoted by lactate and citrate, but phosphate, oxalic acid, and phytol inhibit calcium uptake from food due to complex formation and the production of insoluble salts.

Mineral deficiencies are not uncommon and can have quite a variety of causes—e. g., an unbalanced diet, resorption disturbances, and diseases. *Calcium deficiency* can lead to rickets, osteoporosis, and other disturbances. *Chloride deficiency* is observed as a result of severe Cl^- losses due to vomiting. Due to the low content of iodine in food in many regions of central Europe, *iodine deficiency* is widespread there and can lead to goiter. *Magnesium deficiency* can be caused by digestive disorders or an unbalanced diet—e. g., in alcoholism. Trace element deficiencies often result in a disturbed blood picture—i. e., forms of anemia.

The last column in the table lists some of the functions of minerals. It should be noted that almost all of the **macroelements** in the body function either as *nutrients* or *electrolytes.* Iodine (as a result of its incorporation into iodothyronines) and calcium act as *signaling substances.* Most **trace elements** are *cofactors for proteins,* especially for enzymes. Particularly important in quantitative terms are the *iron proteins* hemoglobin, myoglobin, and the cytochromes (see p. 286), as well as more than 300 different *zinc proteins.*

A. Minerals

Mineral	Content* (g)	Major source	Daily requirement (g)	Functions/Occurrence
Water	35 000-40 000	Drinks Water in solid foods From metabolism 300g	1200 900	Solvent, cellular building block, dielectric, coolant, medium for transport, reaction partner
Macroelements (daily requirement >100 mg)				
Na	100	Table salt	1.1-3.3	Osmoregulation, membrane potential, mineral metabolism
K	150	Vegetables, fruit, cereals	1.9-5.6	Membrane potential, mineral metabolism
Ca	1 300	Milk, milk products	0.8	Bone formation, blood clotting, signal molecule
Mg	20	Green vegetables	0.35	Bone formation, cofactor for enzymes
Cl	100	Table salt	1.7-5.1	Mineral metabolism
P	650	Meat, milk, cereals, vegetables	0.8	Bone formation, energy metabolism, nucleic acid metabolism
S	200	S-containing amino acids (Cys and Met)	0.2	Lipid and carbohydrate metabolism, conjugate formation
Microelements (trace elements)			(mg	
Fe	4-5	Meat, liver, eggs, vegetables, potatoes, cereals	10	Hemoglobin, myoglobin, cytochromes, Fe/S clusters
Zn	2-3	Meat, liver, cereals	15	Zinc enzymes
Mn	0.02	Found in many foodstuffs	2-5	Enzymes
Cu	0.1-0.2	Meat, vegetables, fruit, fish	2-3	Oxidases
Co	<0.01	Meat	Traces	Vitamin B_{12}
Cr	<0.01		0.05-0.2	Not clear
Mo	0.02	Cereals, nuts, legumes	0.15-0.5	Redox enzymes
Se		Vegetables, meat	0.05-0.2	Selenium enzymes
I	0.03	Seafood, iodized salt, drinking water	0.15	Thyroxin
Requirement not known				
F		Drinking water (fluoridated), tea, milk	0.0015-0.004	Bones, dental enamel

Metals   Non-metals

* Content in the body of a 65 kg adult

Lipid-soluble vitamins

Vitamins are essential organic compounds that the animal organism is not capable of forming itself, although it requires them in small amounts for metabolism. Most vitamins are **precursors of coenzymes**; in some cases, they are also precursors of **hormones** or act as **antioxidants**. Vitamin requirements vary from species to species and are influenced by age, sex, and physiological conditions such as pregnancy, breast-feeding, physical exercise, and nutrition.

A. Vitamin supply ●

A healthy diet usually covers average daily vitamin requirements. By contrast, malnutrition, malnourishment (e.g., an unbalanced diet in older people, malnourishment in alcoholics, ready meals), or resorption disturbances lead to an inadequate supply of vitamins from which **hypovitaminosis,** or in extreme cases avitaminosis, can result. Medical treatments that kill the intestinal flora—e.g., antibiotics—can also lead to vitamin deficiencies (K, B_{12}, H) due to the absence of bacterial vitamin synthesis.

Since only a few vitamins can be stored (A, D, E, B_{12}), a lack of vitamins quickly leads to **deficiency diseases.** These often affect the skin, blood cells, and nervous system. The causes of vitamin deficiencies can be treated by improving nutrition and by administering vitamins in tablet form. An overdose of vitamins only leads to **hypervitaminoses,** with toxic symptoms, in the case of vitamins A and D. Normally, excess vitamins are rapidly excreted with the urine.

B. Lipid-soluble vitamins ◑

Vitamins are classified as either lipid-soluble or water-soluble. The lipid-soluble vitamins include vitamins A, D, E, and K, all of which belong to the isoprenoids (see p. 52).

Vitamin A (retinol) is the parent substance of the *retinoids,* which include *retinal* and *retinoic acid.* The retinoids also can be synthesized by cleavage from the provitamin β-carotene. Retinoids are found in meat-containing diets, whereas β-carotene occurs in fruits and vegetables (particularly carrots). **Retinal** is involved in visual processes as the pigment of the chromoprotein *rhodopsin* (see p. 358). **Retinoic acid,** like the steroid hormones, influences the transcription of genes in the cell nucleus. It acts as a differentiation factor in growth and development processes. Vitamin A deficiency can result in *night blindness, visual impairment,* and *growth disturbances.*

Vitamin D (calciol, cholecalciferol) is the precursor of the hormone *calcitriol* (1α,25-dihydroxycholecalciferol; see p. 320). Together with two other hormones (parathyrin and calcitonin), calcitriol regulates the calcium metabolism (see p. 342). Calciol can be synthesized in the skin from 7-dehydrocholesterol, an endogenous steroid, by a photochemical reaction. Vitamin D deficiencies only occur when the skin receives insufficient exposure to ultraviolet light and vitamin D is lacking in the diet. Deficiency is observed in the form of *rickets* in children and *osteomalacia* in adults. In both cases, bone mineralization is disturbed.

Vitamin E (tocopherol) and related compounds only occur in plants (e.g., wheat germ). They contain what is known as a *chroman ring.* In the lipid phase, vitamin E is mainly located in biological membranes, where as an *antioxidant* it protects unsaturated lipids against ROS (see p. 284) and other radicals.

Vitamin K (phylloquinone) and similar substances with modified side chains are involved in carboxylating glutamate residues of coagulation factors in the liver (see p. 290). The form that acts as a cofactor for carboxylase is derived from the vitamin by enzymatic reduction. Vitamin K antagonists (e.g., coumarin derivatives) inhibit this reduction and consequently carboxylation as well. This fact is used to inhibit blood coagulation in *prophylactic treatment against thrombosis.* Vitamin K deficiency occurs only rarely, as the vitamin is formed by bacteria of the intestinal flora.

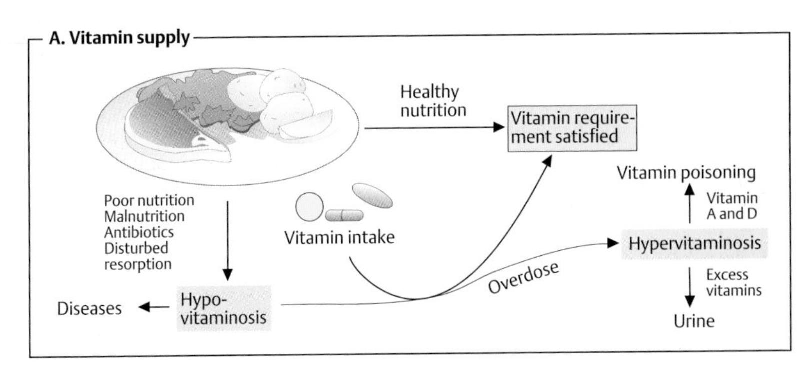
A. Vitamin supply
Healthy nutrition
Vitamin requirement satisfied
Vitamin poisoning
Vitamin A and D
Poor nutrition
Malnutrition
Antibiotics
Disturbed resorption
Vitamin intake
Hypervitaminosis
Overdose
Excess vitamins
Diseases
Hypo-vitaminosis
Urine

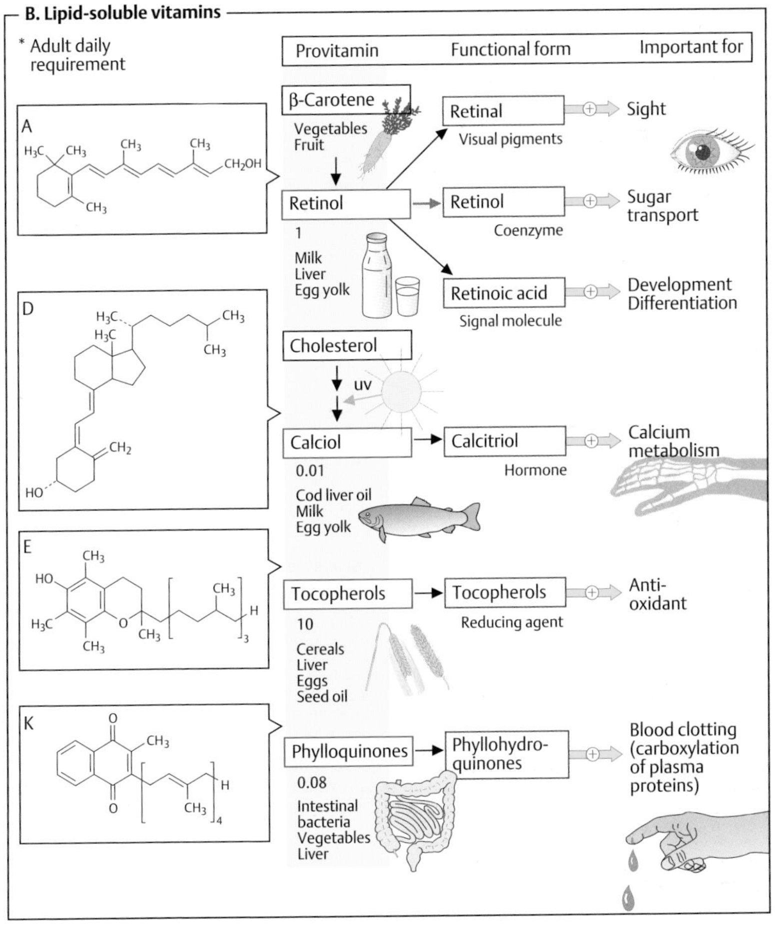
B. Lipid-soluble vitamins
* Adult daily requirement
Provitamin
Functional form
Important for
β-Carotene
Vegetables
Fruit
Retinal
Visual pigments
Sight
Retinol
1
Milk
Liver
Egg yolk
Retinol
Coenzyme
Sugar transport
Retinoic acid
Signal molecule
Development
Differentiation
Cholesterol
uv
Calciol
0.01
Cod liver oil
Milk
Egg yolk
Calcitriol
Hormone
Calcium metabolism
Tocopherols
10
Cereals
Liver
Eggs
Seed oil
Tocopherols
Reducing agent
Anti-oxidant
Phylloquinones
0.08
Intestinal bacteria
Vegetables
Liver
Phyllohydro-quinones
Blood clotting (carboxylation of plasma proteins)
A
D
E
K

Water-soluble vitamins I

The B group of vitamins covers water-soluble vitamins, all of which serve as precursors for coenzymes. Their numbering sequence is not continuous, as many substances that were originally regarded as vitamins were not later confirmed as having vitamin characteristics.

A. Water-soluble vitamins I ◑

Vitamin B_1 (thiamine) contains two heterocyclic rings—a *pyrimidine ring* (a six-membered aromatic ring with two Ns) and a *thiazole ring* (a five-membered aromatic ring with N and S), which are joined by a methylene group. The active form of vitamin B_1 is **thiamine diphosphate** (TPP), which contributes as a coenzyme to the transfer of hydroxyalkyl residues (active aldehyde groups). The most important reactions of this type are oxidative decarboxylation of 2-oxoacids (see p. 134) and the transketolase reaction in the pentose phosphate pathway (see p. 152). Thiamine was the first vitamin to be discovered, around 100 years ago. Vitamin B_1 deficiency leads to *beriberi,* a disease with symptoms that include neurological disturbances, cardiac insufficiency, and muscular atrophy.

Vitamin B_2 is a complex of several vitamins: riboflavin, folate, nicotinate, and pantothenic acid.

Riboflavin (from the Latin *flavus,* yellow) serves in the metabolism as a component of the redox coenzymes flavin mononucleotide (FMN) and flavin adenine dinucleotide (FAD; see p. 104). As prosthetic groups, **FMN** and **FAD** are cofactors for various oxidoreductases (see p. 32). No specific disease due to a deficiency of this vitamin is known.

Folate, the anion of folic acid, is made up of three different components—a *pteridine derivative, 4-aminobenzoate,* and one or more *glutamate* residues. After reduction to tetrahydrofolate (**THF**), folate serves as a coenzyme in the C_1 metabolism (see p. 418). Folate deficiency is relatively common, and leads to disturbances in nucleotide biosynthesis and thus cell proliferation. As the precursors for blood cells divide particularly rapidly, disturbances of the blood picture can occur, with increased amounts of abnormal precursors for megalocytes (*megaloblastic anemia*). Later, general damage ensues as phospholipid synthesis and the amino acid metabolism are affected.

In contrast to animals, microorganisms are able to synthesize folate from their own components. The growth of microorganisms can therefore be inhibited by *sulfonamides,* which competitively inhibit the incorporation of 4-aminobenzoate into folate (see p. 254). Since folate is not synthesized in the animal organism, sulfonamides have no effect on animal metabolism.

Nicotinate and **nicotinamide,** together referred to as "niacin," are required for biosynthesis of the coenzymes nicotinamide adenine dinucleotide (**NAD^+**) and nicotinamide adenine dinucleotide phosphate (**$NADP^+$**). These both serve in energy and nutrient metabolism as carriers of *hydride ions* (see pp. 32, 104). The animal organism is able to convert *tryptophan* into nicotinate, but only with a poor yield. Vitamin deficiency therefore only occurs when nicotinate, nicotinamide, and tryptophan are all simultaneously are lacking in the diet. It manifests in the form of skin damage (*pellagra*), digestive disturbances, and depression.

Pantothenic acid is an acid amide consisting of β-alanine and 2,4-dihydroxy-3,3′-dimethylbutyrate (pantoic acid). It is a precursor of *coenzyme A,* which is required for activation of acyl residues in the lipid metabolism (see pp. 12, 106). *Acyl carrier protein* (ACP; see p. 168) also contains pantothenic acid as part of its prosthetic group. Due to the widespread availability of pantothenic acid in food (Greek *pantothen* = "from everywhere"), deficiency diseases are rare.

Further information

The requirement for vitamins in humans and other animals is the result of mutations in the enzymes involved in biosynthetic coenzymes. As intermediates of coenzyme biosynthesis are available in sufficient amounts in the diet of heterotrophic animals (see p. 112), the lack of endogenous synthesis did not have unfavorable effects for them. Microorganisms and plants whose nutrition is mainly autotrophic have to produce all of these compounds themselves in order to survive.

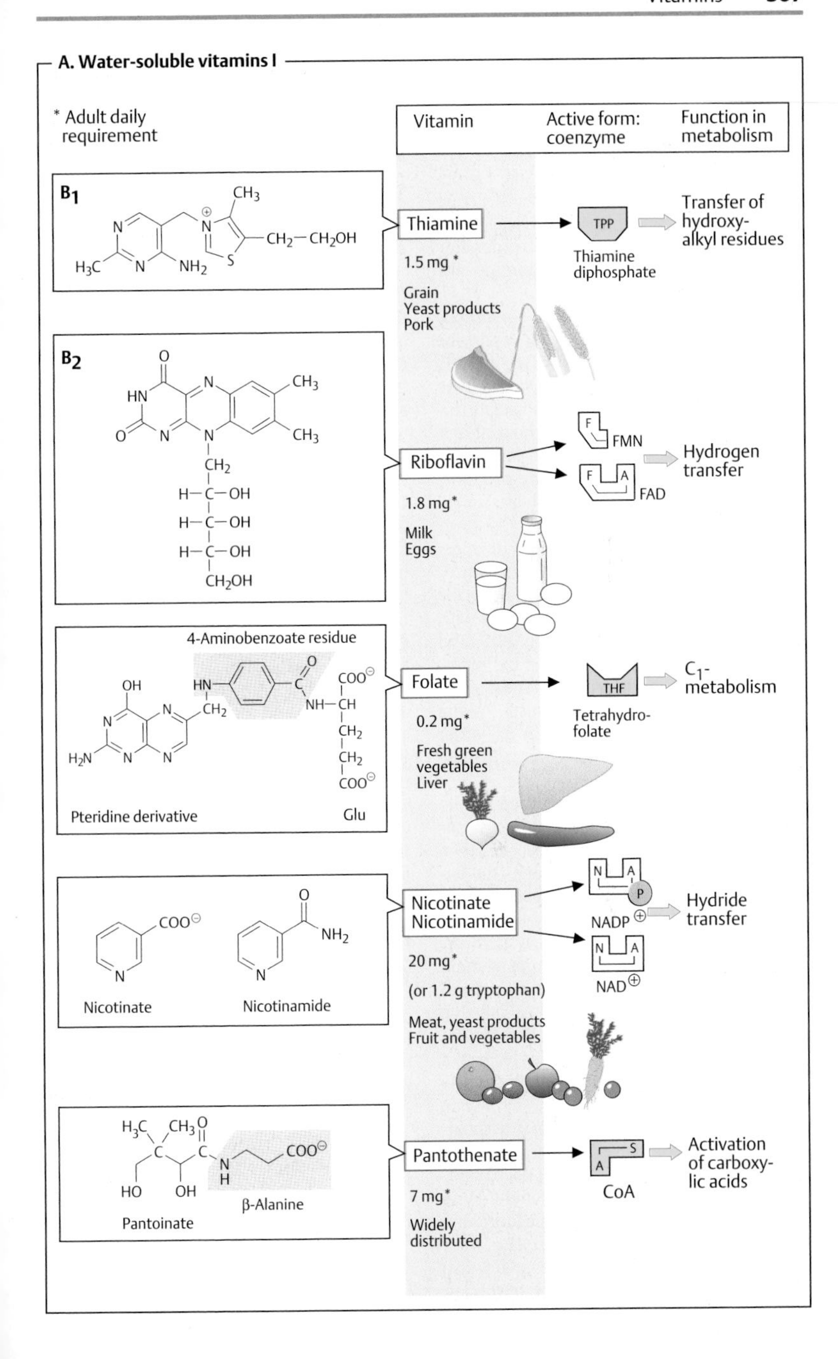
A. Water-soluble vitamins I
* Adult daily requirement
Vitamin
Active form: coenzyme
Function in metabolism
B1
Thiamine
1.5 mg *
Grain
Yeast products
Pork
TPP
Thiamine diphosphate
Transfer of hydroxy-alkyl residues
B2
Riboflavin
1.8 mg*
Milk
Eggs
FMN
FAD
Hydrogen transfer
4-Aminobenzoate residue
Pteridine derivative
Glu
Folate
0.2 mg*
Fresh green vegetables
Liver
THF
Tetrahydro-folate
C1-metabolism
Nicotinate
Nicotinamide
20 mg*
(or 1.2 g tryptophan)
Meat, yeast products
Fruit and vegetables
NADP⊕
NAD⊕
Hydride transfer
Pantoinate
β-Alanine
Pantothenate
7 mg*
Widely distributed
CoA
Activation of carboxy-lic acids

Water-soluble vitamins II

A. Water-soluble vitamins II

Vitamin B_6 consists of three substituted pyridines—**pyridoxal**, **pyridoxol**, and **pyridoxamine.** The illustration shows the structure of pyridoxal, which carries an aldehyde group (–CHO) at C-4. Pyridoxol is the corresponding alcohol ($–CH_2OH$), and pyridoxamine the amine ($–CH_2NH_2$).

The active form of vitamin B_6, **pyridoxal phosphate,** is the most important coenzyme in the amino acid metabolism (see p. 106). Almost all conversion reactions involving amino acids require pyridoxal phosphate, including transaminations, decarboxylations, dehydrogenations, etc. *Glycogen phosphorylase,* the enzyme for glycogen degradation, also contains pyridoxal phosphate as a cofactor. Vitamin B_6 deficiency is rare.

Vitamin B_{12} (cobalamine) is one of the most complex low-molecular-weight substances occurring in nature. The core of the molecule consists of a tetrapyrrol system (*corrin*), with cobalt as the central atom (see p. 108). The vitamin is exclusively synthesized by microorganisms. It is abundant in liver, meat, eggs, and milk, but not in plant products. As the intestinal flora synthesize vitamin B_{12}, strict vegetarians usually also have an adequate supply of the vitamin.

Cobalamine can only be resorbed in the small intestine when the gastric mucosa secretes what is known as *intrinsic factor*—a glycoprotein that binds cobalamine (the *extrinsic factor*) and thereby protects it from degradation. In the blood, the vitamin is bound to a special protein known as *transcobalamin.* The liver is able to store vitamin B_{12} in amounts sufficient to last for several months. Vitamin B_{12} deficiency is usually due to an absence of intrinsic factor and the resulting resorption disturbance. This leads to a disturbance in blood formation known as *pernicious anemia.*

In animal metabolism, derivatives of cobalamine are mainly involved in rearrangement reactions. For example, they act as coenzymes in the conversion of methylmalonyl-CoA to succinyl-CoA (see p. 166), and in the formation of methionine from homocysteine (see p. 418). In prokaryotes, cobalamine derivatives also play a part in the reduction of ribonucleotides.

Vitamin C is L-**ascorbic acid** (chemically: 2-oxogulonolactone). The two hydroxyl groups have acidic properties. By releasing a proton, ascorbic acid therefore turns into its anion, **ascorbate**. Humans, apes, and guinea pigs require vitamin C because they lack the enzyme L-*gulonolactone oxidase (1.1.3.8),* which catalyzes the final step in the conversion of glucose into ascorbate.

Vitamin C is particularly abundant in fresh fruit and vegetables. Many soft drinks and foodstuffs also have synthetic ascorbic acid added to them as an antioxidant and flavor enhancer. Boiling slowly destroys vitamin C. In the body, ascorbic acid serves as a reducing agent in variations reactions (usually hydroxylations). Among the processes involved are *collagen synthesis, tyrosine degradation, catecholamine synthesis,* and *bile acid biosynthesis.* The daily requirement for ascorbic acid is about 60 mg, a comparatively large amount for a vitamin. Even higher doses of the vitamin have a protective effect against infections. However, the biochemical basis for this effect has not yet been explained. Vitamin C deficiency only occurs rarely nowadays; it becomes evident after a few months in the form of *scurvy,* with connective-tissue damage, bleeding, and tooth loss.

Vitamin H (biotin) is present in liver, egg yolk, and other foods; it is also synthesized by the intestinal flora. In the body, biotin is covalently attached via a lysine side chain to enzymes that catalyze carboxylation reactions. Biotin-dependent carboxylases include *pyruvate carboxylase* (see p. 154) and *acetyl-CoA carboxylase* (see p. 162). CO_2 binds, using up ATP, to one of the two N atoms of biotin, from which it is transferred to the acceptor (see p. 108).

Biotin binds with high affinity ($K_d = 10^{-15}$ M) and specificity to *avidin*, a protein found in egg white. Since boiling denatures avidin, biotin deficiency only occurs when egg whites are eaten raw.

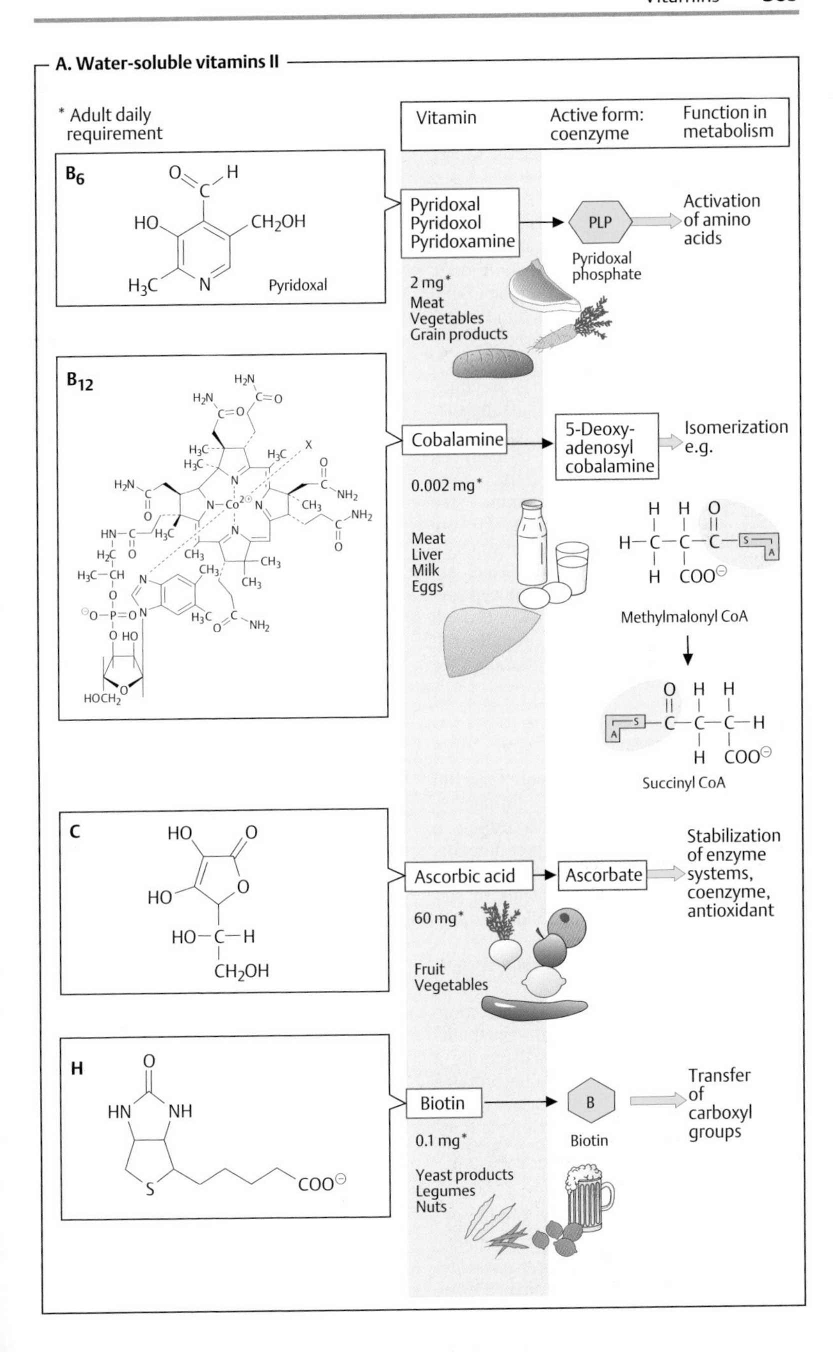
A. Water-soluble vitamins II
* Adult daily requirement
Vitamin
Active form: coenzyme
Function in metabolism
B6
Pyridoxal
Pyridoxal
Pyridoxol
Pyridoxamine
PLP
Pyridoxal phosphate
Activation of amino acids
2 mg*
Meat
Vegetables
Grain products
B12
Cobalamine
5-Deoxy-adenosyl cobalamine
Isomerization e.g.
0.002 mg*
Meat
Liver
Milk
Eggs
Methylmalonyl CoA
Succinyl CoA
C
Ascorbic acid
Ascorbate
Stabilization of enzyme systems, coenzyme, antioxidant
60 mg*
Fruit
Vegetables
H
Biotin
B
Biotin
Transfer of carboxyl groups
0.1 mg*
Yeast products
Legumes
Nuts

Basics

Hormones are *chemical signaling substances.* They are synthesized in specialized cells that are often associated to form *endocrine glands.* Hormones are released into the *blood* and transported with the blood to their *effector organs.* In the organs, the hormones carry out physiological and biochemical regulatory functions. In contrast to endocrine hormones, **tissue hormones** are only active in the immediate vicinity of the cells that secrete them.

The distinctions between hormones and other signaling substances (mediators, neurotransmitters, and growth factors) are fluid. **Mediators** is the term used for signaling substances that do not derive from special hormone-forming cells, but are form by many cell types. They have hormone-like effects in their immediate surroundings. *Histamine* (see p. 352) and *prostaglandins* (see p. 390) are important examples of these substances. **Neurohormones** and **neurotransmitters** are signaling substances that are produced and released by nerve cells (see p. 348). **Growth factors** and **cytokines** mainly promote cell proliferation and cell differentiation (see p. 392).

A. Hormones: overview ●

The animal organism contains more than 100 hormones and hormone-like substances, which can be classified either according to their structure or according to their function. In chemical terms, most hormones are *amino acid derivatives, peptides* or *proteins,* or *steroids.* Hormones regulate the following processes:

- **Growth and differentiation of cells, tissues, and organs**
  These processes include cell proliferation, embryonic development, and sexual differentiation—i. e., processes that require a prolonged time period and involve proteins de novo synthesis. For this reason, mainly steroid hormones which function via transcription regulation are active in this field (see p. 244).
- **Metabolic pathways**
  Metabolic regulation requires rapidly acting mechanisms. Many of the hormones involved therefore regulate *interconversion* of enzymes (see p. 120). The main processes subject to hormonal regulation are the uptake and degradation of storage substances (glycogen, fat), metabolic pathways for biosynthesis and degradation of central metabolites (glucose, fatty acids, etc.), and the supply of metabolic energy.
- **Digestive processes**
  Digestive processes are usually regulated by locally acting peptides (paracrine; see p. 372), but mediators, biogenic amines, and neuropeptides are also involved (see p. 270).
- **Maintenance of ion concentrations (homeostasis)**
  Concentrations of Na^+, K^+, and Cl^- in body fluids, and the physiological variables dependent on these (e. g. blood pressure), are subject to strict regulation. The principal site of action of the hormones involved is the kidneys, where hormones increase or reduce the resorption of ions and recovery of water (see pp. 326–331). The concentrations of Ca^{2+} and phosphate, which form the mineral substance of bone and teeth, are also precisely regulated.

Many hormones influence the above processes only indirectly by regulating the synthesis and release of other hormones (*hormonal hierarchy;* see p. 372).

B. Hormonal regulation system ●

Each hormone is the center of a hormonal regulation system. Specialized glandular cells synthesize the hormone from precursors, store it in many cases, and release it into the bloodstream when needed (**biosynthesis**). For **transport**, the poorly water-soluble lipophilic hormones are bound to plasma proteins known as hormone carriers. To stop the effects of the hormone again, it is inactivated by enzymatic reactions, most of which take place in the liver (**metabolism**). Finally, the hormone and its metabolites are expelled via the excretory system, usually in the kidney (**excretion**). All of these processes affect the concentration of the hormone and thus contribute to regulation of the hormonal signal.

In the effector organs, target cells receive the hormone's message. These cells have hormone receptors for the purpose, which bind the hormone. Binding of a hormone passes information to the cell and triggers a response (**effect**).

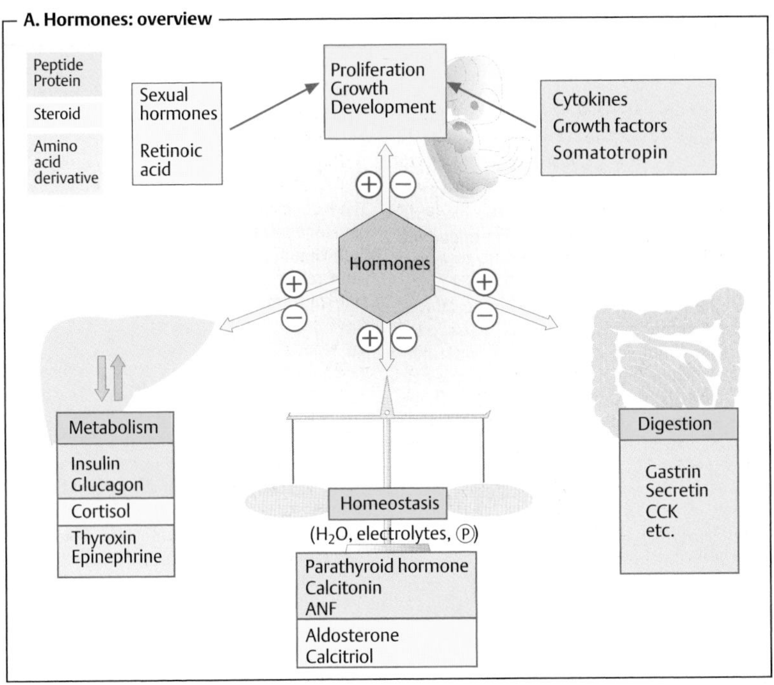
A. Hormones: overview
Peptide
Protein
Steroid
Amino acid derivative
Sexual hormones
Retinoic acid
Proliferation
Growth
Development
Cytokines
Growth factors
Somatotropin
Hormones
Metabolism
Insulin
Glucagon
Cortisol
Thyroxin
Epinephrine
Homeostasis
(H_2O, electrolytes, Ⓟ)
Parathyroid hormone
Calcitonin
ANF
Aldosterone
Calcitriol
Digestion
Gastrin
Secretin
CCK
etc.

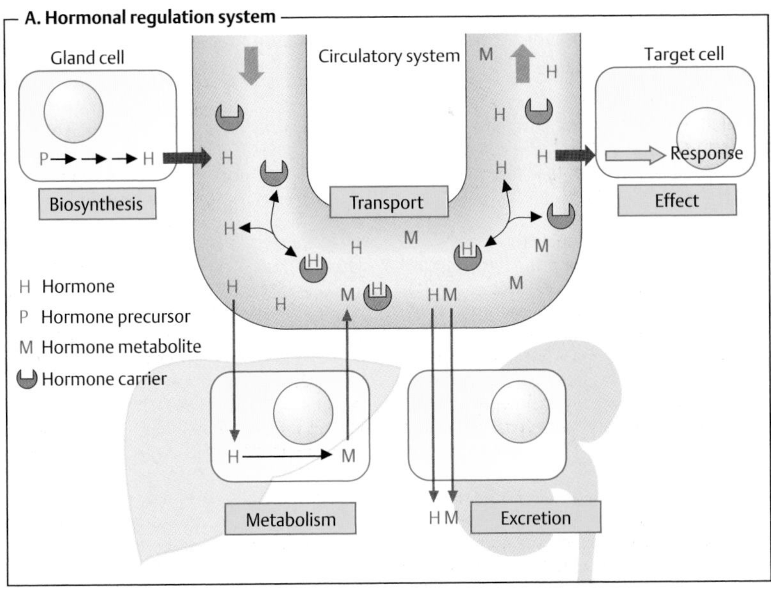
A. Hormonal regulation system
Gland cell
Circulatory system
Target cell
Biosynthesis
Transport
Effect
Response
H Hormone
P Hormone precursor
M Hormone metabolite
Hormone carrier
Metabolism
Excretion

Plasma levels and hormone hierarchy

A. Endocrine, paracrine, and autocrine hormone effects ◑

Hormones transfer signals by migrating from their site of synthesis to their site of action. They are usually transported in the blood. In this case, they are said to have an **endocrine effect** (**1**; example: *insulin*). By contrast, *tissue hormones,* the target cells for which are in the immediate vicinity of the glandular cells that produce them, are said to have a **paracrine effect** (**2**; example: *gastrointestinal tract hormones*). When signal substances also pass effects back to the cells that synthesize them, they are said to have an **autocrine effect** (**3**; example: *prostaglandins*). Autocrine effects are often found in tumor cells (see p. 400), which stimulate their own proliferation in this way.

Insulin, which is formed in the B cells of the pancreas, has both endocrine and paracrine effects. As a hormone with endocrine effects, it regulates glucose and fat metabolism. Via a paracrine mechanism, it inhibits the synthesis and release of *glucagon* from the neighboring A cells.

B. Dynamics of the plasma level ◑

Hormones circulate as signaling substances in the blood at very low concentrations (10^{-12} to between 10^{-7} mol · L^{-1}). These values change periodically in rhythms that depend on the time of day, month, or year, or on physiological cycles.

The first example shows the **circadian** rhythm of the cortisol level. As an activator of gluconeogenesis (see p. 158), cortisol is mainly released in the early morning, when the liver's glycogen stores are declining. During the day, the plasma cortisol level declines.

Many hormones are released into the blood in a spasmodic and irregular manner. In this case, their concentrations change in an **episodic** or **pulsatile** fashion. This applies, for instance, to luteinizing hormone (LH, lutropin).

Concentrations of other hormones are **event-regulated.** For example, the body responds to increased blood sugar levels after meals by releasing *insulin.* Regulation of hormone synthesis, release, and degradation allows the blood concentrations of hormones to be precisely adjusted. This is based either on simple feedback control or on hierarchically structured regulatory systems.

C. Closed-loop feedback control ◑

The biosynthesis and release of *insulin* by the pancreatic B cells (see p. 160) is stimulated by high blood glucose levels (> 5 mM). The insulin released then stimulates increased uptake and utilization of glucose by the cells of the muscle and adipose tissues. As a result, the blood glucose level falls back to its normal value, and further release of insulin stops.

D. Hormone hierarchy ◑

Hormone systems are often linked to each another, giving rise in some cases to a hierarchy of higher-order and lower-order hormones. A particularly important example is the *pituitary–hypothalamic axis,* which is controlled by the central nervous system (CNS).

Nerve cells in the hypothalamus react to stimulatory or inhibitory signals from the CNS by releasing activating or inhibiting factors, which are known as **liberins** (*"releasing hormones"*) and **statins** (*"inhibiting hormones"*). These neurohormones reach the adenohypophysis by short routes through the bloodstream. In the adenohypophysis, they stimulate (liberins) or inhibit (statins) the biosynthesis and release of tropines. **Tropines** (*glandotropic hormones*) in turn stimulate peripheral glands to synthesize glandular hormones. Finally, the **glandular hormone** acts on its target cells in the organism. In addition, it passes effects back to the higher-order hormone systems. This (usually negative) feedback influences the concentrations of the higher-order hormones, creating a feedback loop.

Many steroid hormones are regulated by this type of axis—e. g., thyroxin, cortisol, estradiol, progesterone, and testosterone. In the case of the glucocorticoids, the hypothalamus releases corticotropin-releasing hormone (CRH or corticoliberin, a peptide consisting of 41 amino acids), which in turn releases corticotropin (ACTH, 39 AAs) in the pituitary gland. Corticotropin stimulates synthesis and release of the glandular steroid hormone cortisol in the adrenal cortex.

A. Endocrine, paracrine and autocrine hormone effects

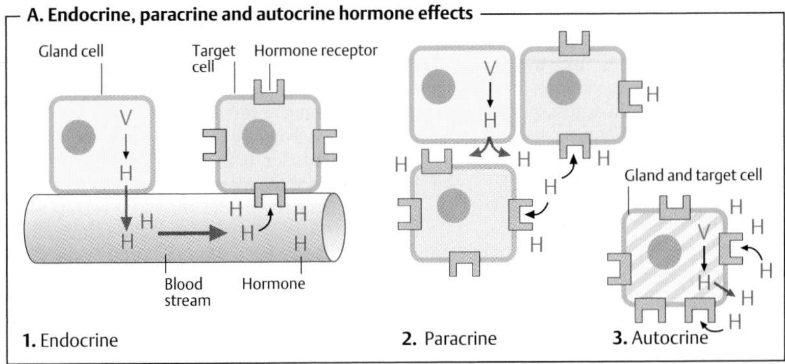

B. Plasma level dynamics

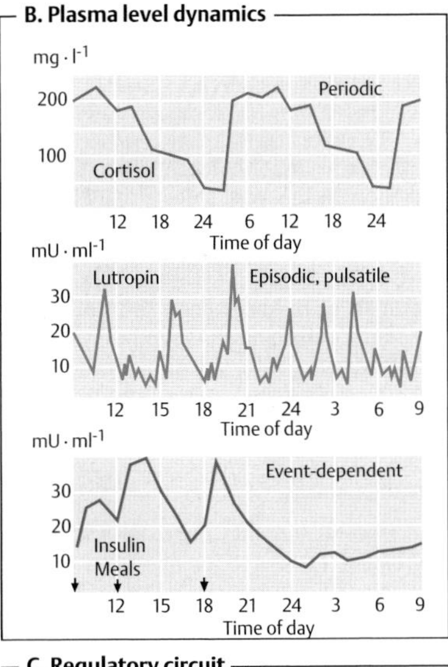

C. Regulatory circuit

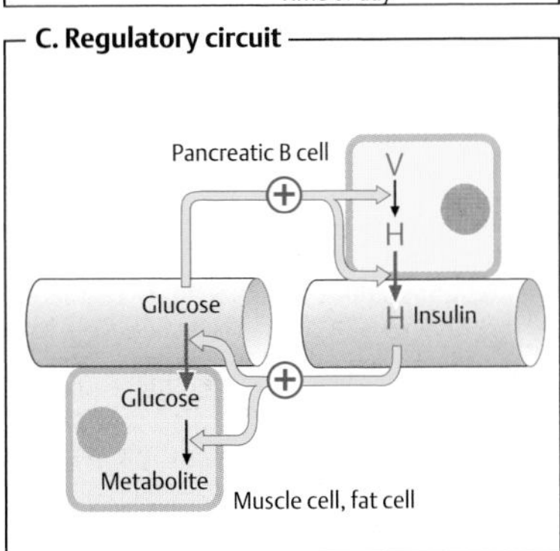

D. Hormone hierarchy

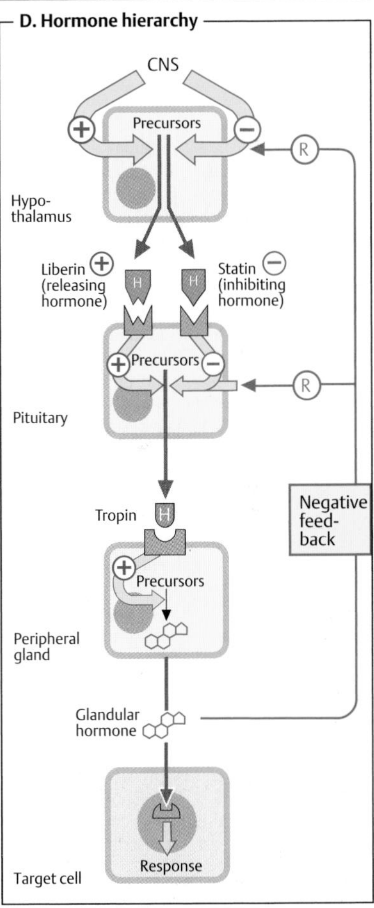

Lipophilic hormones

Classifying hormones into hydrophilic and lipophilic molecules indicates the chemical properties of the two groups of hormones and also reflects differences in their mode of action (see p. 120).

A. Lipophilic hormones ◑

Lipophilic hormones, which include *steroid hormones, iodothyronines,* and *retinoic acid,* are relatively small molecules (300–800 Da) that are poorly soluble in aqueous media. With the exception of the iodothyronines, they are not stored by hormone-forming cells, but are released immediately after being synthesized. During transport in the blood, they are bound to specific carriers. Via intracellular receptors, they mainly act on transcription (see p. 358). Other effects of steroid hormones—e.g., on the immune system—are not based on transcriptional control. Their details have not yet been explained.

Steroid hormones

The most important steroid hormones in vertebrates are listed on p. 57. *Calcitriol* (vitamin D hormone) is also included in this group, although it has a modified steroid structure. The most important steroid hormone in invertebrates is *ecdysone.*

Progesterone is a female sexual steroid belonging to the progestin *(gestagen)* family. It is synthesized in the corpus luteum of the ovaries. The blood level of progesterone varies with the menstrual cycle. The hormone prepares the uterus for a possible pregnancy. Following fertilization, the placenta also starts to synthesize progesterone in order to maintain the pregnant state. The development of the mammary glands is also stimulated by progesterone.

Estradiol is the most important of the *estrogens.* Like progesterone, it is synthesized by the ovaries and, during pregnancy, by the placenta as well. Estradiol controls the menstrual cycle. It promotes proliferation of the uterine mucosa, and is also responsible for the development of the female secondary sexual characteristics (breast, fat distribution, etc.).

Testosterone is the most important of the male sexual steroids (*androgens*). It is synthesized in the Leydig intersitial cells of the testes, and controls the development and functioning of the male gonads. It also determines secondary sexual characteristics in men (muscles, hair, etc.).

Cortisol, the most important *glucocorticoid,* is synthesized by the adrenal cortex. It is involved in regulating protein and carbohydrate metabolism by promoting protein degradation and the conversion of amino acids into glucose. As a result, the blood glucose level rises (see p. 152). Synthetic glucocorticoids (e. g., dexamethasone) are used in drugs due to their anti-inflammatory and immunosuppressant effects.

Aldosterone, a *mineralocorticoid,* is also synthesized in the adrenal gland. In the kidneys, it promotes Na^+ resorption by inducing Na^+/K^+ ATPase and Na^+ channels (see p. 328). At the same time, it leads to increased K^+ excretion. In this way, aldosterone indirectly increases blood pressure.

Calcitriol is a derivative of vitamin D (see p. 364). On exposure to ultraviolet light, a precursor of the hormone can also arise in the skin. Calcitriol itself is synthesized in the kidneys (see p. 330). Calcitriol promotes the resorption of calcium in the intestine and increases the Ca^{2+} level in the blood.

Iodothyronines

The thyroid hormone **thyroxine** (tetraiodothyronine, T_4) and its active form **triiodothyronine** (T_3) are derived from the amino acid *tyrosine.* The iodine atoms at positions 3 and 5 of the two phenol rings are characteristic of them. Post-translational synthesis of thyroxine takes place in the thyroid gland from tyrosine residues of the protein *thyroglobulin,* from which it is proteolytically cleaved before being released. Iodothyronines are the only organic molecules in the animal organism that contain iodine. They increase the basal metabolic rate, partly by regulating mitochondrial ATP synthesis. In addition, they promote embryonic development.

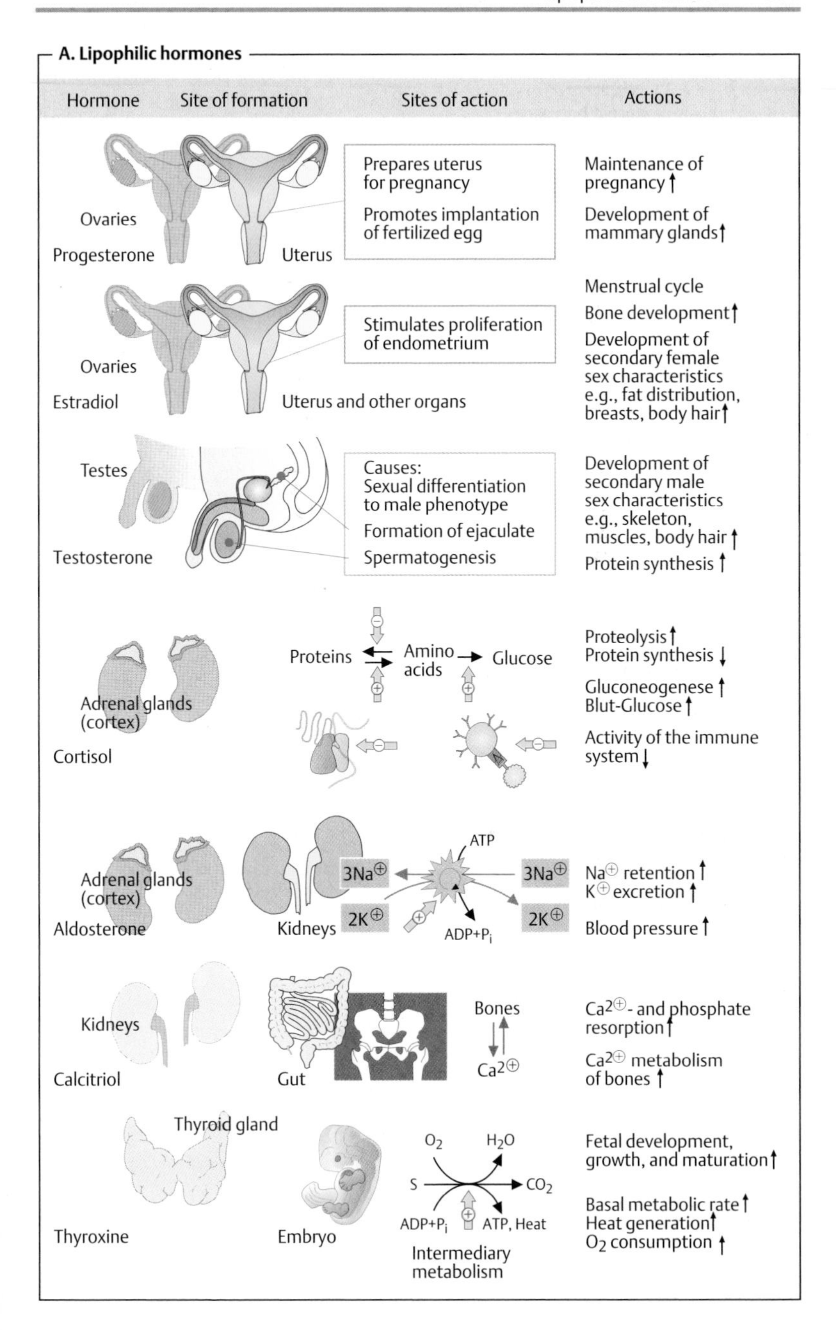
A. Lipophilic hormones
Hormone
Site of formation
Sites of action
Actions
Ovaries
Progesterone
Uterus
Prepares uterus for pregnancy
Promotes implantation of fertilized egg
Maintenance of pregnancy ↑
Development of mammary glands ↑
Ovaries
Estradiol
Uterus and other organs
Stimulates proliferation of endometrium
Menstrual cycle
Bone development ↑
Development of secondary female sex characteristics e.g., fat distribution, breasts, body hair ↑
Testes
Testosterone
Causes:
Sexual differentiation to male phenotype
Formation of ejaculate
Spermatogenesis
Development of secondary male sex characteristics e.g., skeleton, muscles, body hair ↑
Protein synthesis ↑
Adrenal glands (cortex)
Cortisol
Proteins
Amino acids
Glucose
Proteolysis ↑
Protein synthesis ↓
Gluconeogenese ↑
Blut-Glucose ↑
Activity of the immune system ↓
Adrenal glands (cortex)
Aldosterone
Kidneys
ATP
$3Na^{\oplus}$
$2K^{\oplus}$
$ADP+P_i$
$Na^{\oplus}$ retention ↑
$K^{\oplus}$ excretion ↑
Blood pressure ↑
Kidneys
Calcitriol
Gut
Bones
$Ca^{2\oplus}$
$Ca^{2\oplus}$- and phosphate resorption ↑
$Ca^{2\oplus}$ metabolism of bones ↑
Thyroid gland
Thyroxine
Embryo
O_2
H_2O
S
CO_2
$ADP+P_i$
ATP, Heat
Intermediary metabolism
Fetal development, growth, and maturation ↑
Basal metabolic rate ↑
Heat generation ↑
O_2 consumption ↑

Metabolism of steroid hormones

A. Biosynthesis of steroid hormones ○

All steroid hormones are synthesized from **cholesterol.** The gonane core of cholesterol consists of 19 carbon atoms in four rings (A–D). The D ring carries a side chain of eight C atoms (see p. 54).

The cholesterol required for biosynthesis of the steroid hormones is obtained from various sources. It is either taken up as a constituent of LDL lipoproteins (see p. 278) into the hormone-synthesizing glandular cells, or synthesized by glandular cells themselves from acetyl-CoA (see p. 172). Excess cholesterol is stored in the form of fatty acid esters in lipid droplets. Hydrolysis allows rapid mobilization of the cholesterol from this reserve again.

Biosynthetic pathways. Only an overview of the synthesis pathways that lead to the individual hormones is shown here. Further details are given on p. 410.

Among the reactions involved, hydroxylations (H) are particularly numerous. These are catalyzed by specific *monooxygenases* ("hydroxylases") of the cytochrome P450 family (see p. 318). In addition, there are NADPH-dependent and $NADP^+$-dependent *hydrogenations* (Y) and *dehydrogenations* (D), as well as *cleavage* and *isomerization reactions* (S, I). The estrogens have a special place among the steroid hormones, as they are the only ones that contain an aromatic A ring. When this is formed, catalyzed by *aromatase*, the angular methyl group (C-19) is lost.

Pregnenolone is an important intermediate in the biosynthesis of most steroid hormones. It is identical to cholesterol with the exception of a shortened and oxidized side chain. Pregnenolone is produced in three steps by hydroxylation and cleavage in the side chain. Subsequent dehydrogenation of the hydroxyl group at C-3 (b) and shifting of the double bond from C-5 to C-4 results in the gestagen **progesterone.**

With the exception of calcitriol, all steroid hormones are derived from progesterone. Hydroxylations of progesterone at C atoms 17, 21, and 11 lead to the glucocorticoid **cortisol.** Hydroxylation at C-17 is omitted during synthesis of the mineralocorticoid **aldosterone.** Instead, the angular methyl group (C-18) is oxidized to the aldehyde group. During synthesis of the androgen **testosterone** from progesterone, the side chain is completely removed. Aromatization of the A ring, as mentioned above, finally leads to **estradiol.**

On the way to **calcitriol** (vitamin D hormone; see p. 342), another double bond in the B ring of cholesterol is first introduced. Under the influence of UV light on the skin, the B ring is then photochemically cleaved, and the secosteroid *cholecalciferol* arises (vitamin D_3; see p. 364). Two Cyt P450-dependent hydroxylations in the liver and kidneys produce the active vitamin D hormone (see p. 330).

B. Inactivation of steroid hormones ○

The steroid hormones are mainly inactivated in the liver, where they are either reduced or further hydroxylated and then conjugated with *glucuronic acid* or *sulfate* for excretion (see p. 316). The *reduction reactions* attack oxo groups and the double bond in ring A. A combination of several inactivation reactions gives rise to many different steroid metabolites that have lost most of their hormonal activity. Finally, they are excreted with the *urine* and also partly via the *bile.* Evidence of steroids and steroid metabolites in the urine is used to investigate the hormone metabolism.

Further information

Congenital defects in the biosynthesis of steroid hormones can lead to severe developmental disturbances. In the *adrenogenital syndrome* (AGS), which is relatively common, there is usually a defect in *21-hydroxylase,* which is needed for synthesis of cortisol and aldosterone from progesterone. Reduced synthesis of this hormone leads to increased formation of testosterone, resulting in masculinization of female fetuses. With early diagnosis, this condition can be avoided by providing the mother with hormone treatment before birth.

A. Biosynthesis of steroid hormones

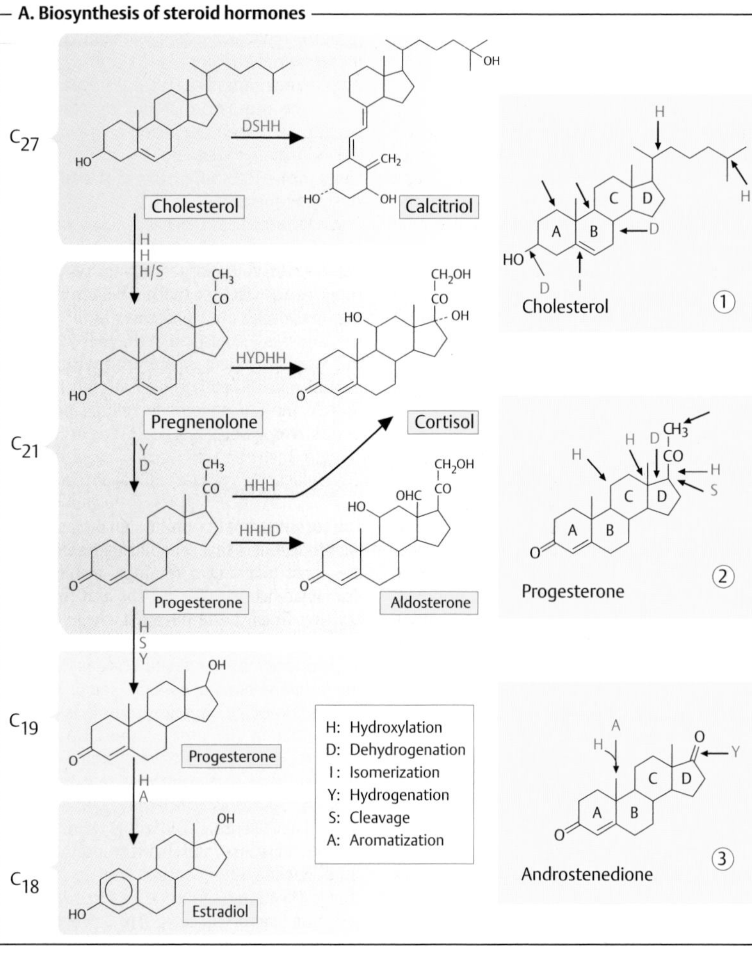

B. Inactivation of steroid hormones

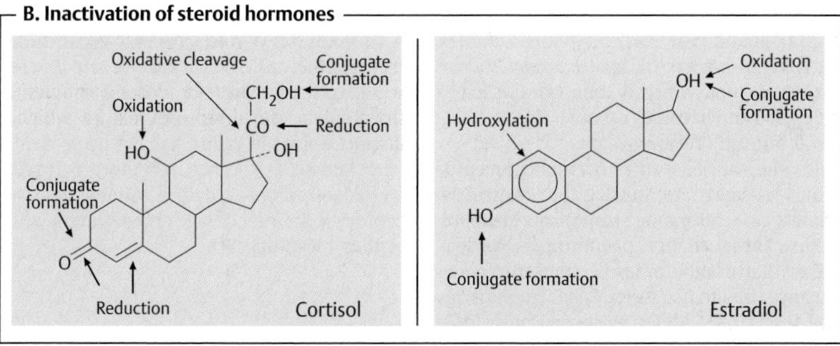

Mechanism of action

A. Mechanism of action of lipophilic hormones ◑

Lipophilic signaling substances include the *steroid hormones, calcitriol,* the *iodothyronines* (T_3 and T_4), and *retinoic acid.* These hormones mainly act in the *nucleus* of the target cells, where they regulate gene transcription in collaboration with their receptors and with the support of additional proteins (known as coactivators and mediators; see p. 244). There are several effects of steroid hormones that are not mediated by transcription control. These alternative pathways for steroid effects have not yet been fully explained.

In the blood, there are a number of transport proteins for lipophilic hormones (see p. 276). Only the free hormone is able to penetrate the membrane and enter the cell. The hormone encounters its receptor in the nucleus (and sometimes also in the cytoplasm).

The **receptors** for lipophilic hormones are rare proteins. They occur in small numbers (10^3–10^4 molecules per cell) and show marked *specificity* and high *affinity* for the hormone (K_d = 10^{-8}–10^{-10} M). After binding to the hormone, the steroid receptors are able to bind as homodimers or heterodimers to *control elements* in the promoters of specific genes, from where they can influence the transcription of the affected genes—i. e., they act as *transcription factors.*

The illustration shows the particularly well-investigated mechanism of action for **cortisol**, which is unusual to the extent that the hormone–receptor complex already arises in the cytoplasm. The free receptor is present in the cytoplasm as a monomer in complex with the chaperone **hsp90** (see p. 232). Binding of cortisol to the complex leads to an *allosteric conformational change* in the receptor, which is then released from the hsp90 and becomes capable of DNA binding as a result of *dimerization.*

In the nucleus, the hormone–receptor complex binds to nucleotide sequences known as **hormone response elements** (HREs). These are short palindromic DNA segments that usually promote transcription as enhancer elements (see p. 244). The illustration shows the HRE for glucocorticoids (GRE; "n" stands for any nucleotide). Each hormone receptor only recognizes its "own" HRE and therefore only influences the transcription of genes containing that HRE. Recognition between the receptor and HRE is based on interaction between the amino acid residues in the DNA-binding domain (**B**) and the relevant bases in the HRE (emphasized in color in the structure illustrated).

As discussed on p. 244, the hormone receptor does not interact directly with the RNA polymerase, but rather—along with other transcription factors—with a coactivator/mediator complex that processes all of the signals and passes them on to the polymerase. In this way, hormonal effects lead within a period of minutes to hours to altered levels of mRNAs for key proteins in cellular processes ("cellular response").

B. Steroid receptors ○

The receptors for lipophilic signaling substances all belong to one *protein superfamily.* They are constructed in a modular fashion from **domains** with various lengths and functions. Starting from the N terminal, these are: the *regulatory domain,* the *DNA-binding domain,* a *nuclear localization sequence* (see p. 228), and the *hormone-binding domain* (see p. 73D).

The homology among receptors is particularly great in the area of the DNA-binding domain. The proteins have cysteine-rich sequences here that coordinatively bind zinc ions (**A**, Cys shown in yellow, Zn^{2+} in light blue). These centers, known as "zinc fingers" or "**zinc clusters**," stabilize the domains and support their dimerization, but do not take part in DNA binding directly. As in other transcription factors (see p. 118), "recognition helices" are responsible for that.

In addition to the receptors mentioned in **A**, the family of steroid receptors also includes the product of the oncogene *erb*-A (see p. 398), the receptor for the environmental toxin *dioxin,* and other proteins for which a distinct hormone ligand has not been identified (known as "orphan receptors"). Several steroid receptors—e. g., the retinoic acid receptor—form functional heterodimers with orphan receptors.

A. Mechanism of action of lipophilic hormones

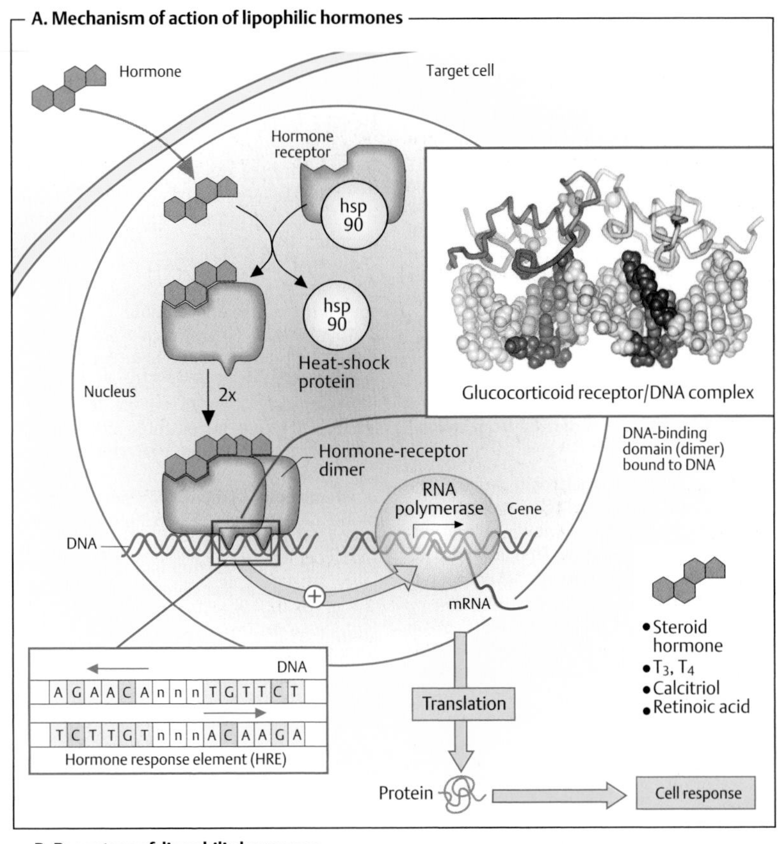

B. Receptors of lipophilic hormones

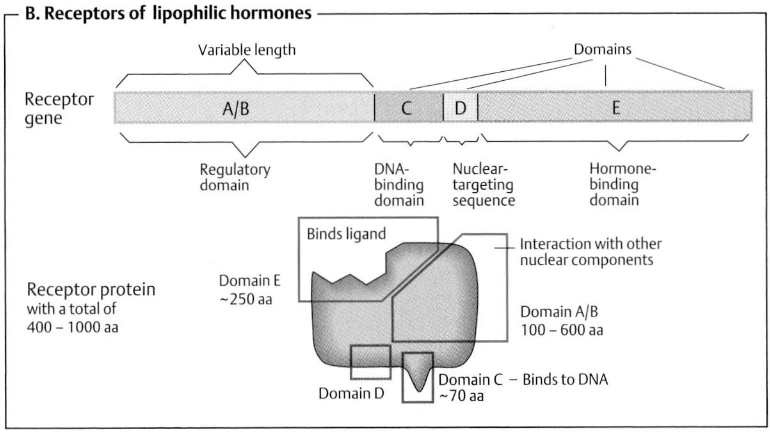

Hydrophilic hormones

The hydrophilic hormones are derived from amino acids, or are peptides and proteins composed of amino acids. Hormones with endocrine effects are synthesized in glandular cells and stored there in vesicles until they are released. As they are easily soluble, they do not need carrier proteins for transport in the blood. They bind on the plasma membrane of the target cells to receptors that pass the hormonal signal on (signal transduction; see p. 384). Several hormones in this group have paracrine effects—i. e., they only act in the immediate vicinity of their site of synthesis (see p. 372).

A. Signaling substances derived from amino acids ◑

Histamine, serotonin, melatonin, and the catecholamines dopa, dopamine, norepinephrine, and epinephrine are known as *"biogenic amines."* They are produced from amino acids by decarboxylation and usually act not only as hormones, but also as neurotransmitters.

Histamine, an important *mediator* (local signaling substance) and *neurotransmitter*, is mainly stored in tissue mast cells and basophilic granulocytes in the blood. It is involved in inflammatory and allergic reactions. "Histamine liberators" such as tissue hormones, type E immunoglobulins (see p. 300), and drugs can release it. Histamine acts via various types of receptor. Binding to H_1 receptors promotes contraction of smooth muscle in the bronchia, and dilates the capillary vessels and increases their permeability. Via H_2 receptors, histamine slows down the heart rate and promotes the formation of HCl in the gastric mucosa. In the brain, histamine acts as a neurotransmitter.

Epinephrine is a hormone synthesized in the adrenal glands from tyrosine (see p. 352). Its release is subject to neuronal control. This "emergency hormone" mainly acts on the blood vessels, heart, and metabolism. It constricts the blood vessels and thereby increases blood pressure (via α_1 and α_2 receptors); it increases cardiac function (via β_2 receptors); it promotes the degradation of glycogen into glucose in the liver and muscles (via β_2 receptors); and it dilates the bronchia (also via β_2 receptors).

B. Examples of peptide hormones and proteohormones ◑

Numerically the largest group of signaling substances, these arise by protein biosynthesis (see p. 382). The smallest peptide hormone, thyroliberin (362 Da), is a tripeptide. Proteohormones can reach masses of more than 20 kDa—e. g., thyrotropin (28 kDa). Similarities in the primary structures of many peptide hormones and proteohormones show that they are related to one another. They probably arose from common predecessors in the course of evolution.

Thyroliberin (thyrotropin-releasing hormone, TRH) is one of the neurohormones of the hypothalamus (see p. 330). It stimulates pituitary gland cells to secrete thyrotropin (TSH). TRH consists of three amino acids, which are modified in characteristic ways (see p. 353).

Thyrotropin (thyroid-stimulating hormone, TSH) and the related hormones *lutropin* (luteinizing hormone, LH) and *follitropin* (follicle-stimulating hormone, FSH) originate in the adenohypophysis. They are all dimeric glycoproteins with masses of around 28 kDa. Thyrotropin stimulates the synthesis and secretion of thyroxin by the thyroid gland.

Insulin (for the structure, see p. 70) is produced and released by the B cells of the pancreas and is released when the glucose level rises. Insulin reduces the blood sugar level by promoting processes that consume glucose—e. g., glycolysis, glycogen synthesis, and conversion of glucose into fatty acids. By contrast, it inhibits gluconeogenesis and glycogen degradation. The transmission of the insulin signal in the target cells is discussed in greater detail on p. 388.

Glucagon, a peptide of 29 amino acids, is a product of the A cells of the pancreas. It is the antagonist of insulin and, like insulin, mainly influences carbohydrate and lipid metabolism. Its effects are each opposite to those of insulin. Glucagon mainly acts via the second messenger cAMP (see p. 384).

A. Signaling substances derived from amino acids

Hormone	Sites of formation	Sites of action	Actions
Histamine	Mast cell (Histamine stores); Basophilic granulocyte	Lungs	Width of bronchi ↓; Capillaries: width ↑, permeability ↑
		Stomach	Gastric acid secretion by parietal cells ↑
Epinephrine	Adrenal glands (medulla)	Heart	Cardiac output ↑; Width of blood vessels ↓; Blood pressure ↑
		Adipose tissue, Liver, Muscle	Metabolism: Glycogenolysis ↑; Blood glucose ↑; Lipolysis ↑

B. Examples of peptide hormones and proteohormones

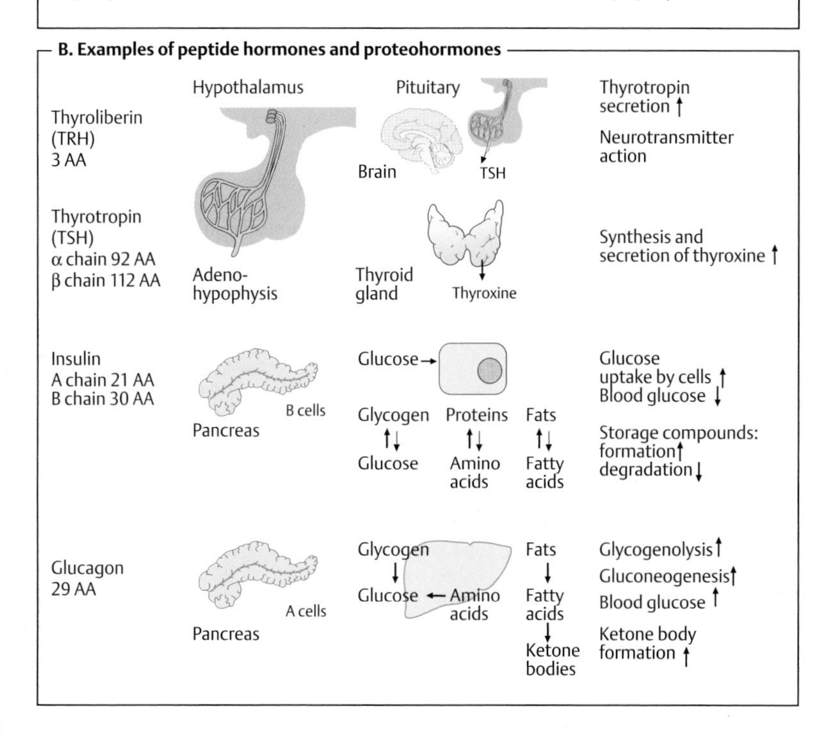

Metabolism of peptide hormones

Hydrophilic hormones and other water-soluble signaling substances have a variety of biosynthetic pathways. Amino acid derivatives arise in special *metabolic pathways* (see p. 352) or through *post-translational modification* (see p. 374). Proteohormones, like all proteins, result from *translation* in the ribosome (see p. 250). Small peptide hormones and neuropeptides, most of which only consist of 3–30 amino acids, are released from precursor proteins by *proteolytic degradation.*

A. Biosynthesis ○

The illustration shows the synthesis and processing of the precursor protein **proopiomelanocortin** (POMC) as an example of the biosynthesis of small peptides with signaling functions. POMC arises in cells of the adenohypophysis, and after processing in the rER and Golgi apparatus, it supplies the opiate-like peptides *met-enkephalin* and *β-endorphin* (implying "opio-"; see p. 352), three *melanocyte-stimulating hormones* (α-, β- and γ-MSH, implying "melano-"), and the glandotropic hormone *corticotropin* (ACTH, implying "-cortin"). Additional products of POMC degradation include two *lipotropins* with catabolic effects in the adipose tissue (β- and γ-LPH).

Some of the peptides mentioned are overlapping in the POMC sequence. For example, additional cleavage of ACTH gives rise to α-MSH and corticotropin-like intermediary peptide (CLIP). Proteolytic degradation of β-LPH provides γ-LPH and β-endorphin. The latter can be further broken down to yield met-enkephalin, while γ-LPH can still give rise to β-MSH (not shown). Due to the numerous derivative products with biological activity that it has, POMC is also known as a **polyprotein**. Which end product is formed and in what amounts depends on the activity of the proteinases in the ER that catalyze the individual cleavages.

The principles underlying protein synthesis and protein maturation (see pp. 230–233) can be summed up once again using the example of POMC:

[1] As a result of *transcription* of the POMC gene and *maturation of the hnRNA,* a mature **mRNA** consisting of some 1100 nucleotides arises, which is modified at both ends (see p. 246). This mRNA codes for prepro-POMC—i. e., a POMC protein that still has a signal peptide for the ER at the N terminus (see p. 230).

[2] **Prepro-POMC** arises through *translation* in the rough endoplasmic reticulum (rER). The growing peptide chain is introduced into the ER with the help of a signal peptide.

[3] Cleavage of the signal peptide and other modifications in the ER (formation of disulfide bonds, glycosylation, phosphorylation) give rise to the mature prohormone ("**pro-POMC**").

[4] The **neuropeptides** and **hormones** mentioned are now formed by *limited proteolysis* and stored in vesicles. Release from these vesicles takes place by exocytosis when needed.

The biosynthesis of peptide hormones and proteohormones, as well as their secretion, is controlled by higher-order regulatory systems (see p. 372). Calcium ions are among the substances involved in this regulation as *second messengers;* an increase in calcium ions stimulates synthesis and secretion.

B. Degradation and inactivation ◑

Degradation of peptide hormones often starts in the blood plasma or on the vascular walls; it is particularly intensive in the kidneys.

Several peptides that contain disulfide bonds (e. g., insulin) can be inactivated by reductive *cleavage of the disulfide bonds* (**1**). Peptides and proteins are also cleaved by *peptidases,* starting from one end of the peptide by *exopeptidases* (**2**), or in the middle of it by proteinases (endopeptidases, **3**). *Proteolysis* gives rise to a variety of hormone fragments, several of which are still biologically active. Some peptide hormones and proteohormones are removed from the blood by binding to their receptors with subsequent *endocytosis* of the hormone–receptor complex (**4**). They are then broken down in the lysosomes. All of the degradation reactions lead to amino acids, which become available to the metabolism again.

A. Biosynthesis

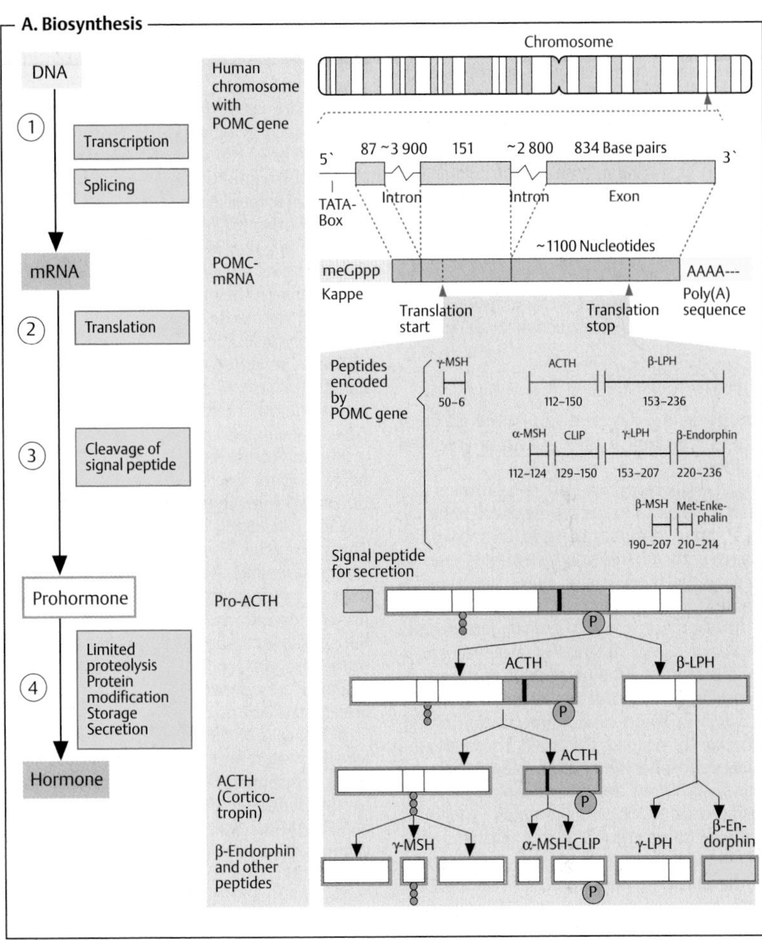

B. Degradation and inactivation

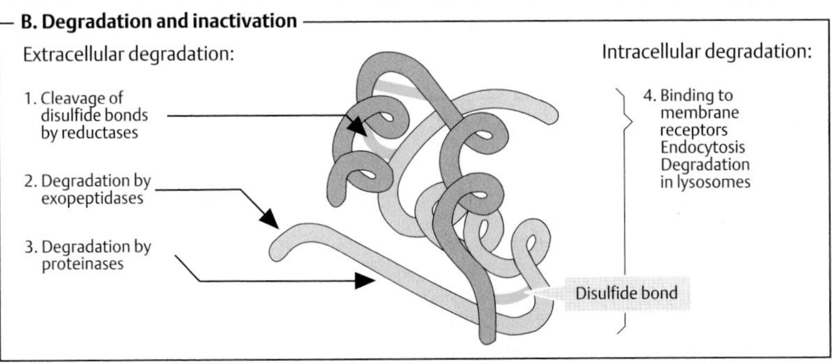

Mechanisms of action

The messages transmitted by hydrophilic signaling substances (see p. 380) are sent to the interior of the cell by *membrane receptors.* These bind the hormone on the outside of the cell and trigger a new second signal on the inside by altering their conformation. In the interior of the cell, this secondary signal influences the activity of enzymes or ion channels. Via further steps, *switching of the metabolism,* changes in the *cytoskeleton,* and activation or inhibition of *transcription factors* can occur *("signal transduction")* can occur.

A. Mechanisms of action ◑

Receptors are classified into three different types according to their structure (see also p. 224):

1. 1-Helix receptors (left) are proteins that span the membrane with only one α-helix. On their inner (cytoplasmic) side, they have domains with *allosterically activatable enzyme activity.* In most cases, these are *tyrosine kinases.*

Insulins (see p. 388), *growth factors,* and *cytokines* (see p. 392), for example, act via 1-helix receptors. Binding of the signaling substance leads to activation of internal kinase activity (in some cases, dimerization of the receptor is needed for this). The activated kinase phosphorylates itself using ATP (*autophosphorylation*), and also phosphorylates tyrosine residues of other proteins (known as *receptor substrates*). Adaptor proteins that recognize the phosphotyrosine residues bind to the phosphorylated proteins (see pp. 388, 392). They pass the signal on to other protein kinases.

2. Ion channels (center). These receptors contain *ligand-gated ion channels.* Binding of the signaling substance opens the channels for ions such as Na^+, K^+, Ca^{2+}, and Cl^-. This mechanism is mainly used by *neurotransmitters* such as *acetylcholine* (nicotinic receptor; see p. 224) and *GABA* (A receptor; see p. 354).

3. 7-Helix receptors (serpentine receptors, right) represent a large group of membrane proteins that transfer the hormone or transmitter signal, with the help of *G proteins* (see below), to *effector proteins* that alter the concentrations of *ions* and *second messengers* (see **B**).

B. Signal transduction by G proteins ◑

G proteins transfer signals from 7-helix receptors to effector proteins (see above). G protein are *heterotrimers* consisting of three different types of subunit (α, β, and γ; see p. 224). The α-subunit can bind GDP or GTP (hence the name "G protein") and has *GTPase activity.* Receptor-coupled G proteins are related to other GTP-binding proteins such as *Ras* (see pp. 388, 398) and *EF-Tu* (see p. 252).

G proteins are divided into several types, depending on their effects. *Stimulatory G proteins* (**G_s**) are widespread. They activate adenylate cyclases (see below) or influence ion channels. *Inhibitory G proteins* (**G_i**) *inhibit* adenylate cyclase. G proteins in the *G_q family* activate another effector enzyme—phospholipase c (see p. 386).

Binding of the signaling substance to a 7-helix receptor alters the receptor conformation in such a way that the corresponding G protein can attach on the inside of the cell. This causes the α-subunit of the G protein to exchange bound GDP for GTP (**1**). The G protein then separates from the receptor and dissociates into an α-subunit and a βγ-unit. Both of these components bind to other membrane proteins and alter their activity; *ion channels* are opened or closed, and *enzymes* are activated or inactivated.

In the case of the β_2-catecholamine receptor (illustrated here), the α-subunit of the G_s protein, by binding to adenylate cyclase, leads to the synthesis of the **second messenger** cAMP. cAMP activates protein kinase A, which in turn activates or inhibits other proteins (**2**; see p. 120).

The βγ-unit of the G protein stimulates a kinase (βARK, not shown), which phosphorylates the receptor. This reduces its affinity for the hormone and leads to binding of the blocking protein *arrestin.* The internal GTPase activity of the α-subunit hydrolyzes the bound GTP to GDP within a period of seconds to minutes, and thereby terminates the action of the G protein on the adenylate cyclase (**3**).

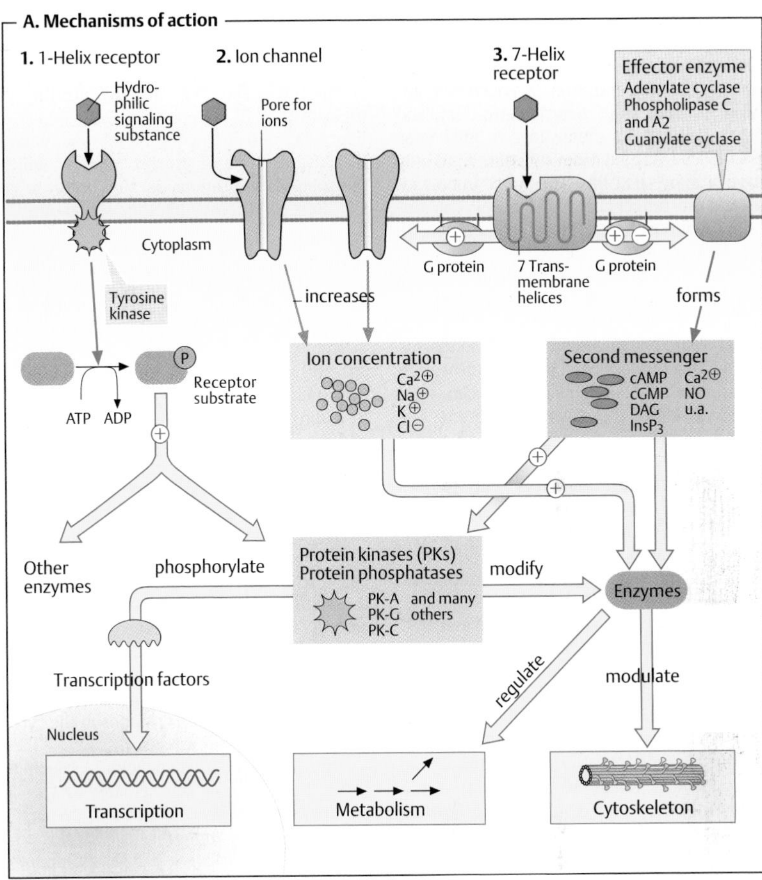
A. Mechanisms of action
1. 1-Helix receptor
Hydro-philic signaling substance
Cytoplasm
Tyrosine kinase
ATP ADP
Receptor substrate
Other enzymes
phosphorylate
Transcription factors
Nucleus
Transcription
2. Ion channel
Pore for ions
increases
Ion concentration
Ca2⊕ Na⊕ K⊕ Cl⊖
Protein kinases (PKs)
Protein phosphatases
PK-A PK-G PK-C and many others
modify
Metabolism
regulate
3. 7-Helix receptor
G protein
7 Trans-membrane helices
G protein
Effector enzyme
Adenylate cyclase
Phospholipase C and A2
Guanylate cyclase
forms
Second messenger
cAMP cGMP DAG InsP3 Ca2⊕ NO u.a.
Enzymes
modulate
Cytoskeleton

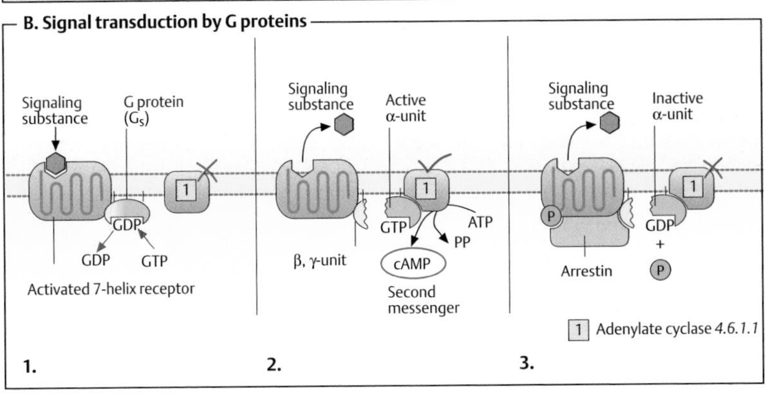
B. Signal transduction by G proteins
Signaling substance
G protein (Gs)
1
GDP
GDP GTP
Activated 7-helix receptor
Signaling substance
Active α-unit
GTP
ATP
PP
cAMP
β, γ-unit
Second messenger
Signaling substance
Inactive α-unit
P
GDP
+
P
Arrestin
1 Adenylate cyclase 4.6.1.1
1.
2.
3.

Second messengers

Second messengers are *intracellular* chemical signals, the concentration of which is regulated by hormones, neurotransmitters, and other extracellular signals (see p. 384). They arise from easily available substrates and only have a short half-life. The most important second messengers are cAMP, cGMP, Ca^{2+}, inositol triphosphate ($InsP_3$), diacylglycerol (DAG), and nitrogen monoxide (NO).

A. Cyclic AMP ◑

Metabolism. The nucleotide **cAMP** (adenosine 3′,5′-cyclic monophosphate) is synthesized by membrane-bound *adenylate cyclases* [1] on the inside of the plasma membrane. The adenylate cyclases are a family of enzymes that cyclize ATP to cAMP by cleaving diphosphate (PP_i). The degradation of cAMP to AMP is catalyzed by *phosphodiesterases* [2], which are inhibited by *methylxanthines* such as caffeine, for example. By contrast, *insulin* activates the esterase and thereby reduces the cAMP level (see p. 388).

Adenylate cyclase activity is regulated by *G proteins* (G_s and G_i), which in turn are controlled by extracellular signals via *7-helix receptors* (see p. 384). Ca^{2+}-calmodulin (see below) also activates specific adenylate cyclases.

Action. cAMP is an allosteric effector of *protein kinase A* (PK-A, [3]). In the inactive state, PK-A is a heterotetramer (C_2R_2), the catalytic subunits of which (C) are blocked by regulatory units (R; autoinhibition). When cAMP binds to the regulatory units, the C units separate from the R units and become enzymatically active. Active PK-A phosphorylates serine and threonine residues of more than 100 different proteins, enzymes, and transcription factors. In addition to cAMP, **cGMP** also acts as a second messenger. It is involved in sight (see p. 358) and in the signal transduction of NO (see p. 388).

B. Inositol 1,4,5-trisphosphate and diacylglycerol ◑

Type G_q G proteins activate *phospholipase C* [4]. This enzyme creates two second messengers from the double-phosphorylated membrane phospholipid *phosphatidylinositol bisphosphate* ($PInsP_2$), i. e., inositol 1,4,5-trisphosphate (**InsP3**), which is soluble, and diacylglycerol (**DAG**). $InsP_3$ migrates to the endoplasmic reticulum (ER), where it opens Ca^{2+} channels that allow Ca^{2+} to flow into the cytoplasm (see **C**). By contrast, DAG, which is lipophilic, remains in the membrane, where it activates type C *protein kinases*, which phosphorylate proteins in the presence of Ca^{2+} ions and thereby pass the signal on.

C. Calcium ions ◑

Calcium level. Ca^{2+} (see p. 342) is a signaling substance. The concentration of Ca^{2+} ions in the cytoplasm is normally very low (10–100 nM), as it is kept down by ATP-driven Ca^{2+} pumps and Na^+/Ca^{2+} exchangers. In addition, many proteins in the cytoplasm and organelles bind calcium and thus act as Ca^{2+} buffers.

Specific signals (e. g., an action potential or second messenger such as $InsP_3$ or cAMP) can trigger a sudden increase in the cytoplasmic Ca^{2+} level to 500–1000 nM by opening Ca^{2+} channels in the plasma membrane or in the membranes of the *endoplasmic* or *sarcoplasmic reticulum. Ryanodine,* a plant substance, acts in this way on a specific channel in the ER. In the cytoplasm, the Ca^{2+} level always only rises very briefly (Ca^{2+} "spikes"), as prolonged high concentrations in the cytoplasm have cytotoxic effects.

Calcium effects. The biochemical effects of Ca^{2+} in the cytoplasm are mediated by special Ca^{2+}-binding proteins (*"calcium sensors"*). These include the *annexins, calmodulin,* and *troponin C* in muscle (see p. 334). **Calmodulin** is a relatively small protein (17 kDa) that occurs in all animal cells. Binding of four Ca^{2+} ions (light blue) converts it into a *regulatory element.* Via a dramatic conformational change (cf. 2a and 2b), Ca^{2+}-calmodulin enters into interaction with other proteins and modulates their properties. Using this mechanism, Ca^{2+} ions regulate the activity of enzymes, ion pumps, and components of the cytoskeleton.

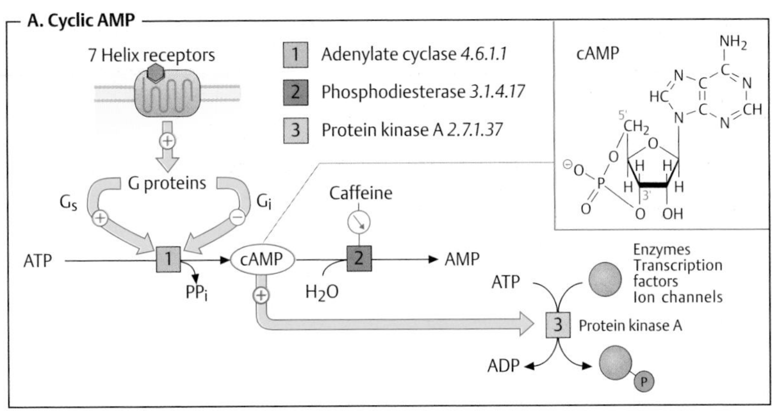
A. Cyclic AMP
7 Helix receptors
1 Adenylate cyclase 4.6.1.1
2 Phosphodiesterase 3.1.4.17
3 Protein kinase A 2.7.1.37
cAMP
G proteins
Gs
Gi
Caffeine
ATP
cAMP
AMP
PPi
H2O
Enzymes
Transcription factors
Ion channels
Protein kinase A
ADP

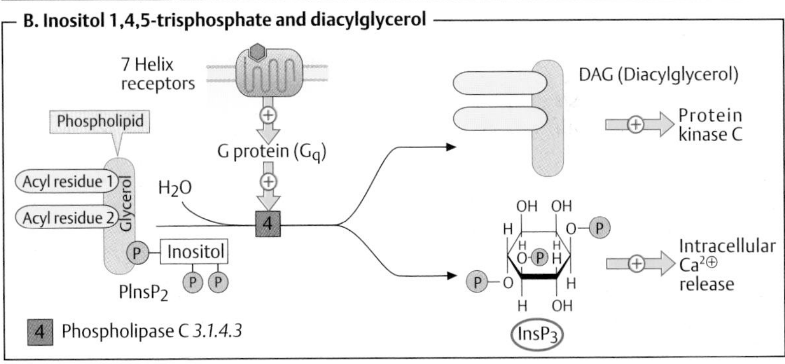
B. Inositol 1,4,5-trisphosphate and diacylglycerol
7 Helix receptors
DAG (Diacylglycerol)
Phospholipid
G protein (Gq)
Protein kinase C
Acyl residue 1
Acyl residue 2
Glycerol
H2O
Inositol
PInsP2
Intracellular Ca2⊕ release
InsP3
4 Phospholipase C 3.1.4.3

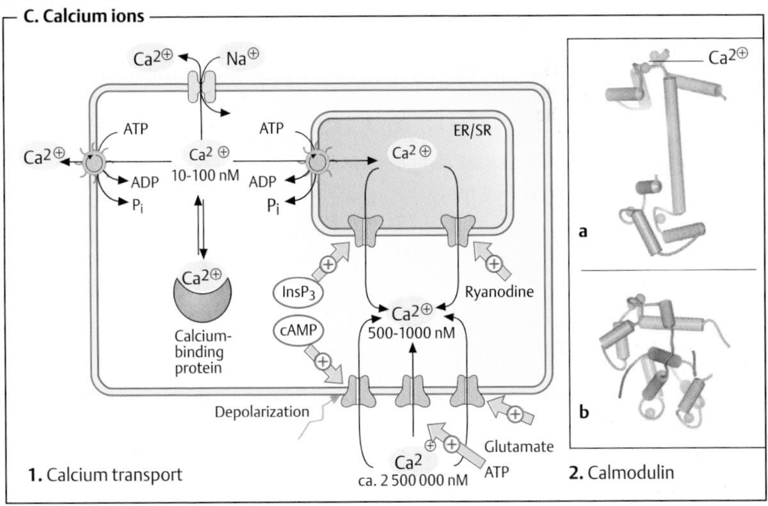
C. Calcium ions
Ca2⊕
Na⊕
ATP
ER/SR
Ca2 ⊕ 10-100 nM
ADP
Pi
Calcium-binding protein
InsP3
cAMP
Ryanodine
Ca2⊕ 500-1000 nM
Depolarization
Glutamate
ATP
Ca2 ca. 2 500 000 nM
1. Calcium transport
2. Calmodulin
a
b

Signal cascades

The signal transduction pathways that mediate the effects of the metabolic hormone **insulin** are of particular medical interest (see **A**). The mediator **nitrogen monoxide** (NO) is also clinically important, as it regulates vascular caliber and thus the body's perfusion with blood (see **B**).

A. Insulin: signal transduction ○

The diverse effects of insulin (see p. 160) are mediated by protein kinases that mutually activate each other in the form of enzyme cascades. At the end of this chain there are kinases that influence gene transcription in the nucleus by phosphorylating target proteins, or promote the uptake of glucose and its conversion into glycogen. The signal transduction pathways involved have not yet been fully explained. They are presented here in a simplified form.

The **insulin receptor** (top) is a dimer with subunits that have activatable tyrosine kinase domains in the interior of the cell (see p. 224). Binding of the hormone increases the tyrosine kinase activity of the receptor, which then phosphorylates itself and other proteins (**receptor substrates**) at various tyrosine residues. **Adaptor proteins**, which conduct the signal further, bind to the phosphotyrosine residues.

The effects of insulin on transcription are shown on the left of the illustration. Adaptor proteins **Grb-2** and **SOS** ("son of sevenless") bind to the phosphorylated **IRS** (insulin-receptor substrate) and activate the G protein **Ras** (named after its gene, the oncogene *ras*; see p. 398). Ras activates the protein kinase **Raf** (another oncogene product). Raf sets in motion a phosphorylation cascade that leads via the kinases **MEK** and **ERK** (also known as MAPK, "mitogen-activated protein kinase") to the phosphorylation of transcription factors in the nucleus.

Some of the effects of insulin on the **carbohydrate metabolism** (right part of the illustration) are possible without protein synthesis. In addition to Grb-2, another dimeric adaptor protein can also bind to phosphorylated IRS. This adaptor protein thereby acquires *phosphatidylinositol-3-kinase* activity (**PI_3K**) and, in the membrane, phosphorylates phospholipids from the phosphatidylinositol group (see p. 50) at position 3. Protein kinase **PDK-1** binds to these reaction products, becoming activated itself and in turn activating protein kinase B (**PK-B**).

This has several effects. In a manner not yet fully understood, PK-B leads to the fusion with the plasma membrane of vesicles that contain the glucose transporter Glut-4. This results in inclusion of Glut-4 in the membrane and thus to increased glucose uptake into the muscles and adipose tissue (see p. 160). In addition, PK-B inhibits glycogen synthase kinase 3 (**GSK-3**) by phosphorylation. As GSK-3 in turn inhibits glycogen synthase by phosphorylation (see p. 120), its inhibition by PK-B leads to *increased* glycogen synthesis. Protein phosphatase-1 (**PP-1**) converts glycogen synthase into its active form by dephosphorylation (see p. 120). PP-1 is also activated by insulin.

B. Nitrogen monoxide (NO) as a mediator ○

Nitrogen monoxide (NO) is a short-lived radical that functions as a locally acting mediator (see p. 370).

In a complex reaction, NO arises from arginine in the endothelial cells of the blood vessels [1]. The trigger for this is Ca^{2+}-calmodulin (see p. 386), which forms when there is an increase in the cytoplasmic Ca^{2+} level.

NO diffuses from the endothelium into the underlying vascular muscle cells, where it leads, as a result of activation of *guanylate cyclase* [2], to the formation of the second messenger **cGMP** (see pp. 358, 384). Finally, by activating a special protein kinase (**PK-G**), cGMP triggers relaxation of the smooth muscle and thus dilation of the vessels. The effects of atrionatriuretic peptide (**ANP**; see p. 328) in reducing blood pressure are also mediated by cGMP-induced vasodilation. In this case, cGMP is formed by the guanylate cyclase activity of the ANP receptor.

Further information

The drug *nitroglycerin* (glyceryl trinitrate), which is used in the treatment of *angina pectoris*, releases NO in the bloodstream and thereby leads to better perfusion of cardiac muscle.

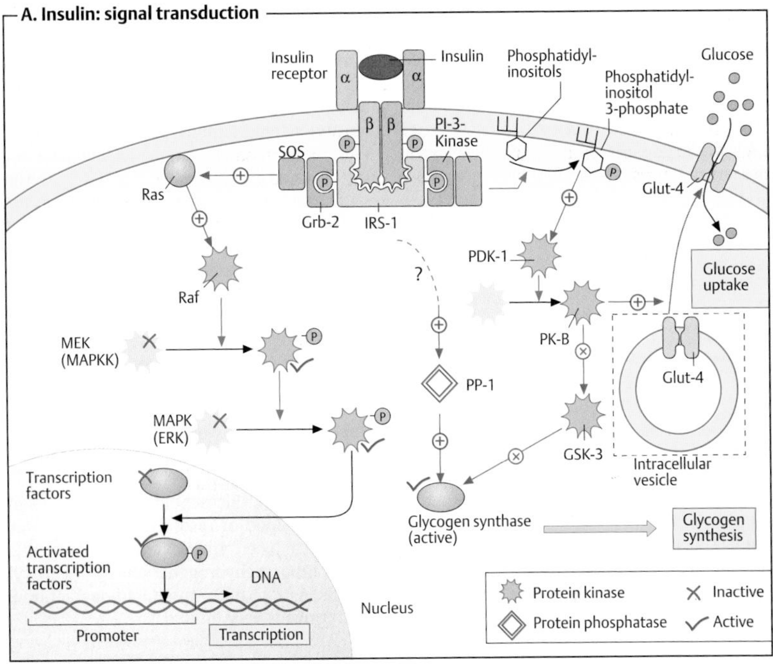
A. Insulin: signal transduction
Insulin receptor
Insulin
α
α
β β
Phosphatidyl-inositols
Phosphatidyl-inositol 3-phosphate
Glucose
PI-3-Kinase
SOS
Ras
Grb-2
IRS-1
Glut-4
Raf
PDK-1
?
Glucose uptake
MEK (MAPKK)
PK-B
PP-1
Glut-4
MAPK (ERK)
GSK-3
Intracellular vesicle
Transcription factors
Glycogen synthase (active)
Glycogen synthesis
Activated transcription factors
DNA
Nucleus
Promoter
Transcription
Protein kinase
Protein phosphatase
Inactive
Active

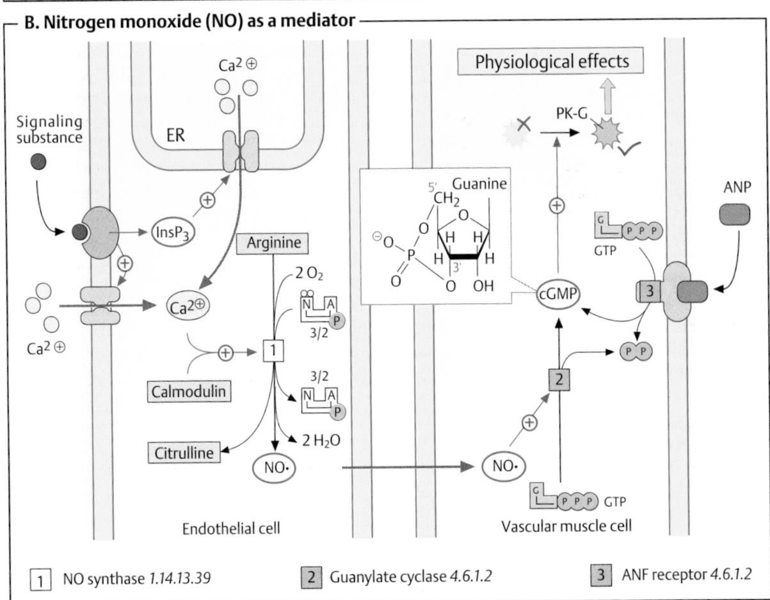
B. Nitrogen monoxide (NO) as a mediator
Physiological effects
Signaling substance
ER
PK-G
Guanine
ANP
InsP3
Arginine
GTP
2 O2
cGMP
Ca2+
3/2
1
Calmodulin
3/2
2
Citrulline
2 H2O
NO·
NO·
GTP
Endothelial cell
Vascular muscle cell
1 NO synthase 1.14.13.39
2 Guanylate cyclase 4.6.1.2
3 ANF receptor 4.6.1.2

Eicosanoids

The eicosanoids are a group of signaling substances that arise from the C-20 fatty acid *arachidonic acid* and therefore usually contain 20 C atoms (Greek *eicosa* = 20). As mediators, they influence a large number of physiological processes (see below). Eicosanoid metabolism is therefore an important drug target. As short-lived substances, eicosanoids only act in the vicinity of their site of synthesis (paracrine effect; see p. 372).

A. Eicosanoids ○

Biosynthesis. Almost all of the body's cells form eicosanoids. Membrane phospholipids that contain the polyunsaturated fatty acid **arachidonic acid** (20:4; see p. 48) provide the starting material.

Initially, *phospholipase* A_2 [1] releases the arachidonate moiety from these phospholipids. The activity of phospholipase A_2 is strictly regulated. It is activated by hormones and other signals via *G proteins*. The arachidonate released is a signaling substance itself. However, its metabolites are even more important.

Two different pathways lead from arachidonate to **prostaglandins, prostacyclins,** and **thromboxanes,** on the one hand, or **leukotrienes** on the other. The key enzyme for the first pathway is *prostaglandin synthase* [2]. Using up O_2, it catalyzes in a two-step reaction the cyclization of arachidonate to prostaglandin H_2, the parent substance for the prostaglandins, prostacyclins, and thromboxanes. Acetylsalicylic acid (aspirin) irreversibly acetylates a serine residue near the active center of prostaglandin synthase, so that access for substrates is blocked (see below).

As a result of the action of *lipoxygenases* [3], hydroxyfatty acids and hydroperoxyfatty acids are formed from arachidonate, from which elimination of water and various conversion reactions give rise to the leukotrienes. The formulae only show one representative from each of the various groups of eicosanoids.

Effects. Eicosanoids act via membrane receptors in the immediate vicinity of their site of synthesis, both on the synthesizing cell itself (*autocrine* action) and on neighboring cells (*paracrine* action). Many of their effects are mediated by the second messengers cAMP and cGMP.

The eicosanoids have a very wide range of physiological effects. As they can stimulate or inhibit smooth-muscle contraction, depending on the substance concerned, they affect blood pressure, respiration, and intestinal and uterine activity, among other properties. In the stomach, prostaglandins inhibit HCl secretion via G_i proteins (see p. 270). At the same time, they promote mucus secretion, which protects the gastric mucosa against the acid. In addition, prostaglandins are involved in bone metabolism and in the activity of the sympathetic nervous system. In the immune system, prostaglandins are important in the inflammatory reaction. Among other things, they attract leukocytes to the site of infection. Eicosanoids are also decisively involved in the development of pain and fever. The thromboxanes promote thrombocyte aggregation and other processes involved in hemostasis (see p. 290).

Metabolism. Eicosanoids are inactivated within a period of seconds to minutes. This takes place by enzymatic reduction of double bonds and dehydrogenation of hydroxyl groups. As a result of this rapid degradation, their range is very limited.

Further information

Acetylsalicylic acid and related non-steroidal anti-inflammatory drugs (NSAIDs) selectively inhibit the *cyclooxygenase activity* of prostaglandin synthase [2] and consequently the synthesis of most eicosanoids. This explains their analgesic, antipyretic, and antirheumatic effects. Frequent side effects of NSAIDs also result from inhibition of eicosanoid synthesis. For example, they impair hemostasis because the synthesis of thromboxanes by thrombocytes is inhibited. In the stomach, NSAIDs increase HCl secretion and at the same time inhibit the formation of protective mucus. Long-term NSAID use can therefore damage the gastric mucosa.

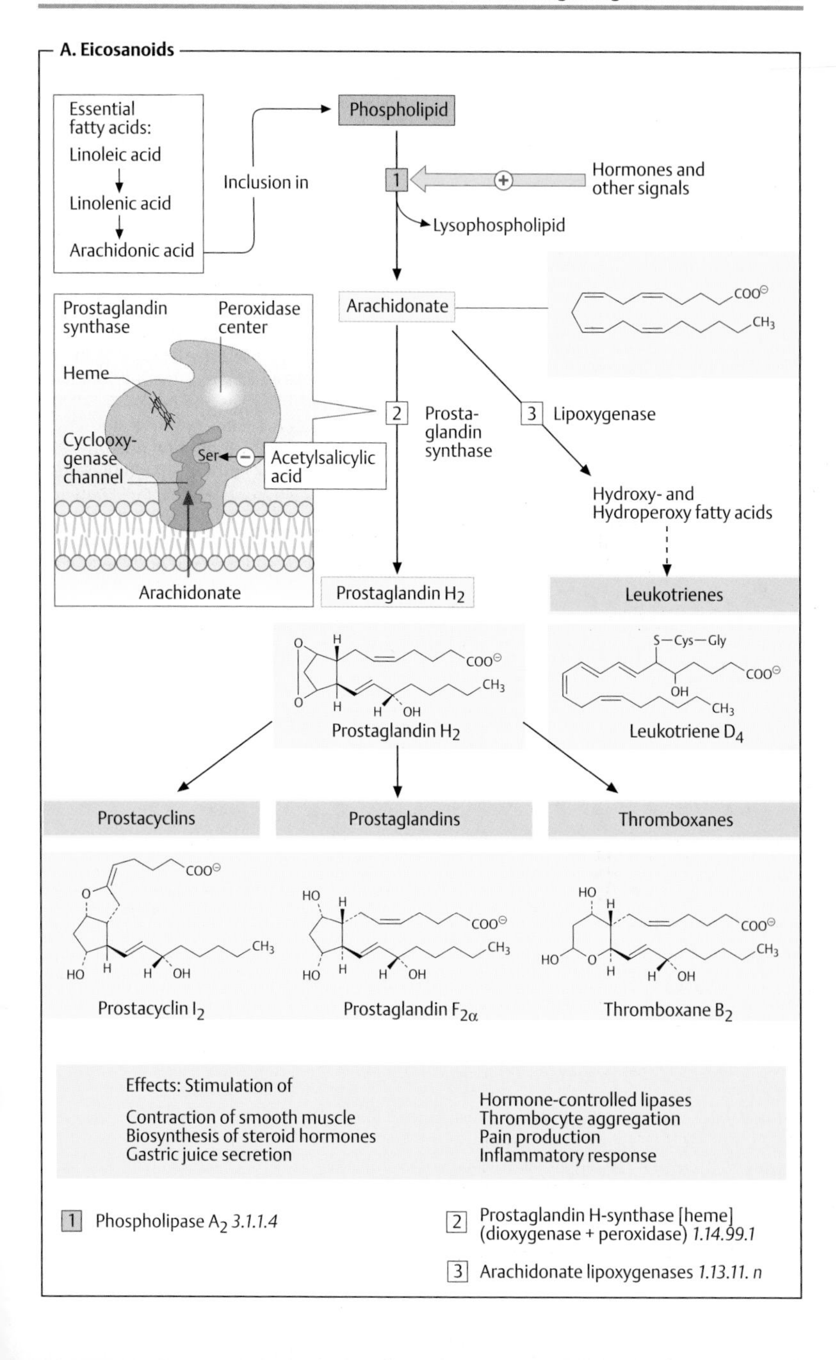
A. Eicosanoids
Essential fatty acids:
Linoleic acid
Linolenic acid
Arachidonic acid
Inclusion in
Phospholipid
1
Hormones and other signals
Lysophospholipid
Arachidonate
COO⊖
CH3
Prostaglandin synthase
Peroxidase center
Heme
Cyclooxygenase channel
Ser
Acetylsalicylic acid
Arachidonate
2
Prostaglandin synthase
3
Lipoxygenase
Hydroxy- and Hydroperoxy fatty acids
Prostaglandin H2
Leukotrienes
S—Cys—Gly
OH
Leukotriene D4
Prostaglandin H2
Prostacyclins
Prostaglandins
Thromboxanes
Prostacyclin I2
Prostaglandin F2α
Thromboxane B2
Effects: Stimulation of
Contraction of smooth muscle
Biosynthesis of steroid hormones
Gastric juice secretion
Hormone-controlled lipases
Thrombocyte aggregation
Pain production
Inflammatory response
1 Phospholipase A2 3.1.1.4
2 Prostaglandin H-synthase [heme] (dioxygenase + peroxidase) 1.14.99.1
3 Arachidonate lipoxygenases 1.13.11. n

Cytokines

A. Cytokines ◑

Cytokines are hormone-like *peptides* and *proteins* with signaling functions, which are synthesized and released by cells of the immune system and other cell types. Their numerous biological functions operate in three areas: they regulate the *development and homeostasis of the immune system;* they control the *hematopoietic system;* and they are involved in *non-specific defense,* influencing inflammatory processes, blood coagulation, and blood pressure. In general, cytokines regulate the growth, differentiation, and survival of cells. They are also involved in regulating apoptosis (see p. 396).

There is an extremely large number of cytokines; only the most important representatives are listed opposite. The cytokines include *interleukins* (IL), *lymphokines, monokines, chemokines, interferons* (IFN), and *colony-stimulating factors* (CSF). Via interleukins, immune cells stimulate the proliferation and activity of other immune cells (see p. 294). Interferons are used medically in the treatment of viral infections and other diseases.

Although cytokines rarely show structural homologies with each other, their effects are often very similar. The cytokines differ from *hormones* (see p. 370) only in certain respects: they are released by many different cells, rather than being secreted by defined glands, and they regulate a wider variety of target cells than the hormones.

B. Signal transduction in the cytokines ○

As peptides or proteins, the cytokines are **hydrophilic signaling substances** that act by binding to *receptors* on the cell surface (see p. 380). Binding of a cytokine to its receptor (**1**) leads via several intermediate steps (**2** –**5**) to the activation of transcription of specific genes (**6**).

In contrast to the receptors for insulin and growth factors (see p. 388), the **cytokine receptors** (with a few exceptions) have *no* tyrosine kinase activity. After binding of cytokine (**1**), they associate with one another to form homodimers, join together with other signal transduction proteins (STPs) to form dimers, or promote dimerization of other STPs (2). Class I cytokine receptors interact with three different STPs (gp130, β_c, and γ_c). The STPs themselves do not bind cytokines, but conduct the signal to tyrosine kinases (**3**). The fact that different cytokines can activate the same STP via their receptors explains the overlapping biological activity of some cytokines.

As an example of the signal transduction pathway in cytokines, the illustration shows the way in which the **IL-6 receptor**, after binding its ligand **IL-6** (**1**), induces the dimerization of the STP **gp130** (**2**). The dimeric gp130 binds cytoplasmic *tyrosine kinases* from the Jak family (“**Janus kinases**,” with two kinase centers) and activates them (**3**). The Janus kinases phosphorylate cytokine receptors, STPs, and various cytoplasmic proteins that conduct the signal further. In addition, they phosphorylate transcription factors known as **STATs** (“**s**ignal **t**ransducers and **a**ctivators of **t**ranscription”). STATs are among the proteins that have an *SH2 domain* and are able to bind phosphotyrosine residues (see p. 388). They therefore bind to cytokine receptors that have been phosphorylated by Janus kinases. When STATs are then also phosphorylated themselves (**4**), they are converted into their active form and become dimers (**5**). After transfer to the nucleus, they bind—along with auxiliary proteins as transcription factors—to the promoters of inducible genes and in this way regulate their transcription (**6**).

The activity of the cytokine receptors is terminated by *protein phosphatases,* which hydrolytically cleave the phosphotyrosine residues. Several cytokine receptors are able to lose their ligand-binding extracellular domain by proteolysis (not shown). The extracellular domain then appears in the blood, where it competes for cytokines. This reduces the effective cytokine concentration.

A. Cytokines

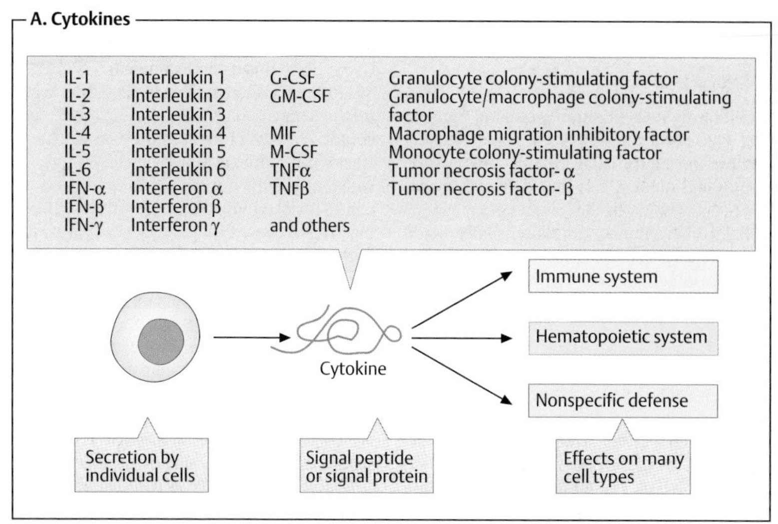

B. Signal transduction in the cytokines

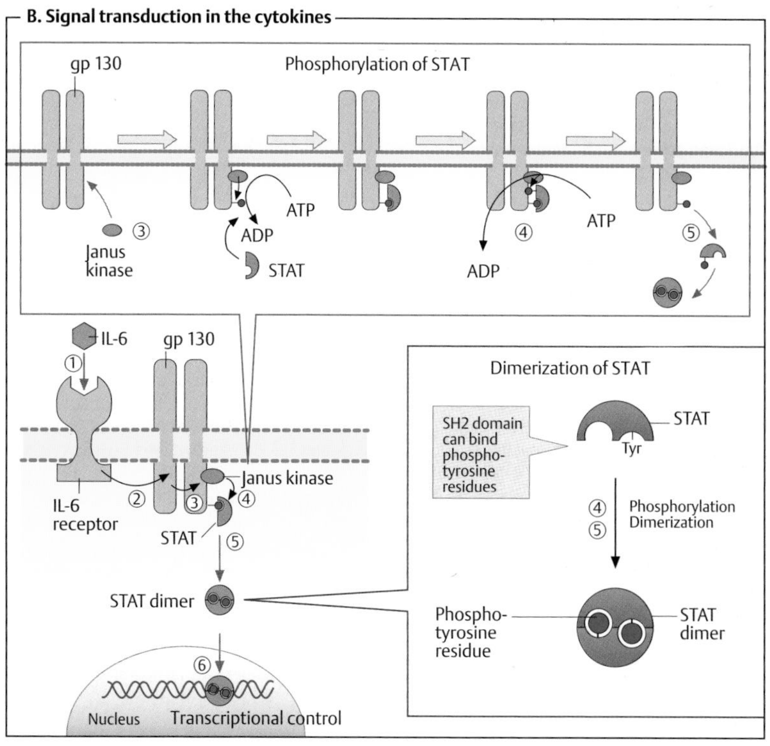

Cell cycle

A. Cell cycle ◑

Proliferating cells undergo a cycle of division (the cell cycle), which lasts approximately 24 hours in mammalian cells in cell culture. The cycle is divided into four different phases (G_1, S, G_2, and M—in that sequence).

Fully differentiated animal cells only divide rarely. These cells are in the so-called **G_0 phase**, in which they can remain permanently. Some G_0 cells return to the **G_1 phase** again under the influence of mitogenic signals (growth factors, cytokines, tumor viruses, etc.), and after crossing a *control point* (G_1 to S), enter a new cycle. DNA is replicated (see p. 240) during the **S phase**, and new chromatin is formed. Particularly remarkable in morphological terms is the actual mitosis (**M phase**), in which the chromosomes separate and two daughter cells are formed. The M and S phases are separated by two segments known as the **G_1** and **G_2** phases (the G stands for "gap"). In the G_1 phase, the duration of which can vary, the cell grows by de novo synthesis of cell components. Together, the G_1, G_0, S, and G_2 phases are referred to as the **interphase**, which alternates in the cell cycle with the short M phase.

B. Control of the cell cycle ○

The progression of the cell cycle is regulated by **interconversion processes**. In each phase, special *Ser/Thr-specific protein kinases* are formed, which are known as cyclin-dependent kinases (**CDKs**). This term is used because they have to bind an activator protein (**cyclin**) in order to become active. At each control point in the cycle, specific CDKs associate with equally phase-specific cyclins. If there are no problems (e.g., DNA damage), the CDK–cyclin complex is activated by phosphorylation and/or dephosphorylation. The activated complex in turn phosphorylates **transcription factors**, which finally lead to the formation of the proteins that are required in the cell cycle phase concerned (enzymes, cytoskeleton components, other CDKs, and cyclins). The activity of the CDK–cyclin complex is then terminated again by proteolytic **cyclin degradation**.

The above outline of cell cycle progression can be examined here in more detail using the G_2–M transition as an example.

Entry of animal cells into mitosis is based on the "mitosis-promoting factor" (**MPF**). MPF consists of **CDK1** (cdc2) and **cyclin B**. The intracellular concentration of cyclin B increases constantly until mitosis starts, and then declines again rapidly (top left). MPF is initially inactive, because CDK1 is phosphorylated and cyclin B is dephosphorylated (top center). The M phase is triggered when a *protein phosphatase* [1] dephosphorylates the CDK while cyclin B is phosphorylated by a *kinase* [2]. In its active form, MPF phosphorylates various proteins that have functions in mitosis—e. g., *histone H1* (see p. 238), components of the cytoskeleton such as the *laminins* in the nuclear membrane, *transcription factors, mitotic spindle* proteins, and various *enzymes*.

When mitosis has been completed, cyclin B is marked with **ubiquitin** and broken down proteolytically by *proteasomes* (see p. 176). Protein phosphatases then regain control and dephosphorylate the proteins involved in mitosis. This returns the cell to the interphase.

Further information

The G_1–S transition (not shown) is particularly important for initiating the cell cycle. It is triggered by the CDK4–cyclin D complex, which by phosphorylating the protein **pRb** releases the transcription factor E2F previously bound to pRb. This activates the transcription of genes needed for DNA replication.

If the DNA is damaged by mutagens or ionizing radiation, the protein **p53** initially delays entry into the S phase. If the DNA repair system (see p. 256) does not succeed in removing the DNA damage, p53 forces the cell into apoptosis (see p. 396). The genes coding for pRb and p53 belong to the **tumor-suppressor genes** (see p. 398). In many tumors (see p. 400), these genes are in fact damaged by mutation.

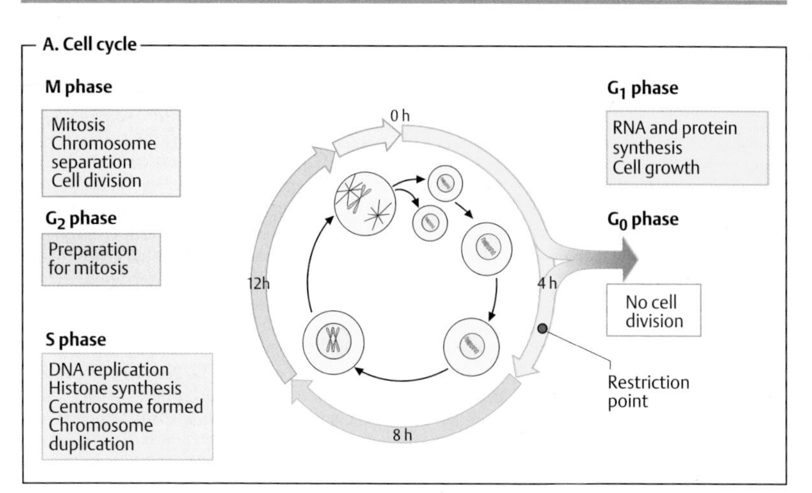
A. Cell cycle
M phase
Mitosis
Chromosome separation
Cell division
G₂ phase
Preparation for mitosis
S phase
DNA replication
Histone synthesis
Centrosome formed
Chromosome duplication
0 h
12h
4 h
8 h
G₁ phase
RNA and protein synthesis
Cell growth
G₀ phase
No cell division
Restriction point

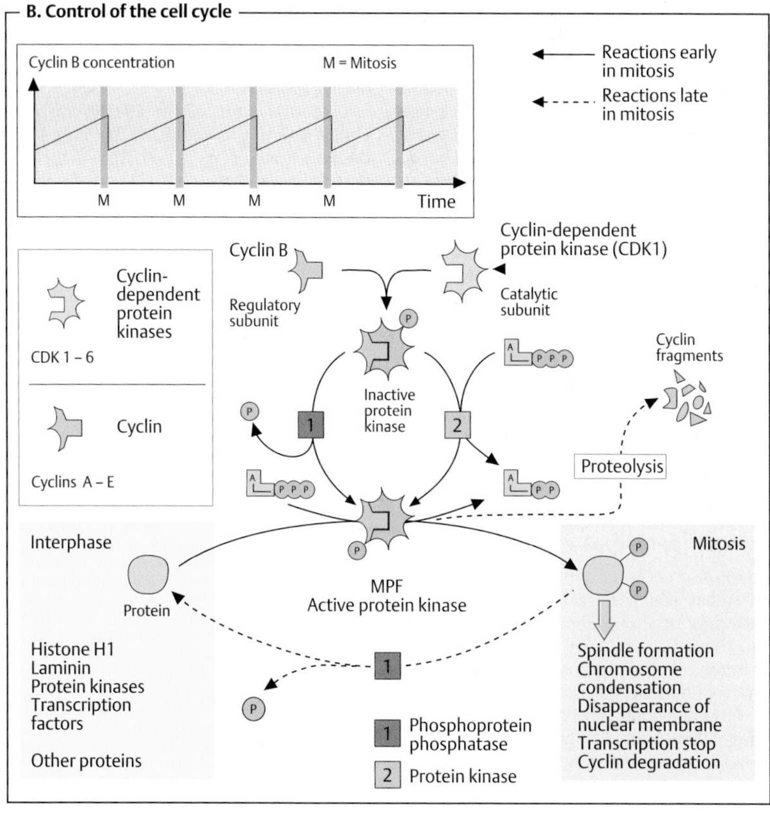
B. Control of the cell cycle
Cyclin B concentration
M = Mitosis
M
M
M
M
Time
Reactions early in mitosis
Reactions late in mitosis
Cyclin-dependent protein kinases
CDK 1 – 6
Cyclin
Cyclins A – E
Cyclin B
Regulatory subunit
Cyclin-dependent protein kinase (CDK1)
Catalytic subunit
Inactive protein kinase
Cyclin fragments
Proteolysis
Interphase
Protein
Histone H1
Laminin
Protein kinases
Transcription factors
Other proteins
MPF
Active protein kinase
Mitosis
Spindle formation
Chromosome condensation
Disappearance of nuclear membrane
Transcription stop
Cyclin degradation
1 Phosphoprotein phosphatase
2 Protein kinase

Apoptosis

A. Cell proliferation and apoptosis ◑

The number of cells in any tissue is mainly regulated by two processes—cell **proliferation** and *physiological cell death*, **apoptosis.** Both of these processes are regulated by stimulatory and inhibitory factors that act in solute form (growth factors and cytokines) or are presented in bound form on the surface of neighboring cells (see below).

Apoptosis is genetically programmed cell death, which leads to "tidy" breakdown and disposal of cells. Morphologically, apoptosis is characterized by changes in the cell membrane (with the formation of small blebs known as "apoptotic bodies"), shrinking of the nucleus, chromatin condensation, and fragmentation of DNA. *Macrophages* and other phagocytic cells recognize apoptotic cells and remove them by phagocytosis without inflammatory phenomena developing.

Cell necrosis (not shown) should be distinguished from apoptosis. In cell necrosis, cell death is usually due to physical or chemical damage. Necrosis leads to swelling and bursting of the damaged cells and often triggers an inflammatory response.

The growth of tissue (or, more precisely, the number of cells) is actually regulated by apoptosis. In addition, apoptosis allows the elimination of unwanted or superfluous cells—e. g., during embryonic development or in the immune system. The contraction of the uterus after birth is also based on apoptosis. Diseased cells are also eliminated by apoptosis—e. g., tumor cells, virus-infected cells, and cells with irreparably damaged DNA. An everyday example of this is the peeling of the skin after sunburn.

B. Regulation of apoptosis ○

Apoptosis can be triggered by a number of different signals that use various transmission pathways. Other signaling pathways prevent apoptosis.

At the center of the apoptotic process lies a group of specialized *cysteine-containing aspartate proteinases* (see p. 176), known as **caspases.** These mutually activate one another, creating an *enzyme cascade* resembling the cascade involved in blood coagulation (see p. 290). Other enzymes in this group, known as **effector caspases**, cleave cell components after being activated—e. g., laminin in the nuclear membrane and snRP proteins (see p. 246)—or activate special DNases which then fragment the nuclear DNA.

An important trigger for apoptosis is known as the **Fas system.** This is used by cytotoxic T cells, for example, which eliminate infected cells in this way (top left). Most of the body's cells have *Fas receptors* (CD 95) on their plasma membrane. If a T cell is activated by contact with an MHC presenting a viral peptide (see p. 296), binding of its *Fas ligands* occurs on the target cell's Fas receptors. Via the mediator protein FADD ("Fas-associated death domain"), this activates *caspase-8* inside the cell, setting in motion the apoptotic process.

Another trigger is provided by **tumor necrosis factor-α** (TNF-α), which acts via a similar protein (TRADD) and supports the endogenous defense system against tumors by inducing apoptosis.

Caspase-8 activates the effector caspases either directly, or indirectly by promoting the **cytochrome c** (see p. 140) from mitochondria. Once in the cytoplasm, cytochrome c binds to and activates the protein Apaf-1 (not shown) and thus triggers the caspase cascade. Apoptotic signals can also come from the cell nucleus. If irreparable DNA damage is present, the **p53 protein** (see p. 394)—the product of a *tumor suppressor gene*—promotes apoptosis and thus helps eliminate the defective cell.

There are also inhibitory factors that oppose the signals that activate apoptosis. These include **bcl-2** and related proteins. The genomes of several viruses include genes for this type of protein. The genes are expressed by the host cell and (to the benefit of the virus) prevent the host cell from being prematurely eliminated by apoptosis.

A. Cell proliferation and apoptosis

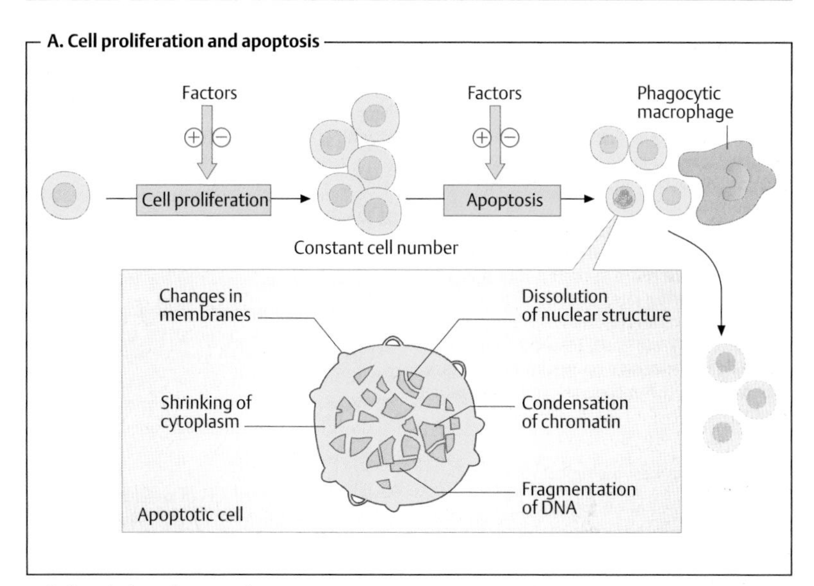

B. Regulation of apoptosis

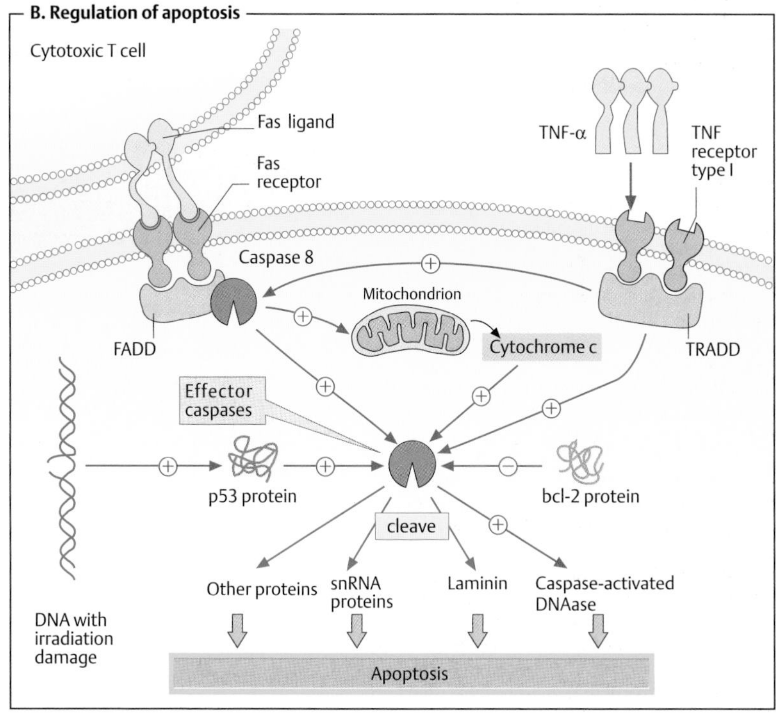

Oncogenes

Oncogenes are cellular genes that can trigger uncontrolled cell proliferation if their sequence is altered or their expression is incorrectly regulated. They were first discovered as *viral (v-) oncogenes* in retroviruses that cause tumors (tumor viruses). Viruses of this type (see p. 404) sometimes incorporate genes from the host cell into their own genome. If these genes are reincorporated into the host DNA again during later infection, tumors can then be caused in rare cases. Although virus-related tumors are rare, research into them has made a decisive contribution to our understanding of oncogenes and their functioning.

A. Proto-oncogenes: biological role ◑

The cellular form of oncogenes (known as c-oncogenes or **proto-oncogenes**) code for proteins involved in controlling growth and differentiation processes. They only become oncogenes if their sequence has been altered by *mutations* (see p. 256), *deletions*, and other processes, or when excessive amounts of the gene products have been produced as a result of *overexpression*.

Overexpression can occur when **amplification** leads to numerous functional copies of the respective gene, or when the gene falls under the influence of a highly active promoter (see p. 244). If the control of oncogene expression by **tumor suppressor genes** (see p. 394) is also disturbed, **transformation** and unregulated proliferation of the cells can occur. A single activated oncogene does not usually lead to a loss of growth control. It only occurs when over the course of time mutations and regulation defects accumulate in one and the same cell. If the immune system does not succeed in eliminating the transformed cell, it can over the course of months or years grow into a macroscopically visible **tumor**.

B. Oncogene products: biochemical functions ◑

A feature common to all oncogenes is the fact that they code for proteins involved in *signal transduction processes*. The genes are designated using three-letter abbreviations that usually indicate the origin of the viral gene and are printed in italics (e. g., *myc* for myelocytomatosis, a viral disease in birds). Oncogene products can be classified into the following groups according to their functions.

1. **Ligands** such as *growth factors* and *cytokines*, which promote cell proliferation.
2. **Membrane receptors** of the 1-helix type with tyrosine kinase activity, which can bind growth factors and hormones (see p. 394).
3. **GTP-binding proteins.** This group includes the G proteins in the strict sense and related proteins such as Ras (see p. 388), the product of the oncogene c-*ras*.
4. **Receptors for lipophilic hormones** mediate the effects of steroid hormones and related signaling substances. They regulate the transcription of specific genes (see p. 378). The products of several oncogenes (e. g., *erbA*) belong to this superfamily of *ligand-controlled transcription factors*.
5. **Nuclear tumor suppressors** inhibit return to the cell cycle in fully differentiated cells. The genes that code for these proteins are referred to as *anti-oncogenes* due to this function. On the role of p53 and pRb, see p. 394.
6. **DNA-binding proteins.** A whole series of oncogenes code for *transcription factors*. Particularly important for cell proliferation are *myc*, as well as *fos* and *jun*. The protein products of the latter two genes form the transcription factor AP-1 as a heterodimer (see p. 244).
7. **Protein kinases** play a central role in intracellular signal transduction. By phosphorylating proteins, they bring about alterations in biological activity that can only be reversed again by the effects of *protein phosphatases*. The interplay between protein phosphorylation by protein kinases and dephosphorylation by protein phosphatases (*interconversion*) serves to regulate the cell cycle (see p. 394) and other important processes. The protein kinase Raf is also involved in the signal transduction of insulin (see p. 388).

A. Proto-oncogenes: biological role

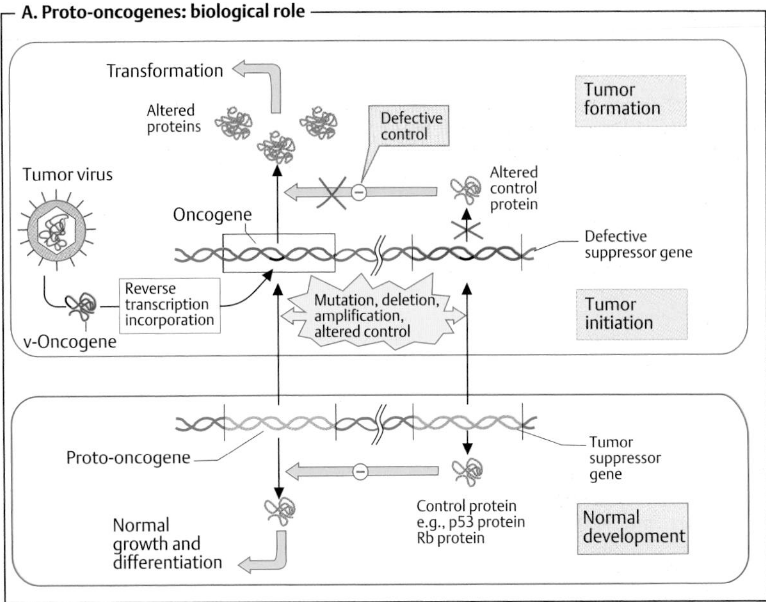

B. Oncogene products: biochemical functions

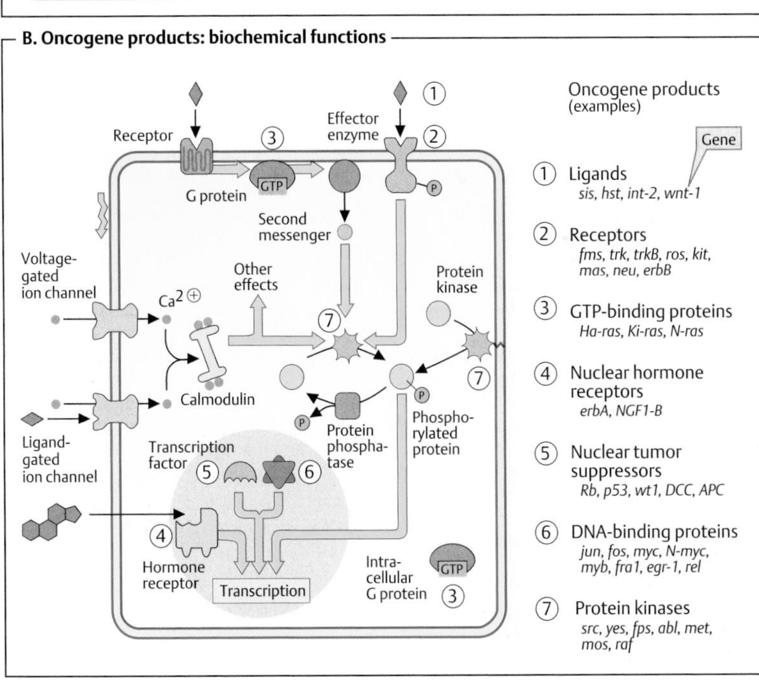

Tumors

A. Division behavior of cells ◑

The body's cells are normally subject to strict "social" control. They only divide until they come into contact with neighboring cells; cell division then ceases due to *contact inhibition.* Exceptions to this rule include embryonic cells, cells of the intestinal epithelium (where the cells are constantly being replaced), cells in the bone marrow (where formation of blood cells takes place), and **tumor cells.** *Uncontrolled cell proliferation* is an important indicator of the presence of a tumor. While normal cells in cell culture only divide 20–60 times, tumor cells are potentially immortal and are not subject to contact inhibition.

In medicine, a distinction is made between *benign* and *malignant tumors.* Benign tumors consist of slowly growing, largely differentiated cells. By contrast, malignant tumors show rapid, invasive growth and tend to form *metastases* (dissemination of daughter lesions). The approximately 100 different types of tumor that exist are responsible for more than 20% of deaths in Europe and North America.

B. Transformation ◑

The transition of a normal cell into a tumor cell is referred to as **transformation.**

Normal cells have all the characteristics of fully differentiated cells specialized for a particular function. Their division is inhibited and they are usually in the G_0 phase of the cell cycle (see p. 394). Their external shape is variable and is determined by a strongly structured cytoskeleton.

In contrast, *tumor cells* divide without inhibition and are often de-differentiated—i. e., they have acquired some of the properties of embryonic cells. The surface of these cells is altered, and this is particularly evident in a disturbance of contact inhibition by neighboring cells. The cytoskeleton of tumor cells is also restructured and often reduced, giving them a rounded shape. The nuclei of tumor cells can be atypical in terms of shape, number, and size.

Tumor markers are clinically important for detecting certain tumors. These are proteins that are formed with increasing frequency by tumor cells (group 1) or are induced by them in other cells (group 2). Group 1 tumor markers include *tumor-associated antigens,* secreted hormones, and enzymes. The table lists a few examples.

The transition from a normal to a transformed state is a process involving several steps.

1. Tumor initiation. Almost every tumor begins with damage to the DNA of an individual cell. The genetic defect is almost always caused by environmental factors. These can include tumor-inducing chemicals (*carcinogens*—e. g., components of tar from tobacco), physical processes (e. g., UV light, X-ray radiation; see p. 256), or in rare cases tumor viruses (see p. 398). Most of the approximately 10^{14} cells in the human body probably suffer this type of DNA damage during the average lifespan, but it is usually repaired again (see p. 256). It is mainly defects in **proto-oncogenes** (see p. 398) that are relevant to tumor initiation; these are the decisive cause of *transformation.* Loss of an **anti-oncogene** (a tumor-suppressor gene) can also contribute to tumor initiation.

2. Tumor promotion is preferential proliferation of a cell damaged by transformation. It is a very slow process that can take many years. Certain substances are able to strongly accelerate it—e. g., *phorbol esters.* These occur in plants (e. g., *Euphorbia* species) and act as activators of protein kinase C (see p. 386).

3. Tumor progression finally leads to a macroscopically visible tumor as a result of growth. When solid tumors of this type exceed a certain size, they form their own vascular network that supplies them with blood (angiogenesis). *Collagenases* (matrix metalloproteinases, MMPs) play a special role in the metastatic process, by loosening surrounding connective tissue and thereby allowing tumor cells to disseminate and enter the bloodstream. New approaches to combating tumors have been aimed at influencing tumor angiogenesis and metastatic processes.

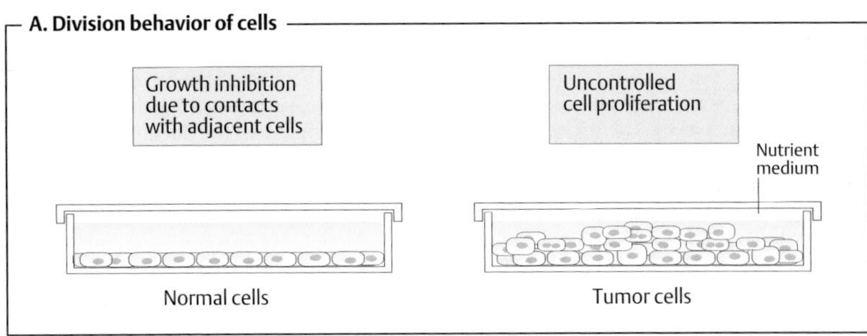
A. Division behavior of cells
Growth inhibition due to contacts with adjacent cells
Uncontrolled cell proliferation
Nutrient medium
Normal cells
Tumor cells

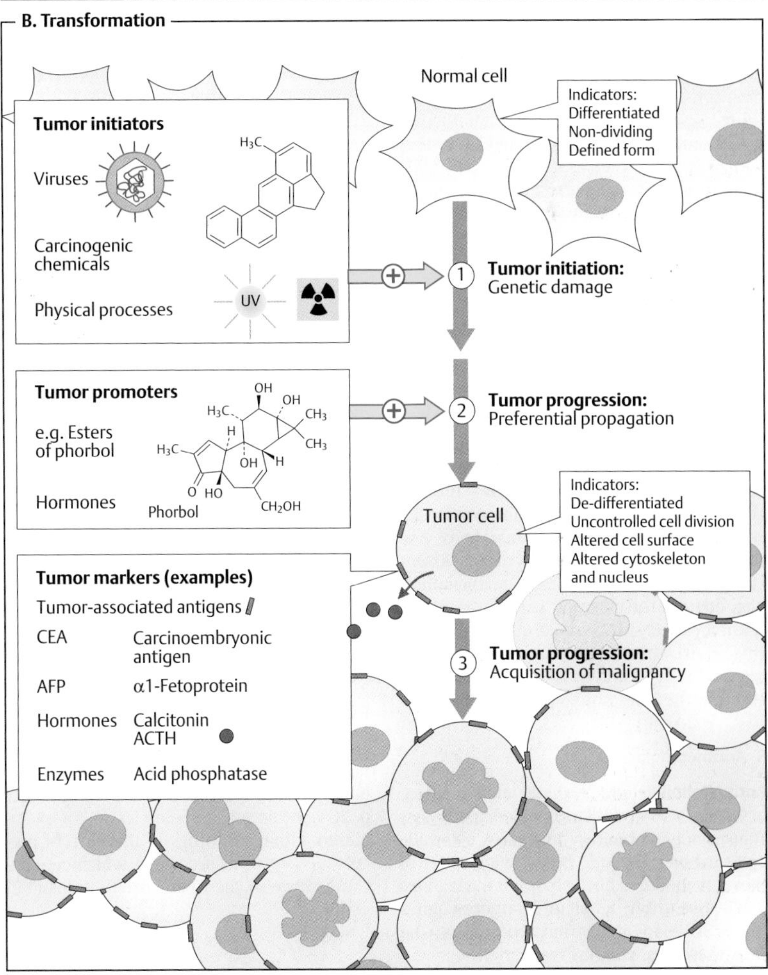
B. Transformation
Normal cell
Indicators:
Differentiated
Non-dividing
Defined form
Tumor initiators
Viruses
Carcinogenic chemicals
Physical processes
UV
1 Tumor initiation:
Genetic damage
Tumor promoters
e.g. Esters of phorbol
Hormones
Phorbol
2 Tumor progression:
Preferential propagation
Tumor cell
Indicators:
De-differentiated
Uncontrolled cell division
Altered cell surface
Altered cytoskeleton and nucleus
Tumor markers (examples)
Tumor-associated antigens
CEA Carcinoembryonic antigen
AFP α1-Fetoprotein
Hormones Calcitonin ACTH
Enzymes Acid phosphatase
3 Tumor progression:
Acquisition of malignancy

Cytostatic drugs

Tumors (see p. 400) arise from degenerated (transformed) cells that grow in an uncontrolled way as a result of genetic defects. Most transformed cells are recognized by the immune system and eliminated (see p. 294). If endogenous defense is not sufficiently effective, rapid tumor growth can occur. Attempts are then made to inhibit growth by physical or chemical treatment.

A frequently used procedure is targeted irradiation with γ-rays, which block cell reproduction due their mutagenic effect (see p. 256). Another approach is to inhibit cell growth by chemotherapy. The growth-inhibiting substances used are known as **cytostatic drugs**. Unfortunately, neither radiotherapy nor chemotherapy act selectively—i. e., they damage normal cells as well, and are therefore often associated with severe side effects.

Most cytostatic agents directly or indirectly inhibit DNA replication in the S phase of the cell cycle (see p. 394). The first group (**A**) lead to chemical changes in cellular DNA that impede transcription and replication. A second group of cytostatic agents (**B**) inhibit the synthesis of DNA precursors.

A. Alkylating agents, anthracyclines ○

Alkylating agents are compounds capable of reacting covalently with DNA bases. If a compound of this type contains *two* reactive groups, intramolecular or intermolecular *crosslinking* of the DNA double helix and "bending" of the double strand occurs. Examples of this type shown here are **cyclophosphamide** and the inorganic complex **cisplatin**. Anthracyclines such as **doxorubicin** (adriamycin) insert themselves non-covalently between the bases and thus lead to local alterations in the DNA structure (see p. 254 B).

B. Antimetabolites ○

Antimetabolites are enzyme inhibitors (see p. 96) that selectively block metabolic pathways. The majority of clinically important cytostatic drugs act on *nucleotide biosynthesis.* Many of these are modified nucleobases or nucleotides that *competitively* inhibit their target enzymes (see p. 96). Many are also incorporated into the DNA, thereby preventing replication.

The cytostatic drugs administered (indicated by a syringe in the illustration) are often not active themselves but are only converted into the actual active agent in the metabolism. This also applies to the adenine analogue **6-mercaptopurine**, which is initially converted to the mononucleotide tIMP (thioinosine monophosphate). Via several intermediate steps, tIMP gives rise to tdGTP, which is incorporated into the DNA and leads to crosslinks and other anomalies in it. The second effective metabolite of 6-mercaptopurine is S-methylated tIMP, an inhibitor of *amidophosphoribosyl transferase* (see p. 188).

Hydroxyurea selectively inhibits *ribonucleotide reductase* (see p. 190). As a radical scavenger, it removes the tyrosine radicals that are indispensable for the functioning of the reductase.

Two other important cytostatic agents target the synthesis of DNA-typical thymine, which takes place at the level of the deoxymononucleotide (see p. 190). The deoxymononucleotide formed by **5-fluorouracil** or the corresponding nucleoside inhibits *thymidylate synthase.* This inhibition is based on the fact that the fluorine atom in the pyrimidine ring cannot be substituted by a methyl group. In addition, the fluorine analogue is also incorporated into the DNA.

Dihydrofolate reductase acts as an auxiliary enzyme for thymidylate synthase. It is involved in the regeneration of the coenzyme N^5,N^{10}-methylene-THF, initially reducing DHF to THF with NADPH as the reductant (see p. 418). The folic acid analogue **methotrexate**l a frequently used cytostatic agent, is an extremely effective competitive inhibitor of dihydrofolate reductase. It leads to the depletion of N^5,N^{10}-methylene-THF in the cells and thus to cessation of DNA synthesis.

Further information

To reduce the side effects of cytostatic agents, new approaches are currently being developed on the basis of **gene therapy** (see p. 264). Attempts are being made, for example, to administer drugs in the form of precursors (known as prodrugs), which only become active in the tumor itself ("tumor targeting").

A. Alkylating agents, anthracyclines

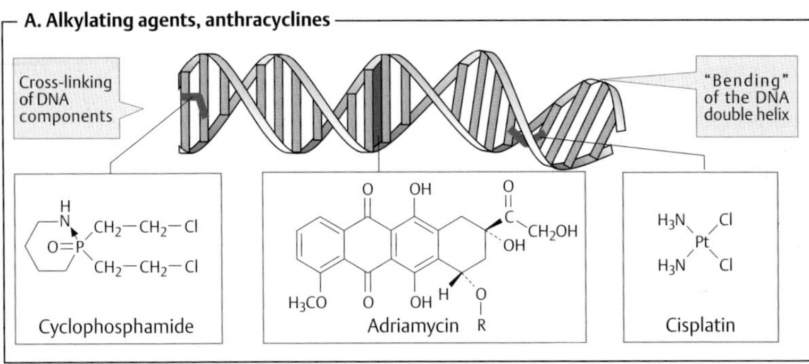

B. Antimetabolites

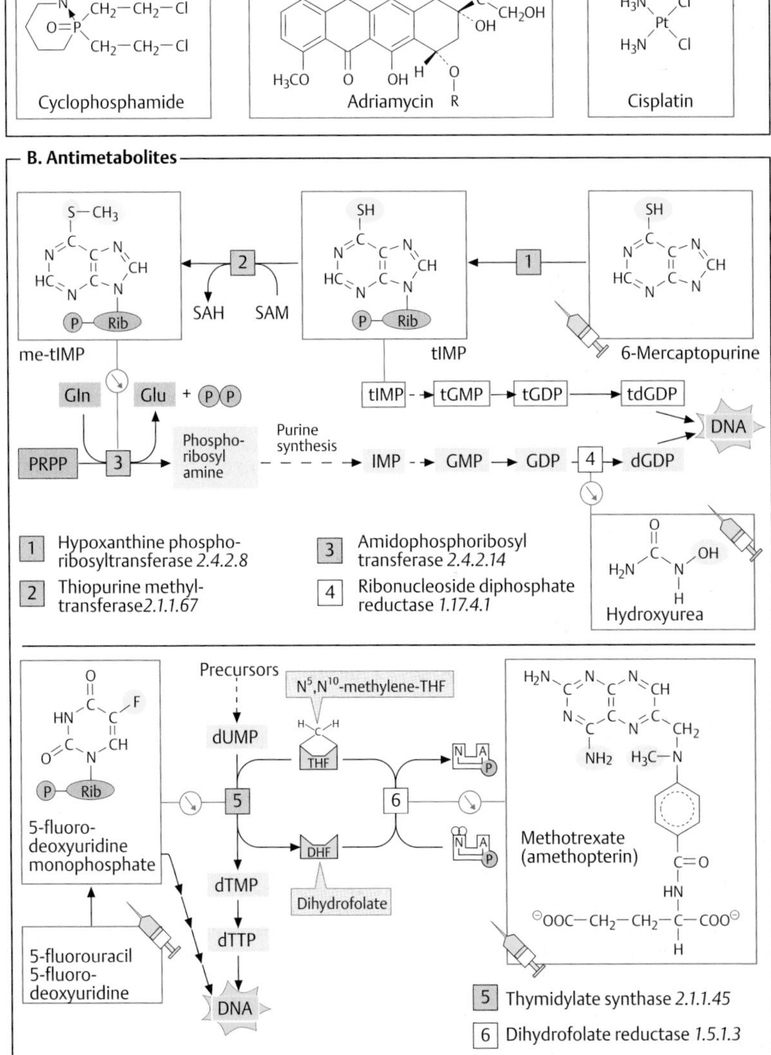

Viruses

Viruses are *parasitic nucleoprotein complexes.* They often consist of only a single nucleic acid molecule (DNA or RNA, never both) and a protein coat. Viruses have *no metabolism of their own,* and can therefore only replicate themselves with the help of host cells. They are therefore not regarded as independent organisms. Viruses that damage the host cell when they replicate are *pathogens.* Diseases caused by viruses include AIDS, rabies, poliomyelitis, measles, German measles, smallpox, influenza, and the common cold.

A. Viruses: examples ◑

Only a few examples from the large number of known viruses are illustrated here. They are all shown on the same scale.

Viruses that only replicate in bacteria are known as **bacteriophages** (or "phages" for short). An example of a phage with a simple structure is **M13.** It consists of a single-stranded DNA molecule (ssDNA) of about 7000 bp with a coat made up of 2700 helically arranged protein subunits. The coat of a virus is referred to as a *capsid,* and the complete structure as a *nucleocapsid.* In genetic engineering, M13 is important as a *vector* for foreign DNA (see p. 258).

The phage **T4** (bottom left), one of the largest viruses known, has a much more complex structure with around 170 000 base pairs (bp) of double-stranded DNA (dsDNA) contained within its "head."

The **tobacco mosaic virus** (center right), a plant pathogen, has a structure similar to that of M13, but contains ssRNA instead of DNA. The **poliovirus**, which causes poliomyelitis, is also an RNA virus. In the **influenza virus**, the pathogen that causes viral flu, the nucleocapsid is additionally surrounded by a *coat* derived from the plasma membrane of the host cell (**C**). The coat carries viral proteins that are involved in the infection process.

B. Capsid of the rhinovirus ○

Rhinoviruses cause the common cold. In these viruses, the capsid is shaped like an *icosahedron*—i. e., an object made up of 20 equilateral triangles. Its surface is formed from three different proteins, which associate with one another to form pentamers and hexamers. In all, 60 protein molecules are involved in the structure of the capsid.

C. Life cycle of HIV ◑

The *human immunodeficiency virus (HIV)* causes the immunodeficiency disease known as **AIDS** (*acquired immune deficiency syndrome*). The structure of this virus is similar to that of the influenza virus (**A**).

The HIV genome consists of two molecules of ssRNA (each 9.2 kb). It is enclosed by a double-layered capsid and a protein-containing coating membrane. HIV mainly infects T helper cells (see p. 294) and can thereby lead to failure of the immune system in the longer term.

During infection (**1**), the virus's coating membrane fuses with the target cell's plasma membrane, and the core of the nucleocapsid enters the cytoplasm (**2**). In the cytoplasm, the viral RNA is initially transcribed into an RNA/DNA hybrid (**3**) and then into dsDNA (**4**). Both of these reactions are catalyzed by *reverse transcriptase,* an enzyme deriving from the virus. The dsDNA formed is integrated into the host cell genome (**5**), where it can remain in an inactive state for a long time.

When viral replication occurs, the DNA segment corresponding to the viral genome is first transcribed by host cell enzymes (**6**). This gives rise not only to viral ssRNA, but also to transcription of mRNAs for precursors of the viral proteins (**7**). These precursors are integrated into the plasma membrane (**8**, **9**) before undergoing proteolytic modification (**10**). The cycle is completed by the release of new virus particles (**11**).

The group of RNA viruses to which HIV belongs are called **retroviruses**, because DNA is produced from RNA in their replication cycle—the reverse of the usual direction of transcription (DNA → RNA).

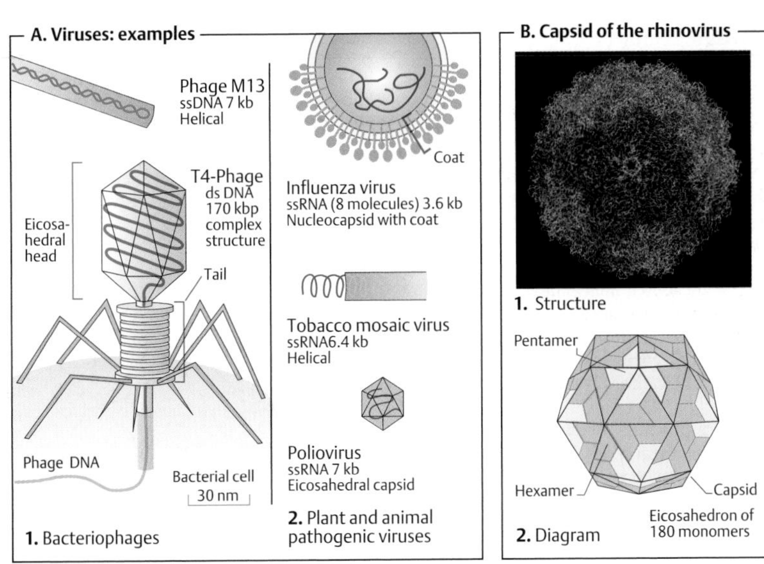
A. Viruses: examples
Phage M13
ssDNA 7 kb
Helical
T4-Phage
ds DNA
170 kbp
complex
structure
Eicosa-
hedral
head
Tail
Phage DNA
Bacterial cell
30 nm
1. Bacteriophages
Coat
Influenza virus
ssRNA (8 molecules) 3.6 kb
Nucleocapsid with coat
Tobacco mosaic virus
ssRNA6.4 kb
Helical
Poliovirus
ssRNA 7 kb
Eicosahedral capsid
2. Plant and animal pathogenic viruses
B. Capsid of the rhinovirus
1. Structure
Pentamer
Hexamer
Capsid
Eicosahedron of 180 monomers
2. Diagram

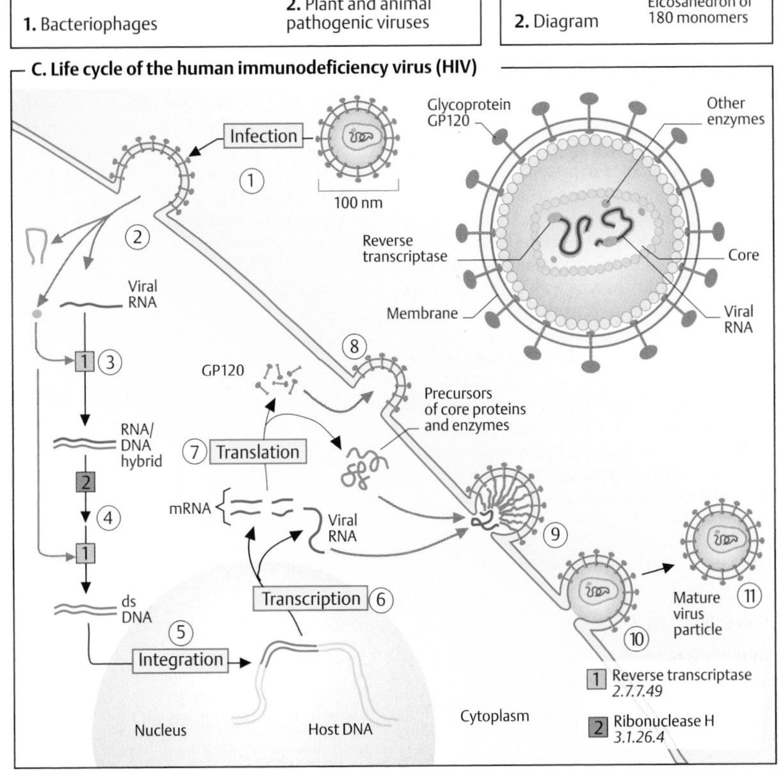
C. Life cycle of the human immunodeficiency virus (HIV)
Infection
1
100 nm
Glycoprotein GP120
Other enzymes
Reverse transcriptase
Core
Membrane
Viral RNA
2
Viral RNA
1 3
RNA/DNA hybrid
2
4
1
ds DNA
5
Integration
8
GP120
Precursors of core proteins and enzymes
7 Translation
mRNA
Viral RNA
Transcription 6
9
10
Mature virus particle
11
Nucleus
Host DNA
Cytoplasm
1 Reverse transcriptase 2.7.7.49
2 Ribonuclease H 3.1.26.4

Metabolic charts

Explanations

The following 13 plates (pp. 407–419) provide a concise schematic overview of the most important metabolic pathways. Explanatory text is deliberately omitted from them.

These "charts":

- Contain details of metabolic pathways that are only shown in outline in the main text for reasons of space. This applies in particular to the synthesis and degradation of the amino acids and nucleotides, and for some aspects of carbohydrate and lipid metabolism.
- Offer a quick overview of a specific pathway, the metabolites that arise in it, and the enzymes involved.
- Can be used for reference purposes and for revising material previously learned.

The most important **intermediates** are shown with numbers in the charts. The corresponding compounds can be identified using the table on the same page.

In addition, at each step the four-figure EC number (see p. 88) for the **enzyme** responsible for a reaction is given in *italics*. The enzyme name and its systematic classification in the system used by the *Enzyme Catalogue* are available in the following **annotated enzyme list** (pp. 420–430), in which all of the enzymes mentioned in this book are listed according to their EC number. The book's index is helpful when looking for a specific enzyme in the text.

In reactions that involve **coenzymes**, the coenzyme names are also given (sometimes in simplified form). Particularly important starting, intermediate, or end products are given with the full name, or as formulae.

Example

On p. 407, the initial step of the dark reactions in plant photosynthesis (in the Calvin cycle) is shown at the top left.

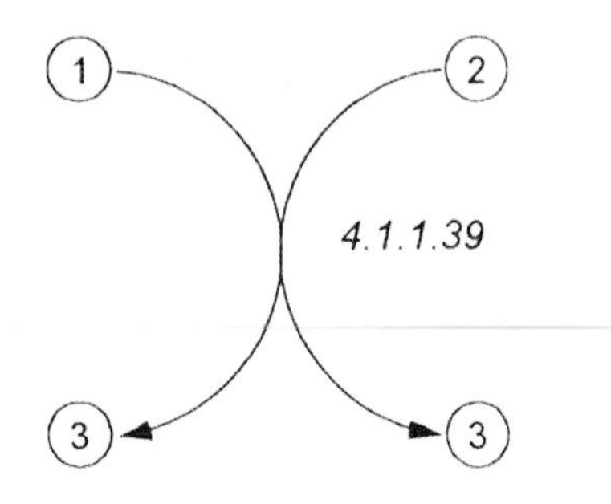

In this reaction, one molecule of ribulose-1,5-bisphosphate (metabolite 1) and one molecule of CO_2 (metabolite 2) give rise to two molecules of 3-phosphoglycerate (metabolite 3). The enzyme responsible has the EC number *4.1.1.39*. The annotated enzyme list shows that this refers to *ribulose bisphosphate carboxylase* ("rubisco" for short). Rubisco belongs to enzyme class 4 (the lyases) and, within that group, to subclass 4.1 (the carboxy-lyases). It contains copper as a cofactor ([Cu]).

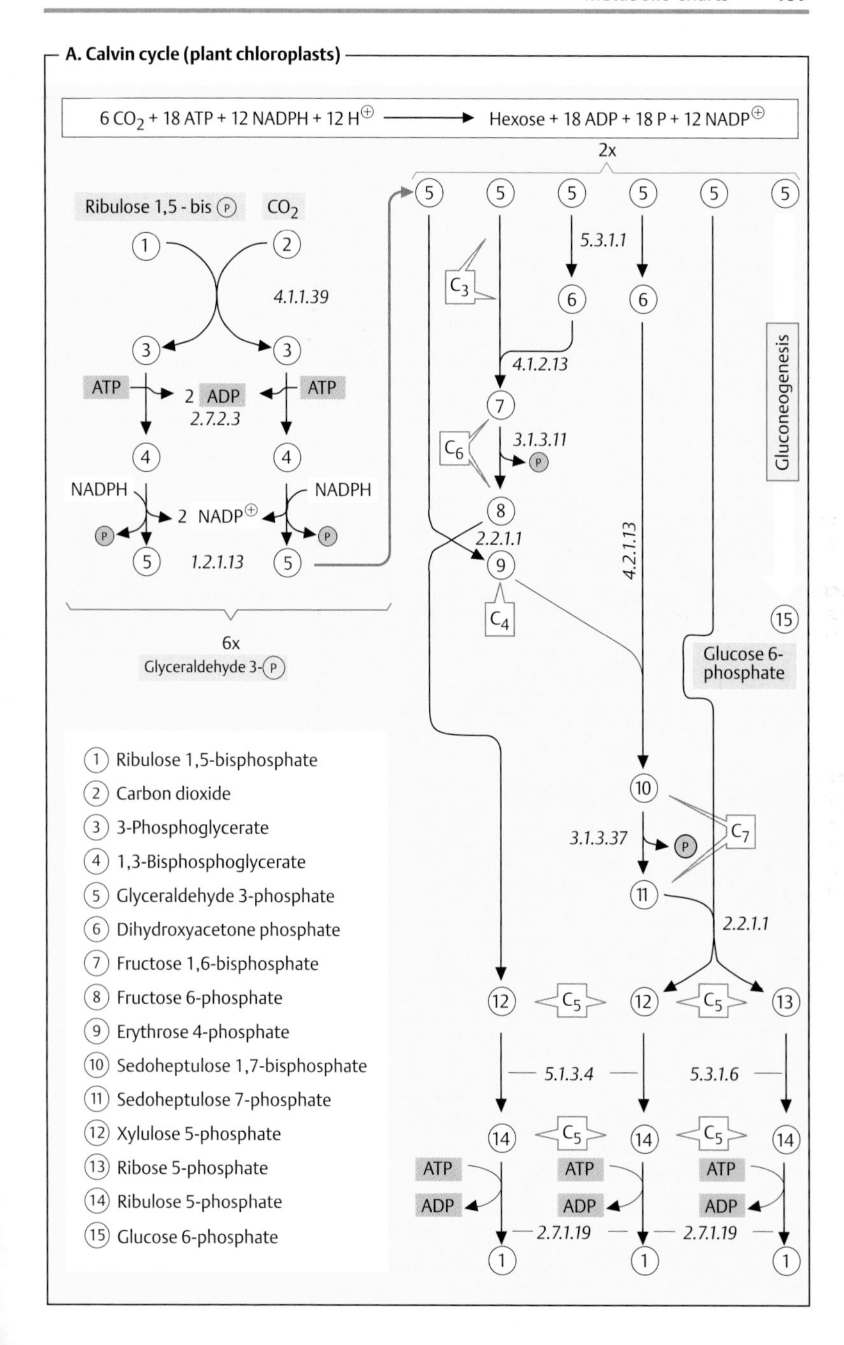
A. Calvin cycle (plant chloroplasts)
6 CO_2 + 18 ATP + 12 NADPH + 12 $H^{\oplus}$ → Hexose + 18 ADP + 18 P + 12 $NADP^{\oplus}$
2x
Ribulose 1,5 - bis (P)
CO_2
4.1.1.39
ATP
2 ADP
ATP
2.7.2.3
NADPH
2 $NADP^{\oplus}$
NADPH
1.2.1.13
6x
Glyceraldehyde 3-(P)
5.3.1.1
C_3
4.1.2.13
3.1.3.11
C_6
2.2.1.1
C_4
4.2.1.13
Gluconeogenesis
Glucose 6-phosphate
3.1.3.37
C_7
2.2.1.1
C_5
C_5
5.1.3.4
5.3.1.6
C_5
C_5
ATP
ADP
ATP
ADP
ATP
ADP
2.7.1.19
2.7.1.19
1 Ribulose 1,5-bisphosphate
2 Carbon dioxide
3 3-Phosphoglycerate
4 1,3-Bisphosphoglycerate
5 Glyceraldehyde 3-phosphate
6 Dihydroxyacetone phosphate
7 Fructose 1,6-bisphosphate
8 Fructose 6-phosphate
9 Erythrose 4-phosphate
10 Sedoheptulose 1,7-bisphosphate
11 Sedoheptulose 7-phosphate
12 Xylulose 5-phosphate
13 Ribose 5-phosphate
14 Ribulose 5-phosphate
15 Glucose 6-phosphate

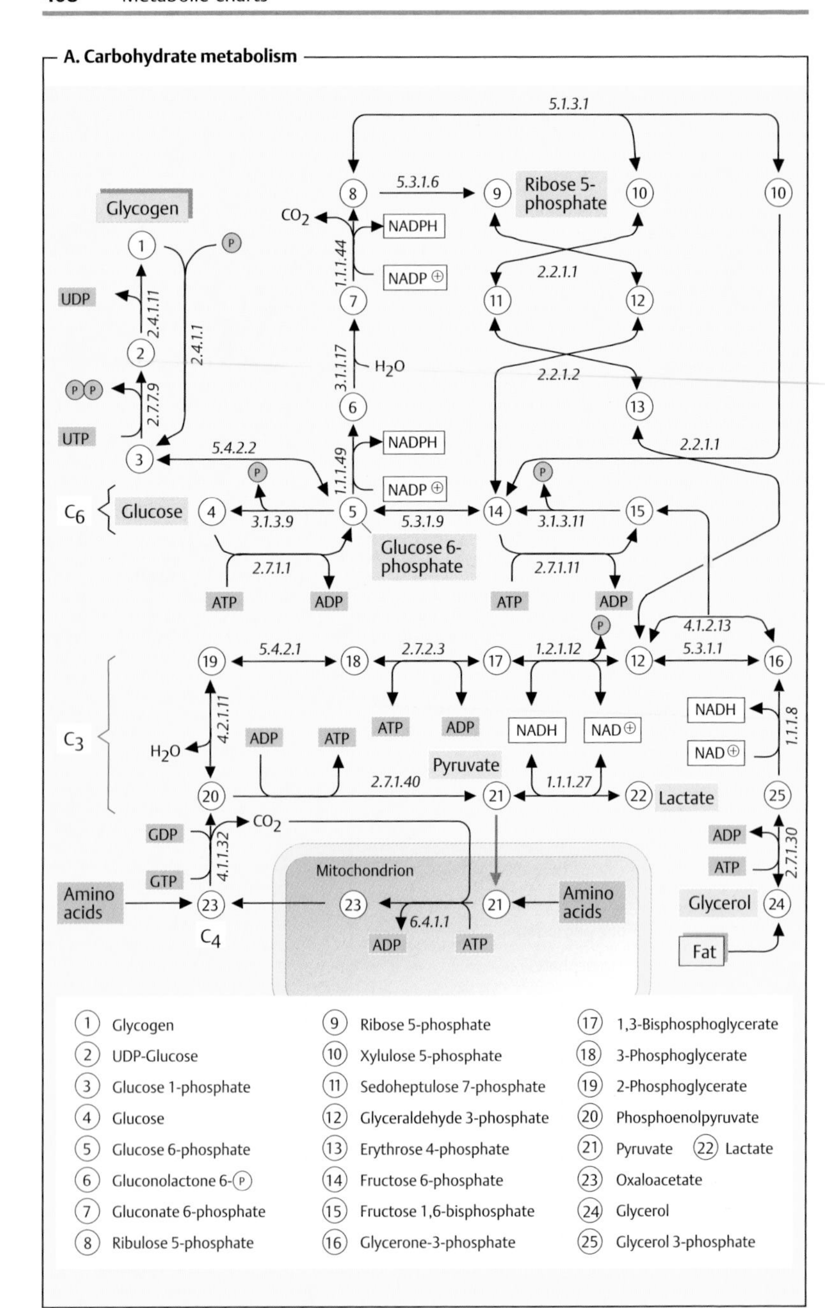
A. Carbohydrate metabolism
5.1.3.1
5.3.1.6
Ribose 5-phosphate
Glycogen
CO2
NADPH
1.1.1.44
NADP ⊕
2.2.1.1
UDP
2.4.1.11
2.4.1.1
3.1.1.17
H2O
2.2.1.2
2.7.7.9
UTP
NADPH
5.4.2.2
2.2.1.1
1.1.1.49
NADP ⊕
C6
Glucose
3.1.3.9
5.3.1.9
3.1.3.11
Glucose 6-phosphate
2.7.1.1
2.7.1.11
ATP
ADP
ATP
ADP
4.1.2.13
5.4.2.1
2.7.2.3
1.2.1.12
5.3.1.1
4.2.1.11
NADH
1.1.1.8
C3
ATP
ADP
NADH
NAD⊕
H2O
ADP
ATP
Pyruvate
NAD⊕
2.7.1.40
1.1.1.27
Lactate
CO2
GDP
ADP
4.1.1.32
2.7.1.30
Mitochondrion
GTP
ATP
Amino acids
Amino acids
Glycerol
6.4.1.1
C4
ADP
ATP
Fat
1 Glycogen
2 UDP-Glucose
3 Glucose 1-phosphate
4 Glucose
5 Glucose 6-phosphate
6 Gluconolactone 6-(P)
7 Gluconate 6-phosphate
8 Ribulose 5-phosphate
9 Ribose 5-phosphate
10 Xylulose 5-phosphate
11 Sedoheptulose 7-phosphate
12 Glyceraldehyde 3-phosphate
13 Erythrose 4-phosphate
14 Fructose 6-phosphate
15 Fructose 1,6-bisphosphate
16 Glycerone-3-phosphate
17 1,3-Bisphosphoglycerate
18 3-Phosphoglycerate
19 2-Phosphoglycerate
20 Phosphoenolpyruvate
21 Pyruvate
22 Lactate
23 Oxaloacetate
24 Glycerol
25 Glycerol 3-phosphate

A. Biosynthesis of fats and membrane lipids

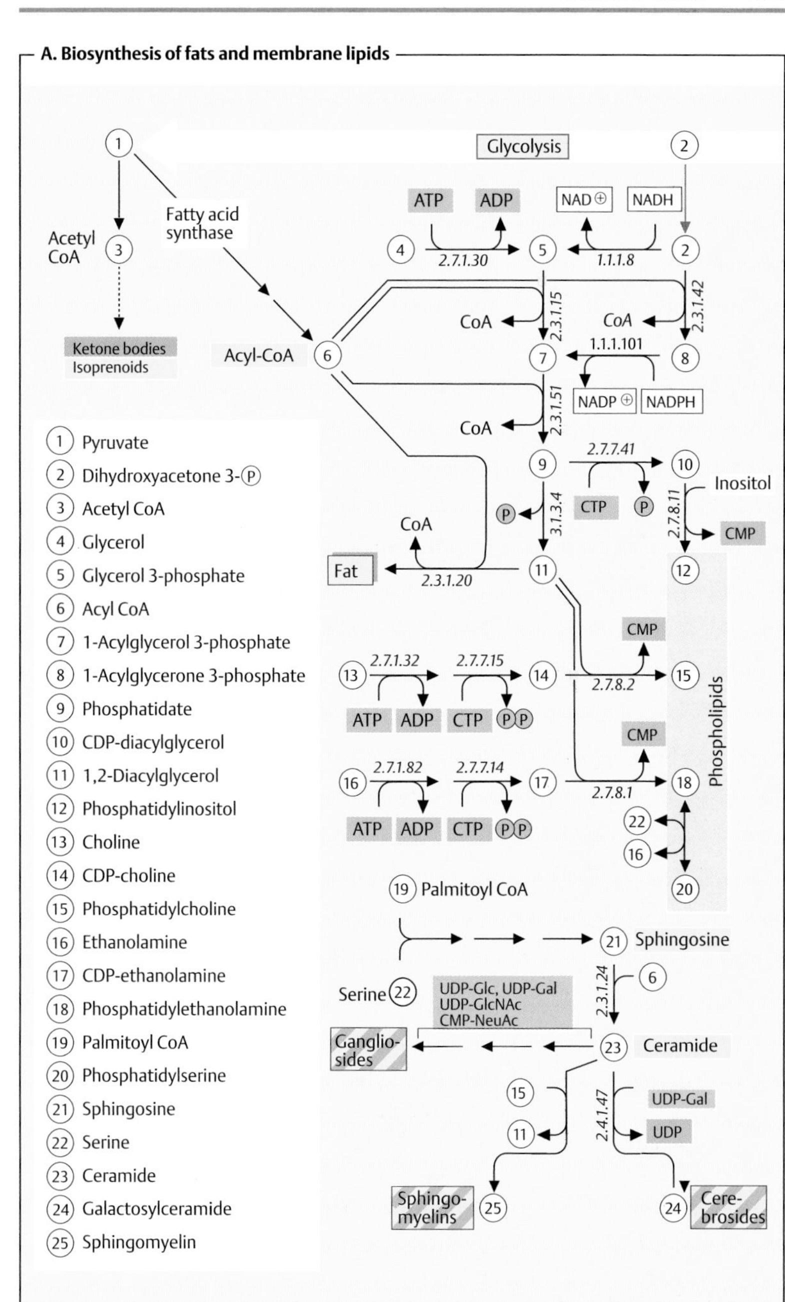

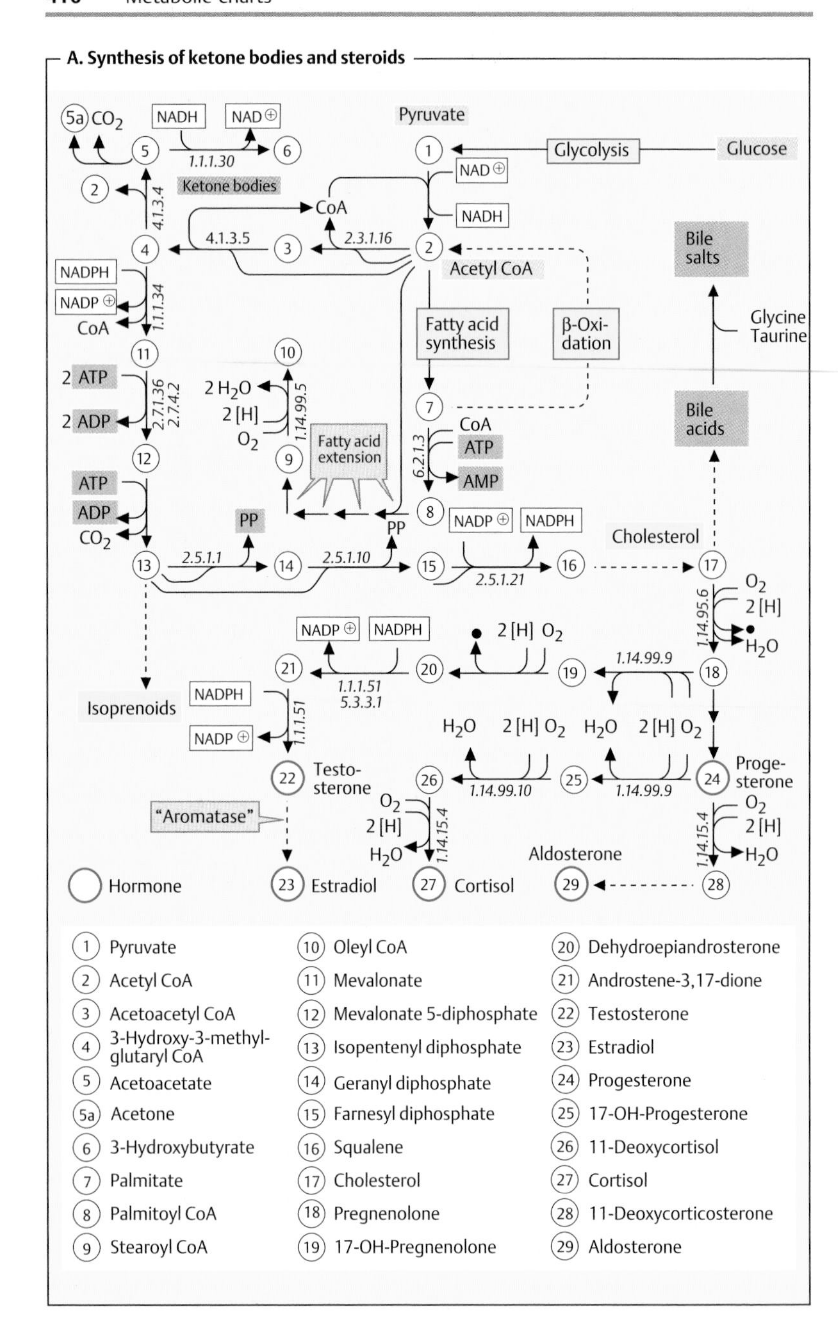
A. Synthesis of ketone bodies and steroids
Pyruvate
Glycolysis
Glucose
NADH
NAD⊕
Ketone bodies
Acetyl CoA
Fatty acid synthesis
β-Oxidation
Bile salts
Glycine
Taurine
Bile acids
Fatty acid extension
Cholesterol
Isoprenoids
Testosterone
Progesterone
"Aromatase"
Aldosterone
Hormone
Estradiol
Cortisol
1 Pyruvate
2 Acetyl CoA
3 Acetoacetyl CoA
4 3-Hydroxy-3-methylglutaryl CoA
5 Acetoacetate
5a Acetone
6 3-Hydroxybutyrate
7 Palmitate
8 Palmitoyl CoA
9 Stearoyl CoA
10 Oleyl CoA
11 Mevalonate
12 Mevalonate 5-diphosphate
13 Isopentenyl diphosphate
14 Geranyl diphosphate
15 Farnesyl diphosphate
16 Squalene
17 Cholesterol
18 Pregnenolone
19 17-OH-Pregnenolone
20 Dehydroepiandrosterone
21 Androstene-3,17-dione
22 Testosterone
23 Estradiol
24 Progesterone
25 17-OH-Progesterone
26 11-Deoxycortisol
27 Cortisol
28 11-Deoxycorticosterone
29 Aldosterone

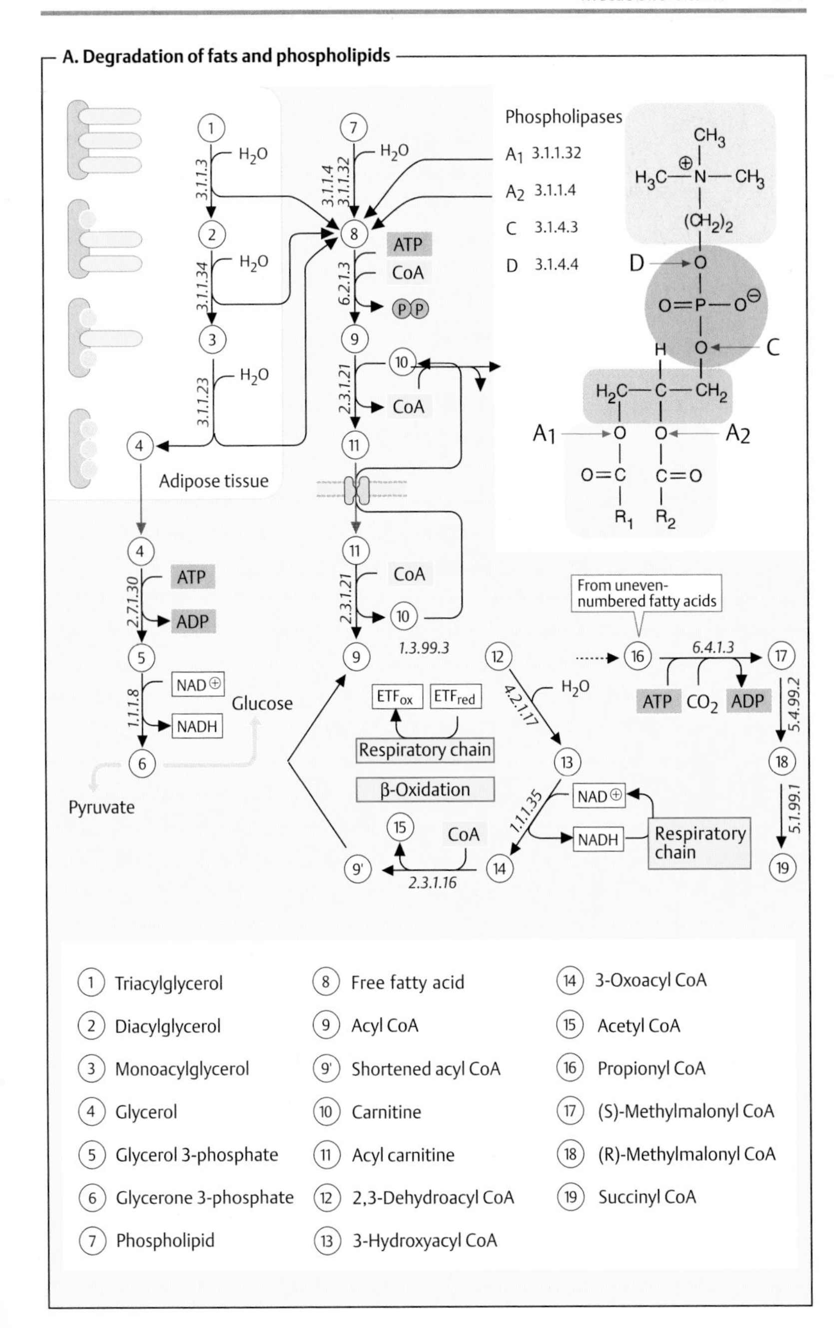
A. Degradation of fats and phospholipids
Adipose tissue
Phospholipases
A1 3.1.1.32
A2 3.1.1.4
C 3.1.4.3
D 3.1.4.4
3.1.1.3
3.1.1.34
3.1.1.23
3.1.1.4
3.1.1.32
6.2.1.3
2.3.1.21
2.7.1.30
1.1.1.8
1.3.99.3
4.2.1.17
1.1.1.35
2.3.1.16
6.4.1.3
5.4.99.2
5.1.99.1
H_2O
ATP
CoA
ADP
NAD⊕
NADH
Glucose
Pyruvate
ETF_{ox}
ETF_{red}
Respiratory chain
β-Oxidation
From uneven-numbered fatty acids
CO_2
1 Triacylglycerol
2 Diacylglycerol
3 Monoacylglycerol
4 Glycerol
5 Glycerol 3-phosphate
6 Glycerone 3-phosphate
7 Phospholipid
8 Free fatty acid
9 Acyl CoA
9' Shortened acyl CoA
10 Carnitine
11 Acyl carnitine
12 2,3-Dehydroacyl CoA
13 3-Hydroxyacyl CoA
14 3-Oxoacyl CoA
15 Acetyl CoA
16 Propionyl CoA
17 (S)-Methylmalonyl CoA
18 (R)-Methylmalonyl CoA
19 Succinyl CoA

A. Biosynthesis of the essential amino acids

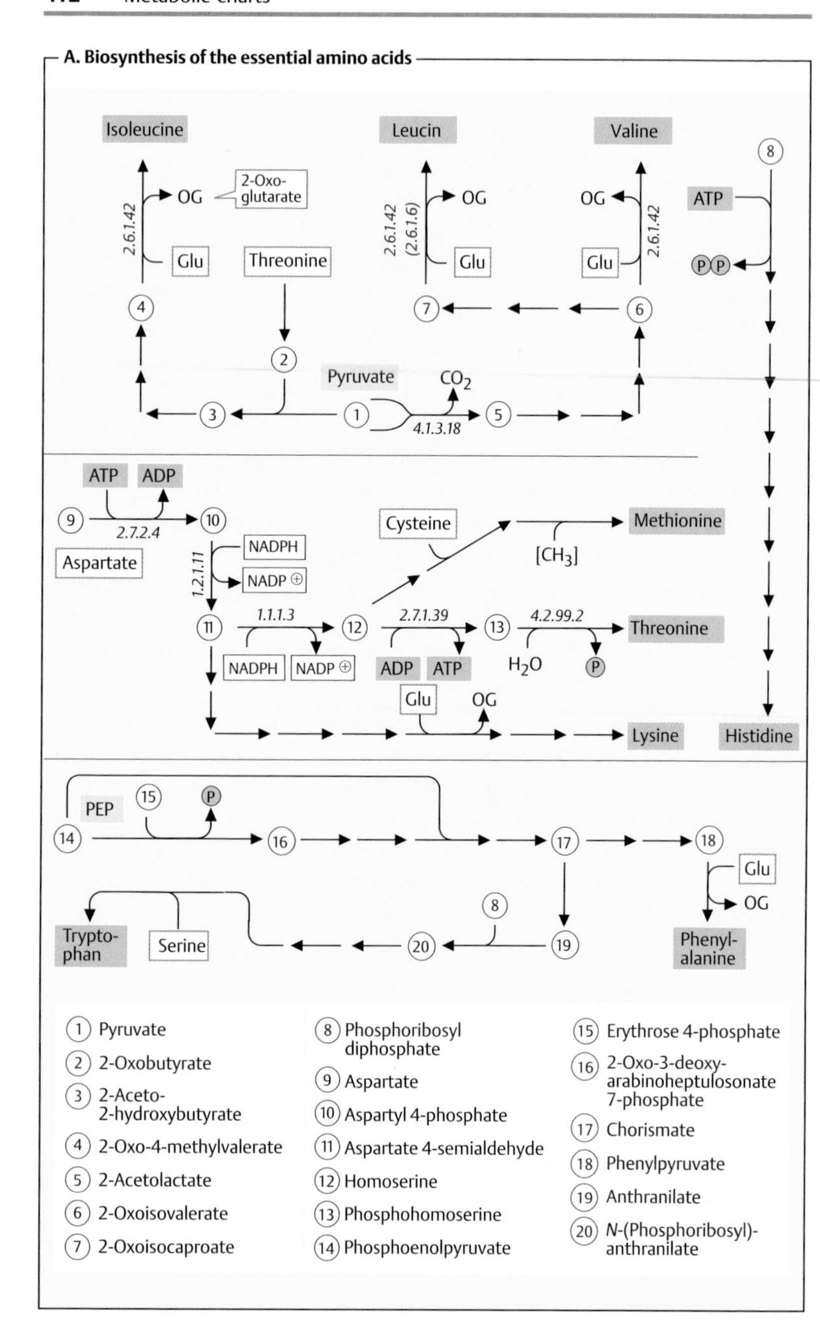

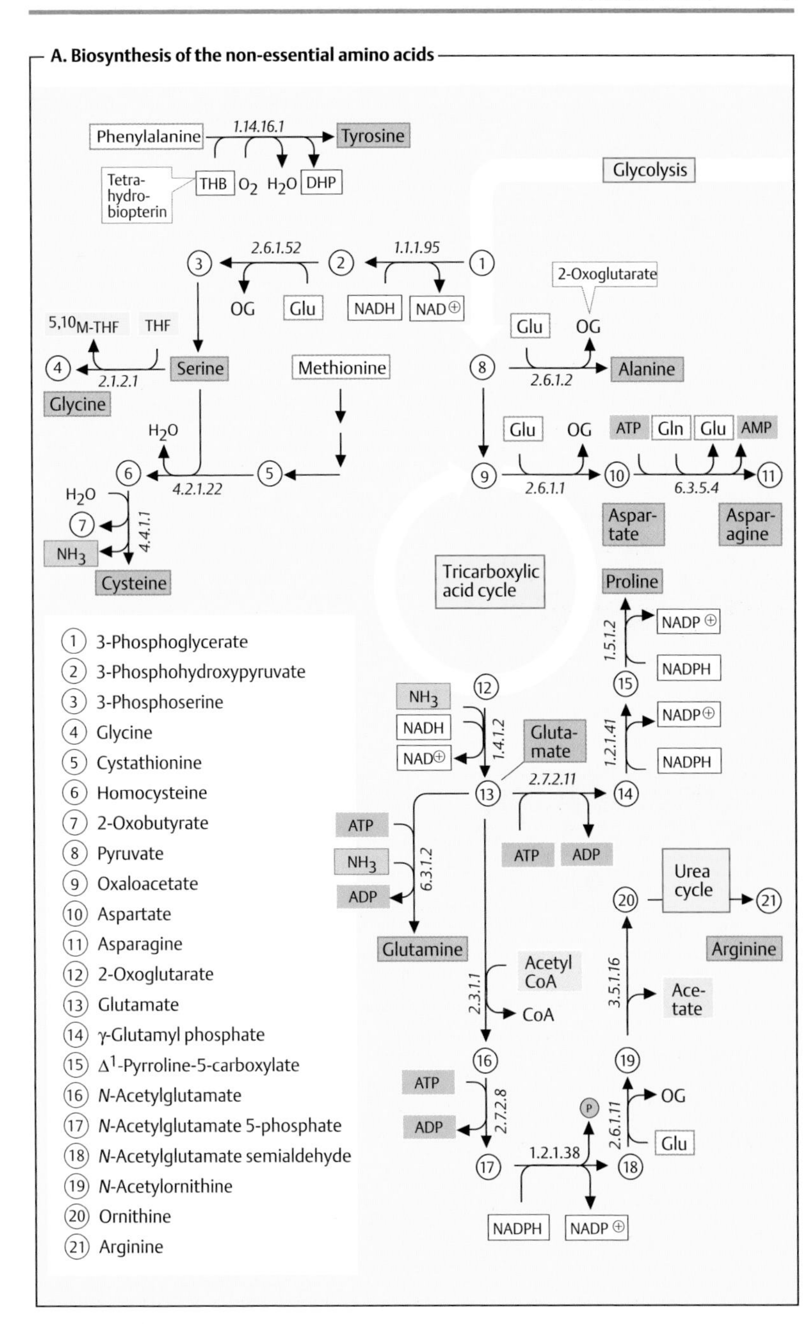
A. Biosynthesis of the non-essential amino acids
Phenylalanine
1.14.16.1
Tyrosine
Tetra-hydro-biopterin
THB
O_2
H_2O
DHP
Glycolysis
2.6.1.52
1.1.1.95
OG
Glu
NADH
NAD⊕
2-Oxoglutarate
5,10M-THF
THF
Glu
OG
Serine
Methionine
Alanine
2.1.2.1
2.6.1.2
Glycine
H_2O
Glu
OG
ATP
Gln
Glu
AMP
4.2.1.22
2.6.1.1
6.3.5.4
H_2O
4.4.1.1
Aspar-tate
Aspar-agine
NH_3
Cysteine
Tricarboxylic acid cycle
Proline
NADP ⊕
1.5.1.2
NADPH
NH_3
NADH
1.4.1.2
Gluta-mate
NADP⊕
1.2.1.41
NADPH
NAD⊕
2.7.2.11
ATP
NH_3
6.3.1.2
ADP
ATP
ADP
Urea cycle
Glutamine
Arginine
Acetyl CoA
2.3.1.1
CoA
3.5.1.16
Ace-tate
ATP
2.7.2.8
ADP
P
2.6.1.11
OG
Glu
1.2.1.38
NADPH
NADP ⊕
1 3-Phosphoglycerate
2 3-Phosphohydroxypyruvate
3 3-Phosphoserine
4 Glycine
5 Cystathionine
6 Homocysteine
7 2-Oxobutyrate
8 Pyruvate
9 Oxaloacetate
10 Aspartate
11 Asparagine
12 2-Oxoglutarate
13 Glutamate
14 γ-Glutamyl phosphate
15 Δ^1-Pyrroline-5-carboxylate
16 N-Acetylglutamate
17 N-Acetylglutamate 5-phosphate
18 N-Acetylglutamate semialdehyde
19 N-Acetylornithine
20 Ornithine
21 Arginine

A. Amino acid degradation I

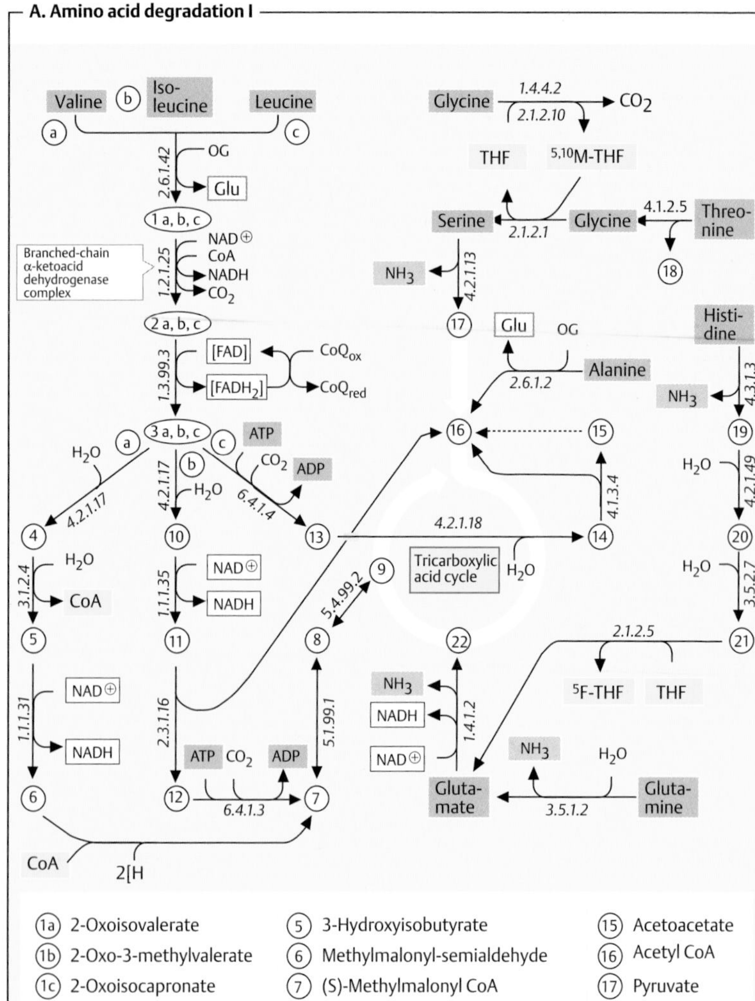

1a 2-Oxoisovalerate
1b 2-Oxo-3-methylvalerate
1c 2-Oxoisocapronate
2a Isobutyryl CoA
2b 2-Methylbutyryl CoA
2c Isovaleryl CoA
3a Methylacrylyl CoA
3b Tiglyl CoA
3c 3-Methylcrotonyl CoA
4 3-Hydroxyisobutyryl CoA
5 3-Hydroxyisobutyrate
6 Methylmalonyl-semialdehyde
7 (S)-Methylmalonyl CoA
8 (R)-Methylmalonyl CoA
9 Succinyl CoA
10 2-Methyl-3-hydroxybutyryl CoA
11 2-Methylacetoacetyl CoA
12 Propionyl CoA
13 3-Methylglutaconyl CoA
14 3-Hydroxy-3-methylglutaryl CoA
15 Acetoacetate
16 Acetyl CoA
17 Pyruvate
18 Acetaldehyde
19 Urocanate
20 Imidazolone-5-propionate
21 *N*-Formiminoglutamate
22 2-Oxoglutarate

A. Amino acid degradation II

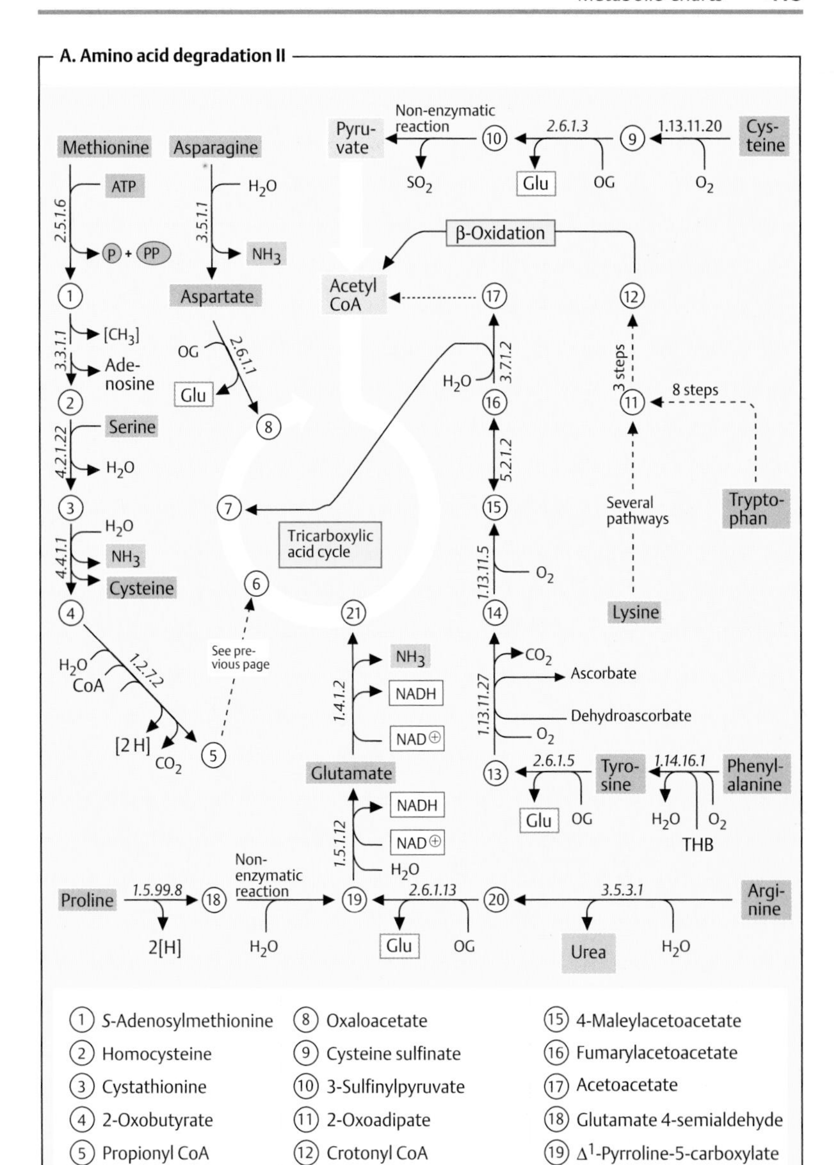

1 S-Adenosylmethionine
2 Homocysteine
3 Cystathionine
4 2-Oxobutyrate
5 Propionyl CoA
6 Succinyl CoA
7 Fumarate
8 Oxaloacetate
9 Cysteine sulfinate
10 3-Sulfinylpyruvate
11 2-Oxoadipate
12 Crotonyl CoA
13 4-Hydroxyphenylpyruvate
14 Homogentisate
15 4-Maleylacetoacetate
16 Fumarylacetoacetate
17 Acetoacetate
18 Glutamate 4-semialdehyde
19 Δ^1-Pyrroline-5-carboxylate
20 Ornithine
21 2-Oxoglutarate

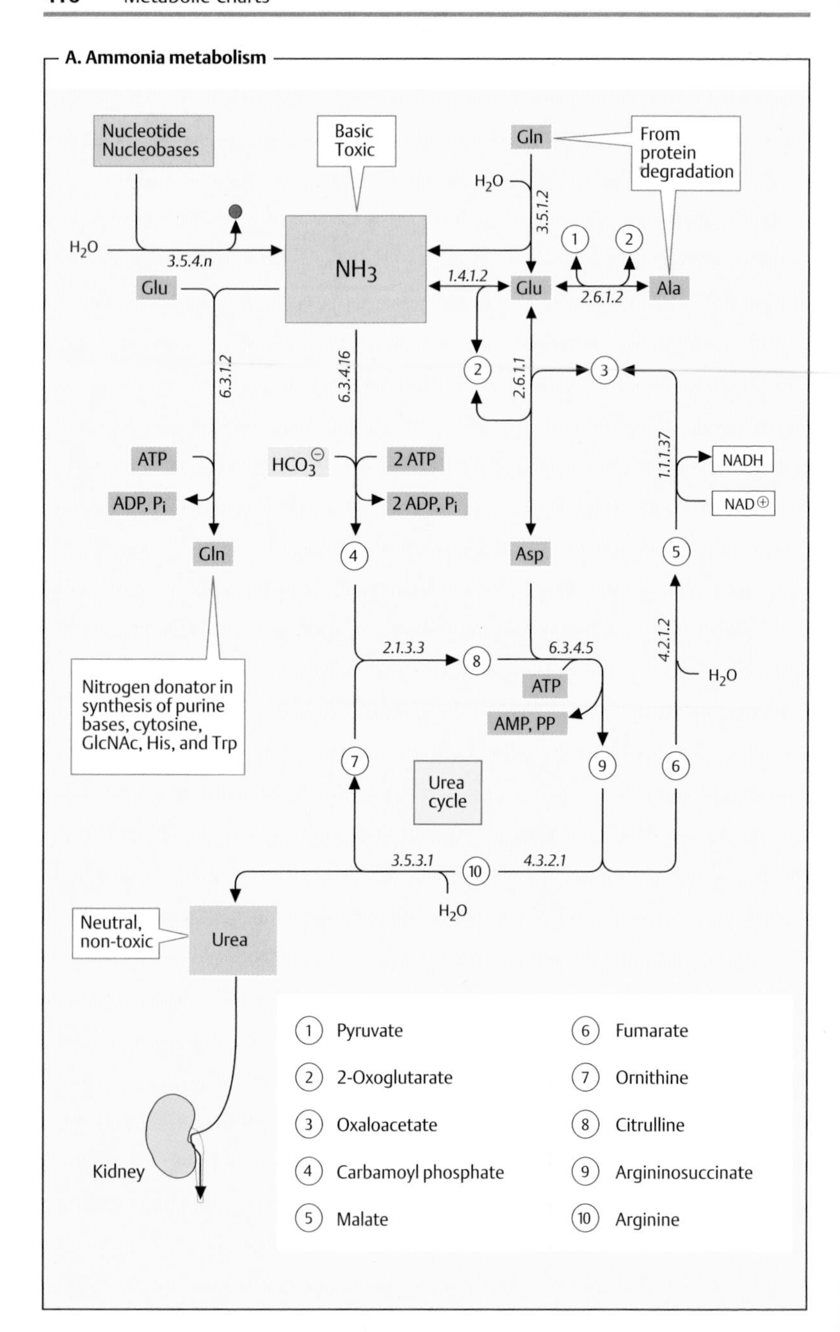
A. Ammonia metabolism
Nucleotide
Nucleobases
Basic
Toxic
Gln
From
protein
degradation
H_2O
3.5.1.2
H_2O
3.5.4.n
NH_3
1
2
1.4.1.2
Glu
Glu
2.6.1.2
Ala
6.3.1.2
6.3.4.16
2
2.6.1.1
3
1.1.1.37
ATP
HCO_3^-
2 ATP
NADH
ADP, P_i
2 ADP, P_i
NAD^+
Gln
4
Asp
5
4.2.1.2
2.1.3.3
8
6.3.4.5
H_2O
ATP
Nitrogen donator in synthesis of purine bases, cytosine, GlcNAc, His, and Trp
AMP, PP
7
9
6
Urea
cycle
3.5.3.1
10
4.3.2.1
H_2O
Neutral,
non-toxic
Urea
Kidney
1 Pyruvate
2 2-Oxoglutarate
3 Oxaloacetate
4 Carbamoyl phosphate
5 Malate
6 Fumarate
7 Ornithine
8 Citrulline
9 Argininosuccinate
10 Arginine

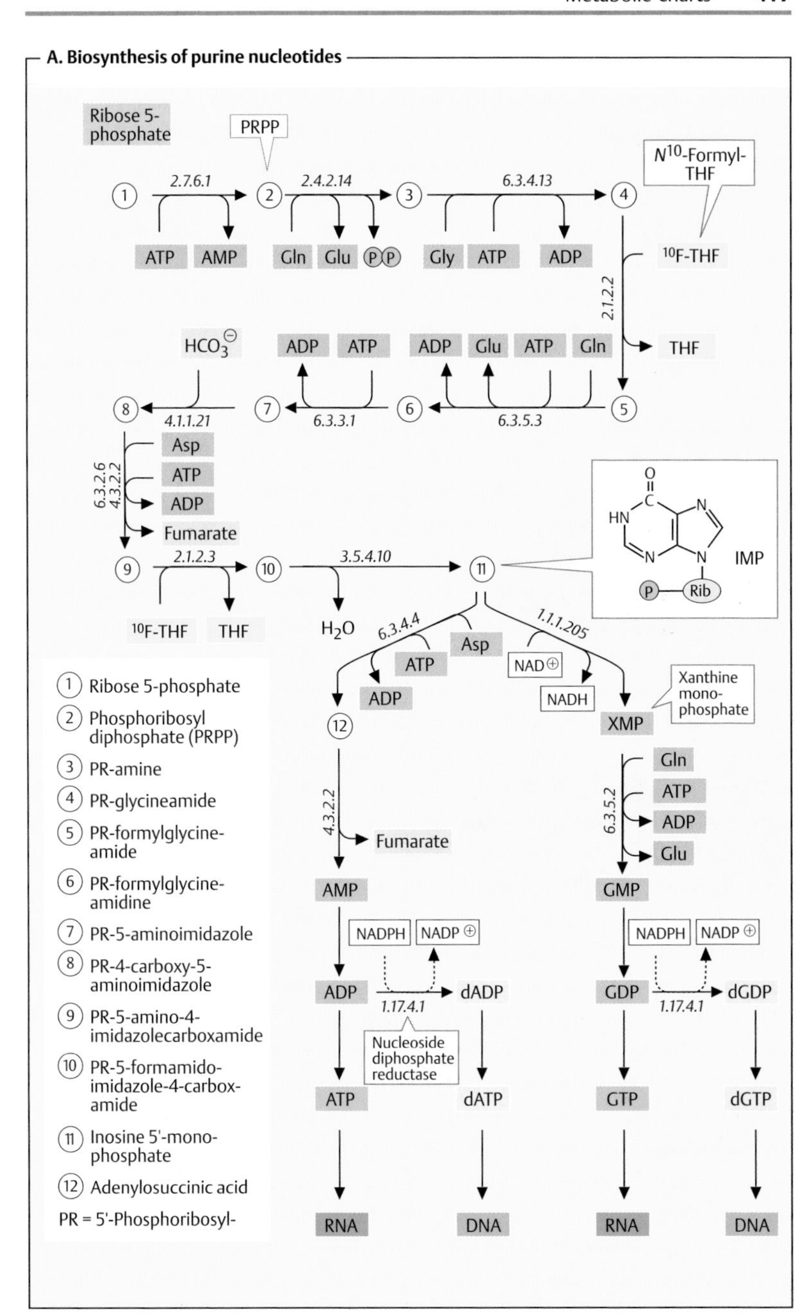
A. Biosynthesis of purine nucleotides
Ribose 5-phosphate
PRPP
N10-Formyl-THF
2.7.6.1
2.4.2.14
6.3.4.13
ATP
AMP
Gln
Glu
Gly
ATP
ADP
10F-THF
2.1.2.2
THF
HCO3
ADP
ATP
ADP
Glu
ATP
Gln
4.1.1.21
6.3.3.1
6.3.5.3
Asp
ATP
ADP
Fumarate
6.3.2.6
4.3.2.2
2.1.2.3
3.5.4.10
IMP
10F-THF
THF
H2O
6.3.4.4
Asp
1.1.1.205
ATP
NAD⊕
ADP
NADH
XMP
Xanthine monophosphate
1 Ribose 5-phosphate
2 Phosphoribosyl diphosphate (PRPP)
3 PR-amine
4 PR-glycineamide
5 PR-formylglycineamide
6 PR-formylglycineamidine
7 PR-5-aminoimidazole
8 PR-4-carboxy-5-aminoimidazole
9 PR-5-amino-4-imidazolecarboxamide
10 PR-5-formamidoimidazole-4-carboxamide
11 Inosine 5'-monophosphate
12 Adenylosuccinic acid
PR = 5'-Phosphoribosyl-
4.3.2.2
Fumarate
Gln
ATP
6.3.5.2
ADP
Glu
AMP
GMP
NADPH
NADP⊕
NADPH
NADP⊕
ADP
dADP
GDP
dGDP
1.17.4.1
1.17.4.1
Nucleoside diphosphate reductase
ATP
dATP
GTP
dGTP
RNA
DNA
RNA
DNA

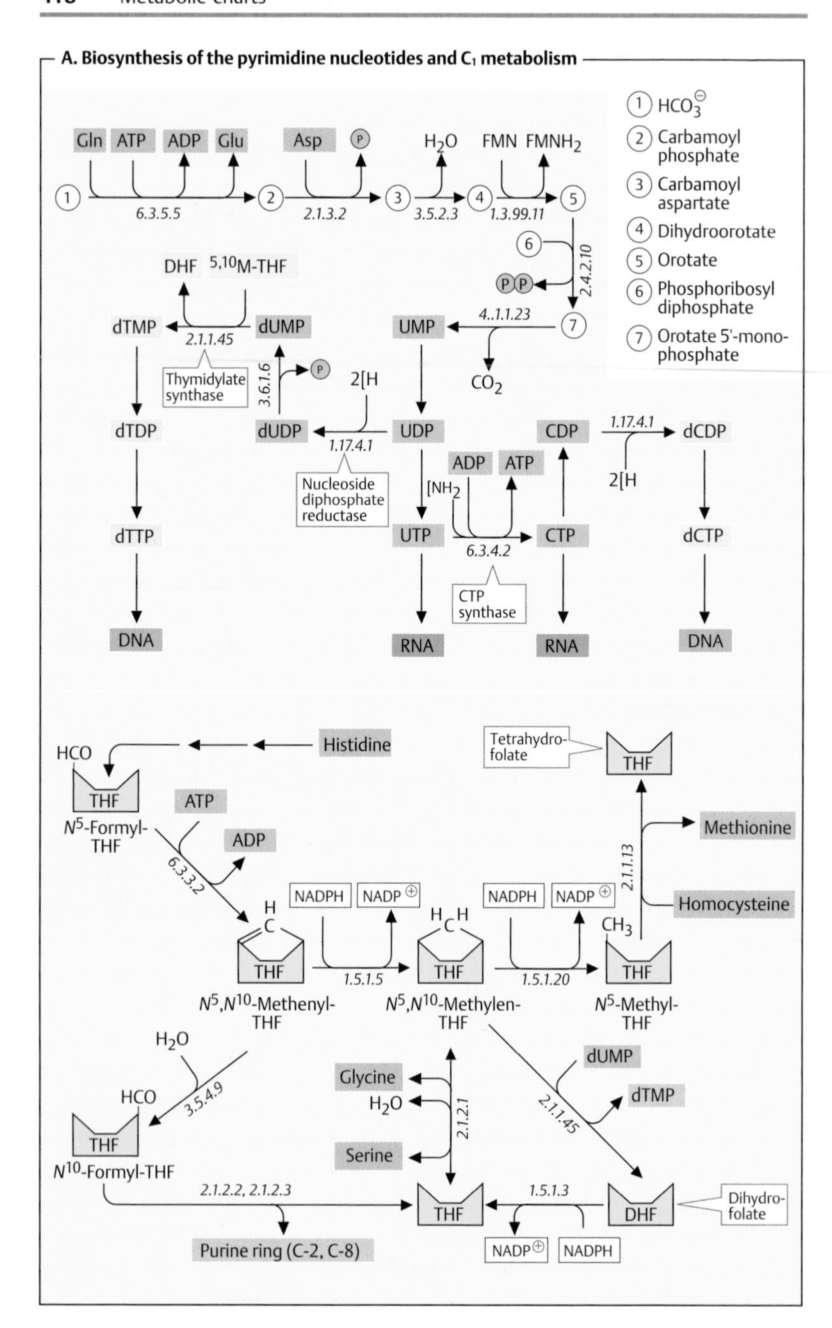
A. Biosynthesis of the pyrimidine nucleotides and C1 metabolism
1 HCO3
2 Carbamoyl phosphate
3 Carbamoyl aspartate
4 Dihydroorotate
5 Orotate
6 Phosphoribosyl diphosphate
7 Orotate 5'-mono-phosphate
Gln ATP ADP Glu
Asp
H2O
FMN FMNH2
6.3.5.5
2.1.3.2
3.5.2.3
1.3.99.11
2.4.2.10
4..1.1.23
DHF 5,10M-THF
dTMP
2.1.1.45
dUMP
UMP
Thymidylate synthase
3.6.1.6
2[H
CO2
dTDP
dUDP
1.17.4.1
UDP
CDP
1.17.4.1
dCDP
Nucleoside diphosphate reductase
ADP ATP
[NH2
2[H
dTTP
UTP
6.3.4.2
CTP
dCTP
CTP synthase
DNA
RNA
RNA
DNA
Histidine
HCO
THF
N5-Formyl-THF
ATP
ADP
6.3.3.2
Tetrahydro-folate
THF
Methionine
2.1.1.13
Homocysteine
NADPH NADP
NADPH NADP
CH3
1.5.1.5
1.5.1.20
N5,N10-Methenyl-THF
N5,N10-Methylen-THF
N5-Methyl-THF
H2O
3.5.4.9
HCO
Glycine
H2O
2.1.2.1
Serine
dUMP
dTMP
2.1.1.45
N10-Formyl-THF
2.1.2.2, 2.1.2.3
1.5.1.3
DHF
Dihydro-folate
Purine ring (C-2, C-8)
NADP NADPH

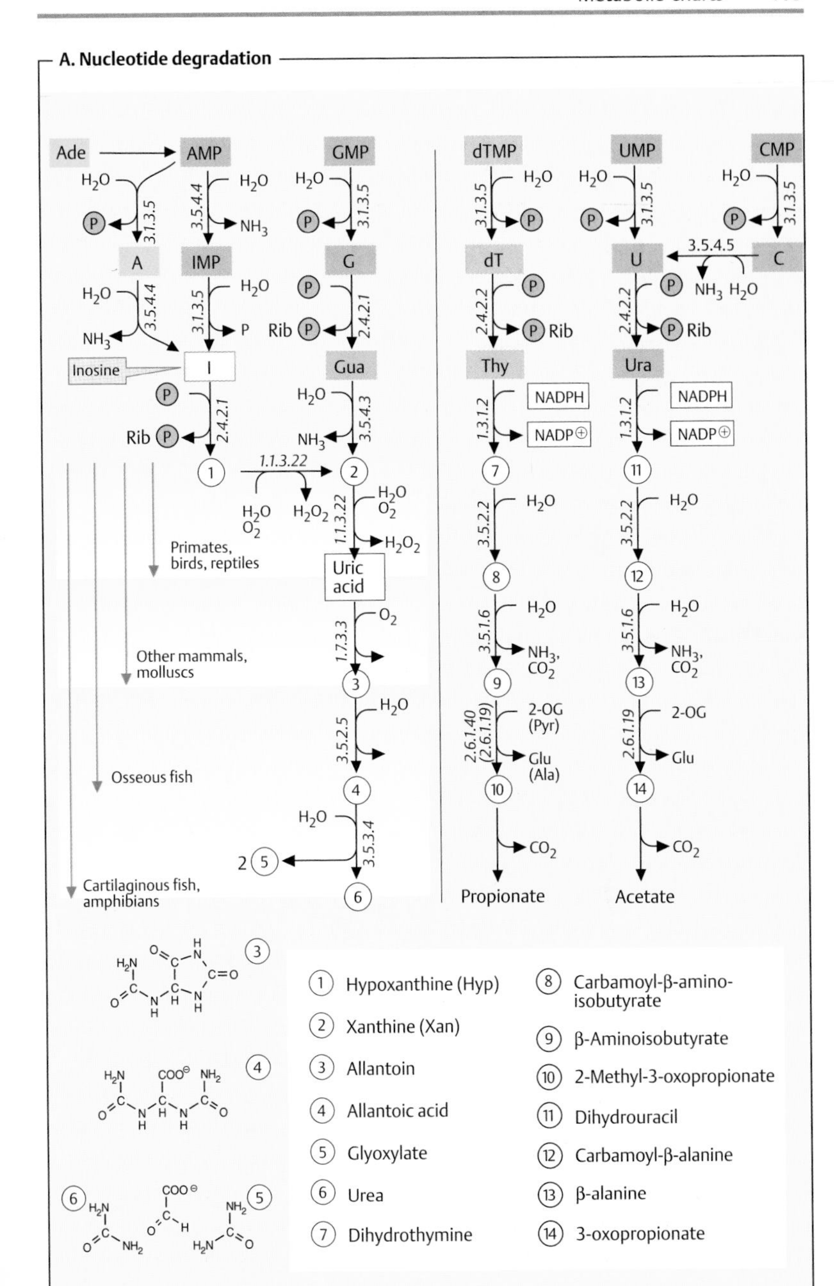
A. Nucleotide degradation
Ade
AMP
GMP
dTMP
UMP
CMP
3.1.3.5
3.5.4.4
A
IMP
G
dT
U
C
3.5.4.5
2.4.2.1
2.4.2.2
Inosine
I
Gua
Thy
Ura
NADPH
NADP⊕
3.5.4.3
1.3.1.2
1.1.3.22
Primates, birds, reptiles
Uric acid
3.5.2.2
Other mammals, molluscs
1.7.3.3
3.5.1.6
3.5.2.5
2-OG
(Pyr)
Glu
(Ala)
2.6.1.40
(2.6.1.19)
2.6.1.19
Osseous fish
3.5.3.4
Cartilaginous fish, amphibians
Propionate
Acetate
1 Hypoxanthine (Hyp)
2 Xanthine (Xan)
3 Allantoin
4 Allantoic acid
5 Glyoxylate
6 Urea
7 Dihydrothymine
8 Carbamoyl-β-amino-isobutyrate
9 β-Aminoisobutyrate
10 2-Methyl-3-oxopropionate
11 Dihydrouracil
12 Carbamoyl-β-alanine
13 β-alanine
14 3-oxopropionate

Annotated enzyme list

Only the enzymes mentioned in this atlas are listed here, from among the more than 2000 enzymes known. The enzyme names are based on the IUBMB's official *Enzyme nomenclature 1992*. The additions shown in round brackets belong to the enzyme name, while prosthetic groups and other cofactors are enclosed in square brackets. Common names of enzyme groups are given in italics, and trivial names are shown in quotation marks.

Class 1: Oxidoreductases (catalyze reduction-oxidation reactions)

Subclass 1.n: What is the electron *donor?*
Sub-subclass 1.n.n: What is the electron *acceptor*?

1.1 **A –CH–OH group is the donor**

1.1.1 **$NAD(P)^+$ is the acceptor** (*dehydrogenases, reductases*)
1.1.1.1 Alcohol dehydrogenase [Zn^{2+}]
1.1.1.3 Homoserine dehydrogenase
1.1.1.8 Glycerol 3-phosphate dehydrogenase (NAD^+)
1.1.1.21 Aldehyde reductase
1.1.1.27 Lactate dehydrogenase
1.1.1.30 3-Hydroxybutyrate dehydrogenase
1.1.1.31 3-Hydroxyisobutyrate dehydrogenase
1.1.1.34 Hydroxymethylglutaryl-CoA reductase (NADPH)
1.1.1.35 3-Hydroxyacyl-CoA dehydrogenase
1.1.1.37 Malate dehydrogenase
1.1.1.40 Malate dehydrogenase (oxaloacetate-decarboxylating, $NADP^+$)–“malic enzyme”
1.1.1.41 Isocitrate dehydrogenase (NAD^+)
1.1.1.42 Isocitrate dehydrogenase ($NADP^+$)
1.1.1.44 Phosphogluconate dehydrogenase (decarboxylating)
1.1.1.49 Glucose 6-phosphate 1-dehydrogenase
1.1.1.51 3(or 17)β-Hydroxysteroid dehydrogenase
1.1.1.95 Phosphoglycerate dehydrogenase
1.1.1.100 3-Oxoacyl-[ACP] reductase
1.1.1.101 Acylglycerone phosphate reductase
1.1.1.105 Retinol dehydrogenase
1.1.1.145 3β-Hydroxy-Δ^5-steroid dehydrogenase
1.1.1.205 IMP dehydrogenase

1.1.3 **Molecular oxygen is the acceptor** (*oxidases*)
1.1.3.4 Glucose oxidase [FAD]
1.1.3.8 L-Gulonolactone oxidase
1.1.3.22 Xanthine oxidase [Fe, Mo, FAD]
1.1.99.5 Glycerol-3-phosphate dehydrogenase (FAD)

1.2 **An aldehyde or keto group is the donor**

1.2.1 **$NAD(P)^+$ is the acceptor** (*dehydrogenases*)
1.2.1.3 Aldehyde dehydrogenase (NAD^+)
1.2.1.11 Aspartate semialdehyde dehydrogenase
1.2.1.12 Glyceraldehyde 3-phosphate dehydrogenase
1.2.1.13 Glyceraldehyde 3-phosphate dehydrogenase ($NADP^+$) (phosphorylating)
1.2.1.24 Succinate semialdehyde dehydrogenase
1.2.1.25 2-Oxoisovalerate dehydrogenase (acylating)
1.2.1.38 *N*-Acetyl-γ-glutamylphosphate reductase
1.2.1.41 Glutamylphosphate reductase

1.2.4 **A disulfide is the acceptor**
1.2.4.1 Pyruvate dehydrogenase (lipoamide) [TPP]
1.2.4.2 Oxoglutarate dehydrogenase (lipoamide) [TPP]

1.2.7 **An Fe/S protein is the acceptor**
1.2.7.2 2-Oxobutyrate synthase

1.3 **A –CH–CH– group is the donor**
1.3.1.10 Enoyl-[ACP] reductase (NADPH)
1.3.1.24 Biliverdin reductase
1.3.1.34 2,4-Dienoyl-CoA reductase
1.3.5.1 Succinate dehydrogenase (ubiquinone) [FAD, Fe_2S_2, Fe_4S_4], “complex II”
1.3.99.3 Acyl-CoA dehydrogenase [FAD]
1.3.99.11 Dihydroorotate dehydrogenase [FMN]

1.4 **A –CH–NH2 group is the donor**
1.4.1.2 Glutamate dehydrogenase
1.4.3.4 Amine oxidase [FAD], “monoamine oxidase (MAO)”
1.4.3.13 Protein lysine 6-oxidase [Cu]
1.4.4.2 Glycine dehydrogenase (decarboxylating) [PLP]

1.5 **A –CH–NH group is the donor**
1.5.1.2 Pyrroline-5-carboxylate reductase
1.5.1.3 Dihydrofolate reductase
1.5.1.5 Methylenetetrahydrofolate dehydrogenase ($NADP^+$)
1.5.1.12 1-Pyrroline-5-carboxylate dehydrogenase
1.5.1.20 Methylenetetrahydrofolate reductase (NADPH) [FAD]
1.5.5.1 Electron-transferring flavoprotein (ETF) dehydrogenase [Fe_4S_4]
1.5.99.8 Proline dehydrogenase [FAD]

1.6 **NAD(P)H is the donor**
16.4.2 Glutathione reductase (NADPH) [FAD]
1.6.4.5 Thioredoxin reductase (NADPH) [FAD]
1.6.5.3 NADH dehydrogenase (ubiquinone) [FAD, Fe_2S_2, Fe_4S_4]–“complex I”

1.8 **A sulfur group is the donor**
1.8.1.4 Dihydrolipoamide dehydrogenase [FAD]

1.9 **A heme group is the donor**
1.9.3.1 Cytochrome *c* oxidase [heme, Cu, Zn] – “cytochrome oxidase,” “complex IV”

1.10 **A diphenol is the donor**
1.10.2.2 Ubiquinol cytochrome *c* reductase [heme, Fe_2S_2]–“complex III”

1.11 **A peroxide is the acceptor** (*peroxidases*)
1.11.1.6 Catalase [heme]
1.11.1.7 Peroxidase [heme]
1.11.1.9 Glutathione peroxidase [Se]
1.11.1.12 Lipid hydroperoxide glutathione peroxidase [Se]

1.13 **Molecular oxygen is incorporated into the electron donor** (*oxygenases*)

1.13.11 **One donor, both O atoms are incorporated** (*dioxygenases*)
1.13.11.5 Homogentisate 1,2-dioxygenase [Fe]
1.13.11.20 Cysteine dioxygenase [Fe]
1.13.11.27 4-Hydroxyphenylpyruvate dioxygenase [ascorbate]
1.13.11.n Arachidonate lipoxygenases

1.14	**Two donors, one O atom is incorporated into both** (*monooxygenases, hydroxylases*)
1.14.11.2	Procollagen proline 4-dioxygenase [Fe, ascorbate]—"proline hydroxylase"
1.14.11.4	Procollagen lysine 5-dioxygenase [Fe, ascorbate]—"lysine hydroxylase"
1.14.13.13	Calcidiol 1-monooxygenase [heme]
1.14.15.4	Steroid 11β-monooxygenase [heme]
1.14.15.6	Cholesterol monooxygenase (side-chain-cleaving) [heme]
1.14.16.1	Phenylalanine 4-monooxygenase [Fe, tetrahydrobiopterin]
1.14.16.2	Tyrosine 3-monooxygenase [Fe, tetrahydrobiopterin]
1.14.17.1	Dopamine β-monooxygenase [Cu]
1.14.99.1	Prostaglandin H-synthase [heme]
1.14.99.3	Heme oxygenase (decyclizing) [heme]
1.14.99.5	Stearoyl-CoA desaturase [heme]
1.14.99.9	Steroid 17α-monooxygenase [heme]
1.14.99.10	Steroid 21-monooxygenase [heme]
1.15	**A superoxide radical is the acceptor**
1.15.1.1	Superoxide dismutase
1.17	**A $-CH_2$ group is the donor**
1.17.4.1	Ribonucleoside diphosphate reductase [Fe]—"ribonucleotide reductase"
1.18	**Reduced ferredoxin is the donor**
1.18.1.2	Ferredoxin-$NADP^+$ reductase [FAD]
1.18.6.1	Nitrogenase [Fe, Mo, Fe_4S_4]

Class 2: Transferases (catalyze the transfer of groups from one molecule to another)

Subclass 2.n: *Which group* is transferred?

2.1	**A C_1 group is transferred**
2.1.1	**A methyl group**
2.1.1.2	Guanidinoacetate *N*-methyltransferase
2.1.1.6	Catechol *O*-methyltransferase
2.1.1.13	5-Methyltetrahydrofolate-homocysteine *S*-methyltransferase
2.1.1.28	Phenylethanolamine *N*-methyltransferase
2.1.1.45	Thymidylate synthase
2.1.1.67	Thiopurine methyltransferase
2.1.2	**A formyl group**
2.1.2.1	Glycine hydroxymethyltransferase [PLP]
2.1.2.2	Phosphoribosylglycinamide formyltransferase
2.1.2.3	Phosphoribosylaminoimidazolecarboxamide formyltransferase
2.1.2.5	Glutamate formiminotransferase [PLP]
2.1.2.10	Aminomethyltransferase
2.1.3	**A carbamoyl group**
2.1.3.2	Aspartate carbamoyltransferase [Zn^{2+}]
2.1.3.3	Ornithine carbamoyltransferase
2.1.4	**An amidino group**
2.1.4.1	Glycine amidinotransferase
2.2	**An aldehyde or ketone residue is transferred**
2.2.1.1	Transketolase [TPP]
2.2.1.2	Transaldolase

2.3 **An acyl group is transferred**

2.3.1 **With acyl-CoA as donor**
2.3.1.1 Amino acid *N*-acetyltransferase
2.3.1.6 Choline *O*-acetyltransferase
2.3.1.12 Dihydrolipoamide acetyltransferase [lipoamide]
2.3.1.15 Glycerol 3-phosphate *O*-acyltransferase
2.3.1.16 Acetyl-CoA acyltransferase
2.3.1.20 Diacylglycerol *O*-acyltransferase
2.3.1.21 Carnitine *O*-palmitoyltransferase
2.3.1.22 Acylglycerol *O*-palmitoyltransferase
2.3.1.24 Sphingosine *N*-acyltransferase
2.3.1.37 5-Aminolevulinate synthase [PLP]
2.3.1.38 [ACP] *S*-acetyltransferase
2.3.1.39 [ACP] *S*-malonyltransferase
2.3.1.41 3-Oxoacyl-[ACP] synthase
2.3.1.42 Glycerone phosphate *O*-acyltransferase
2.3.1.43 Phosphatidylcholine-sterol acyltransferase—"lecithin-cholesterol acyltransferase (LCAT)"
2.3.1.51 Acylgylcerol-3-phosphate *O*-acyltransferase
2.3.1.61 Dihydrolipoamide succinyltransferase
2.3.1.85 Fatty-acid synthase

2.3.2 **An aminoacyl group is transferred**
2.3.2.2 γ-glutamyltransferase
2.3.2.12 Peptidyltransferase (*a ribozyme*)
2.3.2.13 Protein-glutamine γ-glutamyltransferase [Ca]—"fibrin-stabilizing factor"

2.4 **A glycosyl group is transferred**

2.4.1 **A hexose residue**
2.4.1.1 Phosphorylase [PLP]—"glycogen (starch) phosphorylase"
2.4.1.11 Glycogen (starch) synthase
2.4.1.17 Glucuronosyltransferase
2.4.1.18 1,4-α-Glucan branching enzyme
2.4.1.25 4-α-Glucanotransferase
2.4.1.47 *N*-Acylsphingosine galactosyltransferase
2.4.1.119 Protein glycotransferase

2.4.2 **A pentose residue**
2.4.2.7 Adenine phosphoribosyltransferase
2.4.2.8 Hypoxanthine phosphoribosyltransferase
2.4.2.10 Orotate phosphoribosyltransferase
2.4.2.14 Amidophosphoribosyl transferase

2.5 **An alkyl or aryl group is transferred**
2.5.1.1 Dimethylallyltransferase
2.5.1.6 Methionine adenosyltransferase
2.5.1.10 Geranyltransferase
2.5.1.21 Farnesyl diphosphate farnesyltransferase

2.6 **A nitrogen-containing group is transferred**

2.6.1 **An amino group** (*transaminases*)
2.6.1.1 Aspartate transaminase [PLP]—"GOT"
2.6.1.2 Alanine transaminase [PLP]—"GPT"

2.6.1.3 Cysteine transaminase [PLP]
2.6.1.5 Tyrosine transaminase [PLP]
2.6.1.6 Leucine transaminase (PLP]
2.6.1.11 Acetylornithine transaminase [PLP]
2.6.1.13 Ornithine transaminase [PLP]
2.6.1.19 4-Aminobutyrate transaminase [PLP]
2.6.1.42 Branched-chain amino acid transaminase [PLP]
2.6.1.52 Phosphoserine transaminase [PLP]

2.7 A phosphorus-containing group is transferred (*kinases*)

2.7.1 With –CH–OH as acceptor
2.7.1.1 Hexokinase
2.7.1.3 Ketohexokinase
2.7.1.6 Galactokinase
2.7.1.11 6-Phosphofructokinase
2.7.1.19 Phosphoribulokinase
2.7.1.28 Triokinase (triosekinase)
2.7.1.30 Glycerol kinase
2.7.1.32 Choline kinase
2.7.1.36 Mevalonate kinase
2.7.1.37 Protein kinase
2.7.1.38 Phosphorylase kinase
2.7.1.39 Homoserine kinase
2.7.1.40 Pyruvate kinase
2.7.1.67 1-Phosphatidylinositol-4-kinase
2.7.1.68 1-Phosphatidylinositol 4-phosphate kinase
2.7.1.82 Ethanolamine kinase
2.7.1.99 [Pyruvate dehydrogenase] kinase
2.7.1.105 6-Phosphofructo-2-kinase
2.7.1.112 Protein tyrosine kinase

2.7.2 With –CO–OH as acceptor
2.7.2.3 Phosphoglycerate kinase
2.7.2.4 Aspartate kinase
2.7.2.8 Acetylglutamate kinase
2.7.2.11 Glutamate 5-kinase

2.7.3 With a nitrogen-containing group as acceptor
2.7.3.2 Creatine kinase

2.7.4 With a phosphate group as acceptor
2.7.4.2 Phosphomevalonate kinase
2.7.4.3 Adenylate kinase
2.7.4.4 Nucleoside phosphate kinase
2.7.4.6 Nucleoside diphosphate kinase

2.7.6 A diphosphate residue is transferred
2.7.6.1 Ribose phosphate pyrophosphokinase

2.7.7 A nucleotide is transferred
2.7.7.6 DNA-directed RNA polymerase—"RNA polymerase"
2.7.7.7 DNA-directed DNA polymerase—"DNA polymerase"
2.7.7.9 UTP-glucose-l-phosphate uridyltransferase
2.7.7.12 Hexose-1-phosphate uridyltransferase
2.7.7.14 Ethanolamine phosphate cytidyltransferase

2.7.7.15 Choline phosphate cytidyltransferase
2.7.7.41 Phosphatidate cytidyltransferase
2.7.7.49 RNA-directed DNA polymerase—"reverse transcriptase"

2.7.8 Another substituted phosphate is transferred
2.7.8.1 Ethanolaminephosphotransferase
2.7.8.2 Diacylglycerol cholinephosphotransferase
2.7.8.11 CDPdiacylglycerol-inositol 3-phosphatidyltransferase
2.7.8.16 1-Alkyl-2-acetylglycerol cholinephosphotransferase
2.7.8.17 *N*-Acetylglucosaminephosphotransferase

Class 3: Hydrolases (catalyze bond cleavage by hydrolysis)

Subclass 3.n: *What kind of bond* is hydrolyzed?

3.1 An ester bond is hydrolyzed (*esterases*)

3.1.1 In carboxylic acid esters
3.1.1.2 Arylesterase
3.1.1.3 Triacylglycerol lipase
3.1.1.4 Phospholipase A_2
3.1.1.7 Acetylcholinesterase
3.1.1.13 Cholesterol esterase
3.1.1.17 Gluconolactonase
3.1.1.32 Phospholipase A_1
3.1.1.34 Lipoprotein lipase, diacylglycerol lipase

3.1.2 In thioesters 3.1.2.4
3-Hydroxyisobutyryl-CoA hydrolase
3.1.2.14 Acyl-[ACP] hydrolase

3.1.3 In phosphoric acid monoesters (*phosphatases*)
3.1.3.1 Alkaline phosphatase [Zn^{2+}]
3.1.3.2 Acid phosphatase
3.1.3.4 Phosphatidate phosphatase
3.1.3.9 Glucose 6-phosphatase
3.1.3.11 Fructose bisphosphatase
3.1.3.13 Bisphosphoglycerate phosphatase
3.1.3.16 Phosphoprotein phosphatase
3.1.3.37 Sedoheptulose bisphosphatase
3.1.3.43 [Pyruvate dehydrogenase] phosphatase
3.1.3.46 Fructose-2,6-bisphosphate 2-phosphatase
3.1.3.n Polynucleotidases

3.1.4 In phosphoric acid diesters (*phosphodiesterases*)
3.1.4.1 Phosphodiesterase
3.1.4.3 Phospholipase C
3.1.4.4 Phospholipase D
3.1.4.17 3′,5′-cNMP phosphodiesterase
3.1.4.35 3′,5′-cGMP phosphodiesterase
3.1.4.45 *N*-Acetylglucosaminyl phosphodiesterase

3.1.21 In DNA
3.1.21.1 Deoxyribonuclease I
3.1.21.4 Site-specific deoxyribonuclease (type II)—"restriction endonuclease"

3.10.26–7	**In RNA**
3.1.26.4	Ribonuclease H
3.1.27.5	Pancreatic ribonuclease
3.2	**A glycosidic bond is hydrolyzed** (*glycosidases*)
3.2.1	**In *O*-glycosides**
3.2.1.1	α-Amylase
3.2.1.10	Oligo-1,6-glucosidase
3.2.1.17	Lysozyme
3.2.1.18	Neuraminidase
3.2.1.20	α-Glucosidase
3.2.1.23	β-Galactosidase
3.2.1.24	α -Mannosidase
3.2.1.26	β-Fructofuranosidase—"saccharase," "invertase"
3.2.1.28	α,α-Trehalase
3.2.1.33	Amylo-1,6-glucosidase
3.2.1.48	Sucrose α-glucosidase
3.2.1.52	β-*N*-Acetylhexosaminidase
3.2.2.n	Nucleosidases
3.3	**An ether bond is hydrolyzed**
3.3.1.1	Adenosylhomocysteinase
3.4	**A peptide bond is hydrolyzed** (*peptidases*)
3.4.11	**Aminopeptidases** (*N*-terminal exopeptidases)
3.4.11.n	Various aminopeptidases [Zn^{2+}]
3.4.13	**Dipeptidases** (act on dipeptides only)
3.4.13.n	Various dipeptidases [Zn^{2+}]
3.4.15	**Peptidyl dipeptidases** (*C*-terminal exopeptidases, releasing dipeptides)
3.4.15.1	Peptidyl-dipeptidase A [Zn^{2+}]—"angiotensin-converting enzyme (ACE)"
3.4.17	**Carboxypeptidases** (*C*-terminal exopeptidases)
3.4.17.1	Carboxypeptidase A [Zn^{2+}]
3.4.17.2	Carboxypeptidase B [Zn^{2+}]
3.4.17.8	Muramoylpentapeptide carboxypeptidase
3.4.21	**Serine proteinases** (endopeptidases)
3.4.21.1	Chymotrypsin
3.4.21.4	Trypsin
3.4.21.5	Thrombin
3.4.21.6	Coagulation factor Xa—"Stuart–Prower factor"
3.4.21.7	Plasmin
3.4.21.9	Enteropeptidase—"enterokinase"
3.4.21.21	Coagulation factor VIIa—"proconvertin"
3.4.21.22	Coagulation factor IXa—"Christmas factor"
3.4.21.27	Coagulation factor XIa—"plasma thromboplastin antecedent"
3.4.21.34	Plasma kallikrein
3.4.21.35	Tissue kallikrein
3.4.21.36	Elastase
3.4.21.38	Coagulation factor XIIa—"Hageman factor"
3.4.21.43	C3/C5 convertase (complement—classical pathway)
3.4.21.47	C3/C5 convertase (complement—alternative pathway)

3.4.21.68 Plasminogen activator (tissue)–"tissue plasminogen activator (t-PA)"
3.4.21.73 Plasminogen activator (urine)–"urokinase"

3.4.22 **Cysteine proteinases** (*endopeptidases*)
3.4.22.2 Papain

3.4.23 **Aspartate proteinases** (*endopeptidases*)
3.4.23.1 Pepsin A
3.4.23.2 Pepsin B
3.4.23.3 Gastricsin (pepsin C)
3.4.23.4 Chymosin
3.4.23.15 Renin

3.4.24 **Metalloproteinases** (*endopeptidases*)
3.4.24.7 Collagenase

3.4.99 **Other peptidases**
3.4.99.36 Signal peptidase

3.5 **Another amide bond is hydrolyzed** (*amidases*)
3.5.1.1 Asparaginase
3.5.1.2 Glutaminase
3.5.1.16 Acetylornithine deacetylase [Zn^{2+}]
3.5.2.3 Dihydroorotase
3.5.2.7 Imidazolonepropionase
3.5.3.1 Arginase
3.5.4.6 AMP deaminase
3.5.4.9 Methylenetetrahydrofolate cyclohydrolase
3.5.4.10 IMP cyclohydrolase

3.6 **An anhydride bond is hydrolyzed**
3.6.1.6 Nucleoside diphosphatase
3.6.1.32 Myosin ATPase
3.6.1.34 H^+-transporting ATP synthase–"ATP synthase," "complex V"
3.6.1.35 H^+-transporting ATPase
3.6.1.36 H^+/K^+-exchanging ATPase
3.6.1.37 Na^+/K^+-exchanging ATPase–"Na^+/K^+-ATPase"
3.6.1.38 Ca^{2+}-transporting ATPase

3.7 **A C–C bond is hydrolyzed**
3.7.1.2 Fumarylacetoacetase

Class 4: Lyases (cleave or form bonds without oxidative or hydrolytic steps)

Subclass 4.n: *What kind of bond* is formed or cleaved?

4.1 **A C–C bond is formed or cleaved**

4.1.1 **Carboxy-lyases** (*carboxylases, decarboxylases*)
4.1.1.1 Pyruvate decarboxylase [TPP]
4.1.1.15 Glutamate decarboxylase [PLP]
4.1.1.21 Phosphoribosylaminoimidazole carboxylase
4.1.1.23 Orotidine-5′-phosphate decarboxylase
4.1.1.28 Aromatic L-amino acid decarboxylase [PLP]
4.1.1.32 Phosphoenolpyruvate carboxykinase (GTP)
4.1.1.39 Ribulose bisphosphate carboxylase [Cu]–"rubisco"

4.1.2	**Acting on aldehydes or ketones**
4.1.2.5	Threonine aldolase [PLP]
4.1.2.13	Fructose bisphosphate aldolase—“aldolase”
4.1.3.4	Hydroxymethylglutaryl-CoA lyase
4.1.3.5	Hydroxymethylglutaryl-CoA synthase
4.1.3.7	Citrate synthase
4.1.3.8	ATP-citrate lyase
4.1.3.18	Acetolactate synthase [TPP, flavin]
4.1.99	**Other C–C lyases**
4.1.99.3	Deoxyribodipyrimidine photolyase [FAD]—“photolyase”
4.2	**A C–O bond is formed or cleaved**
4.2.1	**Hydrolyases** (*hydratases, dehydratases*)
4.2.1.1	Carbonate dehydratase [Zn^{2+}]—“carbonic anhydrase”
4.2.1.2	Fumarate hydratase—“fumarase”
4.2.1.3	Aconitate hydratase [Fe_4S_4]—“aconitase”
4.2.1.11	Phosphopyruvate hydratase—“enolase”
4.2.1.13	Serine dehydratase
4.2.1.17	Enoyl-CoA hydratase
4.2.1.18	Methylglutaconyl-CoA hydratase
4.2.1.22	Cystathionine β-synthase [PLP]
4.2.1.24	Porphobilinogen synthase
4.2.1.49	Urocanate hydratase
4.2.1.61	3-Hydroxypalmitoyl-[ACP] dehydratase
4.2.1.75	Uroporphyrinogen III synthase
4.2.99	**Other C–O lyases**
4.2.99.2	Threonine synthase [PLP]
4.3	**A C–N bond is formed or cleaved**
4.3.1	**Ammonia lyases** 4.3.1.3
	Histidine ammonia lyase
4.3.1.8	Hydroxymethylbilane synthase
4.3.2	**Amidine lyases** 4.3.2.1
	Argininosuccinate lyase
4.3.2.2	Adenylosuccinate lyase
4.4	**A C–S bond is formed or cleaved**
4.4.1.1	Cystathionine γ-lyase [PLP]
4.6	**A P–O bond is formed or cleaved**
4.6.1.1	Adenylate cyclase
4.6.1.2	Guanylate cyclase

Class 5: Isomerases (catalyze changes within one molecule)

Subclass 5.n: *What kind of isomerization* is taking place?

5.1	**A racemization or epimerization** (*epimerases*)
5.1.3.1	Ribulose phosphate 3-epimerase
5.1.3.2	UDPglucose 4-epimerase
5.1.3.4	L-Ribulose phosphate 4-epimerase
5.1.99.1	Methylmalonyl-CoA epimerase

5.2 **A *cis–trans* isomerization**
5.2.1.2 Maleylacetoacetate isomerase
5.2.1.3 Retinal isomerase
5.2.1.8 Peptidyl proline *cis–trans*-isomerase

5.3 **An intramolecular electron transfer**
5.3.1.1 Triose phosphate isomerase
5.3.1.6 Ribose 5-phosphate isomerase
5.3.1.9 Glucose 6-phosphate isomerase
5.3.3.1 Steroid Δ-isomerase
5.3.3.8 Enoyl-CoA isomerase
5.3.4.1 Protein disulfide isomerase

5.4 **An intramolecular group transfer** (*mutases*)
5.4.2.1 Phosphoglycerate mutase
5.4.2.2 Phosphoglucomutase
5.4.2.4 Bisphosphoglycerate mutase
5.4.99.2 Methylmalonyl-CoA mutase [cobamide]

5.99 **Another kind of isomerization**
5.99.1.2 DNA topoisomerase (type I)—“DNA helicase”
5.99.1.3 DNA topoisomerase (ATP-hydrolyzing, type II)—“DNA gyrase”

Class 6: Ligases (join two molecules with hydrolysis of an “energy-rich” bond)

Subclass 6.n: *What kind of bond* is formed?

6.1 **A C–O bond is formed**
6.1.1.n (Amino acid)-tRNA ligases (*aminoacyl-tRNA synthetases*)

6.2 **A C–S bond is formed**
6.2.1.1 Acetate-CoA ligase
6.2.1.3 Long-chain fatty-acid-CoA ligase
6.2.1.4 Succinate-CoA ligase (GDP-forming)—“thiokinase”

6.3 **A C–N bond is formed**
6.3.1.2 Glutamate-NH_3 ligase—“glutamine synthetase”
6.3.2.6 Phosphoribosylaminoimidazolesuccinocarboxamide synthase
6.3.3.1 Phosphoribosylformylglycinamidine cycloligase
6.3.3.2 5-Formyltetrahydrofolate cycloligase
6.3.4.2 CTP synthase
6.3.4.4 Adenylosuccinate synthase
6.3.4.5 Argininosuccinate synthase
6.3.4.13 Phosphoribosylamine glycine ligase
6.3.4.16 Carbamoylphosphate synthase (NH_3)
6.3.5.2 GMP synthase (glutamine-hydrolyzing)
6.3.5.3 Phosphoribosylformylglycinamidine synthase
6.3.5.4 Asparagine synthase (glutamine-hydrolyzing)
6.3.5.5 Carbamoylphosphate synthase (glutamine-hydrolyzing)

6.4 **A C–C bond is formed**
6.4.1.1 Pyruvate carboxylase [biotin]
6.4.1.2 Acetyl-CoA carboxylase [biotin]
6.4.1.3 Propionyl-CoA carboxylase [biotin]
6.4.1.4 Methylcrotonyl-CoA carboxylase [biotin]

6.5 **A P–O bond is formed**
6.5.1.1 DNA ligase (ATP)

Abbreviations

Abbreviations for amino acids, p. 60
For bases and nucleosides, p. 80
For monosaccharides, p. 38

AA	Amino acid
ACE	Angiotensin-converting enzyme (peptidyl-dipeptidase A)
ACP	Acyl carrier protein
ACTH	Adrenocorticotropic hormone (corticotropin)
ADH	Antidiuretic hormone (adiuretin, vasopressin)
ADP	Adenosine 5′-diphosphate
AIDS	Acquired immunodeficiency syndrome
ALA	5-Aminolevulinic acid
AMP	Adenosine 5′-monophosphate
ANF	Atrial natriuretic factor
ANP	Atrial natriuretic peptide (= ANF)
ATP	Adenosine 5′-triphosphate
AVP	Arginine vasopressin
b	Base
bp	Base pair
BPG	2,3-Bisphosphoglycerate
cAMP	3′,5′-Cyclic AMP
CAP	Catabolite activator protein
CDK	Cyclin-dependent protein kinase (in cell cycle)
cDNA	Complementary DNA
CDP	Cytidine 5′-diphosphate
cGMP	3′,5′-Cyclic GMP
CIA	Chemoluminescence immunoassay
CMP	Cytidine 5′-monophosphate
CoA	Coenzyme A
CoQ	Coenzyme Q (ubiquinone)
CSF	colony-stimulating factor
CTP	Cytidine 5′-triphosphate
d	Deoxy-
Da	Dalton (atomic mass unit)
DAG	Diacylglycerol
dd	Dideoxy-
DH	Dehydrogenase
DNA	Deoxyribonucleic acid
dsDNA	Double-stranded DNA
EA	Ethanolamine
EIA	Enzyme-linked immunoassay
ER	Endoplasmic reticulum
FAD	Flavin adenine dinucleotide
Fd	Ferredoxin
FFA	Free fatty acid
fMet	*N*-formylmethionine
FMN	Flavin mononucleotide
Fp	Flavoprotein (containing FMN or FAD)
GABA	γ-Aminobutyric acid
GDP	Guanosine 5′-diphosphate
Glut	Glucose transporter
GMP	Guanosine 5′-monophosphate
GSH	Reduced glutathione
GSSG	Oxidized glutathione
GTP	Guanosine 5′-triphosphate
h	hour
HAT medium	Medium containing hypoxanthine, aminopterin, and thymidine
Hb	Hemoglobin
HDL	High-density lipoprotein
HIV	Human immunodeficiency virus
HLA	Human leukocyte-associated antigen
HMG-CoA	3-Hydroxy-3-methylglutaryl-CoA
HMP	Hexose monophosphate pathway
hnRNA	Heterogeneous nuclear ribonucleic acid
HPLC	High-performance liquid chromatography
hsp	Heat-shock protein
IDL	Intermediate-density lipoprotein
IF	Intermediary filament
IFN	Interferon
Ig	Immunoglobulin
IL	Interleukin
$InsP_3$	Inositol 1,4,5-trisphosphate
IPTG	Isopropylthiogalactoside
IRS	Insulin-receptor substrate
kDa	Kilodalton (10^3 atomic mass units)
K_m	Michaelis constant

LDH Lactate dehydrogenase
LDL Low-density lipoprotein
M Molarity (mol · L^{-1})
Mab Monoclonal antibody
MAP kinase Mitogen-activated protein kinase
MHC Major histocompatability complex
MPF Maturation-promoting factor
mRNA Messenger ribonucleic acid
N Nucleotide with any base
NAD^+ Oxidized nicotinamide adenine dinucleotide
NADH Reduced nicotinamide adenine dinucleotide
$NADP^+$ Oxidized nicotinamide adenine dinucleotide phosphate
NADPH Reduced nicotinamide adenine dinucleotide phosphate
NeuAc *N*-acetylneuraminic acid
nm Nanometer (10^{-9} m)
ODH 2-Oxoglutarate dehydrogenase
PAGE Polyacrylamide gel electrophoresis
Pan Pantetheine
PAPS Phosphoadenosine phosphosulfate
PCR Polymerase chain reaction
PDH Pyruvate dehydrogenase
PEG Polyethylene glycol
PEP Phosphoenolpyruvate
pH pH value
P_i Inorganic phosphate
PK Protein kinase
PLP Pyridoxal phosphate
PP Protein phosphatase
PPP Pentose phosphate pathway
PQ Plastoquinone
PRPP 5-Phosphoribosyl 1-diphosphate
PS Photosystem
PTH Parathyroid hormone
Q Oxidized coenzyme Q (ubiquinone)
QH_2 Reduced coenzyme Q (ubiquinol)
R Gas constant
rER Rough endoplasmic reticulum
RES Reticuloendothelial system
RFLP Restriction fragment length polymorphism
RIA Radioimmunoassay
RNA Ribonucleic acid
ROS Reactive oxygen species
RP Reversed phase (of silica gel)
rRNA Ribosomal ribonucleic acid
S Svedberg (unit of sedimentation coefficient)
SAH *S*-adenosyl L-homocysteine
SAM *S*-adenosyl L-methionine
SDS Sodium dodecylsulfate
sER Smooth endoplasmic reticulum
sn Stereospecific numbering
snRNA Small nuclear ribonucleic acid
SR Sarcoplasmic reticulum
ssDNA Single-stranded DNA
TBG Thyroxine-binding globulin
THB Tetrahydrobiopterin
THF Tetrahydrofolate
TLC Thin-layer chromatography
TPP Thiamine diphosphate
TRH Thyrotropin-releasing hormone (thyroliberin)
Tris Tris(hydroxymethyl)aminomethane
tRNA Transfer ribonucleic acid
TSH Thyroid-stimulating hormone (thyrotropin)
UDP Uridine 5′-diphosphate
UMP Uridine 5′-monophosphate
UTP Uridine 5′-triphosphate
UV Ultraviolet radiation
V_{max}, V Maximal velocity (of an enzyme)
VLDL Very-low-density lipoprotein

Quantities and units

1. SI base units

Quantity	SI unit	Symbol	Remarks
Length	Meter	m	1 yard (yd) = 0.9144 m 1 inch (in) = 0.0254 m 1 Å = 10^{-10} m = 0.1 nm
Mass	Kilogram	kg	1 pound (lb) = 0.4536 kg
Time	Second	s	
Current strength	Ampere	A	
Temperature	Kelvin	K	°C (degree Celsius) = K – 273.2 Fahrenheit: °C = 5/9 (°F – 32)
Light	Candela	Cd	
Amount of substance	Mol	mol	

2 Derived units

Quantity	Unit	Symbol	Derivation	Remarks
Frequency	Hertz	Hz	s^{-1}	
Volume	Liter	L	$10^{-3} \cdot m^3$	1 U.S. gallon (gal) = 3.785 L
Force	Newton	N	$kg \cdot m \cdot s^{-2}$	
Pressure	Pascal	Pa	$N \cdot m^{-2}$	1 bar = 10^5 Pa 1 mmHg = 133.3 Pa
Energy, work, heat	Joule	J	N · m	1 calorie (cal) = 4.1868 J
Power	Watt	W	$J \cdot s^{-1}$	
Electrical charge	Coulomb	C	A · s	
Voltage	Volt	V	$W \cdot A^{-1}$	
Concentration	Molarity	M	$mol \cdot L^{-1}$	
Molecular mass	Dalton	Da	$1.6605 \cdot 10^{-24}$ g	
Molar mass	–	–	g	
Molecular weight	–	M_r	–	Nondimensional
Reaction rate		v	$mol \cdot s^{-1}$	
Catalytic activity	Katal	kat	$mol \cdot s^{-1}$	1 unit (U) = $1.67 \cdot 10^{-8}$ kat
Specific activity	–	–	kat · (kg enzyme)$^{-1}$	Usually: U · (mg enzyme)$^{-1}$
Sedimentation coefficient	Svedberg	S	10^{-13} s	
Radioactivity	Becquerel	Bq	Decays · s^{-1}	1 curie (Ci) = $3.7 \cdot 10^{10}$ Bq

3 Multiples and fractions

Factor	Prefix	Symbol	Example
10^9	Giga	G	GHz = 10^9 hertz
10^6	Mega	M	MPa = 10^6 pascal
10^3	Kilo	k	kJ = 10^3 joule
10^{-3}	Milli	m	mM = 10–3 mol · L^{-1}
10^{-6}	Micro	μ	μV = 10^{-6} volt
10^{-9}	Nano	n	nkat = 10^{-9} katal
10^{-12}	Pico	p	pm = 10^{-12} meter

4 Important constants

General gas constant, R	R = 8.314 J · $mol^{-1} \cdot K^{-1}$
Loschmidt (Avogadro) number, N (number of particles per mol)	N = $6.0225 \cdot 10^{23}$
Faraday constant F	F = 96480 C · mol^{-1}

Further reading

Textbooks

Alberts B, Bray D, Lewis J, Raff M, Roberts K, Watson JD. The molecular biology of the cell. 4th ed. New York: Garland Science, 2002.
Berg JM, Tymoczko JL, Stryer L. Biochemistry. 5th ed. New York: Freeman, 2002.
Devlin TM, editor. Textbook of biochemistry: with clinical correlations. 5th ed. New York: Wiley-Liss, 2002.
Granner DK, Mayes PA, Rodwell VW. Murray RK. Harper's illustrated biochemistry. 26th ed. New York : McGraw-Hill/Appleton and Lange, 2003.
Lodish H, Darnell J, Baltimore D. Molecular cell biology. 3rd ed. New York: Scientific American Books, New York, 1995.
Mathews CK, van Holde KE, Ahern KG. Biochemistry. 3rd ed. San Francisco: Cummings, 2000.
Nelson DL, Cox MM. Lehninger principles of biochemistry. 3rd ed. New York: Worth, 2000.
Voet D, Voet JG. Biochemistry. 3rd ed. New York: Wiley, 2004.

Reference works

Branden C, Tooze J. Introduction to protein structure. New York: Garland, 1991.
Janeway CA, Travers P, editors. Immunobiology. 5th ed. New York: Garland, 2001.
Michal G, editor. Biochemical pathways: an atlas of biochemistry and molecular biology. New York: Wiley, 1999.
Nature Publishing Group. Encyclopedia of life sciences. http://www.els.net [an Internet encyclopedia with up-to-date overview articles in every field of biochemistry and cell biology].
Webb EC, editor. Enzyme nomenclature 1992. San Diego: International Union of Biochemistry and Molecular Biology/Academic Press, 1992.

Selected periodicals (journals and yearbooks)

Annual Review of Biochemistry. Annual Reviews, Inc., Palo Alto, CA, USA [the most important collection of biochemical reviews].
Current Biology, Current Opinion in Cell Biology, Current Opinion in Structural Biology, and related journals in this series. Current Biology, Ltd. London [short up-to-date reviews].
Trends in Biochemical Sciences. Elsevier Trends Journals, Cambridge, United Kingdom [the "newspaper" for biochemists; official publication of the International Union of Biochemistry and Molecular Biology (IUBMB)].

Source credits

Individual graphic elements in the following plates are based on the following sources, used with the kind permission of the authors and publishers concerned.

Page	Figure	Source
65		Goodsell DS, Trends Biochem Sci 1993; 18: 65–8
201	A	Beckman Instruments, Munich, Bulletin no. DS-555A, p. 6, Fig. 6
203	B	Goodsell DS, Trends Biochem Sci 1991; 16: 203–6, Figs. 1a and 1b
207	C	Stryer L. Biochemistry. New York: Freeman, 1988, p. 945, Fig. 36–47
207	C	Alberts B, et al. The molecular biology of the cell. New York: Garland, 1989, p. 663, Fig. 11–73B and p. 634, Fig. 11–36
279	A, B	Voet D, Voet JG. Biochemistry. New York: Wiley, 1990, p. 305, Fig. 11–45 and p. 306, Fig. 11–47
295	A	Voet D, Voet JG. Biochemistry. New York: Wiley, 1990, p. 1097, Fig. 34–13
297	A	Voet D, Voet JG. Biochemistry. New York: Wiley, 1990, p. 1112, Fig. 34–33
297	A	Janeway CA, Travers P. Immunology. Heidelberg, Germany: Spektrum, 1994, p. 164, Fig. 4.3.c
333	B	Voet D, Voet JG. Biochemistry. New York: Wiley, 1990, p. 1126, Fig. 34–55
335	B	Darnell J, et al. Molecular cell biology. 2nd ed. New York: Freeman, 1990, p. 923, Fig. 23–26

Index

Numbers in *italics* indicate figures.

A

F

G

H

I

J

K

L

N

O

U

V